Handbook of Polymer
Science and Technology

Volume 2:
PERFORMANCE PROPERTIES
OF PLASTICS AND ELASTOMERS

Handbook of Polymer Science and Technology

edited by

Nicholas P. Cheremisinoff

Volume 1 Synthesis and Properties
Volume 2 Performance Properties of Plastics and Elastomers
Volume 3 Applications and Processing Operations
Volume 4 Composites and Specialty Applications

Other Volumes in Preparation

Handbook of Polymer Science and Technology

Volume 2:
PERFORMANCE PROPERTIES
OF PLASTICS AND ELASTOMERS

edited by

Nicholas P. Cheremisinoff

CRC Press
Taylor & Francis Group
Boca Raton London New York

CRC Press is an imprint of the
Taylor & Francis Group, an **informa** business

First published 1989 by MARCEL DEKKER, INC.

Published 2019 by CRC Press
Taylor & Francis Group
6000 Broken Sound Parkway NW, Suite 300
Boca Raton, FL 33487-2742

© 1989 by Taylor & Francis Group, LLC
CRC Press is an imprint of Taylor & Francis Group, an Informa business

First issued in paperback 2019

No claim to original U.S. Government works

ISBN-13: 978-0-367-45101-1 (pbk)
ISBN-13: 978-0-8247-8174-3 (hbk)

Visit the Taylor & Francis Web site at
http://www.taylorandfrancis.com

and the CRC Press Web site at
http://www.crcpress.com

Preface

This volume of the *Handbook of Polymer Science and Technology* focuses on physical, structural, and compositional properties of elastomeric materials and plastics. The book is organized into two sections. Section I, Physicochemical Properties of Elastomers, describes the general properties of synthetic rubbers. The objective of this section is to provide a broad overview of the physical and physicochemical properties of synthetic rubbers that are used in conventional cured applications. Chapters are devoted to the subjects of physical properties of gum and filled elastomers, mechanisms of crystallization, effects and importance of compounding and its relationships to end-use properties, and stability characteristics of synthetic rubber materials. With regard to stability, the subjects of auto-oxidation and ozone resistance are discussed in detail. The proper use of antioxidant and stabilizer systems, as well as curative package selection and compounding practices and systems for achieving desired end-use properties in various consumer-oriented applications such as automotive parts manufacturing, mechanical goods articles, hose and tubing applications, construction articles, and so on, are described. Flammability properties of elastomers are also treated as part of the discussions of general properties. In addition, extension of heat aging properties to high-temperature applications is treated. This section, and indeed the volume, serves as a prelude to later volumes which will discuss generic rubbers and their specific end-use applications and compounding needs. Direct comparisons between rubbers such as EPDM, natural rubber, SBR, and others are reserved for a subsequent volume.

Section II, Properties of Plastics, contains eleven chapters on the properties of major engineering plastics materials. Chapters in this section should be consulted along with Volume 1 for their introductory comparisons of structural properties of certain major plastic materials, including the nylons. Approximately one-third of this section is devoted to the properties of polyethylene because of its wide usage throughout industry and in consumer-oriented applications. This material is again discussed in Volume 3 in terms of its properties for widely practiced extrusion operations. In the present volume, polyethylene as a material is largely discussed in terms of its chemical structure, crystallinity, and resultant high-strength properties, the importance of branching to the polymer's properties, and its thermal stability. The remainder of this section discusses general properties and structural characteristics of polypropylene, polystyrene, and polyurethane-based polymers; general molecular properties of aromatic polymers; and liquid crystal polymers. With the exception of liquid crystal polymers, discussions of specific end-use applications and practices are deferred to a subsequent volume. Because of the unique properties of liquid crystal polymers and the relative new advancements in this area, applications are highlighted.

The efforts of thirty-one specialists from twenty-two institutions and companies are presented in this volume. Each contributor is to be viewed as responsible for his or her statements in the presentation of chapters. Their devotion to contributing time and effort to this work is gratefully acknowledged. Without their efforts this work could not have become a reality. Further thanks are extended to Marcel Dekker, Inc., and to its production staff for the high-quality production of this series.

Nicholas P. Cheremisinoff

Contents

Contributors

Ralph D. Allen   Uniroyal Chemical Company, Inc., Naugatuck, Connecticut

S. Al-Malaika   Department of Chemical Engineering and Applied Chemistry, Aston University, Birmingham, England

Nedhal K. Al-Nidawy   Petroleum Research Center, Jadiriyah, Baghdad, Iraq

Dhoaib Al-Sammerrai   Petroleum Research Center, Jadiriyah, Baghdad, Iraq

A. Apicella   Department of Materials and Production Engineering, University of Naples, Naples, Italy

Jay Bhattacharyya   Gates Rubber Company, Denver, Colorado

Gordon W. Calundann   Celanese Research Company, Summit, New Jersey

Sebastião V. Canevarolo   Departamento de Engenharia de Materials, Universidade Federal de São Carlos, São Carlos-São Paulo, Brazil

Tai-Shung Chung   Celanese Research Company, Summit, New Jersey

Roger L. Clough   Sandia National Laboratories, Albuquerque, New Mexico

M. Dosière   Department of Physical Chemistry and Thermodynamics, University of Mons, Mons, Belgium

Anthony J. East   Celanese Research Company, Summit, New Jersey

Raffaele Gallo   Istituto di Chimica Industriale dell' Universitá di Messina, Messina, Italy

Kenneth T. Gillen   Sandia National Laboratories, Albuquerque, New Mexico

Bruce Hartmann   Polymer Physics Group, Naval Surface Weapons Center, Silver Spring, Maryland

Robert W. Keller   CR Industries, Elgin, Illinois

Joseph P. Kennedy Institute of Polymer Science, The University of Akron, Akron, Ohio

Masakatsu Kochi Institute of Interdisciplinary Research, University of Tokyo, Tokyo, Japan

R. P. Lattimer The BF Goodrich Research and Development Center, Brecksville, Ohio

David F. Lawson Central Research Laboratories, The Firestone Tire & Rubber Company, Akron, Ohio

R. W. Layer The BF Goodrich Research and Development Center, Brecksville, Ohio

Fui-Tseng H. Lee FMC Corporation, Princeton, New Jersey

L. Nicodemo Department of Materials and Production Engineering, University of Naples, Naples, Italy

L. Nicolais Department of Materials and Production Engineering, University of Naples, Naples, Italy

C. K. Rhee The Uniroyal Goodrich Tire Company, Brecksville, Ohio

Febo Severini Instituto di Chimica Industriale dell' Universitá di Messina, Messina, Italy

A. Stevenson Materials Engineering Research Laboratory Ltd., Hertford, England

Ayako Torikai Department of Synthetic Chemistry, Faculty of Engineering, Nagoya University, Nagoya, Japan

Takao Usami Plastics Laboratory, Mitsubishi Petrochemical Company, Ltd., Yokkaichi, Japan

Vittoria Vittoria Istituto di Ricerche su Tecnologia dei Polimeri e Reologia del CNR, Naples, Italy

R. A. Weiss Institute of Materials Science, University of Connecticut, Storrs, Connecticut

Handbook of Polymer Science and Technology

Volume 2:
PERFORMANCE PROPERTIES
OF PLASTICS AND ELASTOMERS

I

PHYSICOCHEMICAL PROPERTIES OF ELASTOMERS

1

Properties and Uses of Triblock Copolymers

Sebastião V. Canevarolo
Universidade Federal de São Carlos
São Carlos-São Paulo, Brazil

Since their introduction in the middle of the 1960s thermoplastic rubbers have steadily increased in their acceptance in the market. Applications of this new class of polymers include shoe soles, adhesives, reinforcement for bitumen, and substitutes for conventional vulcanized rubber products. Their specially designed polymer chain structure allows a rubbery behavior at ambient temperatures as well as a thermoplastic behavior at higher temperatures. So these materials can be processed at melting temperatures as a conventional thermoplastic avoiding the costly and time-consuming vulcanization step common in rubbers.

In the first section a general introduction to the structure and properties of thermoplastic rubbers emphasizes the rheological, optical, and dynamic mechanical properties. An analysis of microphase separation theories is also included.

Thermoplastic rubbers have been shown to have unusual solid-state properties, which must derive from the structure of the melt prior to solidification. In the second section the melt phase is studied in some detail. The molecular architecture of these block copolymers is comprised of hard segments (usually polystyrene) connected by a flexible rubbery chain (polybutadiene or polyisoprene) in a linear or radial structure. Their flow characteristics have been studied and the results correlated with measurements in the solid state. They have been modeled mathematically based on two particular theoretical models. A liquid phase transition was recorded for both models, with appreciable reduction in the apparent activation energy of flow above this temperature. The quality of the domain structure depends on the continuity of the polystyrene phase and has been measured by the stress at yield and by the optical birefringence. A change in response was associated with the melt transition. This melt transition is interpreted as the onset of loss in the long-range order of the domain structure, which is mainly due to the capability of the polystyrene end blocks, which, from this temperature upward, can be removed from their base phase by the attached rubbery chain.

It has therefore been possible to delineate three different processes in the polymer melt in order of ascending temperature: internal stress relaxation, rearrangement of the domain structure, and crosslinking in the rubbery phase. The respective temperature regions are approximately 100°C, 140–186°C (depending on the structure), and greater than 200°C respectively. The transition temperature is also shown to be directly related to the molecular weight of the hard-block segments, supporting the molecular model proposed in the third section.

The fourth section presents some uses and applications of thermoplastic rubbers and the fifth compiles the main conclusions.

INTRODUCTION

Background

Polymer

A *polymer* is a macromolecule which has a simple repeating unit, known as a *mer*. Polymers can be found occurring naturally, as cellulose and natural rubber, or they are produced synthetically as polyethylene, polypropylene, polystyrene, etc. Their physical properties differ so much that their behavior can range from a very flexible and elastic (rubbers) to rigid and strong (plastic) [1].

The rubber industry has developed from the days of Goodyear (1839) to a modern industry based on the knowledge of restricting the rubber chain flow by the formation of chemical crosslinks between chains (vulcanization). The need for incorporating additives dictates the use of several technical steps (mixing and curing) which are costly and time-consuming. Another disadvantage is the impossibility of recycling scrap materials since curing is irreversible. On the other hand, thermoplastic materials can be processed several times because of their ability to flow at high temperatures, but they lack the elasticity common to rubber products.

A new material, which would have the elasticity of the rubber compounds and also could be processed as many times as would be necessary, a thermoplastic rubber, was needed. This was made possible with the development of block copolymers beginning in 1965.

Block Copolymer

Block copolymers, as stated, show the unique behavior of being able to function as conventionally crosslinked elastomers over a certain temperature range (including room temperature), but they soften reversibly at high temperatures, so they can be processed as thermoplastic [2]. These materials are macromolecules comprised of chemically dissimilar, terminally connected segments [3]. According to the components that make up the segments, the thermoplastic rubbers can be classified as shown in Table 1 [4]. As seen in Table 1, the most common type by far is the styrenic in which the thermoplastic blocks are made of polystyrene chemically linked to a rubbery block (polybutadiene or polyisoprene).

A copolymer can be classified according to its structural arrangement as follows:

1. *Random,* in which the disposition of the components in the polymeric chain is irregular.
2. *Graft,* in which branches of one homopolymer are grafted (linked) to the main chain of another homopolymer.
3. *Block,* in which the polymeric chain is built up of blocks of different homopolymer linked by chemical bonds.

The various blocks can be assembled in several ways forming different structures, which can be summarized as follows:

Polymer	Structure
A–B	Diblock or two blocks
A–B–A	Triblock
$(A-B)_{x \geq 3}$	Multiblock or star-shaped block (x = mean functionality of branching); e.g., for $x = 4$,

$$
\begin{array}{ccc}
A & & A \\
\ \searrow & & \swarrow \ \\
& B \quad B & \\
& \times & \\
& B \quad B & \\
\ \swarrow & & \searrow \ \\
A & & A
\end{array}
$$

four-arm star-shaped block

"Tapered block"	The A(B) component concentration varies linearly from 100% at one end of the polymeric chain to 0% at the other end
A–B–C	ABC-type triblock

The letters A, B, and C represent the various homopolymer blocks (or monomer in the tapered block case).

Table 1 Types of Thermoplastic Rubbers

Type	Specific gravity range	Shore hardness range	Market share (%)
Styrenic	0.90–1.14	45A-53D	52
Olefinic	0.89–1.25	60A-60D	18
Polyurethane	1.10–1.34	70A-75D	19
Copolyester	1.13–1.39	35A-72D	6
Polyamide	1.01–1.14	75A-63D	

Source: Ref. 4.

Synthesis of Block Copolymers

There are four processes, depending on the initiator chosen, which yield polymeric chains with blocks well defined in composition, length, and structure [3].

Three-stage sequential addition process. Consider the process for synthesizing styrene–butadiene–styrene (SBS) triblock copolymer. It begins with initiation of styrene polymerization using anionic initiator to form living polystyryl anion, followed by addition of a butadiene monomer to form a living diblock, and finally introduction of a second quantity of styrene to complete the formation of the S–B–S structure. Alkyllithium (preferably *sec*-butyllithium) initiator is used for the polymerization, which is carried out in hydrocarbon solvents (benzene, cyclohexane, etc.) [5, 6]. It is essential that all active hydrogen impurities (water, alcohol, etc.) be scrupulously removed in order to obtain a termination-free system. This polymerization process possesses the considerable virtue of having its molecular weight governed simply by the ratio of monomer to initiator, since every initiator molecule is capable of starting one chain. Another important feature is the ability to achieve a very high degree of monodispersity due to the absence of termination.

$$X- + nA \rightarrow X{\sim}A-$$
$${\sim}A- + mB \ {\sim}AB{\sim}B-$$
$${\sim}AB{\sim}B- + nA \rightarrow {\sim}AB{\sim}BA{\sim}A-$$
$${\sim}AB{\sim}BA{\sim}A- + \text{terminator} \rightarrow A{-}B{-}A$$

Coupling process. This method differs from the previous approach, in that the three-block copolymer is made by coupling two living $A{-}B^{1/2}$ moieties, halfway through the polymerization. In this case the final step is not the addition of monomer but rather the addition of difunctional small molecules (phosgene, dicarboxylic acid esters, etc.) which can couple the diblock chains to form a three-block copolymer:

$$2A{-}B^{1/2} + \cdot X \cdot \rightarrow A{-}B^{1/2}{-}X{-}B^{1/2}{-}A \equiv A{-}B{-}A$$

This approach produces symmetrical end blocks and also avoids possible termination due to impurities in the third monomer addition.

Difunctional initiator process. In this method a difunctional initiator (sodium naphthalene, sodium biphenyl) producing a relatively stable dianionic moiety is capable of growing a living polymer at both ends, first polymerizing the diene monomer, and, while this is still active at both ends, the second monomer is added forming the second block at both ends:

$$-X- + nB \rightarrow -B{\sim}X{\sim}B-$$
$$-B{\sim}X{\sim}B- + nA \rightarrow -A{\sim}AB{\sim}BA{\sim}A- \equiv -A-B-A-$$
$$-A-B-A- + \text{terminator} \rightarrow A-B-A$$

Impurities can deactivate one or both ends of the growing chain, which results in diblock copolymer and homopolymer formation.

Tapered-block process. Owing to the fact that polymerization kinetics very strongly favor the addition of polystyryl anion to butadiene monomer rather than to styrene monomer, and polybutadienyl anion also adds to butadiene more rapidly than to styrene monomer, when a mixture of butadiene and styrene is polymerized a tapered diblock copolymer will be formed. The boundary between the two blocks in this case is very diffuse, and if isoprene is used instead of butadiene the boundary will be even more diffuse. This reduces the incompatibility of the two blocks taking the polymer away from a two-phase system with deterioration of the expected properties [7].

Structure and Properties of Block Copolymers

The special molecular structure of a block copolymer, i.e., a polymeric chain made of various blocks of different components, is bound to induce strong effects on its morphology and properties. In this section special attention is given to this matter with consideration of the domain structure, effects on the glass transition temperature and mechanical properties, equilibrium domain size, and temperature dependence.

Domain Structure

The incompatibility of most high molecular weight polymers, shown, for instance, in their blends, will be retained by the individual low molecular weight blocks that form the copolymer. This incompatibility between two portions of the same chain (the two blocks are chemically linked together) induces phase separation on a microscale which is of the order of approximately a few hundred angstroms [8].

The study of structures on this scale obviously calls for small-angle x-ray scattering (SAXS), small-angle neutron scattering (SANS), and electron microscopy [9, 10]. The first two techniques can determine the structural type and main lattice parameters if an ordered phase separation is present in the copolymer, but for a full assessment, electron microscopy is necessary. By using these techniques, it has been possible to establish that block copolymers containing between 10 and 90% of one block show preferably one structure among five possible structures [11, 12]:

1.   One lamellar structure.
2.   Two hexagonal structures or their inverted structure.
3.   Two cubic structures or their inverted structure.

Lamellar structure. When the ratio between the components is close to one (in fact the content of one component can vary from ~40 to 60%) the diffraction patterns in both scattering analyses (SAXS and SANS) give rise to sharp diffraction lines with a Bragg spacing, typical of a layered structure. This structure can be visualized as a set of plane parallel equidistant sheets. Each sheet results from the superposition of two distinct layers, each layer consisting mainly of one of the block copolymer components. Electron micrographs obtained by orthogonal ultrathin sections of a sample with lamellar structure show a striated structure formed by parallel stripes alternately black (containing the rubbery blocks stained by osmium tetraoxide) and white (containing the polystyrene blocks).

Striated structures will always result from sections of the lamellar structure in planes perpendicular to the planes of the sheets.

Hexagonal structures. The second family of diffraction patterns shows the diffraction lines with a Bragg spacing typical of a two-dimensional hexagonal array. Electron micrographs of stained samples with such a diffraction pattern allow one to distinguish two different structures corresponding to the same family of patterns, one in which white spots hexagonally spaced are immersed in a black background and the other in which black spots hexagonally arranged are immersed in a white background. In both cases micrographs of sections taken perpendicular to the orientation above have shown the same striated structure shown by the lamellar structure. This indicates the presence of a hexagonal cylindrical structure in which one of the phases is segregated in a form of cylinders immersed in a continuous matrix. According to the composition of the cylinder (and matrix) being one or the other component, two possible structures are presented. In the first case, the hexagonal structure, the polystyrene content varies from ~15 to 40% forming cylinders and the dominant rubbery phase fills the space, forming the matrix. When the polystyrene content is increased to ~60–85%, the situation inverts, forming the so-called inverted hexagonal structure, in which polystyrene forms the matrix and the rubber forms cylinders.

Cubic structures. The third possible family of diffraction patterns is characterized by the presence of sharp Bragg spacing lines typical of a cubic structure. Again electron micrographs of stained samples showed two different structures corresponding to the same family. In the first type, with styrene content lower than ~15%, the micrographs in any orientation of sectioning give white spots in a black background indicating a cubic structure. In the second case, for polystyrene content higher than ~85%, the situation is inverted, showing black spots in a white background, i.e., the inverted cubic structure in which the lower rubbery content segregates in spheres in the polystyrene matrix.

Thus far only linear block copolymers have been used as examples for the analysis of their morphologies, but star-shaped structures have also been studied and yield similar morphologies. Recently a wagonwheel type of arrangement was observed when the number of arms is equal to or greater than six [13].

The presence of two incompatible blocks in the same polymer chains is not enough to force a phase separation if the molecular weight is not high enough. Meier in his theoretical calculation [14] predicted that the critical molecular weight for domain formation in the styrene–butadiene block copolymer would be between 5000 and 10,000 when the polybutadiene block molecular weight is of the order of 50,000 (a detailed explanation of this theory will be given later).

Data presented by Holden et al. [15] show that the tensile strength of SBS block copolymers changes from 10.5 kg/cm^2 for (6–81–6) 10^3 molecular weight polymer to 236 kg/cm^2 for (10–53–10) 10^3 molecular weight, i.e., an increase of approximately 20-fold in the tensile strength when the polystyrene block molecular weight is changed only from 6000 to 10,000. Meier [14] interpreted their data as evidence for the onset of domain formation in this molecular weight range. Morton [16] from similar experimental data

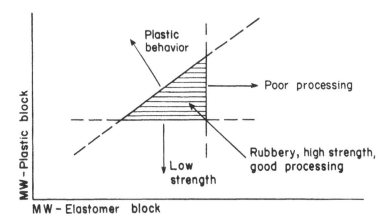

Figure 1 Effect of block molecular weight on properties of thermoplastic rubbers. (From Ref. 18.)

reaches the same conclusion. There is therefore a minimum molecular weight requirement for the end block which at the appropriate styrene content gives optimum rubberiness and strength and leads to a corresponding overall molecular weight in the finished commercial polymer below which satisfactory properties are not obtained [17]. This overall molecular weight determines the melt viscosity of the polymer and hence its processing characteristics.

A favorable balance between processing performance and satisfactory elastomeric response is an essential requirement for thermoplastic rubbers: molecular weights of the blocks of polystyrene and polybutadiene are critical for obtaining this balance. Figure 1 from Bishop and Davison [18] shows the molecular weight relationship diagrammatically: at too low a molecular weight of polystyrene blocks, the thermoplastic rubber has poor tensile strength; at too high a middle-block molecular weight it has poor processibility. Also, nonelastic behavior is obtained if the ratio of end-block to middle-block molecular weights is above a given value [18].

Equilibrium Size of the Domain Structure

As listed in the review of Wilkes and Stein [9] the major rheooptical techniques employed in the study of block copolymers are small-angle x-ray scattering (SAXS), small-angle neutron scattering (SANS), linear birefringence (LB), linear dichroism (LD), and others.

Small-angle x-ray scattering has been one of the most used techniques in which the domain spacing of the periodic arrangement is obtained. A number of papers deal with the basic theory applied to block copolymers [19, 20] and comparison of experimentally measured domain spacing (SAXS) and theoretical predictions [20–22]. From plots of the domain spacing as a function of the total molecular weight the spacing has been shown to increase systematically with total molecular weight [22]. The linear relationship has a slope close to 2/3, which is substantially greater than the power 1/2 expected for the molecular weight dependence of the unperturbed chain dimension. This is a consequence of the chain in the domain space expanding due to the constraint that chains are restricted to their own domains and maintenance of a uniform density by chain adjustments. This chain expansion depends on molecular weight, thus resulting in a power greater than 1/2.

Small-angle neutron scattering (SANS), another powerful technique, has only recently emerged as a tool for the study of block copolymer structures [23, 24]. Hadziioannou et al. [24], studying the chain conformation of two lamellar styrene–isoprene diblock copolymers by SANS, detected that the experimentally measured mean size of the polystyrene blocks falls between those calculated from two extreme models:

1. The Gaussian model, where the blocks can spread laterally far from their anchoring points and display a pronounced degree of interpenetration and entanglement with neighboring chains.
2. The compact cylindrical model, where the chains while disordered locally pack without interpenetration to any appreciable extent.

Measurements using SANS also proved to be more accurate than transmission electron microscopy in measuring domain spacing. Berney et al. [25] drew attention to the consistent discrepancy in the (sphere) sizes of diblock copolymers ($74\% < S\% < 90\%$ wt) measured by these two techniques. The mean radius of the polybutadiene sphere as measured from the electron micrographs was found to be significantly smaller than the SANS values ($R_{EM}/R_{SANS} \simeq 0.77$), a greater discrepancy than can be accounted for by sphere truncation when sectioning the samples for the EM.

Binary blends of di- and triblock copolymers showed that quality of the organization and the sharpness of the interface were not affected by the mixing, which was said to occur on a molecular level [26]. The transition from lamellar to cylindrical structure has been shown to be dependent primarily on the chemical composition of the blend and the transition from lamellar to the single phase, at room temperature, depends primarily on molecular weight. Ternary blends of an A–B diblock copolymer with the A and B homopolymers suggested that the mechanism of domain formation in a ternary system is limited by the relative molecular weight of the homopolymers to those of corresponding blocks of the copolymer for the solubilization of homopolymers into the respective block domain [27].

The morphology of ABC-type triblock copolymer has also been studied. Shibayama et al. [28] synthesized a styrene-(4-vinylbenzyl)–dimethylamine–isoprene ABC-type triblock copolymer and, using SAXS and electron microscopy, analyzed its morphology. Each block chain segregated into respective microdomains, which resulted in the coexistence of three phases in the solid state. The three-phase structure could be directly observed by TEM of stained ultrathin sections (osmium tetraoxide and phosphotungstinic acid for selective staining the isoprene and the amine phase respectively). The typical morphologies presented were first a spherical domain of polyisoprene dispersed in the amine phase, which forms a lamellar structure with the polystyrene phase, and second an alternated trilayer lamellar structure of the three components.

The formation of a domain structure in block copolymers is not only possible in the solid state, as already seen, but also in solution in nonselective (neutral) solvents in concentration as low as 20% wt of polymer [29–31]. It was shown that there are two regimes in the concentration and temperature dependence of the domain size [30]:

1. The equilibrium regime, where the size is thermodynamically controlled, i.e., the greater the segregation power between the two blocks (the higher the concentration or the lower the temperature) the larger the size.
2. The nonequilibrium regime, where the size is controlled kinetically. The average distance between neighboring chemical junction points along the interface or the

average number of block chains per spherical domain cannot follow equilibrium but is fixed more or less to a constant value. Consequently, the domain size decreases with increasing concentration and decreasing temperature simply due to the effects of deswelling and thermal contraction, respectively.

The change from one regime to another as the polymer concentration increases is dependent on the morphology of the cast polymer; that is, for the lamellar structure, close to ~50% wt of polymer (in a different publication ~70% wt is quoted [30]), and for the spherical structure, approximately ~30% wt of polymer.

Effect of Temperature on Domain Structure

Heating phase-separated block copolymer has been found to affect the structure significantly and many researchers have converged on this field. Annealing at low temperatures (<120°C) apparently does not produce great changes [23] in the copolymer structure in contrast to heat treatment at higher temperatures (>180°C) [32–36].

Hadziioannou and Skoulios [32] have shown that an oriented lamellar styrene–isoprene diblock copolymer (M_n 98,000, 54% wt styrene) under heating changed its domain spacing. The spacing increases slightly with temperature in the range from 20°C up to 110°C then decreases to the initial value and then, at about 180°C, begins to increase again with temperature, very rapidly attaining values twice as large as those quoted at room temperature. On further heating above 250°C the x-ray patterns became devoid of any diffraction signal. Two possible reasons were given: the diffraction lines may have gathered in the very low Bragg angle region where they can hardly be detected or the mesomorphic phase may have become molten. On subsequent lowering of the temperature the x-ray patterns remained blank, evidence for the loss of the single-crystalline orientation of the sample. This work was extended to various other di- and triblock copolymers in a further publication [33]. It was found that the disappearance of the lamellar structure on heating proceeded through an irreversible stage where the lamellar thickened considerably. Hadziioannou and Skoulios concluded, contrary to any classical thermodynamic expectation, that the "melting temperature" (~180°C in all cases) did not depend in a decisive way on molecular weight or on structure.

Hashimoto et al. [34] using SAXS of a styrene–isoprene "tapered" block copolymer (M_n 43,000, 47% wt styrene) showed that the diffuse lamellar microdomain structure scattered the x-rays at room temperature corresponding to a domain spacing of 260 nm. On heating the scattering maximum disappeared when the temperature was raised above ~170°C and appeared again, at the same scattering angle, with decreasing temperature. The process was reversible, the transition temperature being much lower than that expected for a normal diblock copolymer (>220°C) due to the broad interface common to tapered block copolymers.

Typical pressure-sensitive adhesives prepared from star-shaped block copolymers of butadiene or isoprene with styrene maintain the domain structure of the block copolymer well beyond the polystyrene domain T_g but form homogeneous melts above a critical temperature [35]. Two different types of behavior were presented for each case. First, the SAXS scattering maximum presented at room temperature by the isoprene-based block copolymer on heating disappeared somewhere between 200 and 230°C. Second, the butadiene-based block copolymer showed that the original scattering maximum tended to disappear with heating but near 180°C a new maximum appeared, which was then stable to temperature at least as high as 230°C. This apparent change in morphology was irreversible; on cooling the spacing remained constant. On reheating the new peak was maintained and "melting" was far from complete.

The effect of heating on the microdomain structures of block copolymers cast from solution was investigated by Fujimura et al. [36]. The radius of the sphere and interdomain distance were observed to increase slightly with increasing temperature from room temperature to 180°C. The original domain structure was concluded to be in a non-equilibrium state in that the number of block polymer molecules per domain in the real system was far less than in the equilibrium state. As the system is heated it tends to approach equilibrium, resulting in an increased number of block polymer molecules per domain and therefore in increased interdomain distance and size of the domain.

Glass Transition Temperature

Meyer and Widmaier [37], working with a triblock copolymer of styrene–isoprene, found that one, two, or even three glass transition temperatures could be detected by DSC. The samples were divided into two sets: the first had different total molecular weights and the same (30% wt) polyisoprene content; for the second, the molecular weight was kept constant (30,000) and the elastomer content varied. The T_g of the homopolystyrene was measured and confirmed to vary asymptotically with the molecular weight, starting from around ~40°C (M_n 2500) and increasing rapidly to ~90.5°C (M_n 32,000) with increasing molecular weight. The variation in the T_g of homopolyisoprene was less pronounced, but nevertheless increased linearly with increasing molecular weight. In the case of the block copolymers the T_g showed a shift (reduction in the T_g of the polystyrene phase and increase in the T_g of the polybutadiene phase) compared to the corresponding homopolymers (of similar molecular weight). The third T_g, intermediate between the other two, was interpreted in terms of the existence of an interphase. For a theoretical evaluation of this T_g, it was assumed that the interphase would be homogeneous and would have, on the average, the same composition as the block copolymer. Three theoretical models elaborated for a binary mixture of homopolymers A and B or for random copolymers were used:

1. Fox's model [38], based on free-volume considerations, which gave

$$\frac{1}{T_g} = \frac{X\mathrm{A}}{T_g(\mathrm{A})} + \frac{X\mathrm{B}}{T_g(\mathrm{B})}$$

   where X is the weight fraction.

2. Dimarzio and Gibbs's [39] entropic theory:

$$T_g = \frac{\phi_\mathrm{A} T_g(\mathrm{A}) + k\phi_\mathrm{B} T_g(\mathrm{B})}{\phi\mathrm{A} + k\phi\mathrm{B}}$$

   where ϕ is the volume fraction and k is a constant (~ 1).

3. Pochan et al. [40], also from thermodynamic considerations, proposed an expression for T_g:

$$\ln T_g = \phi_\mathrm{A} \ln T_g(\mathrm{A}) + \phi_\mathrm{B} \ln T_g(\mathrm{B})$$

Unfortunately the calculated values did not coincide with the experimental temperatures but were considered to be satisfactorily close to them, especially for the Dimarzio and Gibbs model.

A similar approach was employed by Lu et al. [41] using diblock copolymers of styrene–dimethylsiloxane (S-DMS) blended with polystyrene. For both components various samples with different total molecular weight were used. The glass transition temperature was determined by DSC and plots of refractive index as a function of temperature (n vs. T). A single glass transition temperature was observed, in the range of the PS T_g, with a value between that of the polystyrene sample and that of the styrene phase in the copolymer. The T_g of the blends obeyed the Fox model very closely for very large or very small ratios of polystyrene molecular weight to styrene block molecular weight, either ≥ 2.5 or ≤ 0.23.

In a later publication the same group of researchers [42] measured the glass transition temperature in styrene–butadiene di- and triblock copolymers with various molecular weights using refractive index vs. temperature and DSC measurements (these techniques, using SBS, had been the subject of an earlier publication from these authors [43]). All block copolymers investigated having styrene block molecular weight $<22,000$ showed a T_g at least 18 K less than of homopolystyrene with comparable molecular weight. The change in specific heat at the T_g of the microphase was always equal to or greater than that of the homopolystyrene. These results were reported to suggest that some butadiene blocks were mixed into most of the styrene microphase examined, in contrast to earlier results on styrene–dimethylsiloxane (S-DMS) diblock copolymers, where mixing was postulated only at styrene block molecular weight <8200.

Mechanical Properties

The stress–strain properties of SBS block copolymers have been the subject of many publications. Fischer and Henderson [44] showed the reduction in modulus, tensile strength, and elongation at break with increasing temperature (from 25 to 90°C) of SBS with 40% styrene. The stress–strain results indicated that during extension at low strains, much of the load was borne by the continuous phase of polystyrene. Upon further extension the process of yielding occurred with the breakdown of the continuous polystyrene phase and the material became much more flexible and elastomeric. Yielding was accompanied by transfer of a significant fraction of the load from the polystyrene phase to the polybutadiene phase.

The influence of the processing history on the final properties has long since been recognized [45]. The elastic modulus on an extruded SBS block copolymer (30% styrene content) increased in samples in the following sequence: the lowest modulus was obtained across the flow direction of injection moldings, then for compression-molded samples, and finally along the flow direction in injection-molded materials. The modulus also increased in samples prepared at increasing shear rates. It was stated that the material keeps in memory the processing steps to which it had been submitted.

Nandra et al. [46], working with SBS thermoplastic rubbers, found that the tensile strength of injection-molded samples was higher in the transverse direction than in the direction of flow. The explanation given was that the oriented structure (in the flow direction) of the polystyrene phase was reoriented into the stress direction when loading was in the transverse direction. When the specimen was stretched in the transverse direction the strain was readily accommodated by the polybutadiene phase and the reinforcing polystyrene rods were ultimately turned through 90° to reinforce the total structure. Thus at high strain the polystyrene contributed its full strength and the structure was reinforced by the reoriented polystyrene rods giving higher overall strength than in the flow direction.

The addition of polystyrene of high molecular weight to reinforce SBS resins has also been attempted [46, 47]. The hardness of the compound initially reduces with addition of the reinforcing phase (PS), decreasing to a minimum (at ~ 15 phr) and then increasing steadily. The addition of a high molecular weight polystyrene may extract some of the domain polystyrene and destroy the lamellar structure which is partially responsible for the hardness of the SBS elastomer. Hence there is a reduction in hardness. As the polystyrene content increases above 35 phr it dominates the material and leads to an increase in the hardness above the original value [47].

The elastic moduli of an oriented SBS sample with the polystyrene phase as cylinders dispersed in a rubbery matrix can be estimated using the Takayanagi model [48]. The elastic modulus parallel to the orientation direction is given by

$$E_0 = \phi_S E_S + \phi_B E_B$$

where ϕ is the volumetric fraction of the rubbery (B) and glassy (S) phases.

The perpendicular direction, assuming $E_S \gg E_B$, is

$$E'_{90^\circ} = \frac{E_B}{\phi_B}$$

In fact, because the rubbery matrix stretched along $\theta = 90^\circ$ will be prevented from contracting along the direction $\theta = 0^\circ$ by the polystyrene cylinders its true modulus is given by [49]

$$E_{90} = \frac{4}{3} E'_{90} = \frac{4}{3} \frac{E_B}{\phi_B}$$

The restriction on the contraction of the rubber due to the cylinders is reduced to zero at $\theta = 55^\circ$, giving

$$E_{55} = \frac{E_B}{\phi_B} = \frac{3}{4} E_{90}$$

The experimental data agree very well with the theoretical calculations.

Another interesting behavior shown by SBS thermoplastic rubbers is the stress soften-ing which initially was thought to be due to the disruption of the polystyrene continuous phase [50, 51], a conclusion later confirmed by Pedemonte et al. [52]. Their conclusion is supported by two kinds of experiments. The first is stress–strain measurements on speci-mens previously swollen in *n*-heptane, in which the stress-softening effect disappeared as soon as the structure of the material was modified completely by the treatment. The second is direct observation, by electron microscopy, of the structure after the deforma-tion in which the original continuous and regular polystyrene rods aligned in the flow direction of the oriented sample assume a "string of pearls" structure. The stretch causes the formation of a large number of thin ties along the axis of the polystyrene rods deforming the cylinders, and upon further stretching the cylinders break down. For sample with a lamellar structure it is suggested [21, 53], that at the initial stage of deformation, the alternating layers of polystyrene and polybutadiene are stretched apart a little. Increased strain deforms the polystyrene regions into continuous kinks and fractures them into short segments, then finally into small dispersoids embedded in the polybuta-diene matrix. At this point the sample essentially acts as a filled rubber. The same conclusions were reached by Odell and Keller [54] with the introduction of a random rod-breaking theory. Later a more detailed analysis by Séguéla and Prud'homme [55] using small-angle x-ray scattering showed that the deformation mechanism, during stretching of a solvent-cast SBS block copolymer, leads to anisotropy. The polystyrene phase initially randomly rotates to form lamellar grains at an angle close to 20° with respect to the stretching direction in a reversible process. When oriented SBS samples are studied by the same technique [56], the SAXS patterns obtained at various draw ratios indeed indicate that the oriented cylindrical microdomains break and reorient themselves to an angle of 30° to the stretching direction, whatever the original orientation direction. In conclusion, SBS block copolymers at small deformation behave as typical composite materials in which the morphology and properties of the microphase control the mechanical behavior. At high deformation, however, the orientation of the molecules, which originates from the molecular bonding of the constituent polymers forming the microdomains, controls the mechanical behavior and final morphological state, in contrast to physical composites.

On releasing after stretching, the stress-softened sample reversibly recovers its original properties with time and temperature (heals). This suggests that the fragmented system has an excess of free energy relative to the original ordered state owing to [21]:

1. Orientation of the polybutadiene chains, giving rise to decreased entropy.
2. Enormous increase of the interfacial surface area in the fragmented system, giving rise to increased interfacial energy.

Fracture studies on oriented samples with high styrene content (>70% wt) also showed its high anisotropy [57]. These samples are formed by rubber cylinders dissolved in a styrene matrix. When the cracks propagate in a plane perpendicular to the axes of the rubber cylinders or parallel to flow, the fracture is either brittle or ductile accordingly. The brittle fracture surface showed a pattern of alternating rough and smooth bands, characteristic of pure polystyrene. On the other hand, the ductile fracture surface is composed of an array of elongated "chips" arranged practically parallel to the fracture propagation direction.

Infrared Spectroscopy

Infrared spectroscopy has emerged as a powerful tool for studying the chemistry and physics of polymers. Analysis of characteristic group frequencies in spectra allows qualitative and/or quantitative estimates of type and of chemical composition in polymers and copolymers. Structural factors such as branching or crosslinking, steric and conformational order in polymer chains, crystallinity and copolymer sequence distribution are some of the characteristics that can be studied using the IR technique. In the copolymer field infrared methods have been used to investigate domain morphology, changes in domain structure during deformation, and the relationship between domain behavior and physical properties [58].

By using polarized infrared radiation the phenomenon of infrared dichroism can be utilized for an oriented sample. This is a difference in IR absorption by the chemical bond due to its spatial orientation with respect to the polarized beam. Maximum absorption will occur when the polarization plane of the IR beam coincides with the chemical bond and minimum absorption will happen when they are perpendicular [59]. Experimentally the two IR spectra are taken from the oriented sample having its orientation at 0° (parallel, $\parallel$) and 90° (perpendicular, $\perp$) to the plane of polarization of the scanning IR beam (a differential arrangement can be made using two polarizers and the difference spectrum recorded in one run). The two absorptions $A_\parallel$ and $A_\perp$, parallel and perpendicular respectively, at a particular frequency, are measured and their ratio is calculated. This ratio is defined as the dichroic ratio D [59]:

$$D = \frac{A_\parallel}{A_\perp}$$

As the orientation of the chemical bond with respect to the polarized IR beam is of primary importance, the orientation of a particular bond or group of atoms can be detected and evaluated. This is done by the orientation function (f), which is a function of the dichroic ratio as

$$f = \left| \frac{D_0 + 2}{D_0 - 1} \right| \cdot \left| \frac{D - 1}{D + 2} \right|$$

where D_0 is the dichroic ratio for perfect alignment.

The same function can be presented as

$$f = \frac{1}{2}(3 \cos^2 \theta - 1)$$

where θ is the average angle of disorientation. This function has a value of unity for a sample whose elements are completely oriented in the stretch direction, $-\frac{1}{2}$ for the perfect orientation transverse to the stretch direction, and zero for a random orientation.

Using this function, orientation studies in particular portions of the polymeric chain have been done for polethylene [60], polyethylene terephthalate, nylon [61], and polystyrene [62], among others.

In styrene–butadiene copolymers, the above technique cannot be utilized to its full potential because these copolymers show a dichroism effect so weak it was initially thought to be nonexistent [63]. Nevertheless, a number of absorption bands, mainly around 3000 cm^{-1}, associated with the polystyrene phase, were found to be weakly dichroic showing that the polystyrene chains have a preferential orientation normal to the lamellar surface [64].

Knowledge of the ratio between the isomers in the rubbery phase is also very important; this is determined mainly by normal infrared analysis. For sample with a butadiene rubbery phase the following equation can be used to calculate the percentage amount of each isomer in the sample [65]:

$$\text{wt \% isomer} = \frac{A_{\text{isomer}/\epsilon}}{A_{970/86} + A_{909/120} + A_{735/25}}$$

where the absorptions at 970, 909, and 735 cm^{-1} are due to the trans-$1:4$, $1:2$, and cis-$1:4$ isomers respectively. The extinction coefficients ϵ are already provided for each isomer.

In the case of isoprene the characteristic bands are 907, 886, and 864 cm^{-1} for the $1:2$, $3:4$, and $1:4$ isomers. In both cases the presence of various adjacent polystyrene absorption bands interfere and it is necessary to do the calculation for a sample free of polystyrene (obtained during the sample polymerization, removed just before the addition of the styrene monomer) [66].

Gas Permeation Properties

In contrast to the fact that research in the morphology, mechanical properties, and structure–property relationship of block copolymers has been extensive, the transport properties of small molecules has been the subject of few studies.

Odani and co-workers [67] studied the permeation (diffusion, solution, and evaporation) behavior of a series of inert gases (helium, argon, nitrogen, krypton, and xenon) in styrene–butadiene block copolymers. Two morphologies were considered, polystyrene cylinders in a polybutadiene matrix and lamellae. It was shown that for temperatures below the T_g of the polystyrene phase, the gas permeation process (except for helium) is governed primarily by behavior in the polybutadiene matrix. The diffusion behavior in the copolymer was compared with that in homopolybutadiene and the conclusion was that segmental motions in the polybutadiene phase of the copolymer are restricted relative to motions in homopolybutadiene. The reason for the restriction is the chemically joined glassy polystyrene chain. Also, from data on gas sorption in samples of various contents, involving both block copolymers and binary mixtures with homopolystyrene (MW 17,500), it was suggested that the partial mixing of component block chains occurs at the interface between the domains, resulting in rather diffuse domain boundaries. These conclusions were confirmed in a later publication [68].

Theories of Microphase Separation

Since the experimental verification that block copolymers separate to form domain structures, many theories have been developed to explain qualitatively and quantitatively the characteristics of these domains. The simplest way to estimate the periodicity D of the domain structure with respect to the molecular weight of the chain is from the statistical distribution of ideal chains [69]. The root-mean-square end-to-end dimension $\langle \bar{r}^2 \rangle^{1/2}$ for a tetrahedrally bonded chain with n bonds of identical length l is

$$\langle \bar{r}^2 \rangle^{1/2} = \sqrt{2n}\, l$$

in our case

$$n = \frac{\text{molecular weight of the polymeric chain}}{\text{molecular weight of the monomer}} = \frac{M_n}{M_m}$$

so

$$\langle \bar{r}^2 \rangle^{1/2} = \sqrt{2\frac{M_n}{M_m}}\, l = l\sqrt{\frac{2}{M_m}}\, M_n^{1/2}$$

but

$$l\sqrt{\frac{2}{M_m}} = \text{const} = K \quad \text{and} \quad \langle \bar{r}^2 \rangle^{1/2} = R$$

where R is the radius of gyration of the chain

$$R = KM_n^{1/2}$$

By assuming simply the periodicity of the mesophase D equal to the radius of gyration, D can be related to the polymer molecular weight as

$$D - KM_n^{1/2}$$

This principle was first used by Meier [14, 70] for the development of a theory for the formation of spherical domains in A–B block copolymers. This theory established criteria for the formation of domains and their size in terms of molecular and thermodynamic variables. It was shown that the considerable loss in configurational entropy due to the constraints on the spatial placement of chains in a domain structure requires that the critical block copolymer weight required for domain formation is many times greater than is required for phase separation of a simple mixture of the component blocks. The relationship between domain radius R and molecular weight M was given as

$$R = \frac{4}{3}\langle \bar{r}^2 \rangle^{1/2} = \frac{4}{3}\alpha K M^{1/2}$$

where K is a constant and α is the ratio of perturbed to unperturbed chain dimensions. A method to evaluate α was also provided.

The first attempt to develop a microstructural model for the phase-separated states of a triblock copolymer was proposed by Leary and Williams [71–73]. The theory predicted the existence of a first-order phase transition, characterized by a separation temperature T_s. On cooling from the homogeneous melt state the phases separate at T_s forming the two-phase domain structure. Although the model incorporates, with the two pure phases, a mixed region it fails to predict T_s with any accuracy [72]. In a 1985 publication Henderson and Williams [74] extended the previous theory in terms of the interphase profile. In addition to the highly unrealistic original step-function profile, four other profiles were considered (linear, cosine squared, and two cases of hyperbolic tangent). The repeat distance (periodicity) D was found to vary with the total molar volume V. (As the densities of the components are close to one, it is safe to extrapolate the V-dependence of temperature to the M-dependence as

$D\alpha V^{0.75}$ at low V ($D\alpha M^{0.75}$ at low M)

$D\alpha V^{0.5}$ at high V ($D\alpha M^{0.5}$ at high M)

Helfand [75–77] successfully developed a theory of the microdomain structure in systems of block copolymers which has been used as the basis for all further developments. The statistics of the macromolecular conformations were described by the free energy and concentration distribution. In his general theory the mathematical equation involved extensive computer calculations but considerable simplification occurred when, in subsequent publications with Wasserman [78, 79], an approximation was introduced: the interphase width was considered narrow compared to the domain size, creating the narrow interphase approximation theory. With that, the contributions to the free energy could be identified, being mainly interfacial tension, localization of block joints in the interphase, and entropy loss arising from preferential selection of conformations which keep the density uniform. Equations to predict the domain periodicity distance D for a lamellar structure were given, relating to the total molecular weight M_n as

$$D \ \alpha \ M_n^{0.643}$$

The same theory was applied and developed for the other two cases in which the block copolymer separates into spherical [80] and cylindrical [81] domains.

Hashimoto et al. [22] obtained experimentally the periodicity D as a function of total molecular weight for diblock copolymers. Their data, added to those of various other authors, were presented in one plot with a theoretical curve calculated from the theory of Helfand and Meier. The experimental data agree very well with the theoretical prediction in which the spacing D varies with the total molecular weight to the 2/3 power.

The third theory of microphase separation in block copolymers is due to Leibler [82]. He developed a microscopic statistical theory of phase equilibria in noncrystalline block copolymers of type A–B. In particular, the ordered mesophase from a homogeneous melt was studied and a criterion of microphase separation was presented. It was shown that under certain critical conditions a specific unstable mode appears in the homogeneous copolymer melt introducing the microphase separation transition (MST). After the MST a mesophase with the symmetry of a body-centered-cubic (bcc) lattice should appear. For a large range of compositions a bcc mesophase was expected to be a metastable phase. Only two other ordered phases, a hexagonal and a lamellar mesophase, were said to be stable near the MST.

For a lamellar two-block copolymer with composition, f, the periodicity of the mesophase, occurring just below the MST, was expected to be:

$$D = \frac{2\pi}{\sqrt{X^*}} R$$

where $X^* = q^{*2}R^2$ and a diagram of it as a function of composition is presented.

$q^* =$ wave vector at the phase transition

For example, for a 50% composition ($f = 0.5$) the value of X^* is 8.4, so the predicted periodicity is

$$D = \frac{2\pi}{\sqrt{8.4}} R = 2.168 K M_n^{1/2}$$

which is very similar to the behavior predicted by the statistical distribution of ideal chains.

Figure 2 [83] shows the variation of the experimentally measured polystyrene sub-

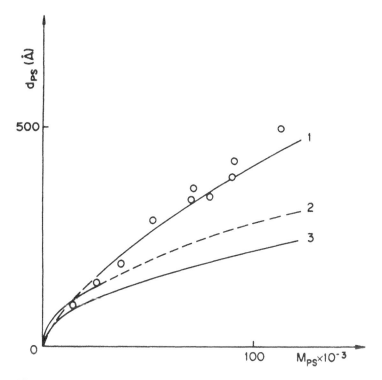

Figure 2 Polystyrene sublayer thickness measured experimentally compared with three major theories: curve 1, Helfand; curve 2, Leibler; curve 3, statistical distribution of ideal chains. (Reprinted with permission from Macromolecules, 1982, *15*, 258. Copyright 1982, American Chemical Society.)

layers in diblock copolymers of styrene–isoprene (SI) as a function of the molecular weight of the corresponding blocks, in comparison with theoretical predictions: curve 1 was calculated from Helfand's theory, curve 2 calculated from Leibler's theory, and curve 3 from the statistical distribution of ideal chains theory.

The thermodynamics of phase separation for star-branched block copolymers $(A–B)_n$ were analyzed by Bauer and Fetters [84] and the results were compared with those for diblock copolymers. The free energy was calculated when the free ends of the B blocks were joined to form the star-branched material. It was shown that the difference in free energy of phase separation between star-shaped and diblock copolymers is given by

$$\Delta G = N_d kT \left| \frac{x-1}{x} \right| \cdot (\ln \phi_B - 0.14)$$

where N_d is the number of diblocks, x is the degree of branching, and ϕ_B is the volume fraction of the lamellar B domains.

Hashimoto [85], on experimental grounds alone, derived a scaling rule relating microdomain size with average molecular weight of binary mixtures of block copolymers. The lamellar identity period (periodicity) D was found to be

$$D \propto \bar{M}_n^a$$

where $2/3 \leq a \leq 4/5$ $(0.667 \leq a \leq 0.8)$. The result was a generalized form of the one used for single-block copolymers, in which $\bar{M}_n$ is the number average molecular weight of the block copolymer mixture.

Rheological Properties of Block Copolymers

General Results

Rheological properties have been used for many years to study and characterize the structure of polymers. Kraus and co-workers [86–88] applied them to block copolymers. Starting in 1965, they studied the effect of introducing one or two long-chain branches in polybutadiene homopolymer to form trichain and tetrachain molecules respectively [86]. At low molecular weights the Newtonian (zero shear) viscosity was lower relative to a linear polymer of the same molecular weight, but at molecular weights exceeding 60,000 (trichain) and 100,000 (tetrachain) the Newtonian viscosity increased rapidly above the corresponding value for a linear polybutadiene. However, non-Newtonian behavior of the branched polymers became more pronounced the higher the molecular weight, so that at moderate to high shear rates, the viscosity of the branched polymer is uniformly lower than that of a linear polymer of similar molecular weight. The explanation given was that at low molecular weights, when the branches of the multichain polymers are not much longer than the entanglement molecular weight (~5600), flow occurs with little coupling between molecules. The principal factor governing viscosity is then the size of the polymer coil, which is smaller for the branched molecules. Ultimately, however, entanglements of branches belonging to different molecules appear to lead to extensive coupling as the branch length is increased.

A second publication [87] reports a study of random and block copolymers of butadiene and styrene. The random copolymer of constant composition along the polymer chain exhibited behavior similar to linear polybutadiene: the flow was Newtonian at low shear stresses and the flow curves for various temperatures could be superimposed. In a random copolymer varying in composition along the polymer chain, non-Newtonian behavior was more pronounced, and temperature–shear rate superposition was not successful, a trend further demonstrated in copolymers with a single long styrene block sequence. With two styrene blocks it was said that below the glass transition temperature of the polystyrene, association of the chains produces a network structure. Further, there was evidence that some of the associations persist at temperatures well above the polystyrene T_g.

A few years later (1971) the various structures for symmetrical linear and star-branched block copolymers of styrene and butadiene were available and Kraus et al. [88] studied the steady flow and dynamic viscosities of B–S–B, (B–S)$_3$, S–B–S, (S–B)$_3$, and (S–B)$_4$ copolymers. It was found that at constant molecular weight and total styrene content viscosities were greater for polymers terminating in styrene blocks, irrespective of branching. Branching decreased the viscosity of both polybutadiene-terminated and polystyrene-terminated block polymers, compared at equal M_w. However, comparisons at equal block lengths showed that the length of the terminal block, not the total molecular weight, governs the viscoelastic behavior of these polymers. This unusual result was attributed to the two-phase domain structure present in these copolymers, which persists to a significant degree in the melt.

Meyer and Widmaier [89] working with S–I and S–I–S (styrene–isoprene) di- and triblock copolymers noticed that they could be linked together by a "linking agent" forming a multiblock structure. Compared with the corresponding linear species, star-shaped block copolymers showed improved properties in adhesives. Coupling introduced chemical junction points which led to a multiblock structure of the linear copolymers and to an increased chain entanglement. As a result, cohesive strength was improved. In star-shaped copolymers, the important parameters are the dimensions of branches and their number.

Sivashinsky et al. [90], using a melt elasticity tester [91], studied the shear stress and relaxation curves of SBS block copolymers. The stress transient following a sudden imposition of shear flow showed a pronounced maximum corresponding to the yield point. Its value was not only a function of the test temperature but also of the solvent(s) from which the sample was cast. THF/MEK consistently gave a slightly higher stress than cyclohexane. Prolonged heat treatment at a temperature higher than the test temperature (7 hr at 150°C and then testing at 119°C) affected the results significantly; the reason given was that the rheological properties could be rationalized by considering heat-induced morphological rearrangement in the sample.

The reduction in the melt viscosity of triblock copolymer blended with a low molecular weight polybutadiene (3000) and an SBR (3600) has also been studied [92]. The latter was much more effective in producing the same effect as the polybutadiene at only a tenth of the quantity (1% SBR = 10% polybutadiene). The reduction in the steady-state viscosity is explained in terms of the plasticization of the interphase. The random copolymer is more compatible than the homopolymer polybutadiene in the interphase styrene–butadiene.

The Arnold–Meier Flow Mechanism

Arnold and Meier [93] in 1970 reported unusual behavior shown by triblock copolymers of styrene–butadiene with different end and midblock sizes. The dynamic viscosity as a function of frequency at several different temperatures exhibited two distinct regions: a high-frequency region where the response was typical of a thermoplastic showing a non-Newtonian flow and a low-frequency region where the viscosity increased with decreasing deformation rate without reaching a steady state. The unusual nature of SBS copolymer was illustrated by comparison with a monodisperse polystyrene with similar total molecular weight. Unlike SBS the viscosity for polystyrene is constant (and equal to the steady-state Newtonian viscosity) at low frequencies and decreases at high frequencies in the characteristic manner of thermoplastic. From the SBS data it was not possible to determine a Newtonian or zero-shear rate viscosity. An explanation for the flow mechanism was then proposed.

Due to the presence of the domain structure, a three-dimensional network will be formed in the block copolymers and their rheology will be dominated by the interplay of processes tending to disrupt and to reform the network and domain systems. The block copolymer can exist in three distinct states which depend on the rate of deformation. The stable at rest or at very low deformation rates will be a state where the molecular network is essentially intact. This state is shown schematically in Figure 3*a* for a semicontinuous domain structure and in Figure 3*b* for dispersed domains. At intermediate deformation rates, the three-dimensional network will be disrupted. Domains and aggregates of many

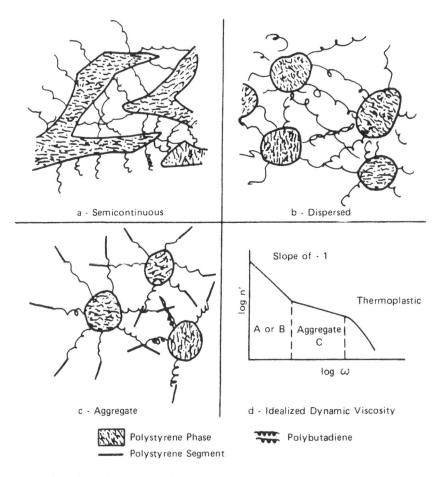

Figure 3 Schematic representation of the block copolymer domain structure performance during flow. (From Ref. 93.)

molecules still occur but they are not linked together to form a three-dimensional network. The system will behave as large star-shaped aggregates as shown in Figure 3*c*. Finally, at high deformation rates, these aggregates will, in turn, be disrupted and the system will behave as an assembly of individual (nonaggregated) molecules. This theory predicts a network response at low shear stress levels (including a yield stress), an intermediate region of shear stresses where the viscosity would be much higher than expected for molecules of the given molecular weight, and finally a region at high shear stresses where the behavior would be typical of ordinary thermoplastic of the same molecular weight. An idealized dynamic viscosity curve based on this theory is shown in Figure 3*d*.

The Melt Transition

Chung and Gale [94] were the first workers (1976) to detect a structure transition in SBS melts. The melt rheological behavior of an SBS block copolymer $(7-43-7)10^3$ was studied using a Weissenberg rheogoniometer. The behavior was highly non-Newtonian with high viscosity and high elasticity (which are characteristic of these copolymers) at 125, 140, and 150°C. The data at these temperatures superposed well, giving a master curve with a constant flow activation energy. However, data at 175°C indicated a marked change in the flow mechanism; at this temperature the flow was Newtonian with negligible elasticity. This was considered to be an indication that the sample went through a structural change from a multiphase structure at low temperatures to a homogeneous structure at some temperature between 150 and 175°C.

Later in 1976 Leblanc [95], working with a star-shaped butadiene–styrene block copolymer, Solprene 415, found that at a critical shear stress (5.5×10^5 dynes/cm^2) a transition occurred in the flow curves, suggesting a change in the melt structure. The flow activation energy (19.2 kcal/mol), deduced from an Arrhenius plot, was said to suggest a mechanism depending on the structure.

Dynamic viscosity and elastic modulus of an S–B–S block copolymer $(7-43-7)10^3$ was used by Gouinlock and Porter [96] to study its rheological behavior as a function of temperature (80–170°C). These linear properties could be time–temperature superposed to yield master curves, each of which exhibited two branches below different (critical) reduced frequencies.

At lower temperatures, the non-Newtonian behavior characteristic of SBS was observed, but, in contrast, Newtonian response occurred at higher temperatures. Thus plots of viscoelastic properties vs. temperature exhibited discontinuities reflecting a narrow transition at about 142°C. The transition could be attributed to a weakening and/or loss of the crosslinking structure due to a sharp increase in phase miscibility and/or the attainment at or above the transition temperature of an easily disruptable phase. This leads to Newtonian behavior observed in the master curve at low reduced frequencies. Gouinlock and Porter suggested that miscibility is at least the major factor since, in the absence of such a phase change, the property changes would be expected to be more gradual. DSC analysis failed to show any thermal events between 125 and 155°C, indicating that heat of mixing associated with phase changes *if any* (this author's comment) is significantly smaller than normally observed.

Chung and Lin [97], using a sample similar to that used two years earlier [94], SBS $(7-43-7)10^3$ block copolymer, repeated their earlier work but with improved accuracy in the temperature recording and with at least 11 temperatures including several in the transition region. The transition was found starting at about 140°C and extending over a narrow transition region from 140 to about 150°C. Time–temperature superposition onto a single master curve was possible only at high reduced frequencies. At low reduced frequencies, two characteristic branches of the master curve were observed. The data at temperatures below the transition region superposed onto the upper branch where the dynamic viscosity n' is a strong function of the frequency, whereas the data at temperatures above the transition region superposed onto the lower branch where n' is independent of frequency (Newtonian behavior). The data at temperatures within the transition region fall between the upper and lower branches. The apparent flow activation energy was found to be constant at about 22.8 kcal/mol below the transition region decreasing to about 17.4 kcal/mol above the transition. Chung and Lin concluded that due to narrowness of the rheological transition, far above the glass transition temperature of the polystyrene domains, added to a viscoelastic behavior being linear only at low frequencies and at temperatures above the transition temperature, an accompanying morphological transition is more likely to be the case rather than a gradual weakening of the polystyrene chains.

This conclusion was later confirmed by electron microscopy [98]. Samples quenched in liquid nitrogen from temperatures above 150°C showed no structure, whereas those quenched from temperatures below 140°C clearly showed a multiphase structure. The melt rheological transition behavior was also shown to be independent of the strain amplitude of the dynamic viscoelastic measurements. This favored the explanation of a morphological transition from a multiphase structure below about 140°C to a single-phase structure above 150°C. Unfortunately no micrograph for that specific condition was shown and the temperatures used for the pair of strain amplitudes were not the same. This casts doubt on the conclusions.

Following the same principles Widmaier and Meyer [99], working with S–I–S triblock copolymer (total molecular weight 45,000; isoprene content 38% w/w), found that above 225°C the dynamic viscosity was independent of frequency up to a critical frequency (Newtonian behavior), whereas for lower temperatures and/or higher frequencies classical non-Newtonian behavior prevailed. All data superposed well on a two-branch master curve. Microscopy and diffraction investigations were also employed by the authors to conclude that two-phase lamellar structure was progressively destroyed, and the transition temperature of 225°C corresponded to the temperature above which complete mixing occurred.

At the end of 1973, the IUPAC Working Party on Structure and Properties of Commercial Polymers started a program on the rheology of thermoplastic elastomers, specifically the A–B–A type. A styrene–butadiene–styrene triblock copolymer, Cariflex TR-1102 $(11–56–11)10^3$, styrene content 28%, was chosen as the representative sample. Fourteen laboratories participated, investigating the effect of the domain structure on the rheological behavior. The final report was prepared by Ghijsels and Raadsen [100] and published in 1980. The main conclusions were:

1. The melt viscosity was much higher than that of otherwise similar random copolymers, especially at low shear rates.
2. No Newtonian behavior was found even at shear rates as low as 10^{-5} sec^{-1} (creep experiments).
3. The viscosity at low shear rates is very sensitive to shear history (generating large scatter in the results from the participating laboratories).
4. In the low shear region (shear stresses <3000 Pa) the complex viscosity is as much as three times the steady-state viscosity at equal values of frequency and shear rate; this implies that the empirical relation of Cox and Merz [101] is not valid for the block copolymer.

5. A residual shear stress is observed in shear stress relaxation experiments; this residual stress is developing during the shear process and its value depends on the previous shear conditions.

The reasons given for this unusual behavior were entirely based on the Arnold–Meier model.

Hashimoto et al. [34] also detected a reversible transition at about 170°C for a styrene–isoprene "tapered" block copolymer (total molecular weight 43,000; styrene content 47%). Its difference from a genuine block copolymer is that in a tapered block copolymer the fraction content of one of the components varies from almost unity at one end of the polymer chain to almost zero at the other end. This mixing of incompatible components in the backbone polymeric chain induces a large interface between the two phases and enhances mixing of unlike units in the domains. The transition temperature, detected by small-angle x-ray scattering, was quoted to be much lower than that of styrene–isoprene diblock copolymers having the same total molecular weight and chemical composition because of the partial mixing of unlike segments.

In 1982 Kraus and Hashimoto [35], using SAXS, worked with two typical pressure-sensitive adhesives prepared from star-branched multichain block copolymers of styrene and butadiene or isoprene $(S-B)_n$ and $(S-I)_n$. They believed that the domain structure was maintained well beyond the T_g of the polystyrene domain, but homogeneous melts were formed above a critical temperature. The isoprene-based copolymer showed that the SAXS maximum present at room temperature persists well above the polystyrene T_g and disappears somewhere between 200 and 230°C. This sample had a high amount of uncoupled diblock copolymer present (26%). For the butadiene-based copolymer the behavior was more complicated, because the original scattering maximum tended to disappear with heating, but a new maximum appeared near 180°C which was then stable to temperatures at least as high as 230°C. The Bragg spacings for the two maxima were different (42 and 28 nm), indicating an irreversible change in morphology; on cooling the spacing remained at 28 nm. On reheating, the new peak was maintained until at least 230°C where "melting" was far from complete.

All papers discussed thus far in this section state categorically that the rheological transition shown by block copolymers is accompanied by a morphological transition from a two-phase to a single-phase structure. Iskandar and Krause [43], working with the same sample as Chung et al. [98], failed to detect a single T_g between those of polystyrene and of polybutadiene to confirm the single-phase morphological structure postulated for the sample after being subjected to the following thermal history: heating the sample to 190°C, keeping at this temperature for 5 min, and then quenching it in liquid nitrogen for 5 min. Chung et al. [98] reported that after this thermal treatment the polymer showed no detectable structure when compared with the two-phase structure observed both in samples receiving no thermal treatment at all and in those quenched from 125°C. Not only were both T_g's (due to the two phases) present in the DSC trace, but also the ΔC_p of the polystyrene phase was much greater than that of a polystyrene homopolymer (0.58 and 0.29 J/g K respectively) indicating that some mixing of butadiene segments into the styrene microphase probably existed. During staining the additional butadiene segments present in the styrene microphase of the quenched sample were probably sufficiently stained to make the polystyrene phase indistinguishable from the polybutadiene phase. Iskandar and Krause concluded that the two-phase structure still existed in the sample SBS $(7-43-7)10^3$ after heating to 190°C (above the transition temperature) and then quenching.

Optical Properties of Block Copolymers

Owing to the fundamental incompatibility of the components, block copolymers segregate and group themselves in distinct and intimately interdispersed microdomains, the size of which is comparable to the dimensions of the macromolecules. When the molecular weight distribution of the blocks is sufficiently sharp, the segregated microdomains aggregate into a regular and periodic arrangements in space. The presence of a regular arrangement involving two phases with different refractive indices induces birefringence.

Birefringence

Birefringence is a bulk optical anisotropy shown by a material having different refractive indices in different directions [102] which can be generated by one or more of the following factors [103]:

1. *Orientation birefringence.* This arises when there is a physical ordering of optically anisotropic elements (e.g., chemical bonds) along with some preferential direction. This can occur in polymers by aligning amorphous or crystalline chains as by an extension or drawing deformation. Intrinsic birefringence is a special case of orientation birefringence and is, by definition, the maximum in orientation birefringence.
2. *Deformation birefringence.* This can occur in any optical system in which by the action of stress (or strain), axial extension, or compression could change the "lattice" spacing, thereby resulting in a refractive index difference along and across the applied deformation axes. Distortion of bond angles and/or bond lengths from equilibrium may also result in a finite birefringence without orientation.
3. *Form birefringence.* This phenomenon arises when the medium contains at least two phases each having a different refractive index, and at least one dimension is small compared with the wavelength of the radiation. This effect depends solely on the volume fractions (ϕ_1, ϕ_2) and the refractive indices (n_1, n_2) of the two microphases. With samples consisting of parallel cylinders, form birefringence is expected to be positive with a value [49, 104] of

$$\Delta n = \frac{\phi_1\phi_2(n_1^2 - n_2^2)^2}{2|(1 + \phi_1)n_2^2 + \phi_2 n_1^2\|\phi_1 n_1^2 + \phi_2 n_2^2|^{1/2}}$$

With samples consisting of parallel layers, it is expected to be negative with a value [49, 64] of

$$\Delta n = \frac{-\phi_1\phi_2(n_1^2 - n_2^2)^2}{2|\phi_1 n_2^2 + \phi_2 n_1^2\|\phi_1 n_1^2 + \phi_2 n_2^2|^{1/2}}$$

When a preferential arrangement of structures is induced by a flowing stream of liquid (or melt) a special case of birefringence arises known as flow birefringence. Wales [105], working with a variety of polymers (high-density polyethylene, low-density polyethylene, polypropylene, polystyrene, and polydimethyl siloxane elastomer), confirmed that the stress-optical coefficient was unique to each class of polymer regardless of its molecular weight, polydispersity, or extrusion temperature.

The stress-optical coefficient is defined as the constant C in the stress-optical law:

$$\Delta n \sin 2X = 2n_{12} = 2C\sigma_{12}$$

where

Δn = birefringence

X = extinction angle

n_{12} = refractive index in the plane 1–2

σ_{12} = stress in the plane 1–2

and the index 1 stands for the flow direction, 2 the direction of the flow gradient, and 3 the neutral direction in a simple shear flow. The extinction angle X is defined as the smaller of the two angles between the shear planes and the vibration planes of the polarizers, which give rise to dark fields. According to the definition, X lies between 0 and 45°.

The validity of the stress-optical law for amorphous polymer melts during steady shear flow is remarkable in that a simple relation bridges the macroscopic stress with the microscopic processes, i.e., the rotation and deformation of the atomic bond.

The independence of the stress-optical coefficient with respect to temperature for triblock copolymers of styrene and butadiene was also recognized by Fischer and Henderson [44]. Birefringence measurements during stress–strain tests indicated that the decrease in modulus and strength of these copolymers with increasing temperature was associated with the decrease in the concentration of elastically effective network chains. Because the stress-optical coefficient is independent of temperature, this decrease in the concentration of elastically effective chains is not due to the onset of flow within the polystyrene regions, at least for temperatures below about 70°C. Rather the decrease was associated with the increased mobility of the polybutadiene chains at higher temperatures, which also leads to an increase in the rate of stress relaxation. Birefringence measured during extension and retraction showed that the stress–strain hysteresis was due to restricted mobility of the polybutadiene chain segments rather than to permanent viscous flow or to a change in the effective network structure of the block copolymers. In other words, the strain accommodation is achieved not by viscous flow, as is usual for thermoplastics, but by changes in the network, principally concerned with the restricted mobility of the polybutadiene blocks. This gives reversibility and more or less complete recovery on removal of the stress, as observed [106].

According to Stein and Tobolsky [107] many processes can contribute to changes in birefringence: chemical reaction, viscous flow or diffusion, crystallization, orientation of crystallites, macroscopic faults, release of stress, and configurational changes. They considered these to be the important processes, but they do not claim the list to be exhaustive.

There may be further complications because of the interactive nature of some of these processes. The relative contribution of different processes will vary from one polymer to another and will be temperature-dependent. One of the ways of examining these phenomena is to measure birefringence during a stress–relaxation test [108].

Qayyum and White [109] followed the birefringence behavior shown by six glassy polymers (polyethylene terephthalate, PET; polymethyl methacrylate, PMMA; polyamide, Trogamid, PA(G); polystyrene, PS; polycarbonate, PC; polyether sulfone, PES) during stress–relaxation and recovery experiments at temperatures below T_g. Of the six polymers tested only two [PET and PA(G)] showed a strong mutual resemblance in the behavior of birefringence under these conditions. The results were discussed with reference to molecular structure, and, although detailed interpretations were not offered, it was observed that the behavior is less complex for polymers that possess their most polarizable groups in the main chain or attached rigidly to it than for those polymers having polarizable side groups with relaxations that do not involve the main chain.

Refractive Index

Refractive index n is defined as the ratio of the velocity of light in vacuum (designated c, and equal approximately to 3×10^8 m/sec) to that in the medium (sample):

$$n = \frac{c}{v_{\text{med}}}$$

The refractive index of vacuum is thus unity; that of air is only very slightly greater ($n_{\text{air}} = 1.000277$). Refractive index values for most materials lie between 1 and 2, although a few have indices somewhat greater than 2.

Refractive indices are constant only for a particular wavelength λ of light and temperature. In most of the cases when no transitions are involved the relation with temperature is linear, being mainly due to thermal expansion. Common values are in the range of $\sim 4.0 \pm 0.6 \times 10^{-4}/°C$. Thus its measurement at different temperatures (keeping a constant λ) may be used to locate transitions.

The glass transition of a polymer occurs at a characteristic temperature T_g in which the glasslike to rubberlike transition occurs over a narrow temperature range. Measurements of the equilibrium refractive index of the polymer at various temperatures below and above the T_g will produce two straight lines intersecting at the transition. The advantage of this technique over others (DSC, for example) is that it produces sharper transitions.

Using this technique Krause et al. [42] measured the glass transition temperatures of the styrene microphase in styrene–butadiene (S–B) diblock and styrene–butadiene–styrene (S–B–S) triblock copolymers with low molecular weight styrene blocks. Differential scanning calorimetry (DSC) measurements were also made. The transitions measured by the refractive index technique were sharp, yielding a unique value, compared with the DSC traces, which were broader up to the 50 K range. The styrene block molecular weights were calculated based on knowledge of the structure (di- or triblock). The styrene content was obtained by NMR (using the procedure of Senn [110] from the aromatic peak at 7.2 ppm and the olefinic peaks at 5.4 ppm) and the total molecular weight of the block copolymer was measured by gel permeation chromatography (GPC). Because GPC gave two distinct calibration curves when using standard polystyrene and polybutadiene homopolymers, the total molecular weight of the block copolymer was calculated taking this into account:

MW = (% S) (MW as PS standard) + (% B) (MW as PB standard)

In all block copolymers investigated having styrene block molecular weights $< 2.2 \times 10^4$, the styrene phase T_g's obtained by both techniques were at least 18 K lower than those of styrene homopolymers with molecular weights comparable to those of the styrene blocks. Also, the change in specific heat at the T_g of the styrene microphase (ΔC_p^s) was always equal to or greater than that of polystyrene homopolymer. These results suggested that some butadiene segments were mixed into the styrene microphase [reducing the T_g and increasing the (ΔC_p^s)].

Transmitted Light Intensity

Very few researchers use the variations in the transmitted light intensity passing through a block copolymer specimen to study structure changes with temperature. Pico and Williams [111] used this technique in plasticized SBS block copolymer films. A good solvent with high boiling point [dipentene (*p*-mentha-1,8-diene)] was used as plasticizer in low-volume concentrations (0.12 ~ 0.30). The clear solutions were trapped in a glass cell and placed in an oven with a transparent window, set at prefixed temperatures. A laser beam was passed through the sample and its intensity at each temperature was recorded with time. Plots of transmittance vs. time at various temperatures were used to detect a temperature range in which a large increase in transmittance was observed (it would be better had they presented their data as plots of transmittance vs. temperature, allowing for the thermal and morphological equilibrium). This opaque-to-clear transition was identified as the separation temperature T_s above which the two microphases mixed. The fact that the transition temperature was found to occur over a range of temperature was attributed to two causes:

1. The sample (Kraton 1101) is a commercial polymer and, as a consequence, very probably polydisperse.
2. Various morphological types could occur (lamellar, inverted spheres, and inverted cylinders) having different T_s, each one of them leading to a range of temperature for the transition to occur.

The rheological behavior of these solutions was also measured in a Weissenberg rheogoniometer and the loss of structure due to raising the temperature above T_s caused a change in the samples' behavior; the Newtonian viscosity at low temperatures was recorded.

Dynamic Mechanical Properties of Block Copolymers

As mentioned earlier, block copolymers comprise two distinct phases at room temperature; this results in complex viscoelastic behavior [112]. Takayanagi and co-workers [48] were the first to analyze the complex modulus of a two-phase material. Using oriented specimens of polyethylene, polypropylene, polyoxymethylene, polyethylene oxide, and polytetrahydrofuran, their complex moduli were determined in directions with reference to the orientation direction. The dynamic tensile modulus E' along the stretched direction (called 0° direction) was found to be lower than that perpendicular to the stretched direction (90° direction) above the temperature of primary dispersion αa for all polymers studied. The α relaxation is attributed to vibrational or reorientational motions within crystals in the polymer [113]. But below the temperature of αa dispersion the E' value in the 0° direction was found to be higher than that of 90° direction as expected from the anisotropy of the crystal modulus. These observations led them to propose a model in which the crystalline C and the amorphous region A were arranged generally in series along the stretched direction and at the same time the C region is more or less continuous along the 90° direction. This assumption was backed by small-angle x-ray scattering analysis in which the drawn sample was shown to be composed of many microfibrils. Those conditions imposed in the model were considered satisfied even when many microfibrils were bound together into a fiber.

This model considered the polymeric material as comprising two phases, a crystalline C and an amorphous A phase, with ϕ_C and ϕ_A volume fractions respectively. The complex tensile modulus of the polymer can be calculated taking into consideration the way that the two phases are connected. Two arrangements were proposed, series and parallel. The complex tensile modulus E^* in series can be written as

$$\frac{1}{E^*} = \frac{\phi_C}{E_C^*} + \frac{\phi_A}{E_A^*}$$

and in parallel as

$$E^* = \phi_C E_C^* + \phi_A E_A^*$$

Figure 4 is the schematic representation of this model.

In another publication Uemura and Takayanagi [114] evaluated the apparent elastic constant G of a two-phase system knowing the moduli of the components. The proposed relationship was

$$G = G_1 \cdot \frac{(7 - 5\nu)G_1 + (8 - 10\nu)G_2 - (7 - 5\nu)(G_1 - G_2)\phi p}{(7 - 5\nu)G_1 + (8 - 10\nu)G_2 + (8 - 10\nu)(G_1 - G_2)\phi p}$$

where G_1 and G_2 are the shear moduli of the medium and the dispersed particles respectively, ν is the Poisson ratio of the medium, and ϕp is the volume fraction of the particles. The result for modulus was extended to dynamic viscoelasticity by the corresponding principle (the shear modulus G is simply replaced by the shear viscosity η; this is valid when the frequency of the applied shear stress is small enough for the inertia force to be ignored and the wavelength in the fluid is sufficiently small compared with the size of the dispersed particles). Experimental verification with dynamic viscoelasticity data for the system of styrene–acrylonitrile copolymer interpolymerized with polybutadiene particles and for the shear viscosity comparisons was carried out for the system linear polyethylene–polybutadiene–1 yielding good agreement. A specific study of the anisotropy in viscoelastic behavior in cold-drawn polyethylenes can be found in various publications [115, 116]; this will not be discussed further here.

The effect of the mechanical structure of the center block on the rheological properties of ABA triblock copolymers with polystyrene end blocks was studied by Futamura and Meinecke [117]. A variety of center structures was used, including polybutadiene, poly-

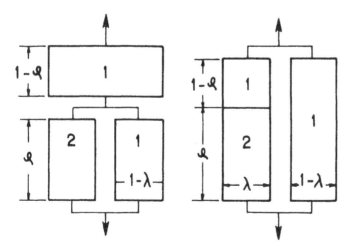

Figure 4   Schematic representation of the Takayanagi model. (From Ref. 48.)

isoprene, ethylene/butene-1 copolymer, ethylene/propylene copolymer, and polydimethyl siloxane. The structure of the center block was said to have a pronounced effect on the rheological properties of these triblock copolymers. However, possibly important variables such as the wide range of total molecular weight and the molecular weight of the polystyrene phase were not taken into account. The characteristic relaxation times of the block copolymer melts increased with increasing incompatibility between the styrene block and the center block (measured in terms of the solubility parameter difference $|\delta_1 - \delta_2|$). The rheological properties of the block copolymers were not influenced significantly by the glass transition or the entanglement molecular weight of the center block.

On the other hand, Annighöfer and Gronski [118], working with I–IS–S poly|isoprene-(isoprene-co-styrene)-styrene| block copolymers with variable interphase thickness (variable wt% of the middle block, IS), found that the glass transition temperature of the hard phase was a function of the size of the middle block, the higher its molecular weight the higher the depression in the polystyrene phase glass transition temperature. Explanation was given in terms of a cooperative softening of the pure hard phase and surface interlayer material. The glass transition of the rubbery phase (isoprene) remained unaffected. The dynamic viscoelastic measurements were carried out using a torsion pendulum.

Stadler and Gronski [119] studied the dependence of the stress relaxation of SBS thermoplastic elastomer on the molecular weight and degree of hydrogenation. The influence of these parameters on the structure of the physical network and the degree of partial mixing at the domain boundaries was investigated by separating the stress–relaxation modulus into a viscoelastic term and an equilibrium network modulus calculated from the relaxation time spectrum. An estimation of the volume fraction of interfacial material and its correlation to the parameters which govern phase separation were also provided.

Kelterborn and Soong [120] used a modified torsion pendulum capable of applying both tensile and torsional deformation to study the dynamic mechanical properties of two SBS block copolymers. Their morphology was altered by large tensile deformation at various temperatures and upon removal of the applied stress, the morphological features of such stretched-and-released systems became time-dependent as the nonequilibrium microstructure reverted to a thermodynamically stable state. This reformation was monitored by DMTA measurements and the results were analyzed using a modified Nielsen model [121] to obtain information on the time-dependent structural state of the samples. These results were then compared with stress–strain curves to provide further insight into the structure breakdown–reformation mechanism. Two competing mechanisms, block pull-out and domain structure fracture, were proposed to explain the experimental observations. During stretching the rigid polystyrene network begins to break down and continues fragmenting with further elongation. At $\sim300\%$ elongation the structure could be identified as two extreme cases. In the first case the polystyrene segments had been pulled out of their domains having a great number of chains mixed with the polybutadiene phase. By contrast, in the second case the polystyrene segments remained in their original domains, but the domains themselves split and fractured. The actual process is a combination of these two extremes and their relative importance depends on system characteristics such as sample composition and type and temperature.

Finally, another interesting field of work is the study of homogeneous block copolymers. This class of materials does not show microphase separation, having only one T_g intermediate between those of the two components. Butadiene–isoprene diblock copolymer and styrene–methylstyrene (MS–S–MS, S–MS–S, MS–S) di- and triblock copolymers are examples of these classes [122, 123].

PROPERTIES OF THERMOPLASTIC RUBBERS

Materials

Table 2 lists a series of commercially available block copolymers with different styrene content and consequently different morphologies [124–126]. Their number average and weight average molecular weight ($\bar{M}_n$ and $\bar{M}_w$ respectively) measured by gel permeation chromatography [127–129] are listed in Table 3. The polydispersity (M_w/M_n), as expected, is low.

Properties of the Molten State

The steady-state flow melt viscosity [130–135] of one of the samples (Cariflex 1102) measured using a Davenport Capillary Rheometer is presented in Figure 5. In the shear rate and temperature range of measurement ($1 < \dot{\gamma} < 1000$ sec^{-1} and $130 < T < 200°C$), the flow behavior of thermoplastic rubbers—Cariflex 1102 is a good example—is mainly non-Newtonian but at high temperatures and low shear rates the curves show a tendency to become Newtonian, as noted by other authors [35, 93, 94, 100]. The flow curves can be reduced to a "master curve" by time–temperature superposition, as shown in Figure 6 using 150°C as reference temperature. Plotting the shift factors (a_T) employed in the

Table 2 Commercially Available Block Copolymers

Elastomer	$S(\%)$	Structure	Morphology	Producer
Cariflex TR-1101	30	Linear SBS	Cylinders of PS	Shell
Cariflex TR-1102	28	Linear SBS	Cylinders of PS	Shell
Cariflex TR-1107	15	Linear SIS	Spheres of PS	Shell
Solprene 415	40	Star-shaped $(S–B)_x$	Lamellar	Phillips
Finaprene 414	40	Star-shaped $(S–B)_x$	Lamellar	Fina
Finaprene 416	30	Star-shaped $(S–B)_x$	Cylinders of PS	Fina

Source: Ref. 163.

Table 3 Number Average and Weight Average Molecular Weight and Polydispersity of Selected Block Copolymers

Elastomer	M_n	M_w	M_w/M_n
Cl101	94,000	115,600	1.23
Cl102	70,860	85,700	1.21
Cl107	116,850	141,400	1.21
S415	125,000	143,700	1.15
F414	93,100	107,000	1.15
F416	99,150	118,000	1.19

Source: Adapted from Ref. 163.

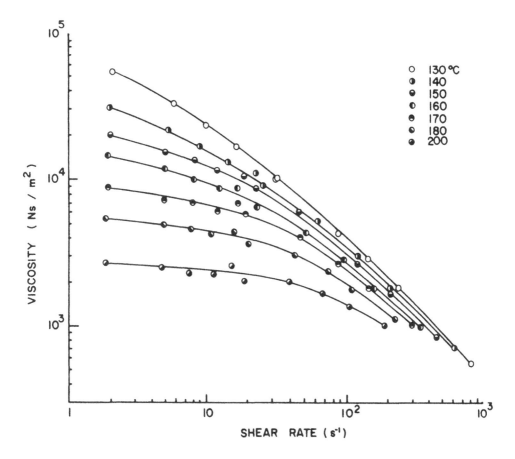

Figure 5 Steady flow melt viscosity curves of Cariflex 1102. (From Ref. 162.)

reduction according to the Arrhenius model yields intersecting straight lines, as shown in Figure 7 (plotted as closed dots).

Another way of studying the rheological properties of a polymeric material is by the melt flow indexer (MFI). The equipment is less expensive than most of the rheometers and so is widely used. Its main difference from a capillary rheometer is that it operates at constant shear stress.

Shenoy and colleagues [136–141] showed theoretically and experimentally that flow curves at various temperatures can be coalesced into a master curve by time–temperature reduction using the inverse of the MFI data as shift factors ($a_T = 1/\text{MFI}$). With this they propose a method to estimate the rheograms of a melt at temperatures relevant to the processing conditions with the use of a master curve, knowing the melt flow index and glass transition temperature of the material.

In other words, the MFI values, which can be measured accurately experimentally, can be substituted for the more questionable shift factors, which are susceptible to error even when obtained by a computer iteration. The unifying approach had been found to be valid for polyolefins and styrene polymers [136], cellulosics [137], engineering thermoplastic [138], polymer composites [139], and polymer blends [140].

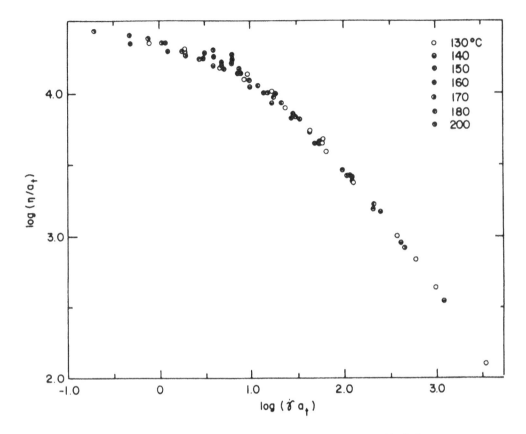

Figure 6 Temperature-reduced steady flow melt viscosity of Cariflex 1102. Reference temperature 150°C. (From Ref. 162.)

The melt flow indices of Cariflex 1102 taken at various temperatures treated according to an Arrhenius model are shown in Figure 7. The same curve also presents the shift factors. The two sets of results overlap within experimental error showing three regions of linear behavior intersecting at ~121 and ~154°C. The calculated apparent flow activation energies are about 34.0 kcal/mol for $T < 121$°C, 23.5 kcal/mol for 121°C $< T < 154$°C, and 16.7 kcal/mol for $T > 154$°C. In this model two possible melt transitions at the intersection of the straight lines are presented. To determine whether they are true transitions another mathematical model is applied to the data. This second mathematical representation is the Vogel [142], Fulcher [143], and Tamman–Hesse [144] equation, abbreviated hereafter by the authors initials VFTH. It states that the variable is affected by the temperature according to its value compared to the theoretical glass transition temperature T_∞:

$$\log(V) = \log A + \frac{C}{2.303(T - T_\infty)} \tag{1}$$

where the variable V can stand for the shift factors a_T or for the inverse of the MFI (1/MFI). In practice T_∞ is chosen as the value that yields the best straight line; the value

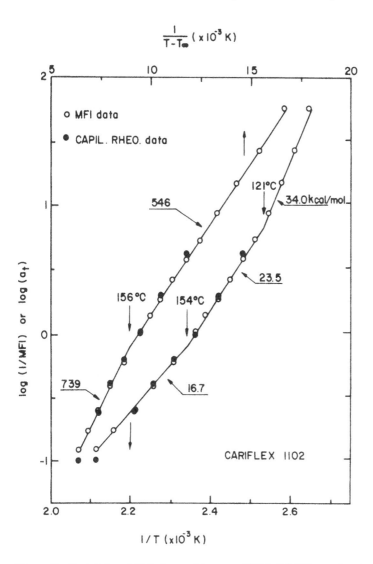

Figure 7 Best fitting shift factors and inverse of MFI (1/MFI) as a function of extrusion temperature according to the Arrhenius equation (1/T) and VFTH equation [1/(T − T_∞)] for Cariflex 1102. (From Ref. 162.)

found for Cariflex 1102 is T_∞ = 318 K (45°C). Figure 7 also shows this representation where, again, both sets of data overlap within the experimental error showing two regions of linear behavior in the range studied with only one melt transition at $T_T \simeq$ 156°C. The C coefficients calculated from the curves are 546 and 739 (in kelvins) below and above T_T respectively.

The same procedure can also be employed for a second sample, Solprene 415, using a 20-kg load for the MFI measurements. The resultant curves for both mathematical models are shown in Figure 8. Again two straight lines intersecting at $T_T \simeq$ 180°C are obtained for both models. The calculated apparent flow activation energies for the Arrhenius model are

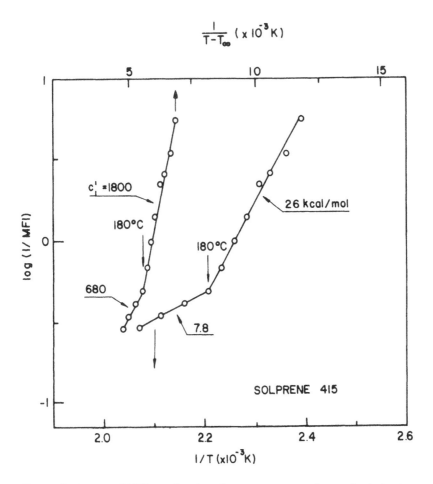

Figure 8 Inverse of MFI as a function of temperature according to the Arrhenius equation $(1/T)$ and VFTH equation $[1/(T - T_\infty)]$ for Solprene 415. (From Ref. 162.)

about 26 kcal/mol for $T < 180°C$ and 7.8 kcal/mol for $T > 180°C$ and the C coefficients 1800 and 680 below and above T_T respectively using $T_\infty = 273$ K (0°C).

Another way of showing the second mathematical representation is using the WLF [145] equation:

$$-\log (a_T) = C° \frac{(T - T_0)}{(T - T_\infty)} \tag{2}$$

where T_0 is a reference temperature and T_∞ is the same temperature as used in the VFTH equation. Figure 9 shows this plot for the two samples, Cariflex 1102 and Solprene 415, both of which give two straight lines intersecting at ∼157 and ∼180°C respectively. The C coefficients calculated from the curves are 4.2 and 5.7 for C1102 and 8.8 and 3.3 for S415 below and above the transition temperature respectively. Due to the reproducibility and consistency of the results in all the mathematical models the remainder of the samples are plotted only as 1/MFI vs. 1/T, according to the Arrhenius equation. The results are

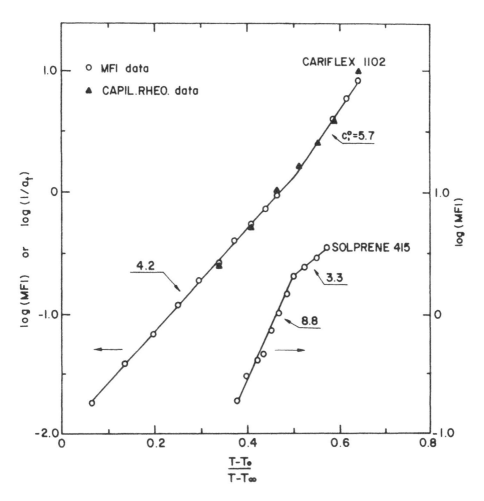

Figure 9 WLF plot of MFI for both samples Cl102 and S415 in the temperature range of $T_g <$ $T < T_g + 100°C$. Inverse of shift factors are also included for Cl102. (From Ref. 162.)

presented in Figure 10. All curves are of two straight lines intersecting at the transition temperature. Table 4 recalls the transition temperatures and apparent flow activation energies below and above T_T for all samples.

The inverse of the MFI (1/MFI) is a direct function of the melt viscosity and behaves as such, being capable of superposition with the shift factors. In fact, for Cariflex 1102 if the reference temperature used during the time–temperature reduction of the flow curves is $T_R = 150°C$ and the MFI measured at a shear stress of 4.5×10^4 Pa (5 kg dead weight), the two sets of results (a_T and 1/MFI) coincide within experimental error. The chosen shear stress is not critical because the activation flow energy is independent of it, as shown by Bartenev [146]. These two sets of data can be treated as one and plotted according to various mathematical models. First, the Arrhenius plot in which three straight lines intersecting at two points is obtained. These two intersections could be considered transitions in the melt state. According to Boyer [147–151] the first (~121°C) could be due to

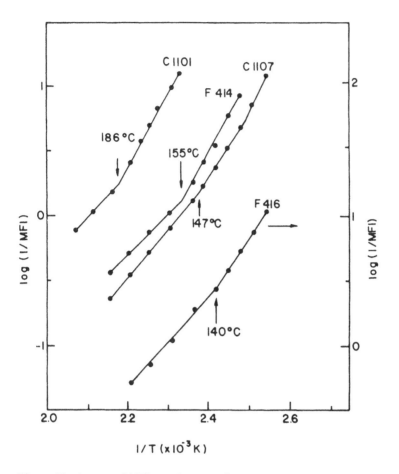

Figure 10 Inverse of MFI as a function of temperature according to the Arrhenius equation for Cariflex 1101, 1107, and Finaprene 414 and 416. (From Ref. 162.)

Table 4 Transition Temperature and Apparent Flow Activation Energy for Selected Block Copolymers

Elastomer	T_T (°C)	$\Delta\epsilon_T < T_T$ (kcal/mol)	$\Delta\epsilon > T_T$ (kcal/mol)
Cl101	186	24.9	14.1
Cl102	155	23.5	16.7
Cl107	147	23.5	16.9
S415	180	26	7.8
F414	155	25.9	14.1
F416	140	22.1	15.3

Source: Adapted from Ref. 163.

Table 5 Vogel Equation *C* Coefficient Values

Sample	T_∞ (K)	C		Ref. 15	
		$T < T_T$	$T > T_T$	T_∞ (K)	C
Cl102	318	546	739	320	463[a]
S415	273	1800	680		
PS				293	1890

[a]Result for $T < T_T$.

the liquid–liquid transition in the PS phase. This argument is refuted by Plazek and others [152–155] owing to its absence when the Vogel (VFTH) model [142–144] is applied. In this case the break in the Arrhenius plot at ~121°C is not present in the Vogel plot. The conclusions, like Plazek's for PS homopolymer [153], is that the break at 121°C is *not* a true transition. On the other hand, the second transition shown by the melt viscosity behavior in the Arrhenius plot is also evident when the Vogel model is applied with a break in the linear behavior at 156 and 180°C for Cariflex 1102 and Solprene 415 respectively. Chung and Gale [94] pointed out earlier the reduction of the apparent flow activation energy for SBS at some temperature above 150°C (with only one experimental point shown above T_T) and this same effect can be observed in some of Arnold and Meier's [93] results. Kraus and Hashimoto [35], working with block copolymer rheology, only show their data as $\log(a_T)$ vs. T (°C), possibly failing to detect the structural changes in melts of block copolymers by study of viscoelasticity. The WLF equation [145], very similar to the Vogel assumption [142], also shows this transition at approximately the same temperature.

The *C* coefficients in the Vogel equation (1) were estimated by the slope of the curves in Figure 8 and are shown in Table 5. As a matter of comparison, data from Holden et al. [15] are also included in Table 5. The data for Cariflex 1102 agree very well with theirs, for both T_∞ and the *C* coefficient, considering that their *C* datum is for temperatures below the transition. The other thermoplastic rubber, Solprene 415, has a high styrene content (40%) with a lamellar morphology and so for temperatures below the transition the polystyrene phase dominates its flow. Temperature dependence is very similar to polystyrene homopolymer, having nearly the same coefficients. Above the transition the flow changes, with a considerable drop in its temperature dependence and consequently in the *C* coefficient. Its T_∞ is lower than for the other elastomer (Cl102) closer to the polystyrene homopolymer datum.

Effect of the Molecular Weight of the Polystyrene End Blocks on the Transition Temperature

With the knowledge of the total average molecular weight of the sample the average molecular weight of the styrene end blocks can be determined with a good degree of certainty if two assumptions are made. First, the polymeric chains are assumed to be symmetrical, i.e., having the polystyrene end blocks with a constant average size in each end of the chain and in the polymer as a whole. Second, in the case of star-shaped samples such as Solprene 415, Finaprene 414, and Finaprene 416, a theoretical four arms per chain is the target, but in commercial samples a minimum of 80% of the chains is assumed to have the target four arms and the rest three arms leading to a mean functionality of branching $\bar{X}$ of

$$\bar{X} = (4 \times 0.8) + (3 \times 0.2) = 3.8$$

Obviously for the linear triblock chains the mean functionality of branching $\bar{X} = 2$. Thus the average number molecular weight of the styrene end blocks can be calculated as follows:

$$M_n\mathrm{PS} = \frac{S\%}{100\%} \cdot \frac{M_n}{\bar{X}} \tag{3}$$

where

    $S\%$ = styrene content (% by weight)

    M_n = total number average molecular weight of the thermoplastic rubber sample (Table 3)

    $\bar{X}$ = mean functionality of branching

The results are shown in Table 6. Also included are the transition temperatures obtained in the earlier section.

Figure 11 is a plot of the transition temperature values T_T of each sample as a function of its styrene block number average molecular weight yielding a straight line. Also included are results reported in the literature, which are summarized in Table 7. In other words the longer the hard end block, the higher the transition temperature. In the literature various authors using a variety of samples with different molecular weights, structures, and morphologies detected a transition in the molten state of thermoplastic rubbers but failed to find a common scaling rule to which all samples could be reduced.

Hadziioannou and Skoulios [32, 33], working with di- and triblock copolymers, found, using SAXS, that a sample of SIS [M_n = 63,000; M_w/M_n = 1.30; S(%) = 50.4] when heated stepwise keeps its lamellar spacing roughly constant up to 210°C when it starts to thicken sharply in an irreversible process. Using the curve of Figure 11 as reference, the calculated transition temperature for this sample would be T_T = 200°C, very close to the value observed experimentally. Another sample SIS–IC [M_n = 50,900; M_w/M_n = 1.25; S(%) = 32.4] showed "a quite different behavior: the position and shape of the Bragg re-

Table 6 Average Molecular Weight of Styrene End Blocks for Selected Block Copolymers

Elastomer	S (%)	$\bar{X}$	M_n PS	T_T (°C)
F416	30	3.8	7,830	140
Cl107	15	2	8,760	147
Cl102	28	2	9,920	155
F414	40	3.8	9,800	155
S415	40	3.8	13,160	180
Cl101	30	2	14,100	186

Source: Adapted from Ref. 163.

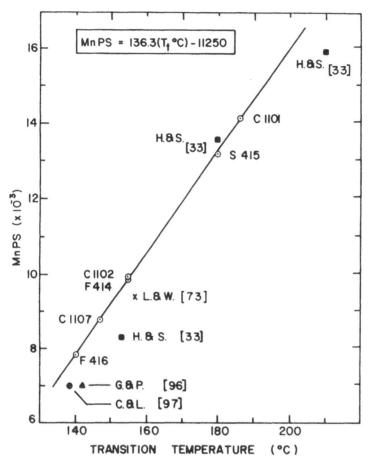

Figure 11 Transition temperature as a function of number average molecular weight of the styrene end blocks in the thermoplastic rubber sample. Also shown are results reported in the literature. (From Ref. 162.)

Table 7 Data Reported in the Literature

M_n PS	T_T (°C)	Ref.
7,000[a]	138.5	97
7,000[a]	142	96
8,250[b]	153	33
9,400[b]	157	73
13,550[b]	180	33
15,900[b]	210	33
14,000[a]	225	99

[a]Nominal value.

[b]Calculated value according to Eq. 2.3.

Source: Adapted from Ref. 163.

flections remained unchanged with temperature and then disappeared suddenly at 153°C'' [33, p. 272]. Using the same logic as above the calculated transition temperature is 143°C. Above T_T, due to the acquired freedom, the sample rearranges to a lower energy structure ultimately losing the long-range order. This was facilitated by its low M_n PS and long exposure time at high temperature ($\sim$18 hr) [33]. In any case, the rearrangement, or the disappearance of the Bragg reflections, would be possible only at T_T or a few degrees above it.

Chung and co-workers [97, 98], working with an experimental triblock copolymer SBS (7–43–7)10^3 with the molecular characterization data provided by the manufacturer (Shell), found that the master curve of the melt rheological behavior at low reduced frequencies is not single, but the results are scattered between two characteristic branches. The upper branch, having a non-Newtonian behavior, comprises mainly the experimental results obtained below T_T. Results above T_T tend to fall closer to the lower branch, which shows Newtonian behavior. Also, the curve of the shift factors employed in the time–temperature superposition forms two intersecting straight lines (at 138.5°C) when plotted in an Arrhenius-type plot. Close results (T_T = 142°C) were found by Gouinlock and Porter [96] using a similar sample, but in this case a clear two-branch master curve is seen. These two values are not far from the calculated datum (T_T = 134°C) and the difference could be attributed to the inaccuracy of the calculated M_n PS, which was not measured by the authors but is a nominal value quoted by the sample's manufacturer.

Widmaier and Meyer [99], working with SIS (14–17–14)10^3, also detected a melt rheological transition at 225°C, interpreting it as the temperature when the two-phase lamellar structure is destroyed to form a complete mixed phase. This temperature is well above our calculated value, which, based on the Mn PS, is $\sim$185°C. An Arrhenius plot of their listed shift factor values [99], as used by Chung [94, 97], has been drawn, but does not show any transition in the temperature range studied, 185–250°C, yielding in fact a very good straight line with a constant calculated apparent activation flow energy of $\sim$28 kcal/mol. Incidentally, the datum at 170°C does not lie on this straight line. The table of shift factors (a_T) presented in their Figure 3 [99] is inverted and should be used as ($1/a_T$).

Gel Content

The gel content formed during heat treatment of SBS–TPR Cariflex 1102 for 30 min measured by soxhlet extraction is shown in Figure 12. The crosslinking process is sharply affected by temperature, being undetectable at temperatures below $\sim$200°C and increasing quickly to 100% gel content at 230°C. After total gelation, infrared analysis failed to reveal any oxidation (no change in absorbance at $\sim$1700 cm^{-1}).

Mechanical Properties of Cariflex 1102

The tensile properties of extruded SBS–TPR are greatly influenced by the thermomechanical history of the samples during preparation. Figure 13 shows the stress–strain curve in the yield region of Cariflex 1102. Generally the curve shows a yield point which is directly affected by the heat treatment temperature. The reduction of the stress after yield is also influenced by the heat treatment: it is quite significant for samples heated below $\sim$140°C, very small when treated at 175°C, and not present at all when the heat treatment temperature is higher than 220°C.

Stress at Yield of SBS–TPR After Heat Treatment

Heat treatment of SBS–TPR tends to increase the stress at yield with increasing temperature and/or decreasing shear rate, as shown in Figure 14. In this case only samples

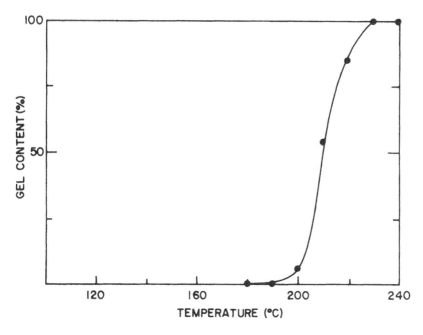

Figure 12 Gel content formed during heat treatment of SBS–TPR Cariflex 1102 (30 min, nitrogen atmosphere). (From Ref. 159.)

extruded at 180°C, arbitrarily chosen, were used and the yield stress for samples heat treated at 220°C was taken at the inflection point in the yield region of the stress–strain curve.

During extrusion of the samples, depending on the applied shear rate, partial or even total destruction of the domain structures occurs, the rate being proportional to the shear rate. In addition, orientation of the chains is induced, increasing with increasing shear rate.

Subsequent heat treatment for these extruded samples can again affect their morphology according to the heat treatment temperature. Temperatures lower than 145°C can induce only internal stress relaxation; however, temperatures higher than this may not only induce stress relaxation but also start to rearrange the domain structure, increasing the order. In the case of very high heat treatment temperatures ($T > 200°C$) crosslinking in the rubbery phase also occurs. All these phenomena contribute to different extents to the performance of the SBS–TPR product. Thus the stress at yield can be used to evaluate the development of the structure of the continuous segregated PS phase.

Optical Properties of Cariflex 1102

Birefringence

Solid block samples of SBS–TPR Cariflex 1102 were obtained by cutting cubes of ~2.5 mm from thick sheets. These sheets were originally made by injection molding or by casting from solutions of THF–MEK or benzene–heptane. The cube X axis was cut parallel to the flow direction in the injected samples and in those cast from solution, parallel to the solvent evaporation direction (perpendicular to the cast evaporation surface). The other two axes during measurements yielded similar results. The birefringence

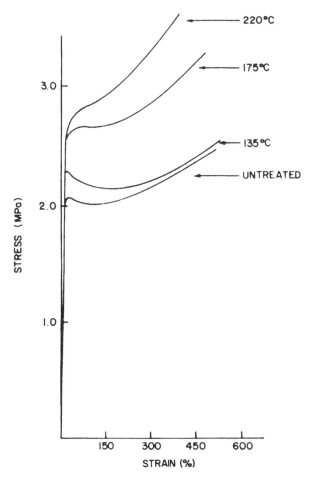

Figure 13 Dependence of the stress–strain curve (yield region only) on heat treatment temperature for extruded SBS–TPR Cariflex 1102 tested at room temperature. (From Ref. 160.)

was measured using both the Ehringhaus tilting compensator [156, 157] and the techniques of Yang et al. [158]. The same procedure was used for the samples after they were heat treated at 200°C for 30 min in vacuum. Table 8 shows the results. In all cases the direction X practically does not show birefringence but its perpendicular directions (Y or Z) do and they are similar. Injection molding gives rise to a much higher initial birefringence compared to the cast-from-solution molding technique. The type of solvent used (THF–MEK or benzene–heptane) is not important for C1102; they yield similar results. The heat treatment increases the birefringence in the samples cast from solution but this effect was not visible in the injection-molded samples.

Any bulk optical anisotropy shown by a noncrystalline material is related to one or more of three factors [103, 104]:

1. Stress (or strain).
2. Low-range molecular orientation.
3. Form, on a scale much larger than molecular sizes where ordered arrangements of similar structural elements occurs, which are, however, small compared with the wavelength of the radiation.

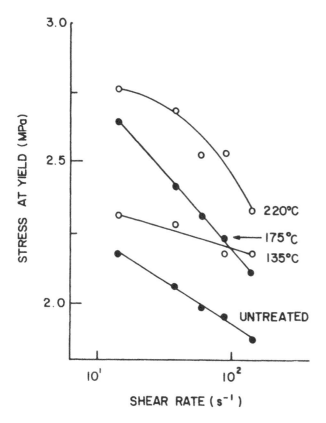

Figure 14 Effect of heat treatment at various temperatures on the stress at yield for extruded SBS–TPR Cariflex 1102 (From Ref. 160.)

Thermoplastic rubbers in a relaxed state show the last two kinds of birefringence owing to the fact that these elastomers are long molecular chains and have a regular domain structure of two segregated phases.

Samples which had been subjected to intense shear during processing and quenched tend to keep the molecular orientation in the solid state. Using an SBS (C1102) with 28% styrene (showing a morphology of styrene cylinders dispersed in a butadiene matrix) as reference, the effects of shear and its direction of application can be studied. Samples processed by injection molding show very high birefringence values ($\sim 0.66 \times 10^{-3}$) when measured across the flow direction and negligible values when measured parallel to

Table 8 Birefringence Measures

Heat treatment	None			30 min at 200°C N$_2$ atmosphere		
	X	Y	Z	X	Y	Z
Injection molded	<0.01	0.66	0.75	<0.01	0.70	0.73
THF/MEK casting	<0.01	0.16	0.16	<0.01	0.23	0.23
Benzene/heptane casting	<0.01	0.16	0.16	<0.01	0.22	0.22

Note: The birefringence data are given in ($\times 10^{-3}$).

the flow direction. This agrees completely with the expected morphology of polystyrene cylinders almost totally oriented in the flow direction. The measured total birefringence is due mainly to form birefringence with a small fraction, adding to that due to the orientation. The calculated birefringence value due solely to form birefringence is $\sim 0.5 \times 10^{-3}$ [49, 161]. Heat treatment (30 min at 200°C) of this highly ordered and oriented arrangement hardly affects the order of the domain structure, maintaining the birefringence almost constant.

On the other hand, if samples are obtained by casting from solution, the polystyrene cylinders set without any preferential orientation direction in the plane perpendicular to the solvent evaporation direction, packing one over the others. This piling produces an order in the parallel direction, which gives rise to birefringence. Common values of birefringence measured across the solvent evaporation direction are $\sim 0.16 \times 10^{-3}$ whatever the pair of solvents used. In fact, the sample C1102, which has a nominal morphology of polystyrene cylinders, always keeps this morphology regardless of preferential dissolution. The styrene content (28%) is right in the middle of the range (15~40%) for developing stable cylinders. In this case, heat treatment (30 min at 200°C) allows the domain structure to rearrange, increasing the birefringence to $\sim 0.23 \times 10^{-3}$, but this is still well below the values shown by shear oriented samples.

Thermooptical Behavior

Figure 15 shows the thermooptical behavior of SBS–TPR Cariflex 1102 developed during cyclic heating and cooling. The regularity of the domain structure gives rise to form birefringence [104] detectable using a polarizing microscope.

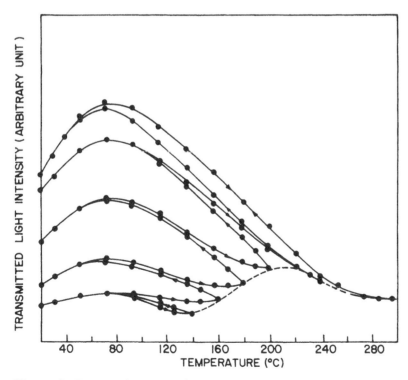

Figure 15 Transmitted polarized light intensity of SBS–TPR Cariflex 1102 subject to cyclic heating–cooling. (From Ref. 159.)

Measurements of the transmitted light intensity between crossed polars can be correlated with birefringence changes via the equation [104]

$$I = I_0 \sin^2 2\alpha \, \sin^2 \frac{\delta}{2} \tag{4}$$

which is applicable to all optically anisotropic specimens, where I = intensity of transmitted light, I_0 = intensity of the polarized light leaving the polarizer, α = angle between the principal axes in the specimen and the vibration direction of the light from the polarizer, and δ = phase difference between the two waves leaving the specimen, giving [159]

$$\left|\frac{\Delta n_i}{\Delta n_f}\right| T = \left|\frac{I_i}{I_f}\right| T \tag{5}$$

Figure 16 shows the birefringence change vs. temperature curve. Above ~140°C this curve begins to show an irreversible increase, which is still present at temperatures as high as ~260°C. The extrapolated curve, presented in the figure as a dashed line, shows the behavior that would occur if only one heating, from room temperature to ~300°C, were employed. The birefringence is still present for temperatures as high as 290°C, as shown in the transmitted light intensity curve which tends asymptotically to a nonzero constant value. This is 40°C higher than noted by Hadziioannou [32, 33] working with styrene–isoprene copolymers. During cooling this increase shows up as an increase in the birefringence at room temperature.

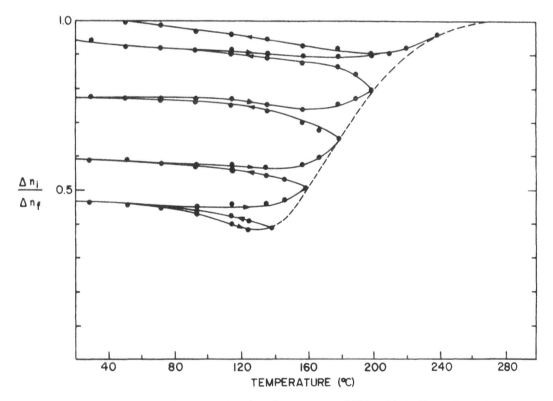

Figure 16 Birefringence change as a function of temperature of SBS–TPR Cariflex 1102 subject to cyclic heating–cooling. (From Ref. 159.)

The polarized transmitted light intensity during heating shows mainly two peaks in the 20~300°C range. The first one, at ~75°C, is due to the glass transition temperature of the polystyrene phase. The second, more interesting, starts at ~145°C, reaches a maximum at ~210°C, and then drops asymptotically to a nonzero constant value. The first peak is reversible but the second is irreversible, giving, on cooling, an increase in the birefringence of the sample at room temperature. This second peak can be further analyzed by a cyclic heating–cooling treatment, in which its development can be seen (see Figure 15). By converting the light intensity into birefringence change, its variation with temperature can be followed. The birefringence doubles its value when the sample is heat treated up to ~300°C. The transformation is a continuous thermoactivated process during which simple feature changes, obviously showing lower activation energy, take place at lower temperatures and more complex rearrangements involving groups of long chains only start to rearrange at higher temperatures. If at any stage during this reorganization the temperature is reduced, the rearrangement is frozen and the birefringence at lower temperatures will be larger than the initial value because of greater order.

Birefringence of Samples Extruded at Constant Shear Stress

Figure 17 shows a log $\Delta n \times 1/T$ plot of the birefringence of extruded SBS–TPR Cariflex 1102. The samples were extruded at constant shear stress at various constant temperatures

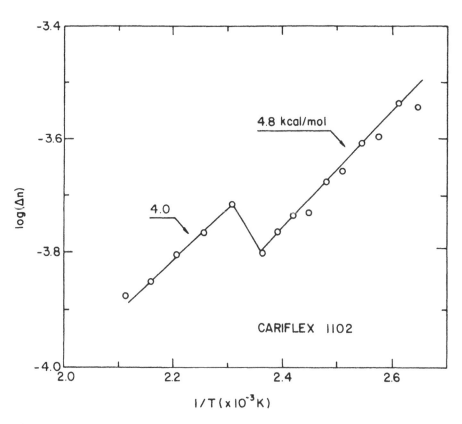

Figure 17 Birefringence of SBS–TPR Cariflex 1102, measured at room temperature, as a function of extrusion temperature. (From Ref. 164.)

and the birefringence was measured at room temperature. As the extrusion temperature increased the birefringence decreased linearly until T_T was reached (between 150 and 160°C) when all results were shifted upward to produce a line approximately parallel to that at the lower temperatures. The difference in the calculated apparent activation energies is very small ($\Delta\epsilon = 0.8$ kcal/mol).

A similar plot can be obtained for SBS–TPR Solprene 415, as presented in Figure 18. The transition lies between 170 and 180°C, and slightly larger values of each particular calculated apparent activation energy and their difference are obtained. (Cariflex 1107 cannot be analyzed in this way because, having a domain structure of polystyrene spheres dispersed in a polybutadiene matrix, it does not show form birefringence.)

Birefringence of MFI extrudates, measured at room temperature, is a superposition of two factors occurring simultaneously. First, birefringence is caused by the molecular orientation. The level of molecular orientation shown at room temperature is dependent on the extrusion temperature: the higher the extrusion temperature, the lower the retained orientation (due to greater relaxation). Thus as the extrusion temperature increases the orientation birefringence contribution reduces. The second factor is the contribution of the form birefringence due to the regular domain structure. Above the transition temperature the flowing melt is formed by the dynamic state of the two instantaneously segregated rich

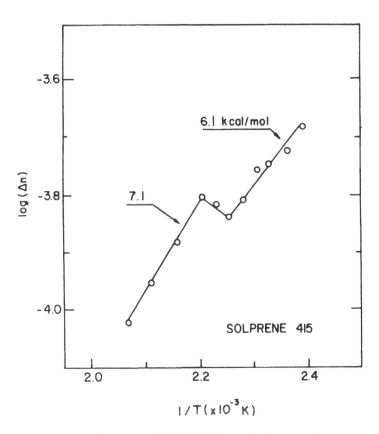

Figure 18 Birefringence of SBS–TPR Solprene 415, measured at room temperature, as a function of extrusion temperature. (From Ref. 162.)

phases with higher mobility (lower flow activation energy) and so during cooling better separation of the two highly incompatible phases can be achieved, higher than that expected by extrapolation from data at temperatures below the transition. In any case this process has a very low temperature activation energy (4~7 kcal/mol in an Arrhenius regime). This effect increases the long-range order and so the form birefringence contribution.

The total birefringence behavior shown by extruded TPR as a function of temperature is therefore as follows: below T_T the birefringence decreases due to the contribution of the orientation; at T_T it undergoes a sudden increase due to the partial contribution of the form birefringence; above T_T it decreases at a smaller rate. This involves contributions from form birefringence, which increases slowly with temperature and orientation and reduces with temperature [162].

Figure 19 shows schematically the contribution of each birefringence (form and orientation) to the total measured birefringence [164].

Birefringence of Samples Extruded at Constant Shear Rate

The birefringence, like the stress at yield, is affected by all the different processing conditions. When using different heat treatment temperatures, the birefringence of SBS–TPR extrudates increases with increasing temperature and/or decreasing shear rate, as

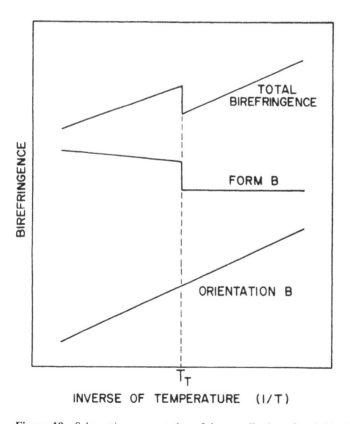

Figure 19 Schematic representation of the contribution of each birefringence (form and orientation) to the total measured birefringence. (From Ref. 164.)

presented in Figure 20 (with samples extruded at 180°C). The greater the shear rate the greater is the forced mixing of the two phases, yielding a structure less ordered and thus reducing the birefringence.

Combined Opticomechanical Behavior

Combining the data from the two previous sections, mechanical and optical properties, as stress at yield and birefringence respectively, one plot can give an overall picture of an extruded SBS–TPR temperature–structure–property relationship. The result is seen in Figure 21. Samples without heat treatment are denoted *untreated* and the others are labeled with the heat treatment temperature to which they were conditioned. The heat treatment increases both the birefringence and the stress at yield but not in a linear way. Instead a steplike behavior is shown where the increase in the stress at yield, for samples heat treated at 135°C, is predominant, forming the stage I. When the heat treatment temperature is increased to 175°C the predominant increase is in the birefringence, stage II, and when 220°C is used the major increase is in the stress at yield, stage III. In the

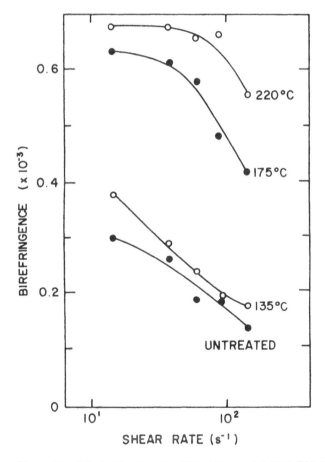

Figure 20 Effect of heat treating (30 min) extruded SBS–TPR Cariflex 1102 at various temperatures on the birefringence. (From Ref. 160.)

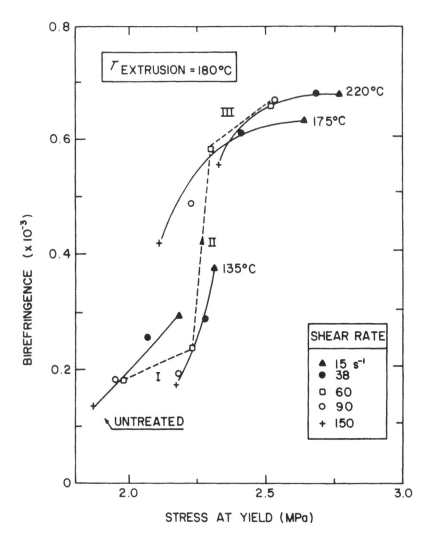

Figure 21 Heat treatment temperature dependence on the birefringence versus stress at yield curve (measured at room temperature) showing the three stages of thermal treatment. (From Ref. 160.)

figure a dashed line is used to follow these three temperature-dependent processes using a set of samples extruded at 60 sec^{-1} chosen arbitrarily. These three stages are internal stress relaxation, rearrangement of the imperfect domain structure, and crosslinking in the rubbery phase. The first and last stages affect principally the stress at yield while rearrangement of the domain structure is more effective in changing the birefringence. Samples subjected to all these three stages do not show the stress reduction after the yield point (Figure 13, upper curves) having a "vulcanized" rubberlike tensile behavior.

THE MODEL

The viscoelastic behavior of block copolymers is not just time-dependent (shear rate and therefore shear stress) as stated by Ghijsels and Raadsen [100], but it is indeed a time–temperature effect. The rubbery chain portions during heating tend to contract due to thermodynamic forces, following the coil model, imposing stress on the PS end portions. This effect can be easily noticed in the form of shrinkage, as high as 15% along the flow direction, when highly oriented samples are heat treated above the T_g of the PS phase. In a static condition and for temperatures below T_T the PS end portions are held firmly inside the PS phase. For temperatures higher than T_T the energy in the PS phase is high enough to allow freedom of movement to the PS chains. Due to the stressing imposed by the rubbery chains, these end portions start to move and in some cases are even removed from their base phase to the rubbery surrounding phase. Thermodynamic incompatibility between the hard and soft segments, which is very high, compels the immediate transfer of this "odd" chain to the nearest PS phase, which could be either the original base phase or a neighboring one. In practice, an equilibrium between these two effects is achieved. The longer the polystyrene end block (the higher the number average molecular weight, M_n), the more entangled the chain is in its original phase and the higher the temperature required to free it. On increasing the temperature above the transition temperature the rubbery chains tend to contract more and more, pulling the polystyrene end block up to a temperature at which crosslinking in the rubbery phase starts to hinder movement. This is the so-called re-arrangement of the domain structure toward a better segregated system, which happens gradually starting at T_T and increasing with temperature up to ~210°C.

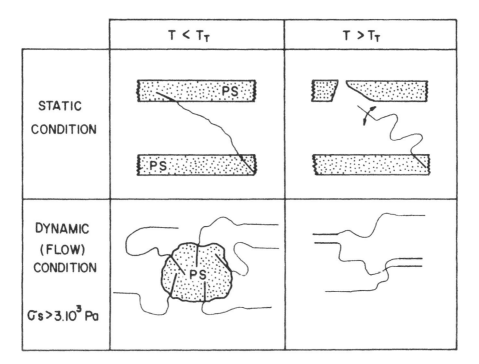

Figure 22 Diagram showing the proposed model (see text for explanation). (From Ref. 162.)

In a dynamic situation where flow is involved the behavior is similar. At temperatures lower than T_T the PS phase in the middle of the star-shaped branched aggregates has not yet obtained energy enough to allow the free movement of the PS chains from it. So the flow occurs by the Arnold and Meier mechanism. For temperatures above T_T the PS end portions can be moved and removed from the center, destroying the stable aggregates, acting as a common thermoplastic [162, 163]. A diagram of this model is shown in Figure 22.

The rearrangement of the domain structure can be accomplished in two ways: temperature (alone) or temperature and shear. The former acts in a broad range (from ~150 to ~240°C for Cariflex 1102); easier moving chains change at lower temperatures while more entangled chains start to rearrange only at higher temperatures [159, 160]. The combination of temperature and shear is much more effective and the rearrangements can occur at the melt transition temperature T_T.

APPLICATIONS OF THERMOPLASTIC RUBBERS

Due to their specially formulated chemical structure thermoplastic rubbers are a unique class of polymers, combining the properties of vulcanized rubber with the processing advantages of conventional thermoplastic. This allows them to soften and flow under heat and pressure and then, on cooling, recover their elastomeric behavior (stretching many times their original length and after release recovering rapidly to their original shape).

Thus articles formed of thermoplastic rubber compounds have physical properties similar to those of vulcanized elastomers which can be manufactured using any thermoplastic processing techniques.

Rubber Bases for Compounds

Injection molding is one of the important techniques to form parts that have high degrees of elasticity and flexibility. Due to the natural anisotropy of these materials, effects of molecular orientation under normal molding conditions are observed. This orientation, a function of the shear employed, is manifested by greater stiffness and modulus in the direction parallel to the flow. This undesirable effect can be lessened to a certain degree, optimizing the mold design. This can be done in compositions with or without processing oil, antioxidant, and UV absorbers, easing the processing and increasing the product performance. Shoe soles for footwear, automotive parts, and flexible molded articles are some of the most common applications.

Thermoplastic rubber can also be molded by extrusion in conditions very similar to those used to process common thermoplastic. Profiles can be made successfully with simple slot or plate dies due to their very low die swells in which the die land length is not critical. Tempered and polished dies produce the best extrudates. The recommended screw design features for processing these rubbers include low compression ratios and rather deep flighted metering sections. The processing temperature range depends on the saturation of the rubbery midblock; unsaturated chains are more susceptible to oxidation needing lower temperature ranges (105 ~ 200°C); saturated midblock chains withstand higher processing temperature ranges (190 ~ 260°C). Some common applications obtained by extrusion are elastic bands, flexible tubes and pipes, film and sheets (heel top pieces for women's shoes can be clicked out of extruded sheets for use in repair work). A new application that has achieved sudden importance is wire and cable insulation and jacketing as a substitute for more expensive polymers.

Blow molding is another processing technique suitable for thermoplastic rubbers. The range of operating temperature is approximately the same as above with preference to the types of machine that produce an intermittent fast drop parison to minimize parison sag [165]. Flexible bottles and containers can be obtained with this technique.

Solution in an appropriate solvent (or mixture of solvents) with addition of a wide range of tackifying resins and plasticizers yields adhesives with varying stiffness, tack, softening point, or open time. The use of aromatic tackifying resins, such as coumarome-indene resin and styrene-based resin produces adhesives with high stiffness and good resistance to creep and heat. On the other hand, use of aliphatic resins such as wood rosin and polyterpene gives softer adhesives with permanent aggressive surface tack. Aliphatic plasticizers such as petroleum-based processing oils or low molecular weight tackifying resins soften the compound and increase its elongation with a minimum effect on the cohesive strength of the adhesive. Aromatic plasticizers associate with the polystyrene phase of the block copolymer which is thus softened. The cohesive strength is thereby drastically reduced. Using a specially developed formulation, solution-based adhesives can be employed as pressure-sensitive laminating or contact adhesives according to need. Some typical applications of the first case are tapes of widely differing nature (heavy duty, insulation, packaging, masking, surgical, general purpose), plasters, labels, carpet tiles, decorative wall covering, and stickers. Laminating adhesives are used as protective and decorative wrapping of books, record sleeves, wall charts, etc. The third case is bonding agent to a wide range of materials such as paper, textiles, leather, rubber, wood, for furniture, building, footwear, and packaging industries.

The coating operation of applying an adhesive layer to substrates can also be done via hot melt. The molten thermoplastic rubber is applied at high speed (up to 300 m/min) to paper or a variety of plastic films such as PVC, polypropylene, polyethylene, polyester, and cellophane.

Blending Agents

Thermoplastic rubber can also be used in low concentration as a blending agent with other polymeric materials to increase impact properties. It can be used as an interphase compatibilizer between polystyrene and polybutadiene in high-impact polystyrene formulations, with high-density polyethylene producing films having exceptional impact and puncture resistance combined with sufficient flexibility to be wound on a reel and to facilitate processing in the rack making, printing, and sealing stages of production. In blends with copolymer polypropylene, the automotive industry uses thermoplastic rubber for bumpers, dashboards, steering wheels, glove compartments, spoilers, air filters, heating and ventilation units, radiator fans, etc. Blends with bitumens have shown an improved performance of the compound in terms of longer fatigue life, reduced permanent deformation, better tensile properties, etc.

In summary, this work attempted to compile scattered information from the triblock copolymer–thermoplastic rubber field, but this chapter is not exhaustive by any means. Rather, evolution in this field is dynamic with everyday innovations.

CONCLUSIONS

Rheological measurements supported by mechanical and optical behavior were used to investigate the molten state of SBS thermoplastic rubbers. Certain solid-state properties of samples with well-defined processing histories were examined, since they are obviously related to the structure of the melt. Transitions in the molten state might be expected to influence the properties of the product and therefore received special attention.

The following conclusions can be drawn:

1. SBS-thermoplastic rubbers showed non-Newtonian flow behavior in the shear rate range studied ($1-1000$ sec^{-1}), which can be reduced to a master curve by time–temperature superposition.

2. The shift factors employed during superposition and the inverse of the melt flow index (MFI) can also be time–temperature superposed.

3. The viscoelastic melt behavior exhibits a transition which is dependent only on the average molecular weight (size) of the polystyrene end blocks. The relationship is linear: the higher the end block size the higher the transition temperature. The melt transition can be detected in plots of two theoretical models, those of Arrhenius and Vogel.

4. The melt transition temperature T_T is interpreted as the temperature above which the polystyrene end block of the polymer chains can be moved within, and removed from, the polystyrene base phase by the attached rubbery chains. (a) In a static state on heating, due to the thermal contraction of the flexible rubbery chains and the high incompatibility between the two chain portions, the polystyrene chains will be pulled to more stable positions, increasing the segregation and thereby causing the rearrangement and sharpening of the domain structure. (b) In a dynamic state the effect is observed as a significant reduction in the apparent flow activation energy. The star-shaped aggregates with a polystyrene phase in the center will not be stable at temperatures above T_T, breaking down and flowing as a thermoplastic.

5. The total birefringence shown by extruded thermoplastic rubbers at increasing temperatures is composed of two parts. There is a contribution which always reduces with increasing temperature resulting from molecular orientation. In addition, there is a contribution from form birefringence associated with the domain structure; this increases suddenly at the melt transition temperature.

6. The stress at yield can also be used to quantify the quality of the polystyrene phase segregation, i.e., how well-developed the domain structure is.

7. The thermosoftening process of styrene–butadiene–styrene thermoplastic rubber during heating can be summarized as occurring in three stages: (a) In the range $70{\sim}95°C$ the polystyrene phase goes through its glass transition temperature, which can be better detected by dynamic analysis (DMTA). Static measurements (DSC and specific volume) are techniques not recommended. (b) Starting at T_T the imperfect initial domain structure begins to rearrange into a well-segregated two-phase structure. This is a thermoactivated process which takes place continuously to approximately $240{\sim}250°C$. (c) Meanwhile starting from $200{\sim}210°C$ crosslinking in the rubbery phase begins to occur, enhancing the modulus and obstructing the flow behavior.

8. Finally, heat treatment of extruded SBS–TPR induces changes with three stages, which can be detected in the birefringence vs. stress at yield plot: (a) internal stress relaxation ($\sim135°C$); (b) rearrangement of the domain structure ($T_T \leq T \leq 240°C$); (c) crosslinking in the rubbery phase ($T > 200°C$).

REFERENCES

1. F. Rodriguez, in *Principles of Polymer Systems*, McGraw-Hill, New York, p. 573 (1983).
2. H. J. M. A. Mieras and E. A. Wilson, *J. IRI, 7*: 72 (1973).
3. A. Noshay and J. E. McGrath, in *Block Copolymers: Overview and Critical Survey*, Academic Press, London, p. 7 (1977).
4. A. Tinker, *Plast. Rub. Int., 11*: 24 (1986).
5. L. J. Fetters and M. Morton, *Macromolecules, 2*: 453 (1969).
6. J. C. Saam, D. J. Gordon, and S. Lindsay, *Macromolecules, 3*: 1 (1970).
7. R. E. Cunningham, M. Auerbach, and W. J. Floyd, *J. Polym. Sci., 16*: 163 (1972).
8. M. J. Folkes and A. Keller, in *The Physics of Glassy Polymers* (R. N. Haward, ed.), Applied Science Publishers, London, p. 548 (1973).
9. G. L. Wilkes and R. S. Stein, *J. Polym. Sci. Polym. Symp., 60*: 121 (1977).
10. A. Skoulios, *J. Polym. Sci. Polym. Symp., 58*: 369 (1977).
11. G. E. Molau, in *Block Polymers*, Plenum Press, New York, p. 102 (1970).
12. B. Gallot, *Pure Appl. Chem., 38*: 1 (1974).
13. D. B. Alward, D. J. Kinning, E. L. Thomas, and L. J. Fetters, *Macromolecules, 19*: 215 (1986); D. J. Kinning, E. L. Thomas, D. B. Alward, L. J. Fetters, and D. L. Handling, Jr., *Macromolecules, 19*: 1288 (1986).
14. D. J. Meier, *J. Polym. Sci. Polym. Symp., 26*: 81 (1969).
15. G. Holden, E. T. Bishop, and N. R. Legge, *J. Polym. Sci. Polym. Symp., 26*: 37 (1969).
16. M. Morton, *J. Polym. Sci. Polym. Symp., 60*: 1 (1977).
17. A. L. Bull, *Plast. Rub. Mat. Appl., 2*: 27 (1977).
18. E. T. Bishop and S. Davison, *J. Polym. Sci. Polym. Symp., 26*: 59 (1969).
19. A. D. Legrand, G. G. Vitale, and D. G. Legrand, *Polym. Eng. Sci., 17*: 598 (1977).
20. T. Hashimoto and H. Kawai, *Polym. Preprints, 20*: 28 (1979).
21. M. Shibayama, M. Fujimura, Y. Tsukahara, T. Hashimoto, and H. Kawai, *Polym. Preprints, 20*: 246 (1979).
22. T. Hashimoto, M. S. Shibayama, and H. Kawai, *Macromolecules, 13*: 1237 (1980).
23. R. W. Richards and J. L. Thomason, *Macromolecules, 16*: 982 (1983).
24. G. Hadziioannou, C. Picot, A. Skoulios, M. L. Ionescu, A. Mathis, R. Duplessix, Y. Gallot, and J. P. Lingelser, *Macromolecules, 15*: 263 (1982).
25. C. V. Berney, R. E. Cohen, and F. S. Bates, *Polymer, 23*: 1222 (1982).
26. G. Hadziioannou and A. Skoulios, *Macromolecules, 15*: 267 (1982).
27. T. Inoue, T. Soen, T. Hashimoto, and H. Kawai, *Macromolecules, 3*: 87 (1970).
28. M. Shibayama, H. Hasegawa, T. Hashimoto, and H. Kawai, *Macromolecules, 15*: 274 (1982).
29. T. Hashimoto, M. Shibayama, and H. Kawai, *Polym. Preprints, 23*: 21 (1982).
30. M. Shibayama, T. Hashimoto, and H. Kawai, *Macromolecules, 16*: 1434 (1983).
31. M. Shibayama, T. Hashimoto, H. Hasegawa, and H. Kawai, *Macromolecules, 16*: 1427 (1983).
32. G. Hadziioannou and A. Skoulios, *Polymer, 21*: 845 (1980).
33. G. Hadziioannou and A. Skoulios, *Macromolecules, 15*: 271 (1982).
34. T. Hashimoto, Y. Tsukahara, and H. Kawai, *J. Polym. Sci. Polym. Lett. Ed., 18*: 585 (1980).
35. G. Kraus and T. Hashimoto, *J. Appl. Polym. Sci., 27*: 1745 (1982).
36. M. Fujimura, H. Hashimoto, K. Kurahashi, T. Hashimoto, and H. Kawai, *Macromolecules, 14*: 1196 (1981).
37. G. C. Meyer and J. M. Widmaier, *J. Polym. Sci. Polym. Phys. Ed., 20*: 389 (1982).
38. T. G. Fox, *Bull. Am. Phys. Soc., 1*: 123 (1956).
39. E. A. Dimarzio and J. H. Gibbs, *J. Polym. Sci., 40*: 121 (1959).
40. J. M. Pochan, C. L. Beatty, and D. F. Pochan, *Polymer, 20*: 879 (1979).
41. Z.-he Lu, S. Krause, and M. Iskandar, *Macromolecules, 15*: 367 (1982).

42. S. Krause, Z.-he Lu, and M. Iskandar, *Macromolecules, 15*: 1076 (1982).
43. M. Iskandar and S. Krause, *J. Polym. Sci. Polym. Phys. Ed., 19*: 1659 (1981).
44. E. Fischer and J. F. Henderson, *J. Polym. Sci. Polym. Symp., 26*: 149 (1969).
45. J. M. Charrier and R. J. P. Ranchoux, *Polym. Eng. Sci., 11*: 381 (1971).
46. D. S. Nandra, D. A. Hemsley, and A. W. Birley, *Plast. Rub. Mat. Appl., 4*: 38 (1979).
47. D. S. Nandra, D. A. Hemsley, and A. W. Birley, *Plast. Rub. Mat. Appl., 2*: 159 (1977).
48. M. Takayanagi, K. Imada, and T. Kajiyama, *J. Polym. Sci. Part C, 15*: 263 (1966).
49. M.J. Folkes and A. Keller, *Polymer, 12*: 222 (1971).
50. S. L. Cooper, D. S. Huh, and W. J. Morris, *Ind. Eng. Chem. Prod. Res. Dev., 7*: 248 (1968).
51. M. Morton, J. E. McGrath, and P. C. Juliano, *J. Polym. Sci., 26*: 99 (1969).
52. E. Pedemonte, A. Turturro, and G. Dondero, *Br. Polym. J., 6*: 277 (1974).
53. J. Diamant and M. Shen, *Polym. Preprints, 20*: 250 (1979).
54. J. A. Odell and A. Keller, *Polym. Eng. Sci., 17*: 544 (1977).
55. R. Séguéla and J. Prud'homme, *Macromolecules, 14*: 197 (1981).
56. T. Pakula, K. Saijo, H. Kawai, and T. Hashimoto, *Macromolecules, 18*: 1294 (1985).
57. A. Weill and R. Pixa, *J. Polym. Sci. Polym. Symp., 58*: 381 (1977).
58. J. C. West and S. L. Cooper, *J. Polym. Sci. Polym. Symp., 60*: 127 (1977).
59. B. E. Read, in *Structure and Properties of Oriented Polymers* (I. M. Ward, ed.), Applied Science Publishers, London, p. 150 (1975).
60. W. Glenz and A. Peterlin, *J. Macromolec. Sci. Phys., B4(3)*: 473 (1970); *J. Polym. Sci., 9*: 1191 (1971).
61. G. W. Urbánczyk, *J. Polym. Sci. Polym. Symp., 58*: 311 (1977).
62. B. Jasse and J. L. Koenig, *J. Polym. Sci. Polym. Phys. Ed., 18*: 731 (1980).
63. M. J. Folkes, A. Keller, and F. P. Scalisi, *Polymer, 12*: 793 (1971).
64. M. J. Folkes and A. Keller, *J. Polym. Sci. Polym. Phys. Ed., 14*: 833 (1976).
65. J. Haslan, H. A. Willis, and D. C. M. Squirrel, in *Identification and Analysis of Plastics*, 2nd ed., Hazell, Watron and Viney, London, p. 441 (1972).
66. T. Hashimoto, N. Nakamura, M. Shibayama, A. Izumi, and H. Kawai, *J. Macromolec. Sci.-Phys., B17(3)*: 389 (1980).
67. H. Odani, K. Taira, N. Nemoto, and M. Kurata, *Polym. Eng. Sci., 17*: 527 (1977).
68. H. Odani, M. Uchikura, K. Taira, and M. Kurata, *J. Macromolec. Sci.-Phys., B17(2)*: 337 (1980).
69. P. J. Flory, *Principles of Polymer Chemistry*, 11th ed., Cornell University Press, Ithaca, N.Y. (1981).
70. D. J. Meier, *Prep. Polym. Colloq. Soc. Polym. Sci. Jpn., 1977*: 83.
71. D. F. Leary and M. C. Williams, *J. Polym. Sci. Polym. Lett., 8*: 335 (1970).
72. D. F. Leary and M. C. Williams, *J. Polym. Sci. Polym. Phys. Ed., 11*: 345 (1973).
73. D. F. Leary and M. C. Williams, *J. Polym. Sci. Polym. Phys. Ed., 12*: 265 (1974).
74. C. P. Henderson and M. C. Williams, *J. Polym. Sci. Polym. Phys. Ed., 23*: 1001 (1985).
75. E. Helfand, *Acc. Chem. Res., 8*: 295 (1975).
76. E. Helfand, *J. Chem. Phys., 62*: 999 (1975).
77. E. Helfand, *Macromolecules, 8*: 552 (1975).
78. E. Helfand and Z. R. Wasserman, *Macromolecules, 9*: 879 (1976).
79. E. Helfand and Z. R. Wasserman, *Polym. Eng. Sci., 17*: 535 (1977).
80. E. Helfand and Z. R. Wasserman, *Macromolecules, 11*: 960 (1978).
81. E. Helfand and Z. R. Wasserman, *Macromolecules, 13*: 994 (1980).
82. L. Leibler, *Macromolecules, 13*: 1602 (1980).
83. G. Hadziioannou and A. Skoulios, *Macromolecules, 15*: 258 (1982).
84. B. J. Bauer and L. J. Fetters, *Macromolecules, 13*: 1027 (1980).
85. T. Hashimoto, *Macromolecules, 15*: 1548 (1982).
86. G. Kraus and J. T. Gruver, *J. Polym. Sci., Part A, 3*: 105 (1965).

87. G. Kraus and J. T. Gruver, *J. Appl. Polym. Sci.*, *11*: 2121 (1967).
88. G. Kraus, F. E. Naylor, and K. W. Rollman, *J. Polym. Sci. Polym. Phys. Ed.*, *9*: 1839 (1971).
89. G. C. Meyer and J. M. Widmaier, *Polym. Eng. Sci.*, *17*: 803 (1977).
90. N. Sivashinsky, T. J. Moon, and D. S. Soong, *J. Macromolec. Sci. Phys.*, *B22*(2): 213 (1983).
91. B. Maxwell and M. Nguyen, *Polym. Eng. Sci.*, *19*: 1140 (1979).
92. N. Sivashinsky and D. S. Soong, *J. Polym. Sci. Polym. Lett. Ed.*, *21*: 459 (1983).
93. K. R. Arnold and D. J. Meier, *J. Appl. Polym. Sci.*, *14*: 427 (1970).
94. C. I. Chung and J. C. Gale, *J. Polym. Sci. Polym. Phys. Ed.*, *14*: 1149 (1976).
95. J. L. Leblanc, *Rheol. Acta*, *15*: 654 (1976).
96. E. V. Gouinlock and R. S. Porter, *Polym. Eng. Sci.*, *17*: 535 (1977).
97. C. I. Chung and M. I. Lin, *J. Polym. Sci. Polym. Phys. Ed.*, *16*: 545 (1978).
98. C. I. Chung, H. L. Griesbach, and L. Young, *J. Polym. Sci. Polym. Phys. Ed.*, *18*: 1237 (1980).
99. J. M. Widmaier and G. C. Meyer, *J. Polym. Sci. Polym. Phys. Ed.*, *18*: 2217 (1980).
100. A. Ghijsels and J. Raadsen, *Pure Appl. Chem.*, *52*: 1359 (1980).
101. W. P. Cox and E. H. Merz, *J. Polym. Sci.*, *28*: 619 (1958).
102. E. M. Slayter, in *Optical Methods in Biology*, Wiley-Interscience, New York, p. 100 (1970).
103. R. S. Stein and G. L. Wilkes, in *Structure and Properties of Oriented Polymers* (I. M. Ward, ed.), Applied Science Publishers, London, p. 57 (1975).
104. M. Born and E. Wolf, in *Principles of Optics*, 2nd ed., Pergamon Press, New York, p. 703 (1964).
105. J. L. S. Wales, in *The Application of Flow Birefringence to Rheological Studies of Polymer Melts*, Delft University Press, Netherlands, p. 111 (1976).
106. A. W. Birley, Unreported work at the Institute of Polymer Technology (IPT), Loughborough University of Technology, Loughborough, Leicestershire, England.
107. R. S. Stein and A. V. Tobolsky, *Text. Res. J.*, *18*: 201 (1948).
108. R. S. Stein and A. V. Tobolsky, *Text. Res. J.*, *18*: 302 (1949).
109. M. M. Qayyum and J. R. White, *J. Appl. Polym. Sci.*, *28*: 2033 (1983).
110. W. L. Senn, *J. Anal. Chem. Acta*, *29*: 505 (1963).
111. E. R. Pico and M. C. Williams, *Polym. Eng. Sci.*, *17*: 573 (1977).
112. N. G. McCrum, B. E. Read, and G. Williams, in *Anelastic and Dielectric Effects in Polymeric Solids*, Wiley, New York, p. 417 (1967).
113. R. C. Rempel, *J. Appl. Phys.*, *28*: 1082 (1957).
114. S. Uemura and M. Takayanagi, *J. Appl. Polym. Sci.*, *10*: 113 (1966).
115. Z. H. Stachurski and I. M. Ward, *J. Polym. Sci. Part A2*, *6*: 1083 (1968).
116. R. W. Gray and N. G. McCrum, *J. Polym. Sci. Part A2*, *7*: 1329 (1969).
117. S. Futamura and E. A. Meinecke, *Polym. Eng. Sci.*, *17*: 563 (1977).
118. F. Annighöfer and W. Gronski, *Colloid. Polym. Sci.*, *261*: 15 (1983).
119. R. Stadler and W. Gronski, *Colloid. Polym. Sci.*, *261*: 215 (1983).
120. J. C. Kelterborn and D. S. Soong, *Polym. Eng. Sci.*, *22*: 654 (1982).
121. L. E. Nielson, *Rheol. Acta*, *13*: 86 (1973).
122. T. Y. Lin and D. S. Soong, *Macromolecules*, *13*: 853 (1980).
123. M. Shen, D. Soong, and D. R. Hansen, *Polym. Eng. Sci.*, *17*: 560 (1977).
124. Shell Elastomers, Technical Manual TR2.1.1, 2nd ed.
125. Shell Elastomers, Technical Manual TR2.1.2, 2nd ed.
126. Shell Elastomers, Technical Manual TR2.1.3, 2nd ed.
127. P. A. Bristow, *Liquid Chromatography in Practice*, hept, England (1976).
128. S. G. Perry, R. Amos, and P. Brewer, *Practical Liquid Chromatography*, Plenum, New York (1972).

129. ASTM D 3593-80 Part 35, p. 882.
130. J. A. Brydron, *Flow Properties of Polymer Melts*, 2nd ed., George Godwin, London (1981).
131. W. Philippoff and F. H. Gaskins, *Trans. Soc. Rheol.*, *2*: 263 (1958).
132. E. B. Bagley, *Trans. Soc. Rheol.*, *5*: 355 (1961).
133. E. B. Bagley, *J. Appl. Phys.*, *28*: 624 (1957).
134. F. N. Cogswell and P. Lamb, *Trans. Plast. Inst.*, *35*: 809 (1967).
135. B. Rabinowitsch, *Z. Phys. Chem.*, *145*: 1 (1929).
136. A. V. Shenoy, S. Chattopadhyay, and V. M. Nadkarni, *Rheol. Acta, 22*: 90 (1983).
137. A. V. Shenoy, D. R. Saini, and V. M. Nadkarni, *J. Appl. Polym. Sci.*, *27*: 4399 (1982).
138. A. V. Shenoy, D. R. Saini, and V. M. Nadkarni, *Rheol. Acta, 22*: 209 (1983).
139. A. V. Shenoy, D. R. Saini, and V. M. Nadkarni, *Polym. Composites, 4*: 53 (1983).
140. A. V. Shenoy, D. R. Saini, and V. M. Nadkarni, *Int. J. Polym. Mat., 10*: 213 (1983).
141. A. V. Shenoy, *Polym. Plast. Technol. Eng.*, *24*: 27 (1985).
142. H. Vogel, *Phys. Z.*, *22*: 645 (1921).
143. G. S. Fulcher, *J. Am. Chem. Soc.*, *8:* 339, 789 (1925).
144. G. Tamman and W. Hesse, *Z. Anorg. Allg. Chem.*, *156*: 245 (1926).
145. M. L. Williams, R. F. Landel, and J. D. Ferry, *J. Am. Chem. Soc.*, *77*: 3701 (1955).
146. G. M. Bartenev, "The Nature and Mechanism of Viscous Flow of the Linear Polymers," Proceedings of the Fifth International Congress on Rheology, Kyoto, vol. 4, p. 313 (1970).
147. R. F. Boyer, in *Encyclopaedia of Polymer Science and Technology,* vol. 2 (suppl.), Interscience, London, p. 745 (1977).
148. R. F. Boyer, *Polym. Eng. Sci.*, *19*: 732 (1979).
149. J. B. Enns, R. F. Boyer, H. Ishida, and J. L. Koenig, *Polym. Eng. Sci.*, *19*: 756 (1979).
150. R. F. Boyer, *J. Polym. Sci. Polym. Phys. Ed.*, *23*: 1, 21 (1985).
151. J. K. Gillham, *Polym. Eng. Sci.*, *19*: 749 (1979).
152. D. J. Plazek, *J. Polym. Sci. Polym. Phys. Ed.*, *20*: 1533 (1982); D. J. Plazek and G. F. Gu, *J. Polym. Sci. Polym. Phys. Ed.*, *20*: 1551 (1982).
153. S. J. Orbon and D. J. Plazek, *J. Polym. Sci. Polym. Phys. Ed.*, *20*: 1575 (1982).
154. J. Chen, C. Kow, L. J. Fetters, and D. J. Plazek, *J. Polym. Sci. Polym. Phys. Ed., 20*: 1565 (1982).
155. J. Chen, C. Kow, L. J. Fetters, and D. J. Plazek, *J. Polym. Sci. Polym. Ed., 23*: 13 (1985).
156. A. Ehringhaus, *Z. Kristallogr. (A)*, *102*: 85 (1939).
157. J. W. Gifford, *Proc. R. Soc. (A)*, *84*: 193 (1910).
158. H. H. Yang, M. P. Chouinard, and W. J. Lingg, *J. Polym. Sci. Polym. Phys. Ed.*, *20*: 981 (1982).
159. A. W. Birley, S. V. Canevarolo, and D. A. Hemsley, *Br. Polym. J.*, *17*: 263 (1985).
160. S. V. Canevarolo, A. W. Birley, and D. A. Hemsley, *Br. Polym. J.*, *18*: 60 (1986).
161. M. J. Folkes (ed.), *Processing, Structure and Properties of Block Copolymers,* Elsevier, London (1985).
162. S. V. Canevarolo, A. W. Birley, and D. A. Hemsley, *Br. Polym. J.*, *18*: 191 (1986).
163. S. V. Canevarolo and A. W. Birley, *Br. Polym. J.*, *19*: 43 (1987).
164. S. V. Canevarolo, Ph.D. Thesis, Loughborough University of Technology, England (1986).
165. *Kraton—Thermoplastic Rubber,* Shell Technical Manual.

2

Crystallization in Elastomers
at Low Temperatures

A. Stevenson
Materials Engineering Research Laboratory Ltd.
Hertford, England

INTRODUCTION

Several elastomer types are capable of crystallization, the principal of which are polyisoprene (natural rubber) and polychloroprene (neoprene). At high strains crystallization can be a desirable feature of elastomers for engineering applications since it can enhance resistance to fracture and fatigue. Its occurrence is related to the level of strain and will be most rapid at the tip of a crack where strains are high. This provides a natural reinforcing mechanism and crystallizing elastomers show better fatigue resistance than other elastomers at high strains. Crystallization can also occur in these elastomers at low temperatures, even without strain, when the consequences (increased modulus and enhanced stress relation rates) can be deleterious for engineering applications. It is important to clearly differentiate between crystallization and glass transition. An elastomer below glass transition will be brittle and may still have an amorphous structure, with a modulus 1000 times that in the rubbery state. The transition to the glassy state occurs when the local temperature is low enough. Crystallization, on the other hand, is a slow process where crystallites grow from within the amorphous matrix over an extended period of time. At equilibrium, less than half the material is crystalline and the elastic modulus increases by a factor of about 100. Rubber is not brittle but tough in this semicrystalline state and shows some characteristics of semicrystalline plastics (e.g., a yield point).

The fact of crystallization in elastomers has been well established for many years [1–9]. This chapter reviews the background literature and theory, concentrating on recent work on elastic modulus effects of direct importance for rubber engineering components exposed to low or subzero temperatures, such as structural bearings for bridges, deep sea oil platforms, or helicopter rotor mounts.

CRYSTALLIZATION PHENOMENA

Crystallization is characterized by several physical phenomena. The rubber crystal unit cell is of higher density (1.0 g/cm^3 for NR [9]) than is amorphous unfilled rubber (0.91 g/cm^3 for NR). Crystallization is thus accompanied by an increase in density (i.e., decrease in volume) of up to 10%. Small volume changes (less than 3% for NR) have been measured by both hydrostatic balance and dilatometric techniques [3, 4]. Crystallite formation also involves the evolution of heat of fusion, and calorimetry has been used by Gorritz and Muller [10] to derive values for the rate and degree of crystallization in NR in broad agreement with volume change data. Differential thermal analysis (DTA) has also been exploited [11, 12], mainly to determine polymer crystal melting temperatures but also (less frequently) to determine crystallization kinetics [13]. Crystallite formation also changes the optical properties of rubber. Treloar [14] and more recently Stein and co-workers [15] have used birefringence and low-angle light-scattering techniques to provide evidence for crystal morphology.

When rubber is held at a fixed strain, crystallization is accompanied by a progressive relaxation of stress—in many cases to and beyond zero. Gent [17–19] made extensive use of both stress relaxation and volume change measurements in investigating the kinetics of crystallization of various vulcanized natural rubbers at different temperatures and strains. This contrasts with other stress changes reported for other polymers [16]. As stress relaxes toward zero the elastic modulus increases dramatically (by up to 2 orders of magnitude). Leitner [20] studied this effect in unvulcanized NR at 0°C and found an approximately linear relation between the increase in Young's modulus and volume change decrease. Subsequent to Leitner's work, direct measurements of elastic modulus were neglected in studies of elastomer crystallization until quite recently [45].

The process of crystallization may be considered in three stages: crystal nucleation, crystal growth, and the equilibrium partially crystalline state. Embryo nuclei may continuously form and grow even in amorphous rubber. However, below a certain critical size they are unstable enough to disappear due to random thermal motions of the molecular chain and may not have measurable macroscopic consequences. Nucleation may occur spontaneously and homogeneously throughout the amorphous phase (most likely at low temperatures) or be "seeded" by foreign surfaces or structural discontinuities. The type and density of nucleation are likely to strongly influence the character of subsequent crystal growth [21].

Above the critical size, nuclei become stable, permitting crystal growth to occur. Crystal growth involves all the physical consequences discussed previously and its kinetics may be characterized in terms of progressive changes in almost any one of them. Avrami [23–25] derived rate relationships for crystals growing uniformly in metals in one, two, or three dimensions. These expressions have been used with some success [17] to describe the kinetics of polymer crystal growth, where the equilibrium state is caused not by neighboring crystals impinging directly on each other (as assumed by Avrami for metals) but by a more complex mechanism of polymer chain restrictions. This may or may not be characterized in terms of spherulites of partially crystalline material behaving in a manner analogous to that of metal grains.

The ultimate degree of crystallinity is below unity and characterizes an equilibrium state, which is a function of the rubber crosslink density [17], temperature [19], and strain [27]. Flory [27, 28] used thermodynamic arguments to derive theoretical expressions for the relaxation of stress due to crystallization (occurring under fixed strain) for the equilibrium degree of crystallinity in terms of strain and temperature, and for the melting temperature as a function of the crystallization temperature. In spite of several somewhat broad assumptions Flory's theory has been shown [17] to be in reasonable agreement with experimental evidence. Various theoretical refinements have been reported [29, 30], although not infrequently the experimental evidence supporting them seems somewhat thin. The main theoretical concepts are discussed in more detail in the following section.

THEORETICAL BACKGROUND

Crystallization becomes theoretically possible whenever the melt free energy exceeds the crystal free energy, enabling a reduction in free energy to result from the crystallization process. Theoretically this may occur whenever $G_{melt} > G_{cryst}$ and will proceed until ΔG is at a minimum. The process may be described thermodynamically:

$$\Delta G = G_{cryst} - G_{amorph}$$
$$= \Delta H - T\,\Delta S \tag{1}$$

where G_{cryst} and G_{amorph} are the Gibbs free energies of the crystalline and amorphous phases respectively, ΔH is the enthalpy of crystallization, and ΔS is the entropy of crystallization.

In practice crystallization in polymer melts and solutions normally occurs only at temperatures lower than Eq. 1 would suggest (i.e., under conditions of some supercooling) and not uniformly over the polymer volume but by growth from small crystal nuclei with large specific areas. Then [22, 32] ΔG in Eq. 1 may be written as

$$\Delta G = \Delta G_c + \Sigma AY \tag{2}$$

where G_c is the bulk free enthalpy change disregarding surface effects, Y is the specific surface free energy, and A is the corresponding surface area.

Nucleation

The equilibrium melting temperature T_m is the characteristic temperature at which all crystallites melt when heated very slowly (i.e., under equilibrium conditions). At temperatures well below this and zero strain (+30°C for NR) primary crystal nucleation is likely to be spontaneous and homogeneous.

Embryo nuclei must, however, grow to a critical size and distribution before any macroscopic effects (such as elastic modulus increase) are observed. There is thus a time lag, called the induction period τ before the observation of any change in properties. Frisch [33] derived a general expression for τ in terms of the statistical distribution of embryo nuclei, and in general the induction period may be written as

$$\tau = \int_{i_{\text{crit}}} \Delta f(i)/I \; di \tag{3}$$

where

$\quad\quad I$ = the nucleation rate (in nuclei per second)

$\quad\quad i$ = the parameter characterizing nucleus size

$\quad\Delta f(i)$ = the difference between the steady-state distribution of nuclei of size i (after long times) and the initial distribution

As a first approximation the nucleation rate may be considered constant throughout the induction period (when nuclei are of subcritical size). This effectively assumes time- and size-dependent crystal surface free energies. Equation 3 may then be simplified to

$$\tau = \frac{N_c}{I} \tag{4}$$

where N_c represents the increase per mole in the number of nuclei of critical size required to produce an observable effect (a modulus increase in the present case).

The classical approach to crystal nucleation was developed primarily by Gibbs [34] and rests on the assumption that fluctuations in the supercooled phase can overcome the energy barrier provided by the crystal surface. The probability of finding a nucleus of given size at constant volume and energy is, according to Boltzmann's law, a function of the entropy change, and may be given by

$$P \propto \exp\left[\frac{\Delta s}{k}\right] \tag{5}$$

At constant pressure and temperature this becomes

$$P \propto \exp\left[\frac{-\Delta G}{kT}\right] \tag{6}$$

On this basis, Turnbill and Fisher [35] used absolute reaction rate theory to derive the following expression for the rate of crystal nucleation in nuclei per mole per second:

$$I = \frac{NkT}{h} \exp\left[-\frac{\Delta G_1 + \Delta G_2}{kT}\right] \tag{7}$$

The term G_1 represents the free enthalpy of crystallization of a nucleus of critical size. Differentiation of Eq. 2 with respect to size for the equilibrium morphology has been shown [22] to lead to the following expression:

$$\Delta G_1 = \frac{32\gamma^2\gamma_e T_m^2}{(h\Delta T\rho_c)^2} \tag{8}$$

where

T = the crystallization temperature

T_m = the equilibrium melting temperature for zero strain

h = the latent heat of fusion (16.4 J/g for NR)

ΔT = the supercooling, $T_m - T$ (the greater the supercooling the smaller the critical nucleus size)

ρ_c = density of crystalline phase

γ, γ_e = surface free energies of the sides and ends of crystals respectively (the greater γ_e is relative to γ, the more fibrous the nucleus structure)

The term G_2 in Eq. 7 describes essentially the mobility of chain segments and has a temperature dependence similar to that of viscosity. G_2 increases rapidly near the glass transition temperature, but at higher temperatures near to T_m Eq. 7 becomes dominated by ΔG_1. A possible (and simple) mathematical form for ΔG_2 is

$$\Delta G_2 = kT \left[\frac{c}{d + T - T_g} \right] \tag{9}$$

where c and d are material constants and T_g is the temperature below which chain segments have insufficient mobility to join the crystal structure. For present purposes this latter temperature is considered equal to the glass transition temperature. Equation 9 is similar in form to the familiar WLF expression, although the latter cannot be expected to apply to a semicrystalline polymer without modification.

Substituting Eqs. 8 and 9 in Eq. 7 yields

$$I = \frac{NkT}{h} \exp - \left[\frac{b^*}{kT} \left(\frac{T_m}{\Delta T} \right)^2 + \frac{c}{d + T - T_g} \right] \tag{10}$$

where b^* is a material constant and the preexponential factor has been ascribed a value of 6.2×10^{12} sec^{-1}.

This leads [45] to the following expression for the induction period:

$$\tau = a \exp \left[\frac{bT_m^2}{T\Delta T^2} + \frac{c}{d + T - T_g} \right] \tag{11}$$

The constant a has absorbed the preexponential factor and depends on the distribution of nuclei required to cause the minimum observable modulus increase. Boltzmann's constant has been absorbed in the constant b. The constants c and d describe the restriction in chain mobility as the crystallization temperature approaches the glass transition.

Equation 4 predicts an infinite induction period at the equilibrium melting temperature T_m. However, the induction period is predicted to become infinite only at ~20°C below T_g.

Growth Rate

The probability at time t of a given point in the rubber matrix never being incorporated in a crystallite, when the latter grow at a local growth rate R, may be expressed by a Poisson probability distribution,

$$P = e^{-R} = 1 - V_c \tag{12}$$

where V_c is the volume fraction of crystalline material.

If the local growth rate R (i.e., that of a single crystal free to grow without restriction) may be expressed by a power law of the form

$$R = Kt^\eta \tag{13}$$

where η and K are constants, then the volume fraction of crystalline material at time t is given by

$$V_c(t) = 1 - e^{-Kt^\eta} \tag{14}$$

Relations of this form were proposed by Avrami [23] to describe crystallization in metals. They have been successfully used to describe crystal growth in polymers by Gent (natural rubber) [17], Mandelkern (NR and polyethylene) [36], Wunderlich (ethylene terephthalate/ethylene sebecate copolymer) [22], and by other authors.

For polymers, the equilibrium volume fraction of crystalline material V_c may be considerably less than unity, so that Eq. 14 becomes

$$V_c(t) = V_c(\infty)(1 - e^{-Kt^\eta}) \tag{15}$$

If the increase in elastic modulus is a simple linear function of V_c at constant strain and temperature, then the elastic modulus at time t may be written as

$$E(t) = E(0) + A \cdot V_c(t) \tag{16}$$

where A is a constant of proportionality, which may depend on both the crystallization temperature and the applied test strain.

Substitution in Eq. 15 yields

$$-\ln\left[\frac{E(\infty) - E(t)}{E(\infty) - E(0)}\right] = Kt^\eta \tag{17}$$

and

$$\log\left\{-\ln\left[\frac{E(\infty) - E(t)}{E(\infty) - E(0)}\right]\right\} = \log K + \eta \log t \tag{18}$$

A plot of t vs. $-\ln[E(\infty) - E(t)]/[E(\infty) - E(0)]$ on logarithmic scales should therefore produce a straight line of slope η and intercept K

Equilibrium Condition

Crystallization becomes possible when the melt free energy exceeds the enthalpy of fusion, and it occurs to minimize the overall Gibbs free energy G (see Eq. 1). For polymer networks this minimum can occur well before all amorphous material is transferred to the crystalline phase. This is caused by restrictions on chain segment mobility imposed by the network. When a chain segment joins a crystallite, there is an effect on the remaining amorphous polymer which tends to make further crystallization less energetically favorable.

This process has been described in entropy terms by Flory [27]. The entropy of a segment joining a crystallite will decrease relative to its amorphous neighbors in the same chain, while this may lead to an increase in entropy for the remaining amorphous part of the chain. The entire problem of determining the equilibrium conditions theoretically may be seen as that of computing the net entropy change in the amorphous phase. Flory provided a solution to this problem for strained polymer networks, assuming that crystal growth proceeds in the direction of strain and that Gaussian chain statistics apply. This leads to the following expression for the equilibrium degree of crystallinity:

$$V_c(\infty) = 1 - \left[\frac{[3/2 - (6/\pi \bar{n})^{1/2}\lambda - (\lambda^2/2 + 1/\lambda)(1/\bar{n})]}{3/2 - (h/R)(1/T_m - 1/T)} \right]^{1/2} \qquad (19)$$

where $\bar{n}$ is the number of chain segments between crosslinks and R is the gas constant.

The number of chain segments $\bar{n}$ between crosslinks may be derived from mechanical stress–strain measurements by means of the following expression (also due to Flory [27]):

$$\sigma_0 = \left(\lambda - \frac{1}{\lambda^2} \right) \frac{\rho}{M_c} \qquad (20)$$

where σ_0 is the initial stress (before any crystallization), ρ is the amorphous rubber density, and M_c is the molecular weight between crosslinks.

EFFECT OF CRYSTALLIZATION ON ELASTIC MODULUS

Experimental Details

The experiments used three different testpiece geometries. Figure 1a shows the tensile testpieces, which were cut from rubber sheet, with gauge length dimensions $2 \times 2 \times 25$ mm. The shear testpieces consisted of four rectangular parallelepipeds of rubber of dimensions $12 \times 20 \times 2$ mm, bonded to outer steel plates and inner T-pieces, which when pulled apart cause each of the four rubber slabs to deform in simple shear. The compression testpieces (Figure 1c) were circular disks of rubber of dimensions 13 mm dia $\times$ 6 mm bonded to thin outer steel plates of the same diameter. They were compressed uniaxially as illustrated causing rubber to bulge at the periphery according to an approximately parabolic profile.

A model elastomer was used to characterize strain and temperature effects chosen for its rapid rate of crystallization. This was a natural rubber (SMR CV60) vulcanized by heating with 1 part per hundred (by weight) of dicumyl peroxide for 60 min at 160°C to produce a relatively lightly crosslinked vulcanizate.

The testpieces were held in a test jig (shown in Figure 2) at a constant prestrain during each test. At regular time intervals small additional displacements were imposed on the testpiece, accurate to 0.1 mm, while a force transducer measured the corresponding change in force. The lower part of the test jig was encased in an insulated Perspex box to form a test cell, and the whole assembly was located in a temperature-controlled cabinet during each test. Test data were recorded remotely with the aid of appropriate electronics.

Measurements were made of the change in force resulting from a small strain increment ($\pm 5\%$) imposed on the testpiece at intervals throughout the crystallization process. The incremental elastic modulus E was defined as the difference in stress divided by the difference in strain as follows:

$$E = \frac{\Delta F/A}{\Delta l/l_0} \qquad (21)$$

where

ΔF = change in force

Δl = displacement increment

A = true area of cross section in the strained state

l_0 = initial unstrained testpiece length

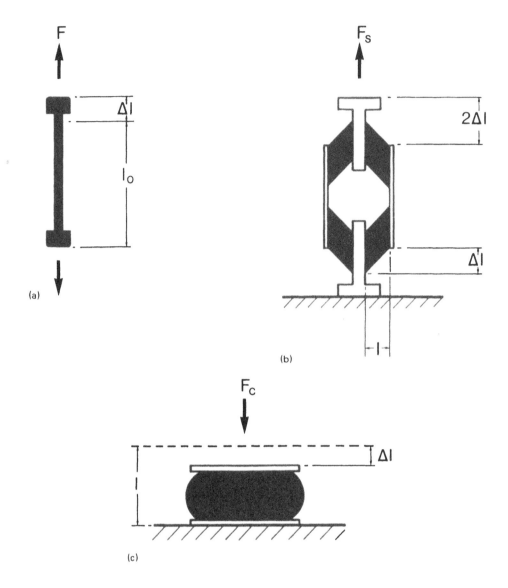

Figure 1 Testpiece geometries: (*a*) tension; (*b*) quadruple shear; (*c*) compression. In each case the rubber part of the testpiece is shown in black.

All measurements were performed without removing the apparatus from the temperature cabinet and without disturbing the insulated test cell encasing the testpieces. The duration of each strain increment was very short (<5 sec) compared to the duration of each experiment (several months) and had no significant effect on the testpieces.

All testpieces were heat treated for 2 hr at 90°C within 24 hr of each experiment. This process reduced variation in the results in a way that suggested the normal existence of a random distribution of nuclei at 22°C. The latter could arise from residual strains from molding the sheets or through random kinetic processes leading to semistable nuclei over long time periods. A similar effect was caused by prestraining tensile testpieces to 300 or

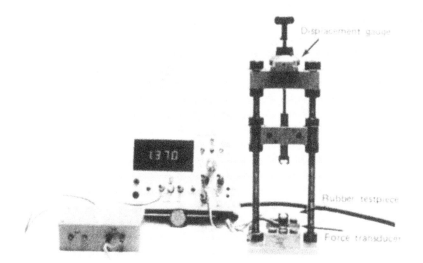

Figure 2 Test arrangement.

500% at 22°C and releasing to zero before starting the test. Subsequent low-temperature crystallization proceeded at a faster initial rate at most prestrains. The importance of careful heat treatment of nominally amorphous rubber testpieces was noted by Gent [17].

It was established by standard thermal conductivity theory [37] (and verified experimentally by means of thermocouples embedded in rubber blocks of various sizes) that the testpieces were within about 5% of thermal equilibrium within 10 min of immersion in the temperature cabinet. The prestrain was imposed on each testpiece at ambient temperature to avoid spontaneous low-temperature nucleation of crystallites occurring before the start of the test. This procedure aims to avoid the growth of "mixed" types of crystallite. In the majority of experiments, induction periods were greater than 30 min.

Effect of Strain

The most noticeable general effect of strain on crystallization is to decrease the induction period before a modulus increase can be measured. The growth rate in fact decreases with increasing strain, and there is no clear effect on the equilibrium degree of crystallinity. These effects have been studied for tensile, compressive, and shear modes of strain and some success has been achieved in unifying the kinetic parameters describing crystallization in these different modes.

Tensile Mode

Typical plots of modulus increase with time are shown in Figure 3 for a series of tensile testpieces held at various prestrains between 10 and 300% extension at −10°C. Each curve is sigmoidal in shape and has several characteristic features.

1. There is a time lag or induction period τ during which the modulus increase does not exceed 10% and is experimentally insignificant.

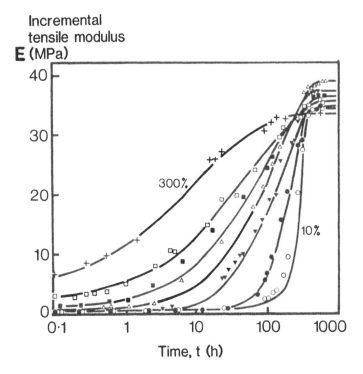

Figure 3 Increase in elastic modulus E with crystallization time t at $-10°C$ for the following tensile strains: $\bigcirc$, 10%; $\bullet$, 50%; $\blacktriangledown$, 100%; $\triangle$, 150%; $\blacksquare$, 200%; $\square$, 250%; $+$, 300%; ———, calculated curves.

2. There is a growth phase characterized by a progressive increase in modulus up to an equilibrium value.
3. There is a characteristic semicrystalline state at equilibrium beyond which no further modulus increase is observed, even after very long time periods. Similar modulus increase curves were obtained at 0, -25, -40, and $-55°C$.

As the tensile strain was increased, the induction period decreased from several days at very low strains and $-10°C$ to a few seconds at high strains above about 200%. Figure 4 illustrates this effect by means of logarithmic plots of nucleation period τ against extension ratio λ for a range of temperatures between 0 and $-55°C$. This figure also shows the discovery of a linear relationship between $\log \tau$ and λ. Extrapolations enable τ at zero strain and λ at elevated temperatures to be derived.

Behavior during the growth phase was characterized by means of Eq. 18. As predicted by the theory, when the parameter $-\ln[E(\infty) - E(t)]/[E(\infty) - E(0)]$ was plotted against crystallization time t on logarithmic scales, straight-line plots were obtained. This was the case at every prestrain tested, as the results of Figure 5 show. The rate of modulus increase during this phase is characterized by the slope of these plots, which has previously been referred to as the Avrami coefficient η. The rate of crystallite growth decreased with increasing strain. The values of these coefficients were used to produce calculated modulus increase curves shown in Figure 3.

The plots of Figure 5 yield values of the crystal growth coefficient η (slope) and the

Nucleation period
τ (h)

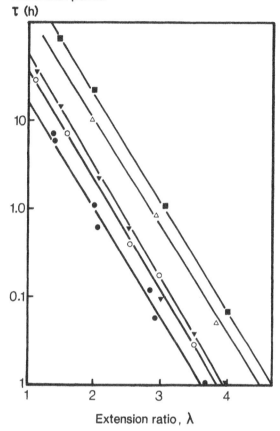

Figure 4 Nucleation period τ from tensile tests of Figure 3 plotted against extension ratio λ for the following crystallization temperatures: △, 0°C; ▼, −10°C; ●, −25°C; ○, −40°C; ■, −55°C.

constant K (intercept), which produce the master plots of Figures 6 and 7 when plotted against extension ratio λ. At zero strain, extrapolation of the curve of Figure 6 yields a value for η of 4.3. The value is below 1 for strains above 140%.

A simple empirical relationship was discovered between η and λ:

$$\eta = \lambda^p \exp[q] \tag{22}$$

This relationship held true for all test temperatures, with values for the material constants p and q of −1.4 and 1.3 respectively. This relationship was also found to hold for other modes of deformation (shear and compression).

The parameter K is subject to greater experimental error than η but appears to follow a smooth curve which intercepts the zero strain axis at $K = 1.4 \times 10^{-11}$. K provides a way of characterizing the early stages of crystallization; it is, however, less directly informative and less accurate than the induction period τ.

The incremental modulus $E(t)$ eventually approached an equilibrium value $E(\infty)$ which at −10°C was about 40 times the initial value. The equilibrium value was not dependent

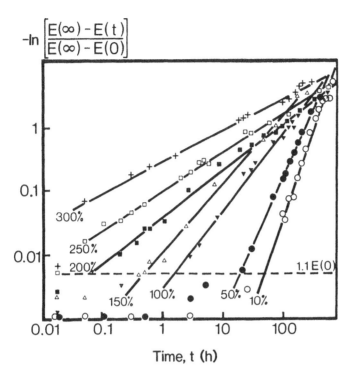

Figure 5 Avrami plots of $-\ln[E(\infty) - E(t)]/[E(\infty) - E(0)]$ against crystallization time t at $-10°C$ for the series of tensile strains of Figure 3. See Figure 3 for key.

on the test strain—a surprising effect in view of the proven influence of strain on crystallite morphology [26], which is contrary to the predictions of Flory's theoretical relation quoted here as Eq. 19.

Shear Mode

The shape of the modulus increase curve in shear was similar to that described for the tensile mode. Figure 8 illustrates a typical set of results for crystallization at $-40°C$ with applied shear prestrains between 20 and 400%. Increasing the shear strain decreased the induction period as it had also done in the tensile mode. The rate of crystallite growth was again successfully described by means of Eq. 17 to the extent that straight-line plots similar to Figure 5 were obtained for every shear strain tested. In most cases there was at least a quasi-equilibrium value of $G(\infty)$, although some experiments were too short to permit definite attainment of equilibrium. Quantitative statements about equilibrium values are therefore subject to some caution. Nevertheless, there was a distinct similarity between the behavior at equilibrium of the shear and tensile modes in the apparent absence of any clear strain dependence.

In simple shear, the shear strain e_s may be related to the angle of deformation θ or to the principal extension ratio λ by

$$e_s = \tan \theta = \lambda - \frac{1}{\lambda} \tag{23}$$

where $\lambda_1 = \lambda$, $\lambda_2 = 1$, $\lambda_3 = 1/\lambda$ and $G = \sigma_s/e_s$, where σ_s is the shear stress.

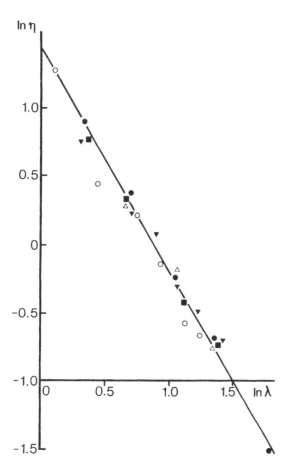

Figure 6 Relationship between crystal growth rate coefficient η and extension ratio λ from Avrami plots of Figure 5 for tensile tests, plotted as ln η vs. ln λ. See Figure 4 for key.

Hooke's law is a better approximation for the deformation of rubber in shear than in either compression or tension. The principal extension ratio can be calculated from the preceding equations for any applied shear deformation [47].

When the induction period τ was plotted against the principal extension ratio, the results were agreed quantitatively with the tensile mode results, as shown by Figure 9. This suggests that a single unique function describes the effect of strain on induction period, which depends not on the mode of deformation but only on the principal strain.

Compressive Mode

Crystallization in compression was associated with increases in modulus, which were qualitatively similar to those described previously for tension and compression. Figure 10 shows a typical set of results for crystallization at $-40°C$ with compressive prestrains between 10 and 70%. There is an induction period followed by a sigmoidal increase to an equilibrium value. Compression differs from the other modes of deformation in that the compression modulus is strongly influenced by shape. For a bonded disk of rubber

$$E_c = E_0(1 + 2S^2) \tag{24}$$

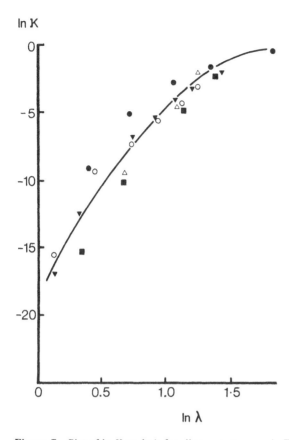

Figure 7 Plot of ln K vs. ln λ for all temperatures as in Figure 4.

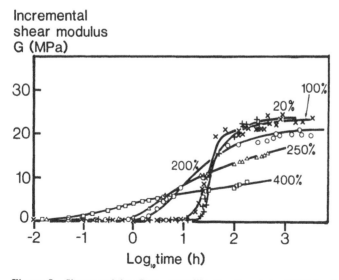

Figure 8 Shear modulus G vs. crystallization time t at $-25°C$ for the following shear prestrains: +, 20%; ×, 100%; ○, 200%; △, 250%; □, 400%.

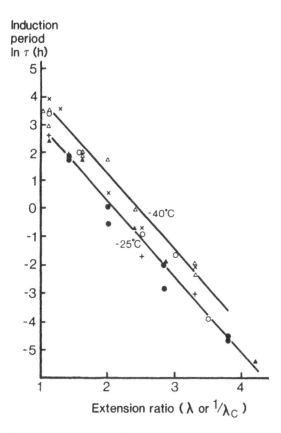

Figure 9 Relationship between nucleation period τ and principal extension ratio for the following cases: tension, ● at $-25°C$ and ○ at $-40°C$; shear, ▲ at $-25°C$ and △ at $-40°C$; compression, + at $-25°C$ and × at $-40°C$.

where E_0 is Young's modulus, S is shape factor ($D/2t$ for a circular disk), D is diameter, and t is the rubber layer thickness.

When a bonded disk is compressed to high strains (such as 70%) the overall stress–strain curve becomes very nonlinear. At this prestrain level the effective shape factor for the small strain cycles used to determine the incremental modulus is increased, resulting in large modulus increases—even without crystallization. Compression of a bonded disk is characterized by a distribution of shear strains adjacent to the outer free surfaces and bulk compression toward the center. The compression modulus is determined by a superposition of these two contributions to stress–strain behavior. The strong shape effect and nonlinearity explain the different initial moduli of Figure 10. Increasing the applied prestrain, however, reduced the time lag or nucleation period, τ, as before.

Figure 9 included data from experiments in compression where induction period was plotted against the reciprocal of the extension ratio $\lambda_c = 1 - \delta l$. It was discovered that the results from all modes of deformation conformed to a single empirical relationship:

$$\tau = e^{\beta\lambda} \tag{25}$$

where

$$\beta = -5/2 \text{ for } \lambda \geq 1$$
$$\beta = +5/2 \text{ for } \lambda < 1$$

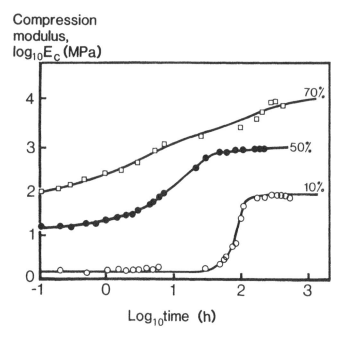

Figure 10 Compression modulus E_c vs. crystallization time t at $-40°C$ for the following compression prestrains: ○, 10%; ●, 50%; □, 70%.

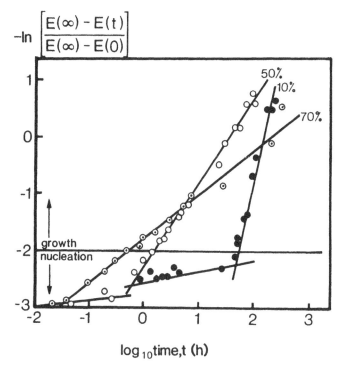

Figure 11 Avrami plots for crystallization in compression at $-40°C$. Tests and key as in Figure 10.

The extrapolated values for τ_0 are the same for compression as for tension and shear. If this result is considered in the more general context of the time-lag analysis of Frisch [33, 44], then it turns out that the same values for constants a, b, c, and d apply for crystallization in compression, with the single exception of a sign change for β [46]. In spite of the large difference in initial compression moduli, the growth phase of crystallization was still adequately described by Eq. 17. This is shown by the linear plots of Figure 11.

In uniaxial compression, the compression ratio λ_c is less than unity and is given by $\lambda_c = 1 - e_c$. If the principal extension ratios along mutually perpendicular axes are represented by λ_1, λ_2, and λ_3, then if the volume of rubber remains a constant, $\lambda_1 \lambda_2 \lambda_3 = 1$. The extension of lateral dimensions corresponding to uniaxial compression ratio λ_c is thus $\lambda_3 = \lambda_2 = \lambda_c^{-0.5}$. However, the crystallization behavior did not scale with this parameter but instead with λ_2^2 (i.e., λ_c^{-1}). In Figure 12, when the rate coefficient η was plotted against λ_c^{-1}, the data coincided with data from other modes of deformation, within experimental uncertainty. It is very interesting that a single relation can be obtained for such different modes of deformation. The two lines drawn on Figure 12 indicate the experimental uncertainty in the values of p and q in Eq. 22.

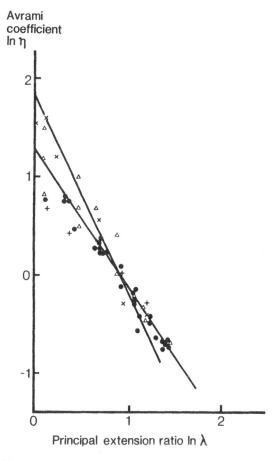

Figure 12 Relationship between crystal growth rate coefficient η for tensile, shear, and compression tests and principal extension ratio λ plotted as $\ln \eta$ vs. $\ln \lambda$. Key as in Figure 9, but with data from tensile tests at all temperatures from Figure 4 plotted as closed circles.

After long exposure periods, equilibrium values for compression modulus were obtained, which did depend strongly on the applied prestrain. This is completely different from the behavior in other modes of deformation where there was no such strain dependence. The strain dependence of the equilibrium modulus can be understood to some extent by means of the equivalent shape factor, as illustrated by Figure 13. It is interesting that shape dependence appeared to persist after crystallization. A fully crystalline material would not be expected to have a shape-dependent elastic modulus, providing further evidence that in elastomers the equilibrium state is one of partial crystallinity with modulus behavior strongly influenced by the amorphous phase.

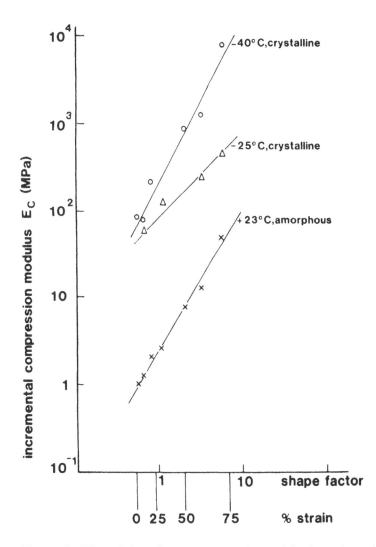

Figure 13 Effect of shape factor on compression modulus in semicrystalline amorphous states. The figure shows compression modulus plotted against equivalent shape factor at the compression strains indicated.

Effect of Temperature

The effect of temperature may be considered in terms of the three stages of the crystallization process, nucleation, growth, and equilibrium. The general shape and nature of the crystallization curve were unaltered by variations in temperature. Even at elevated temperatures there was evidence of similar behavior, although higher strains are required to produce measurable effects and it is clearly easier to obtain reliable data at subzero temperatures. Crystallization phenomena were nonetheless observed in natural rubber at temperatures as high as 100°C. In other work [53], the presence of crystallites at elevated temperatures and high strains has been confirmed by x-ray experiments. Most experiments to characterize the effect of temperature were performed in the range −55 to 0°C.

Experimental results for induction period τ vs. strain at different temperatures produced the series of parallel straight lines shown in Figure 4 for tensile strains. Considering the extrapolated induction period for zero strain at each temperature enabled the effect of temperature to be illustrated more clearly, as in Figure 14. This shape of the curve applies to all strain levels. There was a flat and broad minimum at −25°C for the model natural rubber material. This behavior is consistent with the behavior described by Eq. 11 and the solid line of Figure 14 is a theoretical curve drawn using empirically determined values for parameters a, b, c, and d in Eq. 11. The values in this instance were $a = 0.095$, $b = 12.3$,

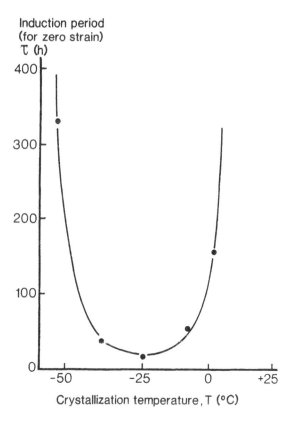

Figure 14 Nucleation period at zero strain τ_0 plotted against temperature T. Experimental ●; solid line is calculated using Eq. 11.

$c = 247$, and $d = 22.5$. Induction period measurements for crystallization in compression showed the relationship between temperature and nucleation period was not changed by the mode of deformation. This was confirmed by shear mode data.

The growth phase of crystallization was successfully characterized by Avrami rate coefficients η in all cases. These were independent of temperature over the complete range studied. This invites the conclusion that, although nucleation is strongly temperature dependent, crystal growth from nuclei is not temperature-dependent. This needs to be understood in the context of the equilibrium condition at the cessation of growth, which did depend strongly on temperature. The elastic modulus at equilibrium increased with decreasing temperature in accord with a possibly linear relationship. Figure 15 illustrates the effect of temperature on the equilibrium modulus for tensile strains. It is possible that the degree of crystallinity depends on temperature but this was not measured directly by x-ray experiments at different temperatures and so this interpretation remains subject to verification.

EFFECT OF CRYSTALLIZATION ON STRESS RELAXATION

During the crystallization tests described, it is also possible to measure absolute stress. This enables stress relaxation to be monitored before, during, and after crystallization.

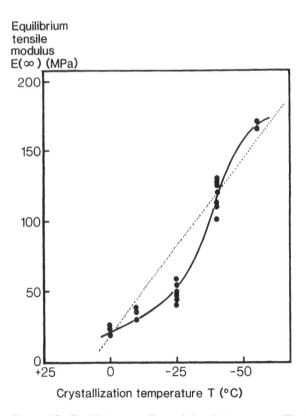

Figure 15 Equilibrium tensile modulus $E(\infty)$ vs. crystallization temperature T. ●, Data for all tensile prestrains are shown; .... averaged empirical trend.

Figure 16 summarizes the behavior for typical tests at 50% prestrain with the model natural rubber material. The very large difference in modulus after crystallization causes curves before and during crystallization to appear accurate when stresses are plotted on the same scale as in fact they are. Amorphous rubber at 23°C has a relatively low stress relaxation rate of 3% per decade of time (3% pd). During crystallization at −25°C this increased to the very high rate of 120% pd. In fact the stress relaxed to zero during this phase and the initial deformation became frozen in. Any further deformations give rise to higher stresses from the new "crystalline" material. After attainment of an equilibrium state of partial crystallinity, a strain increment will again cause stress relaxation, but this time at the intermediate rate of 6% pd.

MECHANICAL BEHAVIOR OF CRYSTALLINE RUBBER

Semicrystalline rubber can to some extent be regarded as a new and different material from amorphous rubber, to which it reverts at temperatures above T_m. The prestrain applied at room temperature becomes frozen-in after crystallization at low temperature, as illustrated in Figure 17. This illustrates a tensile testpiece in its initial amorphous state and

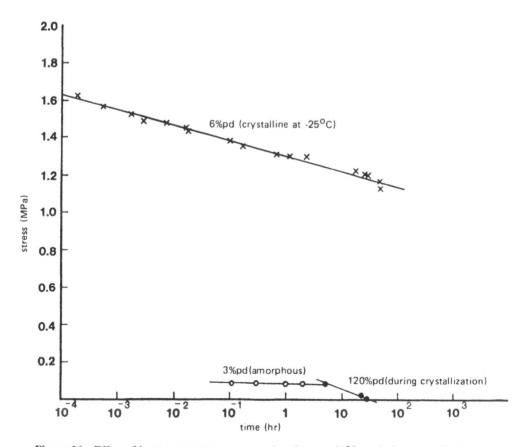

Figure 16 Effect of low temperature on stress relaxation at −25°C: ○, before crystallization; ●, during crystallization; ×, after crystallization.

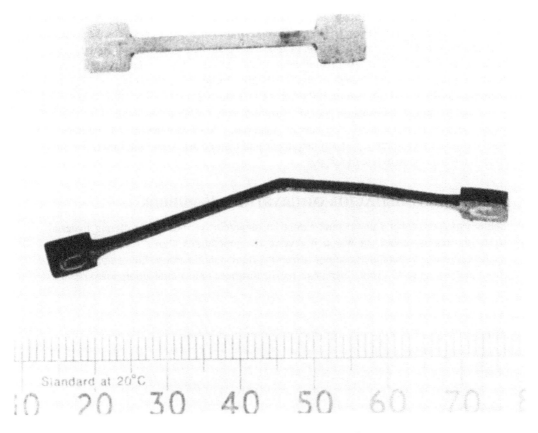

Figure 17 Tensile testpiece before and after crystallization at −25°C.

then after crystallization at 100% and −25°C. In the latter state, the material behaves like a plastic, with high modulus and yield behavior. The central bend shows where the testpiece was flexed and sustained local yielding. After heating to 23°C, this shape disappeared without trace and the testpiece reverted to the shape shown in the upper part of the photograph. A typical force deflection curve for crystalline rubber is seen in Figure 18. This shows an initial small strain region of linear elastic behavior followed by yielding, cold drawing, and eventually fracture.

Small Strain Cycles

Following the attainment of the partially crystalline equilibrium state, repeated small strain cycles were imposed on the rubber to see if this would cause strain softening or crystal melting. These tests were performed using the shear testpieces of Figure 1*b*, after crystallization had occurred at 50% prestrain and −25°C. Figure 19 illustrates the effect of small strain cycles between 50 and 55%. A small reduction in modulus was observed after the first cycle; this softening was less than 7% and even this reduction seemed complete after two cycles. No change was observed after a larger number of cycles of 2.5% amplitude. These results lead to the conclusion that repeated small deformations of crystalline rubber will not cause any substantial softening or crystal melting. This was

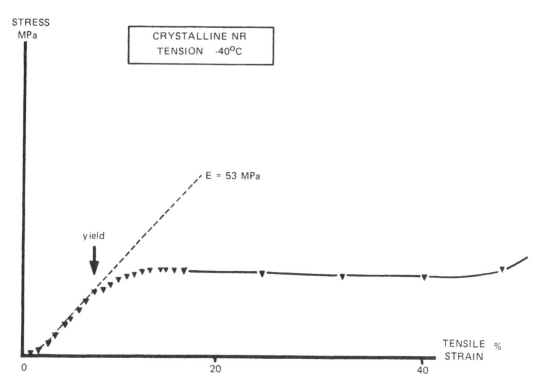

Figure 18 Stress–strain behavior of tensile testpieces after crystallization at −40°C. Yield strain is at 7% and yield stress is 53 MPa.

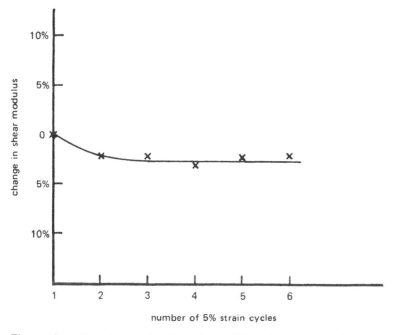

Figure 19 Effect of repeated strain cycles (∓5%) on shear testpiece after crystallization at −25°C.

confirmed in all modes of deformation. Below 7% strain the semicrystalline rubber behaved as a linear elastic material.

Large Strain Cycles—Yield Behavior

Other experiments have shown that if large deformations are applied to rubber after crystallization, then above 7% yielding occurs, followed by a cold-drawing process. This is clearly illustrated for tension by Figure 18. Subsequent (postyield) strain cycles indicate high energy dissipation per cycle and considerable softening. This is illustrated for compression by Figure 20. In this mode also, a yield point occurred at 7%, accompanied by a yield stress of 204 MPa. The results show very high hysteresis per cycle. However, there was no evidence of crystal melting since the modulus remained at least 20 times that of

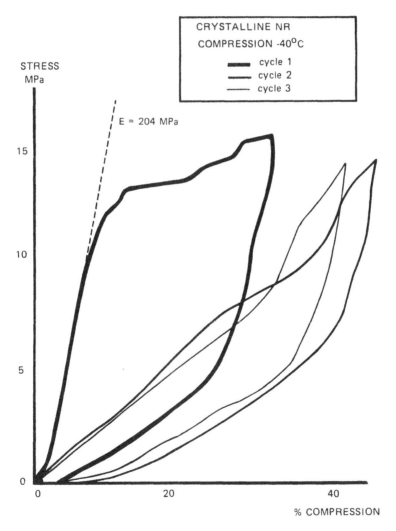

Figure 20 Stress–strain behavior in compression at −40°C. Yield strain is 7% and yield stress is 204 MPa.

noncrystalline rubber at that temperature even after many repeated cycles. The lower two curves of Figure 20 show that even after three cycles softening is virtually complete. It was interesting that although the yield stresses were naturally very different, the yield strains were identical in both modes of deformation.

After a period of several hours at the test temperature there was evidence of recovery after the yield experiments to the initial force–deflection curve, e.g., to cycle 1 of Figure 20. Plastic yielding therefore seemed to break down some of the links in the crystalline network, but not to cause crystal melting. These links then reformed after a further period at the test temperature. Heating the testpieces to melt the crystallites produced testpieces with no trace of permanent damage.

MORPHOLOGY

The morphology of crystalline rubber has been well studied for thin rubber layers using transmission electron microscopy [26]. Strain has been found to have a very strong influence on observable morphological features. In more bulky articles, however, there will be a three-dimensional strain field, which cannot be studied by means of thin layer testpieces and transmission electron microscopy. The three-dimensional nature of the strain field is particularly important in understanding the behavior in compression. An indirect "freeze-fracture" method can provide some information on morphological features. This involves immersing a testpiece to be studied in liquid nitrogen, usually after crystallization has occurred, and then fracturing it by impact. The resulting patterns of the fracture surfaces will reflect any strong morphological features, such as organized planes of crystallites, and can be studied at leisure with a scanning electron microscope. Amorphous control samples were found not to have any regular features, just the typical wave markings of fracture in a brittle glassy polymer below T_g. In contrast, after crystallization in compression, the organization of the markings at the outer edge revealed a clear orientation, which reflected the lines of principal stress during crystallization in compression (see Figure 21). At higher magnifications (Figure 22) ligaments were discovered on another testpiece whose orientation was so far as could be determined the direction that had been the principal strain direction during crystallization. Toward the center of the compression testpiece was a region of very parallel "sheetlike" markings (Figures 23 and 24). This was in a region that had been in hydrostatic compression during crystallization and was strongly suggestive of a platelet structure with cleavage features on a larger scale. The organization of these features on a testpiece crystallized under uniaxial compression is summarized by Figure 25.

The growth of a crystalline phase has been shown [53] to enhance molecular orientation in the remaining amorphous phase. This enhanced orientation combined with some interconnection of the crystallite structure seems the most likely explanation at a morphological level for the very large modulus increases actually observed. Interaction between the crystalline and amorphous phases is also likely to provide a mechanism for limiting the degree of crystallinity at equilibrium (30% for NR). Crystalline growth occurs only with some molecular chain movements which increase the orientation and stored elastic energy in the remaining amorphous phase. This may set up a balancing mechanism that creates an equilibrium degree of crystallinity—which in principle can be different at different temperatures and strains.

Crystallization from rapidly strained melts, strained gels, and strained solutions in which there is considerable molecular extension is generally considered [38–43, 52] to

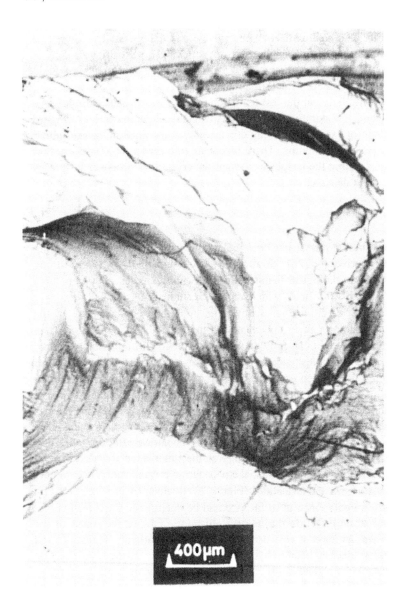

Figure 21 Freeze-fracture surface after crystallization at −40°C and 40% compression. General orientation of ligaments near outer edge of compression testpiece seen at low magnification.

occur with a "chain-extended" morphology. At high prestrains, the crystallization of lightly crosslinked rubber may be similar. There may already be significant chain extension from the initial prestrain. Subsequent crystallization at low temperature will then create a potentially complex final morphology. If crystallization occurs without significant prestrain, a different morphology, of lamellar crystals of the "chain-folded" type, is possible, arranged in a more or less distorted spherulitic pattern, reflecting the overall strain mode and level. In this general way interactions between high strain and low temperature can influence the morphology of crystallization and through this the kinetics and extent of macroscopic changes such as increases in elastic modulus. Low temperatures may also influence nucleation conditions and give preference to large numbers of small (and eventually interconnected) crystallites. An analogy with woven fabric [51] is likely to be more appropriate than one with particulate fillers, due to the very large extent of the modulus effect in some cases. The different modes of deformation are likely to be associated with different geometry strands (fibers, lamellae, platelets, etc.) in the analogy to a woven network.

APPLICATION TO ENGINEERING

The physical behavior described so far has referred to testpieces and to a simple model rubber vulcanizate, chosen to represent accelerated crystallization effects. Crystallization will, however, occur in the same way in engineering components when exposed to low temperatures, and the behavior of these components can be understood by means of the scientific approach described thus far. Quantitative understanding of nucleation, crystal growth kinetics, and equilibrium behavior is indeed the only sound basis for predicting the behavior of components in service. In practice the behavior may be complicated by changes in the strain or temperature conditions during low-temperature exposure. Practical engineering vulcanizates will also have a more detailed chemistry than the model vulcanizate considered so far and this is another factor that needs to be evaluated. This section reviews crystallization phenomena in the context of practical rubber engineering design.

Engineering Components

Bridge bearings are designed to provide a flexible link between a bridge deck and its supporting piers. Figure 26 shows a row of 16 elastomeric bridge bearings supporting the M2 motorway between London and southeast England, after about 20 years service [54]. Figure 27 illustrates the structure of one of these bearings, the rubber being black. Elastomeric bearings are gaining popularity worldwide because of their general lack of maintenance requirements and immunity to seizure due to metal corrosion, to which metallic bearing systems are vulnerable. To function adequately the elastomer must maintain a low shear modulus. In cold climates, some types of rubber vulcanizate will crystallize and stiffen as described in this chapter and the bearing stiffness can increase by an amount unacceptable to the bridge engineer. For this reason there is provision in the appropriate U.K. and U.S. bridge bearing standards, BS 5400.9 and ASTM D4014, for low-temperature tests to exclude unsuitable compounds [59].

A detailed study has been made of crystallization in full-size bridge bearings, supported by the U.K. Department of Transport, and although still not completed for the full range of practical engineering vulcanizates [55], some results have been published for

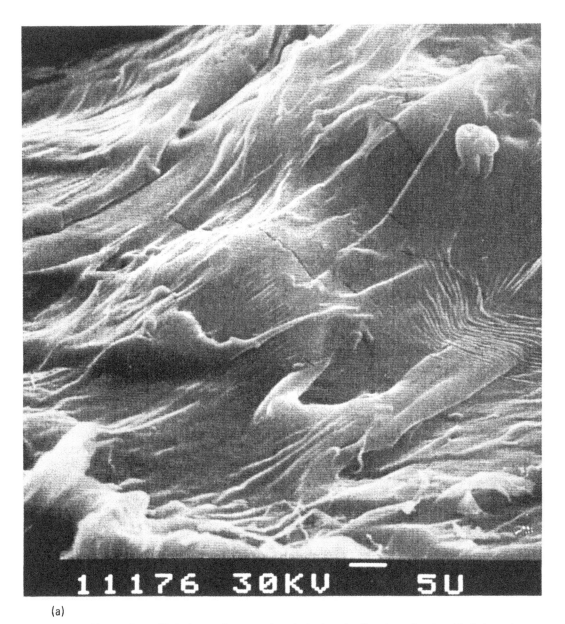

(a)

Figure 22 (*a*) Magnified view of ligaments in principal strain direction. (*b*) Magnified view of typical single ligament in freeze-fracture surface.

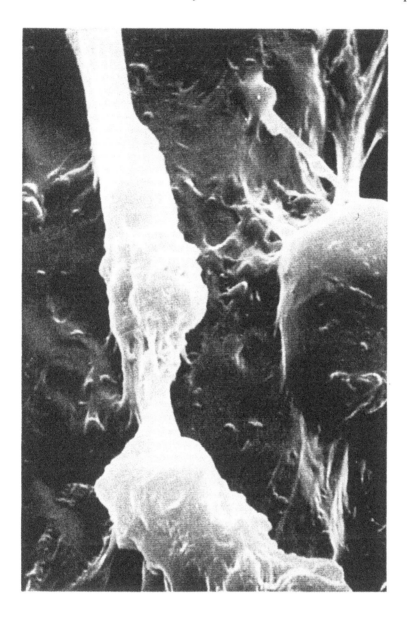

(b)

Figure 23 Fracture surface markings for central region of compression testpiece of Figure 21 crystallized under hydrostatic compression. Shows cleavage steps at low magnification.

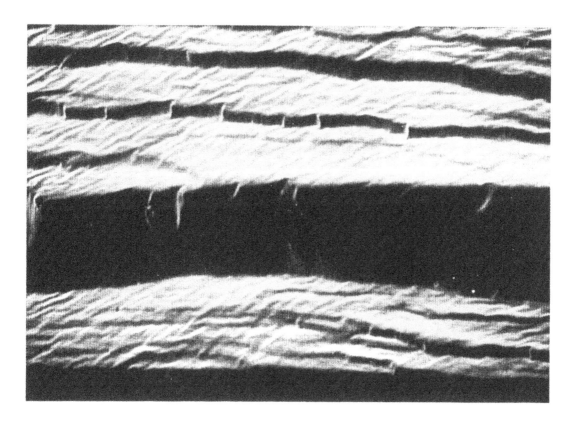

100μm

Figure 24 Magnified view of freeze-fracture surface of central region crystallized in hydrostatic compression. Shows central planes of platelets at high magnification.

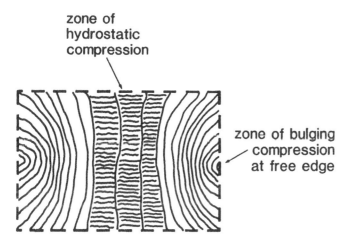

Figure 25 Schematic of orientation of fracture markings for testpieces crystallized at −40°C and 40% compression.

Figure 26 Row of steel-reinforced elastomeric bridge bearings supporting a bridge deck carrying the M2 motorway between London and southeast England.

full-size bridge bearings fabricated using the model elastomer [48]. Figure 28 shows the test arrangement and Figure 29 summarizes some of the results. The bearings were of overall dimensions 432 × 203 × 90 mm and contained six reinforcing steel plates. They were produced for the study by the leading U.K. manufacturer of bridge bearings by normal production techniques. The tests consisted in applying a 25-ton compressive load to a pair of bearings and then superimposing small shear strains of 5% via a centrally located shear plate. An increase in shear stiffness was observed at −21°C from 0.33 kN/mm to about 13 kN/mm. This is consistent with expectations from the laboratory experiments on the same material (see Figure 8). The large size of the bearing did not retard crystallization substantially, since induction periods proved very similar to those of the small testpieces in Figure 8 at the appropriate strain level. The effect of thermal conduction was studied in detail by instrumenting an identical bearing with a three-dimensional array of thermocouples. The results, shown also in Figure 29, demonstrate that the maximum time lag for the center of the bearing, labeled T_1, to cool to the final temperature of the test chamber was 5 hr. This was too short to directly effect the

Figure 27 Bridge bearing after removal from bridge deck sectioned to show internal structure. Rubber parts are black.

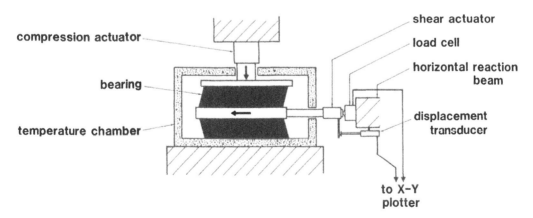

Figure 28 Schematic view of the experimental arrangement for testing full-size bridge bearings at low temperature.

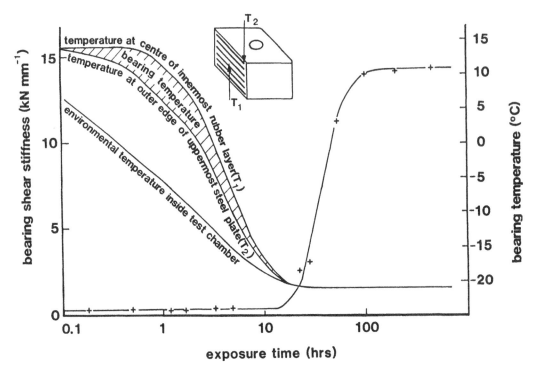

Figure 29 Variation of inner temperature and shear stiffness (+) during exposure to −21°C. T_1 is innermost rubber layer; T_2 is uppermost steel reinforcing layer.

crystallization process, which itself had an induction period of about 30 hr. These tests were performed at nominally constant temperature after the temperature cabinet had cooled to the preset temperature. This took several hours due to the large mass involved. The environment of a bridge bearing in service is likely to be more complex, with temperature varying at very least over daily and seasonal periods. This means that the induction and kinetics of crystallization are likely to be of importance in determining the actual behavior in service. A rise in temperature exceeding 20°C will in most cases be sufficient to cause substantial crystal melting [56].

Structural bearings for deep-water oil platforms have been designed which incorporate flexible elastomeric joints as a means of ensuring low stiffness under lateral forces [56]. One such design is shown in Figure 30, with a single laminated rubber ball joint of about 30-ft diameter located at the base of the tower. Since this platform is designed for hydrocarbon production at water depths of 1000 ft and more, the rubber bearing will be permanently at deep-water temperature. At these locations temperature is about 3°C for the 30-yr life of the platform. This is sufficient time for equilibrium to be reached with any crystallization that may occur in the rubber elements at this depth. Some vulcanizates will crystallize at this temperature and cause modulus increases high enough to be of concern. As a part of the development program for one such platform [58], six candidate elastomers were evaluated for resistance to crystallization at 3°C using both accelerated tests at −25°C and long-term (2-yr) trials at 3°C. Figure 30 illustrates the platform design in this case and indicates the location of the critical elastomeric component.

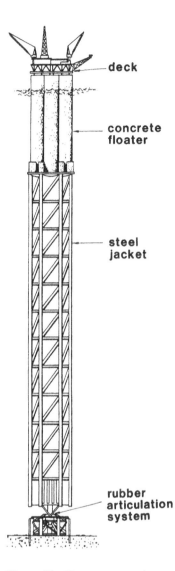

Figure 30 Deep-water gravity tower for hydrocarbon production in deep water. The rubber articulation system is designed to work permanently at 3°C.

Another type of component where low-temperature exposure is possible is the helicopter rotor bearing. In cold climates this can reach −50°C and low modulus in all directions except axial is important for satisfactory function. In this particular case, relatively high-frequency vibrations can create enough internal energy to contribute to internal heat buildup, which will eventually cause crystal melting. There are also many applications in cold climates, such as hoses, fuel tanks, and flexible pipe joints, where low-temperature crystallization may need to be considered as a part of the detailed design of the rubber engineering system.

Test Methods

In general it is essential to perform direct measurements of modulus increase as described in this chapter if quantitative predictions about the performance of engineering components in cold environments are to be made. The strain mode or testpiece size may be varied within some limits, but direct modulus measurements are essential.

There is, however, so far no standard low-temperature modulus test for crystallization of rubber and the standard low-temperature tests that do exist and are included in standard specifications [54] are much less informative and can provide only a qualitative assessment and approximate ranking of different materials. Low-temperature compression set is commonly used. In a study of this specific problem [49, 50], nine engineering vulcanizates were evaluated simultaneously by compression set, hardness, and modulus change at the temperature of interest. Typical results are reproduced in Figure 31 and show clearly that neither hardness nor compression set accurately predicts modulus increase. Low-temperature hardness was generally a much poorer guide than low-temperature compression set. In one case a low-temperature compression set of 65% corresponded to a modulus increase of a factor of between 6 and 14. Low-temperature compression set at best provides a coarse and qualitative indication of the propensity of materials to crystallize at the test temperature. There are also practical difficulties in taking the required measurements at low temperature for both hardness and compression set, which do not occur for a properly designed modulus test where measurements can be made remotely and automatically.

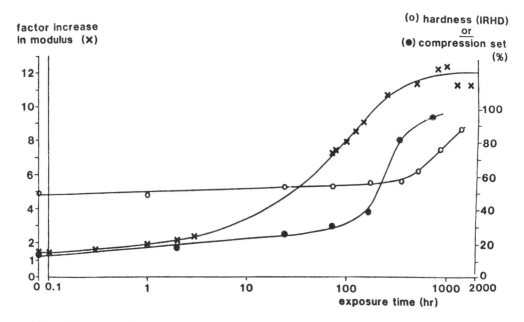

Figure 31 Correlation between modulus increase, compression set, and hardness at −10°C for a natural rubber ''semi-ev'' engineering vulcanizate.

Comparison of Engineering Vulcanizates

The degree and type of crosslinking have an important influence on the amount of crystallization. The use of sulfur vulcanization systems generally increases crystallization resistance as the amount of sulfur and proportion of polysulfidic crosslinks are increased. Low-sulfur compounds of natural rubber consistently show a much greater tendency to crystallize than high-sulfur compounds. Figures 31 and 32 compare the crystallization stiffening of some typical engineering vulcanizates at -10 and $-25°C$. Polychloroprene will generally crystallize at higher temperatures than natural rubber and possesses a higher glass transition temperature T_g. In Figure 32 NR and CR vulcanizates of similar sulfur content and hardness are compared [49]. After 100 hr there was an increase in stiffness exceeding 25 for CR while only about 2 for NR. Different types of polychloroprene will show different degrees of resistance to crystallization, but it is difficult from currently available materials to find a polychloroprene that will stiffen by less than a factor of 8 at $-25°C$. High-sulfur NR compounds can be very resistant to crystallization at this temperature but care needs to be taken in compound design to avoid other undesirable features. It should be emphasized that some natural rubber compounds will stiffen by very much more than this at the same temperature and the detail of the elastomer compound can be more important than the base polymer type.

It is possible to add plasticizers to either NR or CR, and these generally act to depress the T_g and may reduce the stiffening factor near the T_g. However, plasticizers can enhance crystallization slightly by increasing molecular mobility so that if crystallization dominates the behavior their incorporation can be undesirable. Adjustment of the vulcanizing system is the most effective factor for modifying the crystallization behavior of a vulcanizate of any polymer type.

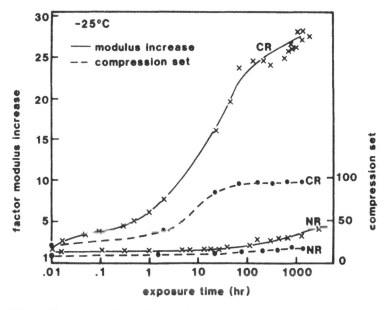

Figure 32 Increase in tensile modulus ($\times$) and low-temperature compression set ($\bullet$) for typical natural rubber and polychloroprene engineering vulcanizates at $-25°C$.

Most engineering compounds contain carbon black. Experiments [56] have shown that the main effect of carbon black is to provide local strain amplification in the rubber phase between the rigid black particles. This does affect the kinetics of crystallization in the same way that applied prestrain has been found to do. It does not, however, seem to affect the equilibrium extent of crystallinity at any temperature. The strain amplification factor can be calculated for a given carbon black content. Equations 22 and 25 can then be used to calculate the crystallization kinetics. No detailed study of the effect on crystallization of different types of black and other fillers has been published, and this remains a potentially interesting line for further research.

Recommendations for Further Research

Most of the published work has been performed on natural rubber. There is a need for systematic and detailed study of a whole range of synthetic rubber types to characterize their low-temperature behavior. It is not entirely clear which materials can crystallize under some conditions and which can never crystallize under any conditions. Natural rubber and polychloroprene are usually considered the crystallizing elastomers, but it is possible that improved crystallizing elastomers could be developed as a new class of materials with improved fatigue and fracture resistance. The key to this work would be a fundamental understanding of the interrelationship between low temperature and strain in causing crystallization. The behavior of polymer blends is particularly poorly understood. The exact mechanisms of crystallization in bulk are still not understood, particularly in the context of the molecular conditions which determine the final degree of crystallinity at equilibrium. Some recent progress has suggested a way that this might be defined theoretically [57], but a full study is required, including x-ray experiments at low temperatures and high strains. The importance of this in the engineering context is set by the need for engineering systems to function in cold climates. Oil exploration in Alaska, for example, has stimulated an interest in flexible bearings, hoses, etc., that will continue to function down to −50°C. All ground transport vehicles rely on a considerable number of elastomeric components from tires and chassis suspension components to hoses and seals. None of the currently available materials will function adequately for long periods at low temperatures. It may well be that there will be in the future a greater stimulus for a better understanding of the low-temperature properties of elastomers.

REFERENCES

1. J. R. Katz, *Chem. Z.*, *49*: 353 (1925).
2. W. Davey, M. Acken, and W. Singer, *Ind. Eng. Chem.*, *24*: 54 (1932).
3. N. Bekkedahl and L. Wood, *Ind. Eng. Chem.*, *33*: 381 (1941).
4. W. Holt and A. McPherson, *J. Res. Nat. Bur. Stand.*, *17*: 657 (1936).
5. L. Wood and N. Bakkedahl, *J. Appl. Phys.*, *17*: 362 (1946).
6. G. Parks, *J. Chem. Phys.*, *4*: 459 (1937).
7. W. Smith and L. Saylar, *J. Res. Nat. Bur. Stand.*, *21*: 257 (1938).
8. J. Goppel and J. Arlman, *Appl. Sci. Res.*, *A1, 3*: 18, 347, 462 (1947).
9. C. Bunn, *Proc. R. Soc.*, *A, 180*: 40 (1942).
10. D. Gorritz and F. Muller, *Knoll. Z.*, *251*: 892 (1973).
11. E. Collins and L. Chandler, *Rub. Chem. Tech.*, *39*: 193 (1966).
12. R. Hammett, R. Wingard, and J. Laud, *J. Rub. Chem. Tech.*, *39*: 206 (1966).
13. W. Cooper and G. Vaughn, *Polymer*, *4*: 329 (1963).
14. L. R. G. Treloar, *Trans. Faraday Soc.*, *43*: 284 (1947).

15. R. S. Stein, *Polym. Eng. Sci., 16*: 152 (1976).
16. P. Calvert and D. Uhlmann, *J. Appl. Phys., 43*: 944 (1972).
17. A. N. Gent, *Trans. Faraday Soc., 50*: 521 (1954).
18. A. N. Gent, *I.R.I. Trans., 30*: 139 (1954).
19. A. N. Gent, *J. Polym. Sci., 18*: 321 (1955).
20. M. Leitner, *Trans. Faraday Soc., 51*: 1015 (1955).
21. E. H. Andrews, *Proc. R. Soc. A, 277*: 562 (1964).
22. B. Wunderlich, *Macromolecular Physics*, Vol. 2, Academic Press, New York (1978).
23. M. Avrami, *J. Chem. Phys., 7*: 1103 (1939).
24. M. Avrami, *J. Chem. Phys., 8*: 212 (1940).
25. M. Avrami, *J. Chem. Phys., 9*: 177 (1941).
26. E. H. Andrews, *Proc. R. Soc. A., 270*: 232 (1962).
27. P. J. Flory, *J. Chem. Phys., 15*: 397, 648 (1947).
28. P. J. Flory, *J. Chem. Phys., 17*: 223 (1949).
29. J. Hoffman and J. Lauritzen, *Knoll. Z. Polym., 231*: 564 (1969).
30. K. J. Smith, *Polym. Eng. Sci., 16*: 168 (1976).
31. A. Keller and M. Machin, *J. Macromol. Sci. B., 1*: 41 (1967).
32. B. Wunderlich, *Macromolecular Physics*, Vol. 1, Academic Press, New York (1976).
33. H. L. Frisch, *J. Chem. Phys., 27*: 90 (1957).
34. J. W. Gibbs, *Trans. Acad. Conn., 3*: 343 (1878).
35. D. Turnbill and J. Fisher, *J. Chem. Phys., 17*: 71 (1949).
36. L. Mandelkern, *Crystallization of Polymers*, McGraw-Hill, New York (1964).
37. H. Carslaw and J. Jaeger, *Conduction of Heat in Solids*, Clarendon Press, Oxford (1959).
38. M. Bruzzone and E. Sorta, *Rub. Chem. Tech., 51*: 207 (1979).
39. A. N. Gent, *J. Polym. Sci., A2–4*: 447 (1966).
40. R. Oono, K. Miyasaka, and K. Ishikawa, *J. Polym. Sci. Phys., 11*: 1477 (1973).
41. D. Greneir and R. E. Prud'homme, *J. Polym. Sci. Phys., 18*: 1655 (1980).
42. C. L. Dugmore, in *Applied Science of Rubber*, Edward Arnold, London, Chap. 7 (1961).
43. A. Keller, M. Hill, and P. Barham, *Coll. Polym. Sci., 258*: 1023 (1980).
44. H. L. Frisch and C. C. Carlier, *J. Chem. Phys., 54*: 4326 (1971).
45. A. Stevenson, *J. Polym. Sci. Polym. Phys. Ed., 21*: 553 (1983).
46. A. Stevenson, *Polymer, 27*: 1211 (1986).
47. A. E. Love, *The Mathematical Theory of Elasticity*, Cambridge University Press, Cambridge (1934).
48. A. Stevenson, Proceedings International Rubber Conference, Sweden, June (1986).
49. A. Stevenson, *Characterisation of Rubber Vulcanizates for Bridge Bearings at Low Service Temperatures*, Technical report to be published by Transport and Roads Research Lab, Crowthorne, Berks (1983).
50. A. Stevenson, *Kautsch. Gummitek. Kunstst., 37*: 105 (1984).
51. J. C. Halpin and J. L. Kardos, *J. Appl. Phys., 43*: 2235 (1972).
52. P. E. Reed, *Proc. R. Soc., Lond. A, 338*: 459 (1974).
53. G. R. Mitchell, *Polymer, 25*: 1562 (1984).
54. A. R. Price and A. Stevenson, Proceedings 2nd World Congress on Joints and Bearings, San Antonio, Texas, October (1986), American Concrete Institute (1987).
55. R. Eyre and A. Stevenson, Transport & Roads Research Lab Report (to be published 1989).
56. A. Stevenson, unpublished work.
57. A. Stevenson and A. G. Thomas (to be published).
58. F. Sedillot and A. Stevenson, *Trans ASME, J. Energy Res. Technol., 105*: 480 (1983).
59. British Standard Specification for Bridge Bearings, BS 5400 pt. 9.2, BSI, London (1983); American Society for Testing Materials, ASTM 04014, ASTM, Philadelphia (1981).

3

Tensile Yield in Crystalline Polymers

Bruce Hartmann
Polymer Physics Group
Naval Surface Weapons Center
Silver Spring, Maryland

INTRODUCTION

Yield is the onset of plastic deformation in a material under an applied load. This is an important process because it determines the practical limit of use more than does the ultimate rupture and because it is the first step in polymer processing by forming, rolling, or drawing. Although the idea of yield is a simple concept, its experimental determination is not easy. Definitions of terms are particularly important since not everyone agrees even on the operational definition of the word *yield*. The presentation here makes use of comparisons with results for metals and ceramics, as well as with amorphous polymers, in order to put the crystalline polymer results in perspective.

The term *crystalline,* as used here, refers to semicrystalline bulk polymer specimens that are also called *polycrystalline.* Bulk polymers are, as a rule, only partly crystalline and are partly amorphous. The degree of crystallinity of a polymer (weight percent of the total polymer that is in the crystalline state) is typically 40%. While polymer crystals have different physical properties in different crystal directions, in a bulk specimen the individual microscopic crystals are oriented at random so that the macroscopic properties are the same in every direction and the physical properties are said to be isotropic. It will be shown that yield properties depend on the degree of crystallinity and the polymer morphology so that the designation of a polymer by a single term, such as polyethylene, does not fully specify the material. There are significant variations among different polyethylenes.

Yield behavior also depends on the test conditions used. Yield varies not only with the test temperature but also with the speed at which the test is made and the magnitude of the test pressure. In some cases, even the gaseous environment (air, nitrogen, oxygen, helium, etc.) can have an effect. Thus yield depends on both material properties and test conditions.

This presentation is organized into three major sections: basic concepts, experimental results, and theoretical interpretations. Basic concepts are important for the understanding of this topic and include definitions of terms used later. Experimental results are given for various common crystalline polymers as functions of temperature, strain rate, and pressure, using polyethylene as the typical example. Finally, theoretical interpretations of the experimental results are considered to the extent that the theories make predictions of practical use in extrapolating and interpolating experimental data and predicting behavior under conditions that have not been measured.

BASIC CONCEPTS

The basic concepts necessary to make proper engineering use of tensile yield measurements are presented in this section. Not only are terms defined, but also limitations of their application and comparisons with related terms are offered in order to give an understanding why the definition is formulated the way it is.

Yield

Yield is defined conceptually as the onset of plastic deformation in a material under an applied load. A plastic deformation is one that remains after the load is removed and is also called a permanent or nonrecoverable deformation. By contrast, at small enough loads, deformation is elastic and is recovered after the load is removed (i.e., the specimen returns to its original length). Yield thus represents the transition from elastic to plastic deformation. As an example, consider a material under an applied tensile load. The length of the specimen will increase, as measured by the elongation or increase in length of the specimen, when the load is applied. Figure 1 is a schematic drawing of a typical tensile load–elongation curve. As the elongation increases, the load at first increases linearly but then increases more slowly and eventually passes through a maximum, where the elongation increases without any increase in load. The peak in the load–elongation curve is the point at which plastic flow becomes dominant and is commonly defined as the *yield point.*

Not all load elongation curves look like Figure 1, which is typical of a material that clearly exhibits plastic deformation prior to rupture. Other materials fracture before reaching a maximum, as shown in Figure 2. Even though no maximum is reached, there may be

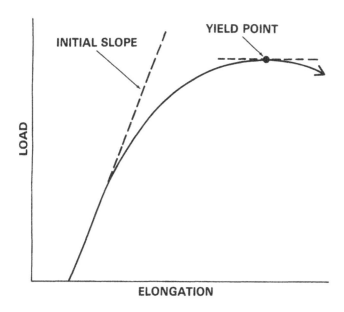

Figure 1   Schematic load–elongation curve for a material with a yield point.

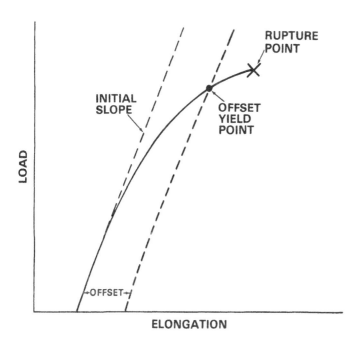

Figure 2   Schematic load–elongation curve for a material with an offset yield point.

plastic deformation and it is usually assumed that yield occurs at some arbitrarily chosen value of strain ϵ (elongation divided by original length). For polymers, the strain at which plastic deformation begins is often taken to be 2% [1], whereas in metals a value of 0.2% is common [2], and in ceramics 0.05% is used [3]. One then draws a line parallel to the initial linear portion of the load–elongation curve but offset by the desired amount. The point where this line intersects the load–elongation curve is called the *offset yield point,* as seen in Figure 2. Use of an offset yield point is common in metals and ceramics where plastic deformation follows immediately after linear elastic deformation. Polymers, however, are more complex and generally exhibit considerable nonlinear elasticity before plastic deformation. Nonlinear elasticity also produces a deviation from the initial linear behavior but is fully recoverable. Since it is not obvious whether the deviation from linearity is caused by plasticity or nonlinearity, yield in polymers is preferably defined as the maximum in the load–elongation curve, if there is a maximum, and as the offset yield point only if there is no maximum.

In some cases, the load rises linearly to fracture with no plastic deformation. The material is then said to be *brittle,* in distinction to the foregoing behavior where plastic deformation occurs and the material is said to be *ductile.* A given material will exhibit ductile or brittle behavior depending on the test conditions. A material that is brittle at low temperature will become ductile at high temperature. The brittle to ductile transition temperature T_{bd} is then an important material characteristic. For polyethylene, T_{bd} is about 125 K [4], depending somewhat on the specific polyethylene and the test conditions. A brittle to ductile transition is also observed in metals and ceramics. For transition metal carbides (TiC, ZrC, VC, NbC, and TaC), it has been pointed out [5] that T_{bd} is about 1500 K while the melting temperatures are about 4500 K so that T_{bd} is about one third of the melting temperature, T_m. In comparing these results with those for polymers, since the melting temperature of polyethylene is 413 K [6], its T_{bd} is also about one third of its melting temperature. These observations suggest that comparisons between different materials should be made at equivalent homologous temperatures, T/T_m. When this is done, polymers and ceramics are both brittle below about $T/T_m = 0.3$ and ductile above this temperature. They differ only in the magnitude of their melting temperatures.

Stress and Strain

The magnitude of a load–elongation curve depends on the dimensions of the specimen being tested and is therefore not a basic material property. Rather than load, the properly normalized variable is stress, the load per unit cross-sectional area of the specimen. Stress σ then has units of pressure and is expressed in pascals (1 MPa = 1 MN/m^2 = 10^7 dyne/cm^2 = 145 psi). Common engineering practice is to calculate σ on the basis of the initial (unloaded) cross-sectional area A_0, so that the engineering (or nominal) stress is given by

$$\sigma(\text{engineering}) = \frac{W}{A_0} \tag{1}$$

where W is the applied load. In a tensile test the cross-sectional area of the specimen decreases as the specimen elongates, and the true stress is given by

$$\sigma(\text{true}) = \frac{W}{A} \tag{2}$$

where A is the instantaneous cross-sectional area at any given elongation. Since the relative change in lateral dimension divided by relative change in length is called Poisson's ratio μ, then σ(true) can be calculated using the formula

$$\sigma(\text{true}) = \frac{\sigma(\text{engineering})}{(1 - \mu\epsilon)^2} \tag{3}$$

Accurate values of σ(true) require a knowledge of μ which is usually not available, and it is often assumed that the volume of the specimen is constant so that $\mu = 0.5$. If, in addition, the strain is small, the ratio of true stress to engineering stress is $1 + \epsilon$. Thus the true stress is somewhat higher than the engineering stress. As an example, for polyethylene at the yield point, σ(true) = 33 MPa while σ(engineering) = 29 MPa [6].

Rather than elongation, the properly normalized variable is *strain,* the elongation divided by the initial length. Strain is therefore dimensionless. Engineering strain is calculated on the basis of the initial length, L_0,

$$\epsilon(\text{engineering}) = \frac{L - L_0}{L_0} \tag{4}$$

where L is the instantaneous length. True strain is defined as

$$\epsilon(\text{true}) = \int_{L_0}^{L} \frac{dL}{L} = \ln\left(\frac{L}{L_0}\right) \tag{5}$$

It is almost universal to use engineering stress and strain when discussing polymers and in this chapter σ and ϵ mean engineering values unless otherwise noted. Since a load–elongation curve can be converted to a stress–strain curve by a simple change of scale without change of shape, Figures 1 and 2 can equally well be considered stress–strain curves because no scales are given. The peak in Figure 1 can then be called the yield point and is specified by a yield stress σ_y and a yield strain ϵ_y. For polyethylene, $\sigma_y = 29$ MPa at room temperature. Other polymers have yield stresses as high as 90 MPa [7] while typical metals vary from 30 to 600 MPa [8] and ceramics are usually in the range from 100 to 800 MPa [5] so there is an overlap in the range of σ_y values for the different types of materials. The yield strain of polyethylene at room temperature is 0.1 and other polymers vary from 0.05 to 0.2, while for ceramics, at about 1500 K, the yield strain is in the range from 0.001 to 0.03 [5].

An alternative definition of yield in terms of true stress and true strain is that the yield point occurs when the slope of the true stress–strain curve equals the true stress. This point can be located using a geometric construction known as Considère's method. A tangent is drawn to the curve from a true strain of -1. The point of tangency is the true yield point. Details of the method are given by Ward [1]. Although this procedure is conceptually attractive, it is more difficult to implement and the results obtained are usually not qualitatively different from the simpler engineering approach [18]. For these reasons, the Considère method is not common and is not used here.

Preyield Behavior

The initial, preyield, stress–strain behavior of a material is of interest here because it is intimately related to the yield behavior. The initial linear portion of the stress–strain curve is a measure of linear elasticity, where stress and strain are linearly proportional, by

Hooke's law, with a constant of proportionality known as the *modulus of elasticity*. In the case of tensile loading, the modulus of elasticity is the tensile modulus or Young's modulus Y (the symbol E is also often used),

$$\sigma = Y\epsilon \tag{6}$$

The stress or strain at which deviation from linearity begins is called the proportional limit. This term is not well defined since it depends on the sensitivity of the test and the degree of deviation considered significant. As mentioned before, deviations from linearity occur as a result of nonlinear elasticity. In the nonlinear region, the modulus is given at any point of the stress–strain curve by

$$Y = \frac{d\sigma}{d\epsilon} \tag{7}$$

and varies with stress (or strain). Y is of interest for a number of reasons. It is a fundamental characteristic of the material that is available from the stress–strain curve used to determine the yield point without further testing. It will be shown that most variables that affect σ_y also affect Y so that results for modulus are directly applicable to yield. As an example, if it is found that the Young's modulus of a crystalline polymer varies with the density of the specimen, then one would expect the yield stress to be similarly affected. For polyethylene at room temperature, $Y = 1.4$ GPa and varies from 1 to 3 GPa for other polymers. In ceramics, Y is commonly in the range from 3 to 30 GPa [3], though diamond is higher. For metals, the range is typically from 30 to 300 GPa [9], though lead is lower.

Another material characteristic that can be obtained from the stress–strain curve is the yield energy, defined as the work (per unit volume) required to produce yield in the polymer. Yield energy E_y is given by the area under the stress–strain curve from zero strain to ϵ_y,

$$E_y = \int_0^{\epsilon_y} \sigma \, d\epsilon \tag{8}$$

Yield energy has units of joules per cubic centimeter (1 J/cm^3 = 0.239 cal/cm^3). In metals this quantity is called *resilience* [8]. For polyethylene at room temperature, E_y = 2.5 J/cm^3.

The final important aspect of preyield behavior is time dependence. Polymers are viscoelastic materials so that their response to mechanical loading is time-dependent. Whereas some materials display instantaneous elastic behavior followed by yield or fracture, polymers typically exhibit delayed elasticity or time dependence. For this reason, the value of the yield stress depends on the rate at which the load is applied to the polymer, as measured by the strain rate, $\dot{\epsilon} = d\epsilon/dt$. The units of $\dot{\epsilon}$ are then reciprocal time (min^{-1}). Since significant variations in yield stress occur only over several decades of reciprocal time, yield stress is usually plotted as a function of log $\dot{\epsilon}$.

Postyield Behavior

Postyield behavior will not be considered here in any detail, but there are some qualitative observations that are relevant. Referring to Figure 1, the stress at high strain is starting to decrease. Beyond this point a dramatic and sudden narrowing of the specimen cross section, called *necking*, may occur as the stress increases. The neck forms at the weakest point along the narrow section of the test specimen and then propagates, at constant stress, along the length of the specimen, a process known as *cold drawing*. Once the neck has traveled the entire length of the narrow section, the stress may start to increase again, a phenomenon known as *strain hardening*. Rupture may occur at a stress higher than σ_y. Since the strength of a polymer is defined as the highest stress it can support, σ_y may or may not be the strength of the polymer depending on the postyield behavior.

The ultimate properties, i.e., those at rupture, are used in defining the *toughness* of a polymer. This is the total work (per unit volume) required to produce rupture. Toughness is given by the area under the stress–strain curve from zero strain up to rupture,

$$\text{toughness} = \int_0^{\epsilon_r} \sigma \, d\epsilon \tag{9}$$

where ϵ_r is the value of the strain at rupture. Toughness is then similar to yield energy.

The postyield behavior of some polymers is marked by *crazing*, which is the formation of multiple microscopic cracks throughout the specimen. When this occurs in an initially transparent polymer, strong light scattering is observed, a phenomenon called *stress whitening*. Occasionally this behavior can even be observed below the yield point [10].

Types of Loading

The most common type of loading used to evaluate polymers is uniaxial tension, but other types of loading are also used and can add insight to the tensile results. The simplest variation of the tensile test is the uniaxial compression test (not to be confused with hydrostatic compression, in which the load is applied from all sides). It is found that compressive stresses are higher than tensile stresses for a given strain. In particular, at the yield point of polyethylene, the compressive yield stress is about 5% higher than the tensile yield stress [11, 12], whereas for polycarbonate the difference is about 14% [12, 13]. This is a bigger effect than seen with metals [12]. The other common type of loading is one that produces yield in shear. The test is usually described as a torsional test. For polyethylene, the shear yield stress τ_y is 13 MPa [14] compared with the tensile yield stress of 29 MPa [6]. For polycarbonate at room temperature, $\sigma_y = 83$ MPa while $\tau_y = 48$ MPa [15]. From basic mechanics it is known that a shear deformation occurs as a shape change at constant volume (at least for small strains). In contrast, under hydrostatic pressure a material undergoes a volume change at constant shape. Since a tensile deformation involves both a shape change and a volume change, a tensile stress can be viewed as a combination of a shear component and a hydrostatic component. For this reason, knowledge of shear yield and the effect of pressure on yield are important for an understanding of tensile yield.

In some cases tensile yield measurements have been made under a superimposed hydrostatic pressure. Polymers that were brittle can become ductile under pressure, and the tensile yield stress increases under pressure. The interest in these measurements is partly because polymers are sometimes processed under pressure and partly because tensile tests involve a pressure component.

Experimental Procedures

The most common experimental procedure used for tensile testing is that described in American Society for Testing and Materials (ASTM) test method ASTM D638, Tensile Properties of Plastics. The specimen has a dumbbell or dogbone shape as shown in Figure 3 for specimens of sufficient thickness (Type I). There are specifications for smaller

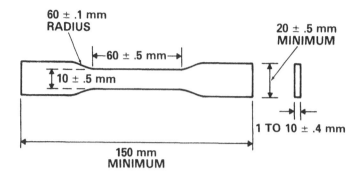

Figure 3 Tension test specimen, ASTM D638, Type I.

specimens, but these are not preferred. The ends of the specimen are wide to provide easy gripping, while the central section is narrower so that elongation will occur preferentially in that region. A strain gage extensometer is attached to the specimen in the narrow region, typically with initial length (or gage length) of 2.54 cm. The ends of the specimen are gripped by heavy, self-aligning metal grips and the load is applied by a Universal Test Machine. (A commonly used machine is manufactured by the Instron Corp.; such a machine is sometimes referred to simply as an Instron.) The top grip is stationary while the bottom grip, attached to a crosshead, moves down. Crosshead speeds available on the typical hydraulic screw-driven machine run from 0.05 to 50 cm/min.

The output of the strain gage extensometer is fed to the x axis of an x–y recorder while the y axis is the applied load from the load cell. Thus the output of the measurement is a continuous load–elongation curve at a given crosshead speed and temperature. When measurements as a function of temperature are desired, an environmental chamber is used. Pressure measurements are more involved and details of the equipment are given in the literature [16]. After the test, the load–elongation curve is converted to a stress–strain curve. Note that the crosshead speed is used only to specify the rate at which the test is run, not to determine the strain. It has been found that calculating strain from the crosshead speed rather than from an extensometer is not accurate. The test is run until, by eye, it is seen that the stress goes through a maximum and then is manually stopped. Thus the test always goes a little past the yield point, making it difficult to make accurate measurements of permanent deformation at yield.

Because of the variations in polymer specimens and test results, a minimum of five replicates should be run under each set of test conditions and the mean and standard deviations reported.

Shapes other than that of Figure 3 can be used, some of which, such as a hollow tube, are described in ASTM D638, and others, such as a cylinder with threaded ends and a narrower central diameter, are described in the literature.

Compressive yield testing uses a flat-ended cylinder for the test specimen, as described in ASTM D695, Compressive Properties of Rigid Plastics. Recall that the sides of the specimen must be unconstrained as compression refers to uniaxial compression. Torsional testing, to determine shear yielding, is described in the literature [17].

Table 1 Tensile Properties of Crystalline Polymers at Room Temperature and a Strain Rate of 2 Min^{-1}

Polymer	σ_y (MPa)	ϵ_y	Y (GPa)	E_y (J/cm^3)	Ref.
Polyethylene	29.3	0.102	1.39	2.47	6
Polypropylene	37.7	0.111	1.60	2.99	18
Poly(4-methylpentene-1)	27.7	0.047	1.61	0.96	10
Polyoxymethylene[a]	86.2		2.69		7
Nylon-6,6	75.7	0.182	2.98	12.50	19
Poly(vinylidene fluoride)	49.6	0.107	1.40	4.15	20
Poly(chlorotrifluoroethylene)	45.6	0.069	1.45	1.95	21
Polytetrafluoroethylene[b]	9.0	0.04	0.41		16

[a]Strain rate $= 0.001$ min^{-1}.
[b]Strain rate $= 0.2$ min^{-1}.

EXPERIMENTAL RESULTS

Experimental yield properties for various common crystalline polymers at room temperature and pressure and a strain rate of 2 min^{-1} are given in Table 1, taken from various sources [6, 7, 10, 16, 18–21]. The polymers are all commercial products. The values in Table 1 are in good agreement with other published values, when the variability of the results with material properties and test conditions is taken into account. For example, the yield stress in a series of polyethylenes varied from 6 to 30 MPa depending on their density and thermal history [22]. For the same series of polymers, Young's modulus varied from 0.2 to 1.3 GPa and was well correlated with the yield stress: the higher the modulus, the higher the yield stress. This is an example of the observation that any variable that affects modulus will also affect yield. This would include density, degree of crystallinity, and thermal history (by changing the morphology).

To examine these polymers in more detail, their behavior as functions of test conditions is considered. The results are presented in order of the test variable considered: temperature, strain rate, pressure, and environment.

Temperature Dependence

As a function of test temperature, stress–strain curves for polyethylene [6] change progressively as shown in Figure 4. As the temperature increases, the yield stress, Young's modulus, and yield energy all decrease, while the yield strain increases. This qualitative behavior is observed over the entire temperature range from the brittle–ductile transition temperature to the melting temperature. Tabular results for polyethylene are given in Table 2.

A plot of yield stress as a function of temperature for polyethylene is shown in Figure 5. Over most of the temperature range shown, the decrease is linear, and σ_y extrapolates to zero near the melting temperature. Some polymers exhibit more curvature at higher temperature but are still roughly linear near room temperature. The rate of change of yield

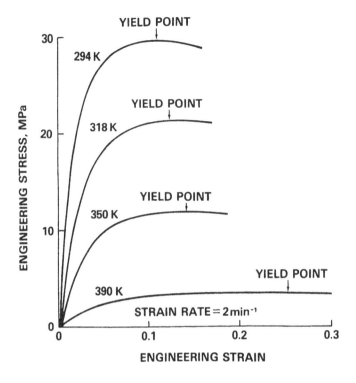

Figure 4 Stress–strain curves for polyethylene.

Table 2 Tensile Properties for Polyethylene

Temperature (K)	Yield stress (MPa)	Yield strain	Young's modulus (GPa)	Yield energy (J/cm³)
294	29.3	0.102	1.39	2.47
302	27.6	0.101	1.18	2.31
310	24.7	0.113	1.07	2.29
318	21.2	0.132	0.07	2.31
326	19.0	0.135	0.57	2.09
334	16.3	0.145	0.45	1.91
342	13.7	0.146	0.39	1.61
350	11.7	0.150	0.30	1.42
358	9.6	0.158	0.24	1.22
366	7.7	0.165	0.17	1.02
374	6.1	0.176	0.13	0.87
390	3.7	0.260	0.06	0.63

Note: Strain rate = 2 min^{-1}.

Source: Ref. 6.

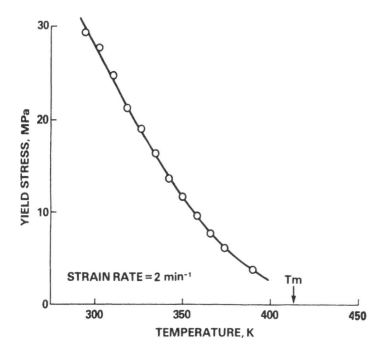

Figure 5 Yield stress vs. temperature for polyethylene.

stress with temperature in the vicinity of room temperature for various crystalline polymers is listed in Table 3. For comparison, the temperature dependence of the shear yield stress of polyethylene is −0.13 MPa/K [17], about one third of the tensile value. The tensile values are not much different from those for amorphous polymers such as polycarbonate [13].

A plot of yield energy vs. temperature for polyethylene is shown in Figure 6. A linear decrease with temperature is observed, extrapolating to zero near the melting temperature. Slopes of yield energy vs. temperature in the vicinity of room temperature for various crystalline polymers are given in Table 3. Recalling that yield energy is the area under the

Table 3 Temperature and Strain Rate Dependence Parameters

Polymer	T_m (K)	$-d\sigma_y/dT$ (MPa/K)[a]	$-dE_y/dT$ (J/cm^3)[a]	$d\sigma_y/d \log \epsilon$ (MPa/dec)[b]	Ref.
Polyethylene	413	0.30	0.021	4.1	6
Polypropylene	437	0.37	0.025	4.3	18
Poly(4-methyl pentene-1)	513	0.29	0.005	3.6	10
Nylon-6,6	538	0.54	0.016	2.5	19
Poly(vinylidene fluoride)	441	0.54	0.031	4.5	20
Poly(chlorotrifluoroethylene)	488	0.42		3.9	21

[a]Strain rate = 2 min^{-1}.

[b]At room temperature.

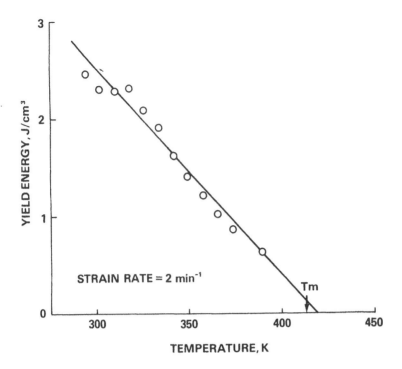

Figure 6 Yield energy vs. temperature for polyethylene.

stress–strain curve, the fact that E_y decreases as the temperature is raised shows that the decrease in yield stress dominates the increase in yield strain.

These results for polyethylene are qualitatively similar to most other polymers, but nylon-6,6 [poly(hexamethylene adipamide)], shows unusual behavior [19]. The yield stress vs. temperature curve (Figure 7) shows the typical decrease with temperature, but the yield strain vs. temperature curve (Figure 8) has a change in slope at 433 K. The yield energy vs. temperature curve (Figure 9) also shows a change at 433 K, contrary to the results for other polymers. Since yield energy is the area under the stress–strain curve, the change in slope of the yield energy is a direct consequence of the yield strain behavior. In particular, the yield stress shows no special behavior at 433 K. It was determined that the change at 433 K is caused by a reversible crystal–crystal transition from a triclinic lattice at room temperature to a pseudo-hexagonal lattice at 433 K. The transition, called the Brill transition, occurs because the *b* lattice constant increases with temperature until the *b* constant equals the *a* constant. In some other polyamides, the *b* constant also increases with temperature, but melting occurs before *b* equals a.

Strain Rate Dependence

As a function of strain rate, yield stress, Young's modulus, and yield energy all increase while yield strain is almost unchanged. This behavior is observed over the entire range of strain rates available. Results for polyethylene are given in Table 4.

A plot of the yield stress as a function of strain rate for polyethylene is shown in Figure 10. Within experimental uncertainty, yield stress is a linear function of log strain rate,

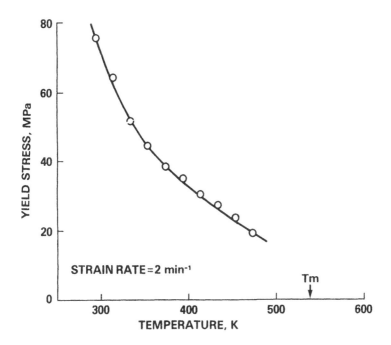

Figure 7   Yield stress vs. temperature for nylon-6,6.

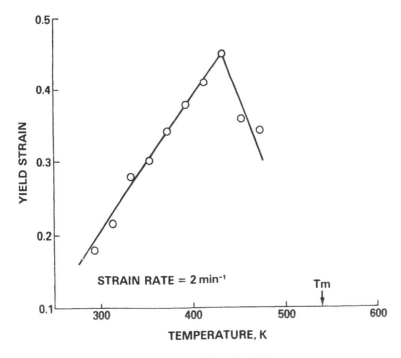

Figure 8   Yield strain vs. temperature for nylon-6,6.

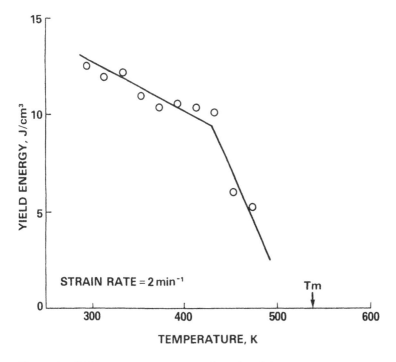

Figure 9 Yield energy vs. temperature for nylon-6,6.

though curvature in such plots is sometimes reported [23] over wide ranges of strain rate. All polymers have qualitatively similar behavior and the rate of change of yield stress with respect to log strain rate is listed in Table 3. This behavior is observed for both crystalline and amorphous polymers. The difference is quantitative only; the values of the slope are lower for crystalline polymers [24]. An even greater slope is observed for the strain rate dependence of yield stress in metals [2, 8] and ceramics [5].

Note that the qualitative strain rate dependence of yield stress, Young's modulus, and yield energy are all opposite to the temperature dependence of these quantities. This can be explained by the viscoelastic nature of polymers and suggests the possibility that time–temperature superposition might be applicable to yield and that yield may be related to creep and other viscoelastic properties. It should be kept in mind, however, that yield is a nonlinear, large-strain property, whereas time–temperature superposition is generally considered in the linear, small-strain region.

Pressure Dependence

As pointed out earlier, the pressure dependence of tensile yield is an important topic because tensile stress has a pressure component and understanding pressure effects is helpful in interpreting tensile yield results. In addition, various processing and forming operations are done under pressure and properties can change significantly under these conditions. For example, some materials that are brittle under atmospheric pressure become ductile under elevated pressure.

As hydrostatic pressure increases, the tensile yield stress and Young's modulus both

Table 4 Strain Rate Dependence of Tensile Properties for Polyethylene

Strain rate (min⁻¹)	Yield stress (MPa)	Yield strain	Young's modulus (GPa)	Yield energy (J/cm³)
		Temperature = 294 K		
0.02	20.5	0.125	0.88	2.12
0.08	23.6	0.112	0.93	2.14
0.2	25.0	0.105	1.09	2.18
0.8	28.3	0.098	1.30	2.20
2.0	29.3	0.102	1.39	2.47
8.0	30.8	0.105	1.51	3.28
		Temperature = 318 K		
0.02	13.4	0.137	0.38	1.48
0.08	15.2	0.133	0.50	1.62
0.2	17.0	0.131	0.61	1.82
0.8	20.3	0.126	0.72	2.14
2.0	21.2	0.132	0.70	2.29
8.0	24.0	0.122	0.77	2.39
		Temperature = 342 K		
0.02	7.9	0.142	0.19	0.89
0.08	9.2	0.142	0.24	1.05
0.2	10.1	0.143	0.26	1.17
0.8	11.7	0.149	0.31	1.43
2.0	13.7	0.146	0.39	1.61
8.0	15.7	0.176	0.40	2.27

Source: Ref. 6.

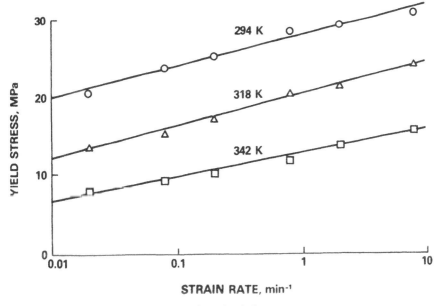

Figure 10 Yield stress vs. strain rate for polyethylene.

increase. Typical results for polyethylene yield stress, at a strain rate of 0.2 min^{-1}, are shown in Figure 11 [25]. A linear pressure dependence is observed. Similarly, the Young's modulus increases with pressure as shown in Figure 12 [25]. Pressure dependence results for various crystalline polymers are summarized in Table 5. The values were taken from a variety of sources [7, 15, 16, 25, 26], tested at differing strain rates, on materials of different density and thermal history. Some of the yield results are for offset yield and some are not. For these reasons, the zero pressure values are not identical to those in Table 2.

Polymers have a high pressure dependence compared with metals because polymers yield at a stress that is very high in relation to their modulus. The ratio of yield stress to bulk modulus is 0.001 to 0.01 for polycrystalline metals, whereas it is 0.02 or higher for polymers [27]. Yield in metals is often said to be independent of hydrostatic pressure, but recent measurements [12] have shown that this is not the case. The effect is relatively small only because the pressures used are small compared to the modulus.

Environment

Gaseous environments can effect the tensile behavior of polymers, usually by causing them to craze at low temperature [28]. Helium is inert, but other gases may produce crazing depending on the test temperature and the partial pressure of the gas. This effect has been observed with nitrogen, argon, oxygen, carbon dioxide, and water vapor.

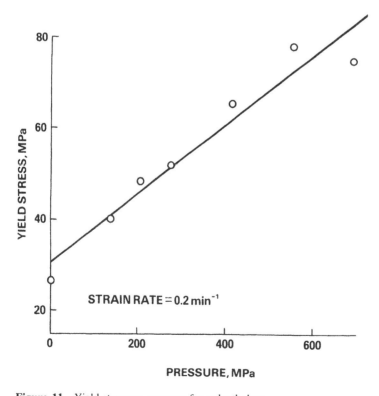

Figure 11   Yield stress vs. pressure for polyethylene.

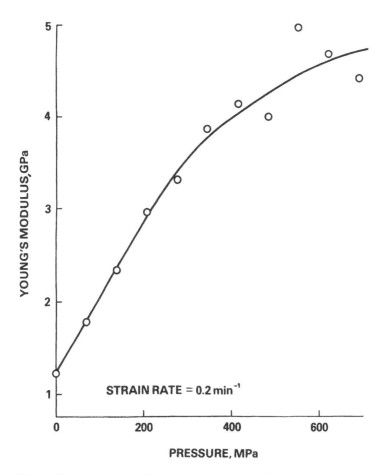

Figure 12 Young's modulus vs. pressure for polyethylene.

Table 5 Pressure-Dependence Parameters

Polymer	σ_y (MPa)	$d\sigma_y/dP$	Y (GPa)	dY/dP	Ref.
Polyethylene	26.1	0.094	1.24	8.7	25
Polypropylene	28.6	0.205	1.52	19.	25
Polyoxymethylene	86.2	0.26	2.69	7.3	7
Nylon-6,6	31.7	0.12	0.50	1.4	26
Polychlorotrifluoroethylene	37.9	0.23	1.16	3.3	15
Polytetrafluoroethylene	9.0	0.08	0.41	3.3	16

Note: All evaluated at room temperature.

Since gaseous environments can have an effect on tensile properties, it is possible that a liquid environment could also. For this reason, one must be careful in making pressure measurements since the pressure-transmitting fluid could interact with the specimen.

THEORETICAL INTERPRETATION

Polymer yield is sometimes interpreted using phenomenological theories, frequently adapted from metals theory, while other theories are specific to polymers. A satisfactory theory must explain the difference between yield in tension and in compression, the existence of shear yield, and the temperature, strain rate, and pressure dependence and their interrelations. This is an area of active development and only an outline is given here.

Rate Theory

The most widely used theory for yield is based on the activated state rate theory of Eyring for viscosity [1]. This theory is used for crystalline and amorphous polymers as well as metals [29] and ceramics [5]. Rate theory assumes that, in the absence of any applied stress, the polymer segments are in potential wells formed by inter- and intramolecular steric hindrances and are separated from one position of stable equilibrium to another by an activation energy (strictly speaking, an enthalpy) ΔH, as seen in Figure 13. Under an applied stress σ, one equilibrium potential is raised by an amount $\sigma \Delta V$ and the other lowered by a like amount, as shown in Figure 13, where ΔV is called the activation volume. Thus the stress biases the activation energy, allowing segmental motion in the direction of the stress. The yield stress σ_y is the value at which the internal viscosity falls to a value such that the applied strain rate is identical to the plastic strain rate predicted by the Eyring equation,

$$\dot{\epsilon} = \dot{\epsilon}_0 \exp\left[\frac{-(\Delta H - \sigma_y \Delta V)}{RT} \right] \tag{10}$$

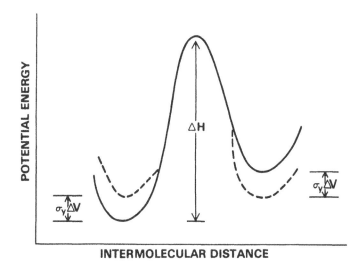

Figure 13 Eyring model for yield.

Table 6 Activation Parameters

Polymer	ΔV^a (nm³)	v (nm³)	$\Delta V/v$	ΔH (kJ/mol)	$\dot{\epsilon}_0$ (min⁻¹)	Ref.
Polyethylene	2.3	0.048	48	84	4×10^{12}	6
Polypropylene	2.1	0.077	27	71	5×10^4	18
Poly(4-methyl pentene-1)	2.6	0.17	15	79	5×10^6	10
Nylon-6,6	3.8	0.33	11			19
Polyvinylidene fluoride	2.1	0.060	35	87	1×10^5	20
Polychlorotrifluoroethylene	2.4	0.090	27	163	5×10^{17}	21

*a*At room temperature.

where $\dot{\epsilon}_0$ is a constant (that includes the concentration of segments, entropy effects, etc.). Solving for σ_y,

$$\sigma_y = \frac{\Delta H}{\Delta V} + \frac{RT}{\Delta V} \ln\left(\frac{\dot{\epsilon}}{\dot{\epsilon}_0}\right) \tag{11}$$

so that σ_y is a linear function of ln strain rate at constant temperature. Behavior of this type is often observed and this fact is usually taken as the justification for the use of Eyring theory. Differentiation of Eq. 11 gives

$$\left(\frac{\partial \sigma_y}{\partial \ln \dot{\epsilon}}\right)_T = \frac{RT}{\Delta V} \tag{12}$$

and activation volume can be determined from a plot of σ_y vs. ln strain rate. Values of ΔV at room temperature for various crystalline polymers are given in Table 6. These values are two or three times smaller than the values for amorphous polymers, which vary from 5 to 20 nm³ [1]. For comparison, the volumes of one polymer segment, v, are also listed where

$$v = \frac{M}{\rho N_A} \tag{13}$$

where M is the molecular weight, ρ is the density, and N_A is Avogadro's number. The ratio of these two numbers is also given in Table 6 and shows that yield involves the cooperative movement of a large number of chain segments (10 to 50).

From a series of plots of σ_y vs. log strain rate at different temperatures, as shown in Figure 10, the value of ΔV at different temperatures can be obtained. ΔV increases as the temperature increases. In polyethylene, for example, ΔV is 2.3 nm³ at 294 K, 2.4 nm³ at 318 K, and 3.6 nm³ at 342 K. Thus the correlated length of chain movement increases as the temperature increases. Similar behavior is observed with glassy polymers but not with metals [29].

The interpretation of yield as resulting from large-scale motion is borne out by the magnitude of the activation energy, which is intermediate between that for the glass transition and for a secondary transition [30]. Values are listed in Table 6. These values were calculated from Eq. 11 using the temperature-dependent ΔV and assuming that ΔH

and $\dot{\epsilon}_0$ are constant. For each pair of temperatures, values of ΔH and $\dot{\epsilon}_0$ were determined. The activation energy from pairs of temperatures tends to decrease as the temperature increases, even becoming negative at high enough temperature, as was found [19] for nylon-6,6, indicating a limitation of this approach. Average values of activation energy are given in Table 6.

In some cases where a plot of σ_y/T vs. log strain rate over a wide range of values is not linear but shows some curvature, the foregoing theory has been extended to include two Eyring processes acting in parallel. The yield stress then consists of two additive terms of the form of Eq. 11. The processes have been interpreted in terms of molecular transitions in the polymer [13, 17]. Thus there is a relation between yield and other measurements, such as dynamic mechanical and dielectric, that are used to study transitions.

This discussion has been for tensile yield at atmospheric pressure. The theory has been extended to include the effect of hydrostatic pressure in the following manner [1]:

$$\dot{\epsilon} = \dot{\epsilon}_0 \exp\left[\frac{-(\Delta H - \tau_y \Delta V + P \Delta \Omega)}{RT}\right] \tag{14}$$

where τ_y is the shear yield stress, P is hydrostatic pressure, ΔV is the shear activation volume, and $\Delta \Omega$ is the pressure (or bulk) activation volume. This analysis fits the data reasonably well.

Internal Energy Theory

It can be assumed that a certain amount of energy is required to thermally induce large-scale cooperative segmental motion over inter- and intramolecular restrictions. Yield energy should then be related to the internal energy change in the polymer. This theory was originally developed for glassy polymers [31–33] and later extended to crystalline polymers [10]. The basis of the theory is that yield energy is related to internal energy U by the relation

$$E_y = \frac{b}{b'} \int_T^{T_g} \left(\frac{\partial U}{\partial T}\right)_P dT \tag{15}$$

where b is the fraction of thermal energy available to overcome the activation energy barrier for flow and b' is the fraction of mechanical energy available to overcome the same barrier. The result can be approximated by

$$E_y = \frac{b}{b'} \bar{\rho} \bar{C}_p (T_g - T) \tag{16}$$

where $\bar{\rho}$ and $\bar{C}_p$ are the average values of density and heat capacity. In the case of crystalline polymers, Eq. 16 becomes [10]

$$E_y = \frac{b}{b'} \bar{\rho} \bar{C}_p (T_m - T) \tag{17}$$

As shown earlier, experimentally E_y is generally found to be a linear function of temperature, extrapolating to zero near the melting temperature. For polyethylene $b/b' = 0.012$, for polypropylene $b/b' = 0.011$, and for poly(4-methyl pentene-1) $b/b' = 0.003$. In the case of nylon-6,6 $b/b' = 0.007$ up to 433 K and $b/b' = 0.055$ above 433 K.

Because directed mechanical energy is more efficient than random thermal motion in causing flow, the value of b/b' is less than 1. For glassy polymers, b/b' varies between

0.02 and 0.03 [32], which is larger than for crystalline polymers, indicating that for crystalline polymers thermal energy is even less effective compared to mechanical energy in surmounting the barrier. An exception is nylon-6,6 above 433 K. In this case b/b' is even greater than for amorphous polymers and indicates a significant increase in the relative effectiveness of thermal energy to produce flow.

Volume-Dependent Theories

It has been suggested [34] that yield stress depends on polymer volume and not on temperature or pressure except as they affect the volume. This idea was based on two observations: Young's modulus is a function of volume and yield stress is a function of Young's modulus. That the Young's modulus of polymers is a function of volume was demonstrated experimentally [35], and similar results were presented for the closely related case of bulk modulus [35]. The volume dependence of the bulk modulus of crystalline polymers also has a theoretical basis and has been derived directly from the intermolecular potential [36]. Thus Young's modulus depends on temperature and pressure only because the volume depends on these variables.

The relation between yield stress and Young's modulus was mentioned earlier with regard to density and thermal treatment differences between different specimens of the same material [22]. Also, for a given specimen, both yield stress and Young's modulus decrease as temperature is raised and both increase as pressure is raised. These observations suggest that yield stress and Young's modulus are related. Since Young's modulus is a function of volume, it follows that yield stress is a function of volume. The temperature and pressure dependence of the yield stress is then a result of the temperature and pressure dependence of the volume. The temperature and pressure derivatives of the yield stress are then not independent but are related by the expression [18]

$$\left(\frac{\partial \sigma_y}{\partial P}\right)_T = -\left(\frac{\beta}{\alpha}\right)\left(\frac{\partial \sigma_y}{\partial T}\right)_P \tag{18}$$

where α is the thermal expansion coefficient and β is the isothermal compressibility (reciprocal of the isothermal bulk modulus). The validity of Eq. 18 has been tested for polypropylene since all of the input data were not available for the same density polyethylene. From Table 3, the temperature derivative of the yield stress is -0.37 MPa/K. The compressibility is 2.87×10^{-4} MPa^{-1} [37] and the thermal expansion coefficient is 5×10^{-4} K^{-1} [38], so that from Eq. 18 the predicted pressure derivative of the yield stress is 0.21, which is in good agreement with the value of 0.20 found experimentally [25]. Since yield stress is a function of volume, it should be plotted vs. volume and not temperature or pressure. Using the preceding values for α and β to calculate volume, the temperature dependence [18] and the pressure dependence [25] of yield stress are well correlated, as shown in Figure 14. The practical significance of this result is that reliable estimates of the pressure dependence of yield stress can be made without having to make these difficult measurements.

While the assumption that yield stress is a function of volume works well for polypropylene and also is reported to be applicable to amorphous polymers such as polymethyl methacrylate and polyvinylchloride [34], results from polycarbonate do not show correlation between temperature and pressure effects [39, 40]. At any specified volume change, changes in temperature affect the yield stress in polycarbonate more than do changes in pressure. It is still possible to find a correlation, but a shifting process must be used. It has

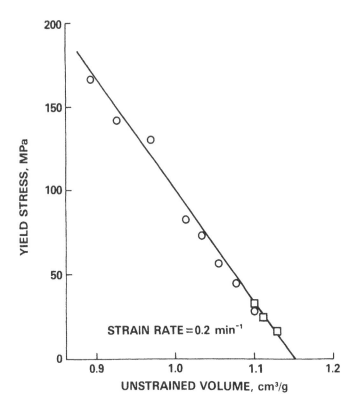

Figure 14 Yield stress vs. unstrained volume for polypropylene (○, pressure dependence; □, temperature dependence).

been suggested [41] that better agreement for polycarbonate would result if the plot were made against the change in fractional free volume rather than against the relative change in total volume. The compression of free volume under pressure is a relatively small proportion of the total compression, whereas the contraction of free volume with decreasing temperature is a relatively large proportion of the total contraction. The free volume changes can be estimated by taking the temperature and pressure coefficients as the differences between the macroscopic coefficients above and below T_g. Phenomena involving molecular mobility are related to free volume, not to total volume. If yield falls in this category, then free volume would be the preferred variable.

It was shown earlier that yield stress and Young's modulus are related, but the form of the relation was not specified. The simplest assumption is that they are directly proportional and that their ratio is a constant. It has been found [42] that the ratio of yield stress to Young's modulus is about the same for numerous amorphous polymers, 0.025 ± 0.010. The values in Table 1 for crystalline polymers also fall in this range. Thus as a rough estimate yield stress is about 1/40 of Young's modulus. This conclusion was reached through a comparison of various polymers at room temperature. It also follows that the yield stress for a given polymer at various temperatures, strain rates, and pressures is a result of the temperature, strain rate, and pressure dependence of Young's modulus. This simple assumption is only partly true. The temperature dependence of the ratio of yield stress to Young's modulus for polyethylene is shown in Figure 15, using data from

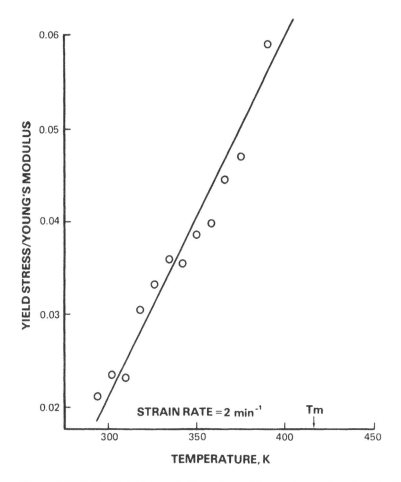

Figure 15 Ratio of yield stress to Young's modulus vs. temperature for polyethylene.

Table 2. The ratio increases with temperature but changes less than either of the individual variables. The strain rate dependence of the ratio (taking data from Table 4) shows considerable scatter but there is a small decrease with increasing strain rate. Similarly, the pressure dependence of the ratio [25] also shows a small decrease with increasing pressure. In summary, some but not all of the temperature, strain rate, and pressure dependence of yield stress can be explained by variations in Young's modulus, and this relationship should be incorporated into the theory of yielding. Similar conclusions have been reached concerning shear yield, where the governing modulus is the shear modulus [43, 44].

Robertson Theory

Robertson [45] pointed out that free volume approaches to yield are principally concerned with intermolecular forces and neglect the equally important intramolecular forces. He assumes that in an amorphous polymer, free rotation can take place about single bonds in the backbone. It is further assumed that there are only two rotational conformations, the trans low-energy state and the cis high-energy state, which is called the flexed state. Stress causes the fraction of flexed bonds to increase from that existing in the glass and leads to yield.

For some orientations of the bonds with respect to the applied stress, the application of a stress will increase the fraction of flexed bonds, whereas for other orientations the fraction will decrease. In the first case this corresponds to an increase in temperature; in the second case, to a decrease. The crux of the theory is the assumption that the rate at which the polymer approaches equilibrium is dependent on temperature, following the WLF equation, so that the rate for the elements that flex under stress is much faster than for the others. This approach was later modified to include the effect of pressure [23].

This theory has the advantage of providing an explanation for the temperature, pressure, and strain rate dependence of yield in amorphous polymers. While the theory has not yet been applied to crystalline polymers, one would assume that an analogous approach would be applicable to that case as well.

SUMMARY

The tensile yield properties of crystalline polymers depend on polymer characteristics such as density, degree of crystallinity, and thermal history as well as test conditions such as temperature, strain rate, and pressure. The behavior of crystalline polymers does not differ qualitatively from that of amorphous polymers or even from metals and ceramics when temperature and pressure are scaled to melting temperature and modulus, respectively.

Because of the general nature of yield, one would expect that the theoretical explanation would not require a specific detailed mechanism applicable to crystalline polymers only. In fact, many of the polymer theories are adapted from theories for metals. Although existing theories are more or less successful in predicting some aspects of yield, none is applicable over all conditions.

REFERENCES

1. I. M. Ward, *Mechanical Properties of Solid Polymers*, 2nd ed., Wiley, New York (1983).
2. I. LeMay, *Principles of Mechanical Metallurgy*, Elsevier, New York (1981).
3. W. D. Kingery, *Introduction to Ceramics*, Wiley, New York (1960).
4. N. Brown and I. M. Ward, *J. Mater. Sci., 18*: 1405 (1983).
5. R. E. Tressler and R. C. Bradt, eds., *Deformation of Ceramic Materials*, vol. 2 (Materials Science Research vol. 18), Plenum, New York (1984).
6. B. Hartmann, G. F. Lee, and R. F. Cole, Jr., *Polym. Eng. Sci., 26*: 554 (1986).
7. D. Sardar, S. V. Radcliffe, and E. Baer, *Polym. Eng. Sci., 8*: 290 (1968).
8. G. E. Dieter, Jr., *Mechanical Metallurgy*, McGraw-Hill, New York (1961).
9. W. Brostow, *Science of Materials*, Krieger, Malabar, Fla. (1985).
10. B. Hartmann and R. F. Cole, Jr., *Polym. Eng. Sci., 23*: 13 (1983).
11. N. Ibrahim, D. M. Shinozaki, and C. M. Sargent, *Mater. Sci. Eng., 30*: 175 (1977).
12. W. A. Spitzig and O. Richmond, *Polym. Eng. Sci., 19*: 1129 (1979).
13. C. Bauwens-Crowet, J.-C. Bauwens, and G. Homès, *J. Mater. Sci., 7*: 176 (1972).
14. S. Rabinowitz, I. M. Ward, and J. S. C. Parry, *J. Mater. Sci., 5*: 29 (1970).
15. J. A. Sauer, *Polym. Eng. Sci., 17*: 150 (1977).
16. J. A. Sauer, D. R. Mears, and K. D. Pae, *Eur. Polym. J., 6*: 1015 (1970).
17. R. W. Truss, P. L. Clarke, R. A. Duckett, and I. M. Ward, *J. Polym. Sci. Phys. Ed., 22*: 191 (1984).

18. B. Hartmann, G. F. Lee, and W. Wong, *Polym. Eng. Sci.*, *27*: 823 (1987).
19. B. Hartmann and R. F. Cole, Jr., *Polym. Eng. Sci.*, *25*: 65 (1985).
20. B. Hartmann, G. Lee, A. Mobin, and J. Lee, unpublished results.
21. B. Hartmann, G. Lee, and Melanie Hicks, unpublished results.
22. P. J. Phillips and J. Patel, *Polym. Eng. Sci.*, *18*: 943 (1978).
23. R. A. Duckett, S. Rabinowitz, and I. M. Ward, *J. Mater. Sci.*, *5*: 909 (1970).
24. J. S. Lazurkin, *J. Polym. Sci.*, *30*: 595 (1958).
25. D. R. Mears, K. D. Pae, and J. A. Sauer, *J. Appl. Phys.*, *40*: 4229 (1969).
26. H. Ll. D. Pugh, E. F. Chandler, L. Holliday, and J. Mann, *Polym. Eng. Sci.*, *11*: 463 (1971).
27. W. Whitney and R. D. Andrews, *J. Polym. Sci. C*, *16*: 2981 (1967).
28. Y. Imai and N. Brown, *Polymer*, *18*: 298 (1977).
29. D. L. Holt, *J. Appl. Polym. Sci.*, *12*: 1653 (1968).
30. R. F. Boyer, *Rub. Chem. Tech.*, *36*: 1303 (1963).
31. J. M. Starita and M. Keaton, *SPE ANTEC*, *29*: 67 (1971).
32. C. W. Macosko and G. J. Brand, *Polym. Eng. Sci.*, *12*: 444 (1972).
33. C. L. Beatty and J. L. Weaver, *Polym. Eng. Sci.*, *18*: 1109 (1978).
34. S. B. Ainbinder, *Mekh. Polimerov*, *3*: 449 (1969).
35. S. B. Ainbinder, *Mekhanika Polim.*, *4*: 986 (1968).
36. M. G. Broadhurst and F. I. Mopsik, *J. Chem. Phys.*, *52*: 3634 (1970).
37. R. W. Warfield, *J. Appl. Chem.*, *17*: 263 (1967).
38. G. N. Foster III, N. Waldman, and R. G. Griskey, *Polym. Eng. Sci.*, *6*: 131 (1966).
39. A. W. Christiansen, E. Baer, and S. V. Radcliffe, *Phil. Mag.*, *24*: 451 (1971).
40. S. V. Radcliffe, in *Deformation and Fracture of High Polymers* (H. H. Kausch, J. A. Hassell, and R. I. Jaffee, eds.), Plenum, New York (1973), pp. 191–209.
41. J. D. Ferry, in *Deformation and Fracture of High Polymers* (H. H. Kausch, J. A. Hassell, and R. I. Jaffee, eds.), Plenum, New York (1973).
42. N. Brown, *Mater. Sci. Eng.*, *8*: 69 (1971).
43. A. S. Argon, *Phil. Mag.*, *28*: 839 (1973).
44. P. B. Bowden and S. Raha, *Phil. Mag.*, *29*: 149 (1974).
45. R. E. Robertson, *J. Chem. Phys.*, *44*: 3950 (1966).

4

Improving the High-Temperature Performance of EPDM

Ralph D. Allen

Uniroyal Chemical Company, Inc.
Naugatuck, Connecticut

BACKGROUND

EPM and EPDM are the ASTM nomenclature for copolymers and terpolymers of ethylene and propylene. The "M" denotes a chemically saturated polymer chain of the polymethylene type, which makes up the repeating units (CH_2), or "vertebrae," in the spine of the polymer. In EPDM, a small amount of a third monomer, a diene, is introduced.

EPM polymers are essentially amorphous or noncrystalline. Also, they are completely saturated (no double bonds). Hence, cross-linking is generally carried out by using a peroxide system. There being no unsaturation in EPM, sulfur vulcanization is not possible.

The structure of the regular, alternating amorphous copolymer of ethylene

$$H_2C{=}CH_2$$

and propylene

$$CH_3{-}\overset{\text{H}}{C}{=}CH_2$$

can be written

$$- (CH_2-CH_2-\overset{\overset{\displaystyle CH_3}{\displaystyle |}}{CH}-CH_2)-_n$$

EPDM rubbers are amorphous terpolymers in which a small amount of a nonconjugated diene is used. The diene introduces unsaturation (double bonds) which are pendant, or in a side chain. The commonly used dienes in the manufacture of EPDM polymers are dicyclopentadiene, ethylidene norbornene and 1,4-hexadiene. The substituent unsaturation provided by the diene permits sulfur vulcanization. EPDM polymers may also be cross-linked by peroxide systems.

The monomer units in EPM and EPDM are randomly distributed. That is, the polymer does not follow a regular alternation sequence of ethylene and propylene units, but contains short runs of both polyethylene and polypropylene interspersed among longer segments of random copolymers.

The three comonomers employed in industry to introduce unsaturation are

dicyclopentadiene (DCPD)

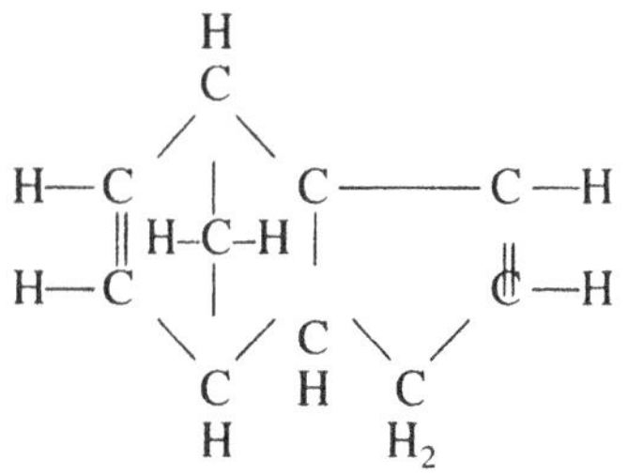

ethylidene norbornene (ENB)

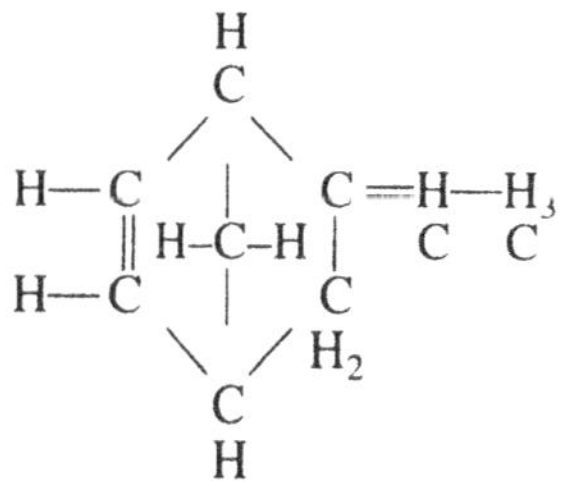

and 1,4-hexadiene (1,4-HD)

$$CH_2{=}CH-CH_2-CH{=}CH-CH_3$$

In all three cases the polymer's double bond is the one illustrated on the left side of the molecule. The most common termonomer employed is ethylidene norbornene because of its ease of incorporation and its greater reactivity toward sulfur vulcanization. The structure of the terpolymer can be illustrated as follows

$$\left[\left[\begin{array}{c} \underset{|}{\overset{|}{H}} \quad \underset{|}{\overset{|}{H}} \\ C - C \\ \underset{|}{\overset{|}{H}} \quad \underset{|}{\overset{|}{H}} \end{array}\right]_x \left[\begin{array}{c} \underset{|}{\overset{|}{H}} \quad \underset{|}{\overset{|}{H}} \\ C - C \\ \underset{|}{\overset{|}{H}} \quad \underset{|}{\overset{|}{CH_3}} \end{array}\right]_y \left[\begin{array}{c} H \quad H \\ C - C \\ \dots \end{array}\right]_z \right]_n$$

ethylene-propylene terpolymer (EPDM)

Increasing the diene content increases the cure rate. The greater degree of unsaturation permits compounds with faster cure rates. Such polymers are more suitable for blending with diene rubbers. Some third monomers are more reactive to sulfur curing systems. That is, ethylidene norbornene is faster than 1,4-hexadiene, which is faster than dicyclopentadiene at equivalent levels.

Because of their structure and composition, EPDM polymers are inherently resistant to ozone and weather. Compounds are suitable for severe outdoor or high ozone applications without need for antiozonants, waxes, or other special additives. EPDM compounds provide good heat and oxidation resistance. No special compounding is required for temperatures up to 100°C. For higher temperatures, the choice of cure systems is important and antioxidants may be required. This will be discussed in considerable detail later in the chapter.

Other inherent properties of EPDM include resistance to polar fluids, such as alcohols, ketones, certain esters, and acetates; resistance to water and aqueous solutions; good low-temperature flexibility; high resilience over a fairly broad temperature range; excellent electrical properties, i.e., high dielectric strength, low power factor, and corona discharge resistance; and radiation resistance.

Its low density (0.865 specific gravity) combined with its ability to accept high levels of extender oils and filler loadings provide the opportunity to make competitive compounds for a variety of applications.

Since becoming commercially available in 1964, EPDM polymers have become widely used in numerous molded and extruded parts. Principle applications include automotive coolant system hoses, tubing, weatherstripping, ignition cable insulation, and isolator mounts. Other uses are found in appliances, agricultural equipment, and the building and construction trades. The most recent important use is in membrane sheeting for commercial building roofing.

Currently there are thirteen manufacturers of EPDM. In 1987, the total worldwide production was 535,000 metric tons.

INTRODUCTION

Various environmental factors must be considered when selecting or designing a rubber compound for a particular application. Although it is usually necessary to consider a combination of requirements, such as weather or ozone resistance, chemical or fluid resistance, low-temperature flexibility, resilience, and abrasion resistance, heat resistance is a more frequently specified requirement than any other, save basic physical properties. In fact, the writers of Specification SAE J200, ASTM D2000, use heat resistance as a basic means of classifying rubber compounds. The first letter in the callout denotes a temperature at which certain minimum requirements for tensile strength, elongation, and hardness changes must be met. Elevated temperatures are simply part of the normal environment in which many rubber parts are expected to function. The increasing complexity of mechanical products, combined with trends toward compactness usually result in the creation of higher operating temperatures. This is particularly true of service temperatures, which can range up to 175°C in today's automobile.

The following environmental factors are important to EPDM (ethylene–propylene–diene monomer): heat resistance, weather or ozone resistance, chemical or fluid resistance, low-temperature flexibility, resilience, and abrasion resistance.

The heat-resistance capabilities of EPDM compounds can be extended by optimizing the polymer and by compounding techniques. The cure system is of prime importance to achieve maximum heat resistance, and as well be shown in this chapter, the level of zinc oxide, the use of chlorosulfonated polyethylene (CSM), and the type and amount of antioxidant also play a part. Parameters that are discussed here include: polymer composition, cure systems, zinc oxide levels, chlorosulfonated polyethylene (CSM), antioxidants, and coagents.

COMPOUND PROPERTIES

In compounding EPDM, three basic types of cure systems may be used. Each gives a different degree of heat resistance, as is commonly known. They are referred to here as (1) normal sulfur or conventional, (2) sulfur donor, and (3) peroxide systems. Within each type, many variations are possible and in common use. Table 1 lists common cure systems for EPDM. Commercial-type compounds based on typical EPDM polymers are used to illustrate comparative differences among these cure systems. To relate their respective serviceable life, retention of elongation after heat aging for 70 and 168 hr were established as measuring points. Using a typical ASTM D2000 requirement of 50% elongation loss as acceptable service performance, the data indicate that after 70 hr heat aging the normal sulfur cure system loses less than 50% elongation at 125°C and 150°C but fails to meet the criterion at the 175°C test temperature. The sulfur donor system provides improved retention at the three temperatures and meets the 50% maximum, even at 175°C. The performance of the peroxide system is obviously outstanding. Table 2 reports typical data on elongation loss.

At the more severe 168-hr aging period, the normal sulfur system fails at 150°C and embrittles at 175°C. The sulfur donor and the peroxide systems provide acceptable retention at 125°C and 150°C but even these systems fail to meet the maximum 50% elongation loss required after the 168-hr aging at 175°C. Table 3 reports elongation loss data after longer agings.

Table 1 Cure Systems for EPDM

Normal Sulfur or Conventional	
MBTS	1.0
TMTD	1.5
ZDMDC	1.0
Sulfur	1.5
Sulfur Donor	
TMTD	3.0
DTDM	2.0
ZDBDC	2.0
ZDMDC	2.0
Sulfur	.5
Peroxide	
Dicumyl peroxide	2.8
Trimethylopropane trimethacrylate	2.0

Key: MBTS, 2,2'-benzothiazole disulfide; TMTD, tetramethylthiuram disulfide; ZDMDC, zinc dimethyl dithiocarbamate; ZDBDC, zinc dibutyl dithiocarbamate; DTDM, 4,4'-dithiodimorpholine.

Table 2 Elongation Loss After Aging for 70 Hours

	Test temperature (°C)		
Cure system	125	150	175
Normal sulfur	−30	−39	−64
Sulfur donor	−14	−18	−40
Peroxide	0	0	−20

Table 3 Elongation Loss After Aging 168 Hours

	Test temperature (°C)		
Cure system	125	150	175
Normal sulfur	−39	−71	B[a]
Sulfur donor	−21	−32	−85
Peroxide	0	−10	−70

[a]B denotes too brittle to test.

POLYMER COMPOSITION

Because there is no unsaturation in the main polymer chain, EPDM polymers are inherently resistant to oxidation. Unsaturation is pendant to the backbone, as illustrated in Figure 1.

It is known that peroxides tend to crosslink polyethylene and primarily cause molecular weight degradation in polypropylene. Landi and Easterbrook studied changes in network structure occurring over periods of time by monitoring stress relaxation of heat-aged, peroxide-cured EPDM compounds [1]. They found that the balance of crosslinking to chain scission is also affected by the type and amount of termonomer, and that this balance of crosslinking to scission relates to the retention of elongation during oxidation.

To illustrate the significance of optimizing the polymer composition, a comparison was made of two EPDM polymers in a formulation utilizing a peroxide cure system. The compound formulation used in this study is given in Table 4.

Table 4   Compound Formulation in Study

Polymer	100.0
CSM	5.0
Zinc oxide	20.0
Antimony trioxide	5.0
N-650 black	60.0
Paraffinic oil	10.0
Antioxidant A	1.0
Trimethylolpropane trimethacrylate	2.0
Dicumyl peroxide (40% active)	7.0
	210.0

Both of the polymers utilize ethylidene norbornene as the third monomer. The typical commercial-grade polymer (R-501) has an iodine number of 10 and an ethylene to propylene ratio of 56:44. By comparison the polymer optimized for heat resistance (HT polymer) has an iodine number of 6 and an ethylene to propylene ratio of 52:48. The two polymers are comparable in molecular weight.

Heat-aging tests were run 14 days at 165°C and physical properties were measured. Data are reported in Table 5. The test data show that the HT polymer has slightly lower tensile strength and greater unaged elongation. The HT polymer provides a significant improvement in heat resistance as judged by elongation change or TE index after aging. The TE index is obtained by multiplying % tensile retained by % elongation retained and dividing by 100.

EFFECTS OF ADDITIVES

Zinc oxide plays a part in the heat resistance of peroxide-cured EPDM compounds. Table 6 reports a compound formulation used to illustrate its effect. Using the HT polymer, levels of 0, 5, and 20 parts of zinc oxide were compared after aging 7 and 14 days at 165°C. Data are reported in Table 7. It is obvious that the use of zinc oxide is necessary to enhance heat resistance, as judged by elongation change or TE index. Five parts of zinc

Table 5   Effect of Termonomer

	R-501	HT polymer
Iodine no. (RI)	10	6
E/P ratio	56/44	52/48
Unaged physical properties[a]		
Hardness, durometer A	69	69
Tensile strength (MPa)	14.3	12.8
Elongation (%)	380	490
Heat aged, 14 days at 165°C		
Hardness change (pts.)	+14	+13
Tensile strength change (%)	−68	−64
Elongation change (%)	−65	−51
TE index	11	18

[a]Specimens cured 30 min at 165°C.

Table 6   Compound Formulation Used to Show Effect of Zinc Oxide

HT polymer	100.0
CSM	5.0
Antimony trioxide	5.0
N-650 black	60.0
Paraffinic oil	10.0
Antioxidant A	1.0
Trimethylolpropane trimethacrylate	2.0
Dicumyl peroxide (40% active)	7.0
Zinc oxide	As shown

Table 7 Effect of Zinc Oxide

	Zinc oxide (parts)		
	0	5	20
Heat aged, 7 days at 165°C			
Hardness change (pts.)	+14	+12	+12
Tensile strength change (%)	−52	−44	−38
Elongation change (%)	−50	−39	−29
TE index	24	34	44
Heat aged, 14 days at 165°C			
Hardness change (pts.)	+18	+15	+13
Tensile strength change (%)	−68	−66	−64
Elongation change (%)	−86	−58	−51
TE index	4	14	18

oxide offers improvement compared to none and 20 parts provides some additional improvement.

Chlorosulfonated polyethylene (CSM) has been reported in other work as improving heat resistance of EPDM [2]. For higher temperature agings the use of five parts has been found to offer benefits. In this work the peroxide-cured HT polymer compound was again employed. The level of antimony trioxide also was varied from 0 to 10 parts as CSM was varied. Table 8 reports the compound formulation prepared to illustrate the effect of CSM on heat aging.

When chlorosulfonated polyethylene is introduced into the compound, hardness increases and elongation decreases compaiatively rapidly during the early stages of exposure to high temperature. This is evident from the results shown after aging for seven days at 165°C. Table 9 illustrates the effect of CSM in the model compound. The beneficial effects become apparent only if long-term tests are conducted. It appears that

Table 8 Compound Formulation Used to Illustrate Effect of CSM

HT polymer	100.0
Zinc oxide	20.0
N-650 black	60.0
Paraffinic oil	10.0
Antioxidant A	1.0
Trimethylolpropane trimethacrylate	2.0
Dicumyl peroxide (40% active)	7.0
Antimony trioxide	As shown
CSM	As shown

Table 9 Effect of CSM

CSM (parts)	0	5	10	10
Sb_2O_3 (parts)	5	5	5	10
Heat aged, 7 days at 165°C				
Hardness change (pts.)	+4	+11	+14	+11
Tensile strength change (%)	−44	−32	−31	−30
Elongation change (%)	−6	−22	−34	−36
TE index	53	53	46	45
Heat aged, 14 days at 165°C				
Hardness change (pts.)	+6	+13	+16	+13
Tensile strength change (%)	−83	−64	−56	−51
Elongation change (%)	−74	−51	−53	−42
TE index	4	18	21	28
Heat aged, 21 days at 165°C				
Hardness change (pts.)	+16	+16	+17	+17
Tensile strength change (%)	−100	−73	−64	−61
Elongation change (%)	−100	−89	−75	−68
TE index	0	3	9	12

chlorosulfonated polyethylene introduces sacrificial sites where oxidation attack is concentrated, thus affording protection to the EPDM polymer. The major benefit is provided by five parts of CSM. However, increasing the level to 10 parts while at the same time increasing antimony trioxide to 10 parts offers some additional improvement. The TE index number becomes more meaningful for making this evaluation.

USE OF ANTIOXIDANTS

As in most other types of rubber, antioxidants can play an important role in the heat resistance of EPDM compounds. However, not all antioxidants are satisfactory for use with a peroxide-cured system. Some are comparatively ineffective; others exert a strong inhibiting effect on the free radical crosslinking mechanism of peroxides. Three of the antioxidants that have been found to be effective are compared here. Table 10 summarizes the ingredients of a model formulation and antioxidants used to demonstrate the effects of antioxidants on heat aging. The formulation used is the same as previously shown. Table 11 reports the experimental results on heat aging obtained.

Among the three antioxidants evaluated, C is more effective than A and A is somewhat better than B in improving heat resistance at 165°C. Again the TE index number assists in making the rating. Judging from the effects on unaged properties, antioxidant A has the most inhibiting effect on the peroxide-cure system; C has the least effect. In fact, the latter is the only one of the three that would be practical for use at the three part level.

USE OF COAGENTS

Coagents are often used in conjunction with peroxides to enhance some property or combination of properties. Some coagents, e.g., sulfur, are known to impair heat resistance. Others provide a beneficial effect. Again using the previously shown formulation, a

Table 10 Model Compound Formulation and Antioxidants Used to Show Effect of Antioxidants

HT polymer	100.0
CSM	5.0
Zinc oxide	20.0
Antimony trioxide	5.0
N-650 black	60.0
Paraffinic oil	10.0
Trimethylolpropane trimethacrylate	2.0
Dicumyl peroxide (40% active)	7.0
Antioxidant	As shown
Antioxidant A	Substituted diphenylamine
Antioxidant B	Polymerized 1,2-dihydro-2,2,4-trimethylquinoline
Antioxidant C	Mixture of alkylated mercaptobenzimidazole and an amine-type antioxidant

Table 11 Effect of Antioxidants

	1	2	3	4	5	6	7
No antioxidant	0						
Antioxidant B		1	3				
Antioxidant A				1	3		
Antioxidant C						1	3
Unaged physical properties[a]							
Hardness, durometer A	77	76	76	71	67	76	77
Tensile strength (MPa)	15.4	14.5	12.9	12.0	7.4	15.1	13.7
Elongation (%)	370	370	460	480	640	390	410
Modulus at 200% (MPa)	7.4	7.2	5.7	4.7	2.5	7.1	7.4
Heat aged, 14 days at 165°C							
Hardness change (pts.)	+13	+8	+11	+13	+12	+10	+7
Tensile strength change (%)	−72	−72	−67	−64	−50	−52	−39
Elongation change (%)	−76	−70	−74	−51	−58	−38	−40
TE index	7	8	9	18	21	30	37

[a]Specimens cured 30′ at 165°C.

comparison was made of five coagents. Table 12 summarizes the compound formulation. The results of the evaluation are reported in Table 13.

The use of trimethylolpropane trimethacrylate, trimethylolpropane acrylate, or triallyl cyanurate increases unaged modulus and tensile strength. After 7, 14, and 21 days aging at 165°C, the trimethylolpropane trimethacrylate provides the best elongation retention. However, tetrahydrofurfuryl methacrylate and 1,2-syndiotactic polybutadiene have good TE index ratings, primarily due to better retention of tensile strength.

Table 12 Compound Formulation Used to Show
Effect of Coagents

HT polymer	100.0
CSM	5.0
Zinc oxide	20.0
Antimony trioxide	5.0
N-650 black	60.0
Paraffinic oil	10.0
Antioxidant A	1.0
Dicumyl peroxide (40% active)	7.0
Coagent	As shown

Table 13 Effect of Coagents

Coagent parts	8	9	10	11	12	13
No coagent	0					
Trimethylolpropane trimethacrylate		2				
Trimethylolpropane acrylate			2			
Triallyl cyanurate				2		
Tetrahydrofurfuryl methacrylate					2	
1,2-syndiotactic polybutadiene						2
Unaged physical properties 30′ at 165°C cure						
Hardness, durometer A	71	73	71	72	67	69
Modulus at 200% (MPa)	2.9	4.8	5.7	8.3	2.6	3.3
Tensile strength (MPa)	9.5	12.1	13.1	13.0	8.5	10.0
Elongation (%)	650	470	420	340	710	590
Heat aged, D865, 7 days at 165°C						
Hardness change (pts.)	+18	+14	+13	+15	+13	+14
Tensile strength change (%)	−32	−29	−33	−21	+4	−27
Elongation change (%)	−32	−23	−29	−29	−45	−27
TE index	46	55	48	56	55	53
Heat aged, D865, 14 days at 165°C						
Hardness change (pts.)	+18	+13	+12	+14	+17	+13
Tensile strength change (%)	−65	−61	−63	−55	−39	−59
Elongation change (%)	−62	−49	−52	−59	−66	−59
TE index	13	20	18	18	20	17
Heat aged, D865, 21 days at 165°C						
Hardness change (pts.)	+18	+19	+15	+17	+18	+15
Tensile strength change (%)	−68	−66	−71	−58	−44	−59
Elongation change (%)	−97	−79	−90	−88	−85	−81
TE index	1	7	3	5	8	8

Based on these findings the formulation for optimum heat resistance would be as reported in Table 14, with data reported in Table 15. The improved retention of properties becomes more significant with longer aging. The results after 21 days at 165°C aging show tensile strength retention comparable to the previously shown data but elongation retention is better.

CURE SYSTEM COMPARISON

While it is believed that in EPDM compounds the best combination of properties for high-temperature service are provided by peroxide cure systems, it is recognized that some

Table 14 Formulation for Optimum Heat Resistance

HT polymer	100.0
CSM	10.0
Zinc oxide	20.0
Antimony trioxide	10.0
N-650 black	60.0
Paraffin oil	10.0
Antioxidant C	3.0
Trimethylolpropane trimethacrylate	2.0
Dicumyl peroxide (40% active)	7.0

Table 15 Heat Aging Data on Optimum Formula

Unaged physical properties[a]	
Hardness, durometer A	77
Tensile strength (MPa)	13.4
Elongation (%)	440
Heat aged, 7 days at 165°C	
Hardness change (pts.)	+10
Tensile strength change (%)	−26
Elongation change (%)	−32
TE index	50
Heat aged, 14 days at 165°C	
Hardness change (pts.)	+11
Tensile strength change (%)	−39
Elongation change (%)	−36
TE index	39
Heat aged, 21 days at 165°C	
Hardness change (pts.)	+12
Tensile strength change (%)	−55
Elongation change (%)	−61
TE index	18

[a]Specimens cured 30′ at 165°C.

applications can be satisfied with sulfur-cured compounds. Also it is recognized that sulfur-cure systems are more widely used and acceptable for production processing.

For high-temperature dry-heat resistance, the use of high levels of thiazole-type (e.g., MBT or MBTS) accelerators has been found to be beneficial. A high-thiazole-cured system is compared here with the sulfur donor system previously discussed and with the optimized peroxide-cured compound using the HT polymer. Because of its lower ter-monomer level, the HT polymer has a slightly slower cure rate when compounded with sulfur systems, but it is considered satisfactory in most processes. Table 16 lists the formulation used and Table 17 reports experimental results. This comparison shows that the high-thiazole system outperforms the sulfur donor system at all aging periods. It is equal to or better than the peroxide-cured compound after the 7- and 14-day aging periods. After the 21-day aging period the peroxide-cured compound is shown to offer superior retention of properties.

To illustrate the range of heat resistance available with EPDM compounds, a comparison of the four cure systems was made with the improved compound utilizing the HT polymer. Long-term heat agings were conducted at standard ASTM test temperatures of 125, 150, and 175°C. The amount of time in hours to reduce the elongation to 20% was used as a measure of elastic decay; 20% was chosen as the point at which embrittlement of the standard test dumbbell would occur. The results are reported in Table 18.

This comparison clearly shows the improvement in heat resistance attainable by selection of polymer, combined with compounding techniques. With the peroxide system improvements are made by use of higher-than-normal levels of zinc oxide, chlorosulfonated polyethylene in combination with antimony trioxide, and the choice of antioxidant and coagent. The advantages of some of these compounding techniques may not be apparent if only short-term heat aging tests are conducted.

A high-thiazole system may be advantageous for some applications.

Application areas where the higher temperature heat resistance capabilities of these EPDM compounds may be of interest are, for example, automotive exhaust hanger straps, fire wall grommets and terminal insulators, hot materials belts, solar energy systems, and steam hose covers. If current trends continue, ever-increasing temperatures and more demanding requirements will necessitate even further advancement in polymer and compounding technology to satisfy future needs.

Table 16   Basic Heat-Resistant Formulation

HT polymer	100.0
Zinc oxide	5.0
N-650 black	60.0
Paraffinic oil	10.0
Zinc stearate	1.5
Antioxidant	1.0
Cure systems	As shown

Table 17 Cure System Comparison

	Sulfur donor	High thiazole
TMTD	3.0	0.8
MBTS		4.0
ZDMDC	2.0	
ZDBDC	2.0	1.5
DTDM	2.0	.8
Sulfur	.5	.7

	Peroxide	Sulfur donor	High thiazole
Unaged physical properties			
Hardness, durometer A	76	72	73
Tensile strength (MPa)	13.5	14.8	12.8
Elongation (%)	400	420	440
Heat aged, 7 days at 165°C			
Hardness change (pts.)	+7	+9	+7
Tensile strength change (%)	−23	−20	+9
Elongation change (%)	−23	−39	−25
TE index	59	49	75
Heat aged, 14 days at 165°C			
Hardness change (pts.)	+7	+14	+9
Tensile strength change (%)	−30	−57	+2
Elongation change (%)	−35	−67	−43
TE index	46	14	67
Heat aged, 21 days at 165°C			
Hardness change (pts.)	+12	+20	+14
Tensile strength change (%)	−39	−68	−31
Elongation change (%)	−60	−98	−82
TE index	24	1	12

Table 18 Elastic Decay Test (hours of heat aging to reduce elongation to 20%)

Cure system	125°C	150°C	175°C
Normal sulfur	1000	350	100
Sulfur donor	1000+	750	140
Peroxide (in 10 I_2 polymer)	1000+	800+	168
Peroxide in HT polymer	1000+	800+	500
High thiazole in HT polymer	1000+	800+	330

GLOSSARY OF INGREDIENTS

Product	Chemical composition	Uniroyal trade name
R-501	EPDM polymer	Royalene 501
HT polymer	EPDM polymer	Royalene 580 HT
MBTS	2,2'-Benzothiazole disulfide	MBTS
TMTD	Tetramethylthiuram disulfide	Tuex
ZDMDC	Zinc dimethyl dithiocarbamate	Methazate
ZDBDC	Zinc dibutyl dithiocarbamate	Butazate
Antioxidant A	Substituted diphenylamine	Naugard 445
Antioxidant B	Polymerized 1,2-dihydro-2,2,4-trimethylaquinoline	Naugard Q
Antioxidant C	Mixture of alkylated mercaptobenzimidazole and an amine-type antioxidant	Naugard 495

The author gratefully acknowledges the assistance of colleagues in the preparation of this paper. The author also thanks Uniroyal Chemical Company, Inc. for its permission to give and publish this paper.

REFERENCES

1. V. R. Landi and E. K. Easterbrook, *Polym. Eng. Sci.*, *18* (15) (1978).
2. E. I. du Pont de Nemours and Co., *Heat Resistance of Nordel*, Bull. No. 6 (1967).

5

Oxidation and Ozonation of Rubber

Robert W. Keller
CR Industries
Elgin, Illinois

INTRODUCTION

Rubber materials find many uses as engineering materials because of their unique combinations of elastic and viscous properties. Rubber compounds can be effectively formulated and designed to provide energy storage, energy dissipation, vibration isolation, or flexible sealing devices where material flexibility is necessary. Many elastomeric polymers are available for use in rubber formulations as well as a wide variety of fillers, extenders, softeners, tackifiers, and vulcanizing systems to provide the finished rubber product with the desired combination of properties. With such a wide variety of compounding materials available to the formulator, it is easy to understand why development of a rubber compound for a specific application can involve considerable designed experimentation and testing. However, if the effects of oxygen and ozone on the rubber product are not taken into account in the formulation and design of the rubber product, a great deal of effort can be wasted and the long-term durability of the product may be in jeopardy.

A few brief examples of what can happen if oxygen and ozone attack in the application are not considered when the rubber product is designed and formulated will help to emphasize the importance of oxidation and ozonation:

1. Automotive and truck tires require several layers of rubber materials to provide low air permeability for the liner, good rigidity for the sidewalls, long-term abrasion resistance for the tread, good wet traction for the tread, good grip in hard braking, resistance to puncture, and recently for overall reduction of rolling resistance to provide good fuel economy. Obviously a great deal of effort is required to produce the desired tire through rubber formulation and tire design. However, failure to include suitable ozone protection can result in catastrophic sidewall cracking before the tire has reached its anticipated life. Also, failure to provide suitable oxygen protection can cause the surface grip characteristics and the wear of the tread to be compromised.

2. Multilayered rubber and steel bushings are used in helicopters and aircraft for isolation of vibration caused by rotors, propellers, and turbines. Considerable effort goes into the design and formulation of the desired composite part to provide the necessary vibration isolation characteristics. If sufficient ozone and oxygen protection is not provided in the design of the vibration isolator, ozone or oxygen attack may cause flex cracking and rupture of the rubber part. This will result in excessive vibration and possible failure of the aircraft that the vibration isolator was mounted on.

3. Elastomeric seals are used in a large variety of applications to retain critical lubricants and exclude detrimental contaminants. In many cases, long-term elastomeric seal life is a key factor in the useful life of engines, gearboxes, power transmissions, and hydraulic cylinders. Oxygen or ozone attack on the elastomeric seal can cause cracking and rupture of the seal elastomer, resulting in fluid leakage or dust ingression, in turn leading to failure of a very expensive engine, gearbox, or transmission.

These examples are intended to display the seriousness of oxygen and ozone attack on rubber articles. Fortunately, a variety of formulating materials are available to rubber chemists to prolong the life of rubber goods exposed to oxygen and ozone. These antidegradant materials are possible due to fundamental work in the chemistry of oxygen and ozone attack on rubber and to organic and organometallic chemistry. This chapter introduces the reader to the chemistry of oxygen and ozone attack on rubber, antioxidant and antiozonant chemistry and mechanisms, and methods of determining oxygen and ozone resistance in the laboratory.

MECHANISM OF RUBBER OXIDATION

Essential to the understanding of rubber oxidation is the understanding of the nature of atmospheric oxygen. The atmospheric oxygen molecule, O_2, demonstrates paramagnetism, which is weak attraction to a magnetic field. Simple Lewis structures could not be internally consistent and explain the paramagnetism of oxygen. However, the relatively simple rules of molecular orbital theory show that the O_2 molecule has two unpaired electrons after bonding. These two unpaired electrons can, for the purposes of understanding oxygen attack on rubber, be considered two free radical species existing in the oxygen molecule. Thus the ability of oxygen to attack unsaturation in hydrocarbon molecules is much easier to understand. Details about the molecular orbital theory as applied to the O_2 molecule are found in most beginning chemistry textbooks.

Bolland and co-workers were the first to understand the oxidative attack on rubber in terms of oxidative attack on low molecular weight hydrocarbon analogs of rubber [1]. The kinetic mechanism proposed by Bolland is given in Figure 1. In Figure 1, the RH represents the rubber hydrocarbon polymer. Steps 2–6 in the mechanism rely on the free radical nature of oxygen to propagate and terminate the free radicals generated in the first stage. Examination of Figure 1 shows that generation of free radicals on the polymer is key to the entire process. Such free radicals can be generated on the polymer through application of heat or by mechanical action causing high shear and rupture of individual polymer chains.

At elevated temperatures, generation of free radicals on the polymer due to carbon–hydrogen bond cleavage or due to carbon–carbon bond cleavage is likely. However, many elastomers are observed to oxidize at relatively low temperatures, say below 60°C, where carbon–hydrogen and carbon–carbon bond cleavage are highly unlikely. Thus it was believed that trace structural impurities in the polymer could account for the oxidation of rubber at relatively mild temperatures. Two separate studies of the low-temperature oxidation of rubber concluded that traces of peroxide were present in the polymer [2, 3]. The ease of oxidation of rubber at low temperature was thus due to the ease of homolytic cleavage of the oxygen–oxygen bonds in the peroxides present in the polymer creating the initiating free radical species necessary for the Bolland oxidation mechanism. Due to the high reactivity of free radicals formed from peroxides, only small traces of such peroxides need be present in the polymer to initiate the low-temperature oxidation of the polymer. Mechanical shear of the polymer during processing and bailing [4] and localized heat

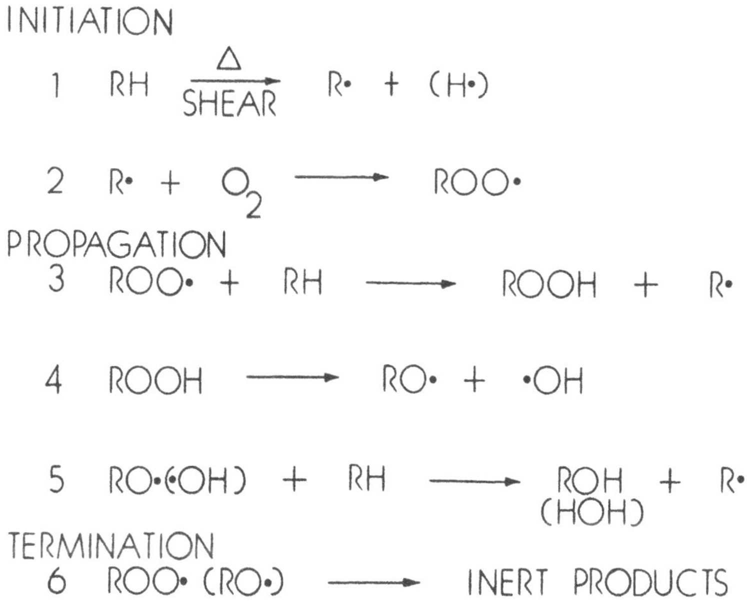

INITIATION

1 $\quad$ RH $\xrightarrow[\text{SHEAR}]{\Delta}$ R• + (H•)

2 $\quad$ R• + O_2 $\longrightarrow$ ROO•

PROPAGATION

3 $\quad$ ROO• + RH $\longrightarrow$ ROOH + R•

4 $\quad$ ROOH $\longrightarrow$ RO• + •OH

5 $\quad$ RO•(•OH) + RH $\longrightarrow$ ROH + R•
$\qquad\qquad\qquad\qquad\qquad\qquad$ (HOH)

TERMINATION

6 $\quad$ ROO• (RO•) $\longrightarrow$ INERT PRODUCTS

Figure 1 Bolland oxidation mechanism.

during drying and packing of the rubber polymer can cause carbon–hydrogen and carbon–carbon bond cleavage. The resultant free radicals formed during processing of the raw rubber will react to form the trace levels of peroxides necessary to account for the ease of rubber oxidation at low temperatures. Therefore, the oxidation of rubber hydrocarbon polymers resembles the Bolland oxidation mechanism for low molecular weight hydrocarbons with the polymer having its own internal source of peroxide initiators present for low-temperature oxidation. In Figure 1, propagation step 4 can be rewritten as initiation step 1 for rubber polymer containing peroxides formed during processing.

It is reasonable to assume that such initiating peroxides are present in even the most carefully prepared raw rubber polymer. Thus it is extremely important to compound rubber for prolonged life in the presence of oxygen through the use of antioxidants and by understanding the effects of pro-oxidants present in the raw polymer or in the rubber compound.

PRO-OXIDANTS

Effects of Polymer Unsaturation

Most of the general-purpose elastomers usually contain high levels of unsaturation in the polymer backbone. Such general-purpose elastomers are natural rubber (NR), styrene–butadiene rubber (SBR), and polybutadiene rubber (BR). The previous examples of general-purpose elastomers include the ASTM designations for the elastomers. It should be noted that the "R" in the ASTM designations does not mean rubber but does refer to the presence of unsaturation in the polymer backbone. Several studies have shown that the unsaturation present in the polymer backbone is a pro-oxidant for the polymers.

Farmer and Sundralingham used model low molecular weight hydrocarbon analogs of unsaturated polymers to study the effects of unsaturation on oxidation [5]. They observed that cyclohexene hydroperoxide was readily formed in a reaction mixture of cyclohexene and oxygen. The ease of hydroperoxide formation was attributed to the ability of the unsaturation present in cyclohexene to stabilize the initial free radical formed in oxygen attack by resonance delocalization. The mechanism they proposed to account for the formation of cyclohexene hydroperoxide is seen in Figure 2. Examination of Figure 2 shows that the first step in the reaction is the abstraction of hydrogen by oxygen from the carbon alpha to the carbon–carbon double bond. The resultant cyclohexene radical is stabilized due to the unsaturation present. The hydroperoxy radical formed by oxygen abstraction of the hydrogen from cyclohexene then recombines with the cyclohexene radical to form the resultant peroxide. Also shown in Figure 2 is the same mechanism for hydroperoxide formation applied to an unsaturated polymer such as polybutadiene. Once again, the ability of the unsaturation to resonance stabilize the radical formed in the initial reaction step allows for relatively easy formation of the hydroperoxide product. Thus unsaturated elastomers such as natural rubber, styrene–butadiene rubber, and polybutadiene can easily react with oxygen to form hydroperoxides. Once formed, these hydroperoxides can then decompose to form radical initiators in the Bolland oxidation scheme. Since oxidative attack according to the Bolland oxidation scheme is a free radical process, only trace amounts of such hydroperoxides need be present to initiate the reaction. Therefore, one would expect unsaturated polymers to be the least resistant to oxidation.

Table 1 lists the various commercially available elastomers in two categories: those resistant to oxidation and those not resistant to oxidation. In general, those elastomers

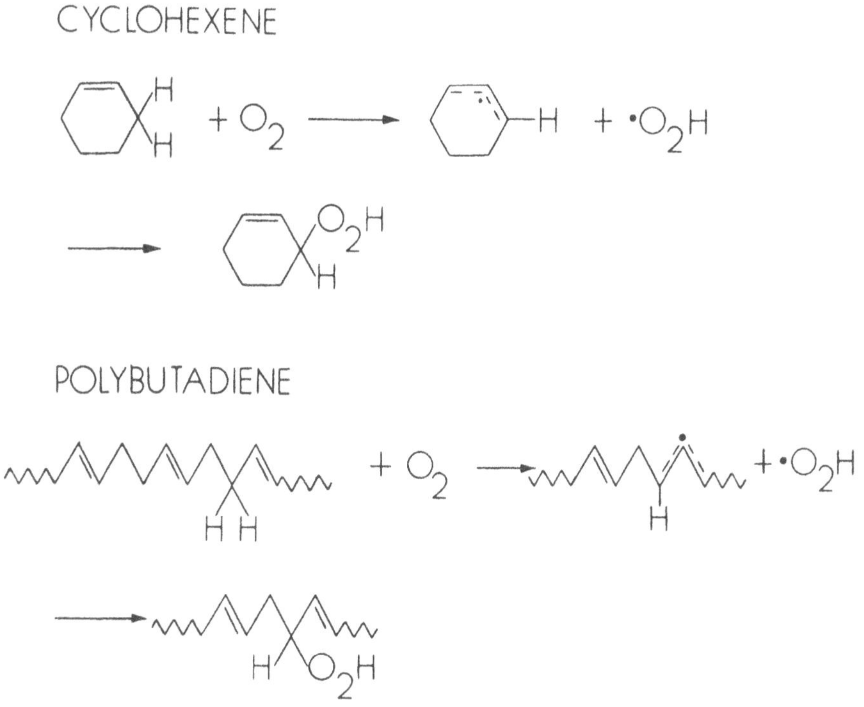

CYCLOHEXENE

POLYBUTADIENE

Figure 2   Oxidation of cyclohexane and poly(butadiene).

Table 1   General Resistance of Elastomers to Oxidative Degradation

Elastomer	ASTM designation
Elastomers resistant to oxidation	
Acrylic	ACM
Chloro-sulfonyl-polyethylene	CSM
Ethylene propylene diene	EPDM
Fluoroelastomers	FKM
Ethylene oxide epichlorohydrin	ECO
Silicones	MQ, VMQ, FVMQ
Polyester urethanes	AU
Polyether urethanes	EU
Butyl rubber	IIR
Halobutyl rubber	BIIR, CIIR
Elastomers not resistant to oxidation	
Natural rubber	NR
Isoprene rubber	IR
Styrene–butadiene rubber	SBR
Poly(butadiene)	BR
Nitrile rubber	NBR
Neoprene	CR

with unsaturation present in the polymer backbone having the letter "R" as the final letter in their ASTM designation are not resistant to oxidation. The only exceptions to the rule regarding unsaturated polymers are butyl and halobutyl elastomers. Butyl and halobutyl elastomers possess unsaturation in the polymer backbone only in sufficient quantities to provide sites for sulfur vulcanization. The residual unsaturation remaining after vulcanization of butyl and halobutyl elastomers is generally so low that these elastomers can be considered oxidation resistant. The reader should keep in mind that, although elastomers are listed in Table 1 as resistant to oxidation, very high temperatures combined with oxygen presence can also cause oxidative degradation of the elastomers resistant to oxidation. The classification of elastomers resistant and not resistant to oxidation in Table 1 is based on observations at temperatures below 100°C.

Effects of Unsaturated Ingredients

Unsaturated organic ingredients present in the elastomer can also have a pro-oxidant effect. Recent studies by Keller and colleagues [6, 7] showed the effects of unsaturated fatty acids on natural rubber oxidation. This work was aimed at a better understanding of the pro-oxidant effects of resinous plant impurities present in natural rubber isolated from the Guayule shrub. Although the Guayule shrub potentially was a commercial source of natural rubber, the plant resins present in the rubber due to isolation caused rapid oxidative softening of the raw rubber and of vulcanized compounds.

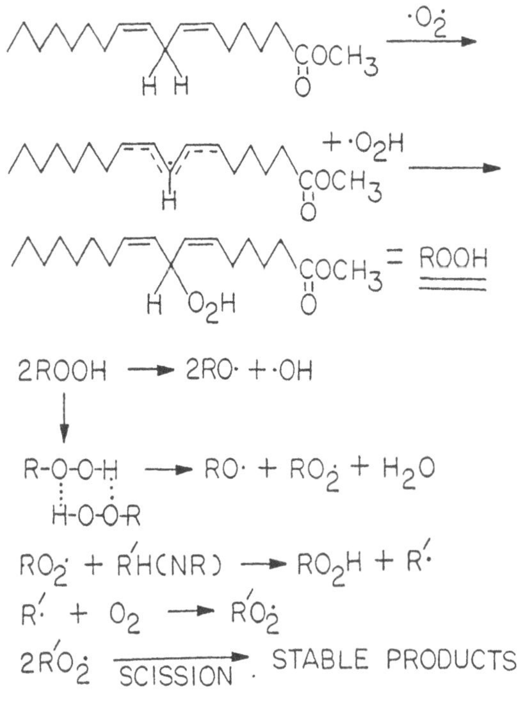

Figure 3 Enhancement of the oxidation of NR by methyl linolenate.

Fatty acids were found to account for roughly 15.5% of the extracted resin by weight. Gas chromatographic analysis of the fatty acids showed high levels of polyunsaturated fatty acids such as linoleic and linolenic acids. Qualitative and further kinetic studies of the effects of the various crude fractions of Guayule resin on natural rubber oxidation showed that the fatty acids present in Guayule resin were responsible for the pro-oxidant effects of the resin on natural rubber. The mechanism proposed to account for these effects is given in Figure 3. In Figure 3, the five carbon-delocalized radical formed during the oxidation of methyl linolenate is more stable than the three carbon-delocalized radical formed during the oxidation of natural rubber. This would make hydroperoxide formation more facile for linoleic acid esters. In turn, decomposition of the linoleic acid ester hydroperoxides would serve as additional initiators for the oxidation chain reaction applied to natural rubber. Thus one should avoid unsaturated and polyunsaturated ingredients when formulating rubber for extended oxidation resistance.

Effects of UV Light

Probably one of the most widespread pro-oxidants for rubber is ultraviolet light since this is present in sunlight and normal incandescent and fluorescent overhead light. Blake and Bruce performed studies where the oxygen absorption rate of natural rubber was studied with and without exposure to UV light [8]. Their studies used a UV transparent quartz cell containing a film of natural rubber. The cell was exposed to oxygen in a closed system where the cell was connected to a sensitive gas manometer so that the oxidation of the natural rubber could be studied with time in terms of volume of oxygen absorbed per unit time. They found that introduction of UV light exposure dramatically increased oxygen absorption, thus demonstrating the pro-oxidant effects of UV light. They also studied the effects of various compounding additives on the absorption rate of oxygen by natural rubber. Results of their studies are presented in Table 2.

Table 2 shows that a common rubber antioxidant, phenyl-β-naphthylamine, actually enhances oxidation of natural rubber with exposure to UV light. Materials which are potent UV light absorbers, such as benzidine and hydroquinone, reduced oxidation with exposure to UV light probably due to their ability to absorb the UV light rendering the natural rubber sample opaque to the detrimental UV light. Ingredients dispersed in the

Table 2 Results of Blake and Bruce on the Oxidation of Rubber Accelerated by UV Light, NR Pale Crepe, 46°C

Additive	Absorption of O_2 (cm^3/hr)
None	0.067
2% Sulfur	0.028
2% Benzidine	0.014
2% Hydroquinone	0.014
2% Phenyl-β-naphthylamine	0.076
5% Zinc oxide	0.010
1% P-33 Carbon black	0.018

natural rubber which cause light scattering and the consequent opacity—sulfur, zinc oxide, and carbon black—also were observed to retard oxidation in the presence of UV light.

The foregoing results demonstrate the importance of UV light as a rubber pro-oxidant. Since UV light is so pervasive in normal environmental exposure to rubber, it is important that formulators be aware of the pro-oxidant effects of UV light and compound rubber materials to maximize resistance to UV light-accelerated oxidation.

MECHANICAL ACTION, OXIDATIVE FLEX CRACKING

Most rubber articles are exposed to repeated mechanical flexing and deformation in service. Static evaluation of oxidation resistance of a rubber material would not necessarily predict the oxidation resistance of the rubber material when exposed to the dynamic flexing and vibration present in the application environment. Studies have shown that flex cracking of rubber materials is accelerated by the presence of oxygen [9–11]. Gent made detailed studies of the chemistry of flex cracking and concluded that environmental flex cracking is caused by oxidative chain scission of the rubber polymer at mechanically induced cracks [11]. Hess and Burgess [12] used photographic and radiographic techniques to study the flex cracking of tires. They concluded that at the natural flaws present in filled rubber, repeated flexing caused reagglomeration of fillers so that a stress concentration point was developed in the rubber compound. This stress concentration point was responsible for crack initiation. Oxidative chain scission of the rubber polymer at the initiated crack then caused rapid crack growth.

The studies cited indicate that oxidative flex cracking occurs in rubber materials exposed to repeated flexing or vibration. The mechanism of oxidative flex cracking involves oxidative chain scission of the rubber polymer at induced cracks. This oxidative chain scission causes rapid crack growth since "fresh" rubber polymer is constantly exposed to oxygen by crack growth and flexing. Since most rubber articles are intended for use in applications where dynamic flexing and vibration are present, methods of compounding for improved oxidative flex crack resistance should be considered in the formulation stage of development. Methods of measuring and rating materials for oxidative flex crack resistance are discussed later.

ANTIOXIDANT TYPES

In general, antioxidants can be divided into two basic chemical types: amines and phenolics. In most rubber systems, amines are more effective in preventing long-term oxidative degradation. However, amine antioxidants usually discolor with aging and may not be the system of choice for light or brightly colored rubber articles where color retention is important. Phenolic antioxidants, in contrast to amine antioxidants, do not discolor on aging but are generally less effective in preventing long-term oxidative degradation. Thus compromises may be necessary in formulating light or brightly colored materials with the use of the generally less effective phenolic antioxidants. If the rubber compound is black, then one can formulate using the more effective amine antioxidants.

The amine antioxidants can be further subdivided into several categories by common chemical types:

1. Secondary diaryl amines: phenyl naphthylamines, substituted diphenylamines, and para-phenylenediamines.

2. Ketone–amine condensates.
3. Aldehyde–amine condensates.
4. Alkyl aryl secondary amines.
5. Primary aryl amines.

Representative chemical structures of the various amine antioxidant types are given in Figures 4 and 5.

Similarly, phenolic antioxidants can also be subdivided by basic chemical types:

1. Hindered phenols.
2. Hindered bisphenols.
3. Hindered thiobisphenols.
4. Polyhydroxy phenols.

Representative chemical structures of the various phenolic antioxidants are given in Figure 6.

In addition, many of the organosulfur compounds and organometallic compounds used in rubber compounds as vulcanization accelerators are known to have antioxidant activity. Mercaptobenzimidazole and its zinc salt have been shown to have antioxidant activity [13]. In addition, mercaptobenzimidazole, when combined with other known antioxidants, has been shown to have a synergistic effect in oxidation prevention when metal ion catalyzed oxidation is prevalent [14, 15]. Also, metal and amine salts of dialkyldithiocarbamates have been shown to have antioxidant action [16].

Choosing the correct combination of antioxidants is specific to the elastomer polymer type as well as the compound formulation and the end use application. For a given

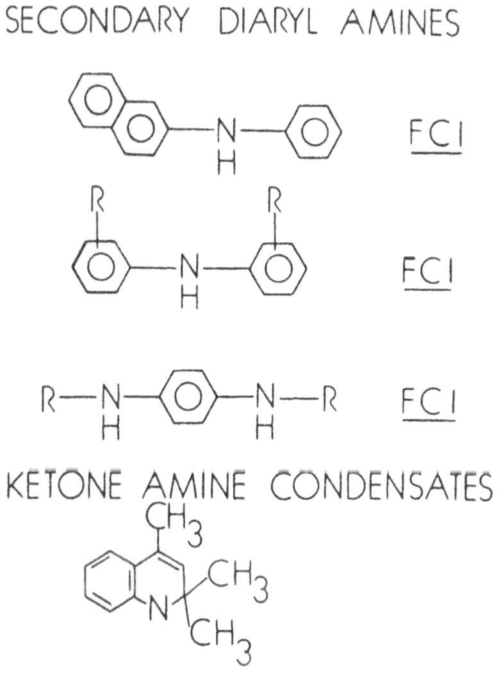

Figure 4 Chemical structures of amine antioxidants.

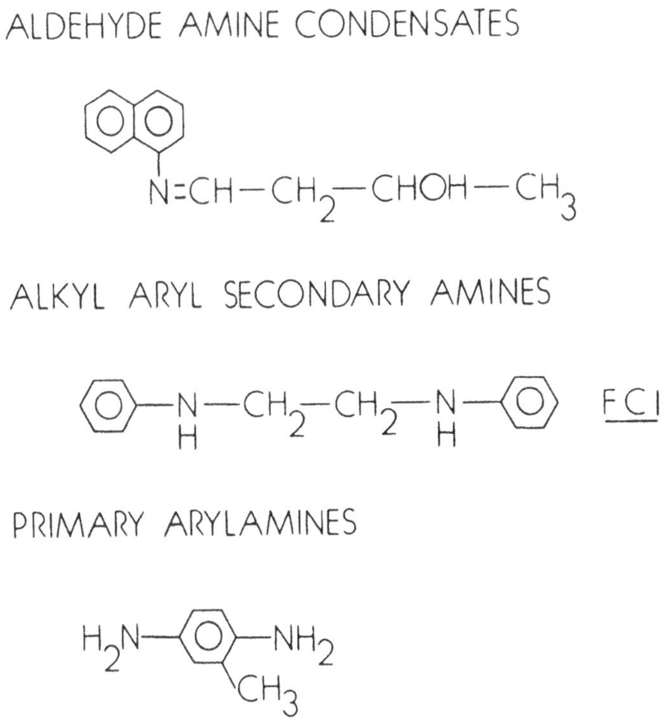

ALDEHYDE AMINE CONDENSATES

$$N=CH-CH_2-CHOH-CH_3$$

ALKYL ARYL SECONDARY AMINES

$$\langle O \rangle - \underset{H}{N} - CH_2 - CH_2 - \underset{H}{N} - \langle O \rangle \quad \underline{F\,C\,I}$$

PRIMARY ARYLAMINES

$$H_2N - \langle O \rangle - NH_2$$
$$CH_3$$

Figure 5 Chemical structures of amine antioxidants.

polymer type, there are antioxidants and antioxidant combinations which have been shown to be the most effective. As mentioned above, color stability of light or brightly colored rubber articles is also a major consideration. Fortunately, much information is available from a variety of sources to assist the rubber formulator in picking the proper antioxidant system for the given polymer, compound, and end use application. Such information will be provided by the polymer manufacturers and the producers of rubber chemicals. Once an antioxidant system is chosen, it is generally necessary to test the effectiveness of the system in accelerated testing to assure that the system performs as expected. Methods of testing are covered later in this chapter.

MECHANISMS OF ANTIOXIDANT ACTION

Four general modes of oxidation inhibition and antioxidant activity are commonly recognized:

1. Absorption of catalytic UV light.
2. Complexation and deactivation of catalytic metals.
3. Decomposition of initiating peroxides.
4. Radical chain reaction quenchers.

As discussed earlier, fillers and metal oxides which provide light scattering and opacity to the compound such as carbon black and zinc oxide provide antioxidant activity in the

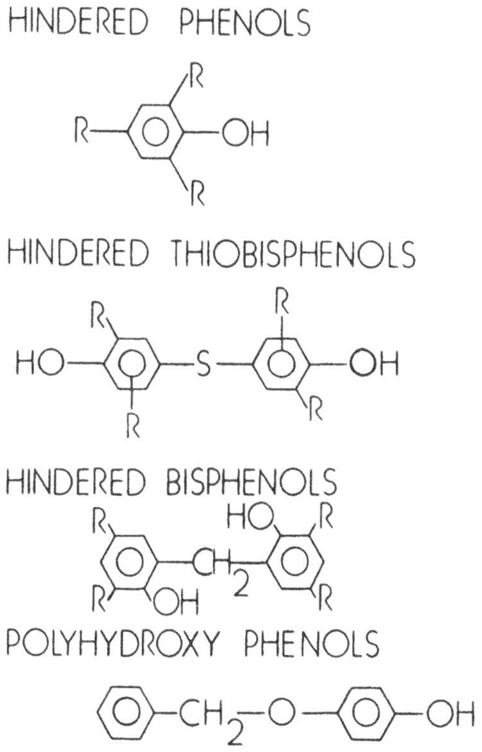

HINDERED PHENOLS

HINDERED THIOBISPHENOLS

HINDERED BISPHENOLS

POLYHYDROXY PHENOLS

Figure 6 Chemical structures of phenolic antioxidants.

presence of UV light by rendering the compound opaque to the UV light. Also, organic compounds soluble in the rubber matrix may prevent absorption of catalytic UV light by their ability to absorb UV radiation, once again rendering the compound opaque to harmful UV light.

Organic compounds capable of forming coordination complexes with metal ions are known to be useful in inhibiting metal ion-activated oxidation. These types of compounds have multiple coordination sites in the individual molecules. The multiple coordination sites wrap themselves around metal ions, forming a "cage" structure which prevents the metal ions from participating in oxidation–reduction reactions, which would activate or accelerate rubber oxidation. EDTA, ethylene diamine tetraacetic acid, and its various sodium salts are classic examples of such coordination compounds.

Materials capable of reacting with peroxides to form nonradical products also function as antioxidants. As discussed earlier, peroxides present in the rubber polymer or in other ingredients present in the rubber compound serve as initiators of the oxidative chain reaction. Only trace amounts of such peroxide initiators need be present to start oxidation. Peroxide decomposers react with the trace peroxide initiators to form nonradical products before such peroxides can initiate rubber oxidation. Mercaptans, thiophenols, and other organic sulfur compounds have been shown to function as antioxidants through a peroxide decomposer mechanism [17]. It has also been suggested that the excellent initial oxidation resistance of compounds containing zinc dialkyl dithiocarbamates is due to a peroxide decomposition mechanism from this common vulcanization accelerator.

The dominant mechanism by which amine and phenolic antioxidants function is one in which the antioxidant creates a relatively stable "trap" for radicals involved in the oxidative chain reaction. The mechanism proposed to account for this behavior is given in Figure 7 [18]. Essentially, this antioxidant mechanism involves reaction of the antioxidant with a peroxy radical, resulting in the abstraction of a labile hydrogen from the antioxidant and formation of an antioxidant radical. The antioxidant radical formed is more stable than the initial peroxy radical involved in oxidation and terminates by reaction with another radical present in the system. Shelton [19] has shown a noticeable deuterium isotope effect in aromatic amine antioxidants, which tends to support this mechanism. In Shelton's work, the labile hydrogens of an aromatic amine antioxidant were replaced with deuterium. Because of the greater atomic mass of deuterium, this replacement would reduce the antioxidant effectiveness if the mechanism in Figure 7 were correct. Such reduction of antioxidant effectiveness was observed by Shelton, confirming the mechanism in Figure 7.

It has also been proposed that aromatic amine and phenolic compounds can form stable pi-electron complexes with peroxy radicals which terminate to form stable nonradical products [20]. Therefore, it would appear that direct hydrogen abstraction, pi-electron complex formation, or both would describe the antioxidant activity of most aromatic amine and aromatic phenolic antioxidants.

In Figure 7, transfer step 3 and termination step 6 are the most important for antioxidant action. These steps remove the free radical functionality from the polymer so that the normal oxidative chain reaction is halted. However, if initiation step 2 and transfer step 4 become dominant in the mechanism, enhanced oxidation can be the result. Addition

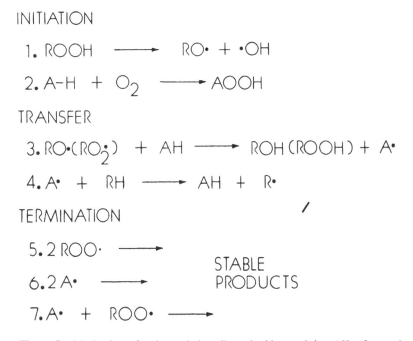

Figure 7 Mechanism of amine and phenolic antioxidant activity. AH refers to the antioxidant.

of excess antioxidant can favor initiation step 2 and transfer step 4. Initiation step 2 would lead to enhanced hydroperoxide formation, resulting in a higher level of potential oxidation initiators present in the rubber compound. Transfer step 4 would cause additional radical formation on the polymer, which could propagate oxidation of the polymer or result in formation of polymer hydroperoxide initiators for the oxidation reaction. Thus it is important that antioxidants be used in small amounts at or near the observed optimum for the rubber compound. Excess antioxidant can have a pro-oxidant effect on the rubber compound.

METHODS OF STUDYING THE OXIDATION RESISTANCE OF RUBBER

Physical Property Testing

The tests most frequently used to study the oxidation resistance of rubber involve the accelerated aging of tensile samples in oxygen-containing atmospheres. Such aging tests can be carried out in a circulating air oven [21], in a high-pressure oxygen bomb [22], or by hanging the tensile samples in a controlled-temperature aluminum aging block [23]. Specific test methods and techniques are given in the cited ASTM procedures. In general, several rubber compounds can be compared for resistance to oxidation reasonably quickly in these types of tests. The resistance to oxidation is quantified by measuring changes in tensile properties such as ultimate tensile strength, ultimate elongation, durometer hardness, and stress at selected percent elongations. Such tests can be run at various times and temperatures to give a relatively comprehensive comparison of the oxidation resistance of various rubber compounds. For elastomers which react with oxygen resulting in crosslinking—generally butadiene-based materials such as polybutadiene, styrene–butadiene rubber, and acrylonitrile–butadiene rubber—the accelerated aging tests would result in increases in tensile stress at a given elongation, increases in durometer hardness, and decreases in ultimate elongation. For elastomers which react with oxygen resulting in chain scission—generally isoprene-based elastomers such as natural rubber and stereoregular polyisoprene rubber—the accelerated aging tests result in decreases in tensile stress at a given elongation, decreases in durometer hardness, and either increases or decreases in ultimate elongation depending on the extent of degradation.

It must be noted that such aging tests are *accelerated* aging tests, which may not directly correlate with the application environment. The ASTM procedures for such aging tests clearly state that such tests are accelerated tests useful for comparison of rubber compounds and may not relate directly to actual long-term behavior. Such tests do provide a relatively quick testing scheme for comparing rubber formulations for oxidation resistance. However, it is generally good practice to run control samples of a known rubber compound for comparison when such accelerated aging tests are performed. In the case of antioxidant system evaluation, the best antioxidant system would be the one that gave the lowest change in properties in the accelerated aging tests. Aging studies in ambient temperature and pressure conditions would generally require excessive time so that compound development would be substantially hindered.

Thermal Analysis

The utility and instrumentation available now in thermal analysis have provided alternative testing procedures for evaluating the oxidation resistance of various rubber compounds. Probably the most useful thermal analysis technique for studying the oxidation resistance of rubber compounds is differential scanning calorimetry (DSC). In DSC, a small sample is placed in an aluminum pan. The pan containing the sample is slowly heated through a range of temperatures along with a similar pan containing no sample. The heating rate can be controlled to several rates from roughly 1°C/min to roughly 40°C/min. The difference in energy required to heat the sample pan and the empty reference pan is amplified and recorded. The reference pan undergoes no transitions over the desired temperature range. Thus any phase transitions occurring in the sample or any chemical reactions occurring in the sample in the selected temperature range would cause a heating rate imbalance between the sample pan and the reference pan. Oxidation is an exothermic process such that a sample undergoing oxidation would require lower energy input to maintain the programed heating rate compared to the empty reference pan. DSC amplifies such differences in energy input and records them as slope changes or peaks and valleys on a recorder trace where the x axis of the trace is temperature and the y axis is the temperature or enthalpy difference between the sample and the reference pans.

Smith and Stephens [24] used DSC to compare antioxidants in polybutadiene rubber and styrene–butadiene rubber. Their measurements involved integration of the primary oxidation exotherm as a measurement of oxidation and oxidation resistance. The areas under the primary oxidation exotherms were used to calculate the enthalpies (ΔH) of oxidation for the various materials tested. A representation of a typical oxidation exotherm and the method used by Smith and Stephens is given in Figure 8. Areas under the oxidation exotherms were converted to enthalpies of oxidation using the equation

$$\Delta H = \frac{K*A*\Delta T_s}{(W*a)} \tag{1}$$

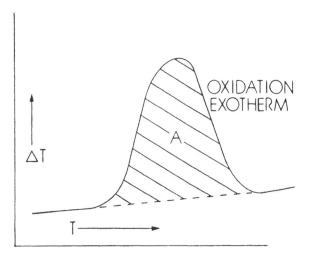

Figure 8 DSC study of oxidation by enthalpy of oxidation method.

where

K = calibration constant of the machine

A = area under oxidation exotherm

ΔT_s = recorder X axis sensitivity in °C/in.

W = sample weight in mg

a = heating rate in °C/min

In their work, an increase in ΔH, the enthalpy of oxidation, obtained from the primary oxidation exotherm would indicate enhanced oxidation, whereas a decrease in ΔH would indicate inhibition of oxidation, or antioxidant effectiveness. Their results for a coagulated styrene–butadiene rubber latex containing various levels of a phosphite antioxidant are shown in Table 3. Close examination of Table 3 shows that there is an optimum level of antioxidant where ΔH is minimized, that is, the greatest oxidation inhibition. Antioxidant levels above or below the optimum level show greater susceptibility to oxidation, manifested by the higher ΔH values. As noted earlier, the mechanism for antioxidant action in Figure 7 indicates that excessive antioxidant would have a pro-oxidant effect. Such a pro-oxidant effect is observed in Table 3 in terms of the increasing ΔH values with increased antioxidant above the optimum level.

Keller and Stephens [7] used integration of successive intervals of primary oxidation exotherms to study the pro-oxidant effects of unsaturated fatty acids present in Guayule resin. The technique involves using the ratio of the area of the oxidation exotherm up to a particular temperature and the total area of the oxidation exotherm. Figure 9 illustrates a typical oxidation exotherm and the method used to determine a rate constant k_{3D} at a given temperature. Rate constants at various temperatures were calculated using the following equations [25]:

$$dH/dT = B*S*X \tag{2}$$

where

dH/dT = enthalpy of oxidation at temperature T

B = measured height of exotherm from interpolated baseline at temperature T

S = recorder sensitivity in mcal/sec/in.

X = calibration constant for the machine

Table 3 Results of Smith and Stephens Using DSC to Study the Oxidation of SBR with Various Levels of Alkylated Phosphite Antioxidant

Antioxidant phr	ΔH (cal/g)
0.22	40.5
0.55	37.0
0.85	36.7
1.16	37.0
1.46	41.0

and

$$k_{3D} = 2(dH/Dt)(3A) * \frac{1 - [1 - (a/A)]^{1/3}}{[1 - (a/A)]^{2/3}} \tag{3}$$

where

k_{3D} = three-dimensional diffusion-controlled rate constant

A = total area of oxidation exotherm

a = fractional area of oxidation exotherm up to temperature T

Typical results of this study are shown in Figure 10. Figure 10 plots the rate constant values as a function of the level of added GFA-OMe, methyl esters of Guayule resin fatty acids, at various temperatures using Eqs. 2 and 3 for interpretation of the primary oxidation exotherms. At all temperatures selected for this study, addition of GFA-OMe caused acceleration of oxidation as indicated by increases in the rate constant values k_{3D}. They observed that the rates of oxidation increased for natural rubber containing increasing levels of the methyl esters of Guayule resin fatty acids, confirming the pro-oxidant effects of these highly unsaturated fatty acids as discussed previously.

Ponce-Velez and Campos-Lopez used three different models of interpretation of oxidation exotherms to study the oxidation of Hevea and Guayule natural rubbers with DSC [25]. The method of Borchardt and Daniels is based on measurements of the heat evolution at a given temperature in the primary exotherm [26]. A typical representation of the natural rubber primary oxidation exotherm and the measurements taken using the Borchardt and Daniels model are given in Figure 8. The following equation is used to calculate rate constants at the various temperatures selected:

$$k_x = \frac{(A/m)^{x-1} * dH/dt}{(A - a)^x} \tag{4}$$

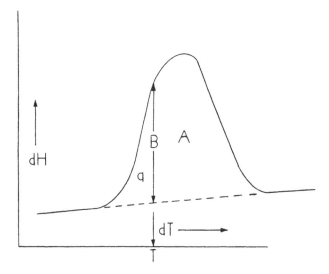

Figure 9 DSC study of oxidation by kinetic rate constant method.

where

k_x = rate constant of reaction order x

dH/dt = heat evolved at time interval dt

A = total area of oxidation exotherm

a = partial area of oxidation exotherm up to time t

m = sample mass

They also utilized the three-dimensional diffusion-controlled model cited in the previous example. Finally, they used a model based on the peak of the oxidation exotherm as a function of heating rate [27, 28]. This final model is based on the observations that the oxidation exotherm maximum will shift to higher temperatures with increased heating rates. The equation used to relate the exotherm maximum, T_m, with heating rate and energy of activation is

$$\frac{d(\log B)}{B(1/T_m)} = \frac{0.457 E_a}{R} \tag{5}$$

where

B = heating rate in K/min

T_m = observed exotherm maximum

E_a = energy of activation

R = universal gas constant in kcal

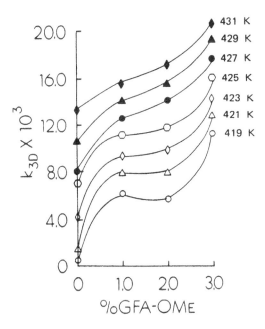

Figure 10 Effects of guayule fatty acid methyl esters (GFA-OMe) on NR oxidation. Rate constants from DSC.

Evaluation of Hevea and Guayule natural rubbers by DSC using these three models showed both materials to be equivalent in oxidation behavior and the activation energies determined using the three models agreed well with the activation energy determined using infrared spectroscopy.

The introduction of relatively inexpensive microcomputers capable of data manipulation required for interpretation of DSC oxidation exotherms makes DSC a very rapid method of evaluating and comparing the oxidation resistance of various rubber compounds.

IR Spectroscopy

One of the most widely used techniques for the study of chemical structure and structure changes in organic materials and polymers is infrared (IR) spectroscopy. IR radiation causes translational and vibrational shifts in chemical bonds and the absorption of the IR radiation by specific chemical structures can be monitored to determine chemical structure and structural changes in organic materials and polymers. IR spectra of the unaged rubber material and of rubber material exposed to oxidative conditions has been used to study structural changes during oxidation and oxidation kinetics.

Kello, Tkac, and Hrivikova used IR spectroscopy to study not only the structural changes occurring in natural rubber during oxidation but also the kinetics of natural rubber oxidation [29, 30]. As oxidation proceeded, they observed increases in carbonyl and hydroxy structures present on the natural rubber polymer. The increases in carbonyl and hydroxy structures did not occur until after a certain induction time during oxidation. By assuming a direct proportionality between the observed induction time for accumulation of hydroxy and carbonyl structures and the rate of oxidation at different temperatures, they determined rates of natural rubber oxidation at different temperatures. The rates of oxidation thus determined at different temperatures were utilized in Arrhenius rate expressions to calculate the energy of activation of rubber oxidation. They found an energy of activation for natural rubber oxidation of 87.9–92.0 kJ/mol, which was in excellent agreement with energy of activation values determined by other methods.

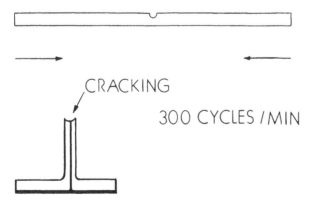

Figure 11 De Mattia test for the flex cracking resistance of rubber.

Mechanical Flex Crack Resistance

The resistance of a rubber compound to oxidative flex cracking is usually determined using a mechanical flexing device designed to flex a sample at a given frequency; any cracks resulting will be formed in a designed flaw in the sample. ASTM D430 describes the use of various dynamic flex devices to evaluate the flex cracking resistance of rubber materials. As an example, the DeMattia flex machine involves a rubber specimen molded with a circular groove perpendicular to the direction of flexing. The specimen is bent at the molded groove at a rate of 300 ± 10 cycles per minute and periodically examined for cracking and crack propagation in the molded groove. A representation of the sample flexing is seen in Figure 11. Several samples are run simultaneously and cracking is judged visually by length, depth, and number of cracks. A material with superior flex crack resistance would yield results with a much greater time of flexing required to produce a given severity of cracking. Antioxidants known to improve flex crack resistance are designated FCI in Figures 4 and 5.

MECHANISM OF OZONE ATTACK ON RUBBER

The ozone attack on unsaturated rubber is basically similar to the ozone attack on model olefins. Criegee and Lohaus [31] originally proposed the mechanism to describe the ozone attack on olefins; this mechanism is summarized in Figures 12 and 13. The initial step in the mechanism involves nucleophilic attack of a resonance form of ozone on the olefin to form product **III**, which is a cyclic ozonide. The initial cyclic ozonide **III** has been isolated [32] and rapidly undergoes rearrangement by heterolytic cleavage of the oxygen–oxygen bond to form a zwitterion **IV**. Product **IV** is quite unstable and will break down to

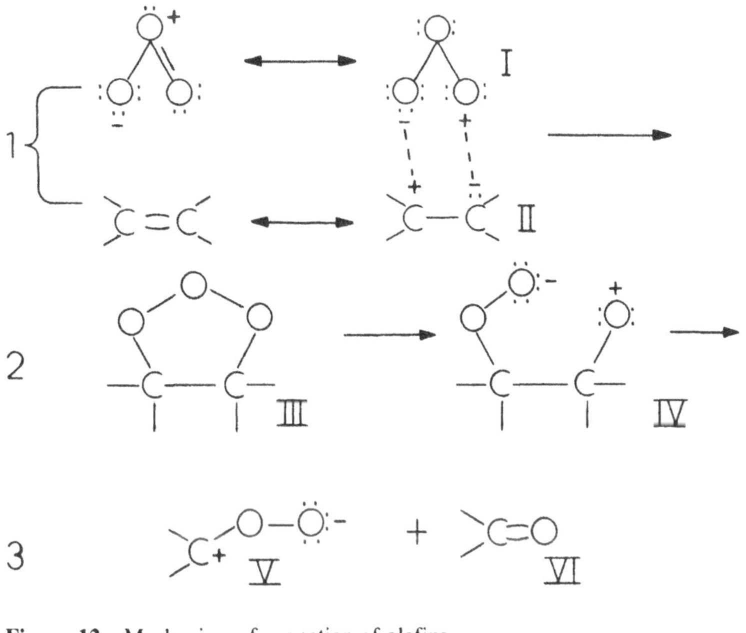

Figure 12 Mechanism of ozonation of olefins.

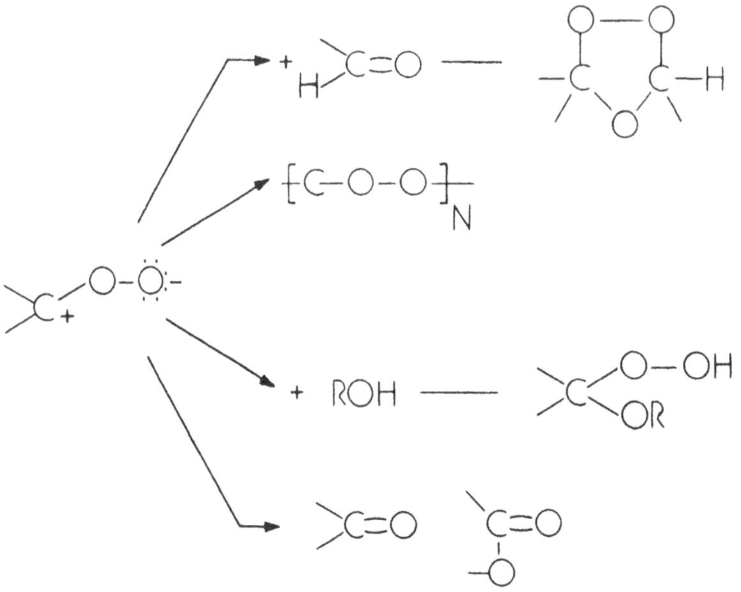

Figure 13 Mechanism of ozonation of olefins.

form another zwitterion **V** and an aldehyde or ketone **VI**. This last zwitterion **V** is the key intermediate in the process. The final zwitterion **V** can then stabilize by four separate processes, as illustrated in Figure 13.

If we regard the initial olefin structure in Figure 12 as that present in an unsaturated polymer, the mechanism can be applied to the ozone attack of rubber. Substitution on the olefin would be expected to follow the general rules for nucleophilic attack:

1. Electron-donating substituents such as alkyl groups would increase the nucleophilic tendency of the olefin and increase reaction rates.
2. Electron-withdrawing substituents such as halogens would decrease the nucleophilic tendency of the olefin and decrease reaction rate.

Thus one would expect that polychloroprene (CR), having chlorine substituents at the residual unsaturation, would be less reactive to ozone than natural rubber (NR), having a methyl group at the residual unsaturation. This type of behavior is generally observed and polychloroprene finds use in many applications where long-term ozone resistance is necessary. Polychloroprene is the only commercial polymer with high levels of unsaturation in the polymer backbone, which is resistant to ozone.

The reaction of ozone with olefins is extremely rapid and has a low activation energy. The fact that ozone attack on olefins occurs very rapidly implies that the attack of ozone on unsaturated rubber occurs at the exposed surfaces and olefins in the interior of the sample would be attacked only after all of the surface olefins had been consumed. Thus ozone attack on unsaturated rubber should be primarily a surface effect.

Various studies have been done which lend support to the mechanism proposed in Figures 12 and 13. Two studies revealed a decrease in polymer molecular weight when unsaturated polymers were exposed to ozone in solution [33, 34]. Allison and Stanley [33] observed decreases in double bonds during the ozonation of styrene–butadiene (SBR) in solution. Others used IR spectroscopy to study the products of ozonation of unsaturated polymers and found evidence to suggest that ozonides, carbonyl compounds, and hydroxy compounds were formed [34, 35]. Therefore, ozonation of unsaturated rubber appears to follow the mechanism proposed in Figures 12 and 13 based on the following experimental results:

1. Consumption of double bonds
2. Decreases in molecular weight
3. Formation of carbonyl compounds
4. Formation of ozonides

ANTIOZONANT TYPES

Because ozonation is primarily a surface effect, it is important to compound using additives designed to provide surface protection. Thus an ideal antiozonant is dispersed into the rubber compound but blooms to the surface after molding. Paraffin waxes provide this type of behavior and create a saturated hydrocarbon ozone barrier after molding. Unfortunately, the wax protective layer is not capable of resisting dynamic flexing. Waxes crack under dynamic flexing and allow ozone to contact and attack the exposed unsaturated rubber underneath the wax surface crack. Thus waxes are used primarily for static ozone protection.

Effective antiozonants for dynamic applications are generally amine compounds and are consequently discoloring in long-term behavior. The four basic types of amine antiozonants are summarized in Figure 14. Only those materials with high to moderate activity in ozone protection are listed in Figure 14. The most effective and most widely

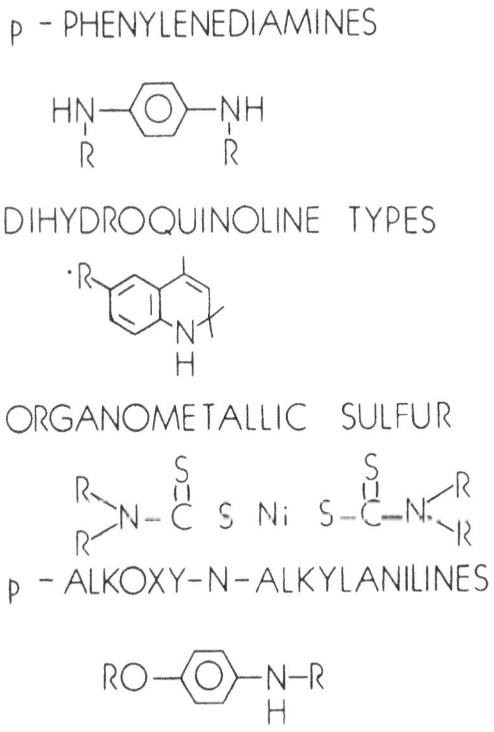

Figure 14 Chemical structures of antiozonants.

used antiozonants are the *p*-phenylene diamine types. These materials are generally used in levels between 0.5 and 2.0 phr in rubber compounds depending on the compound and the application. The *p*-phenylene diamine materials also bloom to the rubber surface to provide protection against surface attack. In many cases a paraffin wax along with a *p*-phenylene diamine are used to provide the best combination of static and dynamic ozone protection.

MECHANISM OF ANTIOZONANT ACTION

Essentially, three theories of antiozonant action have been proposed:

1. The antiozonant blooms to the surface of the rubber compound and reacts preferentially with the incident ozone.
2. The antiozonant blooms to the surface of the rubber and forms a protective film on the surface.
3. The antiozonant reacts with intermediates formed in the ozonation of rubber, preventing chain scission or recombining severed rubber chains.

Andries and co-workers [36] used attenuated total reflectance infrared spectroscopy and scanning electron microscopy to study the surface of carbon black-loaded natural rubber stocks and unreinforced natural rubber stocks. They found that a thick layer of ozonide products built up on the surface of the rubber after exposure to ozone. When the stocks were compounded using an antiozonant, a continuous layer of ozonides resembling the products of ozonation of the antiozonant alone accumulated on the rubber surface. These results indicated that the mechanism of antiozonant action is a combination of the first two theories above: the antiozonant blooms to the surface of the rubber, reacts preferentially with the incident ozone, and forms a protective film of ozonide products, which protects the bulk of the rubber. Since ozone attack on rubber is a surface effect, the buildup of an ozonide film on the surface of the rubber due to the antiozonant action creates a protective shield for the bulk of the rubber.

Lattimer and co-workers [37] used liquid chromatography, field desorption mass spectroscopy, and gas chromatography–mass spectroscopy to study the ozonation of N,N'-di(1-methyl-heptyl)-*p*-phenylenediamine in natural rubber stocks. Their work was key in understanding details of the mechanism of antiozonant action by identifying the products of antiozonant ozonation. Various reaction pathways were proposed to account for the ozonation products observed. All reaction pathways and ozonide products observed were dependent on the initial reaction of the amine antiozonant with ozone to form an initial amine ozonide:

$$R_3N-O-O-O\cdot$$
$$+$$

This work supported the work of Andries and co-workers [36]. Therefore, amine antiozonants appear to function by blooming to the surface of the rubber compound, reacting preferentially with the incident ozone, and forming a protective layer of amine ozonide products, which prevents ozone penetration and attack of the bulk rubber.

METHODS OF STUDYING THE OZONE RESISTANCE OF RUBBER

Since ozone attack on rubber is essentially a surface phenomenon, the test methods involve exposure of the rubber to a specified ozone level at a given temperature for a specified time followed by rating the extent of cracking after testing. Ozone resistance can be determined on flat samples under tensile strain [38], on triangular samples bent around a standard wooden mandril [39], or on flat samples exposed to ozone and dynamic flexing [40]. The most frequently used test involves the triangular samples bent around a specified wooden mandril. The point of the triangular specimen is subjected to the highest static strain when bent. Ozone will cause chain scission of unsaturated rubber polymer followed by crack growth at the point of the triangular specimen due to the static strain. Standard test conditions involve a temperature of 40 ± 1°C at ozone concentration of 50 ± 5 parts per hundred million. A standard test duration is 72 hours. After exposure of the strained triangular sample to ozone, the bent samples are rated for cracking using the following scale:

0 = no cracks visible with the unaided eye or with 7× magnification
1 = tiny cracks visible only with 7× magnification
2 = tiny cracks barely visible with the unaided eye
3 = large cracks readily visible with the unaided eye

It is important to remember that ozone resistance testing requires either static or dynamic strain to produce cracks for rating and comparison. With no strain, the ozone attack will not proceed past the surface of the rubber and the rubber will show no cracking regardless of the test ozone concentration or test duration. Also, rubber products which suffer from ozone attack in the application show degeneration due to cracking under either static or dynamic strain. A tire exposed to high levels of ozone without proper antiozonant protection will display catastrophic sidewall cracking due to both static and dynamic strain after only a few weeks. The standard ozone test procedures are accelerated tests and are used for comparison of formulas and antiozonant packages only.

In general, those elastomers having the final letter designation ''R'' in Table 1—those having unsaturation present in the polymer backbone—are susceptible to ozone attack since ozone attack requires the unsaturated organic structure. Polychloroprene (CR), butyl rubber (IIR), and the halobutyl rubbers (CIIR and BIIR) are the only elastomers with backbone unsaturation which are resistant to ozone attack. In the case of polychloroprene, this is due to the electron-withdrawing nature of the chlorine attached to the unsaturation, which renders the unsaturation resistant to ozone nucleophilic attack. In the case of butyl rubber and the halobutyl rubbers, the unsaturation present exists for the purposes of vulcanization. This unsaturation present in butyl and halobutyl rubbers is generally less than 5%. Therefore, butyl and halobutyl rubbers are resistant to ozone due to the low levels of unsaturation present in the polymer backbone.

REFERENCES

1. J. L. Bolland, *Q. Rev. Chem. Soc.*, *3*: 1 (1949).
2. J. R. Shelton and D. N. Vincent, *J. Am. Chem. Soc.*, *85*: 2433 (1963).
3. L. Bateman, M. Cain, T. Colclough, and J. I. Cunneen, *J. Am. Chem. Soc.*, *1962*: 3570.
4. R. J. Ceresa, *Plast. Inst. Trans. J.*, *28*: 178 (1960).
5. E. H. Farmer and A. Sundralingham, *J. Chem. Soc.*, *147*: 121 (1942).
6. R. W. Keller, D. S. Winkler, and H. L. Stephens, *Rubber Chem. Tech.*, *54*: 115 (1981).
7. R. W. Keller and H. L. Stephens, *Rubber Chem. Tech.*, *55*: 161 (1982).
8. J. T. Blake and P. L. Bruce, *Ind. Eng. Chem.*, *33*: 1198 (1941).
9. A. M. Neal and A. J. Northam, *Rubber Chem. Tech.*, *5*: 90 (1932).
10. H. Winn and J. R. Shelton, *Ind. Eng. Chem.*, *37*: 67 (1945).
11. A. N. Gent, *J. Appl. Polym. Sci.*, *6*: 497 (1962).

12. W. M. Hess and K. A. Burgess, *Rubber Chem. Tech.*, *36*: 754 (1963).
13. J. LeBras, J. C. Danjard, and J. Boucher, *J. Polym. Sci.*, *27*: 529 (1958).
14. A. Haehl, *Rev. Gen. Caoutch.*, *26*: 563 (1949).
15. B. N. Leyland and R. L. Stafford, *Trans. Inst. Rubber Ind.*, *35*: 25 (1959).
16. W. P. Fletcher and S. G. Fogg, *Rubber Age* (*NY*), *84*: 632 (1959).
17. W. L. Hawkins and M. A. Worthington, *J. Polym. Sci. A*, *1*: 3489 (1963).
18. W. L. Hawkins, *Polym. Eng. Sci.*, *5*: 196 (1965).
19. J. R. Shelton, *J. Appl. Polym. Sci.*, *2*: 345 (1959).
20. J. R. Shelton and D. N. Vincent, *J. Am. Chem. Soc.*, *85*: 2443 (1963).
21. ASTM D573-81, *Am. Soc. Test. Mater.*, *Book ASTM Stand. Vol. 9.01.*
22. ASTM D572-81, *Am. Soc. Test. Mater.*, *Book ASTM Stand. Vol. 9.01.*
23. ASTM D865-81, *Am. Soc. Test. Mater.*, *Book ASTM Stand. Vol. 9.01.*
24. R. C. Smith and H. L. Stephens, *J. Elastomers Plast.*, *7*: 156 (1975).
25. M. A. Ponce-Velez and E. Campos-Lopez, *J. Appl. Polym. Sci.*, *22*: 2485 (1978).
26. H. J. Borchardt and F. Daniels, *J. Am. Chem. Soc.*, *79*: 41 (1957).
27. H. E. Kissinger, *J. Res. Natl. Bur. Stand.*, *57*: 217 (1956).
28. T. Ozawa, *J. Therm. Anal.*, *2*: 301 (1970).
29. V. Kello, A. Tkac, and J. Hrivikova, *Rubber Chem. Tech.*, *29*: 1245 (1956).
30. V. Kello, A. Tkac, and J. Hrivikova, *Rubber Chem. Tech.*, *29*: 1255 (1956).
31. R. Criegee and G. Lohaus, *Justus Liebigs Ann. Chem.*, *583*: 6 (1953).
32. R. Criegee and G. Schroeder, *Chem. Ber.*, *93*: 689 (1960).
33. A. R. Allison and I. J. Stanley, *Anal. Chem.*, *24*: 630 (1952).
34. G. Salomon and A. C. van der Schee, *J. Polym. Sci.*, *14*: 181 (1954).
35. F. H. Kendall and J. Mann, *J. Polym. Sci.*, *19*: 503 (1956).
36. J. C. Andries, C. K. Rhee, R. W. Smith, D. B. Ross, and H. E. Diem, *Rubber Chem. Tech.*, *52*: 823 (1979).
37. R. P. Lattimer, E. R. Hooser, H. E. Diem, R. W. Layer, and C. K. Rhee, *Rubber Chem. Tech.*, *53*: 1170 (1980).
38. ASTM D1149-81, *Am. Soc. Test. Mater.*, *Book ASTM Stand. Vol. 9.01.*
39. ASTM D1171-83, *Am. Soc. Test. Mater.*, *Book ASTM Stand. Vol. 9.01.*
40. ASTM D3395-82, *Am. Soc. Test. Mater.*, *Book ASTM Stand. Vol. 9.01.*

6

Techniques for Monitoring Heterogeneous Oxidation of Polymers

Kenneth T. Gillen and Roger L. Clough

Sandia National Laboratories
Albuquerque, New Mexico

An important phenomenon that often occurs during the air aging of polymers is spatially varying (heterogeneous) oxidation. This chapter first presents a general theoretical treatment appropriate to diffusion-limited oxidation, the most common cause of heterogeneous effects. Reviews are then given of a number of experimental techniques which have proven capable of profiling (mapping) spatial variations in oxidation across the relatively

This work was performed at Sandia National Laboratories and supported by the U.S. Department of Energy under contract No. DE-AC04-76DPOP789.

small distances of practical interest (approximately 1 mm or less). Discussion is presented on the uses of the experimental profiles (often in conjunction with theoretical models) for improving materials and predicting degradation behaviors and rates.

INTRODUCTION

During the storage and use of a polymeric material, it may be subjected to various combinations of environmental stresses; examples include elevated temperature, ultraviolet light, high-energy radiation, and mechanical stress. Material degradation or failure can result from the adverse chemical changes caused by such stresses. When air is present during the aging, degradation processes are often dominated by oxidation reactions. Such reactions must be understood in order to understand degradation behaviors and to confidently predict polymer lifetimes. Detailed knowledge of the important oxidative degradation reactions will also aid in the search for improved materials and material formulations.

A major complication which often occurs during attempts to monitor polymer oxidation is the presence of heterogeneous effects. Heterogeneous effects are arbitrarily divided here into two categories, microscopic and macroscopic, dependent on the length scale over which the heterogeneities exist. Since polymers, by their very nature, are heterogeneous on a microscopic level, microscopic oxidation heterogeneities would be anticipated. In crystalline polymers, for example, the amorphous regions are usually much more susceptible to oxidation than the crystalline regions. Of primary interest here are monitoring and understanding macroscopic heterogeneities which occur over regions large enough to average out any microscopic structural and formulational variations. Frequently such macroscopic oxidation effects correspond to differences between the interior and edge regions of materials.

Macroscopic heterogeneities during aging can result from an inhomogeneously processed material. For example, the surface of an extruded material may have enhanced concentrations of impurities (e.g., peroxides, metal species), which make the surface more susceptible to later degradation. Similarly, an incompatible adjacent material may lead to enhanced oxidation effects at the material interface. Heterogeneous oxidation can also arise when an environmental stress interacts nonuniformly with a material. Possibilities include the attenuation of UV light as it passes through a sample and attack near surfaces by ozone or other reactive atmospheric pollutants. The most widespread cause of macroscopic heterogeneous oxidation results from oxygen diffusion-limited effects. This complication occurs whenever the rate of oxygen consumption in a material exceeds the rate at which oxygen from the surrounding atmosphere can be resupplied to the interior by diffusion processes. The importance of this effect will therefore depend on three factors: material geometry, the oxygen consumption rate, and the oxygen permeation rate.

Although diffusion effects in laboratory studies of polymer oxidation can be minimized by using thin samples, this approach has several potential drawbacks. Since oxidation depths may reach only fractions of a millimeter under typical laboratory aging conditions, it may be difficult to prepare sufficiently thin samples. In addition, the aging responses of specially prepared thin samples may not be representative of the bulk material, due either to differences in processing conditions (e.g., compression-molded laboratory samples versus extruded actual material) or to overlooking important degradation effects caused by impurities introduced during processing of the actual material. Moreover, in real situations one is often forced to confront heterogeneous oxidation because this effect can be an important part of the aging process under conditions prevailing in the intended application.

The purpose of this chapter is to discuss some of the experimental techniques that have been used to monitor and understand macroscopic heterogeneous oxidation in polymers and show how data from such techniques might be useful for improving materials and predicting degradation behaviors and rates.

THEORY FOR DIFFUSION-LIMITED OXIDATION

Diffusion-limited oxidation is the predominant cause of heterogeneous degradation. As will be apparent later, it occurs very frequently for polymeric materials aged under typical laboratory conditions. To understand the potential for predictive capabilities of the diffusion-dominated heterogeneous oxidation data to be described below, we must first understand the relationship between oxidation profile shapes and the details of the underlying chemical kinetics.

The kinetics appropriate to the oxidation of polymers has been an intensely studied subject and a number of excellent review articles and books exist [1–5]. Compared to the much simpler oxidation of liquids, many new complications enter for polymers, such as the existence of crystalline and amorphous regions and the influence of the restricted mobility of the solid phase. Even so, it is usually assumed that a small number of general primary and secondary reactions are most important to the oxidation of polymers. In the interest of simplifying the present discussion, analyses are confined to these critical consensus reactions. For an unstabilized polymer the following reactions are appropriate:

Initiation:

$$R \xrightarrow{k_i} R\cdot$$

Propagation:

$$R\cdot + O_2 \xrightarrow{k_2} RO_2\cdot$$

$$RO_2\cdot + RH \xrightarrow{k_3'} RO_2H + R\cdot$$

Bimolecular termination:

$$2R\cdot \xrightarrow{k_4} \text{Products}$$

$$R\cdot + RO_2\cdot \xrightarrow{k_5} \text{Products}$$

$$2RO_2\cdot \xrightarrow{k_6} \text{Products}$$

Branching reactions, entailing the breakdown of ROOH species formed in the third reaction to give more free radical species, need to be added to this simplified mechanism whenever they are important (often at high temperatures or for extended time periods). Assuming steady-state concentrations of all free radical species plus long kinetic chain lengths (e.g., many propagation cycles compared to termination rates and little oxygen formed from the last reaction compared to oxygen consumed by the second reaction), standard kinetic analysis yields for the oxygen consumption rate [2]

$$\frac{d[O_2]}{dt} = \frac{k_3 k_2 [O_2] k_i^{0.5}}{[2k_4 k_3^2 + 2k_6 k_2^2 [O_2]^2 + 2k_5 k_2 k_3 [O_2]]^{0.5}} \tag{1}$$

where $k_3 = k_3'[RH]$. With the usual assumption relating the termination rate constants ($k_5^2 = 4k_4 k_6$), Eq. 1 simplifies to

$$\frac{d[O_2]}{dt} = \frac{C_{1b}[O_2]}{1 + C_{2b}[O_2]} \tag{2}$$

where

$$C_{1b} = \frac{k_2 k_i^{0.5}}{(2k_4)^{0.5}} \tag{3}$$

$$C_{2b} = \frac{k_6^{0.5} k_2}{k_4^{0.5} k_3} \tag{4}$$

For a well-stabilized polymer, the three bimolecular termination reactions shown above will be replaced in the simplest scenario by the following two termination reactions:

$$RO_2 \cdot + AH \xrightarrow{k_7'} \text{Products}$$

$$R \cdot + AH \xrightarrow{k_8'} \text{Products}$$

where AH stands for an antioxidant (free-radical scavenger). Assuming the antioxidant concentration is large enough not to be substantially depleted by the termination reactions, these reactions are pseudo first order (unimolecular). In fact, aside from reactions involving stabilizers, unimolecular termination kinetics might be expected to be favored more in bulk polymers than in liquids. Radicals are formed in pairs, and the greatly decreased diffusional mobility characteristic of solid polymers retards the movement of paired radicals away from each other. This increases the probability of termination involving radicals derived from the geminate pair. Although termination involving either the geminate pair or radicals derived from the geminate pair is bimolecular from a mechanistic standpoint, it will give rise to kinetics which are unimolecular (e.g., oxidation will depend linearly on the initiation rate). With unimolecular termination reactions substituted for the three bimolecular termination reactions given earlier, it is easy to show that kinetic analysis gives

$$\frac{d[O_2]}{dt} = \frac{C_{1u}[O_2]}{1 + C_{2u}[O_2]} \tag{5}$$

where

$$C_{1u} = \frac{k_2 k_i}{k_8} \tag{6}$$

$$C_{2u} = \frac{k_2 k_7}{k_8 (k_3 + k_7)} \tag{7}$$

and $k_7 = k_7'[AH]$, $k_8 = k_8'[AH]$.

As seen from Eqs. 2 and 5, the general form of the oxygen consumption rate is identical for the two kinetic schemes given, a convenient circumstance for the analysis that follows. There are, however, major differences in the meanings of the constants C_{1j}

and C_{2j} ($j = b$ or u), as is clear from Eqs. 3, 4, 6, and 7. The most important difference is the square root dependence of C_{1b} on the initiation rate constant k_i compared to a first-order dependence of C_{1u}; the profound implications of this difference on experimental data will become clear later.

Using the bimolecular kinetic scheme result given in Eq. 2 and assuming the oxygen consumption and permeation rates are independent of time, Cunliffe and Davis [6] showed how it is possible to derive the shapes of heterogeneous oxidation profiles for slabs of material of thickness L when oxygen diffusion-limited effects are important. Their steady-state results depend on two parameters, α and β, which are given by

$$\alpha = \frac{C_1 L^2}{D} \tag{8}$$

$$\beta = C_2 Sp = C_2 [O_2]_e \tag{9}$$

where C_1 and C_2 are the constants in Eqs. 3 and 4, D and S are the diffusion and solubility parameters for oxygen in the material, p is the oxygen partial pressure of the surrounding atmosphere, and $[O_2]_e$ denotes the oxygen concentration at the edge of the sample. Since the unimolecular scheme leads to the same functional form for the oxygen consumption rate, the profile shapes generated in terms of α and β will be identical, except that the appropriate values of C_1 and C_2 come from Eqs. 6 and 7. Figures 1–3 show theoretical profiles of normalized oxidation and the corresponding steady-state oxygen concentrations for a slab of material exposed to oxygen on both sides. P denotes the percentage of the distance from one sample edge to the opposite edge. Results for various values of α at

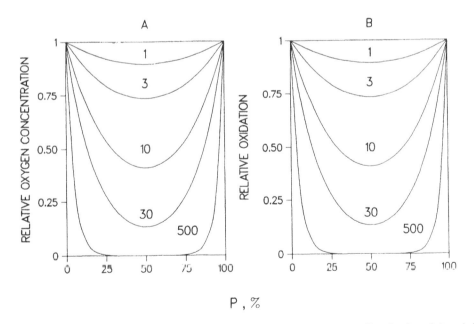

Figure 1 Theoretical oxygen concentration (A) and oxidation (B) profiles for $\beta = 0.1$ and the various values of α indicated. The parameters α and β are defined in Eqs. 8 and 9. P represents the percentage of the distance from one oxygen-exposed surface of the sample to the opposite oxygen-exposed surface. (Curves calculated using approach of Ref. 6.)

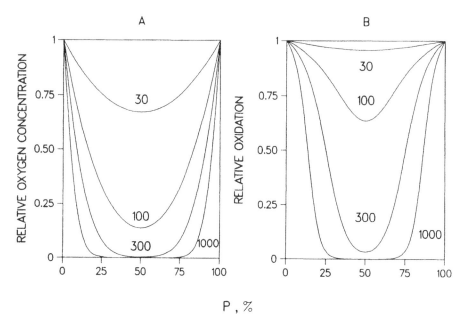

Figure 2 Same as Figure 1 except β = 10.

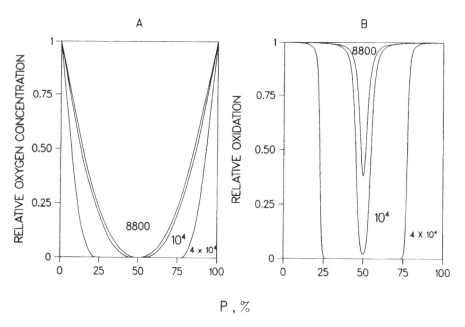

Figure 3 Same as Figure 1 except β = 1000.

a β value of 0.1 are shown in Figure 1. When β ≪ 1, the oxidation will be proportional to the steady-state oxygen concentration, as is clear from Figure 1, and the oxidation profiles will be U-shaped. When β is small and the sample is thick enough (α large) such that little oxidation occurs in the center, the dropoff in the oxidation from the edge is exponential. For large values of β, oxidation is not sensitive to the steady-state oxygen concentration until the latter has dropped substantially, as seen in Figure 3 where β = 1000. The theory then predicts sharp boundaries (cliff-shaped) between oxidized and nonoxidized regions. As illustrated later in this chapter, rather than approaching either limit, most air aging experimental results suggest that intermediate values of β may be appropriate. Predicted curves under an intermediate condition (β = 10) are shown in Figure 2. Since P_{ox}, the oxygen permeation rate, is equal to $D \times S$, it is easy to show from Eqs. 8 and 9 together with either Eq. 2 or Eq. 5 that the rate of oxygen consumption at the edge (the true oxygen consumption rate of the material, unaffected by diffusion effects) is given by

$$\frac{d[O_2]}{dt} = \frac{\alpha p P_{ox}}{L^2(\beta + 1)} \tag{10}$$

Thus if the shape of an oxidation profile can be used to estimate α and β, Eq. 10 allows estimates of the oxygen consumption rate to be made from known values of the oxygen permeation constant or vice-versa.

If one examines the integrated theoretical oxidation versus sample thickness, a linear relationship will hold at small thicknesses (diffusion-limited effects absent) and an asymptotic limit will be reached when the thickness becomes large enough for the oxidation to drop to zero in the center of the sample. The transition from the linear region to the asymptotic value will depend on the value of β, as shown in Figure 4. Since many

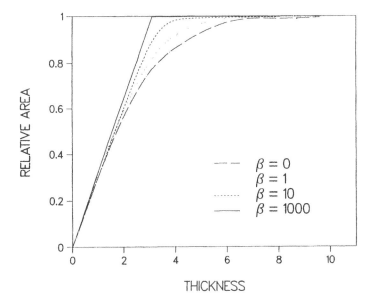

Figure 4 Theoretical curves for the integrated oxidation area normalized to its infinite thickness-limiting value as a function of sample thickness for the indicated values of the theoretical parameter β (Eq. 9).

experimental studies examine the thickness dependence of parameters related to the integrated oxidation, the shape of the experimental curves can in principle be used to estimate β. In practice, however, the limited dependence of the curves on β implies that accurate determination of β from such an approach is difficult.

Better estimates of β can often be made from experiments designed to follow the dependence of the oxidation on the oxygen pressure surrounding the sample. Figure 5 shows predicted theoretical increases in the oxidation at the edge of a sample (or throughout if no diffusion limitation exists) versus the ratio of oxygen pressure to a given reference oxygen pressure. As shown in the figure, the results depend strongly on β_{REF}, the value of β under reference oxygen pressure conditions. It is not surprising that large effects occur when β is small and small effects occur when β is large.

The discussion of the theoretical profiles to this point has avoided the distinction between bimolecular and unimolecular kinetic schemes, since the theoretical shapes in terms of α and β are identical. For the two schemes, however, the dependence of α on the initiation rate constant k_i is quite different with $\alpha \sim k_i^{0.5}$ for the bimolecular and $\alpha \sim k_i$ for the unimolecular. When experiments can be run where k_i can be changed without changing any other rate constants (e.g., under isothermal conditions), evidence relative to the appropriate scheme is potentially available. Two methods of quantitatively controlling k_i under isothermal conditions involve changing the radiation dose rate for gamma-initiated oxidation and changing the concentration of an oxidation-initiating agent. For the oxidation at the edge of a sample (or for the whole sample in the absence of diffusion effects), the bimolecular scheme predicts that the oxidation rate per unit time is proportional to $k_i^{0.5}$, and therefore the rate per integrated radiation dose is proportional to $k_i^{-0.5}$. The

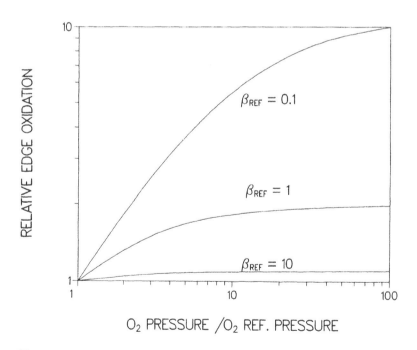

Figure 5 Theoretical predictions of the change in the relative oxidation at the edge of a material versus the increase in oxygen pressure surrounding the material at the three indicated values of β.

Table 1 Values of Exponent n in $d[O_2]/dt \sim k_i^n$

Limit	Rate/unit time		Rate/unit dose	
	Unimolecular	Bimolecular	Unimolecular	Bimolecular
High O_2 pressure or thin samples (no diffusion effects) or at sample edge	$n = 1$	$n = 0.5$	$n = 0$	$n = -0.5$
Asymptote for thick samples	$n = 0.5$	$n = 0.25$	$n = -0.5$	$n = -0.75$

unimolecular scheme predicts a dependence on k_i^1 (rate/unit time) and k_i^0 (rate/unit dose). In the opposite limit, where the sample is thick enough so that minimal oxidation occurs in the center of the sample (the asymptotic limit of Figure 4), the unimolecular scheme leads to $k_i^{0.5}$ (rate/unit time) and $k_i^{-0.5}$ (rate/unit dose), whereas the bimolecular gives $k_i^{0.25}$ (rate/unit time) and $k_i^{-0.75}$ (rate/unit dose). These important distinctions are summarized in Table 1.

Even though it is based on relatively simple bimolecular and unimolecular kinetic schemes, the theory just described offers a convenient framework for discussion of some of the experimental results that follow. In many cases, adding reactions to the schemes described will not significantly alter the resulting profile shapes or their dependence on thickness, pressure, or initiation rate. It should be noted, however, that caution must be used when applying the foregoing steady-state theory (oxygen consumption and permeation assumed independent of time) to experimental data that may not be consistent with these assumptions. For heat aging in particular, oxygen consumption may increase dramatically with time (often after an induction period). In such cases, application of steady-state theories may be possible only in the earlier stages of aging, when the consumption may be relatively constant. Another complication that can become important in the later stages of heat aging is a dramatic decrease in the permeation constant of oxygen for regions of the sample which have become much harder due to the aging. Both effects (a decrease in oxygen permeation rate and an increase in oxygen consumption rate) will lower the oxidative penetration distance, enhancing the oxidative dropoff rate at the sample edge [7]. Thus in the later stages of oxidative heat-aging studies, very rapid dropoffs in oxidation rates near sample edges might be anticipated. If, during aging, significant changes occur in the oxygen permeation and/or oxygen consumption rates, such changes would have to be accounted for before kinetic information could be extracted from theoretical fits to experimental profile data.

EXPERIMENTAL TECHNIQUES FOR MONITORING HETEROGENEOUS OXIDATION

Infrared Spectroscopy

In the early 1960s researchers [8] recognized that the detailed chemical information available from infrared spectroscopy [9], coupled with its ability to make measurements on thin film samples, made it an ideal technique to monitor heterogeneous oxidation effects. Any oxidation-sensitive infrared peak that can be monitored either as a function of the sample thickness or for microtomed slices as a function of cross-sectional position will yield information on oxidation heterogeneities. Most of the infrared studies to date have concentrated on the carbonyl region ($\sim$1720 cm^{-1}) of polyolefin materials (e.g., polyethylene, polypropylene), since carbonyl peaks are often absent for these materials when new, and since these peaks are characterized by high extinction coefficients, resulting in high sensitivities. Since the carbonyl region typically represents a superposition of a number of oxidized product peaks (e.g., ketones, aldehydes, esters, acids) at slightly different wavelengths and with differing extinction coefficients, extraction of quantitative information will often require simplifying assumptions. In most cases, either the maximum height of the hybrid carbonyl peak or its area is chosen for studies designed to follow the oxidation of the material. This assumption will yield a parameter quantitatively related to the amount of oxidation, either if one peak dominates the carbonyl region or if the relative mix of products contributing to the carbonyl peak remains approximately constant versus both aging time and geometric position in the sample.

Giberson [8] used infrared techniques to monitor diffusion-limited oxidation effects for gamma-irradiated polyethylene. Figure 6 is a plot of some of his carbonyl optical density data versus sample thickness for samples given a total dose of 1 MGy (1 megagray, 100 Mrad) at three different dose rates. Comparing the shapes of the curves in Figure 6 with those in Figure 4 implies that β is intermediate to small ($\sim$1 or less). This would imply that oxidation profiles for this material would be U-shaped (see Figures 1–3). Since the initiation rate k_i is linearly related to the dose rate, the dose-rate dependent data shown in Figure 6 allow us to choose between the unimolecular and bimolecular kinetic limiting models discussed earlier. The results in the thin sample limit imply that the oxidation per

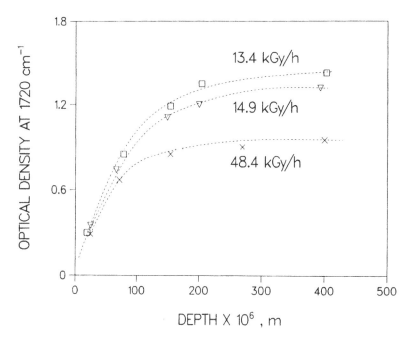

Figure 6 Carbonyl infrared optical density versus sample thickness for low-density polyethylene samples aged with 1 MGy gamma radiation dose at the three indicated dose rates. (Data from Ref. 8.)

unit dose has little if any dependence on the initiation rate, immediately implying unimolecular kinetics (i.e., $n \sim 0$), by the criterion given in Table 1. In the opposite thick sample limit, Table 1 shows that $n = -0.5$ would be expected for unimolecular kinetics (vs. $n = -0.75$ for bimolecular). The actual data in Figure 6 indicate that the experimental $n \sim -0.3$. A likely reason for this discrepancy is the possibility that the ambient temperature was slightly higher under the high-dose-rate exposure conditions. This would increase the oxygen permeation constant, leading to slightly deeper oxidative penetration, which would be consistent with the results.

A recent study by Papet and co-workers [10] obtained detailed carbonyl profiles of radiation-aged additive-free, low-density polyethylene by examining 20-μm slices obtained using a microtome equipped with glass knives. Figure 7 shows their experimental results (squares) plotted versus depth into the sample for a sheet of material aged in air to 1 MGy at 2.5 kGy/hr and 20°C. The dashed curve through the data points represents a theoretical fit of the data using $\beta = 10$.

Figure 8 shows depth-dependent carbonyl absorption data (squares) given by Moisan [11] for a polyolefin material after aging for six days in air at 100°C. Since the carbonyl (oxidation) has an approximately exponential drop with distance into the sample, analysis based on the preceding diffusion-limited theoretical discussions would imply a small value for β (i.e., less than ~0.3). The solid curve represents a theoretical fit to the data using $\beta = 0.1$. Again, caution is necessary when applying steady-state theory to these data, since oxygen permeation and consumption may change significantly with time.

Sack and co-workers [12] used infrared techniques to study the mechanism of copper-

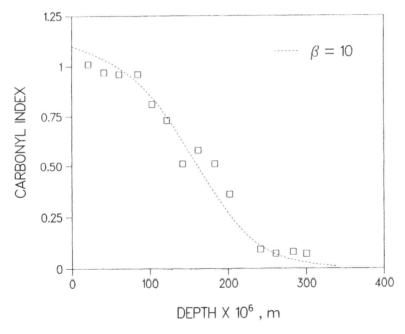

Figure 7 Carbonyl index (relative infrared peak height) versus depth away from air-exposed surface for microtomed low-density polyethylene samples after irradiation of 0.6-mm-thick sheet in air at 20°C to 1-MGy total dose. (Data from Ref. 10.) Curve represents theoretical fit to data using β = 10.

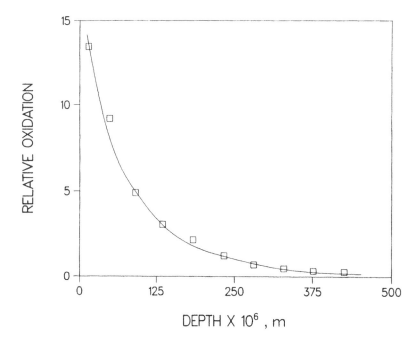

Figure 8 Relative oxidation as determined from the carbonyl absorbance versus depth away from air-exposed surface of polyolefin material after aging for 6 days at 100°C. (Data from Ref. 11.) Curve represents theoretical fit to data using $\beta = 0.1$.

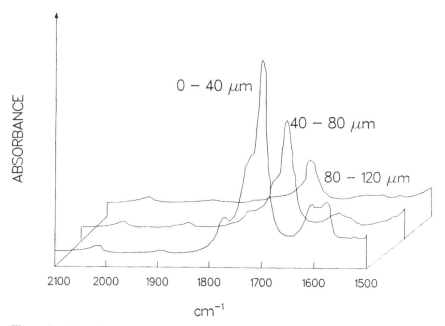

Figure 9 Infrared spectra for a low-density polyethylene material after oxidizing for 24 hr at 110°C in contact with copper foil. The three spectra correspond to regions of the polyethylene at different distances from the copper foil (the distance span for each spectrum is indicated). Data from Ref. 12.)

catalyzed thermal oxidation of low-density polyethylene. This degradation mechanism, which was first studied in great detail by workers at Bell Laboratories [13, 14], can lead to heterogeneous oxidation whenever a polyolefin material is aged adjacent to copper or a copper-containing material. Sack and co-workers obtained infrared absorbance plots for three polyethylene films of varying thicknesses (40–120 μm) after aging for 24 hr at 110°C [12]. During the aging, one side of each film was exposed to the air while the other was in contact with copper foil. By suitable spectral subtraction, spectra corresponding to the three depth intervals 0–40, 40–80, and 80–120 μm were obtained, as shown in Figure 9, demonstrating strongly enhanced oxidation in the region nearest the copper plate. It is clear from these spectra that the carbonyl region (1700–1800 cm^{-1}) contains the super-position of a number of unresolved peaks. The additional peaks appearing near 1600 cm^{-1} are attributed to carboxylate species, which are connected with the diffusion of copper ions into the polymer bulk. By microtoming samples after air aging at 110°C, oxidation profiles were obtained for a 1-mm-thick polyethylene sheet in contact with copper foil. Figure 10 shows data after 100 hr of aging where the absorbance at 1715 cm^{-1} relative to a reference polyethylene absorbance at 2020 cm^{-1} is plotted versus distance away from the metal foil. In the figure, the metal–polymer interface is located at 0, whereas the polymer–air interface is at 1000. Two opposing effects of the increasing copper concentration as the interface is approached account for the maximum in oxidation observed some distance from the interface [12]. First the increasing copper concentration acceler-

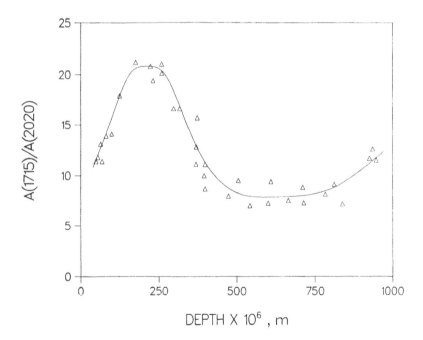

Figure 10 Relative carbonyl infrared peak heights (ratio of 1715 cm^{-1} peak to 2020 cm^{-1} reference peak) versus depth away from copper foil for a 1-mm-thick low-density polyethylene material aged in air for 100 hr at 110°C. Data were obtained on sections which were microtomed parallel to the surface after aging. The copper–polymer and polymer–air interfaces are located at 0 and 1000 respectively. (Data from Ref. 12.)

ates the oxidation (copper ions catalytically break down hydroperoxides to peroxy radicals), but when it exceeds a critical level of about 550 ppm, the copper ions take part in chain-breaking reactions (e.g., they react with peroxy radicals), inhibiting the oxidation.

Numerous additional studies have used infrared techniques for profiling oxidative degradation. Davis and co-workers [6, 15] obtained carbonyl profiles by using microtomed slices from naturally and artificially UV-aged low-density polyethylene. A similar approach was used by Terselius and co-workers [16] for thermally oxidized, high-density polyethylene and by Trotignon and co-workers [17] for low-density polyethylene exposed to pile radiation. A slightly modified sample preparation approach was used by Muller [18] on photooxidized polyethylene sheets. A milling machine was used to obtain shavings from successive 10-μm thickness layers. The shavings from each layer were then melted into films for infrared analysis.

Attenuated Total Reflection Spectroscopy

Besides standard transmission techniques, another infrared approach that can be very sensitive to changes in oxidation near the surfaces of samples involves attenuated total reflection (ATR) spectroscopy. This technique monitors the infrared spectra of a sample surface placed in intimate contact with a reflection element [19]. By varying the reflection elements and the angle of incidence, one can obtain infrared spectra corresponding to various penetration depths into the surface of the sample (typically from 0.1 to ~ 5 μm).

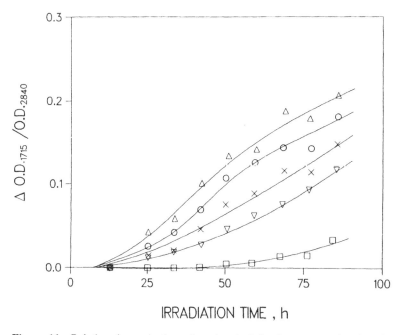

Figure 11 Relative change in the carbonyl optical density versus aging time for a photooxidized polypropylene film. The bottom curve (squares) comes from transmission infrared techniques. The other four curves come from ATR techniques using different reflection elements and incidence angles and correspond to the following penetration depths: 0.385 μm (triangles), 0.432 μm (circles), 0.835 μm ($\times$), and 1.17 μm (upside-down triangles). (Data from Ref. 20.)

Although this represents a fairly small range of accessible depths, in some instances oxidation heterogeneities will occur over such ranges. For instance, Carlsson and Wiles [20] used Ge and KRS-5 reflection elements, each at two different angles of incidence, to obtain ATR spectra of photooxidized polypropylene films at four different penetration depths. Analysis of the absorptions at 3400 and 1715 cm^{-1} allowed them to monitor the time and depth dependence of ROOH and carbonyl concentrations respectively. The carbonyl absorption results versus aging time for the four different ATR depth conditions (ranging from 0.385 to 1.17 μm) and for standard transmission infrared are shown in Figure 11. The results clearly indicate a rapid dropoff in oxidation away from the surface. Figure 12 gives our estimates, based on the data of Figure 11, for the carbonyl profiles after 50 and 85 hr of aging. ATR techniques have also been used to probe near-surface photooxidative heterogeneities of polystyrene [21] and near-surface ozonation of high-density polyethylene [22].

Modulus Profiling

A recently developed technique, which may turn out to be the most universally applicable method for following heterogeneous oxidation of polymers above their glass transition temperatures (e.g., elastomers), involves modulus profiling. This technique [23] allows one to quickly and easily obtain more than 20 quantitative tensile compliance measurements per millimeter of sample cross section. The importance of the technique (and the name) comes from the fact that the inverse of the tensile compliance ($1/D$) is closely related to the very commonly measured tensile modulus, which in turn is very sensitive to

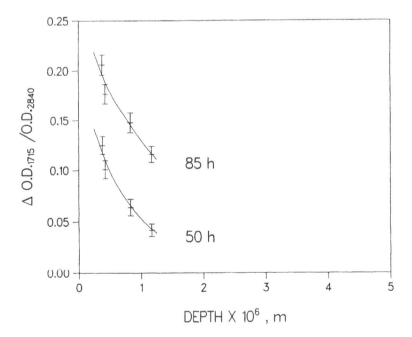

Figure 12 Relative change in carbonyl optical density versus depth away from air-exposed surface of photooxidized polypropylene film samples after 50 and 85 hr of aging. Data are based on ATR penetration depths and come from the data plotted in Figure 11.

processing and aging environments. Figure 13 is a schematic of the apparatus, which is based on extensive but simple modifications of a Perkin-Elmer TMS-1 thermomechanical analyzer. The apparatus allows one to quantitatively monitor versus time and load conditions the penetration into a sample of a specially prepared, paraboloidally shaped indentor tip. A small (typically 0.2 g) contact mass M_C is used to load the probe into the sample at time 0, followed by the addition of a larger mass (typically 1–5 g) at time t_1, giving a total mass of M_M. The difference in penetration d between time t_1 and $2t_1$ is the experimental quantity of interest. For a paraboloidally tipped probe with a focal point to vertex distance of $R/2$, linear viscoelasticity leads to the expression [23]

$$\frac{1}{D(t_1)} = \frac{9g(M_M^{2/3} - M_C^{2/3})^{3/2}}{16R^{1/2}d^{3/2}} \tag{11}$$

where g is the gravitational constant and $D(t_1)$ is the tensile compliance at time t_1 (typically 30 sec). The sample holder (Figure 13 detail) consists of a specially designed vise, constructed of a hollowed-out plastic cylinder having adjustable opposing screws contacting aluminum plates. Cross-sectioned samples (three of rectangular cross section

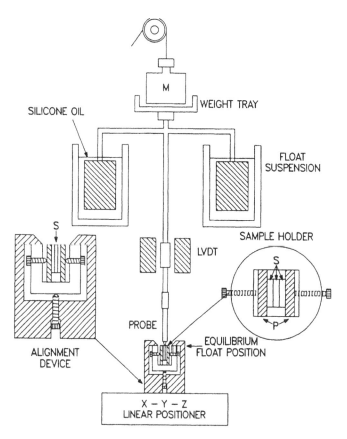

Figure 13 Schematic diagram of modulus profiling apparatus. The detailed top view of the sample holder shows three samples labeled with an S held between metal plates (P). The detail to the left shows a side view of the sample holder held in the alignment device. (From Ref. 23.)

are shown in the figure) are held together in a sandwich arrangement between the plates, after which the entire vise assembly is metallographically polished to a smooth, flat surface. The profile obtained consists of a series of penetration measurements taken across the cross-sectional surface of the sample. The accuracy (within 10% of conventional modulus measurement techniques), reproducibility (typically better than $\pm 5\%$), and linearity ($1/D$ independent of M_M for values ranging from 1 to 100 g) of the instrument have been demonstrated on a variety of elastomeric materials [23].

Data taken on a 1.5-mm-thick Viton material after gamma-radiation aging in air at 70°C serve to illustrate the power of modulus profiling for unraveling complex aging behaviors [24]. Figure 14 shows ultimate tensile elongation and tensile strength data obtained from stress–strain curves for this material versus dose rate and total radiation dose. The ultimate properties appear to degrade in opposite directions at high (elongation drops rapidly, tensile strength slowly) and low (elongation constant, tensile strength drops rapidly) dose rates. Thus the high-dose-rate samples become hard and brittle, whereas those aged at low dose rates become softer and are more easily stretched. Modulus profiling results for samples aged to 0.77 MGy at the three indicated dose rates are presented in Figure 15. The unaged material had a flat profile, indicated by the horizontal line. At the highest dose rate, diffusion-limited oxidation effects lead to an anaerobic (oxygen-starved) region in the interior of the sample. In this region the radiation crosslinks the material, leading to a large increase in the modulus. As the sample edges are approached, oxidative scission becomes important, resulting in a substantial drop in modulus relative to the unaged material. For the sample aged at the intermediate dose rate, diffusion-limited oxidation effects are still present but more moderate since the slower rate of oxidation allows further

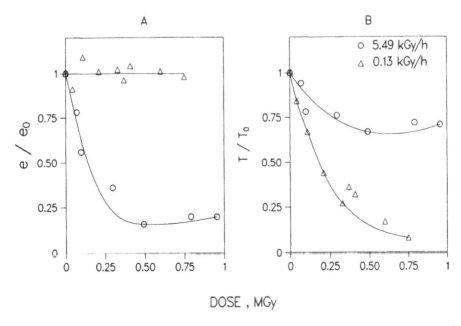

Figure 14 Changes in ultimate tensile properties versus dose for a Viton material gamma-irradiated in air at 70°C and the indicated dose rates. (A) Reduced elongation (e/e_0); (B) reduced tensile strength (T/T_0). (Data from Ref. 24.)

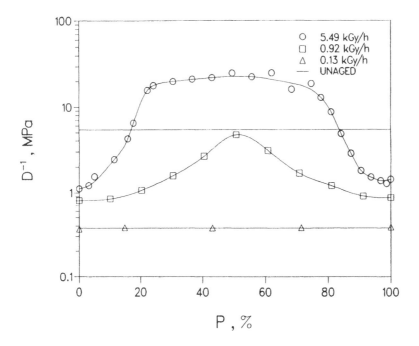

Figure 15 Modulus profiles for a 1.5-mm-thick Viton material after gamma-irradiation aging in air at 70°C and the indicated dose rates. The unaged material had a flat profile indicated by the horizontal line. *P* gives the percentage of the distance from one air-exposed surface to the opposite air-exposed surface. (From Ref. 23.)

oxidative penetration into the sample. At the lowest dose rate, diffusion effects disappear and homogeneous oxidative scission is observed. From these results one can immediately understand the reasons for the unusual tensile results of Figure 14 in terms of diffusion-dominated degradation behaviors. Additional information is available from the modulus profiling results at the outer edges of the samples. Since the edge results are unaffected by diffusion anomalies, the progression to lower values for the edge modulus as the dose rate is lowered immediately implicates an additional dose-rate-dependent mechanism. This could involve bimolecular termination kinetics or a time-dependent step in the chemical oxidation mechanism (e.g., peroxide decomposition leading to chain branching as a rate-determining step).

The modulus profiling technique also yields heat-aging data with a richness of detail largely unavailable until recently. Figure 16 shows, for example, results for 1.4-mm-thick nitrile rubber sample aged at 120°C. The unaged profile drops slightly on one side, indicating a slight undercure near that edge, perhaps due to slight thermal gradients present during the cure of the compression-molded sheets. At early aging times, there is only a slight hint of heterogeneous oxidation, whereas such effects become very important later. Again, this could be indicative of an increase in oxygen consumption rate with time and/or an ever-decreasing oxygen permeation rate as the sample edges harden with time. Since modeling of oven-aging data versus temperature is an important method for estimating (extrapolating) material lifetimes by means of the Arrhenius technique, monitoring

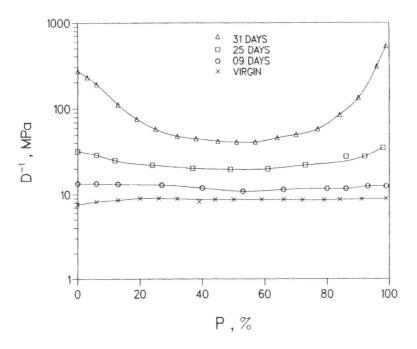

Figure 16 Modulus profiles for a 1.4-mm-thick nitrile rubber after 120°C oven-aging for the indicated time periods. *P* gives the percentage of the distance from one air-exposed surface to the opposite air-exposed surface. (From Ref. 23.)

modulus profiles versus time and temperature should prove extremely helpful in developing rigorously defendable models.

To compare the detailed information available from the modulus profiling technique with theoretical oxidation profiles such as those shown above, relationships must be derived between changes in modulus and oxidation. This is not a trivial task since modulus values will generally change both in the presence and absence of oxidation reactions. This is clear from the Viton results (Figure 15) where crosslinking in the absence of oxygen raises the modulus in the center of the sample, while oxidative scission lowers the modulus at the edges. Fairly detailed models for the relationship between the modulus and combinations of crosslinking and scission would be necessary before comparisons with the theoretical oxidation profiles could be attempted. This is a generic problem for many of the techniques that yield profiles; the property profiled may have a complicated or unknown relationship with the amount of oxidation. Nonetheless, in such instances, the technique may provide important profiling information which is complementary to profiles more directly related to the extent of oxidation. Modulus profiling, for example, provides profiles of mechanical property changes, information which often has the most direct relevance to understanding the environmental limitations of a material.

Though the foregoing discussion relates to one particular apparatus used for profiling mechanical properties across the cross-sectioned surface of polymer samples, other instrumentation of this type can also be employed to obtain such profiles of relative hardness or modulus. For instance, a Knoop hardness instrument was used to obtain heterogeneous oxidation profiles in UV-exposed samples of polypropylene [24]. This instrument is particularly suitable for profiling hard (high-modulus) materials.

Density Profiling

Density profiling utilizes a density gradient column to monitor the density of successive slices cut across a sample [25]. It depends on the observation that oxidation reactions typically lead to readily measurable changes (usually increases) in polymer density. Although there are a number of other methods available for measuring density, only the density gradient column approach has the capability to obtain accurate measurements on the extremely small samples necessary for achieving sufficient resolution for profiling across small distances. Unfortunately, artifacts associated with the measurement technique make its application to some materials difficult. The most important problem occurs when the material being investigated absorbs significant amounts of the liquid(s) used to make the gradient column. This effect, which will result in swelling of the material, can lead to an incorrect estimate of density. Since absorption of water by most polyolefins is small, density profiling has been successfully applied to these materials using saltwater columns. Figure 17 shows results [25] obtained from the density profiling technique for an unaged 0.8-mm-thick chemically crosslinked polyethylene (CLPE) material and for samples radiation aged in air to approximately 1.15 MGy at the three indicated dose rates. A straight horizontal line is drawn to represent the unaged material, since the variation in density across unaged samples was found to be less than ± 0.001 g/cm^3. For aged

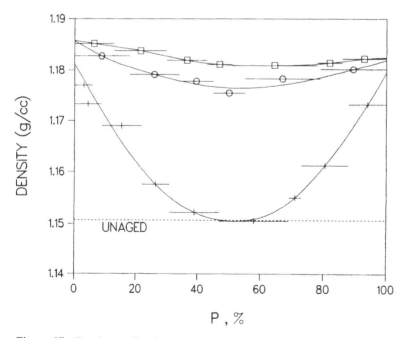

Figure 17 Density profiles for an unaged sample of chemically crosslinked polyethylene and for three samples after gamma-irradiation in air to approximately the same total dose at 43°C. Squares, 1.14 MGy at 0.175 kGy/hr; circles, 1.16 MGy at 0.64 kGy/hr; crosses, 1.19 MGy at 8.9 kGy/hr. P gives the percentage of the distance from one air-exposed surface to the opposite air-exposed surface. (From Ref. 25.)

samples, the density data for the series of slices taken across the sample from one air-exposed surface to the opposite air-exposed surface are plotted as a series of horizontal bars. Each bar spans the region from which the slice was taken, measured as a percentage of the distance from one side of the sample to the other. Under the high-dose-rate aging conditions, the oxidation is extremely heterogeneous with substantial oxidation (density increase) near both air-exposed surfaces and essentially no oxidation in the middle of the sample. In the lower dose rate experiments, the oxidation rate slows down, allowing more time for diffusion processes to replenish the oxygen used up in the sample interior. This leads to decreased heterogeneity and eventually to a homogeneously oxidized material. Since the results at the sample edges (diffusion-limited effects absent) indicate that a factor of 50 drop in dose rate (k_i) has little effect on the change in edge density, unimolecular kinetics (see Table 1) and therefore the absence of "chemical" dose-rate effects (arising from a time-dependent step in the oxidation chemistry) are indicated.

In many instances, including the CLPE material profiled in Figure 17, the changes in density for aged samples appear to have an approximately linear dependence on time. Thus if the oxygen consumption rate is approximately constant with time, the density changes would be expected to be linearly related to the amount of oxidation [25]. This would allow profiles of density changes to be compared directly with the theoretical oxidation profiles. Assuming such a linearity exists, Figure 18 plots the experimental

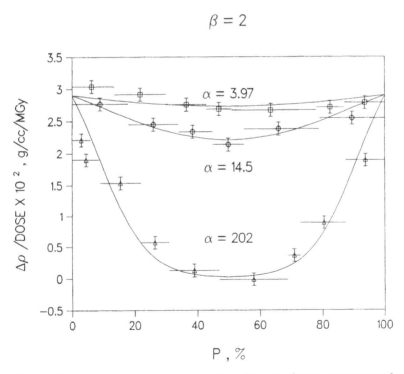

Figure 18 Crosses give experimental profiles of density change per megagray for CLPE using the density profile data from Figure 17. Width of crosses span region of sample from which the slice was taken, measured as a percentage of the distance from one side of the sample to the other; height of crosses gives the estimated experimental uncertainty. Solid curves are theoretical fits using $\beta = 2$ and the indicated values of α.

changes in density per megagray as crosses (vertical lines give estimated uncertainties) and fits these data with the solid theoretical curves from the unimolecular diffusion model. The parameter β, which is independent of the dose rate, is equal to 2 for all three curves. Since unimolecular kinetics is indicated, the fitting procedure requires that the parameter α be linearly related to k_i, the initiation (dose) rate. The theoretical fits to the data are excellent and, as with a number of other theoretical fits of data discussed in previous sections, we note that an intermediate value of β appears appropriate. If values of the oxygen consumption and permeation constants are available, the derived values of α and β can be independently confirmed (see Eq. 10). As mentioned earlier, if the oxygen consumption and permeation rates turn out to have a significant dependence on aging time, the theoretical fits shown in Figure 18 would be inappropriate, since the theory was derived assuming nonvarying consumption and permeation rates. It should also be noted that the use of the unimolecular model (discussed in the theoretical section) forces the edge values of the theoretical curves to be independent of dose rate. If the bimolecular model had been used, the edge value would have dropped by a factor of more than 7 times going from the low dose rate to the high dose rate $[(890/17.5)^{0.5} = 7.13]$. Thus clear evidence for unimolecular behavior exists for this material.

Another interesting example of the information available from density profiling involves oven-aging results for an ethylene–propylene rubber (EPR) material [25]. Mechanical property data covering a temperature range of 100–170°C had non-Arrhenius behavior at both high and low temperatures. Even though diffusion-limited oxidation effects could be expected to give non-Arrhenius behavior at high temperatures, such effects should become less important at the lower temperatures. This was confirmed by density profile data, yet a second heterogeneous mechanism was found. Figure 19 shows profiles

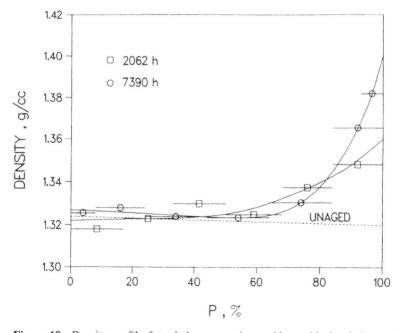

Figure 19 Density profile for ethylene–propylene rubber cable insulation material which had been heat aged in air at 100°C for the indicated times. *P* gives the percentage of the distance from one air-exposed surface to the opposite air-exposed surface. (From Ref. 25.)

for samples aged at the lowest temperature compared to the unaged profile. The dramatic density increases near the inside of the insulation are due to the same copper-catalyzed oxidation mechanism mentioned earlier for the low-density polyethylene material [12]. Although the EPR samples were removed from their copper conductors before aging, copper salts had already diffused into the insulation during the high-temperature extrusion manufacturing process. Enhanced copper concentrations near the inside surface of the insulation were confirmed using emission spectroscopy and microprobe analysis. Data from density profiling also showed that copper-catalyzed oxidation can be an important mechanism leading to dose-rate effects under ambient temperature radiation-aging conditions [26].

Another interesting aspect of the density data shown in Figure 19 is their sensitivity to the early stages of degradation. Comparing the mechanical property and density profile data shows that clear and significant changes appear in the density profiles before noticeable changes are apparent in the mechanical properties [25].

Gel Permeation Chromatography

For soluble materials, gel permeation chromatography (GPC) can be used to obtain molecular weight distributions versus aging time, which in turn can be used to estimate scission and crosslinking yields. Bowmer and co-workers [27], in a study of polystyrene which was gamma-irradiated in air, applied this approach in a manner that led to profiles of scission and crosslinking yields. After aging, successive layers (minimum thickness of ~0.1 mm) were removed from irradiated material by wet grinding on a lapping wheel.

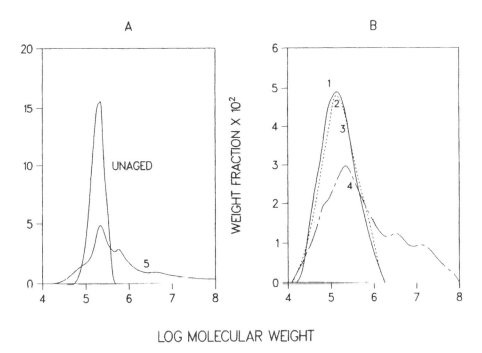

Figure 20 Gel permeation chromatograms for polystyrene after gamma-irradiation of a 3.2-mm-thick bar in air for 1 MGy. Curve 1, surface scrapings spanning approximately 0–5% of cross section; other curves represent scrapings spanning 5–10% (2), 10–20% (3), 20–30% (4), and 30–50% (5). (Data from Ref. 27.)

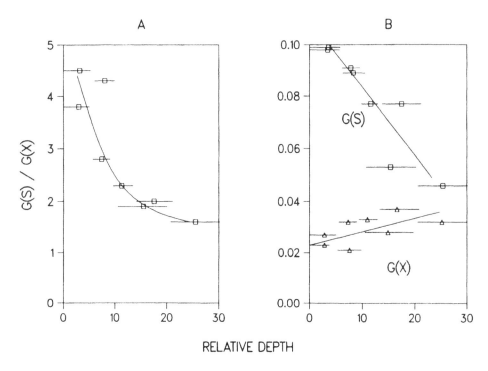

Figure 21 Scission yield $G(S)$ and crosslinking yield $G(X)$ versus relative depth (percentage of cross section) obtained from analysis of gel permeation chromatograms like those seen in Figure 20. (A) $G(S)/G(X)$ versus relative depth. (B) $G(S)$ and $G(X)$ versus relative depth. (From Ref. 27.)

The polymer removed from a given layer was recovered by filtering the washing water, dissolving the grindings in chloroform, filtering to remove particles of the abrasive used for grinding, evaporating the solution, and drying. For polystyrene aged to 1 MGy in air, Figure 20 gives GPC molecular weight distribution curves versus depth into the sample. Comparing these curves to that of the molded, unirradiated sheet (also shown in the figure), one sees that scission predominates near the sample surface, whereas crosslinking becomes much more important relative to scission further into the sample. These molecular weight distributions, upon analysis, led to profiles for the crosslinking and scission yields, $G(X)$ and $G(S)$, as shown in Figure 21 [27]. Diffusion-limited oxidation is again the mechanism responsible for these results. Dickens and co-workers [28, 29] used GPC in a similar way to obtain estimates of the number of bonds broken as a function of depth into plates of thermally and photolytically aged poly(methyl methacrylate) (PMMA) samples. Pastuska and co-workers [30] also used GPC to obtain molecular weight profiles for naturally and artificially UV-aged PMMA.

Nuclear Magnetic Resonance Self-Diffusion Coefficients

Through the use of pulsed-field gradient techniques [31], proton nuclear magnetic resonance (NMR) can be used to measure self-diffusion constants (D_s) as low as $\sim 10^{-14}$ m^2/sec. Recently Belousova et al. [32] used stimulated echo methods [33] together with the pulsed-field gradient technique to profile self-diffusion coefficients for low- and high-density polyethylene samples which had been gamma-radiation aged in air. Measurements were run on each microtomed layer at 160°C, a temperature which is unfortunately high enough to raise concerns about the possibility of further degradation during the measurements. Some of the resulting profiles for the low-density polyethylene material after various total doses are shown in Figure 22. Unaged material had a flat profile with $D_s \sim 10^{-12}$ m²/sec. For well-characterized polyethylene samples, D_s is related to the inverse square of the molecular weight [34]. Since radiation in a vacuum crosslinks polyethylene, the anticipated molecular weight increase should lead to a drop in D_s, an expectation confirmed by the 1-MGy vacuum data in Figure 22. For the samples aged in air, the dominance of oxidative scission over crosslinking near the sample surfaces leads to a drop in molecular weight, as evidenced by increasing values of D_s. Further into the samples, where oxidation is reduced due to diffusion-limited effects, D_s begins to drop and eventually approaches a constant value at depths where anaerobic aging occurs. Comparing the self-diffusion profile shapes with the theoretical oxidation profiles would be difficult, first requiring a conversion from a D_s profile to a molecular weight profile, followed by a further conversion to an oxidative scission profile. Although $D_s \sim MW^{-2}$ for well-characterized (narrow molecular weight distribution) samples, deviations from this relationship would be expected for commercial samples of polyethylene undergoing combinations of crosslinking and scission. We saw earlier (Figure 20) that the radiation-induced crosslinking and scission occurring in the polystyrene material of narrow initial molecular weight distribution led to dispersed and complicated distributions for the aged materials.

Even though comparisons of the D_s profile shapes with theory are difficult, the dose-rate dependencies at the sample edges and temperature dependencies of the profiles lead to

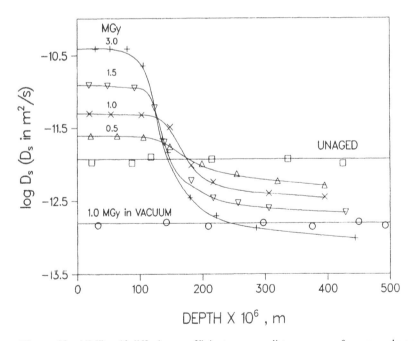

Figure 22 NMR self-diffusion coefficients versus distance away from sample surface for low-density polyethylene samples after gamma-irradiation in air or vacuum at 0.6 Gy/sec for the indicated total doses. (From Ref. 32.)

interesting conclusions. Figure 23A shows D_s profiles for the low-density polyethylene material aged to 0.5 MGy at 0.6 and 0.2 Gy/sec. Since the value of D_s at the edge of the sample (no oxygen diffusion-limited effects) is approximately independent of dose rate, unimolecular kinetics is again indicated (see Table 1). If bimolecular kinetics was appropriate, the increase in value of the D_s at the edge would be 73% greater at the lower dose rate $[(0.6/0.2)^{0.5} = 1.73]$. The region where the oxidation drops off also increases at the lower dose rate by approximately the 73% expected for the unimolecular model (~32% would be expected for the bimolecular case). Figure 23B gives data for samples aged to 0.2 MGy at 0.6 Gy/sec and two different temperatures. Since the edge results for the two temperatures are approximately the same, we have evidence that the oxygen consumption per Gy is independent of temperature for these data. This implies that the increased oxidative penetration at the higher temperature comes predominantly from the increased oxygen permeation constant at the higher temperature. Knowing that the penetration distance is proportional to the square root of the permeation constant then allows us to use the data shown to estimate that the activation energy for oxygen permeation is ~6 kcal/mol, which is in reasonable agreement with most literature values for low-density polyethylene.

Miscellaneous Methods Sensitive to Scission and Crosslinking

Besides modulus profiling and GPC and NMR D_s measurements, several other methods are sensitive to the scission and crosslinking which occur during aging of polymeric

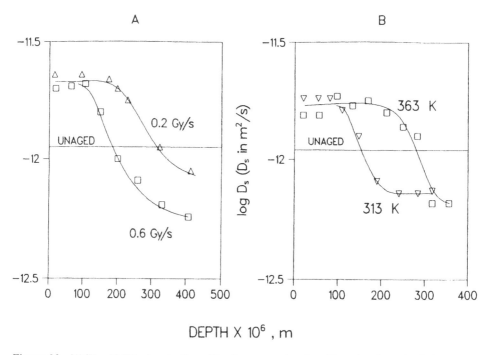

Figure 23 NMR self-diffusion depth profiles for gamma-irradiated low-density polyethylene materials. (A) 0.5 MGy at the indicated dose rates. (B) 0.2 MGy at 0.6 Gy/sec and the two indicated temperatures. (Data from Ref. 32.)

materials. By microtoming or milling successive layers, a number of these techniques have been used to qualitatively profile oxidative heterogeneities. Wilski and co-workers [35, 36] pioneered the use of viscosity profiles to show the importance of diffusion-limited oxidation effects for gamma-irradiated polyethylene, polypropylene, and PVC. Viscosity measurements were made on aged samples after sectioning and dissolving in suitable solvents. Bowmer et al. [27] used limiting viscosity numbers obtained from analysis of solution viscosity results on their radiation-aged polystyrene samples to show that scission becomes more important relative to crosslinking as the sample surface is approached. Solution viscosity techniques cannot be applied to insoluble (gelled) material. A gel fraction profile (resolution of 0.3 mm) was obtained versus dose rate for gamma-irradiated polyethylene by Kuriyama et al. [37]. Although this technique distinguishes between soluble and insoluble material, it cannot be applied to initially crosslinked material or to a material that never gels.

Chemiluminescence

For a few materials, polypropylene in particular, chemiluminescence (CL) techniques have proven useful for following oxidation processes [38, 39]. It is thought that CL occurs due to the decay to the ground state of an excited ketone product formed during the reaction of a peroxy radical (ROO·). Yoshii et al. [40] used CL techniques carried out at 80°C to generate profiles of commercial-grade copolypropylene (6% ethylene units) samples which had been irradiated in air either with gamma rays or with an electron beam. CL counts were first measured on a 1-mm-thick aged film. After scraping away ~50 μm of surface layer from both sides of the sample, a new CL measurement was taken; this procedure was repeated until the CL intensity stopped dropping. Figure 24 plots the

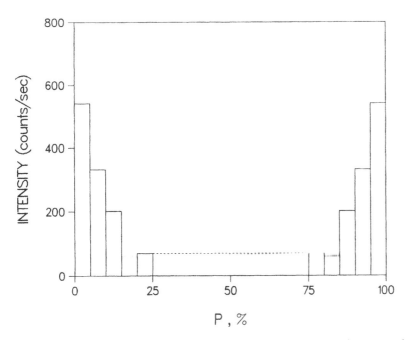

Figure 24 Chemiluminescence profile for a polypropylene material after gamma-irradiation in air to 0.05 MGy at 2 kGy/hr. (Data from Ref. 40.)

decrease in CL counts found versus position across the sample for a gamma dose of 0.05 MGy at 2 kGy/hr. The authors conclude that diffusion-limited oxidation is responsible for the heterogeneous character of their data. If the amount of oxidation is assumed to be linearly related to the CL intensity, we note that the U-shaped profiles found in this study imply a low to intermediate value for β. This assumption might be reasonable for a sample aged at one dose rate, if little CL intensity is lost during the time span required to successively remove sufficient layers to profile the material. However, the transitory nature of the CL intensity would have to be quantitatively understood before meaningful conclusions could be attempted for CL intensity data versus dose (or initiation) rate.

Crystallinity Profiles—Digital Scanning Calorimetry and X-Ray Crystallography

For crystalline polymers, the amount and type of crystallinity will depend both on processing conditions and on aging environments. Since both differential scanning calorimetry (DSC) and x-ray crystallography techniques [9] are sensitive to sample crystallinity and can be done on small samples (e.g., thin microtomed slices), these techniques have been used to profile crystalline polymers [10, 17]. Figure 25 shows DSC crystalline melting point (temperature at the maximum of the endotherm of fusion) profiles from Papet et al. [10] for the same radiation-aged low-density polyethylene material discussed earlier (in the infrared section). The first thing to note is the drop in melting point near the surface for the unaged material, implying a lowering in crystallinity as the edge is approached. By integrating the fusion endotherms, the authors estimated that the amorphous fraction for the bulk unaged material was ~73%; this fraction was found to increase

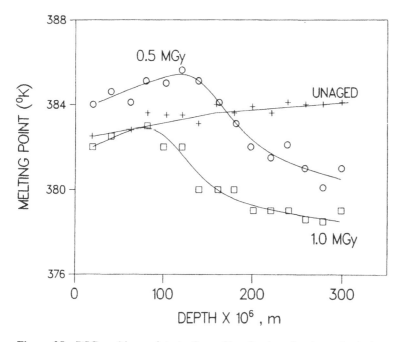

Figure 25 DSC melting point depth profiles for low-density polyethylene material gamma-irradiated in air to the indicated total doses. (Data from Ref. 10.)

substantially (to 80–85%) for the outer 50 μm of the sample. The observation of enhanced amorphous content near the surfaces of molded crystalline polymers has been observed previously [41] and may be the primary cause of near-surface oxidative anomalies observed in various studies [6, 17]. Figure 25 shows that aging leads to complicated changes in the melting point profiles. Qualitative interpretation of the changes can be made by comparing the infrared (Figure 7), melting point (Figure 25), and crystalline fraction profiles for the same material [10]. At lower total doses and near the surface, oxidation-induced crystallization occurs; crystal destruction occurs in the oxygen-depleted interior of the sample and for higher doses near the surface.

It should be noted that the density of crystalline polymers is sensitive to the amount of crystallinity; for this reason, density measurements have also been used to map crystallinity [41]. Since polymer oxidation (including noncrystalline materials such as rubbers) usually leads to density increases, caution is advised whenever density data are used to follow crystallinity changes caused by aging.

Oxidative Induction Time Profiles from Differential Thermal Analysis

Differential thermal analysis (DTA) monitors the energy change associated with reactions or transitions in a material by following the difference in temperature between the material and a reference material as the two materials are subjected to identical environmental conditions [9]. It is closely related to the more commonly used DSC technique discussed in the previous section and can be similarly used to profile crystalline melting temperatures and percentage crystallinity. Kramer and Koppelmann [42] recently showed that such techniques could also be used to profile the remaining oxidative stability of a material. They used long-term isothermal DTA to obtain the oxidative induction time (break from baseline indicating onset of exothermic reaction) versus cross-sectional position for polybutylene and crosslinked polyethylene hot water pipes. Figure 26 shows oxidative induction time profiles (isothermal DTA carried out at 200°C) for unaged and field-aged 5.6-mm wall thickness polybutylene hot water pipe. Field aging causes a rapid drop in the oxidative stability, especially near the inside of the pipe (left side of profile). This was explained by the large amount of stabilizer that had been extracted during long-term contact with the hot flowing water.

Optical Profiling

In many cases the occurrence of heterogeneous oxidation may result in differential changes in the polymer sample which can be observed by optical examination. Where applicable, this can provide a rapid and simple means by which qualitative or semiquantitative information may be obtained on heterogeneous effects. Some polymeric materials, or common additives which they contain, change color under degradation-inducing conditions. In certain cases the nature of the color change depends strongly on the occurrence of oxidizing versus nonoxidizing conditions. For example, when commercial samples of a clear polyurethane elastomer were subjected to gamma irradiation by the authors, and then cut in cross section, the interior region turned dark brown, whereas the regions near the sample surfaces were light yellow. These two regions clearly defined the depth of significant oxidation in the material under the environmental conditions employed. Seguchi and Arakawa [43] conducted a study of the depth of oxidation of polyethylene samples subjected to ionizing radiation at a variety of dose rates and oxygen partial pressures by monitoring color changes in cross-sectioned samples. In this case, using clear samples

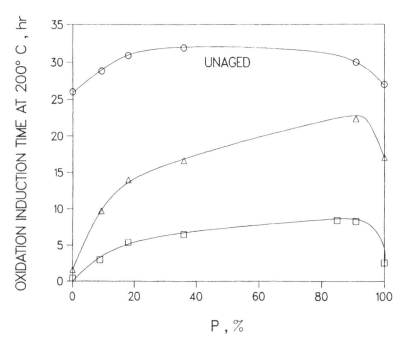

Figure 26 Oxidation induction time profiles for polybutylene hot water pipe. Aged data is for field exposures of 15 months at 90°C (triangles), followed by an additional 3 months at 110°C (squares). Inside surface of 5.6-mm-thick pipe (exposed to hot water) is at zero on horizontal scale. (Data from Ref. 42.)

containing diphenyl-*p*-phenylenediamine (DPPD), the DPPD turned dark red-brown in oxidized regions but exhibited no color change in unoxidized (interior) regions. Many organic dyes are bleached by oxidation reactions. Incorporation of such materials into the formulation of transparent plastics could provide a general means of identifying heterogeneous oxidation, as long as the dye does not interfere with the oxidation depth. An example of dye interference with oxidation might occur in a photodegradation environment if the dye absorbed (and thereby attenuated) the incident light.

Changes in optical properties can also be monitored in a more quantitative way, using visible or UV spectroscopy. For example, Bowmer and co-workers [27] found that polystyrene yellows when subjected to gamma irradiation (i.e., the strong aromatic absorption in the UV shifts progressively to longer wavelength with absorbed dose). The color change is much more pronounced when oxygen is present, and this fact was used in examining the oxidation profile of 3-mm-thick polystyrene bars subjected to gamma irradiation in air. This was accomplished by grinding away successive layers from the polystyrene bars, dissolving the grindings in an organic solvent, and measuring the absorption at 325 nm as a function of depth in the sample.

Polishing/Reflectance Technique

A somewhat more general optical technique, which does not depend on color changes, exists for rapid identification of heterogeneous oxidation in a qualitative or semiquantitative mode [44]. In this method, a sample of a degraded polymer is cut in cross section, and the cross-sectional surface is subjected to polishing using standard metallographic techniques. Rubbers may be conveniently held for polishing using a tiny vise or by potting in epoxy resin. The polished surface is examined under an optical microscope, with the sample held at an angle such that the illuminating light is reflected directly into the scope from the sample surface. Oxidized and nonoxidized regions of the sample typically degrade in very different ways. Very often these two regions of the sample acquire distinctly different mechanical properties—in particular different hardness. These regions of different hardness take on different lusters when polished. As a result, the oxidized and nonoxidized regions show up as bands of different reflectivity upon examination. The technique works equally well for transparent or opaque materials, and it has been successfully applied to estimating the depth of significant oxidation, upon degradation, in a wide variety of commercial rubbers [44]. The technique corresponds directly to depth-dependent mechanical property differences resulting from the heterogeneous oxidation; it works best in materials for which such property differences are large and is most distinct in highly degraded samples. Figure 27 provides an example of an ethylene–propylene rubber material for which oxidized and nonoxidized regions are clearly visible by the reflectance technique.

Profiling of Low Molecular Weight Components

Low molecular weight compounds in polymeric materials, obtained by solvent extraction of the sample, may be monitored as a function of degradation [45]. This can be done in terms of either the appearance of degradation products or the disappearance of compounds initially present in the formulation. By extraction of successive sections proceeding toward the sample interior, it is possible to obtain profiles of the degradation. Matsumoto [46] followed the disappearance of antioxidants in stacks of thin polyethylene films which were separated and extracted following thermal oxidation. Antioxidant content was determined using liquid chromatography. A similar approach was used on chloroprene rubber materials, which were aged in air-circulating ovens, sectioned, extracted, and then submitted to gas chromatographic analysis [7]. Antioxidant (AH) profiles are complicated by the fact that the AH may disappear both by oxidative chemistry and by evaporative loss from the sample. Moreover, depending on the temperature and time scale of the experi-

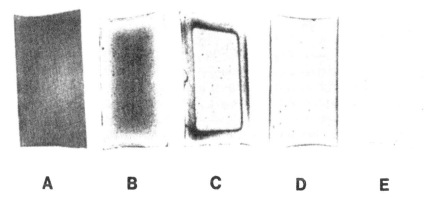

A　　　**B**　　　**C**　　　**D**　　　**E**

Figure 27　Metallographic polishing results for EPR samples which had been gamma-irradiated at 70°C. (A) Unaged material; (B) 1.65 MGy in air at 6.7 kGy/hr; (C) 2.97 MGy in air at 6.7 kGy/hr; (D) 1.75 MGy in air at 1.1 kGy/hr; (E) 2.53 MGy in vacuum at 11 kGy/hr. (From Ref. 44.)

ment and the diffusion coefficient of the AH in the polymer, the low molecular weight AH species may itself be able to migrate through the bulk sample at an appreciable rate, and this may complicate the interpretation of any profile that might arise due to the heterogeneous nature of the oxidation. By adapting the mathematics of heat flow in a solid, Calvert and Billingham [47, 48] derived a theoretical model applicable to the evaporative (i.e., nonreactive) loss of additives from polymers. The model contains three parameters, the solubility of the additive, its volatilization rate at the polymer surface, and its diffusion coefficient in the bulk of the polymer. After aging a material whose AH concentration was initially independent of spatial location in the sample, the AH concentration could remain independent or become dependent on spatial location, depending on the relative importance of the three model parameters. It would be difficult to generalize the model to include reactive loss (in addition to evaporative loss) of the AH, especially in the presence of diffusion-limited oxidation effects.

Miscellaneous Profiling Techniques

It should be clear that almost any technique capable of following the oxidation of polymers can in principle be used to profile heterogeneous oxidation. However, since most heterogeneous oxidation mechanisms are operative over fractions of a millimeter, those techniques which are sensitive enough to probe small amounts of material are the best candidates. Certain techniques (e.g., modulus profiling) may allow direct profiles to be obtained by sequentially moving across the cross section of a sample, avoiding the often difficult and tedious option of microtoming, milling, or scraping the sequential samples necessary for nondirect techniques. Depending on the polymer and its aging environment, there are a number of high-resolution modern spectroscopic techniques for surfaces which should be capable of directly profiling oxidative degradation. Two examples include micro-Raman spectroscopy and the recently introduced but increasingly widespread micro-infrared approach. In a study of the copper-catalyzed oxidation of radiation-aged ethylene–propylene rubber cable insulation material, microprobe analysis techniques were used directly on a cross section of the insulation to show the presence of enhanced copper concentrations near the copper conductors [26]. This enhanced copper concentration was also confirmed by profiling concentric rings of material using emission spectroscopy. Sack and co-workers [12] used neutron activation analysis of microtomed LDPE slices to show the presence after thermal aging of order-of-magnitude enhancements in copper content for material aged near the copper foil.

It might be possible to use chemical reagents to treat cross-sectioned surfaces of oxidized samples to render them more amenable to profiling analysis, since the small concentrations of oxygen incorporated into polymers as a result of oxidation can often be difficult to detect by surface techniques. In particular, there are a number of chemical reagents that react with oxygen-containing organic functional groups to yield a metal precipitate (or a metal-containing compound). For example, both Tollen's reagent and Benedict's solution react with aldehydes [49] to give insoluble precipitates (metallic silver and cuprous oxide, respectively). The resulting metal species could be quantitatively profiled with much more sensitivity by such techniques as electron microprobe [50].

High-Pressure Oxygen Techniques

For many years, researchers have attempted to use elevated oxygen pressures as a means of reducing or eliminating diffusion-limited oxidation effects for accelerated aging exposures. For instance, Seguchi and co-workers [51] pioneered the use of high oxygen pressures under high radiation dose rate conditions as a means of eliminating diffusion-limited oxidation effects at high dose rates and therefore rapidly simulating much longer low dose rate aging exposures. Similarly, Faulkner [52] recently applied high-pressure oxygen techniques to thermal aging of thick stabilized polypropylene in order to speed up sample embrittlement times, thereby allowing accelerated aging to be conducted at temperatures closer to the service temperatures.

From the theory and profile results summarized in earlier sections, it is obvious that a number of potential problems can occur when oxygen overpressure techniques are used. First of all, when β is small to intermediate under air-aging conditions, Figure 5 shows that raising the oxygen pressure will lead to large increases in oxidation at the sample edges. This is an obvious result since a low value of β implies oxidation linearly proportional to oxygen pressure (see Figure 1). In such instances, increasing the oxygen pressure to eliminate diffusion-limited oxidation effects will eventually result in uniform oxidation throughout the sample, but at a level higher than would occur under air-aging conditions. Only when evidence exists that either the edge oxidation or the overall oxidation (small enough dimensions such that diffusion effects are unimportant) is insensitive to oxygen pressure (0.2 atm and higher) can high oxygen pressures be rigorously used to eliminate physical diffusion anomalies in thicker samples. But even in these instances, it can be inappropriate for radiation-aging situations to use the homogeneously oxidized results obtained at high radiation dose rates to make predictions at much lower dose rates, since lack of data versus dose rate will overlook the existence of any important chemical dose-rate effects [53, 54].

APPLICATIONS OF PROFILING RESULTS

It should be clear from the preceding examples that heterogeneous oxidation effects are a very common occurrence when polymers are aged in air. The information gained from oxidation profiling techniques is invaluable for monitoring and understanding such effects. It is worthwhile to briefly summarize some of the uses which can be made of oxidation profile data. The utility of the results depends to a large extent on the techniques used and the extent of the data. Even limited or qualitative data (e.g., Figure 27) will often be sufficient for recognizing the existence of diffusion-limited oxidation or material incompatibility effects. If material oxidation information is desired in the absence of physical diffusion-limited oxidation complications, reduction of such anomalies can sometimes be accomplished by either using lower thickness samples or, in some cases (see previous section), by raising the oxygen pressure. When material incompatibilities are discovered, one of the two interacting materials can be either replaced or modified to eliminate such problems.

If more detailed, semiquantitative profiling results are obtained, additional information is available. For instance, it may be possible to recognize the existence of nonuniform material processing (Figure 16), a common precursor of nonuniform aging effects. Processing parameters can then be adjusted to minimize any such spatial variations in material properties. Unusual macroscopic material property data (Figure 14) may be immediately comprehensible from profiling results (Figure 15).

As oxidation profile information becomes more quantitative and detailed, greater predictive capabilities are possible. For instance, by determining the dependence of the oxidation rate on the initiation rate (k_i), extrapolations can be made to experimentally inaccessible k_i's. This dependence can be ascertained by following the oxidation at the sample surfaces (no diffusion limitations) as a function of k_i. It can also be determined using macroscopic techniques as a function of k_i in the low-thickness or high-oxygen-pressure limits (see Table 1). In radiation environments, as an example, the edge dependence for a Viton material was used to recognize the existence of "chemical" dose-rate effects (Figure 15). Although bimolecular termination reactions can lead to chemical dose-rate effects (oxidative degradation rates per Gy, which increase as the dose rate is lowered), other chemical mechanisms can also be responsible for such effects. For instance, recent studies of a polyvinylchloride material at dose rates low enough to eliminate diffusion-limited oxidation effects found both unimolecular termination and chemical dose-rate effects caused by the slow, rate-determining breakdown of the hydroperoxides formed during the oxidative propagation kinetic steps (see "Theory for Diffusion-Limited Oxidation" section). Modeling of these effects led to the capability of predicting material responses in radiation environments for extended (tens of years) time periods [53].

By comparing detailed quantitative experimental profiles with theoretically modeled profiles, kinetic and physical parameter data are available, as are consistency tests of the various theories. Profile shapes versus the initiation rate (Figures 17 and 18) not only allow choices to be made between competing theories (e.g., unimolecular versus bimolecular termination), but also allow estimates to be made of the theoretical parameters (e.g., α and β). By using the theoretical relationship (Eq. 10) relating α, β, and the oxygen permeation rate to the oxygen consumption rate, measurement of either one of the latter two quantities can be used to estimate the other quantity. If both quantities are measured, internal consistency of the experimental results with the theory can be checked. In such instances, detailed kinetic rate constant information may be available. In using experimental profiles for comparison with theory, it is critical that a known relationship exist between the experimental parameter and the amount of oxidation, the simplest being a linear relationship. An additional complication in such comparisons occurs whenever the oxygen consumption and/or permeation rates depend on aging time, since this would require modifications of the time-independent theory.

If sufficient data can be collected such that confidence in the appropriateness of a theoretical model exists, it should be possible to use either estimated or measured values of oxygen consumption and permeation rates as a means of directly estimating when important diffusion effects are present. For instance, if the theory presented were found to be appropriate, the following expression [6, 53] could be used to estimate the maximum sample thickness L_{max} for which the integrated oxidation across the sample exceeds ~90% of what completely uniform oxidation would give:

$$L_{max} = \left(\frac{8pP_{ox}}{\Phi}\right)^{0.5} \tag{12}$$

where Φ is the oxygen consumption rate and p and P_{ox} have the same meaning as in Eqs. 9 and 10. For aging simulations intending to model homogeneous oxidation, such expressions could offer significant shortcuts in choosing experimental conditions sufficient to eliminate diffusion-limited oxidation effects.

It should be apparent that further development and use of experimental profiling techniques in conjunction with a better theoretical understanding of the profile shapes will be useful for improving polymeric materials and more reliably predicting their lifetimes.

ACKNOWLEDGMENT

The authors would like to thank N. J. Dhooge for able assistance in preparing the figures for this chapter.

REFERENCES

1. J. F. Rabek, in *Chemical Kinetics, Vol. 14, Degradation of Polymers* (C. H. Bamford and C. F. H. Tipper, eds.), Elsevier, Amsterdam, p. 425 (1975).
2. Y. Kamiya and E. Niki, in *Aspects of Degradation and Stabilization of Polymers* (H. H. G. Jellinek, ed.), Elsevier, Amsterdam, p. 79 (1978).
3. T. Kelen, *Polymer Degradation*, Van Nostrand Reinhold, New York (1983).
4. S. S. Stivala, J. Kimura, and L. Reich, in *Degradation and Stabilization of Polymers*, Vol. 1 (H. H. G. Jellinek, ed.), Elsevier, Amsterdam, p. 1 (1983).
5. N. Grassie and G. Scott, *Polymer Degradation and Stabilisation*, Cambridge University Press, Cambridge (1985).
6. A. V. Cunliffe and A. Davis, *Polym. Degrad. Stab.*, *4*: 17 (1982).
7. R. L. Clough and K. T. Gillen, *Polym. Mater. Sci. Eng.*, *58* (1988).
8. R. C. Giberson, *J. Phys. Chem.*, *66*: 463 (1962).
9. J. F. Rabek, *Experimental Methods in Polymer Chemistry*, Wiley, Chichester (1980).
10. G. Papet, L. Jirackova, and J. Verdu, *Radiat. Phys. Chem.*, *29*: 65 (1987).
11. J. Y. Moisan, in *Polymer Permeability* (J. Comyn, ed.), Elsevier Applied Science Publishers, London, p. 119 (1985).
12. S. Sack, S. Schar, and E. Steger, *Polym. Degrad. Stab.*, *7*: 193 (1984).
13. M. G. Chan and D. L. Allara, *J. Colloid Interface Sci.*, *47*: 697 (1974).
14. M. G. Chan and D. L. Allara, *Polym. Eng. Sci.*, *4*: 12 (1974).
15. G. C. Furneaux, K. J. Ledbury, and A. Davis, *Polym. Degrad. Stab.*, *3*: 431 (1980–81).
16. B. Terselius, U. W. Gedde, and J. F. Jansson, *Polym. Eng. Sci.*, *22*: 422 (1982).
17. J. P. Trotignon, J. Verdu, and R. Roques, "The Role of the Skin-Core Structure in the Ageing of Semi-Crystalline Polymers," Morphology of Polymers, Proceedings, Europhysics Conference on Macromolecular Physics, 17th Meeting (B. Sedlacek, ed.), Walter de Gruyter and Co., Berlin, p. 297 (1986).
18. K. Muller, *Angew. Makromol. Chem.*, *114*: 69 (1983).
19. D. J. Carlsson and D. M. Wiles, *Can. J. Chem.*, *48*: 2397 (1970).
20. D. J. Carlsson and D. M. Wiles, *Macromolecules*, *4*: 174 (1970).
21. S. Curran, S. Siggia, R. Porter, and E. Otocka, *Org. Coat. Plast. Chem.*, *42*: 760 (1980).
22. N. M. Emanuel, *Polym. Sci. USSR*, *27*: 1505 (1985). Translation of *Vysokomol. Soyed.*, *A27*: 1347 (1985).
23. K. T. Gillen, R. L. Clough, and C. A. Quintana, *Polym. Degrad. Stab.*, *17*: 31 (1987).
24. R. L. Clough and K. T. Gillen, in *Polymer Stabilization and Degradation*, ACS Symposium Series no. 280 (P. P. Klemchuk, ed.), American Chemical Society, Washington, D.C., p. 411 (1984).
25. K. T. Gillen, R. L. Clough, and N. J. Dhooge, *Polymer*, *27*: 225 (1986).
26. K. T. Gillen and R. L. Clough, *Radiat. Phys. Chem.*, *22*: 537 (1983).
27. T. N. Bowmer, L. K. Cohen, J. H. O'Donnell, and D. J. Winzor, *J. Appl. Polym. Sci.*, *24*: 425 (1979).
28. B. Dickens, J. W. Martin, and D. Waksman, *Polymer*, *25*: 706 (1984).
29. B. Dickens, J. W. Martin, and D. Waksman, *Polym. Degrad. Stab.*, *15*: 265 (1986).
30. G. Pastuska, V. Lehmann, and U. J. Berlin, *Kunststoffe*, *76*: 59 (1986).
31. E. O. Stejskal and J. E. Tanner, *J. Chem. Phys.*, *42*: 288 (1965).

32. M. V. Belousova, V. D. Skirda, O. E. Zgadzai, A. I. Maklakov, I. V. Potapova, B. S. Romanov, and D. D. Rumyanthev, *Acta Polym., 36*: 557 (1985).

33. J. E. Tanner, *J. Chem. Phys., 52*: 2523 (1970).

34. M. Tirrell, *Rubber Chem. Tech., 57*: 523 (1984).

35. H. Wilski, E. Gaube, and S. Rosinger, *Kerntechnik, 5*: 281 (1963).

36. H. Wilski, *Kunststoffe, 53*: 862 (1963).

37. I. Kuriyama, N. Hayakawa, Y. Nakase, J. Ogura, H. Yagyu, and K. Kasai, *IEEE Trans. Electr. Insul., EI-14*: 272 (1979).

38. G. D. Mendenhall, *Agnew Chem. Int. Ed. Engl., 16*: 225 (1977).

39. G. D. Mendenhall, H. K. Agarwal, J. M. Cooke, and T. S. Dziemianowicz, in *Polymer Stabilization and Degradation*, ACS Symposium Series No. 280 (P. P. Klemchuk, ed.), American Chemical Society, Washington, D.C., p. 373 (1985).

40. F. Yoshii, T. Sasaki, K. Makuuchi, and N. Tamura, *J. Appl. Polym. Sci., 31*: 1343 (1986).

41. M. R. Kamal and F. H. Moy, *J. Appl. Polym. Sci., 28*: 1787 (1983).

42. E. Kramer and J. Koppelmann, *Polym. Degrad. Stab., 14*: 333 (1986).

43. T. Seguchi and K. Arakawa, *Japanese Atomic Energy Research Institute Publication JAERI-M 9671* (1981). In Japanese

44. R. L. Clough, K. T. Gillen, and C. A. Quintana, *J. Polym. Sci. Polym. Chem. Ed., 23*: 359 (1985).

45. R. D. Prottas, M. T. Shaw, and J. F. Johnson, *Trans. Am. Nucl. Soc., 46*: 365 (1984).

46. S. Matsumoto, *J. Polym. Sci. Polym. Chem. Ed., 21*: 557 (1983).

47. P. D. Calvert and N. C. Billingham, *J. Appl. Polym. Sci., 24*: 357 (1979).

48. N. C. Billingham and P. D. Calvert, in *Developments in Polymer Stabilisation*, Vol. 3 (G. Scott, ed.), Applied Science Publishers, London, p. 139 (1980).

49. R. Shriner, R. Fuson, and D. Curtin, *The Systematic Identification of Organic Compounds*, 5th ed., Wiley, New York, pp. 117, 173 (1964).

50. J. I. Goldstein, D. E. Newbury, P. Echlin, D. C. Joy, C. Fiora, and E. Lifshin, *Scanning Electron Microscopy and X-Ray Microanalysis*, Plenum, New York (1981).

51. T. Seguchi, S. Hashimoto, K. Arakawa, N. Hayakawa, W. Kawakami, and I. Kuriyama, *Radiat. Phys. Chem., 17*: 195 (1981).

52. D. L. Faulkner, *J. Appl. Polym. Sci., 31*: 2129 (1986).

53. K. T. Gillen and R. L. Clough, *J. Polym. Sci. Polym. Chem. Ed., 23*: 2683 (1985).

54. R. L. Clough and K. T. Gillen, *J. Polym. Sci. Polym. Chem. Ed., 19*: 2041 (1981).

7

Flammability of Elastomeric Materials

David F. Lawson
Central Research Laboratories
The Firestone Tire & Rubber Company
Akron, Ohio

INTRODUCTION

Elastomers fit and serve numerous functions very well. Some of the largest volume applications of rubber—e.g., tires, gaskets, seals, footwear—do not normally require flame resistance. However, elastomers, like all organic substances, can burn, and some of their applications require that special attention be paid to their flammability properties and to the selection of flame-resistant materials. It is the intent of this chapter to summarize the major flammability characteristics of various elastomers, the means and mechanisms that have been used for inhibiting flammability, reference sources for selected applications and formulations for flame-resistant elastomers, tests for flammability, and issues of smoke and combustion toxicity. This is done to enable the reader to better assess elastomer flammability when selecting materials for a specific application.

Typically, the flammability of a material is expressed in terms of its ease of ignition, its surface flame spread, and the amounts of heat, fuel, smoke, and toxic gases that are produced during burning, as well as the rate at which these quantities are generated. A number of fire response tests have been designed to obtain relative numerical values for these characteristics under strictly specified test conditions. The numerical flame spread ratings, burning extents, burning rates, and oxygen indices found in this chapter are the results of controlled laboratory tests. These results are not intended to represent hazards presented by any material under actual fire conditions. The reader must determine whether a given material satisfies the requirements of a specific application.

ELASTOMER PYROLYSIS AND COMBUSTION

The combustion of an organic polymer involves chemical and physical processes which occur both within the polymer and in the surrounding gas phase. Fabris and Sommer [1, 2] have summarized the combustion models of Buck [3], Van Krevelen [4], and others, using Scheme 1.

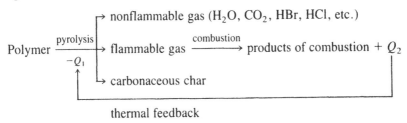

Scheme 1 Processes of combustion of an organic polymer. Q_1, Q_2 = heat. (From Ref. 1 by permission of the Division of Rubber Chemistry, Inc., American Chemical Society, Akron, Ohio.)

Several interdependent factors control burning. For example, the nature and rate of formation of pyrolyzates (which comprise the flammable gas) determine to a large extent heat output Q_2 and rate of burning, while the pyrolyzates themselves are affected greatly by the heat input Q_1 and the composition of the polymer. Since it has several stages, the combustion process can be interrupted and studied from various points. Prior studies of thermal decomposition, combustion, and inhibition of elastomers and various other polymers have been reviewed [1, 2, 5–9]. The book by Cullis and Hirschler [5] provides a comprehensive review of polymer combustion in general, particularly the mechanisms of thermal decomposition, combustion, and inhibition. The review articles by Fabris and Sommer [1] and Lawson [9] are directed specifically toward elastomers. The following discussion will take up separately the topics of polymer decomposition, or pyrolysis, and combustion for several major classes of elastomer.

Hydrocarbon Rubbers and Nitrile Rubber

This section considers ethylene–propylene copolymers (EPM) and terpolymers (EPDM); polyisobutylene (IM) and butyl rubber (IIR); natural rubber and synthetic polyisoprenes (NR and IR); polybutadiene (BR); and copolymers of butadiene with styrene (SBR). Because of certain similarities in their decomposition and burning characteristics, butadiene–acrylonitrile copolymers (NBR) are considered with this group as well. Fabris and Sommer [1] have reviewed aspects of the thermal degradation of these elastomers.

In the pyrolysis of ethylene–propylene copolymers Wall and Straus [10] found that the thermal stability of the polymers decreased as the content of propylene units increased. Thermal stability was still further reduced by quaternary carbon atoms in the polymer chain. Volatilization is nearly complete at 440°C for polyethylene, 400°C for polypropylene, and 350°C for polyisobutylene after heating for 30 min (Figure 1) [11].

An analysis of the volatiles from the thermal degradation of polyethylene, polypropylene, and polyisobutylene revealed an increasing tendency to revert to monomer in the above order at 500 and 800°C. At 1200°C, however, the order appears to become reversed (Table 1) [12–14].

The relative stabilities of various butadiene and isoprene polymers are compared in Figure 2. Thermal degradation in air starts between 280 and 340°C, with the polyisoprenes occupying the lower end of this range. In all cases the entire sample was

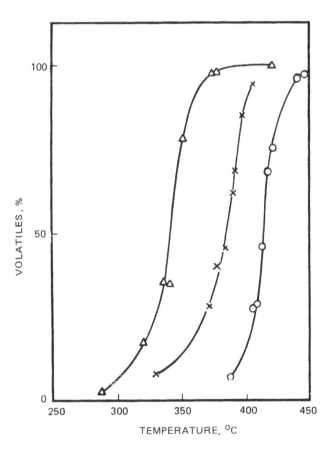

Figure 1 Pyrolysis of polyethylene (○), polypropylene (×), and polyisobutylene (△). Heating time for each experiment 30 min. (Data from Ref. 11. From Ref. 1 by permission of the Division of Rubber Chemistry, Inc., American Chemical Society, Akron, Ohio.)

Table 1 Monomer Yields from Pyrolysis of
Polyethylene (PE), Polypropylene (PP), and
Polyisobutylene (IM)

Temperature (°C)	% Monomer		
	PE	PP	IM
500		0.4	35–38
800	5.5	18	65–73
1200	26	13–18	11–15

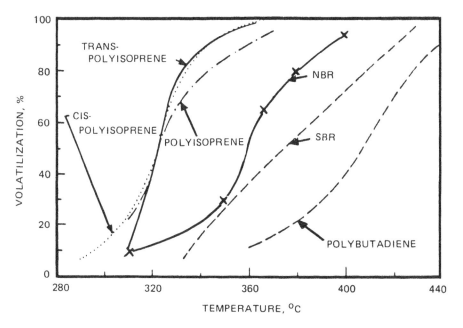

Figure 2 Relative thermal stabilities of diene rubbers. Ordinate is percentage volatilized in 30 min at the indicated temperature. (From Ref. 2.)

consumed below 450°C [15–17]. Even though the butadiene polymers appear more stable than the polyisoprenes, the differences may be slight at the temperature range of ignition and burning. The surface temperature at which volatiles were emitted during thermal decomposition of a variety of rubbers at rates sufficient for ignition by a pilot flame varied between 325 and 500°C [18]. This temperature range is broad and very dependent on the specific experimental conditions. The lower limit corresponds to the onset of thermal degradation as indicated in Figure 2. Autoignition took place at a much higher surface temperature, 620–670°C.

Thermal degradation of elastomers takes place at moderate temperatures and is accompanied by the evolution of volatile low molecular weight hydrocarbons. These are

Table 2 Heats of Combustion of Polymers

Polymer	$-\Delta H$ (kJ/g)	Polymer	$-\Delta H$ (kJ/g)
Rubber	40	Cotton	17
Polyisobutylene	47	Polyformaldehyde	17
Polypropylene	46	Poly(vinyl chloride)	18
Polystyrene	41	Poly(methyl methacrylate)	25
ABS	36	Polyester resin	18

capable of generating large quantities of heat on burning. A comparison of specific heats of combustion for hydrocarbon polymers and polymers also containing hetero-elements is given in Table 2.

Several groups have attempted to link classical condensed-phase pyrolysis studies with actual combustion conditions. Tkac has used electron spin resonance (ESR) to detect volatile free radicals evolved from polymers during pyrolyses at 513–623 K under high vacuum [19]. By trapping the radicals in a matrix at 78 K, energy transfer and oxidation to diamagnetic products were minimized. Substituted alkyl, vinyl, and allyl diradicals were detected in the volatile pyrolyzates from natural rubber, *cis*-polyisoprene, polypropylene, and polystyrene. Other polymers and radicals were also detected in the residues. It was suggested that the rate of reactive radical formation is a primary determinant of the burning behavior. Smith and Youren found that the products of pyrolysis of ethylene–propylene–diene terpolymer at 770–870 K were those predicted to arise from 1:5, 1:5:9, and 1:3 H-transfer and some beta-scission of polymer radicals formed initially by random scission along the backbone [20]. At higher temperatures (>900 K) secondary products arise from the cracking of high molecular weight pyrolyzates. Further fragmentation, cyclization, and aromatization occur at still higher temperatures (>1000 K). Butyl rubber was found to form isobutylene at 770–870 K, primarily via stepwise unimolecular elimination from randomly broken polymer chains.

Polydienes also undergo random chain scission, followed by H-transfer and elimination to form low molecular weight products. However, extensive crosslinking and cyclization via free-radical addition occur during pyrolysis, and the solid polymers are subject to oxidation at pyrolysis temperatures [21, 22]. Thermal cyclization reactions of polydienes have been reviewed [23].

In research by Stuetz and co-workers [24, 25], "polymer flammability" was regarded as a variable property, dependent on the intrinsic combustibility of the polymer, as well as extrinsic factors such as orientation and oxygen supply. "Intrinsic combustibility," as determined from the minimum oxygen concentration required for self-sustained combustion, is characteristic of polymer structure and was found to correlate with thermooxidative stability. Flame quenching was found to be related to the rate of diffusion of oxygen into the product effluent stream in polyethylene and polypropylene. Exothermic reactions of oxygen at the polymer surface were implicated in the formation of combustion fuels [24].

Secondary cool flames (FFS) occurring during the combustion of poly(1-butene), poly(4-methyl-1-pentene), and polypropylene were found by Baillet, Delfosse, and Lucquin [26] to be due to reactions of oxygen with tertiary C–H in the polymers. Crosslinking of EPM had the effect of decreasing the availability of tertiary C–H and inhibiting FFS. Brauman reported that although surface oxidation occurs during thermal decomposition in air, its effect on the rate of burning is unimportant once steady-state burning is achieved [27]. Using a vertical, top-burning, driven-rod configuration, she found the pyrolytic regression rates for polystyrene and a polyester thermoset to be independent of the atmosphere. However, the rate of attainment of steady-state was oxygen dependent. It may be concluded that reactive C–H units, such as tertiary and allylic structures and branch points, appear to provide sites for exothermic oxidation, which drives fuel formation. However, other factors such as the burning configuration and physical state of the polymer also influence the burning process.

When rubber is heated excessively under adiabatic conditions in the presence of air, exothermic processes occur, which can lead to rapid self-heating of the mass and eventually to self-ignition. For example, in experiments where SBR was heated above 300°C, Emmons showed that exothermic processes dominate and the sample was capable of self-ignition if it was insulated [28]. Spontaneous ignition would not occur if the rubber was not heated beyond 220°C (Figure 3).

Brauman has also shown that condensed-phase charring correlates with reduced fuel production, lower burning rates, and reduced smoke formation [29]. Polybutadiene, IR, and high vinyl polybutadiene were characterized as high-volatile-loss, low-char polymers relative to polymers containing acrylonitrile, halogen, or main-chain aromatic groups (Figure 4). Polychloroprene (CR) forms a char which is stable below 623 K in nitrogen. Although the elemental composition of chars seems to be important, Brauman was unable to find definitive structural criteria for chars relating to fire behavior [30].

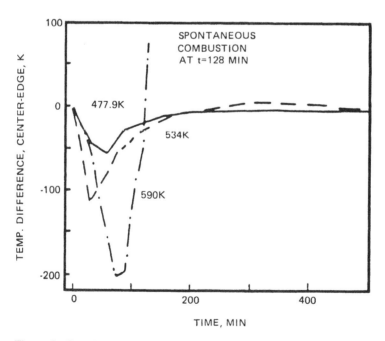

Figure 3 Experimental heating curves of SBR blocks at three furnace temperatures, showing temperature differential between center and edge of block, and exotherm at higher temperatures [28]. (Adapted from Ref. 28.)

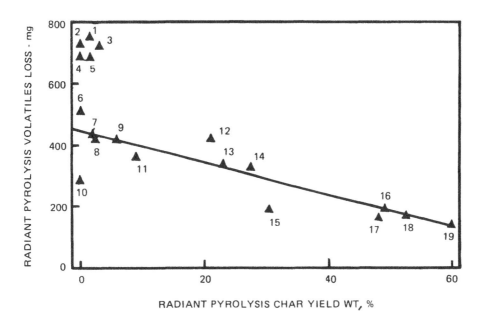

Figure 4 Relationship between char yield and fuel production during radiant pyrolysis of various polymers: (1) high-vinyl polybutadiene; (2) foamed polyurethane; (3) poly(cis-1,4-butadiene); (4) polyisoprene; (5) unsaturated thermoset polyester A; (6) unsaturated thermoset polyester B; (7) unsaturated thermoset polyester C; (8) impact polystyrene; (9) polydivinylbenzene; (10) polypropylene; (11) polyethylene terephthalate; (12) acrylonitrile–vinyl acetate copolymer; (13) polycarbonate; (14) polyphenylene oxide; (15) polyethersulfone; (16) poly(amide-imide); (17) polyimide A; (18) polyphenylene sulfide; (19) polyimide B15. (From Ref. 29 by permission of Technomic Publishing Co., Lancaster, PA.)

Carbon-Chain Elastomers with Noncarbon Substituents

The pyrolysis and combustion of these polymers have been reviewed by Fabris and Sommer [1]:

1. *Polychloroprene.*—Polychloroprene, as a consequence of its chlorine content (40%) and its tendency to form substantial amounts of crosslinked char, inherently burns less readily than the hydrocarbon rubbers. This superiority in flame resistance has made polychloroprene the material of choice in many applications demanding reduced flammability [31].

The thermal degradation of polychloroprene was studied by a number of researchers [32–34]. Dynamic thermogravimetric analysis of chloroprene rubber in air showed the main weight loss between 290 and 400°C, during which nearly complete dehydrochlorination takes place [33–35]. An exotherm found in the same temperature region was interpreted as oxidation [34] and crosslinking [36] (Figure 5).

Cured CR gum by itself has an oxygen index of 26–31 (see section on Fire Testing for a description of oxygen index, OI, and its significance). Fillers are usually incorporated, and their choice is critical for the retention of flame-retardant behavior.

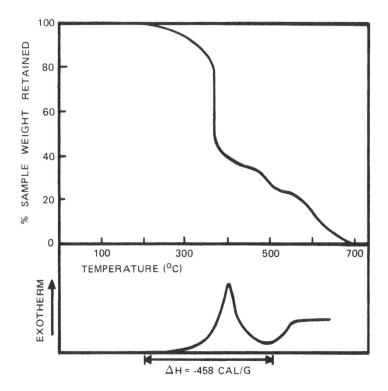

Figure 5   TGA and DSC curves for raw polychloroprene (CR). Heating rate 50°C/min. Top: TGA in air; bottom: DSC in air. (From Ref. 1 by permission of the Division of Rubber Chemistry, Inc., American Chemical Society, Akron, Ohio.)

Hilado and Casey reported that chlorinated elastomers, including CR, have a greater tendency to char than do IR, SBR, NBR, and EPDM [37]. Fabris and Sommer continue [1]:

2.   *Chlorosulfonated polyethylene.*—Treatment of polyethylene with a mixture of sulfur dioxide and chlorine introduces some sulfonyl chloride groups into the chain. A typical elastomer contains about 25–40% chlorine and 1–2% sulfonyl chloride groups. Polymers with chlorine contents of 35% or higher are difficult to ignite under atmospheric conditions even though they burn and char while in contact with a flame. The complex, stepwise degradation pattern shown in the thermogravimetric traces (Figure 6) is probably due to the loss of sulfur and hydrogen chloride, followed by main-chain degradation [38].

3.   *Chlorinated polyethylene.*—Chlorinated polyethylene is available in two grades that differ in chlorine content (36 and 45%, Dow Chemical Co.) [39]. The polymer is generally dehydrochlorinated to the level of unsaturation desired for curing by heating with zinc oxide.

Thermal degradation patterns of the raw polymers are very similar to those of chlorosulfonated polyethylene (Figure 6) or poly(vinyl chloride) [40]. The dehydrochlorination rates in nitrogen or oxygen at 150 and 180°C were found to be considerably faster for the polymer with the higher chlorine content [41]. The effect appears more pro-

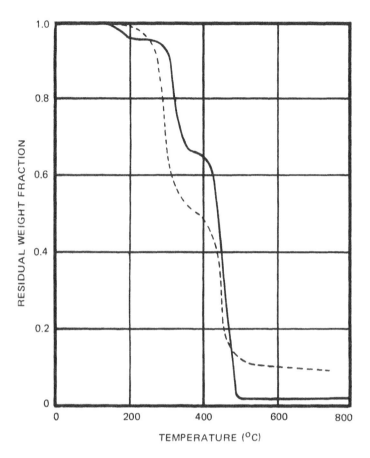

RESIDUAL WEIGHT FRACTION

TEMPERATURE (°C)

Figure 6 TGA trace for chlorosulfonated polyethylene (solid curve) and chlorinated polyethylene (dashed curve). (Data from Refs. 38 and 40. From Ref. 2.)

nounced in the inert (N_2) atmosphere (Figure 7). Compounding with antimony oxide causes a short induction period in the thermal degradation process, followed by a fast dehydrochlorination (Figure 8).

The high chlorine content of these elastomers is generally sufficient to pass all but the most severe flammability tests. As with all polymers whose reduced flammability derives from their capability to release hydrogen chloride, carbonate fillers will tend to reduce the flame resistance.

4. *Fluorocarbon elastomers.*—Highly fluorinated elastomers (e.g., hexafluoropropylene–vinylidene fluoride copolymers) are known for their relatively high resistance to thermal oxidation [42, 43]. This resistance is primarily due to: (1) the extreme stability of the carbon–fluorine bond and (2) the shielding of the carbon–carbon bond by the fluorine atoms from attack by oxygen.

Figure 9 shows the weight loss of fluorocarbon polymers on heating [44]. Lowering the content of chlorotrifluoroethylene in its copolymers with vinylidene fluoride (Kel-F) slightly shifts the thermal decomposition curves toward higher temperatures. The chlorine-free vinylidene fluoride–hexafluoropropylene copolymers (Viton A, Fluorel)

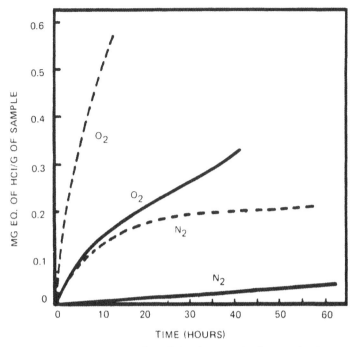

Figure 7 Dehydrochlorination of chlorinated polyethylenes in nitrogen and oxygen atmospheres at 150°C. Chlorine content: 35.2% (solid curves); 45.2% (dashed curves). (From Ref. 1 by permission of the Division of Rubber Chemistry, Inc., American Chemical Society, Akron, Ohio.)

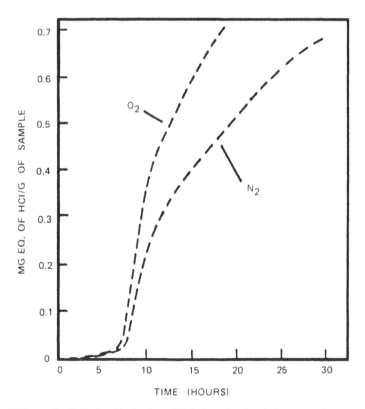

Figure 8 Dehydrochlorination of chlorinated polyethylene containing 5 phr Sb_2O_3. (From Ref. 1 by permission of the Division of Rubber Chemistry, Inc., American Chemical Society, Akron, Ohio.)

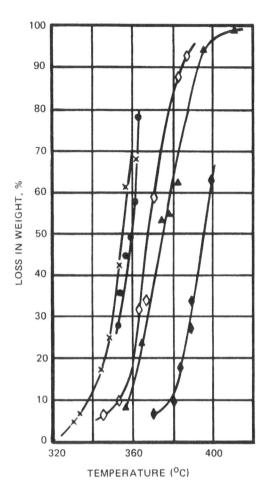

Figure 9 Weight loss of various fluorine-containing polymers after heating under vacuum (each point represents a fresh sample heated for 2 hr): polychlorotrifluoroethylene (×); copolymers of chlorotrifluoroethylene and vinylidene fluoride with mole ratios of 64:36 (●), 45:55 (◇), and 23:73 (▲); and copolymer of hexafluoropropylene and vinylidene fluoride (◆). (Data from Ref. 45. From Ref. 2.)

show the highest stability. The latter elastomers remain virtually unaffected when exposed to temperatures of 120–150°C over extended periods of time and in 28 days at 232°C lose only 10% of their original tensile strength.

Fluoroelastomers are reported not to support combustion [45 47]. Lead oxide cured hexafluoropropylene tetrafluoroethylene copolymer (Viton B) has an oxygen index of 100. The oxygen index values are extremely sensitive to compounding variables. Curing with MgO lowers the OI to 55, while a lead oxide-cured, calcium carbonate-filled compound had an OI of only 42 [48]. Evaluations at NASA showed Fluorel L-3206-6 to have excellent resistance to ignition under the most severe conditions encountered in space exploration [43, 49]. The materials also meet the specific requirements for low odor and toxicity of off-gases [50, 51].

Differing opinions have been offered about the effects of low molecular weight fluorocarbons and HF as components of the gaseous combustion products.

Heterochain Elastomers

Polyurethanes. Polyurethanes are extremely versatile materials, made into solid elastomers, coatings, adhesives, fibers, and flexible foams. The main emphasis in flammability has been focused on cellular materials, due to their use in large-volume applications such as mattresses, seating, and rug and carpet underlay. This is a vast field of research, and only a brief summary can be given here. For more comprehensive treatments of the field, the reader should consult reviews by Papa [52], Babiec and co-workers [53], Tilley and co-workers [54], and Fabris [51]. Fabris and Sommer have described the thermal degradation of polyurethane elastomers [1]:

> Polyurethane elastomers contain a variety of different chemical bonds. In addition, the thermal decomposition pattern is affected by additives and by the presence of catalysts, and can be extremely complex. About two-thirds of a flexible polyurethane foam consists of polyol, in the majority of cases either poly(propylene ether) diol or triol. The pyrolysis products from this portion of the polymer can be expected to be identical to those obtained from polypropylene oxide rubbers (Table 3) [55]. Rate of fuel production from polyether triol and the corresponding urethane foams under dynamic conditions is shown in the thermograms of Figure 10.

> Hileman and co-workers [56] analyzed the products of pyrolysis in inert atmosphere of a commercial, flexible polyether urethane foam. Secondary products derived from the isocyanate portion of the urethane during pyrolysis in nitrogen at 200–300°C were collected as a yellow distillate. Further pyrolysis of this distillate at higher temperature (800–1000°C) gave a variety of nitrogen-containing compounds (acetonitrile, acrylonitrile, benzonitrile, pyridine, and hydrogen cyanide) [57]. Similar results were obtained by Paabo and Comeford [58] and Napier and Wong [59] in oxidizing atmospheres. Little, if any, char formation is generally observed during pyrolysis or combustion of flexible polyurethane foams.

Table 3 Compounds Identified in the Pyrolysis
Gases of Isotactic Poly(propylene Oxide)

Ethylene
Ethane
Acetaldehyde (major)
Propylene oxide
Acetone
1-Hydroxypropylene oxide-1,2
Isopropanol
Methyl ethyl ketone
Methyl isopropyl ether
Ethyl isopropyl ether
Dipropyl ether

Source: Ref. 55.

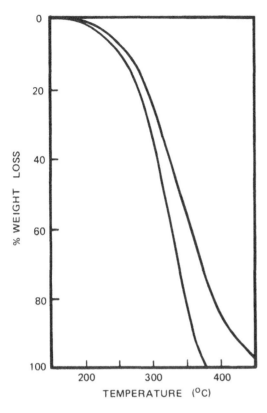

Figure 10 TGA of poly(propylene ether) triol, MW 3000 (left) and polyether urethane foam derived from it (right). Atmosphere: air; heating rate: 10°C/min. (From Ref. 1 by permission of the Division of Rubber Chemistry, Inc., American Chemical Society, Akron, Ohio.)

Thermal oxidative degradation of urethanes made from poly(propylene ether) and toluene diisocyanate in air was studied by Ingham and Rapp [60]. The authors concluded that degradation of the polypropylene oxide chain in nitrogen takes place between 250 and 320°C, predominantly by an unzipping mechanism. Backus and co-workers [61] arrived at similar conclusions from thermal analysis in air. Again, a major weight loss was apparent at 250–350°C, which can be interpreted as degradation and volatilization of polyether.

Voorhees et al. [62] proposed a decomposition mechanism for a TDI–PPO polyurethane foam which involves both elimination of polyol and cracking at the urethane linkage via a cyclic transition state (Figure 11). Additional products are formed by further decomposition of these intermediates. Other studies on the pyrolysis of polyurethanes have been reported [63, 64].

Eaves and Keen [65] showed that when flexible foam was heated to 250–350°C, a self-propagating degradation reaction occurred in the foam interior which could result in ignition upon reaching the surface. The stability of the foam toward heating increased with the age of the foam. A model for the smoldering combustion of polyurethane foam was proposed by Rogers et al. (Scheme 2) [66]. In the absence of air, tar and gases are formed by pyrolysis steps 2 and 4. In air, the cell structure essential to smoldering is

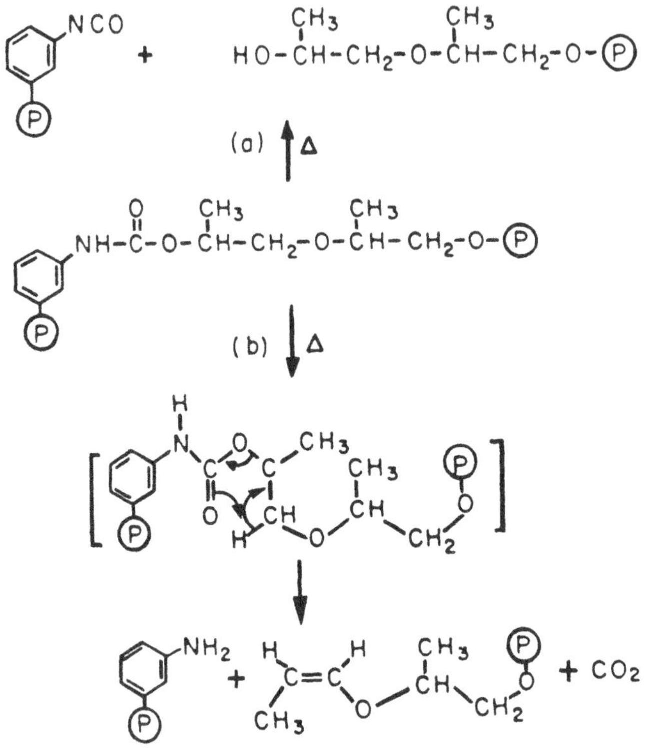

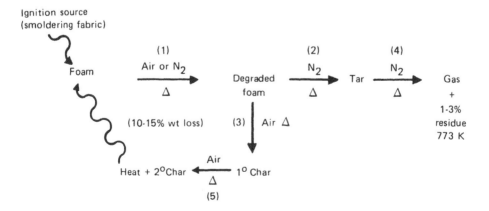

Figure 11 Proposed initiation of polyurethane thermal decomposition [62]. (From Ref. 9 by permission of the Division of Rubber Chemistry, Inc., American Chemical Society, Akron, Ohio.)

Scheme 2 Model for the smoldering of polyurethane foam. (Data from Ref. 66. From Ref. 9 by permission of the Division of Rubber Chemistry, Inc., American Chemical Society, Akron, Ohio.)

retained as a primary char forms (step 3). If sufficiently exothermic, heat from the condensed-phase combustion of the primary char (step 5) drives the smolder wave. Mechanisms of smoldering in other materials have been reviewed [5, 67].

Continuing with the summaries of Fabris and Sommer [1]:

2. *Epichlorohydrin elastomers.*—Polyepichlorohydrin (ECO) elastomers are available as amorphous homopolymers as well as copolymers with ethylene oxide, and have been used wherever the combination of low-temperature flexibility, ozone resistance, and oil stability is required. The chlorine content of 38% is responsible for the rubber's inherent flame resistance, which has been rated as slightly inferior to that of CR [68].

3. *Silicone elastomers.*—The main advantages of silicone rubbers are the wide temperature range of their serviceability (-110 to $310°C$) and their excellent chemical and weathering resistance. Improvements in thermal stability and flame retardance were achieved by material replacement of the methyl groups in the original poly(dimethylsiloxane) with phenyl [69] or vinyl groups [70], or by incorporation of carborane units into the polymer chain [71, 72].

Thermogravimetric analysis of vulcanized silicone rubbers showed that decomposition started at about $300°C$ and was virtually complete at $400°C$. The residue was approximately 55% of the original sample weight and consisted essentially of silicic acid.

Silicone rubbers do ignite, but their combustion is characterized by a high production of char, with no flaming drip. The average char obtained on exposing a selected silicone elastomer to an arc jet was 90% of the original polymer weight [73]. The combustion of low molecular weight silicones was studied by Lipowitz [74]. Flame-resistant silicone compounds have been described in which the flammability is observed to increase with increasing humidity, especially in the case of phenyl-substituted silicones [75]. Halogenated organic compounds, metal oxides, dibutyltin dichloride, and platinum or platinum compounds have all been claimed to further reduce the burning tendency of silicone rubbers [77–85].

4. *Polyphosphazenes.*—Poly(organophosphazenes) are a relatively new class of high-performance polymers with a number of specialty applications [86]. The inorganic —P=N— backbone contributes a high degree of thermal stability and flame resistance to these polymers, depending on the structure of the ester [87]. Best known are phosphazene polymers in which the phosphorus carries aryloxy-(I) or fluoroalkoxy-(II) substituents:

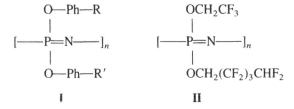

The flame and smoke properties of two series of mixed poly(aryloxyphosphazene) elastomers (I with R and R' alkyl or alkoxy) were reported by Quinn and Dieck [88]. They found OI's between 23 and 27, with OI depressed by higher alkyl groups or the presence of alkoxy groups on the aromatic rings.

The use of fillers has a remarkable effect on further increasing the flame resistance, resulting in oxygen indices of 30 to 40 and higher, and reducing smoke generation [87–

91]. When burned, the nonhalogenated esters are reported to form combustion gases of relatively low toxicity [92].

Fabris and Sommer [1] continue:

5. *Nitroso and triazine elastomers.*—Nitroso and triazine elastomers were developed and investigated primarily by the U.S. Air Force and NASA as specialty materials for spacecraft and high-velocity aircraft [45]. The nitroso rubbers (III) are copolymers of trifluoronitrosomethane with tetrafluoroethylene or trifluoroethylene:

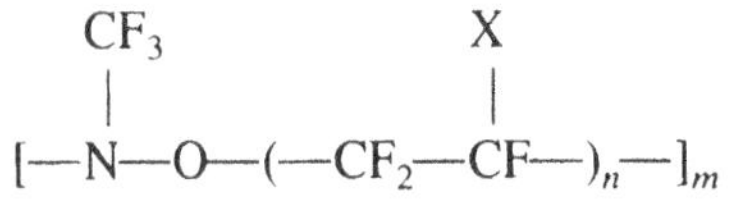

$$
\begin{array}{cc}
CF_3 & X \\
| & | \\
[-N-O-(-CF_2-CF-)_n-]_m
\end{array}
$$

III

where X = H or F. The copolymer with X = F is the more stable, undergoing a major weight loss in the region of 240–280°C [93, 94]. These polymers showed high resistance to chemical attack, but were curable only with polyamines to give relatively weak vulcanizates.

Carboxynitroso rubbers (IV) are made by the copolymerization of trifluoronitrosomethane with tetrafluoroethylene and perfluoronitrosobutyric acid or its esters:

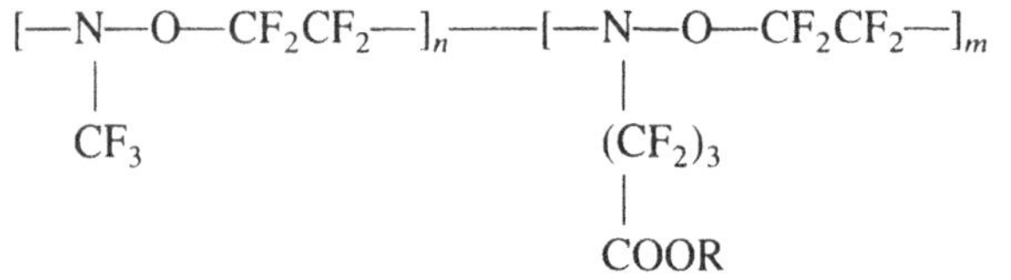

$$
\begin{array}{cc}
[-N-O-CF_2CF_2-]_n \text{———} [-N-O-CF_2CF_2-]_m \\
| \qquad\qquad\qquad\qquad | \\
CF_3 \qquad\qquad\qquad (CF_2)_3 \\
\qquad\qquad\qquad\qquad | \\
\qquad\qquad\qquad\qquad COOR
\end{array}
$$

IV

where R = H or alkyl, $n = 99$, and $m = 1$. Unlike the nitroso rubbers, they are curable with metal oxides, epoxides, and salts of fluorocarboxylic and other organic acids. Like the nitroso rubbers, the carboxynitroso rubbers also decompose at relatively low temperatures. However, the volatiles resist ignition, and the elastomers were extremely resistant to burning even in oxygen-enriched, hyperbaric atmospheres. The toxicity of the volatiles has been questioned.

Triazine elastomers (V) have been described by Graham [95] and by Dorfman and co-workers [96]. The materials consist of triazine rings separated by perfluorinated polymethylene groups and carrying only perfluoroalkyl substituents:

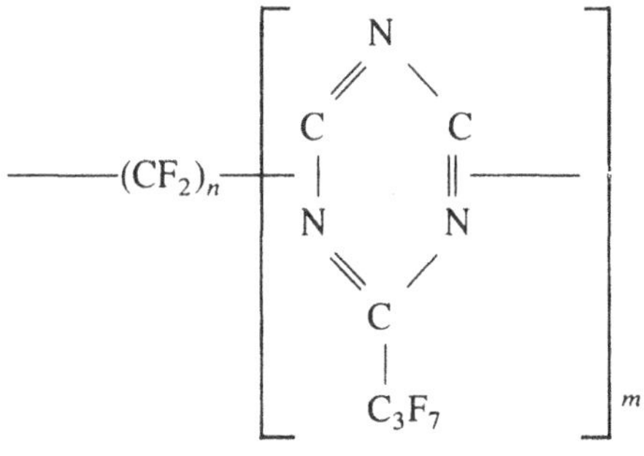

$$
-(CF_2)_n-
\begin{bmatrix}
& N & \\
& \diagup\diagdown & \\
C & & C \\
| & & \| \\
N & & N \\
& \diagdown\diagup & \\
& C & \\
& | & \\
& C_3F_7 &
\end{bmatrix}_m
$$

V

These polymers have outstanding thermal stability and retain their elastomeric character after exposure to 260–430°C. Elastomers III–V must be considered highly experimental at this time.

INHIBITION

Most nonhalogenated hydrocarbon-based elastomers can be ignited and will burn with much heat unless the pyrolysis–combustion cycle is interrupted [1]. Culverhouse suggested eight steps in a hypothetical pyrolysis–combustion cycle at which the process may be inhibited [97]:

1. Heat input to the polymer
2. Temperature rise
3. Emission of dispersed-phase products
4. Formation of a combustible mixture with oxygen
5. Ignition
6. Flaming
7. Flame spread
8. Afterglow and solid-phase combustion

Most of the carbon-chain elastomers use similar inhibitor systems. Table 4 lists a number of flame-retardant additives used most often in elastomers. Inhibition is achieved by modification of either condensed-phase or dispersed-phase processes. Research has shed light on mechanisms by which several of the more important flame-retardant systems operate.

Alumina Trihydrate [Al(OH)$_3$, ATH] and Other Fillers

At high loadings (> 20 wt %), mass dilution and/or endothermal dehydration is thought to slow the buildup of heat, favoring char-forming processes and reducing flammability and smoke generation in polydienes [98]. Although the effect is predominantly physical, evidence has been found for catalytic activity as well [99]. Antia, Cullis, and Hirschler [100] and Hirschler and Tsika [101] have reported a synergism between Al(III) oxide, as well as its mono- and trihydrates, and decabromobiphenyl (DBBP) in acrylonitrile–butadiene–styrene polymer (ABS). The process is largely confined to the condensed phase, where the alumina compound apparently catalyzes the degradation of DBBP and release of HBr. Chalabi, Cullis, and Hirschler [102, 103] have also reported that, at less

Table 4 Flame-Retardant Additives Commonly Used in Elastomers

Organic	Inorganic
Chlorinated paraffins	Al(OH)$_3$, Mg(OH)$_2$, MgCO$_3$
Polychlorinated alicyclics	Sb$_2$O$_3$, Fe$_2$O$_3$, ZnO
Brominated aromatics, particularly ethers	Clay, silica, CaCO$_3$
Organohalophosphates	Zn borate
Organophosphates and phosphites	NH$_4$ phosphates, halides

than 20 phr, pyrogenic silica is superior to $Al(OH)_3$ for smoke suppression in polystyrene. A "skin" is formed, comprised of silica particles bonded by crosslinked styrene oligomers. The effect is related to the surface area of the filler, with 10- to 20-fold increases in the amount of carbonaceous residue observed.

Calcium and magnesium carbonate and zinc oxide are frequently used as fillers in all types of polymers, especially elastomers. In halogenated systems, these fillers can depress OI by absorbing hydrogen halides, which function as flame inhibitors. Clays, silicas, and other minerals are also commonly used as flame-retardant fillers in elastomers.

Antimony Oxide and Other Metal Compounds

Antimony oxide (Sb_2O_3) is widely used in elastomers, chiefly in combination with halogen-containing compounds. It synergizes with the halogens in gas phase reactions by forming a volatile Sb trihalide (reaction 1), but also engages in solid phase reactions.

$$Sb_2O_3 + 6HX \rightarrow 2SbX_3 + 3H_2O, \qquad X = Cl, Br \tag{1}$$

Mechanistic studies of antimony–halogen chemistry have been reviewed elsewhere [104, 105]. Ohe, Takiyama, and Takamoto report that a high degree of synergism exists between Sb_2O_3 and chlorine in CR or poly(vinyl chloride) (PVC), at ratios of 1.30–2.35 mol $Sb_2O_3/100$ mol Cl [106]. Aging effects have been observed by Clough in EPDM compounds containing Sb_2O_3 and chlorinated paraffin waxes [107]. Flammability increased as 50% of the Sb_2O_3 and 20% of the Cl were lost after aging at 383 K for four months. No appreciable losses on aging were detected in EPDM which contained cyclic chlorinated hydrocarbons and Sb_2O_3, due presumably to the greater stability of the cyclic compounds toward HCl loss. Chlorosulfonated polyethylene (CSPE) lost both Sb and Cl but became less flammable due to the concurrent loss of other volatile flammable substances. Brauman used Sb_2O_3 intentionally to enhance the formation of char in styrene polymers containing chloromethyl groups [108]. Although antimony compounds generally increase smoke in rubbers containing chlorinated paraffins, they have variable effects on smoke in other systems.

Several metal compounds synergize with halogens for flame retardation and smoke suppression. Ferrocene [109] and other compounds of iron [110] have been used in PVC and rubber compounds to enhance the effectiveness of halogens. Whelan has shown that the oxygen index of a 50:50 NBR/PVC is steadily increased with increasing concentrations of Fe_2O_3 in NBR containing chlorinated paraffin (Tables 5 and 6) [111]. Lecomte,

Table 5 Effect of Fe_2O_3 on the Oxygen Index of 50:50 NBR/PVC

Fe_2O_3 (pph)	Halogen/metal		OI
	pph	Atomic ratio	
0	28/0		29.3
1.3	28/0.909	48	31.3
2.5	28/1.75	25	32.8
5.0	28/3.50	13	35.9

Source: Ref. 111.

Bert, Michel, and Guyot [112] proposed a mechanism to explain flame retardation and smoke suppression by ferrocene (Cp_2Fe) in PVC, which should also be relevant to Fe_2O_3 and other iron compounds. $FeCl_2$ is formed at temperatures below 300°C and is believed to crosslink the polymer and suppress benzene formation during heating by decreasing the length and mobility of polyene sequences, possibly through transfer of the triplet state energy of activated pi-systems. Also observed below 300°C are the Lewis acids ferricenium chloride and ferric chloride, which can catalyze dehydrochlorination [113], while alpha-Fe_2O_3 and gamma-$FeO(OH)$, observed above 500°C, are thought to be oxidation catalysts.

Molybdenum oxides and salts also function as flame retardants and smoke suppressants in halogen-containing elastomers [114, 115]. Unlike Sb_2O_3, MoO_3 is retained largely in the condensed phase [116], where studies in PVC indicate it may be involved in precombustion processes, either by suppressing benzene formation [117] or by catalyzing "reductive coupling" of the degrading polymer to promote extensive intermolecular crosslinking to give char [118].

Kroenke discusses the mechanisms of a wide variety of metal additives which function as smoke suppressants in PVC [119]. Positive synergistic interactions have been established for Fe(III) oxide and Mo(VI) oxide [120] and for a number of metal chelates [121] used with halogens in polyolefins.

From thermal analyses of polyolefin compounds, Cullis and Hirschler report that metal chelates and organic chlorine compounds enhanced carbonization in excess of their mass contribution [122]. They classified the metal chelates by function: (1) pro-oxidants, e.g., $Co(acac)_2$ and $Zn(acac)_2$, which favor interchain crosslinking, and (2) antioxidants, e.g., $Ni(acac)_2$ and copper stearate, which favor intrachain cleavage. Both processes reportedly lead to char formation.

Halogens

The use of halogens to impart flame resistance to elastomers is well established [1]. Polychloroprene rubber (CR), chlorinated polyethylene (CPE), and CSPE rubber (CSPE) are selected for flame-retardant uses largely because chlorine comprises much of their

Table 6 Comparative Effects of Fe_2O_3 and Sb_2O_3 on Oxygen Index of NBR Containing 16 pph of Chlorinated Paraffin

Nitrile content	Metal oxide (pph)	Halogen/metal		OI
		pph	Atomic ratio	
Medium		11.2/0		21.5
	Sb_2O_3 (10)	11.2/8.3	4.6	23.3
	Fe_2O_3 (4.2)	11.2/2.94	6.0	24.1
High		11.2/0		22.5
	Sb_2O_3 (10)	11.2/8.3	4.6	25.0
	Fe_2O_3 (4.2)	11.2/2.94	6.0	27.8
Medium	No additives			19.3
High	No additives			22.5

Source: Ref. 111.

structures. Enhanced charring is commonly observed with polymers containing halogens in their structure or as additives, especially in combination with the metal oxides noted above, but the predominant mode of flame-retardant action is centered in gas phase processes. Thermal decomposition releases HCl or HBr, which can inhibit flames by radical transfer with reactive species in the propagation reactions, e.g., Eqs. 2 and 3 [8].

$$H\cdot + HBr \rightleftharpoons H_2 + Br\cdot \tag{2}$$

$$OH\cdot + HBr \rightleftharpoons H_2O + Br\cdot \tag{3}$$

Other mechanisms have been proposed, e.g., by Larsen and Ludwig, who related flame retardation by halogenated organic additives to a mass effect, in which the heat required for vaporization is not supplied by the diminished heat of combustion of the halogen compound [123]. Tkac also reported that decabromodiphenyl oxide (DBDPO) reduces the concentration of free radicals in both the solid phase and the volatile products of various polymers, including natural rubber [124]. He noted similar effects for Sb_2O_3 and triphenyl phosphate.

Phosphorus Compounds

Organic phosphates and other compounds of phosphorus are used to flame retard elastomers, and their greatest use is found in heterochain polymers, especially polyurethanes. Although gas phase inhibition reactions can occur, flame-retardant activity is centered extensively in the condensed phase. Dehydration takes place with the formation of polyphosphates and carbonaceous char. A vast literature deals with the flame retardation of polyurethanes, especially flexible foams. Readers are directed to other reviews for details [1, 2, 51–54]. Various combinations of phosphorus and halogen compounds are commonly used as additives. Several of these are summarized in Table 7. The tendency of polyurethane foam to smolder is reportedly improved by some phosphate additives, e.g., VI, but worsened by others, e.g., VII [66].

$$
\begin{array}{ccc}
ClCH_2 & CH_2OP(O)(OC_2H_4Cl)_2 \\
\diagdown\ \diagup \\
C & & (BrCH_2CHBrCH_2O)_3P{=}O \\
\diagup\ \diagdown \\
ClCH_2 & CH_2OP(O)(OC_2H_4Cl)_2 \\
\end{array}
$$

| VI | VII |

Miscellaneous

Melamine or urea additives and blends with various copolymers have contributed to reductions in the flammability of polyurethanes [125–128]. Some approaches for other polymers include structure modifications through postreactions, copolymerization, and surface treatments, and the use of intumescent systems. Several of these are described briefly next.

Postreactions

Fabris and Sommer [1] described several postreactions:

NR has been swollen with trichlorobromomethane and treated with high energy irradiation or a radical initiator [129, 130]. The process leads to the addition of trichloromethyl

Table 7 Some Commercially Available Phosphorus and Halogen Flame-Retardant Additives for Flexible Foams

Chemical or trade name	Structure
Tris(2-Chloroethyl) phosphate	$PO(OCH_2CH_2Cl)_3$
Tris(2-Chloropropyl) phosphate	$PO[OCH_2CH(CH_3)Cl]_3$
Tris(2,3-Dibromopropyl) phosphate	$PO(OCH_2CHBrCH_2Br)_3$
Tris-2-(1,3-Dichloropropyl) phosphate	$PO[OCH(CH_2Cl)_2]_3$

Phosgard C-22-R

$$ClCH_2CH_2OP(O)OCH\overset{\displaystyle CH_3}{\underset{\displaystyle CH_2CH_2Cl}{|}}\left[P(O)OCH\overset{\displaystyle CH_3}{\underset{\displaystyle OCH_2CH_2Cl}{|}}\right]_n P(O)(OCH_2CH_2Cl)_2$$

2XC-20

$$(CH_2Cl)_2—C—[CH_2OPO(OCH_2CH_2Cl)_2]_2$$

Fyrol 99

$$ClCH_2CH_2O\left[OCH_2CH_2OP(O)\underset{\displaystyle ClCH_2CH_2O}{|}\right]_n OCH_2CH_2Cl$$

and bromine across the double bond, giving polymers which retained their high elasticity even at high halogen contents. In a similar reaction CCl_4 was added to 1,4-polybutadiene in the presence of $FeCl_3$ catalyst [131]. The reaction produced transparent polymers with reduced flammability. In contrast to these, the addition of chlorine or bromine to polydiene rubbers leads to flame-retardant, but nonelastic, materials.

Flame-resistant EPDM is reportedly obtained by halogenation [132] or by grafting with monomers containing halogens and/or phosphorus [133, 134].

Copolymerization

The patent literature lists various comonomers which are claimed to impart flame retardation to rubbers. These include bis(2,3-dibromopropyl)fumarate [135], allyl or vinyl phosphonates [136, 137], monochlorostyrene [138], halogen-substituted olefins [139], vinylidene chloride [140], and the Diels–Alder adduct of allyl alcohol and hexachlorocyclopentadiene [141]. More recently, elastomeric copolymers have been synthesized from dienes and halophenyl acrylates [142, 143], vinylbenzyl derivatives [144, 145], and combinations of acrylates and vinylidene chloride [146, 147].

Surface Treatments

Polydienes have been reported to exhibit improved surface flammability properties when chlorinated on the surface with *t*-butyl hypochlorite in alcohol [148] or with thionyl chloride [149]. Fluorocarbon elastomers have been used as coatings on polyurethanes [150].

Intumescent Systems

The concept of intumescence—in which expansion and charring take place upon heating to develop a heat-resistant protective outer layer—was originally used with coatings, but it has been exploited in elastomers. Early posttreatments of CR sponge used melamine–formaldehyde resins, which had an intumescent effect [151]. More recently, intumescent systems have been claimed for polydienes which employ benzenesulfonic hydrazide [152], or combinations of sodium silicate–phenolic resin [153] or melamine–hexamethylenetetramine nitrate [154]. Polyolefins have been treated with intumescent systems based on cyclic nitrogen compounds [155, 156], or on combinations of pentaerythritol, a phosphate source, and melamine [157–159].

Formulations for Flame-Retardant Rubber Compounds

Specific formulations for flammability-modified elastomeric compounds may be found in the original references cited above. Selected examples of some formulations may also be found in the reviews by Fabris and Sommer [1, 2] and Lawson [9]. It should be noted that the incorporation of additives or fillers may change the mechanical properties, processability, resistance to weathering or aging, and/or cost of the rubber compound.

Applications

The review of Fabris and Sommer contains an excellent survey of a wide variety of applications in which flame-resistant elastomers are used [1].

FIRE TESTING

The most reliable predictive tests for flammability are those that model the conditions of the potential fire system. Since most fire tests measure very specific properties, perfor-

Table 8 Flammability Tests Commonly Used with Elastomers

Test	Application	Reference
DOC FF-4-72 (Cigarette Test)	Mattresses and upholstery	163
DOC FF-1-70 (Methenamine Pill Test)	Carpet backing and rug underlay	164
DOT MVSS-302	Automobile interior	165
ASTM-E-84 (Steiner Tunnel Test)	Carpet backing and rug underlay	166
ASTM E-162 (Radiant Panel Test)	Interior of mass transit cars	167
IEEE 383 (Ribbon Burning Test)	Wire and cable	168
ASTM D-1692 (Horizontal Flame Test)[a]	Cellular materials	169
DMS 1510A (Vertical Flame Test)	Aircraft interiors	170
MIL R 200 92 G (Hot-Bolt Test)	Shipboard mattress and rubber sheets	171
USBM Schedule 2G	Coal mine conveyors	172
ASTM D-2863 (Oxygen Index Test)	Laboratory research test	173
UL-94	Laboratory-scale vertical/horizontal burn	174
UL-44	Rubber insulated wire and cables	175
ASTM E-108 (UL-790)	Roof covering materials	176
Factory Mutual Roofing	Roof covering systems	177
ASTM E-648 (Flooring Radiant Panel)	Floor covering systems for corridor and exits	178
UFAC Cigarette Test	Upholstered furniture	179
ASTM E-662 (NBS Smoke Density Chamber)	Visible smoke generation	180

[a]ASTM D-1692 has been discontinued from the official ASTM handbook.

mance results in several tests must be considered together to assess the potential fire behavior of a material or system [160]. Both large- and small-scale tests have been used for elastomers, although few, if any, were originally designed specifically to test elastomeric materials. Many of these have been discussed in earlier reviews [1, 2, 5, 160–162], and they are summarized in Table 8. The reader is directed to the individual references for more details [163–180]. General features of some of the tests have been summarized by Fabris and Sommer [1]:

The Cigarette Test (DOC FF-4-72) is used to test flammability of household furnishings and mattresses. It determines the capability of an upholstery material or an assembly to withstand ignition by contact with a lighted cigarette. The test is passed if the length of charred material on the upholstered surface is not more than 50 mm in any direction from the nearest point of the cigarette. [163]

In the Pill Test (DOC FF-1-70) a sample is held flat on a horizontal asbestos cement surface by a metal frame with a 203.2 mm diameter hole in the center. A methenamine tablet is then placed in the middle of the exposed sample surface and ignited. The specimen passes the test if the charred portion does not extend within 25.4 mm of the edge of the circular hole at any point. [164]

In Motor Vehicle Safety Standard No. 302, a sample measuring 101.6 × 355.6 mm is mounted horizontally in a test chamber. One end is exposed to a vertical burner flame. The test is passed if the rate of burning of the sample is less than 100 mm/min. [165]

The Steiner Tunnel Test (ASTM-E-84) is a widely used test for surface flame spread. A 7.62 m long and 0.50 m wide sample is mounted, face down, on the ceiling of the tunnel. The specimen is ignited at one end of the tunnel (the fire end) by two gas burners, while air at an average velocity of 75 m/min is blown through the tunnel in the direction of flame travel. Flame spread classification is determined relative to asbestos cement board = 0 and red oak flooring = 100. Fuel contribution and smoke density by light obscuration are determined on a similar scale. [166]

The National Bureau of Standards Radiant-Panel Test (ASTM-E-162) employs a vertically mounted, porous refractory panel maintained at 670 ± 4°C as the radiant heat source. A specimen (150 × 460 mm) is supported in front of this panel in an inclined (30°) position such that the upper edge is closest to the panel. Ignition of the sample is forced by a pilot flame near its upper edge, and the flame travels downward. A constant air flow of 200 mm/s opposite to the direction of flame front travel is maintained. The temperature rise above 204°C, recorded by thermocouples placed at the exhaust stack, is taken as a measure of heat contribution. Smoke is determined gravimetrically by filtration of airborne particles. [167]

The Ribbon Burning Test (IEEE 383) measures flame propagation on a 2.4 m long vertically hanging cable while it is exposed on its lower end to a flame of 871–899°C for 20 min. In order to meet the specifications, the flame shall not travel the entire length of the cable. [168]

In DMS 1510A, a vertically mounted specimen (305 × 115 mm) is exposed at the lower end for 60 s to a flame with a minimum temperature of 843°C. Time and extent of burning after removal of the flame are recorded. [170]

The Hot-Bolt Test (MIL R 200 92 G) calls for contact of a sample of cellular material for 2 h with a steel bolt initially heated to 800 ± 10°C. At 1 and 2 min intervals, a Bunsen burner flame is held 50 mm above the center of the test piece. Any ignition or smoldering is reported. [171]

In USBM Schedule 2G, conveyor belt samples (150 × 13 mm) are exposed for 1 min to the flame of an upright Bunsen burner. Burning must cease within 1 min and afterglow stop within 3 min after removal of the flame. [172]

The Oxygen Index Test (ASTM D-2863) has been employed widely for laboratory screening. It is both simple and reproducible, and has been used to compare relative flammabilities. An 80 mm long rod or strip of material is placed in a metered, flowing column of oxygen and nitrogen, and ignited at the upper end with a pilot flame in the fashion of a candle. The Oxygen Index (OI or LOI) is defined as the minimum volume percent of oxygen in an oxygen–nitrogen mixture which will just permit the sample to burn in this manner [173]. Under test conditions the predominant mode of heating is conduction from the flame; convective and radiant heating are minimized. As a result, OI results sometimes fail to correlate with other burning tests. Melting and dripping away from the flame front may give anomalously high OI results. [182]

By taking OI measurements at two or more different environmental temperatures, a more informative picture of the flammability response may be gained. OI results included in this chapter refer to measurements made near normal room temperature.

UL-94 is a small-scale ignition/burn test commonly used with plastics, and it can be used with insulation for wire and cables. In the test, a material in either vertical or horizontal orientation is exposed to an ignition source for specified lengths of time. The ease of ignition, length of burn, self-extinguishment, presence of flaming drips, and other burning behaviors are recorded to establish classifications of 94HB, 94V-0, 94V-1, or 94V-2, depending on the results [174].

One of the tests, ASTM E-108, is well established but is seeing increasing use with elastomers due to the greatly increased application of elastomeric single-ply membrane roofing. In it, a roof-covering system is given a class A, B, or C rating, depending on its effectiveness against severe, moderate, or light fire exposure. Test decks of the roof covering, measuring from 1 × 1.3 m to 1 × 4 m, are subjected to ignition in a draft by an intermittent flame, a steady flame, and a burning brand, according to the format of the respective class rating system [176].

Other tests modeling specific end uses have been employed. The UFAC cigarette test is related to a voluntary standard for upholstered furniture. It involves the application of lighted cigarettes in the crevices of fabric-covered furniture mockups [179]. Woolley and co-workers have reported that the covering fabric is a critical factor in the ignition of polyurethane foams with candles or cigarettes [181]. The Flooring Radiant Panel (ASTM E-648) [178] was designed to measure heat requirements (critical radiant flux) for the propagation of fires involving floor covering systems in public-use corridors and exits. In studies of critical radiant flux (CRF) of carpets with latex foam underlayments, Anolick, Cook, and Stewart found that padded carpets required less heat than unpadded for combustion, that systems with CR foam paddings required more heat than those with SBR foam, and that other flammability tests should also be used with the Flooring Radiant Panel, since factors in addition to CRF are involved in the fires [183]. Hilado and co-workers used a laboratory test to examine the flash-fire potentials of a variety of materials [184], and other tests have been reported which simulate fires involving cellular insulations for rooms and compartments [185] and pipes [186], mattresses [187], theater seats [188], and various transportation modes including aircraft, buses, and railcars [188, 189]. A recently described flammability test for belting used in mines rates candidates according to flame spread rate, heat release rate, and critical ignition flux [190].

The NBS Smoke Density Chamber (ASTM E-662) is widely accepted for the measurement of smoke generation [191]. In it a sample is either burned or thermally degraded in a closed chamber, and the smoke produced is measured with a photoelectric cell as the time-dependent attenuation of a vertical beam of light. Results are reported as maximum specific optical density, D_m, time to $D = 16$, and time to 90% of D_m.

Excellent compilations and discussions of worldwide tests and standards are found in books by Troitzsch [192] and Cullis and Hirschler [5].

SMOKE AND COMBUSTION TOXICITY

The major lethal hazards of uncontrolled or accidental fires involving any materials, including elastomers, are heat, oxygen depletion, and toxic gases [193]. In certain fires, visible smoke may also accompany or precede the development of these hazards, hampering escape from a fire location until the ambient temperatures and atmospheres become untenable [194]. All of these factors, in addition to those of ignition, heat development, flame spread, and extinguishment, are of concern in efforts to improve fire safety.

Visible Smoke

The formation of and methods for measurement and reduction of visible smoke from various polymers have been reviewed recently by Lawson [194], Sutker [195], and Hirschler [196, 197]. Visible smoke from burning polymers results from incomplete combustion. Smoke may consist of liquids (e.g., mists from pyrolyzates) or solids (e.g., soot, sublimed components, or, particularly with silicones, solid oxides). Carbonaceous soot is most commonly encountered in the flaming combustion of organic polymers, and it is attributable to processes outlined in Scheme 3 [198].

Most of the polydiene rubbers tend to form much visible smoke readily on burning, since their pyrolyzates are aromatic and polyenic compounds which proceed rather directly to soot. However, visible smoke formation is not an intrinsic material property, and observed smoke levels can vary greatly with extrinsic factors, such as heat flux [199] oxidant supply, burning configurations [200], air movement, and so on. At least 11 different tests are used to measure smoke. The National Bureau of Standards Smoke Density Chamber (ASTM E-662, NFPA Method 258-1982) is receiving widest acceptance in the United States at this writing. Details of this and other smoke tests have been described elsewhere [1, 194, 201].

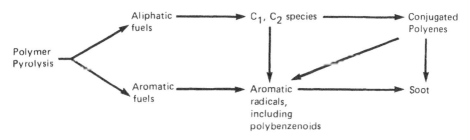

Scheme 3 Production of carbonaceous soot from the flaming combustion of organic polymers. (From Ref. 198 by permission of The Plastics and Rubber Institute, London, England.)

Several hypothetical materials approaches have been proposed to limit visible smoke production from polymers involved in fire [194]:

1. Enhanced polymer stability
2. Reduction of the concentrations of aromatic and polyenic components of pyrolysates evolved as fuels
3. Formation of char and retention of combustible substrates in the solid phase
4. Reducing burning rates to allow fuel/oxidant stoichiometries favorable to clean burning
5. Enhanced oxidation of combustion products
6. Alteration of light transmission characteristics or nucleation of soot

The first four of these approaches are most consistent with the overall improvement of flammability properties and a number of practical systems have employed them. The subject has been reviewed in greater depth elsewhere [194].

The polydienes and most other elastomers do not of themselves have an exceptional degree of thermal stability which would limit their combustion. Fluoroelastomers and silicones may exhibit this to a certain extent, as do some polyphosphazenes, I and II. Extremely stable polyphosphazene elastomers with high fire resistance and low smoke generation have been obtained when the substituents are aryloxy or perfluoroalkoxy [87, 202]. The aryloxy derivatives have been used in cushioning and insulating foams and wire coverings. Highly flame-resistant foams have also been obtained with polyimides. Flexible rather than elastomeric, they take advantage of highly stable polyimide linkages such as VIII [89].

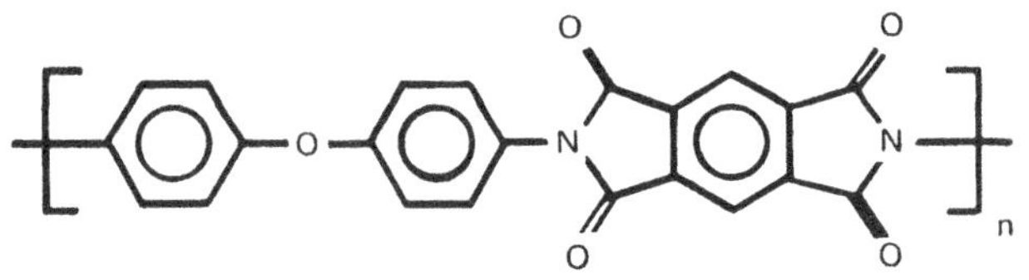

VIII

Polyurethane foams also reportedly produce less smoke on burning when the backbone contains isocyanurate groups, but the resulting foams tend to be rigid and friable. It is difficult to build units of high thermal stability and still maintain rubbery properties since these structures contribute to stiffness. Alternatively, polymer structures with less tendency to form smoke may also be selected to substitute for or blend with smokier polymers. For example, a 55% ethylene/45% vinyl acetate (EVA) copolymer was blended 4:1 with chlorinated polyethylene and substituted for 100% CSPE in wire insulation [90]. Although somewhat more flammable, the blend produced 70% less visible smoke and also less CO and HCl on burning than CSPE. In another example, a copolymer prepared by grafting vinylidene chloride to a diene–acrylate elastomer had good flame resistance and low-temperature flexibility, as well as about half the ASTM-E662 smoke of an unmodified polybutadiene [146]. Other diene/acrylate/vinylidene chloride interpolymers and graft copolymers have also shown promising low-smoke properties [147]. Low smoke generation and improved weatherability were reported in PVC compositions containing an acrylic elastomer [203], in polyesters and polyamides containing silicone resins [204], and in poly(carbonate-co-siloxane) multiblock polymers [205].

Visible smoke generation from polydienes [206] and polyurethanes [207] is generally increased with increasing quantities of halogenated flame-retardant additives. However, other additives can reduce visible smoke formation. The best known of these is alumina trihydrate (ATH), which has exhibited both smoke-suppressant and flame-retardant effects in a wide variety of elastomers when used at relatively high loadings [194]. The mechanism was discussed previously. ATH has also been incorporated into flexible polyurethane foams through use of a halogenated latex binder [208], and ATH is used in flame-retardant polyurea–urethane foams developed and licensed by Mobay Chemical Company [209]. By use of coupling agents such as silanes and surface modifications such as fatty acids, high loadings of ATH may be incorporated into polyolefins for use in electrical insulation [210]. Other fillers reported to have noteworthy smoke-suppressant activity include $Mg(OH)_2$ [98, 211], magnesium oxychloride [212], and $MgCO_3$ [213]. Tables 9–11 contain low-smoke formulations for several elastomers. Of this group, EPDM filled with $MgCO_3$ exhibited the lowest smoke under flaming conditions. The data also suggest the influence of polymer structures on observed smoke levels.

Table 9   Effect of Plasticizers, Additives, and Fillers on Flammability and Smoke Generation in CSPE

	A	B	C	D	E
CSPE[a]	100	100	100	100	100
MoO_3	7.5	7.5	7.5	7.5	7.5
Arizona 208[b]	25				
Triactyl trimellitate		25		25	25
Butyl oleate			25		
MgO/PER-200[c]		4/3		4/3	4/3
Whitetex clay[d]	55		55		
$Al(OH)_3$		55			
Buca clay[d]				55	
Catalpo clay[d]					55
ASTM E-662					
Flaming					
D_{mc}[e]	332	319	280	219	264
tD_{100} (sec)[f]	138	188	122	162	194
Nonflaming					
D_{mc}	520	334	473	302	152
tD_{100} (sec)	158	240	138	197	256
ASTM D-2863					
Oxygen index (%)	30	34	30	34	34

[a]Chlorosulfonated polyethylene, Hypalon trademark of E.I. duPont de Nemours Co., litharge cure [213].

[b]Tall oil fatty ester, Arizona Chemical Co.

[c]Pentaerythritol.

[d]Freeport Kaolin Co.

[e]Maximum specific optical density, corrected.

[f]Time to $D = 100$.

Source: Ref. 213.

Table 10 Effect of Additives on Flammability of CR (NA-22 Cure)

	A	B	C	D	E
CR (Neoprene W)	100	100	100	100	100
Suprex clay	90		90	90	
Burgess KE					20
Precipitated $CaCO_3$		50			
$Al(OH)_3$		73			80
Sundex 790[a]	15				
TCP[b]		15	15		
DOS[c]				15	
DFR 121[d]					12
ASTM E-662					
Flaming					
D_{mc}	692	479	639	647	375
tD_{16} (sec)	30	67	39	46	59
Nonflaming					
D_{mc}	773	693	630	607	256
tD_{16} (sec)	101	121	109	105	105
ASTM D-2863					
Oxygen index (%)	37	35	39	39	52

[a]Sun Petroleum Products Co.
[b]Tricesyl phosphate.
[c]Dioctyl succinate.
[d]Ferrocene-based F. R. Plasticizer, Syntex Corp.
Source: Ref. 213.

As noted previously, much of the smoke-suppressant activity of hydrated metal oxide fillers at high loadings appears to be associated with endothermal effects of dehydration. However, Hirschler reports evidence for specific smoke-suppressant interactions in a number of metal oxide–polymer combinations [214] and Kroenke described many systems effective in PVC [119], which may have uses in halogen-containing elastomer compounds as well. Metal oxides reported to be useful as smoke inhibitors on their own in elastomeric systems include Al_2O_3, ATH, MgO, CrO_3 [215], and $Mg(OH)_2$. Some metal oxides active in the presence of halogens include Fe_2O_3 [110], ZnO [216], and MoO_3 [114–116].

Kroenke has reported that several polymers, including CR, NBR, CSPE, and epichlorohydrin elastomer (ECO), have lower smoke on burning when formulated with a sulfate glass [217] or with melamine molybdate [218]. The latter may function at least in part by intumescence. Intumescent fillers have been used in SBR [219] and EPDM [155, 156], with accompanying reductions in smoke generation. Other intumescent systems were discussed briefly in the section on "Inhibition."

CR or fluoroelastomers have been used as binders with halogenated additives, Sb_2O_3, and ammonium polyphosphate to impregnate polyurethane foams for flame resistance and smoke suppression [220]. DuPont's Vonar interliner, a CR latex foam heavily loaded with flame-retardant fillers and additives, is employed as a sacrificial wrap around more

Table 11 Effect of Additives on Flammability and Smoke Emission by Peroxide-Cured EPDM and Ethylene–Acrylic Copolymers

	A	B	C	D	E	F
EPDM	100	100	100			
E-A (Vamac N123)[a]				100	100	100
$Al(OH)_3$	200	150	150	120	120	100
$Mg(OH)_2$		50				20
$MgCO_3$			50			
Sunpar 2280[b]	32	10	10			
Santicizer 148[c]		12.5	12.5			7.5
Kemgard 911A[d]			3		3	
ASTM E-662						
Flaming						
D_{mc}	354	91	80	201	245	163
tD_{16} (sec)	138	510	540	103	125	395
Nonflaming						
D_{mc}	548	569	582	285	255	301
tD_{16} (sec)	188	130	137	177	183	203
ASTM D-2863						
Oxygen index (%)	26	26.5	27.4	36.3	39.7	33.7

[a]E.I. duPont de Nemours Co.
[b]Sun Petroleum Products Co.
[c]Monsanto Corp.
[d]Sherwin-Williams Co.
Source: Ref. 213.

flammable CR or polyurethane cushions, insulating the cushion from fire and reducing smoke output [221].

Combustion Toxicity

The uncontrolled burning of any material poses a potential toxicity hazard due to the gases and hot air produced and the depletion of oxygen from the surrounding air. The combustion toxicity of elastomers appears to be comparable to that of most plastics and natural products. However, polymer combustion toxicology in general is presently the subject of some controversy. Although most of the details are outside the scope of this review, the controversy centers around which, if any, of the current laboratory protocols is appropriate for determining and quantifying the fire toxicity hazard of a given material.

Six of the major laboratory combustion toxicology protocols are listed in Table 12 [222]. Differences among these methods include chamber configuration, mode of combustion, endpoint (t_i = time to incapacitation, t_d = time to death, LA_{50} or LC_{50} = area or concentration for 50% lethality, etc.), species and number of subjects, type of exposure, and the postexposure observation period. For example, the PITT test [226], which can be used to follow either lethality or sensory irritation effects, uses a relatively slow (20°C/min) thermal degradation of the material and only four mice. On the other hand, the NBS test [225] uses a larger number of rats for statistical reasons and a more rapid rate of

Table 12 Summary of Principal Laboratory Methods for Combustion Toxicology

Method	Combustion device	Furnace temperature	Air flow	Animals/ no. per test	Exposure mode	Toxicity measurements	Reference
DIN	Movable annular tube furnace	Fixed, 473–873 K	Dynamic	Rats, at least 5, usually 20	Head-only or whole body	LC_{50} (30 min + 14 day)	223
FAA	Tube furnace	Fixed, 898 K	Static recirculating	Rats, 3; at least 3 tests	Whole body	t_i and t_d^a	224
NBS	Cone calorimeter furnace	Fixed, 25 K below and above autoignition temperature	Static	Rats, 6	Head-only	LC_{50} (30 min + 14 day)	225
Radiant heat	Radiant heat furnace	Fixed heat fluxes up to 5 W/cm^2	Static	Rats, 6	Head-only	LA_{50}, t_i and gross respiratory tract pathology	222
PITT	Tube furnace	Ramped to 873 K above 0.2% weight less temperature	Dynamic	Mice, 4	Head-only	RD_{50}, LC_{50} (30 min + 10 min), SI, asphyxiation range, histopathology, LT_{50}	226
USF	Tube furnace	Fixed or ramped, 473–1073 K	Static or dynamic	Mice, 4; at least 2 tests	Whole body	t_i and t_d	227

^{a}FAA method modified for fixed or ramped heating, flaming or nonflaming combustion, and determination of LC_{50}.

Source: Ref. 222.

pyrolysis. However, neither of these tests models the conditions of a developing fire, under which the concept of combustion toxicity is most meaningful. Steady-state combustion, as in the German DIN test [223], may be closest to those conditions. Although precise distinctions of toxicity hazards may not be achievable with any of these protocols, they may be of use in identifying materials with unusually toxic combustion products [228]. Information concerning the total fire hazard, in which ignitability, flame spread, heat release, and smoke generation are also considered, may be forthcoming from current studies which use fire modeling and toxic responses in species other than rats and mice.

A study of structural fires showed that CO and acrolein were the most hazardous gases commonly encountered by firefighters [229]. Although combinations of respiratory irritants, NO_x, HCl, and acrolein, were important in selected fires, exposures to HCN, CO_2, and oxygen deficiency were not found to be significant in the study.

A prison fire in Tennessee was believed to involve an SBR foam covered with neoprene–nylon and repaired in places with polyurethane foam. At the time, news reports of HCN poisoning implicated the polyurethane, but a later study concluded that the elevated HCN blood levels were sublethal, and CO was the major asphyxiant [230]. Studies with monkeys showed evidence for HCN and CO poisoning when polyurethane foam was pyrolyzed at 1173 K but not at lower temperatures, 873 and 573 K [231]. An irritating yellow smoke containing neither HCN nor TDI was formed at these temperatures. Delayed lung inflammations have been observed in monkeys and humans exposed under similar conditions [232].

HCN is produced abundantly by nitrile polymers [233, 234] and, in lesser amounts, by polyurethanes at 1173 K, but only 12% of the available nitrogen in polyacrylonitrile is converted to HCN at 993 K [234]. HCN is also formed in large quantities by nitrogenous natural products and in small quantities even by hydrocarbon materials [233]. Laboratory exposures with rats showed no significant differences in combustion toxicity between several selected polyurethane foams and a polyester fiberfill material [235].

A number of synthetic construction and furnishing materials, including elastomers, have been reported to present no greater toxic hazard than common food items which might become involved in a kitchen fire, and less hazard than wood under preignition and preflashover conditions [236, 237]. However, high levels of irritating smoke were reported for synthetic materials used in mines, including elastomeric foams and belts and diesel fuels [238], and widely varying combustion toxicities were reported for candidate electrical cable insulation materials [239].

Both PVC and CR cable insulations generate HCl in a similar manner upon pyrolysis, but the amount and rate of formation are moderated by the presence of compounding ingredients [240]. Figure 12 shows the effect of $CaCO_3$ on the generation of HCl from pyrolysis of CR, in comparison with unfilled PVC. Much research and controversy has taken place over the role of HCl and acid gases in combustion toxicity. Hinderer concluded that toxicology data did not show PVC to be an unusually hazardous material [241].

In addition to HCl and hydrocarbons, small quantities of organochlorine compounds have been reported from the combustion of PVC and poly(tetrachloroisoprene) [242], and many products attributable to plasticizers have been identified [240]. Alternative cable jacket materials, such as silicone rubbers, may avoid HCl generation and afford less CO [243], but even here acute respiratory distress in firefighters has been reported, owing to other products of combustion [244]. Both the degradation temperature and the incorpora-

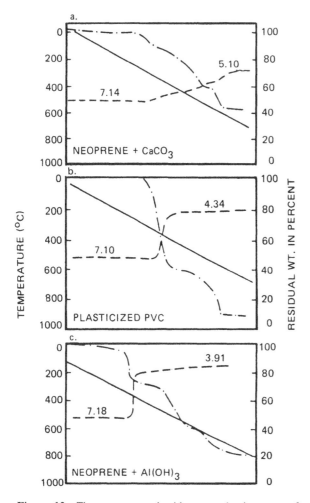

Figure 12 Thermograms and acid–gas production curves for CR and PVC insulation compounds heated at 20°C/min [240]. Solid curves, temperature; dashed curves, change in pH of water trap by off-gases; dot-dash curve, residual weight of compound. (Adapted from Ref. 240.)

tion of flame-retardant additives have been found to have variable effects on combustion toxicity results [245].

The hydrocarbon combustion products from a smoldering foam rubber identified as SBR were similar to those obtained on pyrolysis at 673 K [234]. These included (in decreasing order) isoprene, styrene, limonene, toluene, benzene, and styrene derivatives and oligomers. Hydrocarbons detected under flaming conditions were apparently formed at 673–773 K. Additive effects of hydrocarbons and HCN were speculated to contribute to incapacitations in combustion atmospheres [234]. Polycyclic aromatic hydrocarbons may be adsorbed and carried by soot particles in the dense smoke resulting from the burning of old tires in the open air.

Some compounding approaches are known for limiting the evolution of toxic gases, such as the use of $CaCO_3$ for HCl as noted above [240], $Ca(OH)_2$ for CO [246], and CaO with polysulfide elastomers for HCN [247]. Certain polymer structures generate less toxic atmospheres than others under given conditions of burning, e.g., polyimides [248] and nonhalogenated polyphosphazenes [92]. These results, together with the flame resistance and low smoke generation of these polymers, has made them candidates for high-risk submarine, shipboard, and aircraft applications.

The ubiquity of CO and the wide variability of fire conditions and exposure situations suggest that the problem of combustion toxicity will not be controlled simply through developments in materials alone. Moreover, the unselective substitution of materials strictly on the basis of the potential combustion gases, e.g., nonhalogenated elastomers for halogenated, may intensify other combustion parameters such as ignitability or heat release, resulting in questionable gains in overall fire safety [245].

CONCLUSIONS

Most of the flame-retardant technology now used with elastomers has been known for some time. This has made it possible to select or formulate elastomeric materials for a wide variety of applications requiring flame resistance, in which satisfactory performance is achieved. Advances have been made in recent years in the understanding of elastomer combustion and inhibition mechanisms, in the measurement and control of visible smoke generation from elastomers, and in the toxic effects of combustion atmospheres. By integrating existing technology with recent developments and the likely advances of the future, elastomeric materials will continue to receive acceptance for a wide variety of applications.

ACKNOWLEDGMENTS

The author acknowledges with indebtedness the excellent review article by Fabris and Sommer in *Rubber Chemistry and Technology* [1]. Much of the background discussion in various parts of this chapter, especially in the sections "Elastomer Pyrolysis and Combustion," "Inhibition," and "Fire Testing," is drawn from their work, and they have generously offered comments on this manuscript. Most of this chapter relies on a more recent review by the author, parts of which are reproduced by permission of the publisher [9]. The author is most grateful to The Firestone Tire and Rubber Co. for support and permission to publish this work.

REFERENCES

1. H. J. Fabris and J. G. Sommer, *Rubber Chem. Technol., 50*: 523 (1977).
2. H. J. Fabris and J. G. Sommer, in *Flame Retardancy of Polymeric Materials*, Vol. 2 (W. C. Kuryla and A. J. Papa, eds.), Marcel Dekker, New York, Chap. 3 (1973).
3. G. S. Buck, Jr., in *Kirk-Othmer Encyclopedia of Chemical Technology*, Vol. 6, Wiley, New York, p. 543 (1963).
4. D. W. van Krevelen, *Polymer, 16*: 615 (1975).
5. C. F. Cullis and M. M. Hirschler, *The Combustion of Organic Polymers*, Clarendon Press, Oxford (1981).
6. P. C. Warren, in *Polymer Stabilization* (W. L. Hawkins, ed.), Wiley-Interscience, New York, Chap. 7 (1972).
7. T. H. Rogers, Jr., and R. E. Fruzzetti, in *Flame-Retardant Polymeric Materials*, Vol. 1 (M. Lewin, S. M. Atlas, and E. M. Pearce, eds.), Plenum, New York, Chap. 5 (1975).

8. J. W. Hastie, *J. Res. Nat. Bur. Stand. Sect. A*, *77*: 733 (1973).
9. D. F. Lawson, *Rubber Chem. Technol.*, *59*: 455 (1986).
10. L. A. Wall and S. Straus, *J. Polym. Sci.*, *44*: 313 (1960).
11. S. L. Madorsky, *Thermal Degradation of Organic Polymers*, Interscience, New York (1964).
12. S. L. Madorsky, *SPE J.*, *17*: 665 (1961).
13. S. L. Madorsky and S. Straus, *Soc. Chem. Ind. Monogr.*, *13*: 60 (1961).
14. S. Straus and S. L. Madorsky, *J. Res. Nat. Bur. Stand. Sect. A*, *66*: 401 (1962).
15. S. L. Madorsky, S. Straus, D. Thompson, and T. Williamson, *J. Res. Nat. Bur. Stand.*, *42*: 499 (1949).
16. S. L. Madorsky and S. Straus, *J. Res. Nat. Bur. Stand.*, *61*: 77 (1958).
17. S. Straus and S. L. Madorsky, *J. Res. Nat. Bur. Stand.*, *50*: 165 (1950).
18. W. K. Smith and J. B. King, *J. Fire Flammability*, *1*: 282 (1970).
19. A. Tkac, *J. Polym. Sci. Polym. Symp.*, *57*: 109 (1977).
20. D. A. Smith and J. W. Youren, *Br. Polym. J.*, *8*: 101 (1976).
21. C. F. Cullis and H. S. Laver, *Eur. Polym. J.*, *14*: 571 (1978).
22. M. A. Golub, *J. Polym. Sci. Polym. Lett. Ed.*, *15*: 369 (1977).
23. D. N. Schulz, S. R. Turner, and M. A. Golub, *Rubber Chem. Technol.*, *55*: 809 (1982).
24. D. E. Stuetz, A. H. DiEdwardo, F. Zitomer, and B. P. Barnes, *J. Polym. Sci. Polym. Chem. Ed.*, *18*: 967 (1980).
25. D. E. Stuetz, A. H. DiEdwardo, F. Zitomer, and B. P. Barnes, *J. Polym. Sci. Polym. Chem. Ed.*, *18*: 987 (1980).
26. C. Baillet, L. Delfosse, and M. Lucquin, *Eur. Polym. J.*, *17*: 791 (1981).
27. S. K. Brauman, *J. Polym. Sci. Polym. Chem. Ed.*, *15*: 1507 (1977).
28. H. W. Emmons, *J. Appl. Polym. Sci.*, *26*: 2447 (1981).
29. S. K. Brauman, *J. Fire Retard. Chem.*, *6*: 249 (1979).
30. S. K. Brauman, *J. Fire Retard. Chem.*, *6*: 266 (1979).
31. K. Oplustil, duPont Technical Bulletin SD-113, "Factors Affecting Flame Spread Rating of Neoprene Vulcanizates" (1968).
32. G. S. Skinner and J. H. McNeal, *Ind. Eng. Chem.*, *40*: 2303 (1948).
33. C. Torii, K. Hoshii, and S. Isshiki, *J. Soc. Rubber Ind. Jpn.*, *29*: 3 (1956); *Chem. Abstr.*, *50*: 17511a (1956).
34. C. W. Stewart, R. L. Dawson, and P. R. Johnson, *Rubber Chem. Technol.*, *48*: 132 (1975).
35. B. W. Mapperley and P. R. Sewell, *Eur. Polym. J.*, *9*: 1255 (1973).
36. R. F. Schwenker, Jr., and L. R. Beck, Jr., *Text. Res. J.*, *30*: 624 (1975).
37. C. J. Hilado and C. J. Casey, *J. Fire Flammability*, *10*: 227 (1979).
38. D. A. Smith, *Kautsch. Gummi Kunstst.*, *19*: 477 (1966).
39. Dow Chemical Co., Plastics Department, Tech. Bulletin, "The Use of Chlorinated Polyethylene for Fire-Retardant Sponge Underlay for Carpeting."
40. F. H. Winslow, L. D. Low, and W. Matreyek, *Am. Chem. Soc., Div. Org. Coat. Plast. Chem. Pap. 31*(1): 124 (1971).
41. I. A. Abu-Isa, *J. Polym. Sci. Part A-1*, *10*: 881 (1972).
42. P. Thiery, *Fireproofing*, American Elsevier, New York, (1970) Chaps. 4 and 5.
43. D. E. Supkis, NASA General Working Paper, "Description and Application of Fluorel L-3203-6"; also available as MSC-01275, NASA, Manned Spacecraft Center, Houston, Texas.
44. J. E. Smariga and G. T. Castino, *Archit. Eng. News*, *13*: 34 (March 1970).
45. D. G. Sauers, NASA Conference on Materials for Improved Fire Safety, NASA Manned Spacecraft Center, Houston, Texas, p. 17-1 (May 1970).
46. A. L. Moran, duPont, Elastomers Chemical Dept. Report No. 59-4, "Viton B" (1959).
47. Minnesota Mining and Manufacturing Co., Tech. Information Y-TTDC (Fluorel), "Fluorel Elastomers," p. 9.
48. P. R. Johnson, R. Pariser, and J. J. McEvoy, *Rubber Age*, *107*(5): 29 (1975).

49. D. A. Stivers, *SAMPE Tech. Conf. Proc., 16*: 363 (1970).
50. A. R. Gregory, *J. Appl. Polym. Sci., 17*: 2909 (1973).
51. H. J. Fabris, *Adv. Urethane Sci. Technol., 4*: 89 (1976).
52. A. J. Papa, in *Flame Retardancy of Polymeric Materials*, Vol. 3 (W. C. Kuryla and A. P. Papa, eds.), Marcel Dekker, New York, Chap. 1 (1975).
53. J. S. Babiec, Jr., J. J. Pitts, W. L. Ridenhour, and R. J. Turley, *Plast. Technol., 21*(7): 47 (1975).
54. P. N. Tilley, H. G. Nadeau, H. E. Reymore, P. H. Waszeciak, and A. A. R. Sayigh, *J. Cell. Plast., 4*: 56 (1968).
55. S. L. Madorsky and S. Straus, *J. Polym. Sci., 36*: 183 (1959).
56. F. D. Hileman, K. J. Voorhees, L. H. Wojcik, M. M. Birky, P. W. Ryan, and I. N. Einhorn, presented at a meeting of the Rubber Division, American Chemical Society, Philadelphia (October 15–18, 1974). Abs. *Rubber Chem. Technol., 48*: 343 (1975).
57. W. D. Woolley, *Br. Polym. J., 4*: 45 (1972).
58. M. Paabo and J. J. Comeford, NTIS Technical Report AD-763, 327 (July 1973).
59. D. H. Napier and T. W. Wong, *Br. Polym. J., 4*: 45 (1972).
60. J. D. Ingham and N. S. Rapp, *J. Polym. Sci. Part A, 2*: 689 (1964).
61. J. K. Backus, D. L. Bernard, W. C. Darr, and J. H. Saunders, *J. Appl. Polym. Sci., 12*: 1053 (1968).
62. K. J. Voorhees, F. D. Hileman, I. N. Einhorn, and J. H. Futtrell, *J. Polym. Sci. Polym. Chem. Ed., 16*: 213 (1978).
63. T. J. Ohlemiller, J. Bellan, and F. E. Rogers, *Combust. Flame, 36*: 197 (1979).
64. N. Grassie, M. Zulfiqar, and M. I. Guy, *J. Polym. Sci. Polym. Chem. Ed., 18*: 265 (1980).
65. D. E. Eaves and O. V. Keen, *Br. Polym. J., 8*: 41 (1976).
66. F. E. Rogers, T. J. Ohlemiller, A. Kurtz, and M. Summerfield, *J. Fire Flammability, 9*: 5 (1978).
67. D. D. Drysdale, *Fire Prev. Sci. Technol., 23*: 18 (1980).
68. W. D. Willis, L. O. Amberg, A. E. Robinson, and E. J. Vandenberg, *Rubber World, 153*(1): 88 (1965).
69. E. Kay, Royal Aircraft Establishment Tech. Report 69025 (March 1969).
70. V. L. Zakutinskii, I. G. Blyakher, M. V. Koloskova, F. S. Tolstukhina, and T. B. Sulimov, *Kauch. Rezina, 27*(10): 18 (1968).
71. R. E. Kesting, K. F. Jackson, E. B. Klusmann, and F. J. Gerhart, Chemical Systems, Inc., Santa Ana, Calif., Tech. Report AD-704-812 (March 1970).
72. M. J. Shaw and A. V. Tobolsky, *Polym. Eng. Sci., 10*: 225 (1970).
73. N. S. Vojvodich, *J. Macromol. Sci.-Chem., A3*: 367 (1969).
74. J. Lipowitz, *Am. Chem. Soc. Div. Org. Coat. Plast. Chem. Pap. 36*(2): 456 (1976).
75. T. L. Laur and L. B. Guy, *Rubber Age, 102*(12): 63 (1970).
76. Hooker Chemical Corp., Preliminary Data Sheet No. 349, "Dechlorane 604" (August 1970).
77. A. E. Pepe, Germ. Pat. 1,936,345 (1970).
78. Dow Corning Corp., Br. Pat. 1,161,052 (1969).
79. R. A. Compton, Can. Pat. 768,223 (1967); Dow.
80. B. D. Karstedt, U.S. Pat. 3,539,530 (1970); General Electric.
81. R. I. Smith and B. D. Karstedt, U.S. Pat. 3,840,492 (1974); General Electric.
82. C. W. Pfeifer and W. J. Bobear, U.S. Pat. 3,711,520 (1973); General Electric.
83. M. Hatanaka, R. Mahekawa, and H. Maruyama, U.S. Pat. 3,862,082 (1975); Toshiba Silicone Kabushiki Kaisha.
84. A. Pepa, U.S. Pat. 3,838,089 (1974); Stauffer Chemical Company.
85. S. B. Smith, U.S. Pat. 3,923,705 (1975); Dow.
86. R. E. Singler, N. S. Schneider, and G. L. Hagnauer, *Polym. Eng. Sci., 15*: 321 (1975).
87. E. J. Quinn and R. L. Dieck, *J. Cell. Plast., 13*: 96 (1977).

88. E. J. Quinn and R. L. Dieck, *J. Fire Flammability, 7*: 5 (1976); *J. Fire Flammability, 7*: 358 (1976).
89. J. Gagliani, R. Lee, U. A. K. Sorathia, and A. L. Wilcoxson, NASA Contract Report, NASA-CR-160576, SR79-R-4674-38 (1980).
90. J. M. Guilhaumou and P. Reumont, Eur. Pat. Appl. 17579 (1980); Thomson-Brandt.
91. A. E. Oberster and D. F. Lawson, U.S. Pat. 4,064,095 (1977); Firestone Tire & Rubber Co.
92. K. Sebata, J. H. Magill, and Y. C. Alarie, *J. Fire Flammability, 9*: 50 (1978).
93. W. W. Wright, *Soc. Chem. Ind. Monogr., 13*: 248 (1961).
94. A. R. Schultz, K. Noll, and G. A. Morneau, *J. Polym. Sci., 62*: 211 (1962).
95. T. L. Graham, *Rubber Age, 101*(8): 43 (1969).
96. E. Dorfman, W. E. Emerson, R. J. Gruber, A. A. Lemper, and T. L. Graham, *SAMP Tech. Conf. Proc., 15*: 413 (1970).
97. D. Culverhouse, *Rubber World,* 18 (April 1983).
98. D. F. Lawson, E. L. Kay, and D. T. Roberts, *Rubber Chem. Technol., 48*: 124 (1975).
99. C. F. Cullis and M. M. Hirschler, in *The Combustion of Organic Polymers*, Clarendon Press, Oxford, pp. 240–241 (1981).
100. F. K. Antia, C. F. Cullis, and M. M. Hirschler, *Eur. Polym. J., 17*: 451 (1981).
101. M. M. Hirschler and O. Tsika, *Eur. Polym. J., 19*: 375 (1983).
102. R. Chalabi and C. F. Cullis, *Eur. Polym. J., 18*: 1067 (1982).
103. R. Chalabi, C. F. Cullis, and M. M. Hirschler, *Eur. Polym. J., 19*: 461 (1983).
104. J. J. Pitts, in *Flame Retardancy of Polymeric Materials*, Vol. 1 (W. C. Kuryla and A. J. Papa, eds.), Marcel Dekker, New York, pp. 133–194 (1973).
105. C. F. Cullis and M. M. Hirschler, in *The Combustion of Organic Polymers*, Clarendon Press, Oxford, pp. 276–291 (1981).
106. H. Ohe, S. Takiyama, and S. Takamoto, *Text. Res. J., 48*: 303 (1978).
107. R. L. Clough, *J. Polym. Sci. Polym. Chem. Ed., 21*: 767 (1983).
108. S. K. Brauman, *J. Polym. Sci. Polym. Chem. Ed., 17*: 1129 (1979).
109. J. J. Kracklauer, U.S. Pat. 4,049,618 (1977); Syntex Corporation.
110. D. Florence, U.S. Pat. 3,993,607 (1976); Armstrong Cork.
111. W. P. Whelan, *J. Fire Retard. Chem., 6*: 206 (1979).
112. L. Lecomte, M. Bert, A. Michel, and A. Guyot, *J. Macromol. Sci. Chem., A11*: 1467 (1977).
113. D. F. Lawson, *J. Appl. Polym. Sci., 20*: 2183 (1976).
114. F. W. Moore, C. J. Hallada, and H. F. Barry, U.S. Pat. 3,956,231 (1976); Amax, Inc.
115. D. F. Lawson, *Res. Disclosure, 160*: 7 (1977).
116. F. W. Moore, *SPE Tech. Pap., 23*: 414 (1977).
117. W. H. Starnes and D. Edelson, *Macromolecules, 12*: 797 (1979).
118. R. P. Lattimer and W. J. Kroenke, *J. Appl. Polym. Sci., 26*: 1191 (1981).
119. W. J. Kroenke, *J. Appl. Polym. Sci., 26*: 1167 (1981).
120. C. F. Cullis, M. M. Hirschler, and T. R. Thevaranjan, *Eur. Polym. J., 20*: 841 (1984).
121. C. F. Cullis, A. M. M. Gad, and M. M. Hirschler, *Eur. Polym. J., 20*: 707 (1984).
122. C. F. Cullis and M. M. Hirschler, *Eur. Polym. J., 20*: 53 (1984).
123. E. R. Larsen and R. B. Ludwig, *J. Fire Flammability, 10*: 69 (1979).
124. A. Tkac, *Nehorlavost Polym. Mater., 6* (1976); *Chem. Abstr., 87*: 168680S.
125. J. P. Brown and J. Hynds, Br. Pat. 75 24098 (1978); Imperial Chemical Ind., Ltd.
126. N. K. Kalfoglou, *J. Appl. Polym. Sci., 26*: 823 (1981).
127. J. Hunter and S. R. Abbott, Br. Pat. Appl. 2,122,232 A1 (1984); British Vita PLC.
128. R. N. Tracalossi, W. V. Greenhouse, and M. S. Buchanan, U.S. Pat. 4,385,131 (1983).
129. E. G. Cockbain, T. D. Pendle, E. G. Pole, and D. T. Turner, in *Proceedings of the 4th Rubber Technology Conference* (T. H. Messenger, ed.), Institution of the Rubber Industry, London, p. 498 (1963).
130. E. G. Cockbain, T. D. Pendle, and D. T. Turner, *Chem. Ind.* (London), 318 (1960).

131. H. Rosin and M. Asscher, *J. Appl. Polym. Sci., 13*: 1721 (1969).
132. G. M. Ronkin, *Kauch. Rezina, No. 12*, 17 (1978); *Chem. Abstr., 90*: 88505k.
133. R. Muroi and K. Sato, Jpn. Pat. 77 130,888 (1977); Toyoda Gosei Co., Ltd.
134. M. M. Guseinov, B. Y. Trifel, D. E. Mishiev, G. M. Arjbov, S. S. Shchoegol, T. N. Malinova, and N. I. Dadasheva, U.S.S.R. Pat. 595,343 (1978); Y. G. Mamedaliev Institute of Petrochemical Processes, Sumgait; *Chem. Abstr., 88*: 154070b.
135. W. Cummings, U.S. Pat. 3,260,772 (1966); U.S. Rubber Co.
136. Borden Co., "Development of Fire-Retardant Elastomers" (May 30, 1960); NTIS Tech. Report AD-466131.
137. B. F. Goodrich Co., Belg. Pat. 720,954 (1969).
138. Dow Chemical Co., Technical Bulletin, "Monochlorostyrene in Fire-Retardant Foam Rubber."
139. Standard Brands Co., Belg. Pat. 729,226 (1969).
140. Polymer Corp., Australian Pat. 32,843/68 (1969).
141. R. G. Bauer, R. W. Kavchok, and J. M. O'Connor, *Am. Chem. Soc. Div. Org. Coat. Plast. Chem. Pap. 36*(2): 715 (1976).
142. A. H. Weinstein, *Rubber Chem. Technol., 54*: 767 (1981).
143. K. Adachi, I. Watanabe, N. Sawatari, and K. Saito, Jpn. Pat. 79 107,996 (1979); Fujitsu Ltd.
144. A. G. Altenau and D. F. Lawson, *Res. Disclosure, 171*: 5 (1978).
145. J. Spanswick, U.S. Pat. 4,101,510 (1978); Standard Oil Co. (Ind.).
146. D. F. Lawson, R. A. Hayes, and A. G. Altenau, U.S. Pat. 4,383,071 (1983); Firestone Tire and Rubber Co.
147. D. F. Lawson, *Res. Disclosure, 261*: 56 (1986).
148. Y. Jya, Y. Wada, M. Aonuma, T. Kobayashi, and K. Inokuchi, U.S. Pat. 3,968,316 (1976); Nippon Zeon Co., Ltd.
149. Zaidan Hojin Kagakuhin Kensa Kyokai, Jpn. Pat. 82 96836 (1982).
150. J. T. Howarth, A. A. Massucco, K. R. G. Sheth, U.S. Pat. Appl. 657,998 (1976); U.S. National Aeronautics and Space Administration.
151. F. C. Weissert, U.S. Pat. 2,880,183 (1959); Firestone Tire & Rubber Co.
152. J. Rucinski, L. Slusarski, W. Rzymski, and W. Przybl, Pol. Pat. 100,348 (1979); Politechnika Lodzka; *Chem. Abstr., 91*: 158863X.
153. R. L. Langer, U.S. Pat. 4,273,879 (1981); Minnesota Mining and Mfg. Co.
154. E. Vasatko, J. Kavalek, and V. Karasek, Germ. Pat. 2,855,691 (1979); Statni Vyzkumny Ustav Materialu.
155. G. Bertelli, R. Locatelli, and P. Roma, Proceedings, International Rubber Conf., Venice, p. 924 (1979).
156. G. Bertelli, R. Locatelli, and P. Roma, *Ind. Gomma, 24*(9): 52 (1980).
157. H. W. Bost, U.S. Pat. 4,026,810 (1977); Phillips Petroleum Co.
158. Caltop S.A., Germ. Pat. 2,704,274 (1977).
159. K. Kobayashi and H. Matsuo, Jpn. Pat. 78 64,259 (1976); Shinko Kasei Kogyo. Co., Ltd.
160. A. F. Robertson, in *Behavior of Polymeric Materials in Fire* (E. L. Schaeffer, ed.), Am. Soc. Test. Mater. Spec. Tech. Publ. No. 816, p. 3 (1983).
161. National Materials Advisory Board, NAS, *Fire Safety Aspects of Polymeric Materials*, Vol. II. *Test Methods, Specifications and Standards*, Technomic Publishing Company, Westport, Conn. (1979).
162. A. H. Landrock, *Handbook of Plastics Flammability and Combustion Toxicology*, Noyes Publications, Park Ridge, N.J. (1983).
163. U.S. Department of Commerce, "Flammability Standard for Mattresses," DOC F 4-72, Fed. Reg. 37 (110), 11362 (June 7, 1972).
164. U.S. Department of Commerce, "Standard for Surface Flammability of Carpets and Rugs," DOC FF 1-70, Fed. Reg. 35 (74), 6211 (April 16, 1970).

165. U.S. Fed. Highway Adm., MVSS 302, Fed. Reg. 36 (5), 289 (January 8, 1971).
166. Am. Soc. Test. Mater., ASTM-E-84-70, ASTM Std. 14, p. 472 (November 1972).
167. Am. Soc. Test. Mater., ASTM-E-162-67, ASTM Std. 14, p. 558 (November 1972).
168. Institute of Electrical and Electronic Engineers, "IEEE Standard for Type Test of Class IE Electrical Cables, Field Splices and Connections for Nuclear Power Generating Stations," IEEE Std. 383 (April 1974).
169. Am. Soc. Test. Mater., ASTM-E-1692-73, ASTM Std. 27, 521 (July 1973).
170. McDonnell-Douglas Co., Douglas Material Specification DMS 1510A (July 9, 1970).
171. Military Specification, MIL-R-0020092G (Ships), Class 3 (January 4, 1972).
172. "Electric Motor-Driven Mine Equipment and Accessories," U.S. Department of Interior Bureau of Mines, Washington, Schedule 2G (1968).
173. Am. Soc. Test. Mater., ASTM D2863-70, ASTM Std. 27, 727 (1972).
174. Underwriters Laboratories, Inc., "Standard for Tests for Flammability of Plastics Materials for Parts in Devices and Appliances," Std. UL94, 3rd ed. (January 1980).
175. Underwriters Laboratories, Inc., "Rubber Insulated Wires and Cables," Std. UL44 (July 1968).
176. Am. Soc. Test. Mater., ASTM E-108-80a, ASTM Std. 18, 927 (1982).
177. Factory Mutual Research Corp., Norwood, Mass.
178. Am. Soc. Test. Mater., ASTM E-648-78, ASTM Std. 18, 1328 (1982).
179. Upholstered Furniture Action Council, Box 2436, High Point, N.C.
180. Am. Soc. Test. Mater., ASTM E-662-79, ASTM Std. 18, 1362 (1982).
181. W. D. Woolley, S. A. Ames, A. I. Pitt, and K. Buckland, *Fire Saf. J.*, 2: 39 (1980).
182. C. P. Fenimore and F. J. Martin, *Combust. Flame, 10*: 135 (1966); F. J. Martin, *Combust. Flame, 12*: 125 (1968); J. L. Isaacs, *J. Fire Flammability, 1*: 36 (1970).
183. C. Anolick, G. S. Cook, and C. W. Stewart, *Rubber Chem. Technol.*, 52: 871 (1979).
184. C. J. Hilado, H. J. Cumming, and R. M. Murphy, *J. Elastomers Plast., 11*: 239 (1979).
185. W. M. Widenor, NASA Tech. Memo. 78523, Compil. Present. Pap., Conf. Fire Resist. Mater. (Firemen), N79-12029, p. 479 (1978).
186. T. Fritz, *Proc. SPI Annu. Tech./Mktg. Conf.*, 27: 346 (1982).
187. R. M. Murch, *J. Consum. Prod. Flammability, 8*: 3 (1981).
188. R. H. Morford, *J. Fire Flammability, 8*: 279 (1977).
189. G. L. Nelson, A. L. Bridgman, W. J. O'Conell, and J. B. Williams, *J. Fire Flammability, 8*: 262 (1977).
190. J. M. Kuchta, M. J. Sapko, F. J. Perzak, and K. E. Mura, *Fire Technol., 17*: 120 (1981).
191. D. Gross, J. J. Loftus, and A. F. Robertson, ASTM Spec. Tech. Publ. No. 422, p. 166 (1967).
192. J. Troitzsch, *International Plastics Flammability Handbook,* Carl Hanser Verlag, Munich (English ed. distributed by Macmillan, New York, 1983).
193. J. B. Terrill, R. R. Montgomery, and C. F. Reinhardt, *Science, 200*: 1343 (1978).
194. D. F. Lawson, in *Flame-Retardant Polymeric Materials,* Vol. 3 (M. Lewin, S. B. Atlas, and E. M. Pearce, eds.), Plenum, New York, Chap. 2 (1982).
195. B. J. Sutker, in *Behavior of Polymeric Materials in Fire* (E. L. Schaeffer, ed.), Am. Soc. Test. Mater. Spec. Tech. Publ. No. 816, p. 78 (1983).
196. M. M. Hirschler, *J. Fire Sci., 3*: 380 (1985).
197. M. M. Hirschler, *J. Fire Sci., 4*: 42 (1986).
198. A. M. Calcraft, R. J. S. Green, and T. S. McRoberts, *Plast. Polym., 42*: 200 (1974).
199. P. G. Edgerley and K. Pettett, *Fire Mater., 2*: 11 (1978).
200. L. H. Breden and M. Meisters, *J. Fire Flammability, 7*: 234 (1976).
201. C. J. Hilado and R. M. Murphy, *J. Elastomers Plast., 12*: 79 (1980).
202. D. F. Lawson and T. C. Cheng, *Fire Res. (Lausanne), 1*: 223 (1978).
203. W. E. Garrison, Jr., U.S. Pat. 4,121,016 (1978); E. I. DuPont de Nemours and Co.
204. A. G. Moody and R. J. Penneck, Germ. Pat. 2,832,893 (1979); Raychem, Ltd.

205. A. Factor, K. N. Sannes, and A. M. Colley, *J. Fire Flammability, 12*: 101 (1981).
206. D. F. Lawson, *Org. Coat. Plast. Chem. Preprints, 43*: 171 (1980).
207. G. F. Baumann and J. F. Szabat, *Adv. Urethane Sci. Technol., 4*: 212 (1975).
208. P. V. Bonsignore, *J. Cell. Plast., 15*: 163 (1979).
209. A. S. Wood, *Modern Plast.*, 60 (May 1984).
210. *Modern Plast., 60*: 56 (March 1983).
211. M. Moseman and J. D. Ingham, *Rubber Chem. Technol., 51*: 970 (1978).
212. D. F. Lawson and E. L. Kay, U.S. Pat. 4,059,560 (1977); Firestone Tire and Rubber Co.
213. R. E. Fenwick, *Plast. Rubber News*, 31 (March 1983).
214. M. M. Hirschler, *Polymer, 25*: 405 (1984).
215. A. Zaopo, Br. Pat. 2,108,130 A1 (1983); Soc. Cavi Pirelli, S.p.A.
216. M. I. Jacobs, Eur. Pat. Appl. 4175 (1979); Uniroyal, Inc.
217. W. J. Kroenke, U.S. Pat. 4,371,655 (1983); B. F. Goodrich Co.
218. W. J. Kroenke, U.S. Pat. 4,129,540 (1978); B. F. Goodrich Co.
219. D. F. Lawson, *Res. Disclosure, 160*: 10 (1977).
220. K. R. Sidman and P. Monaghan, U.S. Pat. Appl. 710,798 (1978); United States National Aeronautics and Space Administration.
221. C. Anolick, L. R. Perkins, C. W. Stewart, J. R. Galloway, and W. Hellyer, in *Ann. Convention, American Hospital Association*, Atlanta, Ga. (1977).
222. H. L. Kaplan, A. L. Grand, and G. E. Hartzell, *Combustion Toxicology. Principles and Test Methods*, Technomic Publishing Company, Lancaster, Pa. (1983).
223. H. Klimisch, H. W. M. Hollander, and J. Thyssen, *J. Combust. Toxicol., 7*: 243 (1980).
224. C. R. Crane, D. C. Sanders, B. R. Endecott, J. K. Abbott, and P. W. Smith, Report No. FAA-AM-77-9, Department of Transportation, Federal Aviation Administration, Office of Aviation Medicine, Washington (March 1977).
225. U.S. National Bureau of Standards, NBSIR 82-2532 (June 1982).
226. Y. C. Alarie and R. C. Anderson, *Toxicol. Appl. Pharmacol., 51*: 341 (1979).
227. C. J. Hilado and J. E. Schneider, *J. Combust. Toxicol., 6*: 91 (1979).
228. National Materials Advisory Board, NAS, *Fire Safety Aspects of Polymeric Materials*, Vol. 3. *Smoke and Toxicity* (*Combustion Toxicology of Polymers*), Technomic Publishing Co., Westport, Conn. (1978).
229. W. A. Burgess, R. D. Treitman, and A. Gold, "Air Contaminants in Structural Fire Fighting," Final Report, NFPCA Grant 7x008, SPI Grant: Fire Ground Survey of Air Contaminants (March 1979).
230. M. M. Birky, M. Paabo, and J. E. Brown, *Fire Saf. J., 2*: 17 (1980).
231. D. A. Purser and P. Grimshaw, *Fire Mater., 8*: 10 (1984).
232. D. A. Purser and P. Buckley, *Med. Sci. Law, 23*: 142 (1983).
233. I. Yamamoto, R. Eyanagi, J. Io, F. Morita, M. Uchida, T. Kagoshima, K. Hayashi, Y. Kondo, and S. Urano, *Daiichi Yakka Daigaku Kenkyu Nempo, 6*: 11 (1975); *Chem. Abstr., 88*: 99952p.
234. W. D. Woolley, S. A. Ames, and P. J. Fardell, *Fire Mater., 3*: 110 (1979).
235. B. C. Levin, M. Paabo, M. L. Fultz, C. Bailey, and W. Yin, Nat. Bur. of Stand. Internal Rep. 83-2791 (1983); NTIS Tech. Rep PB 84-140 227.
236. C. J. Hilado and P. A. Huttlinger, *Fire Technol., 17*: 177 (1981).
237. C. J. Hilado and P. A. Huttlinger, *J. Elastomers. Plast., 13*: 177 (1981).
238. P. Florschuetz, *Freiberg. Forschungsh., A614*: 115 (1979); *Chem. Abstr., 92*: 202841k.
239. R. R. Raje, J. J. Sciarra, and L. Greenberg, *J. Combust. Toxicol., 8*: 45 (1981).
240. N. J. Alvares, A. E. Lipska-Quinn, and H. K. Hasegawa, in *Behavior of Polymeric Materials in Fire* (E. L. Schaffer, ed.), Am. Soc. Test. Mater. Spec. Tech. Publ. 816, p. 46 (1983).
241. R. K. Hinderer, *J. Fire Sci., 2*: 82 (1984).
242. E. S. Lahaniatis, D. Bieniek, L. Vollner, and T. Korte, *Chemosphere, 10*: 935 (1981).

243. H. J. Einbrodt and H. Jesse, *GAK Gummitek. Asbest. Kunstst.*, *36*: 648 (1983).

244. T. J. Smith, A. W. Musk, A. Gold, and P. Roto, *Int. Arch. Occup. Environ. Health, 41*: 139 (1978).

245. C. Herpol, *Fire Mater.*, *7*: 193 (1983).

246. D. F. Lawson, U.S. Pat. 4,361,668 (1982); Firestone Tire and Rubber Co.

247. G. L. Deets and S. P. Nemphos, U.S. Pat. 4,141,931 (1979); Monsanto Corp.

248. C. J. Hilado, A. M. Machado, and D. P. Brauer, *Proc. West. Pharmacol. Soc.*, *22*: 201 (1979).

8

Antiozonant Protection of Rubber Compounds

R. P. Lattimer and R. W. Layer
The BF Goodrich Research and Development Center
Brecksville, Ohio

C. K. Rhee
The Uniroyal Goodrich Tire Company
Brecksville, Ohio

INTRODUCTION

The term *antiozonant,* in its broadest sense, denotes any additive that protects rubber against ozone deterioration. Most frequently, the protective effect results from a reaction with ozone, in which case the term used is *chemical antiozonant.* Ozone is generated naturally by electrical discharge and also by solar radiation in the stratosphere [1]. These sources produce ground-level ozone concentrations of 1–5 parts per hundred million (pphm). In urban environments, however, ozone reaches much higher levels, up to ~25 pphm, due to the ultraviolet photolysis of pollutants [1]:

$$NO_2 \xrightarrow{h\nu} NO + O$$
$$O_2 + O + M \rightarrow O_3 + M$$

Ozone levels are dependent on the time of day and also the season. Levels reach a maximum in late morning and drop to nearly zero at night. Summer months show considerably higher ozone levels than winter months. Indoors, ozone is formed by fluorescent lighting.

Only a few parts per hundred million of ozone in air can cause rubber cracking, which may destroy the usefulness of elastomer products. Ozone will attack any elastomer with backbone unsaturation. Degradation results from the reaction of ozone with rubber double bonds. Unstretched rubber reacts with ozone but is not cracked. In this case, since only the double bonds at the surface are attacked, the degradation is confined to a thin surface layer (~0.5 μm thick). Occasional "frosting" (a white or gray bloomlike appearance) may occur in transparent unstretched rubbers, but elongation is required to induce the characteristic ozone cracking. As the stressed polymer chains cleave under ozone attack, new high-stress surface is exposed. The localized continuation of this process results in visible cracking, which is always perpendicular to the applied stress.

Ozone cracking was first recognized as a serious problem after World War II. Tires on mothballed military vehicles were found to be so badly cracked that they were unserviceable. A government-sponsored research program led to the discovery of chemicals that prevented ozone cracking when added to rubber compounds. Commercial antiozonants have been available since the early 1950s. Since that time, the ozone degradation problem has worsened as atmospheric ozone concentrations have gradually increased, especially in urban industrial areas.

CHEMISTRY OF OZONE ATTACK

Ozone cracking is a physicochemical phenomenon, and many factors are involved in explaining the effect of ozone attack on elastomers [2–7]. From a chemical point of view, ozone attack on olefinic double bonds causes chain scission and the formation of the decomposition products seen in Figure 1 [8, 9]. These chemical reactions are believed to

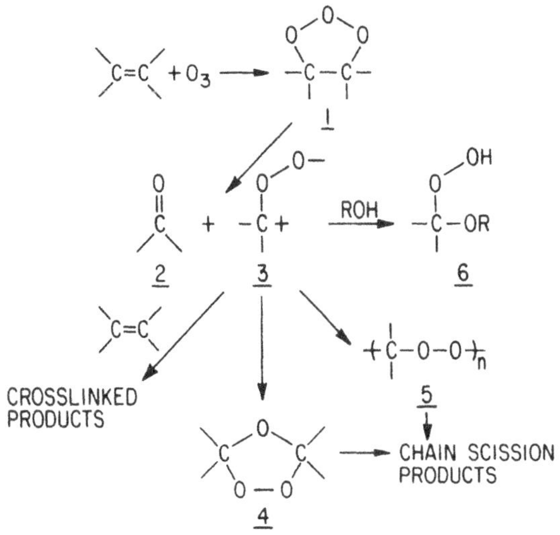

Figure 1 Diene rubber–ozone reaction scheme.

be similar for both small olefins and unsaturated rubbers. The first step is the formation of a relatively unstable primary ozonide (or molozonide, **1**), which cleaves to an aldehyde or ketone **2** and a carbonyl oxide (or zwitterion) **3**. Subsequent recombination of **2** and **3** produces a secondary ozonide (or just ozonide) **4**. In small olefins, ozonide formation is generally a facile process. In stretched rubber, however, ozonide formation is more difficult, since the cleaved intermediates **2** and **3** may be forcefully separated to relieve the stress [10].

Interestingly, both crosslinking and chain scission products may form during rubber ozonation. Rubbers containing trisubstituted double bonds (e.g., polyisoprene, IR, and butyl rubber, IIR) are more prone to yield chain scission products [11]. Several pathways can lead to chain scission. Ozonides are reasonably stable in neutral environments, but they will decompose readily under the influence of heat or various reducing agents to yield such chain scission products as aldehydes, ketones, acids, and alcohols. Polymeric peroxides **5** may be formed initially from the carbonyl oxide, but these are unstable and will eventually decompose to yield chain scission products. The rate of chain scission is increased in the presence of active hydrogen (e.g., water), probably due to reaction with carbonyl oxides to form reactive hydroperoxides **6**.

Crosslinking products may also be formed, especially with rubbers containing disubstituted double bonds (e.g., polybutadiene, BR, and styrene–butadiene rubber, SBR) [11]. It is proposed that this is due to attack of carbonyl oxides, in their biradical form, on the rubber double bonds.

Typical rate constants for the reaction of ozone with various polymers in solution are listed in Table 1 [5]. It is clear that typical diene rubbers (polyisoprene and polybutadiene) have rate constants several orders of magnitude greater than polymers having a saturated backbone (polyolefins). Other unsaturated elastomers having high reaction rates with ozone include styrene–butadiene (SBR) and acrylonitrile–butadiene (NBR) rubbers. Polychloroprene (CR) is less reactive than other diene rubbers (Table 2), and it is therefore inherently more resistant to attack by ozone. The physical aspects of ozone cracking are clearly seen in the case of the predominantly saturated butyl rubber (IIR), a copolymer of isobutylene and isoprene containing ~1–3% isoprene. Although the chemistry is the same as for its diene rubber counterpart, IR, butyl rubber is much more resistant to ozone cracking.

Table 1 Rate Constants of Polymer–Ozone Reaction in Solution at 20°C

Polymer	k_1 (L/mol-sec)
cis-Polyisoprene (IR)	110,000
cis-Polybutadiene (BR)	60,000
Polystyrene	0.3
Polypropylene	0.08
Polyethylene	0.046
Polyisobutene	0.012

Source: Adapted from Ref. 5.

Table 2 Rates of Cut Growth for Various Vulcanizates

Polymer	Rate (mm/min)	
	No DEHA	DEHA[a]
Natural rubber (NR)	0.22	0.26
SBR (25%-styrene)	0.37	0.40
NBR (40%-acrylonitrile)	0.04	0.20
Butyl rubber (IIR)	0.02	0.24
Chloroprene rubber (CR)	~0.01	0.05

[a]With 25 phr diethylhexyladipate.
Source: Adapted from Ref. 17.

PHYSICAL FACTORS IN OZONE ATTACK

As mentioned earlier, unsaturated elastomers must be stretched for ozone cracking to occur. Typically, ozone cracking initiates at sites of high stress (flaws) on the rubber surface. Thus in general the rubber article should be designed to minimize potential sites of high elongation such as raised lettering. Similarly, clean molds should be used to reduce the incidence of surface flaws.

Ozone attack leads to chain scission and the formation and propagation of cracks. As crack growth proceeds, fresh surfaces are continuously exposed for further ozone attack. Degradation continues until failure relieves the inherent stress. It has been calculated that only ~1% of all the ozone that reacts with the rubber is responsible for the formation of cracks [12].

The study of ozone crack growth has been an active area of research since the early 1960s [13–21]. Early work [13] showed that ozone crack growth proceeds at a linear rate for typical elastomers and is proportional to the concentration of ozone. Rates of cut growth for various vulcanizates, both with and without the plasticizer diethylhexyladipate, are listed in Table 2 [17]. The rates in NBR and IIR are roughly an order of magnitude less than those for NR and SBR. This is not due to low unsaturation, but rather because of high internal hysteresis. In other studies [20, 21] it was shown that the rate of crack growth at relatively low temperatures (near T_g) is dependent on segmental mobility, but at higher temperatures (far from T_g) segmental mobility is not an important factor. Also, at higher temperatures the rate is not strongly dependent on the degree of unsaturation of the elastomer. In the presence of an added plasticizer (which increases chain mobility), the cut growth rates for several elastomers (NR, SBR, NBR, and IIR) are rather similar. Note in Table 2 that polychloroprene, despite its high degree of unsaturation, has unusual resistance. This is attributed to the reduced reactivity of the double bond due to the electronegative chlorine atom.

A key result of the early crack growth studies was the "critical stress" effect, i.e., no crack growth occurs unless a specific stress value is exceeded. Some critical stress values for typical elastomers are listed in Table 3 [14]. In practical terms these stress values correspond to threshold tensile strains of ~3–5%, depending on stiffness. It has been found that critical stress values are largely unchanged by temperature, plasticization, and

Table 3 Critical Stress Values

Polymer	S_2 (kg/cm²)	
	20°C	50°C
Natural rubber (NR)	1.23	1.18
SBR (25%-styrene)	1.06	
NBR (40%-acrylonitrile)	0.97	1.13
Butyl rubber (IIR)	0.91	0.68
Chloroprene rubber (CR)	2.30	

Source: Adapted from Ref. 14.

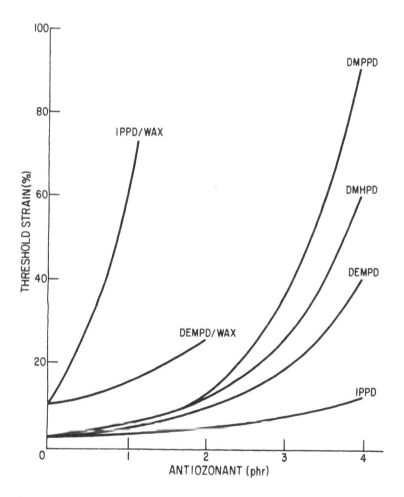

Figure 2 Enhancement of threshold strain of natural rubber by antiozonants. (See Table 6 for abbreviations of antiozonants. Ozone concentration = 25 pphm. Wax, when added, is 1 phr of the microcrystalline type.) (Adapted from Ref. 4.)

ozone concentration. Polychloroprene has a higher critical stress value than other diene rubbers, consistent with its reduced reactivity to ozone.

It has been shown that effective antiozonants have two important functions [15, 18]: (1) all antiozonants decrease the rate of crack growth, and (2) some antiozonants (particularly *N,N'*-dialkyl-*p*-phenylenediamines, *p*-PDAs) increase the critical stress or threshold strain (Figure 2). Some antiozonants reduce the number and/or size of cracks as well as the growth rate. For example, *N*-alkyl-*N'*-aryl-*p*-PDAs reduce the number of cracks, while 6-ethoxy-2,2,4-trimethyl-1,2-dihydroquinoline (a non-*p*-PDA antiozonant) produces a large number of fine cracks, which increase stress relaxation. Recent morphological studies have shown that subsurface ozone cracking is complex and highly branched [22]. Local stresses, as well as local concentrations of antiozonants, are important in determining the number and extent of crack growth. Since ozone reacts much more rapidly with the antiozonant than the rubber, the physical properties of the ozone–antiozonant reaction products must also account for some of these differences.

DESIRABLE PROPERTIES OF ANTIOZONANTS

Some desirable properties of an antiozonant additive are as follows:

1. A *physical* antiozonant must provide an effective barrier against the penetration of ozone at the rubber surface. In this regard, the barrier should be continuous at the surface, unreactive and impenetrable to ozone, and capable of renewing itself if damaged (as by abrasion). Flexibility (extensibility) under dynamic stress conditions is also desirable. A *chemical* antiozonant, on the other hand, must first be extremely reactive with ozone. A compound that is not reactive enough will not protect adequately. An antiozonant must not be too reactive with oxygen, however, or it will not persist long enough in the rubber to afford long-term ozone protection.

2. The antiozonant should possess adequate *solubility* and *diffusivity* characteristics. Since ozone attack is a surface phenomenon, the antiozonant must migrate to the surface of the rubber to provide protection. Poor solubility in rubber may result in a problem with excessive bloom, or else the loading level obtainable may be insufficient for long-term protection. The diffusion rate to the exposed surface must be high enough to meet the incoming ozone flux but low enough to ensure a reservoir in the rubber bulk during the useful lifetime of the article. Because antiozonants must diffuse and replenish themselves on the surface, it is not possible to have an effective "bound antiozonant."

3. The antiozonant should have no adverse effects on the rubber *processing* characteristics (mixing, fabrication, vulcanization, physical properties). In general purpose (diene) rubbers, this implies that the antiozonant must be compatible with sulfur curing systems. The antiozonant should not be destroyed during vulcanization.

4. The antiozonant should be effective under both *static* and *dynamic* conditions over a wide range of extension and temperature conditions.

5. The antiozonant should *persist* in the rubber over its entire life cycle. It should be resistant to loss via oxidation, vaporization, or extraction by water or other solvents.

6. For non-carbon black–filled rubbers, the antiozonant must be *nondiscoloring* and *nonstaining*.

7. The antiozonant should have a *low toxicity* and should be nonmutagenic.

8.   The antiozonant should be acceptable *economically*. That is, it should have a low manufacturing cost and be usable at low bulk concentration levels.

Rubber is protected against ozone attack by the addition of physical and/or chemical antiozonants. Hydrocarbon waxes are the most common type of physical antiozonant, and *p*-phenylenediamine derivatives are the prevalent chemical antiozonants. These two categories are discussed separately.

HYDROCARBON WAXES

Waxes are derived from petroleum and are of two common types, *paraffin* and *microcrystalline* [23–26]. Paraffin waxes have lower carbon numbers, a higher proportion of straight-chain hydrocarbons, and lower melting points compared to the microcrystallines. Typical carbon numbers are $n = 20–50$ for paraffin waxes and $n = 30–70$ for microcrystallines. If a wax is present in a vulcanizate at a level exceeding its solubility, some of it will migrate to the rubber surface where it can form a *physical barrier* to prevent the penetration of ozone. Waxes, of course, are essentially unreactive to ozone, so that there is no appreciable element of chemical protection. Commercial waxes are usually *blends* of paraffin and microcrystalline waxes. The solubility and mobility of various waxes are affected by the polymer type, the filler types and loading, the state of cure, and the time and temperature of storage. Each type of commercial wax has particular solubility and migration characteristics that can be matched to the expected operating environment of the vulcanizate. The various components of typical blends will migrate efficiently over a temperature range of ~10–60°C (Figure 3). Blended waxes have the advantage that the various oligomers will provide an effective bloom over a wider temperature range than

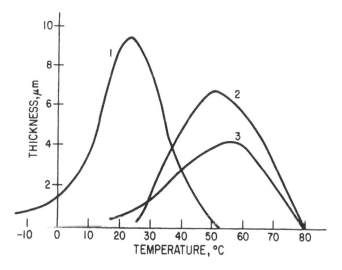

Figure 3   Thickness of wax bloom as a function of temperature. Rubber is a butadiene/α-methylstyrene copolymer. Time of storage = 30 days. Curve 1, paraffin wax; 2, microcrystalline wax; 3, blended wax. (Adapted from Ref. 26.)

will a paraffin or microcrystalline wax alone. At higher temperatures waxes become more soluble in the rubber so that less bloom formation occurs.

Some advantages of waxes are (1) they are of relatively low cost, (2) they are nonstaining, and (3) they generally have no adverse effects on rubber processing/vulcanization. Waxes act as internal lubricants in rubber and increase the scorch safety somewhat. Unfortunately, waxes have a number of shortcomings. First, they are ineffective under dynamic stress conditions. This is likely due to a lack of adhesion between the wax film and the rubber and to the inextensibility of the wax bloom. Second, their protection capability is highly dependent on exposure temperature. Protection is difficult to achieve at both very low (<10°C) and very high (>50°C) temperatures. Third, waxes may easily be lost during storage or use by embrittlement, scuffing, or delamination. Because of these limitations, chemical antiozonants must be used (often in combination with waxes) in most diene rubber applications.

CHEMICAL ANTIOZONANTS

Many compounds have been reported in the literature to be chemical antiozonants, and nearly all contain nitrogen. Compound classes include derivatives of 2,2,4-trimethyl-1,2-dihydroquinoline, N-substituted ureas or thioureas, substituted pyrroles, and nickel or zinc dithiocarbamate salts. The most effective antiozonants, however, are derivatives of p-phenylenediamine (p-PDA):

R—NH—⟨O⟩—NH—R'

The commercial materials are grouped into three classes: N,N'-dialkyl-p-PDAs, N-alkyl-N'-aryl-p-PDAs, and N,N'-diaryl-p-PDAs. The N,N'-dialkyl-p-PDAs (where the alkyl group may be 1-methylheptyl, 1-ethyl-3-methylpentyl, 1,4-dimethylpentyl, or cyclohexyl) are the most effective in terms of their reactivity to ozone [27–29]. These derivatives increase the critical stress required for the initiation of crack growth, and they also reduce the rate of crack growth significantly [18]. The sec-alkyl group is most active, for reasons that are not yet completely clear. The drawbacks of these derivatives are (1) their rapid destruction by oxygen and consequent shorter useful lifetimes, (2) their activity as vulcanization accelerators and hence increased scorchiness, (3) their tendencies to cause dark red or purple discoloration, and (4) their liquid nature and hence difficulty in handling. The dialkyl-p-PDAs are seldom used alone in rubber compounds, although they can be used effectively when blended with alkyl/aryl-p-PDAs.

The N-alkyl-N'-aryl-p-PDAs (where the aryl is phenyl and the alkyl may be isopropyl, sec-butyl, cyclohexyl, 1-methylheptyl, or 1,3-dimethylbutyl) are the most widely used p-PDAs. These derivatives reduce the rate of crack growth and also the number of cracks, but in general-purpose rubbers the critical stress is not appreciably raised unless the compound is combined with a wax or a dialkyl type. It is interesting that their behavior is synergistic with a wax under static conditions to a much greater extent than with dialkyl derivatives (Figure 2) [4]. The alkyl/aryl-p-PDAs are in general excellent antiozonants, particularly in dynamic environments. These derivatives are destroyed only slowly by oxygen and increase the scorchiness of the stock only slightly. In addition to their antiozonant ability, they are also very effective as antifatigue (mechanostabilization) agents and antioxidants. Since they are low-melting solids, they are easily handled. A principal problem in selected applications is that these derivatives (like the dialkyls) are staining.

The *N,N'*-diaryl-*p*-PDAs (where the aryl group may be phenyl, tolyl, or xylyl) are only moderately active antiozonants, which can only be used at low concentrations (generally less than 2 phr) because of their poor solubility. However, they have minimal scorching effects, are the most stable toward oxygen, and stain the least. Their main advantage is high resistance to loss by consumption and vaporization [2], and in combination with more reactive antiozonants they can offer a degree of increased protection in such long-term applications as radial passenger tires.

Very few chemical antiozonants outside the class of *p*-PDAs have much commercial importance. One of the few exceptions to this rule is 6-ethoxy-2,2,4-trimethyl-1,2-di-hydroquinoline, one of the first commercial antiozonants. It should be clear from the foregoing discussion that the principal objection to *p*-PDA antiozonants is their staining characteristics. The lack of suitable alternative antiozonants for light-colored diene rubber articles is one of the major outstanding problems in rubber technology. A notable exception is chloroprene rubber, which has more natural ozone resistance than other diene rubbers. For example, pentaerythritol acetal derivatives have been shown to be effective nonstaining antiozonants for CR [30]. These derivatives are reactive enough with ozone to protect CR vulcanizates, but they are not effective in other general-purpose diene rubbers.

For natural rubber, two compound classes which have some promise as nonstaining antiozonants are substituted thioureas and metal dithiocarbamates [7]. Tributylthiourea has been used as an NR antiozonant, but its activity is considerably less than that of *p*-PDAs and in addition it is very scorchy [4]. Zinc dialkyl dithiocarbamates reduce the rate of crack growth [18], but their scorch resistances are prohibitively low. A few other classes of compounds have been reported to react rapidly with ozone, such as phosphines and stibines. These materials also exhibit antiozonant activity when swollen into rubber vulcanizates after the cure. However, they cannot be used practically for two reasons: (1) they react readily with oxygen, and (2) they are destroyed during vulcanization [31]. Toxicity rules out selenium and tellurium compounds, which have also been reported to have antiozonant activity [4].

MECHANISM OF ACTION OF CHEMICAL ANTIOZONANTS

Several theories have appeared in the literature regarding the mechanism of protection by *p*-PDA antiozonants. The *scavenger* theory states that the antiozonant diffuses to the surface and preferentially reacts with ozone, with the result that the rubber is not attacked until the antiozonant is exhausted [28, 32–33]. The *protective film* theory is similar, except that the ozone–antiozonant reaction products form a film on the surface that prevents attack [32]. The *relinking* theory states that the antiozonant prevents scission of the ozonized rubber or recombines severed double bonds [17]. A fourth theory states that the antiozonant reacts with the ozonized rubber or carbonyl oxide (**3**, Figure 1) to give a low molecular weight, inert *self-healing film* on the surface [3].

The literature suggests that more than one mechanism may be operative for a given antiozonant and that different mechanisms may be applicable to different types of anti-ozonants. All of the evidence, however, indicates that the scavenger mechanism is the most important one. All antiozonants react with ozone at a much higher rate than does the rubber which they protect. For example, selected antiozonants have been shown to react preferentially with ozone in rubber films as well as in solution (Table 4) [5, 33]. Anti-ozonant–ozone reaction rates are typically one to two orders of magnitude higher than the rates for diene rubbers. Surface spectroscopy also supports preferential attack of ozone on

Table 4 Rate Constants of Ozone Reaction in Solution at 20°C

Compound	$k \times 10^{-5}$ (L/mol-sec)
N,N'-diisopentyl-p-PDA	80
N,N'-di-n-octyl-p-PDA	70
N-phenyl-N'-isopropyl-p-PDA	70
N,N,N'-tributylthiourea	20
cis-Polyisoprene (IR)	4.4
cis-Polybutadiene (BR)	0.6

Source: Adapted from Ref. 5.

the antiozonant [34, 35]. The ozonation products of two common antiozonants have been characterized in detail [36, 37]. Among the numerous products isolated are nitroso and nitro compounds, nitrones, and amides. The extremely high reactivity of p-phenylene-diamines, compared to other amines, with ozone has been shown to be due to their unique ability to react free radically with ozone [29]. Other evidence shows that antiozonants diffuse to the rubber surface rapidly enough to account for protection against ozone attack [38].

Although all antiozonants must react rapidly with ozone, not all highly reactive materials are antiozonants. Something else in addition to the scavenging effect is required. The protective film theory contends that ozonized products, to a considerable extent, prevent ozone from reaching the rubber. There is visual and microscopic evidence for formation of a protective film on the rubber surface [35]. Spectroscopic characterization has shown that this film consists of unreacted antiozonant and many of the same components observed in ozonized liquid antiozonant [34–36]. The components of the film are polar and tend not to diffuse back into the rubber bulk. The surface film evidently acts as a secondary scavenger, as well as a partial physical barrier, for ozone. It has been assumed that the physical properties of these films play an important role in the functioning of an antiozonant. Differences in these properties have been used to explain why N,N'-tri-substituted and N,N'-tetrasubstituted p-phenylenediamines are not as effective as their N,N'-disubstituted counterparts, even though all of these derivatives scavenge ozone at about the same rate.

The relinking [17] and self-healing film [3] theories require chemical interaction between the antiozonant, or ozonized antiozonant, and the rubber or ozonized rubber. The evidence for these interactions is sparse in the literature [39, 40]. The products of the ozone–antiozonant reaction are soluble in acetone [39]. Thus if only the scavenger and protective film mechanisms are operative, no nitrogen from the antiozonant should be left in the rubber after ozonation and subsequent acetone extraction. Nitrogen analyses of extracted rubber showed, however, that some of the nitrogen was unextractable; this nitrogen was presumably attached chemically to the rubber network [39]. Experiments showing that antiozonants react readily with aldehydes led to the suggestion that the antiozonant may work by relinking aldehydic end groups of rubber chains that have been broken by ozonolysis [40]. More recent work has shed doubt on these conclusions, however [41, 42]. While it was confirmed that vulcanization and ozone aging can lead to attachment of p-PDA fragments to the rubber matrix, mechanisms other than antiozonant action can cause this unextractable nitrogen effect. In particular, it has been proposed that the principal reason for nitrogen becoming bound during aging is the trapping of macro-alkyl radicals by aromatic amine-derived species (e.g., nitrones and nitroxides) [42].

In summary, the scavenger model is believed to be the principal mechanism of anti-ozonant action. Ozone–antiozonant reaction products form a surface film that provides additional protection against ozone. Although some evidence indicates that chemical interactions can occur between the antiozonant and rubber, these effects are secondary. Although the evidence for these conclusions has been gathered almost exclusively from experiments with *p*-PDA derivatives, recent work suggests similar conclusions for penta-erythritol-based compounds [30].

MANUFACTURE AND PRODUCTION

The *N,N'*-dialkyl-*p*-PDAs are manufactured by reductively alkylating *p*-PDA with ke-tones. Alternatively, these compounds can be prepared from the ketone and *p*-nitroaniline with catalytic hydrogenation. The *N*-alkyl-*N'*-aryl-*p*-PDAs are made by reductively alkyl-ating *p*-nitro-, *p*-nitroso-, or *p*-aminodiphenylamine with ketones. *p*-Nitrodiphenylamine is made by condensing *p*-chloronitrobenzene with aniline using a cuprous salt as the catalyst. *p*-Nitrosodiphenylamine is made from diphenylamine and nitrosyl chloride in the presence of excess HCl. *N,N'*-Diaryl-*p*-PDAs are made by condensing aniline or its derivatives with hydroquinone at elevated temperatures using an acid catalyst.

The United States annual production of substituted *p*-PDAs (Table 5) [43–45] is related to automobile production. A list of commercial antiozonants is given in Table 6 [43, 46].

ALTERNATIVES TO ANTIOZONANTS

Any diene rubber article subjected to flexing, bending, or folding requires protection against ozone. Chemical antiozonants, primarily *p*-PDAs, are used in general-purpose commercial applications (mainly tires, hoses, flat belts, and transmission belts) where staining and discoloration are not serious problems. Waxes act as antiozonants for diene rubbers under conditions requiring little or no flexing, such as tie-down straps. Under these conditions, waxes are more effective than the *p*-PDA antiozonants [24]. However,

Table 5 U.S. Production of Phenylenediamine Antiozonants (metric tons)

	1970	1973	1976	1978	1980	1981	1983	1984
Total substituted *p*-PDAs[a]	27,353	32,564	32,559	32,732	24,716	28,091	29,441	26,283
N,N'-di(1,4-dimethyl pentyl)-*p*-PDA	2,421	2,744			2,260	2,338		
U.S. factory auto sales, thousands[b]	6,547	9,658	8,498	9,165	6,375[c]	6,253[c]	6,781	7,773

[a]Listed in Table 6.

[b]Adapted from Ref. 44.

[c]Adapted from Ref. 45.

Source: Adapted from Ref. 43, except where otherwise stated.

Table 6 U.S. Manufacturers of Commercial Antiozonants

p-Phenylenediamine	Manufacturer	Trade name	Price[a] ($/kg)	MP (°C)
N,N'-di(1,4-dimethylpentyl) (DMPPD)	Monsanto	Santoflex 77	6.33	Liquid
	Universal O.P.	UOP 788		
	Uniroyal	Flexzone 4L		
	American Cyanamid	Cyzone DH		
	Eastman	Eastozone 33		
	Mobay	Vulkanox 4030		
N,N'-di(1-ethyl-3-methylpentyl) (DEMPD)	Uniroyal	Flexzone 8L	6.77	Liquid
	Universal O.P.	UOP 88		
	R.T. Vanderbilt	Antozite 2		
N,N'-di(1-methylheptyl) (DMHPD)	Universal O.P.	UOP 288	6.28	Low MP
	R.T. Vanderbilt	Antozite 1		
N-cyclohexyl-N'-phenyl (CHPD)	Uniroyal	Flexzone 6H	7.01	110
	Mobay	Antioxidant 4010		
	Universal O.P.	UOP 36		
N-(1,3-dimethylbutyl)-N'-phenyl (HPPD)	Goodyear	Wingstay 300	6.64	48–50
	Monsanto	Santoflex 13		
	Uniroyal	Flexzone 7L/7F		
	Universal O.P.	UOP 588		
	R.T. Vanderbilt	Antozite 67		
	Mobay	Vulcanox 4020		
	Pennwalt	Anto$_3$ "E"		
	Akron Chemical	Antiozonant PD-2		
	Vulnax	Permanax 6PPD		
N'-isopropyl-N'-phenyl (IPPD)	Uniroyal	Flexzone 3C	7.08	70–78
	Monsanto	Santoflex IP		
	Mobay	Vulkanox 4010 NA		
	American Cyanamid	Cyzone IP		
	Universal O.P.	UOP 388		
	Akron Chemical	Antiozonant PD-1		
	Pennwalt	Anto$_3$ "H"		
	Vulnax	Permanax IPPD		
N,N'-mixed ditolyl (DTPD)	Goodyear	Wingstay 100	5.69	90–150
	Goodyear	Wingstay 200		60
N,N'-diphenyl (DPPD)	R.T. Vanderbilt	AgeRite DPPD	9.37	144–153
	Uniroyal	J-Z-F		
	Vulnax	Permanax DPPD		
N,N'-di(2-naphthyl) (DNPD)	R.T. Vanderbilt	AgeRite White	12.43	224–230
	Mobay	Vulkanox DNP		
N,N'-dicyclohexyl (DCHPD)	Universal O.P.	UOP 26		106

[a]Adapted from Ref. 46.

Source: Adapted from Ref. 43.

in applications involving dynamic stress where discoloration cannot be tolerated, alternatives to the traditional antiozonants must be used. Typical applications include white tire sidewalls, gaskets, weatherstripping, gloves, and sporting goods. The two principal alternatives to antiozonants are ozone-resistant elastomers and blends of ozone-resistant elastomers and diene rubbers.

Ozone-Resistant Elastomers

Ozone-resistant elastomers which have no unsaturation are an excellent choice when their physical properties suit the application, for example, polyacrylics, polysulfides, silicones, polyesters, and chlorosulfonated polyethylene [47]. Such polymers are also used where high ozone concentrations are encountered. Elastomers with pendant, but not backbone, unsaturation are similarly ozone-resistant. Elastomers of this type are the ethylene–propylene–diene (EPDM) rubbers, which possess a weathering resistance that is not dependent on environmentally sensitive stabilizers. The impact of these ozone-resistant elastomers has been more selective than general, however, since producers are reluctant to forfeit such properties as resilience, resistance to physical degradation (via stress relaxation, tear, and abrasion), and ease of processing (including green strength, building tack, and cure rate) [7]. Another important factor is that most of the ozone-resistant elastomers are considerably more costly than traditional diene rubbers.

Other elastomers, such as butyl rubber (IIR) with low double-bond content, are fairly resistant to ozone. As unsaturation increases, ozone resistance decreases. Butyl and nitrile rubbers have naturally low crack growth rates in their unplasticized states (Table 2). Chemical antiozonants are frequently used to enhance ozone resistance in these elastomers. Chloroprene rubber (CR) is also quite ozone-resistant. In CR the N,N'-diaryl-p-PDAs are more effective than the N,N'-dialkyl derivatives, a reversal of the normal order [48]. Other antiozonants for CR include nickel dibutyldithiocarbamate, 5-norbornene-2-methanol derivatives, and pentaerythritol acetals.

Blends

Blends of diene and backbone-saturated rubbers are frequently used in applications where discoloration by chemical antiozonants cannot be tolerated, yet where cost is still a primary consideration (e.g., white sidewalls of tires). Disadvantages are that physical properties have to be compromised and usually the two rubbers differ greatly in their rates of vulcanization. Usually at least a 25% replacement by the ozone-resistant rubber is needed for an appreciable enhancement in ozone protection [7]. It is necessary for the two rubbers to exist as separate domains after mixing. Then, as microcracks form and grow on the surface, they encounter a region of ozone-resistant rubber, which relieves the local stress and stops the crack growth [5]. Perhaps the first such blend was the addition of more than 30 phr EPDM to SBR [49]. Halogenated butyl rubbers, chlorosulfonated polyethylene, poly(vinyl acctate), and poly(vinyl chloride) have also been used [5, 50].

TESTING

Laboratory Tests

Since antiozonants are affected by most compounding ingredients, each new rubber compound requires the development of a cost-effective antiozonant system. Outdoor as well as accelerated ozone-chamber tests are available [51–56]. Laboratory tests involve exposing a statically or dynamically elongated test sample to ozone and measuring the time to crack formation, the severity of cracking, or the decay of 100% modulus with time. Cracking is affected by ozone concentration and flow rate, temperature, humidity, sample shape, and type of strain (static, dynamic, or both). Cracking is accelerated by increasing ozone concentration. Normal test concentrations are 10–25 pphm, but they can be as high as 50 pphm in accelerated tests. Ozone concentrations should not exceed 200 pphm, since at this level the system cannot be protected and the antiozonant cannot be replenished by migration to the surface. Associated with concentration is the flow rate of the ozone stream. An increase in the flow increases the diffusion of the ozone onto the surface of the rubber and accelerates cracking. Elevated temperatures increase the reaction rate and affect the migration rate of the antiozonant. Above 70°C ozone decomposes, and therefore testing is usually carried out at 30–50°C. Humidity can accelerate ozone cracking, and the maximum recommended value is ~65%.

Strain affects both the number and size of cracks. Low strain (10–20%) causes fewer but larger cracks; higher strain gives numerous small cracks. Testing is usually done at low strain. Most antiozonants are used in applications where flexing is involved, and dynamic tests are needed. However, since actual field conditions involve a combination of static and dynamic stress, an intermittent static–dynamic ozone exposure often provides a more realistic test regimen. In Figure 4 it can be seen that static, dynamic, and intermittent testing can give results which differ greatly.

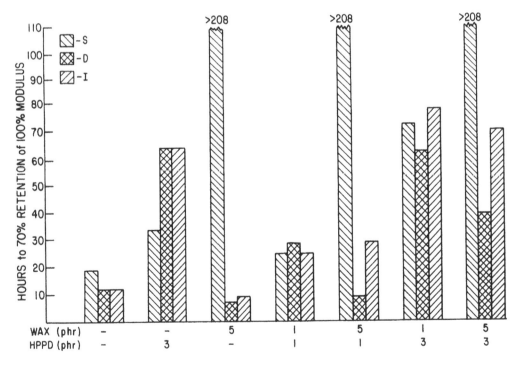

Figure 4 Effects of wax/p-PDA combinations on ozone resistance under varied laboratory test conditions. Rubber is an NR tread compound. Wax is of the blended type. HPPD is N-(1,3-dimethylbutyl)-N'-phenyl-p-PDA. S, static; D, dynamic; I, intermittent. (Adapted from Ref. 24.)

Field Tests

In the laboratory one can study in a controlled environment the various degradative forces that act on a rubber compound (oxygen, ozone, heat, fatigue, and others). During its useful lifetime, however, a rubber article sees a variety of degradative forces in a continuing but lower level than is experienced in the laboratory. Thus laboratory tests are best used as screening procedures for new additives and compounds. Only field testing of rubber articles can provide a true judgment of the protective capability of a particular formulation. For example, extensive field testing of tires is necessary to establish optimum formulations for the various components. In actual tire testing of antidegradants, the degree of cracking is rated. Crack rating is a difficult process, and arbitrary (subjective) systems are used. Nevertheless, good correlations between laboratory and road tire tests have been reported for different antiozonants at various concentrations [57]. In Figure 5 it is clear that increasing levels of antiozonant provide better crack protection in tire tests.

HEALTH AND SAFETY FACTORS

The first *p*-PDA antiozonants were low molecular weight N,N'-dialkyl-*p*-PDAs, which caused skin irritations. Current higher molecular weight N,N'-dialkyl or N-alkyl-N'-aryl derivatives are not primary skin irritants. A notable exception is N-isopropyl-N'-phenyl-*p*-

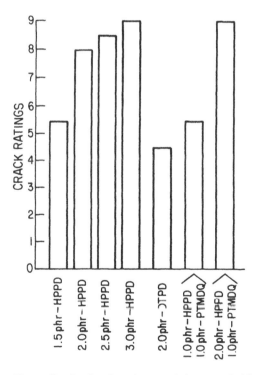

Figure 5 Crack ratings from road tire tests. Rubber is an SBR/BR sidewall compound. Distance driven is 62,000 km. A rating of 10 = no cracks. See Table 6 for abbreviations of antiozonants; PTMDQ = polymerized 2,2,4-trimethyl-1,2-dihydroquinoline. (Adapted from Ref. 57.)

PDA, which causes dermatitis. However, since some individuals are more sensitive than others, antiozonants should always be handled with care [58]. When skin contact does occur, the affected area should be washed with mild soap and water. In case of eye contact, flush well with water. Inhalation of rubber chemicals should be avoided, and respiratory equipment should be used in dusty areas.

USES AND FORMULATIONS

Chemical antiozonants are routinely used to protect diene rubbers (NR, IR, BR, SBR, NBR) against atmospheric ozone for extended periods of time. Large volumes are used in tire, belt, and hose applications. The N-alkyl-N'-aryl-p-PDAs have largely displaced the N,N'-dialkyl-p-PDAs as the materials of choice since they are less scorchy, more persistent, and more easily handled. The N,N'-diaryl-p-PDAs are more effective antiozonants for chloroprene rubber (CR). In addition, these derivatives provide the best bin cure protection. (Bin cure is the tendency for compounded CR to vulcanize prematurely during storage.)

Antiozonants (p-PDAs) are added to raw rubber stocks during mixing at 1–5 phr concentrations and, most frequently, at 2–4 phr. Too little antiozonant increases the severity of cracking. On the other hand, there is a maximum (normally about 5 phr) above which no further protection is obtained [57]. Paraffin waxes (1–2 phr) may also be added to rubber formulations containing p-PDAs for increased static ozone protection [24] (see Figure 4). Caution should be used in these formulations, however, since wax may cause a significant *decrease* in ozone protection in a dynamic environment. In general, wax levels above 1.5–2.0 phr should be used only after careful testing to assure that the dynamic ozone protection is not reduced.

Certain fillers may absorb or oxidize p-PDA antiozonants, slightly increasing the antiozonant requirement. Absorption increases with acidity of the filler. Similarly, as the oxygen content of the carbon black increases, more antiozonant is required. These effects are strongest for N,N'-dialkyl derivatives and less so for the more stable, less basic N-alkyl-N'-aryl-p-PDAs.

The p-PDA antiozonants are compatible with sulfur-curing systems. However, N,N'-dialkyl-p-PDAs and to a lesser extent N-alkyl-N'-aryl-p-PDAs increase scorchiness and cure rate. An effective solution when this is a problem is to use a prevulcanization inhibitor (PVI), several of which are available commercially. Another approach is to add the quinone diimine form of the antiozonant to the rubber recipe [7]. This will reduce back to the parent p-PDA with the application of heat during cure. Unfortunately, p-PDAs are not compatible with peroxide-curing systems [59].

Extender oils and plasticizers increase the ozone cracking of unprotected rubbers. This is due to increased chain mobility and a more facile separation of the ozonized rubber fragments [17, 60]. Although extender oils increase the antiozonant requirement for butyl rubbers [61], little or no additional antiozonant is required for SBR and NR vulcanizates [62].

General-purpose antioxidants extend the service life of antiozonants and reduce the overall cost of the rubber protective system. For example, the mixture of polymerized 2,2,4-trimethyl-1,2-dihydroquinoline and an N-alkyl-N'-aryl-p-PDA has been shown to be an effective long-lasting antidegradant combination (see Figure 5) [57].

Water can leach antiozonants and shorten the life of the product [63]. Leaching of N,N'-dialkyl-p-PDAs decreases with increasing size of the alkyl group and becomes negligible for groups containing more than seven carbon atoms. *N*-alkyl-*N'*-aryl-*p*-PDAs are less easily extracted than the dialkyl derivatives of similar molecular weight [64]. Oil-resistant nitrile rubbers (NBR) are used in applications involving contact with petroleum products. Under these conditions, the *p*-PDA antiozonants are leached from the rubber. Poly(vinyl chloride) may be used for ozone protection in these applications [50].

REFERENCES

1. Air Quality Criteria for Photochemical Oxidants, Natl. Air Pollution Admin. Publ. No. AP-63, U.S. Government Printing Office, Washington (1970).
2. J. C. Ambelang, R. H. Kline, O. M. Lorenz, C. R. Parks, and C. Wadelin, *Rubber Chem. Technol.*, *36*: 1497 (1963).
3. L. D. Loan, R. W. Murray, and P. R. Story, *J. Inst. Rubber Ind.*, *2*: 73 (1968).
4. P. M. Lewis, *NR Technol.*, *1*: 1 (1972).
5. S. D. Razumovskii and G. E. Zaikov, *Dev. Polym. Stab.*, *6*: 239 (1983).
6. C. K. Rhee, R. P. Lattimer, and R. W. Layer, in *Encyclopedia of Polymer Science and Engineering*, Vol. 2 (H. Mark, ed.), Wiley, New York (1985).
7. P. M. Lewis, *Polym. Degrad. Stab.*, *15*: 33 (1986).
8. R. Criegee, *Angew. Chem. Int. Ed. Eng.*, *14*: 745 (1975).
9. R. L. Kuczkowski, *Acc. Chem. Res.*, *16*: 42 (1983).
10. A. S. Kuzminsky, *Dev. Polym. Stab.*, *4*: 71 (1981).
11. K. W. Ho, *J. Polym. Sci. Part A*, *24*: 2467 (1986).
12. S. D. Razumovsky, V. V. Podmasteriyev, and G. Zaikov, *Polym. Degrad. Stab.*, *16*: 317 (1986).
13. M. Braden and A. N. Gent, *J. Appl. Polym. Sci.*, *3*: 90 (1960).
14. M. Braden and A. N. Gent, *J. Appl. Polym. Sci.*, *3*: 100 (1960).
15. M. Braden and A. N. Gent, *Trans. Inst. Rubber Ind.*, *37*: 88 (1961).
16. E. H. Andrews and M. Braden, *J. Polym. Sci.*, *55*: 787 (1961).
17. M. Braden and A. N. Gent, *Rubber Chem. Technol.*, *35*: 200 (1962).
18. M. Braden and A. N. Gent, *J. Appl. Polym. Sci.*, *6*: 449 (1962).
19. E. H. Andrews and M. Braden, *J. Appl. Polym. Sci.*, *7*: 1003 (1963).
20. A. N. Gent and J. E. McGrath, *J. Polym. Sci. Part A*, *3*: 1473 (1965).
21. A. N. Gent and H. Hirakawa, *J. Polym. Sci. Part A*, *5*: 157 (1967).
22. L. L. Ban, M. J. Doyle, and G. R. Smith, *Rubber Chem. Technol.*, *59*: 176 (1986).
23. F. Jowett, *Elastomerics, 111*(9): 48 (1979).
24. D. A. Lederer and M. A. Fath, *Rubber Chem. Technol.*, *54*: 415 (1981).
25. F. Jowett, *Rubber World, 188*(2): 24 (1983).
26. R. M. Mavrina, L. G. Angert, I. G. Anisimov, and A. V. Melikhova, *Kauch. Rezina, 31*(12): 29 (1972).
27. R. F. Shaw, Z. T. Ossefort, and W. J. Touhey, *Rubber World, 130*: 636 (1954).
28. W. L. Cox, *Rubber Chem. Technol.*, *32*: 364 (1959).
29. R. W. Layer, *Rubber Chem. Technol.*, *39*: 1584 (1966).
30. D. Brück, H. Konigshofen, and L. Ruetz, *Rubber Chem. Technol.*, *58*: 728 (1985).
31. G. Bertrand and E. Leleu, *Rubber World, 192*(1): 32 (1985).
32. E. R. Erickson, R. A. Berntsen, E. L. Hill, and P. Kusy, *Rubber Chem. Technol.*, *32*: 1062 (1959).
33. S. D. Razumovskii and L. S. Batashova, *Rubber Chem. Technol.*, *43*: 1340 (1970).
34. J. C. Andries, D. B. Ross, and H. E. Diem, *Rubber Chem. Technol.*, *48*: 41 (1975).
35. J. C. Andries, C. K. Rhee, R. W. Smith, D. B. Ross, and H. E. Diem, *Rubber Chem. Technol.*, *52*: 823 (1979).

36. R. P. Lattimer, E. R. Hooser, H. E. Diem, R. W. Layer, and C. K. Rhee, *Rubber Chem. Technol., 53*: 1170 (1980).
37. R. P. Lattimer, E. R. Hooser, R. W. Layer, and C. K. Rhee, *Rubber Chem. Technol., 56*: 431 (1983).
38. G. J. Lake, *Rubber Chem. Technol., 43*: 1230 (1970).
39. O. Lorenz and C. R. Parks, *Rubber Chem. Technol., 36*: 194 (1963).
40. O. Lorenz and C. R. Parks, *Rubber Chem. Technol., 36*: 201 (1963).
41. R. P. Lattimer, R. W. Layer, and C. K. Rhee, *Rubber Chem. Technol., 57*: 1023 (1984).
42. R. P. Lattimer, J. Gianelos, H. E. Diem, R. W. Layer, and C. K. Rhee, *Rubber Chem. Technol., 59*: 263 (1986).
43. Synthetic Organic Chemicals, U.S. Production and Sales, 1970-1985, U.S. Government Printing Office, Washington (1985).
44. U.S. Department of Commerce, Business Statistics, U.S. Government Printing Office, Washington (1985).
45. *The World Almanac Book of Facts 1986*, Newspaper Enterprise Association, New York (1986).
46. "Compounding Ingredient Price List," *Rubber World, 194*(6): 49 (1986).
47. *The Vanderbilt Rubber Handbook* (R. O. Babbit, ed.), R. T. Vanderbilt Co., Norwalk, Conn. (1978).
48. D. H. Geschwind, W. F. Gruber, and J. Kalil, *Rubber Age, 99*(11): 69 (1967).
49. Z. T. Ossefort and E. W. Bergstrom, *Rubber Age, 101*(9): 47 (1969).
50. J. R. Dunn, *Dev. Polym. Stab., 4*: 223 (1981).
51. A. Hartmann and F. Glander, *Rubber Chem. Technol., 29*: 166 (1956).
52. J. Crabtree and A. R. Kemp, *Ind. Eng. Chem. Anal. Ed., 18*: 769 (1946).
53. J. R. Beatty and A. E. Juve, *Rubber World, 131*(2): 232 (1954).
54. A. G. Veith, "Report on Interlaboratory Ozone Test Program of ASTM Committee D-11, Subcommittee XV on Life Tests for Rubber Products, 1957," Am. Soc. Test. Mater. Spec. Tech. Publ. No. 229, 113 (1958).
55. K. M. Davies and D. G. Lloyd, *Dev. Polym. Stab., 4*: 111 (1981).
56. *1980 Annual Book of ASTM Standards*, American Society for Testing and Materials, Philadelphia, Parts 37 and 38 (1980).
57. L. A. Walker and J. J. Luecken, *Elastomerics, 1980*(5): 36.
58. W. E. McCormick, *Rubber Chem. Technol., 45*: 627 (1972).
59. Z. T. Ossefort, *Rubber Chem. Technol., 32*: 1088 (1959).
60. D. J. Buckley, *Rubber Chem. Technol., 32*: 1475 (1959).
61. D. C. Edwards and E. B. Storey, *Rubber Chem. Technol., 28*: 1096 (1955).
62. W. L. Cox, *Rubber World, 140*: 88 (1959).
63. G. R. Browning and R. R. Barnhart, *Rubber Chem. Technol., 44*: 1441 (1971).
64. E. J. Lotos and A. K. Sparks, *Rubber Chem. Technol., 42*: 1471 (1969).

9

Antioxidants and Stabilizers for Hydrocarbon Polymers: Past, Present, and Future

S. Al-Malaika

Aston University
Birmingham, England

INTRODUCTION

Product performance of most organic polymers is severely curtailed by extreme conditions and environments prevalent during manufacture and use. The combination of oxidative processes (to which industrial polymers are subjected during high-temperature processing and under subsequent aggressive service conditions) and physical effects (such as compatibility and volatility) are responsible for the undesirable alterations in mechanical properties and ultimate failure of manufactured polymer articles.

The diversity of commercial polymers and their technical applications and the challenging requirements for the stability of these polymers under different environmental conditions (e.g., high temperature, stress, UV light, ozone, high-energy irradiation) increase the demand for more efficient stabilizers and antioxidants having optimum technical properties. Most hydrocarbon polymers cannot be processed and used without antioxidants.

Although antioxidants and stabilizers are normally used at very low concentrations (in the range of 0.1–0.5% w/w), they have a key position in compounding ingredients.

The term antioxidant goes back to the early work of Moureu and Dufraisse, who in 1921 introduced the term "antioxygen" to describe compounds that act catalytically by retarding oxidation. Different terminologies have been used for antioxidants in different polymer matrices: antidegradants, antifatigue agents, and antiozonants are used in rubber technology, whereas the plastics industry prefers the use of melt, heat, or light stabilizers. The terms primary and secondary antioxidants are also often used but the distinction is arbitrary and offers little to the understanding of their mechanisms of action. In this chapter the term antioxidant is used to describe all chemical agents that inhibit the polymer matrix from attack by oxidants; they are classified (or identified) according to their mode of action in the oxidative process.

The development of antioxidants and stabilizers began as an empirical science more than 100 years ago but real scientific and technological progress began relatively recently when an understanding of the basic underlying mechanisms of polymer degradation began to emerge. This has made it possible to ascribe specific chemical and physical functions to antioxidants and stabilizers according to their mode of action in the oxidative process. An important aspect in the development of our understanding of the role of antioxidants is the realization that their effectiveness depends not only on the chemical inhibition process but also on physical factors such as solubility of antioxidants and their compatibility with the polymer, diffusion and migration phenomena within the polymer matrix, volatility and extractability of antioxidants. As polymers are being used progressively in more demanding engineering and medical applications in which they have to withstand increasingly aggressive environments, the importance of the physical processes (mainly related to the loss of antioxidants from the polymer matrix) becomes evident to such an extent that in some cases its contribution to the antioxidant activity would exceed that of the function of the chemical structure of the antioxidant.

EARLY THEORIES OF AUTOOXIDATION AND ANTIOXIDANT ACTION

Deterioration of rubber was recognized as early as 1800, although the process of rubber oxidation was unknown until Hofmann [1], in 1861, showed that oxygen is the primary cause of rubber deterioration under ambient conditions. The existence of peroxides in simple oxidizable substrates was a major discovery at the turn of the century [2].

The work of Moureu and Dufraisse in the 1920s proved to be a cornerstone in the development of concepts of autooxidation and antioxidant action. Oxidizable compounds were now considered to compete effectively with the substrate for oxygen, and compounds that could retard oxidation catalytically were described as "antioxygens." In the mid-1920s Moureu and Dufraisse proposed a theory (Scheme 1) that accounts for the role of antioxygens as inhibitors which act by destroying active peroxides, which are the primary cause of oxidative degradation [3]. Although their general mechanism did not account for the nature of the reactive intermediate, and hence is now only of historical interest, it was very important at the time because it recognized the significance of peroxides in autooxidation. Subsequent progress in the study of reactive intermediates involved in autooxidation led Bäckstrom [4] to propose kinetic chain theory (see Scheme 1) in which alkyl peroxyl radicals were considered to be chain-propagating intermediates (e.g., during photooxidation of benzaldehyde to benzoic acid; Scheme 1b).

Bäckstrom's theory was elaborated in the late 1940s by researchers at the British Rubber Producer Research Association (BRPRA), who consolidated a basic autooxidation theory [5]. This was the turning point in autooxidation studies and all modern developments in this area are still based on the pioneering work carried out at BRPRA. According

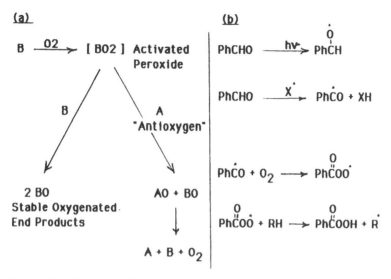

(a)

B $\xrightarrow{\text{O2}}$ [BO2] Activated Peroxide

B

A
"Antioxygen"

2 BO
Stable Oxygenated.
End Products

AO + BO

A + B + O$_2$

(b)

PhCHO $\xrightarrow{h\nu}$ PhĊH (O)

PhCHO $\xrightarrow{\dot{X}}$ PhĊO + XH

PhĊO + O$_2$ $\longrightarrow$ PhC(=O)OO·

PhC(=O)OŎ + RH $\longrightarrow$ PhC(=O)OOH + Ṙ

Scheme 1 Moureu and Dufraisse's catalytic "antioxygen" mechanism (a), and Backstrom's free radical chain theory of autooxidation (b).

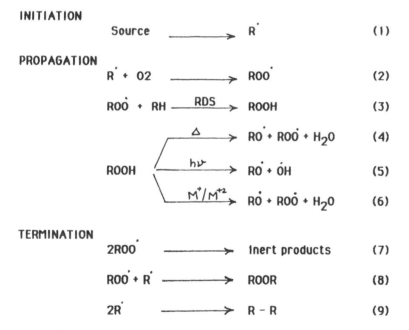

INITIATION

Source $\longrightarrow$ R· (1)

PROPAGATION

R· + O2 $\longrightarrow$ ROO· (2)

ROŎ + RH $\xrightarrow{\text{RDS}}$ ROOH (3)

ROOH $\xrightarrow{\Delta}$ RO· + ROO· + H$_2$O (4)

$\xrightarrow{h\nu}$ RO· + ȮH (5)

$\xrightarrow{M^+/M^{+2}}$ RO· + ROŎ + H$_2$O (6)

TERMINATION

2ROO· $\longrightarrow$ Inert products (7)

ROŎ + R· $\longrightarrow$ ROOR (8)

2R· $\longrightarrow$ R – R (9)

Scheme 2 Radical mechanism of hydrocarbon autooxidation.

to their chain reaction theory, a complex set of elementary reaction steps for hydrocarbon autooxidation was shown as a series of initiation, propagation, and termination steps (Scheme 2). An important result of the work of Bolland, Bateman, and co-workers (at BRPRA) relates to the importance of oxygen concentration. Under atmospheric pressure and conditions where there is a constant supply of free radicals, reaction 3 (Scheme 2) is rate-determining and alkylperoxyl radicals are by far the most important species present, hence termination occurs almost exclusively by bimolecular reactions of alkylperoxyl radicals (reaction 7). At low oxygen pressures, however, termination proceeds mainly by reactions of alkyl radicals (reactions 8 and 9) [6]. A more detailed account of the history of antioxidant and stabilizers has been reviewed recently [7].

POLYMER DEGRADATION AND PROCESSING

The free radical chain reaction model which was originally developed by researchers of the BRPRA [5] for low molecular weight olefins was later successfully used for a range of polymers [8, 9]. Similar radical chain processes occur during the processing operation employed to convert thermoplastic polymers to finished products since this involves not only high temperature and the presence of a small amount of oxygen, but also conditions of high shear during the melting of the polymer. Unlike low molecular weight compounds, polymers undergo bond scission in their backbone when subjected to shear.

Mechanical scission of carbon–carbon bonds in the polymer backbone was first inferred from rheological changes in polymers during shearing by Kauzman and Eyring [10] over 45 years ago. The prior thermal history of polymers will therefore, to a major extent, determine their subsequent service performance [11, 12]. The damage sustained during processing of polymers varies depending on the nature of the polymer, although reactions that occur during processing of most polymers are generally similar. For example, low-

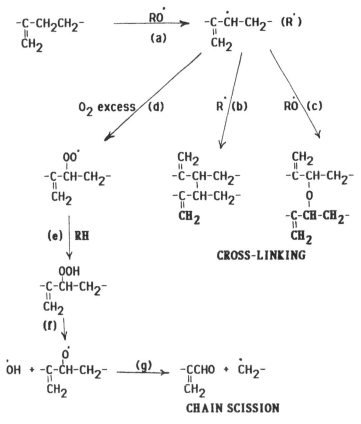

Scheme 3 Chemical changes occurring during processing of LDPE.

density polyethylene (LDPE) undergoes molecular enlargement [13] (Scheme 3), whereas polypropylene (PP) undergoes chain scission [14] (Scheme 4). In these polymers both macroalkyl and macroalkoxyl radicals occur as intermediates during polymer processing. Reactions of macroalkyl radicals with oxygen give rise to hydroperoxides, which are potent sensitizers for polymer degradation both during processing and in use [15]. Similarly, polyvinyl chloride (PVC) undergoes chain scission reactions with the formation of macroradicals followed by two competing reactions (see Scheme 5): HCl elimination

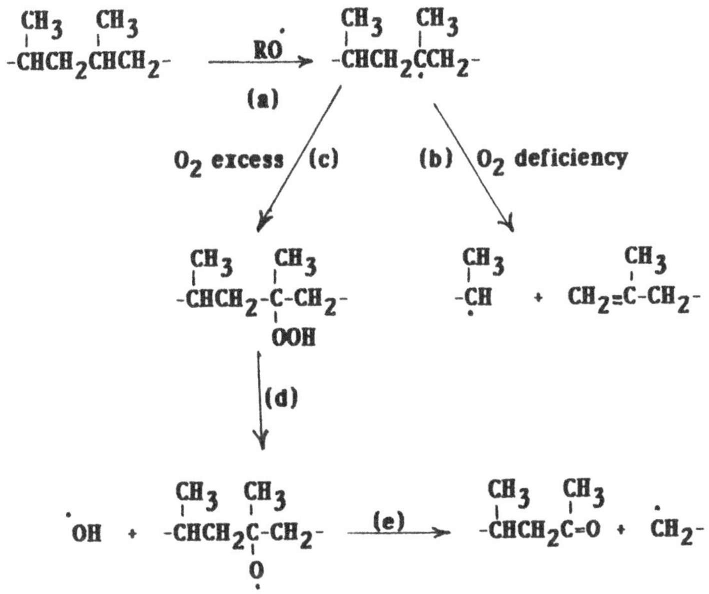

Scheme 4   Chain scission in PP during processing.

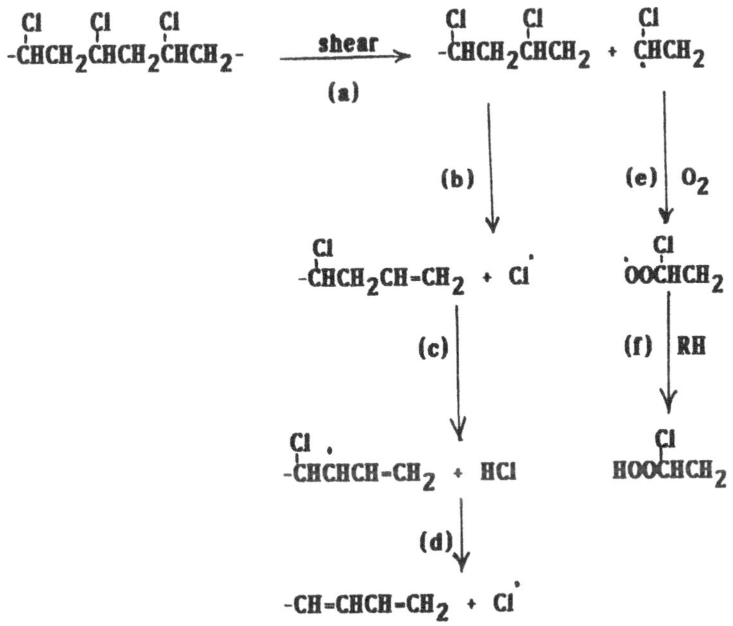

Scheme 5   Chemical changes occurring during processing of PVC.

(reactions b and c) and peroxidation (reactions e and f) [16]. Degradation of PVC is normally accompanied by discoloration due to the development of conjugation in the polymer chain (see reaction d in Scheme 5).

MODERN THEORIES OF MECHANISMS OF ANTIOXIDANT ACTION

The free radical chain reactions (see Scheme 2) proposed earlier by BRPRA form the basis of current models of autooxidative degradation processes. These reactions are now viewed in terms of two interrelated oxidative cycles, the first of which involves the formation of macroalkyl and alkyl peroxyl radicals, whereas the second involves the formation of radical generators as in Scheme 6 [17, Chap. 7]. It is now generally accepted [18–20] that the formation of hydroperoxides in the chain sequence (see Scheme 6) is the most important radical initiator in the autooxidation process of hydrocarbon polymers. Further, the nature of the initiation step in which macroalkyl radicals are initially formed influences the overall oxidation rate of the substrate.

Chemical agents which can effectively interfere with the autooxidation cycles above can act as antioxidants and protect the deterioration of polymers during processing and in use. Accordingly, Scott classified antioxidants [8] according to the mechanism by which they interfere with the process of autooxidation and proposed two distinct classes: chain-breaking antioxidants (CB), which interrupt the main oxidative cycle by removing the main propagating radicals (R˙ and ROO˙); and preventive antioxidants, which interfere with the second cycle by preventing the introduction of chain-initiating radicals into the polymer system (see Scheme 6). The most important preventive mechanism, both theoretically and practically, is hydroperoxide decomposition (PD) in a process that does not involve the formation of free radicals.

Chain-Breaking Antioxidants

Historically, the chain-breaking mechanism was the first to be investigated since it embraces the traditional rubber antioxidants, phenols and aromatic amines. Chain-breaking

CB-A Chain breaking (electron acceptor)
CB-D Chain breaking (electron donor)
PD   Preventive (peroxide decomposing)
UVA   Uv absorbers
Q   Quenchers
MD   Metal deactivators

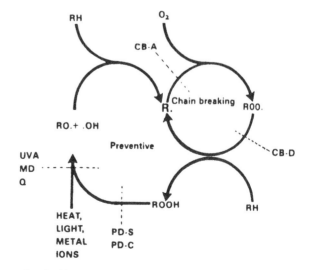

Scheme 6 Oxidative degradation processes and antioxidant mechanisms.

antioxidants were further subdivided to distinguish between electron (and proton) donors: chain-breaking donors (CB-D), which reduce ROO˙ to ROOH, and electron acceptors, chain-breaking acceptors (CB-A), which oxidize or spin trap R˙ in the absence of oxygen [8]. At ambient oxygen pressure ROO˙ are normally present in large excess over R˙. Consequently, termination occurs either by coupling of alkylperoxyl radicals (reaction 7, Scheme 2) or by their reaction with the antioxidant, AH (see reaction 1 below); in the latter case a CB-D process is favored over a CB-A. At low oxygen pressure, however, e.g., inside a screw extruder, termination through alkyl radicals (reactions 8 and 9, Scheme 2) become increasingly important [6].

$$ROO˙ + AH \rightarrow ROOH + A˙ \qquad (1)$$

Antioxidants which can function by both CB-D and CB-A mechanisms have a clear advantage over those operating by a single mechanism since in many antioxidant processes both alkyl and alkylperoxyl radicals are formed to some degree, as will be discussed later.

Chain-Breaking Donor Mechanism

The most important commercial examples of antioxidants which fall in this class are the hindered phenols and aromatic amines. It was known as early as 1950 [21–23] that the function of CB-D antioxidants depends on their reaction rate with ROO˙ and the reactivity of the generated antioxidant radical (e.g., phenoxyl, A˙, from hindered phenols, see reaction 1). Furthermore, the stoichiometry of the reaction of CB-D antioxidants was found [24] to be two, i.e., two peroxyl radicals are deactivated by one antioxidant molecule. It was becoming apparent too that transformation products of the initial antioxidant, e.g., hindered phenol, can greatly influence the autooxidation reaction and that products formed by their further reactions were complex [24]. The detailed chemistry of these reactions has been discussed elsewhere [8, 24]. Important features for the antioxidant reactions of a typical hindered phenol, 2,6-di-tert-butyl-4-methyl phenol (BHT), are summarized in Scheme 7. BHT-derived transformation products have both advantages and disadvantages in melt and thermal stabilization of polymers. In polyolefins, for example, ethylene bisphenol (Scheme 7, **6**) is as effective an antioxidant as the starting phenol and the stilbene quinone **7** is effective in oxygen-deficient atmosphere, while the peroxydienones **1** and **3** are potential prooxidants, particularly at high temperatures [25, Chap. 6].

Less systematic studies have been conducted on the oxidation of aryl amines, although it was generally found that the same principles for phenols apply. Scheme 8 summarizes the transformation products of a diphenyl amine [26, 27]. Similar to hindered phenols, secondary oxidation products of the highly reactive aminyl radical (**II**) give products which themselves are powerful antioxidants, e.g., **III** and **IV** [26]. However, a disadvantage of arylamines is that they cause considerable discoloration of the polymer to which they are added due to the formation of extensively conjugated quinonoid oxidation products (e.g, **V**).

Chain-Breaking Acceptor Mechanism

It was shown earlier that alkyl radicals predominate in an autooxidizing system under conditions of limited oxygen supply, e.g., during processing of polymer in a screw extruder, or when the rate of initiation is high compared with the rate of diffusion of oxygen into the polymer, e.g., during UV-initiated oxidation or mechanically induced

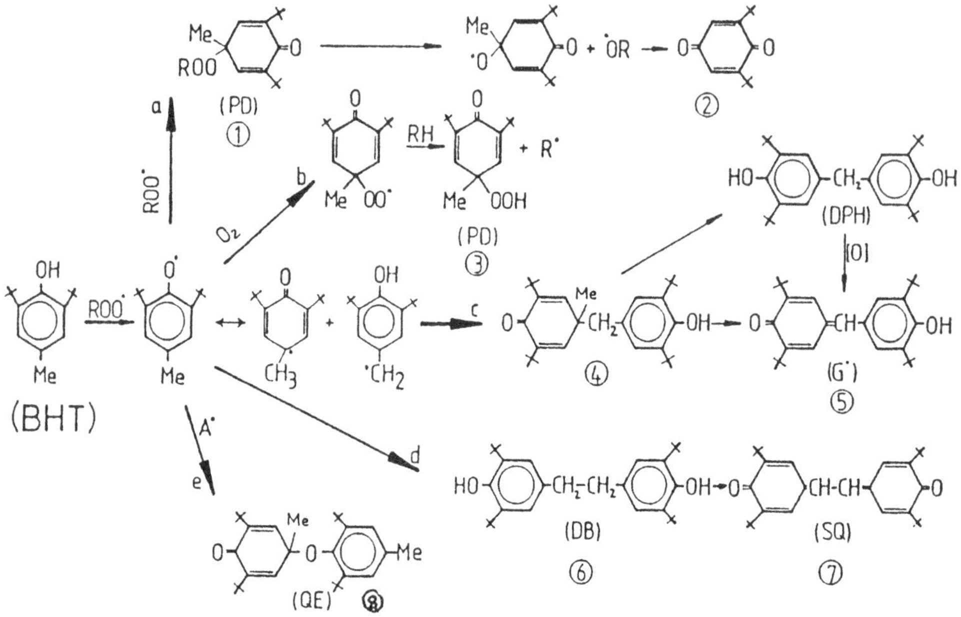

Scheme 7 Chemistry of oxidation of BHT.

oxidation. It is known that in solid polymers alkyl radicals play a more important part in termination processes than in liquid hydrocarbons, due to the low solubility of oxygen [28, 29] and its lower rate of diffusion [30] in polymers than in liquid hydrocarbons. Removal of alkyl radicals by CB-A antioxidants therefore plays an important part in uninhibited and inhibited oxidations.

The molecular requirements for an effective CB-A antioxidant become evident from a consideration of the transition state of a typical phenolic antioxidant (reaction 2 below) [8]. Since the transition state involves a partial transfer of an electron to the aromatic nucleus, electron-releasing and electron-delocalizing substituents reduce the energy of the transition state and increase antioxidant activity [8]. This requirement of CB-A antioxidants (oxidizing agents) is the same as for polymerization inhibitors which function in the absence of oxygen [8].

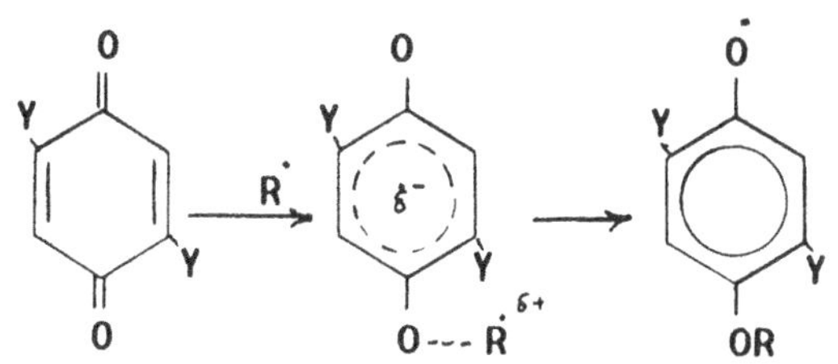

$$(2)$$

Catalytic (Redox) Chain-Breaking Antioxidants

Catalytic deactivation of peroxidic intermediates was one of the earliest theories of antioxidant action put forward by Moureu and Dufraisse [3] in the mid-1920s, although this

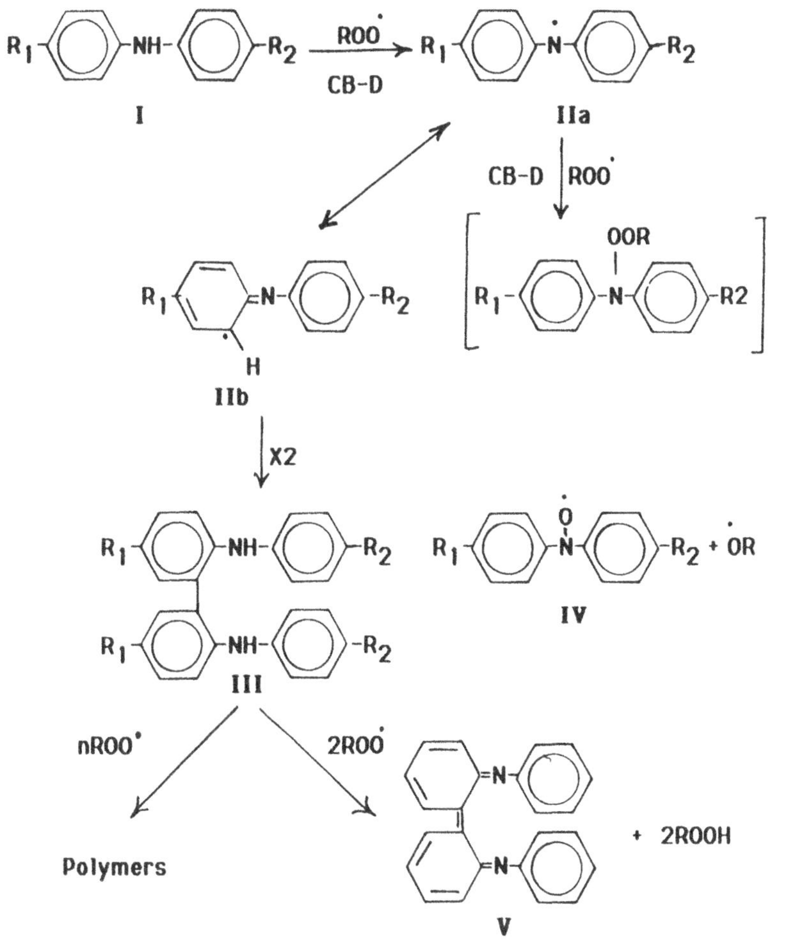

Scheme 8 Chemistry of oxidation of arylamines.

was largely concerned with peroxidolytic antioxidants. The mechanism of catalytic chain-breaking antioxidants, however, remained unknown until 1971 when Scott [31] suggested that some metal ions—notably copper—might function as catalytic chain-breaking antioxidants by a redox mechanism involving alternating oxidation of alkyl and reduction of alkylperoxyl radicals. Nine years later, results from the same laboratories [32] showed the formation of conjugated unsaturation in linseed oil-based paint media (with brown coloration) during copper acetate-catalyzed photooxidation. Meanwhile, in 1978 Bolsman and co-workers [33] proposed a catalytic radical scavenging mechanism in order to explain the high efficiency of nitroxyl radicals in inhibiting autooxidizing hydrocarbon substrates. Many other catalytic antioxidant systems have since become known [17, Chap. 6]; stable phenoxyl and nitroxyl radicals [34] for melt stabilization of polyolefins, stable nitroxyls as antifatigue agents for rubber [35] and as photostabilizers for polyolefins [17, Chap. 7; 34], iodine atoms and soluble copper as melt stabilizers for PP [36], and some transition metal ions for the photostabilization of paint films [32].

The mechanism of catalytic chain-breaking antioxidants as melt and photo stabilizers for polyolefins is illustrated here using two recent examples. The first is based on the stable free radical Galvinoxyl (**G**), which was shown to be an effective melt stabilizer for PP [36]. The mechanism of antioxidant action of this compound has been studied in detail [37] and was shown to involve a cyclical regenerative reaction involving alternating CB-A and CB-D antioxidant steps. It was shown that there is a rapid formation of hydro-galvinoxyl (HG; see reaction 3) and a simultaneous formation of unsaturation in the polymer (when the shearing forces are at their highest) and that the concentrations of G and HG alternate reciprocally as the conditions in the mixer alter (Figure 1). The total concentration of the redox couple (G/HG) remained essentially unchanged during the initial stage of processing when the macroalkyl radical and oxygen concentrations are both high. Galvinoxyl is ultimately destroyed by oxidation to 2,6-di-tert-butylbenzoquinone, BQ [37]. It was also shown that galvinoxyl is relatively ineffective in an air atmosphere, whereas HG is highly effective. The mechanism proposed to account for this behavior is summarized in Scheme 9. The central feature is the CB-A–CB-D cycle in which the G is reversibly reduced and reoxidized with initially little loss of the redox couple from the system.

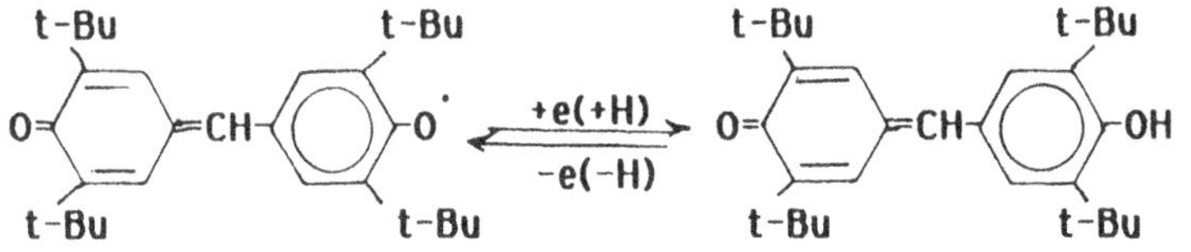

$$\text{(G)} \xrightleftharpoons[-e(-H)]{+e(+H)} \text{(HG)} \qquad (3)$$

The second example is based on stable nitroxyl radicals derived from hindered piperidine light stabilizers; the best known commercial example is Tinuvin 770 (**I**). The UV stabilizing role of this class of compounds is unique since it has been shown [38, 39] that it does not act by the normal photostabilizing mechanisms: It is an ineffective UV screen, does not quench singlet oxygen or triplet carbonyls, and does not catalyze the decomposition of hydroperoxides. Furthermore, they were shown to rapidly disappear during the very early stages of photooxidation with the formation of the derived nitroxyl radical, the concentration of which reaches to a very low stationary level at later stages of photooxidation [17, Chap. 7; 38; 40; 41] and persists at this low level throughout the lifetime of the polymer (see Figure 2).

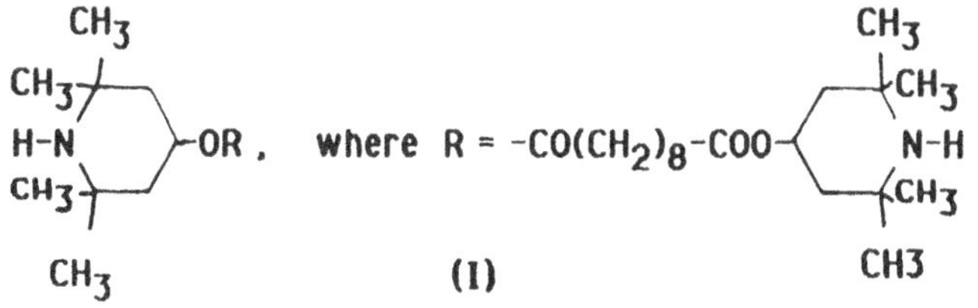

where R = -CO(CH$_2$)$_8$-COO-

(I)

Tinuvin 770 and related hindered piperidines have been shown [15] to be ineffective as heat and melt stabilizers for polyolefins since, unlike aromatic amines, aliphatic amines are not efficient CB-D antioxidants. Moreover, although hindered piperidines do not catalytically decompose hydroperoxides, they do react with them slowly during process-

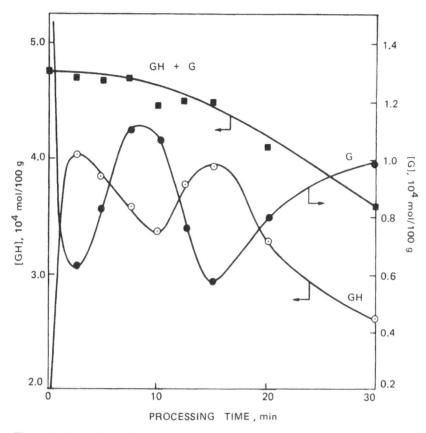

Figure 1 The effect of processing time (200°C) on the concentration of galvinoxyl (G) and hydrogalvinoxyl (HG) during processing of PP initially containing 4.75×10^{-4} mol/100 g of G. (After Ref. 38.)

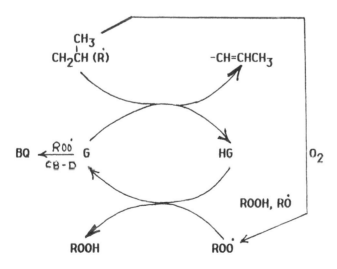

Scheme 9 Mechanism of antioxidant of galvinoxyl (G) during melt stabilization of PP.

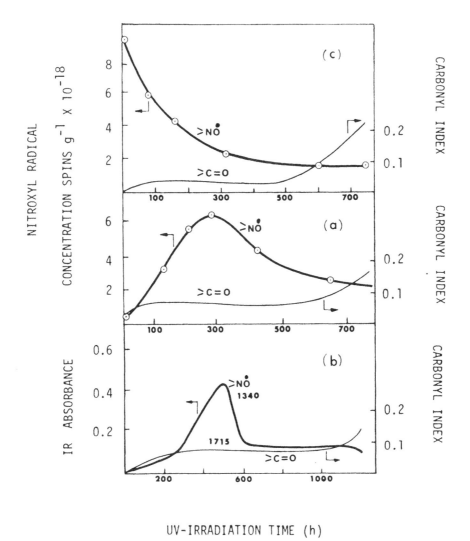

Figure 2 Effect of Tinuvin 770 (**I**) on the photooxidation of (*a*) PP and (*b*) methylcyclohexane. Concentration of 1 in (*a*) is 3×10^{-4} mol/100 g and in (*b*) is 1×10^{-2} *M*. The growth of the corresponding nitroxyl (>NO) and buildup of carbonyl groups (>C = 0) is shown. The effect of added nitroxyl with the structure **I**, R = OH (concentration 6×10^{-4} mol/100 g) on the photooxidation of PP (*c*) is also shown (numbers on curves are vibrational frequencies in cm^{-1}). (After Ref. 17.)

ing [39] to form the corresponding nitroxyl radical in a stoichiometric radical generating process (see reaction 4) with associated thermal degradation of the polymer. This accounts not only for their lack of stabilizing activity during melt processing but also for the rapid initial formation of carbonyl during photooxidation (see Figure 2). The initial high rate of oxidation is even more pronounced during the photooxidation of methylcyclohexane containing Tinuvin 770 under conditions where no prior thermal oxidation had taken place (Figure 2). This is additional evidence for the importance of oxidation in the overall mechanism of hindered piperidines.

$$\underset{R'}{\overset{}{\bigg\langle}}\!\!\!\!\!\bigg\rangle N\text{-}H + ROOH \xrightarrow{\ ROO^{\bullet}\ } R'\!\!\!\!\bigg\langle\!\!\!\!\bigg\rangle \overset{\bullet}{N} + H_2O + R\overset{\bullet}{O}$$

(ROOH)

$$\downarrow ROO^{\bullet}$$

(4)

$$R'\!\!\!\!\bigg\langle\!\!\!\!\bigg\rangle N\text{-}\overset{\bullet}{O} + R\overset{\bullet}{O}$$

Investigations by a number of research groups [30, 34, 38, 41, 42] have led to the conclusion that the nitroxyl radical is able to compete with oxygen for alkyl radicals in a CB-A process to give alkyl hydroxylamine (>NOR). Furthermore, it has been suggested [43] that nitroxyl radicals associate with hydroperoxides through hydrogen bonding at room temperature and that the consequent proximity of nitroxyl radicals to the alkyl radicals generated from the photolytically unstable hydroperoxides improves the trapping efficiency of the former.

It is now generally accepted [34, 40] that macroalkylhydroxylamine (>NOR) is formed as an intermediate (or transient) under photooxidative conditions, and two different mechanisms have been proposed for the regeneration of the nitroxyl from the alkylhydroxylamine. The first involves the reaction of an alkylperoxyl radical with the alkylhydroxylamine (Scheme 10, reaction e) [43, 44]. This reaction must be able to compete with the substrate (Scheme 10, reation e vs. f) for the alkylperoxyl radicals. The second assumes

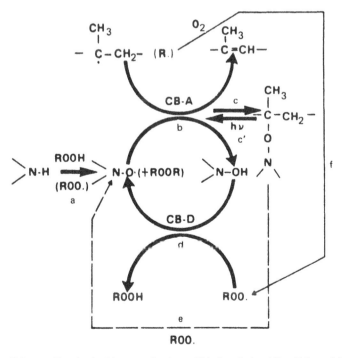

Scheme 10 Antioxidant mechanism of hindered piperidine light stabilizers.

[39, 45] that the formation of the alkylhydroxylamine is inherently reversible, particularly in the presence of light (see Scheme 10, reactions c and ċ), and that the ultimate termination step is the oxidation of the macroalkyl radical by nitroxyl (Scheme 10, reaction b). In the presence of oxygen, secondary, and particularly tertiary alkylhydroxylamines (but not the primary analogs, which are not antioxidants, since they are relatively stable [46]) are rapidly converted to the nitroxyl radicals [40] with the formation of oxidized products of the alkyl radicals. At low oxygen concentrations, on the other hand, irreversible formation of hydroxylamine and unsaturation will predominate (Scheme 10, reaction b) [39]. It was found [45] that free hydroxylamines are efficient CB antioxidants but differ from nitroxyl radicals in that they react with alkylperoxyl radicals, in a CB-D mechanism, to generate nitroxyl radicals (see Scheme 10, reaction d). Although the exact nature of the nitroxyl regeneration step still remains to be resolved, it almost certainly must involve the oxidation of free hydroxylamine, alkylhydroxylamine, or both, by the peroxides formed in the cycle; the key step remains the reaction of macroalkyl radicals with the nitroxyl in a CB-A process. Subsequently, the reason for the very high effectiveness of hindered amines and their oxidation products at low concentrations in the polymer is most probably due to the high $[P\dot{P}]/[PPO\dot{O}]$ ratio in solid polymers (e.g., PP) when compared to $[\dot{R}]/[RO\dot{O}]$ ratio in liquid hydrocarbon models [47]. For a CB-A antioxidant to operate effectively in the latter system, it would therefore have to be several orders of magnitude more reactive to alkyl radicals than oxygen, although clearly this is not the case when the substrate is a polymer.

Both nitroxyls and their derived hydroxylamines are good UV stabilizers [34, 39]. The overall high efficiency of nitroxyl radicals as UV stabilizers in polyolefins is therefore due to the complementary nature of the donor- and acceptor-antioxidant mechanisms involved. Scheme 10 summarizes the regenerative cycle involving nitroxyl radicals formed from hindered piperidines and emphasizes the redox antioxidant function of the CB-A–CB-D combination during photooxidation of PP. In this respect the behavior of nitroxyl radicals in polyolefins during their antioxidant function is remarkably similar to that described earlier for galvinoxyl (Scheme 9).

Preventive Antioxidants

Hydroperoxides are the most important product of the cyclical autooxidation process since they are the main source of further radicals in the system. Hydroperoxides are both thermally and photolytically unstable, giving rise to $RO\cdot$ and $\cdot OH$ radicals (see reactions 4–6, Scheme 2). Hydroxyl radicals react rapidly with hydrocarbons to give alkyl radicals, which feed back into the main autooxidation cycle, thus maintaining the chain reaction. Agents that decompose hydroperoxides by a process which does not give rise to free radicals (peroxide decomposers, PD) or which in some way stabilize hydroperoxides (e.g., by absorbing UV irradiation, UVA, or by deactivating transition metal ions, MD) prevent the reinitiation of the chain reaction and are classified generally as preventive antioxidants (see Scheme 6) [8].

The two main classes of peroxidolytic antioxidants are the phosphite esters [25, Chap. 1] and sulfur-containing compounds [48]. The simple trialkyl phosphates **II**, table 1) act by a stoichiometric peroxidolytic mechanism (PD-S) and are not effective as the cyclic catechol phosphites (**III**), which decompose hydroperoxides catalytically, PD-C [25, Chap. 1]. The derived phosphate (**IV**), which also acts by a catalytic mechanism, is also effective. These and many sulfur-containing compounds, which are known to act by catalytically removing hydroperoxides in a nonradical process, have been shown to be

Table 1 Typical Peroxide Decomposers

Structure	Number	Common name or code
(RO)$_3$P	II	Phosphate
	III	Cyclic catechol phosphite
	IV	Derived phosphate
(ROCOCH$_2$CH$_2$)$_2$S	V	Thiodipropionate ester
	VI	MDRC
	VII	MDRP
	VIII	MRX

effective melt stabilizers for some hydrocarbon polymers such as PP [15, 48, 49]. Examples of such sulfur-containing compounds include monosulfide of dipropionate ester (**V**), metal complexes of dithioic acids, of which metal dithiocarbamates (MDRC, **VI**), dithiophosphates (MDRP, **VII**), and xanthates (MRX, **VIII**) are typical examples. Extensive mechanistic studies on sulfur-containing compounds [48, 50–52] have revealed that compounds of this class undergo a complex series of oxidation reactions involving intermediate free radicals to give sulfur acids which are catalysts for the decomposition of hydroperoxides. For example, Scheme 11 gives a summary of the most important features of the antioxidant action of the thiodipropionate ester. Besides the sulfur acids, which are primarily responsible for imparting long-term thermal oxidative stability to hydrocarbon polymers, intermediate sulfinic acid and sulfinyl radical have also been implicated in trapping mechanism involving macroalkyl radicals [53].

For peroxidolytic antioxidants to be good photostabilizers they, and their transformation products, must be photolytically stable so that the combined effect of the parent compound and its products leads to an effective photoantioxidant. Evidently not all peroxidolytic antioxidants fulfill this dual role. The monosulfide (**V**), for example, is not

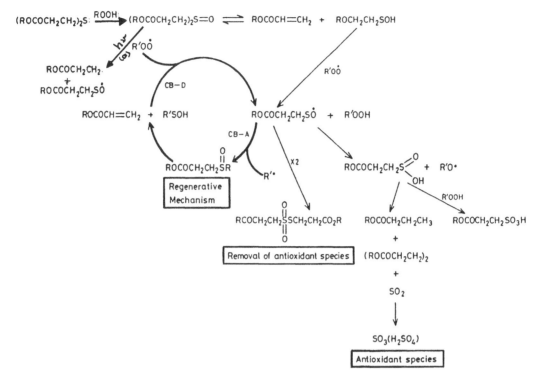

Scheme 11 Mechanism of antioxidant action of thiodipropionate esters.

useful as photoantioxidant (although it is a good thermal and melt stabilizer) because the intermediate sulfoxide is very sensitive to UV light and undergoes photolysis to initiate free radicals (Scheme 11, reaction a) [54]. In contrast to the simple sulfides, metal dithiolates (e.g., M = Ni) are much more effective UV stabilizers due to their much higher UV stability and their ability to liberate the ionic catalysts for peroxide decomposition over a much longer period of time [48]. All metal dithiolate antioxidants show two distinct phases of antioxidant activity [50, 55–57]:

1. An initial chain-breaking activity.
2. Peroxide-decomposing activity, which appears during the later stages of oxidation and particularly at high ratios of peroxide to dithiolate metal complex.

Contributions of each phase varies depending on the nature of the metal complex. Peroxide decomposition involves a catalytic nonradical process (PD-C) triggered by acids formed by oxidative breakdown of the original sulfur-containing metal complex. The metal complex itself therefore is not the true peroxidolytic antioxidant but acts as a precursor and a reservoir for the effective catalytic peroxide decomposers.

In the case of nickel complexes of dithioic acids (e.g., NiDRP), peroxide decomposition was shown to take place in a three-stage process (Figure 3), an initial fast stage followed by a secondary induction period leading to a third catalytic stage [55]. Product analysis of reactions of NiDRP with a model hydroperoxide (e.g., cumene hydroperoxide, CHP) at elevated temperatures using ^{31}P NMR, IR, and TLC have shown that the first

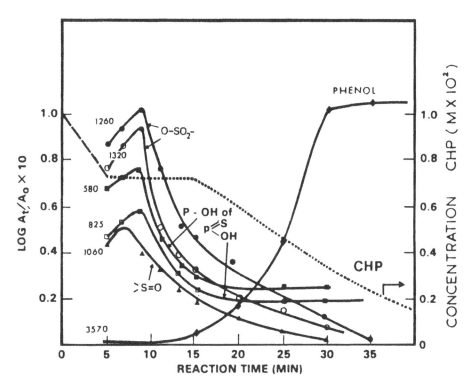

Figure 3 Formation and decay of products formed in the reaction between cumene hydroperoxide (CHP) and nickel dibutyl dithiophosphate (NiDBP) at molar ratio [CHP]/[NiDBP] = 100 in chlorobenzene at 110°C. Numbers on curves are vibrational frequencies in cm^{-1}. The decay of CHP is superimposed. (After Ref. 58.)

rapid stage is associated with the formation of disulfides (Scheme 12), which are slowly converted to nickel-free oxidation products of which the sulfonic acid (DBDiSO) is the most important since it readily loses SO_2 to give thionophosphoric acid, which can be seen from Figure 3 (and from ^{31}P NMR results [58]) to be a relatively stable end product in the system. Figure 3 also shows that although CHP is decomposed from the beginning of the reaction, phenol, the product of ionic decomposition of CHP, does not start to be formed until after 10 min, when the intermediate sulfonic acid begins to break down [56]. This clearly identifies the antioxidant activity with the sulfur acids formed from NiDRP by hydroperoxide oxidation.

Although it has been known for a long time that in the case of sulfur-containing antioxidants, the effective peroxidolytic catalysts were not the parent compounds but their oxidation products [55, 59, 60], the realization that melt processing operations exert profound effect on the light stability of polymer articles containing them is relatively recent [61]. The main reason for this is the fact that earlier mechanistic studies were mainly conducted on model compounds or hydrocarbon oils.

The evidence suggests that the primary functions of metal dithiolates, e.g., nickel complexes, is to act as light-stable reservoirs for oxidatively derived antioxidants. Since these species are partially formed during processing in the presence of a trace of oxygen,

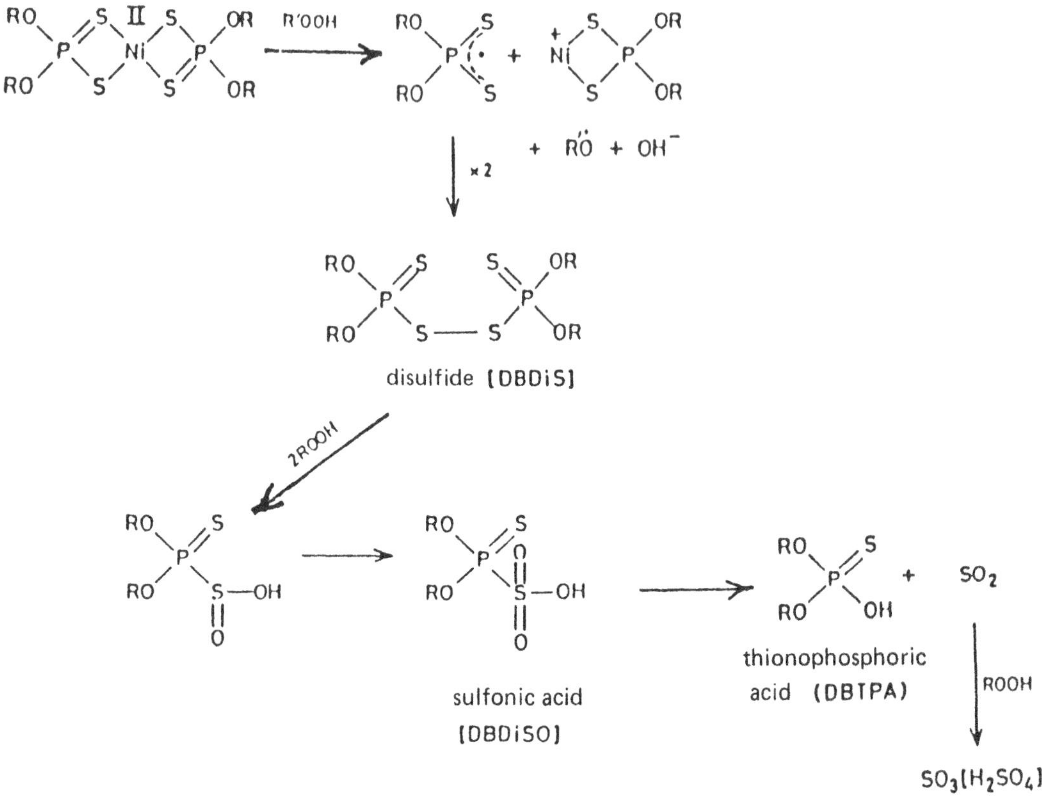

Scheme 12 Antioxidant mechanism of nickel dithiophosphate.

the metal dithiolates are all very effective melt stabilizers for polyolefins and no hydroperoxides are formed at the end of the processing operation. Consequently, the UV-stabilizing effectiveness of the dithiolate group of antioxidants is less sensitive to the effects of the processing operation than are the nonsulfur UV absorbers, e.g., 2-hydroxy-4-octyloxy benzophenone (HOBP) [62]. Nevertheless, under severe processing conditions, some reduction in activity does occur due to partial conversion of the nickel complex to the disulfide [63] (see Scheme 12 and Figure 4). However, the disulfide, e.g., thiophosphoryl disulfide, which is the primary oxidation product of the corresponding nickel dithiophosphate, actually improves in performance as a UV stabilizer when subjected to the same severe processing (Figure 4). The inset of Figure 4 compares the effect of processing (e.g., under mild conditions) on the UV-stabilizing activity of the thiophosphoryl disulfide (DiPDS) in PE and PP [the hatched area indicates the extent of stabilization afforded by the disulfide-containing polymer samples (PP and PE) compared to their corresponding unstabilized polymer controls]. Clearly a higher level of stabilizing activity is observed in PP than in PE. This difference in activity is believed to be a reflection of the relative ease of oxidation of PP compared to PE, hence higher peroxide concentration level in the more oxidizable polymer [63].

The behavior of metal dithiolates during processing was subsequently exploited [64, 65] in the development of very effective stabilization systems in which a process of controlled oxidation is adopted during processing of the polymer in the presence of sulfur-based antioxidants. Examination of Table 2 shows that a combination of NiDRP and the

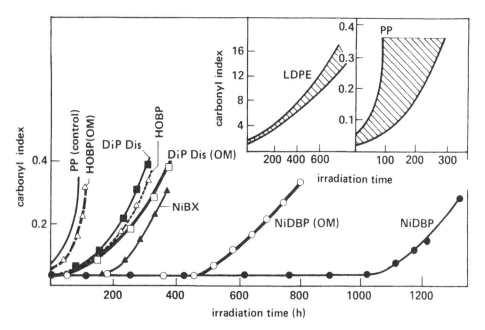

Figure 4 Effect of mixing procedure (processing under mild conditions, closed mixer, CM, and under oxidative conditions, open mixer, OM, of an internal mixer, a RAPRA torque rheometer) on photooxidative stability of PP-containing additives (2.5×10^{-4} mol/100 g). Inset compares photostabilizing effectiveness of mildly processed (CM) DiPDS in PE and PP. (After Ref. 65.)

corresponding disulfide (DRDS) with a UV-absorber (e.g., HOBP) gives a synergistic stabilizer system whose activity is greatly augmented by oxidative processing. Figure 5 compares the UV-stabilizing effectiveness of this oxidatively processed stabilizing system (called here COPS) with that of a number of commercially based synergistic systems and also with the same system when processed under the normal mild conditions (called here CMPS). It is important to mention here that oxidative processing is not normally used by polymer manufacturers since it adversely affects the UV stability of the product (refer to Table 2 for the effect of oxidative processing on the control) due to the formation of sensitizing groups and especially hydroperoxides. In the present case, however, the peroxides are used beneficially to convert the antioxidant precursor to the effective anti-

Table 2 Effect of Oxidative Processing on the Activity of Synergistic Combination of Antioxidants Containing Dithiophosphates

		Embrittlement time (hr)	
Stabilizer	Concentration (g/100 g)	Normally processed	Oxidatively processed[a]
Control (no additive)		90	65
DODS	0.1		
NiDBP	0.1	2100	4900
HOBP	0.2		

[a]As described in Ref. 65.

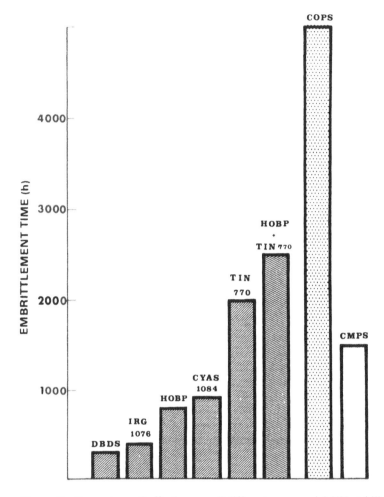

Figure 5 Comparison of effectiveness of different commercial UV stabilizers with oxidatively processed, under controlled conditions, polypropylene-containing sulfur antioxidants (DBDS + NiDBP + HOBP, referred to as COPS), and with the same system but processed mildly in a closed mixer (referred to as CMPS). Total concentration in each case is 0.4%. (After Ref. 66.)

oxidant during the processing operation. This is a unique exploitation of the idea of generating active antioxidants in situ and should be of greater interest in commercial applications.

TIME-CONTROLLED STABILIZATION

Metal dithiolate antioxidants exhibit a general characteristic feature upon UV exposure. Typically, an induction period is observed which is proportional to concentration and which ends very sharply upon the consumption of the metal complex. The end of the induction period marks the start of photooxidation, which proceeds at a rate that is primarily a function of the concentration and the nature of the free metal ion produced [17, Chap. 7; 52]. While most metal chelates are generally found to be good UV stabilizers for polyolefins [48], iron metal chelates have a dual function as UV stabilizers at one concentration and as UV activators at another [66, 67]. This phenomenon has been exploited in the design of time-controlled plastic films with predetermined lifetimes. The complementary stabilizer–degradant roles exhibited by iron complexes have made possible the commercial use of these additives, e.g., FeDRC (see Table 1, **VI**) to increase the yields of agricultural crops.

The basic requirements for photodegradable plastic mulching films for agricultural purposes [67] are that the plastic material should retain its strength throughout its entire service lifetime and should rapidly lose its mechanical strength shortly before cropping, so that it is easily disintegrated under the mechanical action of harvesting machines. The ideal behavior of an antioxidant–photoactivator system therefore is that it should impart a limited, but predefined, induction period before the physical properties of the film begin to change, followed by a process of fast accelerated oxidation, leading to rapid mechanical failure of the mulching film. The role of iron complexes in time-controlled photostabilization has been studied in some detail [66–68].

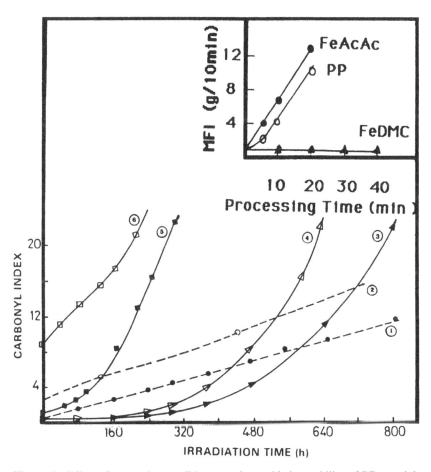

Figure 6 Effect of processing conditions on photooxidative stability of PE-containing additives (1×10^{-3} mol/100 g): (1) control processed for 10 min in closed mixer (10, CM); (2) control processed for 10 min in open mixer (10, OM); (3) FeDMC (10, CM); (4) FeDMC (30, OM); (5) FeAcAc (10, CM); (6) FeAcAc (30, OM). Inset shows the effect of FeDMC and FeAcAc (2.5×10^{-4} mol/100 g) on the melt flow index (MFI) of PP processed in closed mixer of RAPRA torque rheometer at 180°C.

Transition metal carboxylates and iron complexes of oxygen-chelate ligands, e.g., iron acetyl acetonate, FeAcAc [66, 69], were shown to satisfy only the second requirement of the ideal photosensitizer; i.e., they cause rapid and autoaccelerated photooxidation of polyolefins under all concentrations and also cause considerable oxidation of the polymer during processing as reflected from high melt flow index values (see Figure 6 inset). Iron complexes of dithiocarbamates, on the other hand, act more like the ideal antioxidant photoactivator [52, 66]. They stabilize the polymer during processing (Figure 6 inset) at all concentrations and, at the same time, cause rapid photooxidation in a concentration-dependent autoaccelerated oxidation process during the subsequent UV exposure (Figure 6). At very low concentration, FeDRC is much less photostable than the Ni or Co analogs. At higher concentrations it undergoes a role reversal and becomes a short-term stabilizer [66, 69]. The change from acting as photoantioxidant (which is partly due to reaction a in Scheme 13 and partly due to screening and quenching of excited species) to photopro-oxidant at the end of the induction period (Figure 6, curves 3 and 4) is due to the concomitant increase in concentration of ionic iron by photolysis of the complex. Photo-sensitization of the polymer substrate by the metal complex therefore starts at the end of the induction period due to the formation of ferric carboxylate [52, 67], which is produced by replacement of the dithiocarbamate ligands present in the polymer. In commercial practice, a two-component combination of nickel (to control the length of the induction period) and iron dithiocarbamate (to effect rapid oxidation) is used (Figure 7).

At appropriate concentrations of the stabilizer–activator combination therefore both

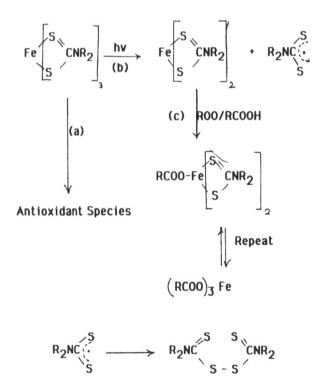

Scheme 13 Mechanism of antioxidant photoactivator, FeDRC.

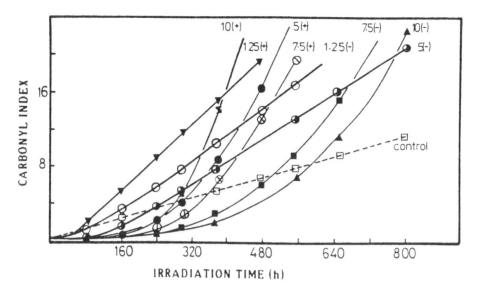

Figure 7 Effect of concentration of FeDMC on photooxidation of PE in the presence (+) and absence (−) of NiDEC (2.5×10^{-5} mol/100 g). Numbers on curves are concentrations of FeDMC (mol/100 g $\times$ 10^{-4}). (After Ref. 68.)

the service lifetime of the film (which is proportional to the length of the induction period) and the rate of the catalyzed photooxidative degradation of the matrix can be precisely controlled. The correct choice of induction period and the rate of the activation process are dictated by the application of the manufactured product.

FACTORS CONTROLLING ANTIOXIDANT EFFECTIVENESS

Additive volatility, diffusivity, solubility, and morphology, orientation, and surface-to-volume ratio of the polymer sample are some of the physical parameters that can be as important as the intrinsic activity of the antioxidant (dictated by its chemistry and its transformation products) in determining the effectiveness of antioxidants and stabilizers. Therefore, determining the efficiency of antioxidants in polymers under practical conditions is a first step toward establishing the real effectiveness of antioxidants. Several complementary technological tests are available to ascertain physical and chemical characteristics of antioxidants; each of these tests has its own limitations. For example, oxygen absorption experiments (a ''closed test'') are good indicators of intrinsic activity of antioxidants in model systems, which the antioxidant is completely soluble and cannot be physically lost. They are, however, poor predictors of antioxidant activity in polymer artifacts under aggressive service conditions due to the dominating influence of solubility and volatility. Accelerated oven aging tests (an ''open test'' normally in an air stream), on the other hand, tend to accentuate the loss of antioxidants from the surface by volatilization. BHT (see Table 3 for structure) is a good example to use here. BHT, which is one of the most efficient antioxidants known for liquid hydrocarbons (from oxygen absorption), was found to be virtually useless (because of high volatility) during air (flowing) oven test of plastics and rubbers [17, Chap. 6].

Table 3 Typical Hindered Phenol Antioxidants

	Common or commercial name
R = H	BHT
R = $CH_2COOC_{18}H_{37}$	Irganox 1076
R = $(-CH_2COOCH_2)_4C$	Irganox 1010
R = (with trimethylbenzene structure)	Ethyl 330
R = (isocyanurate ring structure)	Good-Rite 3114
R = (Topanol structure)	Topanol CA

The loss of antioxidants from the surface of the plastic through "blooming" (where the additive migrates to the surface and is deposited on the surface) and volatilization is a consequence of antioxidant solubility in the polymer [70]. Further, antioxidant substantivity (ability to remain in the substrate) [25, Chap. 6] under high-temperature processing and exposure to air in service are affected by diffusion characteristics of the antioxidant from the bulk to the surface [70]. The problem of additive loss from the polymer depends on the temperature and the nature (gas, liquid, solid) of the surrounding medium, and is most serious in plastic articles with high surface-to-volume ratio, e.g., in films, coatings, and fibers.

The failure of typical rubber antioxidants such as BHT to protect thermoplastics in an accelerated air oven as an aging test (although they were good melt stabilizers) was soon to lead to the development of higher molecular weight antioxidants such as Irganox's 1076 and 1010, Ethyl 330, Good-Rite 3114, and Topanol CA (see Table 3). A limitation to this approach, however, is that the performance of antioxidants may actually be reduced on a weight basis, because the weight ratio of the functional group to the inert residue is steadily reduced in the antioxidant [71, 72]. This problem of functional group dilution was thought to be overcome by incorporating an antioxidant group into a polymer as a conventional vinyl or condensation polymer so that the antioxidant function is repeated at short intervals along the polymer chain. However, it was soon found [73] that antioxidants of high molecular mass made in this way are again not very effective due to their limited miscibility with commercial polymers.

Over the last two decades the trend has been toward the development of oligomeric additives of sufficient solubility in the host polymer to perform their function. In the 1960s Thomas and co-workers [74] had conducted pioneering and systematic work for the production of oligomeric antioxidants for high-temperature rubbers. In the 1970s a number of commercial oligomeric light stabilizers which are much less readily lost from polymers under aggressive environments were developed (e.g., Chimasorb 944, **IX**). It was later suggested [25, Chap. 6], however, that such oligomeric antioxidants (which are more substantive at high temperatures), can still be lost by solvent leaching.

Vogl and co-workers [75] have synthesized polymerizable and polymeric UV stabilizers of several types which generally contain both a vinyl group and a UV absorber function. To increase effectiveness, more than one UV absorber group was incorporated in the same molecule and these molecules were homopolymerized and copolymerized with monomers such as styrene and methyl methacrylate.

Another approach was developed by Scott and co-workers in the 1970s. This involved the use of additives that are capable of reacting with polymers to produce polymer-bound functions, which can both improve additive performance and modify polymer properties [25, Chap. 6; 76]. The general approach is to mechanochemically synthesize bound antioxidants, highly concentrated bound antioxidants, or highly concentrated bound antioxidant adduct (master batches), which can be used as conventional additives for polymers during normal compounding procedures. This has provided a potential solution to the problem of loss by volatilization, since such antioxidants can be removed only by breaking chemical bonds. Moreover, these were found to be molecularly dispersed along the polymer chain and are, by definition, infinitely soluble [77]. The principle of mechanochemical reactions of these reactive additives in polymers to form bound antioxidants was

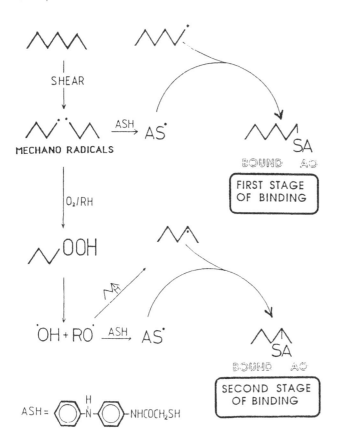

Scheme 14 Mechanism of thiol adduct formation in saturated hydrocarbon polymers.

based on the work of Watson and co-workers in the 1950s. During their studies on rubber mastication these workers [78] showed that macroalkyl radicals produced during mastication (a shearing process) could be utilized to initiate polymerization of vinyl monomers with the formation of block copolymers.

Thiol-containing antioxidants were used and high binding was achieved especially in rubbers [79] and unsaturated thermoplastic polymers such as ABS [80] and, more recently, polyolefins, e.g., PP and PE [25, Chap. 6; 81]. In the case of the saturated polymers, it was found that most of the adduct is formed during the first minute of processing, when the applied torque in the mixer is high. This stage involves the formation of macroalkyl radicals by mechanical scission of the polymer chain (Scheme 14). The second stage utilized the chain-end hydroperoxides as radical generators (see Scheme 14). More recent work [81] has shown that the highly effective hindered piperidine light stabilizers can be prepared in adduct master batch form in PP. Their use in unstabilized PP imparts high UV stability. In this respect they are similar to the oligomeric additives but without their disadvantage of leaching out by solvents.

PROSPECTS FOR THE FUTURE

The century-old trial-and-error approach in dealing with hydrocarbon oxidation has gradually developed into a well-established scientific discipline. The present state of the art of stabilization technology is the result of unabated research and progress in the theories and models of autooxidation and antioxidant function. In practice, these achievements have transformed plastics from a meager existence to a viable and progressive industry with unlimited scope and potential. Undoubtedly, newer markets and challenges will emerge. There remain, however, several unsolved problems that require greater efforts by the scientific community.

During the course of the development of antioxidants and mechanisms of antioxidant action, a large amount of invaluable information and data has accumulated. A period of consolidation is timely, to examine, reevaluate, and reappraise the fundamentals of this experimental science and to invest more time and effort in developing an integrated theoretical model based on the current state of the art in stabilization technology and design of antioxidants with the aim of achieving more accurate prediction of lifetimes of manufactured articles, with minimum accelerated and long-term weathering tests.

At present, accurate prediction of the service lifetime of manufactured articles is impossible. Further studies and detailed information are needed to establish the effect of applied stress on thermo- and photooxidative stabilities. Although we already know that stressed polymers undergo random chain scission leading to mechanoinitiated radicals, with deleterious effects on the oxidative stability of the melt during processing (at high torque), only a limited amount of information is available on the effect of deformation on oxidative stability (and lifetime) of fabricated articles, especially under aggressive environments.

The problems of additive solubility, compatibility, and the mechanics of physical loss of antioxidants from polymers have been clearly defined but there is still the need to establish some structure-solubility/migration correlation of antioxidants in polymers. In polyolefins, further work is needed on the distribution of antioxidants to achieve maximum performance. A consequence of the micro heterogeneous environment of semi-crystalline polymers is that oxygen solubility and diffusion rate are critical and affect the efficiency of the catalytic function of chain-breaking antioxidants. By increasing the oxidation potential of the antioxidant, or reducing oxygen diffusion, the efficiency of chain-breaking antioxidants to capture alkyl radicals is enhanced. The presence of high levels of molecularly dispersed polymer-bound antioxidant moieties should overcome inherent problems of heterogeneity and limited mobility.

It has been suggested that piperidines are efficient because of associative reactions with primary oxidation products, i.e., hydroperoxides. This viewpoint is interesting and may have wider implications for PD antioxidants, especially the sulfoxide, where these have been shown to form hydrogen-bonded complexes with hydroperoxides in solution; this may also explain the exceptional activity of some sulfur compounds and their oxidation products as photoantioxidants.

Migration of additives into the human environment is a major concern for food packaging, body implants, and prosthetic replacements which remain in contact with body fluids for many years. In the case of implants and prosthetics, problems of durability and biocompatibility must be solved for further advance in organ replacement surgery. Polymer-bound reactive antioxidants may contribute to solving some of these problems. It must be emphasized, however, that by virtue of their antioxidant function, most antioxidants and stabilizers are transformed to some other products during processing and service lifetime of the host polymer. Consequently, the safety aspects (toxicological) of fabricated articles throughout their entire life span must be assessed. Much more research is needed to furnish the necessary data for objective assessments by the consumer world.

Durability of thin films of polymers used in the microelectronic industry is a new challenge. Improvement in oxidative stability of conducting polymers is as important as improvements in their mechanical and electrical properties. The utilization of antioxidants (or degradants) in applications requiring controlled stabilization (or degradation), e.g., in agriculture, is another potential area of growth with interdisciplinary activity involving stabilizer design and agricultural sciences.

REFERENCES

1. A. W. Hofmann, *J. Chem. Soc., 13*: 87 (1861).
2. C. Engler and J. Weissberg, *Bericht, 33*: 1090, 1097, 1109 (1900).
3. C. Moureu and C. Dufraisse, *Chem. Rev., 3*: 113 (1926–27).
4. H. L. J. Bäckstrom, *J. Am. Chem. Soc., 49*: 1460 (1929); *Z. Phys. Chem. B, 25*: 99 (1934).
5. J. L. Bolland, *Q. Rev., 3*: 1 (1949); L. Bateman, *Q. Rev., 8*: 147 (1954).
6. L. Bateman and A. L. Morris, *Trans. Faraday Soc., 49*: 1026 (1953).
7. S. Al-Malaika, in *History of Polymeric Composites* (R. B. Seymour and R. D. Deanin, eds.), VNU Science Press, The Netherlands, p. 223 (1987).
8. G. Scott, in *Atmospheric Oxidation and Antioxidants*, Elsevier, London, Chaps. 4 and 5 (1965).
9. D. Barnard, L. Bateman, and J. F. Smith, in *Chemistry and Physics of Rubber-Like Substances* (L. Bateman, ed.), Applied Science Publishers, London, p. 593 (1963).
10. W. Kauzman and H. Eyring, *J. Am. Chem. Soc., 62*: 3113 (1940).
11. G. Scott, in *Stabilization and Degradation of Polymers* (D. L. Allara and W. L. Hawkins, eds.), American Chemical Society, Washington (1978).
12. S. Al-Malaika and G. Scott, *Eur. Polym. J., 16*: 709 (1980).
13. K. B. Chakraborty and G. Scott, *Eur. Polym. J., 13*: 73 (1977).
14. K. B. Chakraborty and G. Scott, *Polym. Deg. Stab., 1*: 37 (1978).
15. G. Scott, *Pure Appl. Chem., 52*: 365 (1980).
16. B. B. Cooray and G. Scott, in *Developments in Polymer Stabilisation*, Vol. 2 (G. Scott, ed.), Applied Science Publishers, London, Chap. 2 (1980).
17. S. Al-Malaika and G. Scott, in *Degradation and Stabilisation of Polyolefins* (N. S. Allen, ed.), Applied Science Publishers, London (1983).
18. D. J. Carlsson and D. M. Wiles, *J. Macromol. Sci. Macromol. Chem., C14*: 65, 155 (1976).
19. A. Garton, D. J. Carlsson, and D. M. Wiles, in *Developments in Polymer Photochemistry*, Vol. 1 (N. S. Allen, ed.), Applied Science Publishers, London, Chap. 4 (1980).
20. G. Scott, *ACS Symp. Ser., 25*: 340 (1976).
21. F. Bickel and E. C. Kooyman, *J. Chem. Soc., 22*: 5 (1956).
22. D. S. Davis, H. L. Goldsmith, A. K. Gupta, and G. R. Lester, *J. Chem. Soc., 1956*: 4926.
23. J. I. Wasson and W. M. Smith, *Ind. Eng. Chem., 45*: 197 (1953).
24. J. Pospisil, in *Developments in Polymer Stabilisation*, Vol. 1 (G. Scott, ed.), Applied Science Publishers, London, Chap. 1 (1979).
25. G. Scott, in *Developments in Polymer Stabilisation*, Vol. 4 (G. Scott, ed.), Applied Science Publishers, London (1981).
26. J. Pospisil, in *Developments in Polymer Stabilisation*, Vol. 7 (G. Scott, ed.), Elsevier Applied Science Publishers, London, Chap. 1 (1984).
27. N. Grassie and G. Scott, in *Polymer Degradation and Stabilisation*, Cambridge University Press, Cambridge, Chap. 5 (1985).

28. A. S. Michael and H. J. Bixler, *J. Polym. Sci.*, *50*: 393 (1961).
29. N. M. Emmanuel, E. T. Denisov, and Z. K. Maizus, in *Liquid Phase Oxidation of Hydrocarbons*, Plenum, New York (1967).
30. E. T. Denisov, in *Developments in Polymer Stabilisation*, Vol. 3 (G. Scott, ed.), Applied Science Publishers, London, Chap. 2 (1980).
31. G. Scott, *Br. Polym. J.*, *3*: 24 (1971).
32. F. Rasti and G. Scott, *Studies in Conservation*, *25*: 145 (1980).
33. T. A. B. M. Bolsman, A. P. Block, and J. A. G. Frijns, *Rec. Trav. Chim. Pays-Bas*, *97*: 310, 313, 320 (1978).
34. S. Al-Malaika, E. O. Omikorede, and G. Scott, *Polym. Comm.*, *27*: 173 (1986); *J. Appl. Polym. Sci.*, *33*: 703 (1987).
35. A. A. Katbab and G. Scott, *Eur. Polym. J.*, *17*: 599 (1981).
36. T. J. Henman, in *Developments in Polymer Stabilisation*, Vol. 1 (G. Scott, ed.), Applied Science Publishers, London, Chap. 2 (1979).
37. R. Bagheri, K. B. Chakraborty, and G. Scott, *Polym. Deg. Stab.*, *5*: 145 (1983).
38. V. Ya. Shlyapintokh and V. B. Ivanov, in *Developments in Polymer Stabilisation*, Vol. 5 (G. Scott, ed.), Applied Science Publishers, London, Chap. 3 (1982).
39. R. Bagheri, K. B. Chakraborty, and G. Scott, *Polym. Deg. Stab.*, *4*: 1 (1982).
40. D. J. Carlsson and D. M. Wiles, in *Developments in Polymer Stabilisation*, Vol. 1 (G. Scott, ed.), Applied Science Publishers, London, Chap. 7 (1979).
41. F. Tudos, G. Balint, and T. Kelen, in *Developments in Polymer Stabilisation*, Vol. 6 (G. Scott, ed.), Applied Science Publishers, London, Chap. 4 (1983).
42. D. J. Carlsson, D. W. Grattan, and D. M. Wiles, *J. Appl. Polym. Sci.*, *22*: 2217 (1978).
43. Y. B. Shilov and E. T. Denisov, *Vysokomol. Soedin.*, *A16*: 2313 (1979).
44. G. A. Kovtun, A. Alexandrov, and V. A. Goluev, *Izv. Akad. Nauk. SSSR Ser. Khim.*, *1974*: 2197.
45. G. Scott, in *Developments in Polymer Stabilisation*, Vol. 7 (G. Scott, ed.), Elsevier Applied Science Publishers, London, Chap. 2 (1984).
46. H. Berger, T. A. B. M. Bolsman, and D. M. Brouwer, in *Developments in Polymer Stabilisation*, Vol. 6 (G. Scott, ed.), Applied Science Publishers, London, Chap. 1 (1983).
47. E. T. Denisov, in *Developments in Polymer Stabilisation*, Vol. 5 (G. Scott, ed.), Applied Science Publishers, London, Chap. 2 (1982).
48. S. Al-Malaika, K. B. Chakraborty, and G. Scott, in *Developments in Polymer Stabilisation*, Vol. 6 (G. Scott, ed.), Applied Science Publishers, London, Chap. 3 (1983).
49. S. Al-Malaika, P. Huczkowski, and G. Scott, *Polym. Comm.*, *25*: 1006 (1984).
50. S. K. Ivanov, in *Developments in Polymer Stabilisation*, Vol. 3 (G. Scott, ed.), Applied Science Publishers, London, Chap. 3 (1980).
51. J. R. Shelton, in *Developments in Polymer Stabilisation*, Vol. 4 (G. Scott, ed.), Applied Science Publishers, London, Chap. 2 (1981).
52. S. Al-Malaika, A. Marogi, and G. Scott, *J. Appl. Polym. Sci.* (in press).
53. C. Armstrong and G. Scott, *J. Chem. Soc.*, *1971*: 1747; C. Armstrong, M. A. Plant, and G. Scott, *Eur. Polym. J.*, *11*: 161 (1975).
54. J. R. Shelton and K. E. Davis, *Int. J. Sulfur Chem.*, *8*: 217 (1973).
55. S. Al-Malaika and G. Scott, *Eur. Polym. J.*, *16*: 503 (1980).
56. S. Al-Malaika and G. Scott, *Polym. Comm.*, *23*: 1711 (1982).
57. S. Al-Malaika, A. Marogi, and G. Scott, *J. Appl. Polym. Sci.*, *30*: 789 (1985).
58. S. Al-Malaika, M. Coker, and G. Scott, unpublished work.
59. D. Barnard, L. Bateman, M. E. Cain, T. Colclough, and J. I. Cuneen, *J. Chem. Soc.*, *1961*: 5337.
60. J. D. Holdsworth, G. Scott, and D. Williams, *J. Chem. Soc.*, *1964*: 4692.
61. V. Hutson and G. Scott, *Eur. Polym. J.*, *10*: 45 (1974).
62. K. B. Chakraborty and G. Scott, *Eur. Polym. J.*, *13*: 1007 (1977).

63. S. Al-Malaika and G. Scott, *Eur. Polym. J., 19*(3): 241 (1983).

64. S. Al-Malaika, *Br. Polym. J., 16*: 301 (1984).

65. G. Scott and S. Al-Malaika, Br. Pat. 2,117,779 (1983).

66. S. Al-Malaika, A. Marogi, and G. Scott, *J. Appl. Polym. Sci., 31*: 685 (1986).

67. D. Gilead and G. Scott, in *Developments in Polymer Stabilisation,* Vol. 5 (G. Scott, ed.), Applied Science Publishers, London, Chap. 4 (1982).

68. S. Al-Malaika, A. Marogi, and G. Scott, *Polym. Deg. Stab., 18*: 89 (1987).

69. M. U. Amin and G. Scott, *Eur. Polym. J., 10*: 1019 (1974).

70. N. C. Billingham and P. D. Calvert, in *Developments in Polymer Stabilisation,* Vol. 3 (G. Scott, ed.), Applied Science Publishers, London, Chap. 3 (1980).

71. M. A. Plant and G. Scott, *Eur. Polym. J., 7*: 1173 (1971).

72. S. Al-Malaika, P. Desai, and G. Scott, *Plast. Rubber Processes Appl., 5*: 15 (1985).

73. B. W. Evans and G. Scott, *Eur. Polym. J., 10*: 453 (1974).

74. D. K. Thomas, in *Developments in Polymer Stabilisation,* Vol. 1 (G. Scott, ed.), Applied Science Publishers, London, Chap. 4 (1979).

75. S. Fu, A. Gupta, A. C. Albertsson, and O. Vogl, in *New Trends in the Photochemistry of Polymers,* (N. S. Allen and J. Rabek, eds.), Elsevier Applied Science Publishers, London (1985).

76. S. Al-Malaika, in *Chemical Reactions on Polymers* (J. L. Benham and J. F. Kinstle), American Chemical Society, Symposium Series 364, ACS, Washington (1988), p. 409.

77. G. Scott, in *Developments in Polymer Stabilisation,* Vol. 8 (G. Scott, ed.), Applied Science Publishers, London, in press.

78. G. Ayrey, C. G. Moore, and W. F. Watson, *J. Polym. Sci., 19*: 1 (1956).

79. G. Scott and S. M. Tavakoli, *Polym. Deg. Stab., 4*: 333 (1982).

80. G. Scott and E. Setudeh, *Polym. Deg. Stab., 5*: 1 (1983).

81. S. Al-Malaika, A. Ibrahim, and G. Scott, unpublished work.

10

Curatives for Castable Urethane Elastomers

Fui-Tseng H. Lee
FMC Corporation
Princeton, New Jersey

INTRODUCTION

Curatives, or curing agents, are chemical compounds that can react with functional groups in a prepolymer and bring necessary polymerization to completion. As a result, a liquid or a low melting solid prepolymer is transformed to a hard polymer with a corresponding increase in molecular weights and an enhancement of physical properties [1–3].

Curatives are necessary components in the preparation of castable urethane elastomers especially by the prepolymer process. By employing various curatives and diisocyanate-terminated prepolymers, elastomers having high strength and elasticity, good load-bearing capacity, and resilience can be prepared. Urethane elastomers are also characterized by high tear strength; resistance to oil, oxygen, and ozone; good low-temperature properties; and exceptional abrasion resistance [4–6]. Both amine and hydroxy compounds are used as curatives for the preparation of urethane elastomers. The resulting polymers are either polyurethane–urea or polyurethane elastomers. Both of these elastomer types are usually referred to as urethane elastomers.

In the prepolymer process the same compound can be applied at a different stage to function as a chain extender as well as a curative. The main function of a chain extender is to extend the polymeric chain, but it may or may not bring the polymerization to a completion. Curatives, on the other hand, should not only extend the polymeric chain but also cure the polymer. Thus, by comparison, curatives are used at a much later stage of a polymerization process and are used in small quantities than chain extenders in a typical urethane elastomer formulation. The difference between curatives and chain extenders is illustrated by the following simple scheme:

$$\text{Low MW polymer} \xrightarrow{\text{chain extender}} \text{Higher MW polymer} \xrightarrow[\text{(chain extender)}]{\text{curative}} \text{Product}$$

In the early urethane elastomer literature curatives were customarily referred to as chain extenders. Only in recent years has the term "curative" been used in various urethane elastomer publications and commercial brochures [7–9]. Castable urethane elastomers can also be prepared by the one-shot process in which all the ingredients are mixed and cast in a mold. In this process the chain extender also serves as a curative.

Since the functional groups in urethane prepolymers are isocyanates, any active hydrogen-containing compound is potentially a curative. However, due to various limitations in the processing and general requirements on the final elastomer properties, only certain compounds are suitable as curatives. Aromatic diamines and aliphatic or aromatic hydroxy compounds constitute the majority of curatives used today. For some applications, alkyl hydroxylamines are occasionally used. The aliphatic diamines are too reactive as curatives for cast urethane to be of any importance [4]. Most hydroxy-containing curatives are bifunctional, although a few triols are sometimes used especially when higher crosslinking is needed to optimize certain elastomer physical properties such as compression set.

Commercially available amine and hydroxy curatives are listed in Tables 1 and 2, respectively. Some additional diamine curatives which have been investigated or cited in the patent literature are given in Table 3. The chemical formulas of the two general types of curatives are shown in Schemes 1 and 2.

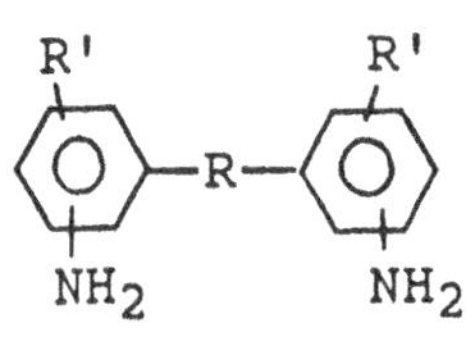

R = CH₂
R′ = H, Cl, or COOCH₃

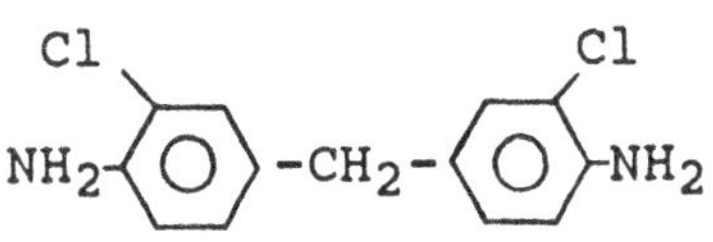

MOCA

Scheme 1 Amine curatives.

Table 1 Diamine Curatives for Castable Urethane Elastomers

Common or trade name	Chemical name	Equivalent weight	Physical properties					Ref.
			Form	MP (°C)	Specific gravity at melt	Moisture content (%)		
MOCA, Curene, MBCA, Cyanaset M	4,4'-Methylene bis(2-chloro-aniline)	133.5	Light yellow granules	100–102	1.26	0.2 (max)		22, 54, 63–66
MDA	4,4'-Methylene dianiline	63.5	Amber solid	92–93	1.05			22, 62, 68
Caytur 21	Methylene dianiline sodium chloride complex dispersed in dioctyl phthalate	217	White liquid		1.13	0.08		9, 76
Cyanacure	1,2-Bis(2'-aminophenylthio)-ethane	138	Light tan flakes	72–77	1.2			54, 77
Polacure	Trimethyleneglycol-di-*p*-aminobenzoate	157	Solid	125–128	1.14			23, 72, 78
Polycure 1000	Methylene bis(methyl-anthranilate)	155	Yellow granules	126–143		<2		71, 79–80
Ethacure 300	3,5-Dimethylthio-2,4 and 2,6-toluenediamine	107	Clear amber liquid		1.21	<0.08		73, 82
Conacure AH-40	Mixture of *m*-phylene di-amine and cumene	133.3	Amber liquid	18	1.02	<0.5		18
Caytur 7		67	Liquid					22
Baytec 1604	3,5-Diamine-4-chloro-iso-butylbenzoate	121	Dark brown flakes	86–90				74

Trademarks: MCCA, Du Pont; Caytur, Uniroyal Chemical; Cyanaset and Cyanacure, American Cyanamid; Polacure, Polaroid; Curene, Anderson Development; Polycure, PTM&W Industries; Conacure, Conap; Ethacure, Ethyl Corp.; Baytec 1604, Mobay Corp.

Table 2 Diol and Triol Curatives for Castable Urethane Elastomers

Common or trade name	Chemical name	Chemical structure	Equivalent weight	Physical properties			Ref.
				Form melting range (°C)	Specific gravity at 20°C or at melt	Moisture content (%)	
BDO	1,4-Butanediol	$HOCH_2CH_2CH_2CH_2OH$	45.1	Clear liquid	1.015	<0.2	8, 18, 83, 93
TMP	Trimethylolpropane	$CH_3C(CH_2OH)_3$	40	Solid 30			8, 47
EG	Ethylene glycol	$HOCH_2CH_2OH$	31	Liquid			47
HQEE	1,4-Bis(2-hydroxyethyoxy)-benzene	OCH_2CH_2OH / $-OCH_2CH_2OH$	99.1	96–100	1.15		87, 95–96
HER	Hydroquinone di(β-hydroxy-ethyl)ether Resorcinal di(β-hydroxy-ethyl)ether	$-OCH_2CH_2OH$	99.1	89	1.16	0.1 max	88, 95–96
Conacure AH-50			90	Liquid	1.15		18
TIPA	Triisopropanolamine	$N(iso\text{-}C_3H_7OH)_3$	64	Liquid	1.15		85
Isonol 93	Triol		90.4	Liquid	1.14		84
Pluracol TP440	Triol		141	Liquid	1.03	0.05 max	86

Trademarks: Isonol, Upjohn Co.; Pluracol, BASF Wyandotte Corp.; HER, Anderson Development; Conacure, Conap. TIPA is a product of Dow Chemical.

Table 3 Less Commonly Known Curatives

Chemical name	Ref.
2,6-Dichloro-*p*-phenylenediamine	55
N,N-Bis(2-hydroxypropyl)phenylamine	56
1-Chloro-2,6-diaminobenzene	57
1-Methyl-4-chloro-3,5-diaminobenzene	57
1,4-Dichloro-3,5-diaminobenzene	57
Methyl-3,5-diaminobenzoate	58
Isobutyl-3,5-diaminobenzoate	58
(2-Ethylhexyl)-3,5-diaminobenzoate	58
Isobutyl-3,5-diamino-4-chlorobenzoate	59
4,4′-Diamino-3,3′-bis-ethoxycarbonyl diphenylmethane	59
1,5-Diaminoaphthalene	60
Ethyl-4-methyl-3,5-diaminobenzoate	60
3,3′,5,5′-Tetrachlorobenzidine	61
4,4′-Methylene-bis(2,3-dichloroaniline)	61
4,4′-Methylene-bis(2,6-dibromoaniline)	61

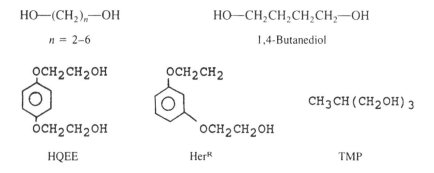

$HO—(CH_2)_n—OH$

$n = 2–6$

$HO—CH_2CH_2CH_2CH_2—OH$

1,4-Butanediol

$CH_3CH(CH_2OH)_3$

HQEE   Her^R   TMP

Scheme 2 Hydroxy curatives.

Ever since the liquid castable technique was first introduced in the early 1950s [10, 11], the two most frequently used curatives are 4,4′-methylene bis(2-chloroaniline), MOCA, and 1,4-butanediol. In the production of high-hardness urethane elastomers (80 Shore A to 50 Shore D), 75–80% of curatives used are aromatic diamines, primarily MOCA. The remaining are hydroxy compounds, mainly 1,4-butanediol. For medium-hardness elastomers (70 to 50 Shore A), diols and triols as well as MOCA mixed with diols or/and triols are used. For low- to very-low-hardness elastomers (50 to 20 Shore A) diol curatives mixed with a plasticizer are employed.

Although MOCA has been shown to induce cancer in laboratory animal tests, there seems to be no direct evidence of a carcinogenic effect on humans during many years of extensive use in the industry [12]. Nevertheless, concerns over MOCA's potential toxicity to humans and the possibility of its being regulated by EPA have stimulated a great deal of research to find a replacement [13]. As a result, several new curatives of diamine or

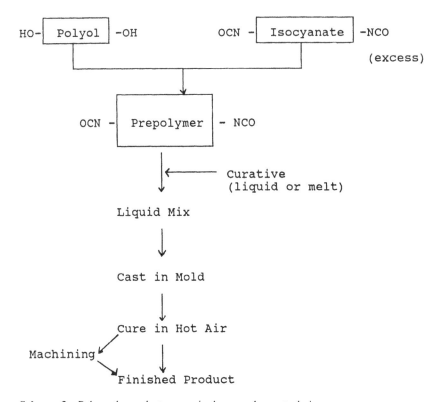

Scheme 3 Polyurethane elastomers via the prepolymer technique.

hydroxy type became available commercially in 1970s and early 1980s [14]. Although these new curatives are generally much safer chemicals than MOCA, none of them has the versatility or is economically competitive to successfully replace MOCA as a general-purpose curative. These new curatives are discussed along with MOCA and other curatives in the "Review of Commercial Curatives" section.

With regard to the MOCA toxicity issue, the Polyurethane Manufacturing Association (PMA) and the Occupational Safety and Health Administration (OSHA) have worked closely together for many years and have developed a mutually acceptable standard which largely reflects the recommended practices in industry [7, 15–17]. It is therefore reasonable to expect that MOCA will remain the most important curative for castable urethane elastomers.

Before a detailed discussion of how each of the commercial curative performs in various castable urethane formulations, it will be useful to review briefly the manufacturing process, the urethane chemistry, and the structure–property relationship of castable urethane elastomer.

MANUFACTURING PROCESS

Urethane elastomers are made from three basic ingredients: a long-chain polyester or polyether polyol, a diisocyanate, and a chain extender or curative. Both the one-shot and the prepolymer process have been used [4–6]. The majority of castable urethanes produced today are made by the prepolymer process.

In this process the predried long-chain polyol is first reacted with a 20–40% excess of diisocyanate to obtain an isocyanate-terminated prepolymer, which is usually a viscous liquid or low-melting solid. The exact amount of diisocyanate needed is dependent on the desired NCO content (expressed as %NCO) in the resulting prepolymer. Storage-stable prepolymers with %NCO ranging from 2 to 10% are available commercially [18]. In the subsequent step, the degassed liquid or molten prepolymer is mixed and cured with a curative or a mixture of curatives, which has previously been melted and degassed. The liquid mix is then poured into a heated mold. As indicated in Scheme 3, additional curing in hot air is required to complete the process. Moreover, for the products to reach their final physical strength, they are often postcured at elevated temperatures and conditioned at ambient temperature for several hours or several days, depending on the formulation.

URETHANE CHEMISTRY

The basic building blocks in urethane elastomers may include polyether, polyester, urethane(carbamate), urea, and other minor but important linkages such as allophanate and biuret. The four principal chemical reactions which usually take place during the curing of a prepolymer are shown in the first four chemical equations that follow. In addition, if water is present as an impurity in the formulation, a fifth reaction can also occur. In these equations, Ar and Ar′ represent aromatic ring or aromatic moities and R represents an aliphatic group.

Urethane formation:

$$2OCN\text{—}Ar\text{—}NCO + HO\text{—}R\text{—}OH \rightarrow$$
$$OCN\text{—}Ar\text{—}NHCOO\text{—}R\text{—}OOCNH\text{—}Ar\text{—}NCO \qquad (1)$$

Urea formation:

$$OCN\text{—}Ar\text{—}NCO + H_2N\text{—}Ar'\text{—}NH_2 \rightarrow OCN\text{—}Ar\text{—}NHCONH\text{—}Ar\text{—}NH_2 \quad (2)$$

Allophanate formation:

$$\begin{array}{cc} H & H \\ | & | \end{array}$$
$$\text{—}OOCN\text{—}Ar\text{—}NCOOR\text{—}O\text{—} + Ar'NCO \rightarrow \text{—}OOCNH\text{—}Ar\text{—}NCOO\text{—}R\text{—}O\text{—}$$
$$|$$
$$CONH\text{—}Ar' \quad (3)$$

Biuret formation:

$$\text{—}CONH\text{—}Ar\text{—}NHCONH\text{—}Ar\text{—}NH\text{—} + Ar'NCO \rightarrow$$
$$\text{—}CONH\text{—}Ar\text{—}NH\text{—}CO\text{—}N\text{—}ArNH\text{—}$$
$$|$$
$$CONHAr'$$
$$(4)$$

Reaction with water:

$$OCN\text{—}Ar\text{—}NCO + H\text{—}O\text{—}H \rightarrow OCN\text{—}Ar\text{—}NH_2 + CO_2 \qquad (5a)$$

$$OCN\text{—}Ar\text{—}NH_2 + OCN\text{—}Ar\text{—}NCO \rightarrow OCN\text{—}Ar\text{—}NHCONH\text{—}Ar\text{—}NCO \quad (5b)$$

In the preparation of castable urethanes, the formation of urethane linkages is largely accomplished during the manufacturing of prepolymers which involves reacting polyether or polyester polyols with an excess diisocyanates (Eq. 1). Either tolylenediisocyanate (TDI), 2,4-isomer or 80:20 mixture of 2,4- and 2,6-isomers, or 4,4'-methylenediphenyl-diisocyanate (MDI) can be employed. When a hydroxy curative is used additional urethane linkages are formed during curing of a prepolymer. In this case all the principal linkages in the hard segments in the resulting elastomers will be urethanes (carbamates) and the secondary linkage will be allophanates (Eq. 3). On the other hand, if a diamine curative is used, urea groups are formed and the principal linkages in the resulting elastomers will be ureas and urethanes. Subsequently, the urea group may react with a carbamate to give biuret as the secondary linkage. The urea and the biuret formation are shown in Eqs. 2 and 4, respectively.

Water reacts readily with isocyanates to give amines and a gaseous by-product, CO_2, as shown in Eq. 5. This by-product can cause bubble formation in the resulting elastomer and weaken its physical property. Therefore, the removal of water from the castable system is of utmost importance [5].

As expected, the extent of each of the reactions occurring during the preparation of urethane elastomers is determined by many factors including compositional, such as the chemical reactivity of polyols, diisocyanates, and curatives, as well as processing conditions such as processing temperature. Some of the more important factors which control these chemical reactions are briefly discussed in the following sections.

Relative Reaction Rate

During the curing of urethane elastomers usually more than one of the previously mentioned reactions will take place simultaneously; therefore, the relative rates of these reactions have an important bearing in determining the final structure and the property of the resulting polymer. The rate of reaction and the activation energy of several diisocyanates with different active hydrogen compounds including (1) a polyester polyol, (2) water, (3) diphenyl urea, (4) 3,3'-dichlorobenzidine, and (5) *p*-phenylene dibutylcarbamate were determined [19, 20]. Data show that reactions between diisocyanate and amine or hydroxy functions are considerably faster and require less activation energy than those between diisocyanates and urea functions. This is true for a highly reactive 2,6-tolyene-diisocyanate (2,6-TDI) or a much less reactive alkydiisocyanate, 1,6-hexamethylene di-isocyanate (HMDI). The reactions between diisocyanates and urethanes (carbamate), which lead to allophanate formation, are by far more difficult and require a higher reaction temperature. Thus the order of relative reaction rates or relative reactivity for these compounds toward NCO function is as follows:

$$—NH_2 > —OH > H_2O \gg —NHCONH— > —NHCOOR—$$

As indicated in this reactivity sequence, aromatic diamine curatives are more reactive than hydroxy curatives; as a result, MOCA will give a much shorter pot life and cure time than 1,4-butanediol would in the same prepolymer system.

Several polyester and polyether polyols which contain only primary hydroxy groups are found to have a similar reaction rate constant in reacting with a diisocyanate [19]. A slightly lower reaction rate is shown for low molecular weight diols and those polyols containing secondary hydroxy functions. As a result the polypropylene glycols, which have approximately 96% secondary hydroxy groups, are expected to be slower in urethane formation. The reactivity of this type of polyether polyols is often improved by capping with ethylene oxides, which produce primary hydroxy groups.

Effect of Temperature

Liquid castable urethanes are mostly cured at 85–110°C, although some formulations can be cured at room temperature when a very reactive curative is used. As temperature increases, all reaction rates tend to converge. The effect is that secondary or crosslinking reactions are favored, especially the biuret formation, resulting in a marked improvement in compression set but poorer high-temperature properties. This is due to the fact that allophanate and biuret dissociate at elevated temperatures leading to the formation of a more linear polymer with a significant loss of practically all the properties [21]. As the velocity constants of various reactions are changing at different rates with temperature, it is necessary to have a close control of temperature to obtain elastomers of consistent quality.

The presence of a weak catalyst such as adipic or other organic acid can also regulate some of the urethane reactions and therefore is occasionally used. These catalysts tend to mildly accelerate the chain-extending reaction but retard the crosslinking reactions [19, 22]. Both metal ions and bases which accelerate all isocyanate reactions are usually not used for the preparation of urethane elastomers.

Curative Stoichiometry

The amounts of curative used have a significant control over the formation of allophanate and biuret, which link two polymeric chains and thus increase crosslinking. Since some degree of crosslinking between long polymer chains is desirable for producing polymers with a better balanced physical property, curatives are usually used at a slightly less than stoichiometric amount (85–95% theory). Elastomers produced by using this range of curatives usually have the best combination of properties. Use of less curative will leave more isocyanate functions available for reacting with urethane or urea leading to the formation of a more rigid polymer with poorer high-temperature characteristics. A higher curative stoichiometry, on the other hand, will increase polymer chain length and result in a softer polymer with reduced modulus and other properties. As mentioned earlier, the allophanate and biuret bonds are not stable at elevated temperatures; therefore it is not desirable to have too many of these secondary linkings.

The amount of a curative required for a particular formation can be calculated as

$$\frac{\text{wt. of prepolymer} \times \%\text{NCO} \times \text{eq. wt. of curative} \times \% \text{ theory}}{4200}$$

Diisocyanates

Aromatic diisocyanates which are commonly used in the preparation of urethane prepolymers are tolylenediisocyanate (TDI) and 4,4′-methylene diphenyl diisocyanate (MDI). TDI diisocyanates are available as a pure isomer (2,4- or 2,6-isomer) or as a mixture (80 : 20 mix of 2,4- and 2,6-isomers). Aliphatic diisocyanates such as 1,6-hexamethylene diisocyanate (HDI) and others are much less important. 1,5-Naphthalene diisocyanate (NDI), which is widely used in Europe, is rarely used in the United States.

The isocyanate groups attached to diphenyl methane and naphthalene aromatic rings are considerably more reactive than those attached to tolulene in TDI [19]. The NCO group at the 4-position, which is less hindered, is two to three times more reactive than the one at the 2-position at the reaction temperature of 100°C. During the preparation of prepolymers the more reactive NCO groups, as expected, react first, leaving behind the less reactive NCO groups. The reactivity of these less reactive NCO functions is further reduced in the prepolymer because of increased steric hindrance. The NCO groups in a prepolymer are no longer monomeric but are polymeric in nature. This is true for either symmetrical diisocyanates such as 2,6-TDI, MDI, and NDI or nonsymmetrical diisocyanates such as 2,4-TDI. It should be noted that these less reactive NCO groups are the ones that react with curatives during the curing of prepolymers.

Studies [19, 20] show that the monomeric isocyanates at 2- or 4-positions are four times faster in forming urethanes with primary hydroxy groups than those NCO groups in the polymeric isocyanates at the reaction temperature of 60°C. Therefore, the reduction in reaction rates when 50% of isocyanates had reacted is quite significant. The difference in relative reaction rates between monomeric and polymeric isocyanates becomes somewhat greater—six times instead of four—when secondary hydroxy groups are involved in the reaction. Interestingly, there is no difference in the reaction rate for allophanate and urethane formations between monomeric and polymeric isocyanates, even though the rate of formation for urethane is about 20 times faster than allophanate. It is also interesting to note that the frequently used 2,4-TDI diisocyanate, although considerably more reactive than the 2,6-isomer toward hydroxy functions, is less reactive toward a urea group. This means that prepolymers prepared from 80 : 20-TDI will give more crosslinked elastomers than those made from 2,4-TDI. For this reason, certain curatives are better to be used with prepolymers which are derived from 80 : 20-TDI rather than from 2,4-TDI for achieving optimum elastomeric properties through a slightly higher degree of crosslinking [23].

In addition, the prepolymers prepared from 2,4-TDI in general are less reactive than those derived from 2,6-TDI. This difference in the NCO reactivity often can be a determining factor for the selection of a suitable curative. Most of the TDI-based prepolymers used today are 2,4-TDI and are best cured with MOCA. On the other hand, the more reactive 2,6-TDI–based prepolymers are more suitable for some of the newer, more hindered and less reactive diamine curatives [23].

STRUCTURE–PROPERTY RELATIONSHIPS

It is well recognized that urethane elastomers are $(AB)_n$-type block copolymers consisting of chemically different ''soft'' and ''hard'' segments (blocks) [24]. The primary polymer chain can be depicted as in Scheme 4.

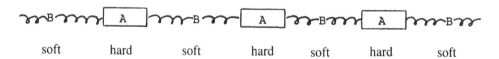

soft   hard   soft   hard   soft   hard   soft

A = hard segment, curative + diisocyanate
B = soft segment, polyols + diisocyanate

Scheme 4

Evidence indicates that during the polymerization and the solidification of urethane pre-polymers, phase separation occurs due to the basic thermodynamic incompatibility of the soft and the hard segments [25]. This leads to the formation of hard and soft domains. However, the topology of the block copolymer molecule imposes restrictions on this segregation and consequently some degrees of phase mixing exit in urethane elastomers.

Recent studies on the morphology of various urethane and urethane–urea elastomers by differential screening calorimetry, transmission electron microscopy, and x-ray diffraction show that the hard segments tend to associate with each other through aromatic π-electron attraction and through hydrogen bonding due to the polar nature of the urethane or urea moieties [26–30]. As a result the hard segments cluster or aggregate into hard domains, which are dispersed within the rubbery matrix of the soft segments, and reinforce the matrix by functioning as pseudo-crosslinkers. Infrared analysis has provided direct evidence for the existence of hydrogen bondings between the ether groups in the soft segments and the —NH— groups in urethane hard segments [31]. The hydrogen bonding increases the glass transition temperature of the soft segment and reduces T_g of the hard segment, leading to improvements in hardness and tensile strength [32]. Urea groups, which are more polar than urethanes because of the presence of the two NH groups, are more effective in providing physical crosslinking between polymeric chains and thus in improving physical properties.

Many investigations [32–39] have been carried out to study the structure–property relationships of urethane elastomers as related to the nature of the soft and hard segments. Results show that the hard segments control the hardness, modulus, and tear strength and the soft segments provide flexibility, low-temperature properties, and solvent and oil resistances. The overall structure–property relationship of urethane elastomers is basically governed by five factors: molecular weight, degree of crosslinking, degree of crystallinity, amount of hydrogen bonding, and rigidity of chain units. These factors in turn are dependent on many compositional variables such as the type and the amount of long-chain polyol, diisocyanate, and curative as well as the processing conditions such as reaction temperature and curing time. Curatives play a major role in determining the final elastomer properties through their contributions to the nature and the amounts of hard segments. The hard segment content is also determined by the isocyanate content (%NCO) of the pre-polymer.

In the following sections, some details on how the nature of the soft and hard segments influence the elastomer properties are briefly reviewed.

Soft Segments

Polyester Versus Polyether Polyol

The long-chain polyether or polyester polyols which make up the soft segments in urethane polymers are either low-melting or amorphous materials with low glass transition temperatures (−50 to −80°C). Polyester polyols, which are the condensation products of adipic acid and ethylene glycol or 1,4-butanediol, and have an average molecular weight of 1500–3000, are used mostly with MDI diisocyanate. These types of prepolymers are usually cured with hydroxy curatives. Polyether polyols, which are condensation adducts of propylene oxide (or propylene oxide and end-capped with ethylene oxide) using either 1,4-butanediol or ethylene glycol as an initiator and having a molecular weight of 1000–

3000, are reacted mainly with TDI. The resulting prepolymers are generally cured with diamine curatives.

Many studies have been carried out to compare the elastomer properties to the type of polyols used (40–47). In general, elastomers containing the polyester soft segments have a higher physical strength and a better solvent and chemical resistance than comparable polyether-based products. The polyethers, on the other hand, offer better flexibility, hydrolytic stability, and superior low-temperature properties due to their lower glass transition temperatures. The stronger physical properties observed on the polyester urethane elastomers are due to the more polar nature of polyester groups and their greater ability to form hydrogen bondings with urethane hard segments. Their greater tendency to undergo crystallization on extension is also a contributing factor. The difference between polyester and polyether urethanes can be somewhat reduced by using polyethers of lower molecular weight to increase the polarity. Currently the polyether polyols constitute the majority of soft segments used in the urethane elastomers, mainly because they are more resistant to hydrolysis and are less expensive.

Among the polyethers, poly(tetramethylene)ether glycol (PTMG) is the most frequently used. This polyol is shown to contain the optimum number of carbon atoms and offer the best balanced properties compared to other polyethers. This is due to the regularity of its chain structure and its ability to crystallize upon extension.

The presence of a side chain such as a methyl substitution in either polyether or polyester polyols resulted in the reduction of both tensile and modulus significantly in either MDI- or TDI-based elastomers. The use of a diamine curative instead of a diol, thus introducing the more polar urea linkages into the polymer backbone and increasing the polar group content, can counteract the effect introduced by the methyl group [40, 41].

Molecular Weights of Polyols

Polyols with molecular weights in the range of 600–3000 are generally used. Polyols of molecular weight less than 600 will give hard elastomers which have little flexibility and poor strength. The optimum molecular weight depends on the particular composition required [48]. In general, the main consequences of increasing the molecular weight of the soft segments for a given overall molar ratio of polyol to isocyanate plus curative are reduction in modulus and hardness and improvement in tear, tensile strength, and elongation at break. This is due to increase in flexibility and a small reduction in hard-segment interaction. These effects are illustrated in Figure 1. The data obtained are from elastomers based on TDI–PTMG–MOCA [49].

The greater tendency for the soft segments to cold harden or slowly crystallize on standing with increasing molecular weight is also attributed to the observed variation in elastomer properties [25, 47]. The degree of cold hardening can be reduced through the use of co-polyols, which introduce some structural irregularity to the system. However, some degree of cold hardening is desirable in urethane elastomers since the ability of the flexible polyester segments to crystallize upon extension of the elastomer is shown to increase the strength of the material [44].

Hard Segments

The hard segments in urethane elastomers are essentially low molecular weight polyurethanes or polyurethane–ureas. The properties of the hard blocks determine the inter-

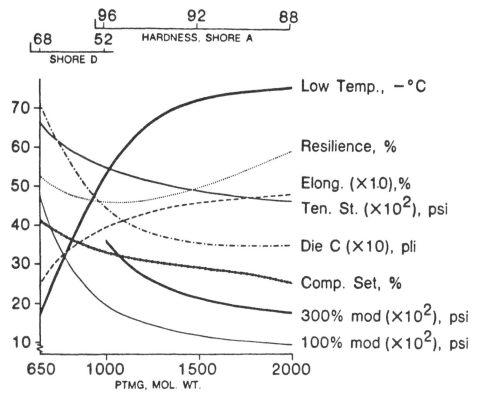

Figure 1 Effect of polyether polyol molecular weight on elastomer properties. Elastomer composition: TDI/PTMG at 2 : 1 NCO/OH; MOCA at 95% theory. (From Ref. 49.)

chain interactions in the elastomers to a large extent and thus determine the network "structure" in these materials.

Studies have shown that bulky, more rigid aromatic diisocyanates having a symmetrical molecular structure result in elastomers which are higher in hardness and modulus [41, 47]. Diisocyanates with a greater molecular flexibility or with a spatial separation by a methyl side chain tend to produce softer and more elastic polymers. Thus the substitution of methyl group(s) in *p*-phenylene diisocyanate or in 4,4′-methylene diphenyl diisocyanate results in general decrease in tensile strength, modulus, tear, and hardness. Similarly, asymmetric molecules such as 2,4-TDI in comparison to the more symmetrical 2,6-TDI produce elastomers of lower strength. The use of diamine curatives can restore some of the symmetry and also increase the aromatic ring contents which bring rigidity and improvement of certain physical properties. Increasing the concentration of the aromatic ring, however, has a somewhat similar effect to increasing the urethane content, which is a greater modulus and hardness along with a moderate increase in tear strength and moderate loss in elongation; but unlike hydrogen bondings through the polar groups, this does not improve toughness or tensile strength. Variations in diisocyanate structures are found to have little influence on low-temperature properties of urethane elastomers when compared to other changes.

Degree of Crosslinking

Crosslinking in urethane elastomers can be achieved by either chemical or physical means [4–6]. The use of curatives at less than 100% of theory, triol curatives, branched polyols, or polymeric isocyanates can provide chemical (or primary) crosslinking, while increasing polar groups can improve physical (or secondary) crosslinking through hydrogen bonding and other intermolecular forces.

Although some degree of crosslinking is desirable for better overall elastomer property, the relationship between crosslinking and the properties of urethane elastomer is less straightforward than in other polymers. This is because urethane elastomers rely on intermolecular forces for their high modulus properties and increased chemical crosslinking tends to reduce the orientation of the polymer chain and the possibility of hydrogen bonding [45, 50, 51].

The disruptive effect that chemical crosslinking has on the properties of an elastomer was examined in an MDI–polyester system [46]. It was found that increasing a triol to diol curative ratio resulted in generally decreased physical properties except compression set and volume swelling toward dimethylacetamide. The effects of increasing the NCO/OH ratio are similar. Usually, a considerable improvement in elastomeric properties is observed when the NCO/OH ratio is increased from 0.9 to 1.0. Further increase of this ratio to 1.3, however, results in the reduction of elastomer strength except volume swelling and compression set, which remain constant.

Curatives: Diamine Versus Hydroxy

Aromatic diamines, because of their higher reactivity toward isocyanate functions, are used only with the less reactive and less symmetrical TDI-based prepolymers, especially those derived from 2,4-TDI. MOCA and other diamines are too reactive for MDI-based prepolymers to have sufficient pot life to allow for casting. The diamine-cured TDI elastomers are characterized by high mechanical and/or physical strength. The less reactive hydroxy curatives can be used with either TDI- or MDI-terminated prepolymers. However, when TDI prepolymers are cured with hydroxy curatives the resulting elastomers have much lower hardness and physical strength. This is mainly due to the difference in the nature of the hard segments in the resulting elastomers. The presence of highly polar urea groups and the additional rigid, π-electron–containing aromatic rings in the diamine-cured elastomers allow their polymer chains to interact more effectively than hydroxy-cured elastomers. On the other hand, when hydroxy curatives are used with the more reactive, highly symmetrical, and rigid MDI-based prepolymers, elastomers with properties which are comparable to MOCA–TDI systems can be obtained [52].

A general comparison of the elastomer physical properties obtained from four MOCA–TDI and two 1,4-butanediol–MDI systems are presented in Table 4 [53, 54]. As can be seen, with the exception of the percentage elongation, other properties either improved or stayed constant as the isocyanate contents (hardness contents) are increased. Hardness, modulus, tensile strength, and tear are among those properties showing significant improvements, whereas compression set and rebound are unaffected, especially with MOCA–TDI systems. It is noteworthy that among these six systems, the elastomers containing polyester soft segments have properties which are considerably better than the TDI–polyether–MOCA or MDI–polyether–1,4-butanediol systems. For example, the tear property is greatly enhanced in the polyester-type elastomer.

A more specific comparison of diamine and hydroxy curatives is given in Table 5 in

Table 4 Range of Properties Obtainable from MOCA- and Diol-Cured Elastomers[a]

Elastomer property	MOCA TDI– polyether[b]	MOCA TDI– polyester[c]	1,4-BD MDI– polyether[b,d]
% NCO, range	2.8–9.5	2.3–6.7	6.4–9.6
Hardness			
Durometer A	80	71	83–95
Durometer D	73	71	
Modulus, psi			
100%	400–4700	300–3600	980–1800
300%	625	500–7800	1780–3800
Tensile strength, psi	3000–9000	5050–9050	5900–5500
Elongation, %	800–210	725–340	565–375
Tear strength			
D470, pli	70–110	125–510	130–125
Die C, pli		260–830	
Resilience (rebound), %	45	32–33	65–35
Compression set, Method B	45	34–38	32–35
NBS abrasion index, % of standard	110–400		220
Compression modulus, psi			
At 10% compression		145–1870	425–825
At 25% compression		370–3850	1125–2250
Specific gravity	1.08–1.21	1.25–1.30	1.09–1.14

[a]Not including diol-cured TDI–polyether- or polyester-based elastomers of medium to low hardness or high hardness MDI–aromatic diol-cured elastomers.
[b]From Ref. 53, Adiprene L 42 and L 315, MOCA at 100 and 95% of theory, respectively.
[c]From Ref. 54, Cyanaprene A-7QM and D-7, Cyanaset M (MOCA) at 95% of theory.
[d]From Ref. 53, Adiprene M 483 and M467, 1,4-butanediol at 100% of theory.
Trademarks: Adiprene, Uniroyal Chemical Company; MOCA, Du Pont; Cyanaprene and Cyanaset, American Cyanamid.

which the physical properties of elastomers prepared from a same TDI–polyester prepolymer and cured with MOCA and 1,4-butanediol, respectively, are summarized [18]. It is evident that, with the exception of compression set, MOCA-cured elastomers are superior in all other properties, particularly the hardness, modulus, and tear. Consequently, diols such as 1,4-butanediol are used with TDI prepolymers only when softer products with medium strength are desired. Products of such nature are useful in a variety of printing and roller-coating applications. [8].

As mentioned earlier, comparable elastomer properties can be obtained from either TDI–MOCA or MDI–diol systems. This is illustrated in Table 6, in which the physical properties of polyether urethane elastomers prepared from three systems are compared [52]. These systems are MOCA–TDI, MDI–1,4-butanediol, and MDI–1,4-bis(beta-hydroxyethoxy) hydroquinone (HQEE). With the exception of higher hardness shown by the two MDI–diols systems, a higher tear strength in MDI–HQEE, and a slightly better compression set with MOCA–TDI, other properties are practically the same. Although the MDI–diols systems can produce elastomers with equal or even better physical strength than TDI–MOCA systems, they are generally considered to be the more difficult systems to process

Table 5 Comparison of Urethane Elastomers Cured with Diamine and Polyol Curatives

	Curative	
	Conacure AH-50	MOCA
Parts per 100 prepolymer[a]	6.7	9.5
Processing conditions		
Prepolymer temp., °C	100	100
Curative temp., °C	25	120
Pot life, min	100	6
Demold time at 100°C, hr	4	0.5
Physical properties		
Hardness, Shore A	56	84
Tensile strength, psi	5800[b]	8600[c]
Modulus, psi		
100%	250	800
300%	500	1280
Elongation, %	550	700
Tear strength, pli (Die C)	190	500
Compression set, % (method B)	8	26
Cured specific gravity	1.23	1.25

[a]Conathane RN-3038 (3.2% NCO, TDI/polyester) and Conacure AH-50, a liquid triol curative, are the products of Conap, Inc. (Ref. 18).
[b]Die C.
[c]Die D.

Table 6 Comparison of Elastomeric Properties of TDI–MOCA- and MDI–Diol-Cured Polyether Urethanes

	Formulation[a]		
Elastomer property	TDI–MOCA	MDI–BDO[b]	MDI–HQEE[c]
Hardness, Shore A	90	95	95
Modulus, psi			
100%	1100	1100	1300
300%	2200	2800	1970
Tensile strength, psi	5500	6110	6100
Elongation, %	430	390	690
Split tear, pli	90	90	190
Bashore rebound, %	42	40	55
Compression set, %	25	30	30

[a]Curative % theory: MOCA = 90, diols = 95.
[b]BDO, 1,4-butanediol.
[c]HQEE, 1,4-bis(2-hydroxyethoxy)benzene.
Source: Ref. 52.

and are used less frequently for producing high-hardness urethane elastomers [52]. The advantages and the disadvantages of these two approaches are further elaborated in the next section.

REVIEW OF COMMERCIAL CURATIVES

In the following sections the processing and the performance characteristics of many commercially available curatives are discussed and whenever possible are compared with those of MOCA.

Diamine Curatives

Commercially important diamine curatives along with their typical physical properties are presented in Table 1. Aromatic diamines which have been investigated but have little commercial application are listed in Table 3 [55–61].

Both unsubstituted and substituted aromatic diamines have been used for casting urethane elastomers. With one or two exceptions, they are mostly the derivatives of biphenyl or diphenyl methane. Unsubstituted aromatic amines such as methylene dianiline (MDA) and *m*-phenylene diamine (MPD) are usually too reactive to have sufficient pot life for hand batch mixing and the use of a dispensing machine is required [4–6]. Substitutes such as chlorine atoms and methyl or alkyl carboxylate groups in the aromatic ring(s) can reduce the chemical reactivity of the amine functions through the electron-withdrawing effect and/or steric hindrance, to the extent that a reasonable pot life can be obtained. Some of the newer diamine curatives with more bulky substituents are considerably less reactive than MOCA. As a result, they are more sensitive to the 2,4- and 2,6-TDI isomer ratios used in the preparation of the prepolymer. For optimum performance, the less reactive diamine curatives are best used with the more reactive prepolymers derived from 2,6-TDI or 80 : 20-TDI [23].

In an earlier study seven aromatic diamine curatives which are the derivatives of benzene, biphenyl, and diphenyl methane with substituents such as Cl, OCH_3, and CH_3 were compared [62]. Data indicate that among these diamines MOCA imparts a better balance of tensile, tear, modulus, and hardness properties to the resulting elastomers. Results also show that ortho-methyl and chlorine substituents in these diamines did not cause reduction in elastomer properties as were observed with diisocyanates having methyl substituents [47]. Benzidine, which compared favorably with MOCA, is no longer being used because it is a cancer-suspect agent.

MOCA

4,4′-Methylene bis(2-chloroaniline) or 3,3′-dichloro-4,4′-diamino diphenyl methane, commonly known as MOCA, was first introduced as a curative for TDI–polyether prepolymers (Adiprene L) by du Pont in late 1950 [62–64]. Since then it has become the most used general purpose curative for the production of TDI-based high-performance urethane elastomers. It is used for both polyether- and polyester-based urethanes. Because of concerns over its toxicity, MOCA is no longer manufactured in the United States. The material produced offshore is marketed under various common names or trademarks which include MBCA, MBOCA, Bis-amine A, Curene 422, and Cyanaset M [7, 54, 65, 66].

Typical physical properties of MOCA are listed in Table 7. MOCA is a solid with

Table 7 Typical Properties of 4,4'-Methylene bis(2-Chloroaniline)[a]

Molecular weight	267.2
Amine equivalent	133.5
Physical form	Expanded pellets or granules
Color	Light tan
Odor	Slight, aromatic amine odor
Specific gravity	
Solid, 24°C	1.44
Melt, 107°C	1.26
Melting range, °C	100–102
Moisture content, %	0.2% max
Storage stability	Excellent, slightly hydroscopic in un-opened fiber drums
Solubility	Soluble in ketones, esters, and aromatic hydrocarbons

[a]Known commercially as MOCA, MBCA, Bis-amine A (Palmer, Davis, Seika, Inc.); Cyanaset M (American Cyanamid); Curene 442 (Anderson Development Co.), etc.

melting range of 100–109°C. In the processing of liquid castables, MOCA is usually handled as a liquid at 120°C. MOCA is the most favored curative mainly because of its excellent balance of pot life, cure rate, elastomer properties, and overall handling ease [65]. It is relatively easy to melt MOCA. Its tendency to supercool, which permits MOCA to be processed at temperatures below its melting point, is also an advantage. In addition, MOCA is very soluble in various prepolymers and will not easily crystallize out upon mixing with these materials.

When MOCA is used to cure Adiprene L-100, a TDI–polyether prepolymer (%NCO = 4.1 ± 0.2), the resultant elastomers show a good balance of properties, for example, hardness: 90 Shore A; tensile strength: 4500 psi; 100% modulus: 1100 psi; 300% modulus: 2100 psi; elongation: 450%; tear strength: 75 pli by D-470 or 400 pli by Die C; compression set: 27%, abrasion resistance: 175 NBS index [22]. In addition, good low-temperature properties and good resistance to oil, solvent, oxidation, and ozone are obtained. The range of elastomeric properties obtainable from MOCA and TDI–polyether or polyester prepolymers has already been discussed (see Table 4). MOCA-cured urethane elastomers have many applications such as industrial tires, mining equipment parts, wheels, pump impellers, and gears [4–6].

MOCA stoichiometry. In comparison with other curatives, MOCA is generally known as being more "forgiving," meaning that the elastomer properties are not overly sensitive to the stoichiometry (% theory) of MOCA. Therefore, product quality control with MOCA-cured products is relatively easy. The effect of varying MOCA content (from 90 to 110% theory) in a TDI–polyether-based elastomer is illustrated in Figure 2 [49]. These results are similar to an earlier study carried out on an Adiprene L-100–MOCA system with a much wider range of MOCA stoichiometry (40–120% theory) [22]. As can be seen from Figure 2, hardness is not affected by curative stoichiometry. The highest tensile strength and good compression set are obtained with 90–95% theory of MOCA.

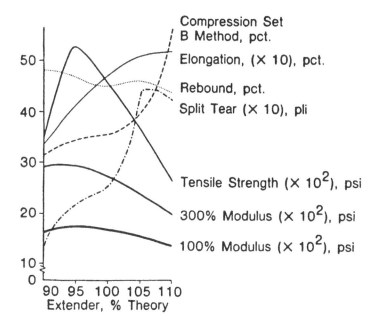

Figure 2 Ninety-five Shore A elastomer properties versus MOCA curative stoichiometry. Prepolymer: TDI-20 Polymeg 1000. (From Ref. 49.)

Increasing MOCA reduces the availability of NCO function for crosslinking and thus more linear elastomers with lower tensile strength and higher ultimate elongation are formed. The compression set, which can be taken as an inverse measure of crosslinking, increases rapidly as the MOCA concentration increases from 90% theory to 110%. Best tear resistance is obtained from 103 to 110% theory of MOCA; tear strength decreases at lower stoichiometry. The earlier study on Adiprene L-100 shows that good abrasion resistance was obtained with 90–95% theory of MOCA, and flex life (DeMattia flex test, not nickel) improved markedly at concentrations of 105–110% stoichiometry. MOCA stoichiometry is shown to have little effect on the hardness property. The effect of MOCA stoichiometry on elastomer properties based on TDI–polyester systems is also quite forgiving, as indicated by the data in Table 8 [54]. The prepolymer used is Cyanaprene A-9, which has an isocyanate content (%NCO = 4.2 ± 2) similar to that of Adiprene L-100.

Processing conditions and elastomer properties. In processing MOCA, it is typically heated to 110–120°C for melting. It is then mixed with a previously degassed liquid or molten prepolymer at 100°C and poured into a mold for curing at this temperature for 1–3 hr. For mixing, either the hand mix or dispensing machine is used. The pot life is generally within 1–15 min and solidification takes place within 5 45 min, depending on the grade of prepolymer (type and % NCO). MOCA is stable at 120°C for up to 48 hr. Thermal decomposition by MOCA is exothermic and self-propagating; therefore, heating MOCA above 140°C even for a short time should be avoided [5].

As mentioned earlier, factors such as mixing, curing temperature, and curing time have a pronounced effect on elastomer properties. The curing reaction begins the moment MOCA is mixed with the prepolymer; it continues throughout the fluid stage and beyond the point of solidification until the elastomer is completely cured. Since the reactions

Table 8 Effect of Cyanaset M Concentration on the Physical Properties of
TDI–Polyester Elastomers

% Stoichiometry	85	90	95	100
Cyanaset M curative, pph, to use with 4.2% NCO prepolymer	11.1	11.7	12.4	13.0
Hardness, Shore A	91	91	91	92
Modulus, psi				
At 100% elongation	980	1010	1015	1020
At 300% elongation	1870	1960	2030	2040
Tensile strength, psi	6300	6500	6600	6650
Elongation, %	520	540	545	560
Tear strength, pli (kn/m)				
Die C	505	540	550	540
Split	290	300	320	310
Compression set, %	26	28	34	40
Bashore rebound, %	31	32	32	31
Compression modulus, psi				
At 10% compression	435	450	460	450
At 25% compression	1090	1120	1120	1100

Note: Prepolymer: Cyanaprene A-9, 4.2 ± 0.2 % NCO. Prepolymer temperature, 100°C; curative temperature, 110–115°C; curing, 100°C/45 min; postcure, 100°C/16 hr; potlife, 6–8 min; gel time, 8–12 min. Cyanaset and Cyanaprene are trademarks of American Cyanamid. *Source:* Ref. 54.

between curatives and prepolymers are exothermic and urethanes are thermal insulators, the rise in mass temperature can be significant and can be influenced by the temperature of the prepolymer and curing agent; the mold material and configuration and the mold temperature; and the environmental or oven temperature. The size of the casting has an effect also, since small castings can transfer heat to the surroundings and may not reach as high a temperature as large castings.

The temperature at which the curing reactions takes place has an important effect on elastomer properties. The formation of biuret branching, like urea formation, is also promoted by an increase in temperature; consequently, different properties can be obtained from a same system by varying the reaction temperature. In general, the effect of increasing the temperature is similar to decreasing MOCA concentration. Thus compression set improves while tear and tensile properties decrease in value. With TDI–polyether–MOCA elastomers, highly consistent properties are obtained when the mixing temperature is held within the range of 70–100°C, since the maximum reaction temperature then rarely exceeds 116°C. At reaction temperatures above 116°C the amount of biuret branching formed reaches a concentration at which the orderly structure of the polymer is reduced and the softer, more flexible, and weaker elastomers are produced. It should be pointed out that for castable urethane elastomers curing usually continues after the pieces are removed from the mold, and postcuring at room temperature for several days is needed to reach ultimate properties.

Curing reactions between MOCA and prepolymers can also be accelerated by the presence of a small amount of carboxylic acid such as adipic acid so that shorter demold time can be allowed. As expected, the catalytic reaction will also shorten pot life. Studies show that 0.20 phr of adipic acid reduces pot life from 10 min to less than 5 min [22]. However, no further reduction in pot life or curing time is obtained when the concentrations of this acid are greater than 0.30 phr. The final properties of elastomers are shown to be comparable for the catalyzed and the uncatalyzed systems. The main advantage of the catalyzed system is earlier demolding, which allows faster mold turnover and significantly reduces the number of molds required to produce a given number of molded parts.

Diamine Curative Blends

Aromatic diamines are sometimes used as binary or ternary blends [22]. Such mixtures may be useful because they can be handled at lower temperatures and used to control rate and gel characteristics or to modify properties. Examples of such mixtures are blends of MOCA and methylene dianiline (MDA) for increasing reactivity (compared to MOCA alone) without significantly affecting elastomer properties. A eutectic mixture of *m*-phenylene diamine and curmene diamine (Caytur 7) is a liquid at temperatures above 65°F (18°C). This permits the room-temperature mixing of this curative mixture. Physical properties of elastomers made from Adiprene L-100 and various diamine blends, as shown in Table 9, are quite similar except compression set, which is higher for the blends compared to when they are used alone. A more significant difference in elastomer physical properties is observed, however, when ternary diamine blends are used. They yield much softer products than are obtained from single diamine or from binary blends. This is believed to be caused by a significant decrease in the amounts of intermolecular bonding, due to irregularities placed in the polymer chain by the three diamines [22, 24, 25]. MOCA can also be blended with diols or triols to produce a medium-hardness elastomer with moderate physical and mechanical strength [18].

4,4'-Methylenedianiline (MDA)

As expected, without the chlorine substitution at the orthoposition to the amine function, MDA is much more reactive than MOCA so that a dispensing machine is needed to handle formulations containing MDA. The physical properties of elastomers obtained from these two curatives are, however, very similar except that MDA tends to give elastomers with a slightly lower hardness, by two to three Shore A units [62, 67]. Elastomers with compression set of 50 or less are reportedly obtained from MDA and prepolymers based on a PTMG polyol and a mixture of diisocyanates, methylene-bis(4-cyclohexylisocyanate), and 80 : 20-TDI [68].

In some applications, MDA is mixed with a less reactive curative, such as MOCA, glycols, polyols, inert diluents, or plasticizers to increase the pot life. For example, Andur C-141, which is a mixture of MDA and an inert diluent, is recommended for use with prepolymers made from the less reactive aliphatic diisocyanates [69]. The elastomer properties using a mixture of MDA and MOCA as a curing agent to cure Adiprene L-100 prepolymer have already been mentioned (Table 9).

The use of MDA in the past had been limited to special applications due to difficult processing conditions; however, in view of the health hazards associated with MOCA, the use of MDA is expected to increase. It should be noted that MDA is on the NIOSH list of suspected carcinogens (a subfile of the NIOSH toxic substance list PB-2 44334, June 1975).

Table 9 Comparison of Elastomer Properties Using Diamine Curative Blends[a]

Curatives					
MOCA	12.5	8.7		7.2	7.5
Methylenedianiline (MDA)		2.9			1.85
Caytur 7			6.2	2.5	
m-Phenylenediamine (MPD)					1.00
Mol ratio, MOCA/MDA/Caytur 7/MPD	100:0:0:0	70:30:0:0	0:0:100:0	60:0:40:0	60:20:0:20
Mixing and curing					
Mix temp., °C	100	85	70	85	100
Cure, hr/°C	3/100	3/100	1/100	3/100	3/100
Pot life at mix temp., min	12	2	30–40 sec	3	7
Average demolding time, min	20	8		10	12
Physical property					
Hardness, Shore A	90	90	90	78	82
Modulus, psi					
100%	1125	1000	975	375	500
300%	2200	1875	1675	850	1050
Tensile strength, psi	4650	4350	5000	2850	4600
Elongation at break, %	430	470	480	470	450
Tear strength, split, lb/in.	85	84	113	71	50
Abrasion resistance NBS Index, %	160	147	175	133	88
Compression set (ASTM D624 Method B, 22 hr at 70°C, %)	32	31	54	46	26

[a]Parts per 100 Adiprene L-100 prepolymer. Adiprene and Caytur 7 (a blend of MPD and cumenediamine) are trademarks of Uniroyal Chemical Co., Inc.
Source: Ref. 22.

Recent Developments in Diamine Curatives

As discussed earlier, search for a MOCA replacement in recent years has intensified because of concerns over its potential health hazards to humans. As a result, during the past few years several new diamine curatives became available commercially. These non-MOCA diamine curatives are shown to be relatively safe chemicals, without the health hazards believed to be associated with MOCA [13–14]. This is especially true with the esters of aminobenzoic acids. Esters of *p*-aminobenzoic acid have long been used as local anesthetic and substantial toxicological characterization is reported in the literature [13]. Two of the new curatives, Polycure 1000 and Polacure 740M, are derivatives of amino benzoic acids.

The use of these diamines generally does not require revising of fabrication processes but some modifications in formulations and handling conditions are necessary since their physical properties (melting point, equivalent weight, specific gravity, etc.; see Table 1) and reactivities are different from those of MOCA. In general, these non-MOCA curatives require a narrower set of processing parameters such as curative stoichiometry, curing temperature, and postcuring conditions. In addition, they are generally more sensitive to 2,4- and 2,6-TDI isomer content used during the preparation of the prepolymers [23]. For these reasons, and the fact that they are more expensive than MOCA, their applications are limited to some specific applications. The chemical structures of the non-MOCA curatives are shown in Scheme 5.

For these non-MOCA curatives, generally 85% theory of curative is used rather than 90–95% so that a better balance of physical properties can be obtained through a higher

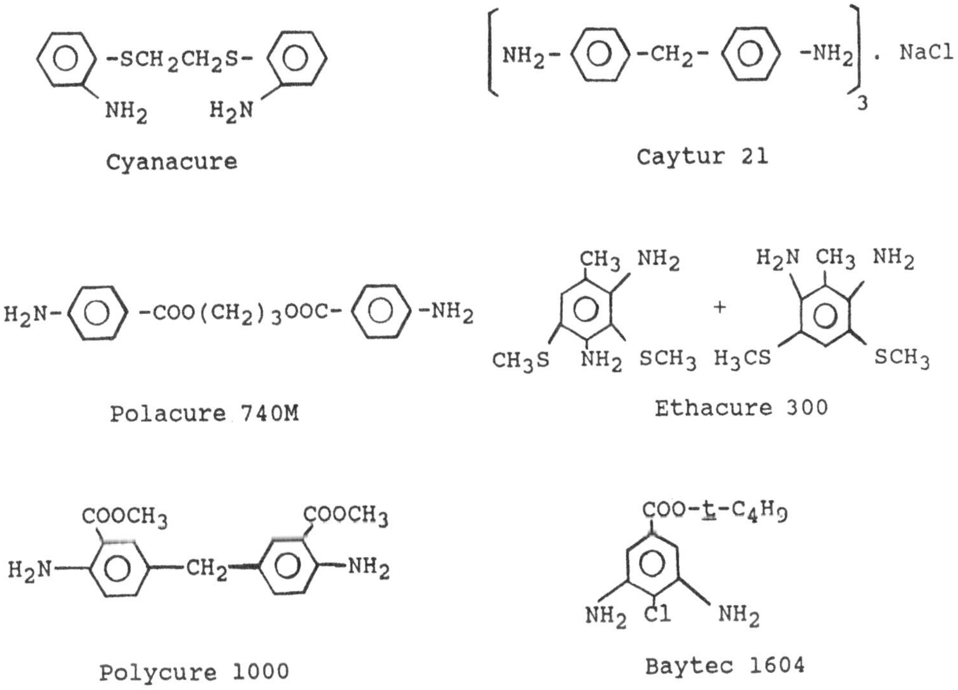

Scheme 5

degree of biuret crosslinking. The physical properties of several elastomers prepared from Vibrathane B835 or Vibrathane B839 (both are TDI–polyether prepolymers having approximately 4% NCO) and cured with MOCA, Polacure, Cyanacure, or Polycure are compared in Table 10 [70, 71]. Except for Polacure, which gives a better tear strength, other non-MOCA systems resulted in a slightly lower tear property than MOCA. However, other properties are quite comparable. For a further comparison among some of the new curatives, Conathane RN 1521, a TDI–polyether prepolymer with 6.3 % NCO, is cured with Polacure, Polycure, Cyanacure, and Caytur 21, respectively [18]. Results, as given in Table 10, show that both Polacure and Polycure tend to produce elastomers with a higher hardness than either Cyanacure or Caytur 21. In terms of other properties, Polacure gives the highest tensile strength (9900 psi) and tear (750 pli) followed by Polycure, Caytur 21, and Cyanacure. Since these curatives are different in both physical and chemical properties, it is to be expected that the processing conditions to achieve optimum performance for each curative would be different. Information regarding their optimum processing conditions for general or specific applications is available from the product literature [9, 54, 72–74].

In the following discussion, some of the important characteristics of these new diamine curatives are covered.

Table 10 Comparison of MOCA and New Diamine Curatives

| Property | Curative[a] | | | | |
	MOCA	Cyanacure	Polacure	MOCA	Polycure
% theory	95	85	85	95	95
Parts per 100	12.5	11.6	13.1		
Melting temp., °C	110	85	115	115	115
Mixing and curing					
Mixing, °C	83	83	76	82	82
Pot life, min	4–5	4–5	7	4	6
Cure temp., °C, 1 hr	100	83	100		
Physical properties					
Hardness, Shore A	90	90	90		
Modulus, psi					
At 100%	1110	1010	1000	2230	1990
At 300%	2180	1820	2130	4320	5020
Tensile strength, psi	5850	4200	6380	5830	5860
Elongation, %	440	480	460	370	340
Tear strength, pli					
Die C (D624)	440	400	510	640	540
Split (D470)	90	80	80		
Compression set (method B), %	34	33	38		
Bashore rebound, %	43	45	42		

[a]Prepolymer Vibrathane B835 (3.94–4.35% NCO, TDI–polyether) was used to compare MOCA, Cyanacure and Polacure. MOCA and Polycure were compared using Vibrathane B839.
Trademarks: MOCA, Du Pont; Vibrathane, Uniroyal Chemical; Cyanacure, American Cyanamid; Polacure, Polaroid; Polycure, PTM&W Industries.
Source: Refs. 70, 71.

Caytur 21

Caytur 21 is a 50:50 mixture of methylene dianiline salt complex and dioctyl phthalate [9, 75, 76]. The coordination complex, formed from 3 mol MDA and 1 mol NaCl, renders the normally very reactive MDA inactive at temperatures under 60–70°C. However, when heated above 100°C the complex breaks down, freeing the MDA, which then rapidly cures isocyanate prepolymers. The main advantages provided by this curative are longer pot life and rapid cure, and it can produce elastomers with a wide range of hardness (79 Shore A to 62 Shore D) and with other physical properties similar to those MOCA-cured elastomers if higher isocyanate-containing prepolymers (by about 2.3%) are used for Caytur 21. The only exception is the compression set. Elastomers cured with Caytur 21 will inherently have higher compression set properties because of the presence of a plasticizer. This is especially true for systems using prepolymers of higher isocyanate content, which require higher concentrations of Caytur 21. The compression set can be optimized to a large extent by lowering curative levels to 70–80% of theory, higher curing temperatures (130°C), and postcuring at 70–100°C for a longer period of time [9]. Comparisons of elastomer properties obtained from Caytur 21 and other non-MOCA curatives have already been discussed (Table 10).

In terms of dynamic properties, several elastomer systems based on Adiprene–diamines or MDI–1,4-butanediol are compared. Data show that Caytur 21 performed equally well or better than MOCA, followed by Polycure, Cyanacure, and finally diol-cured elastomers. In addition, Adiprene L-700–Caytur 21 elastomers as measured on a Rheovibron dynamic tester at 100 Hz gives a constant modulus with increasing temperature (20–140°C) while the MDI–diol system showed significant drop in dynamic modulus in this temperature range.

Caytur 21 is recommended for the curing of Adiprene L-42, L-83, and others in this series which, however, do not include the well-known Adiprene L-100 or L-167. The latter two are most often cured with MOCA. The combination of Caytur 21 and these two prepolymers is not recommended because of their higher curing temperatures, slower curing, poor large-part curability, and higher cost. However, comparable elastomer physical properties can be obtained from Caytur 21 and MOCA from different prepolymers [9].

Cyanacure

The chemical composition for Cyanacure is 1,2-bis(2′-aminophenylthio)ethane [54, 77]. It has an equivalent weight of 138 and a melting range of 72–77°C, which is considerably lower than that of MOCA (110°C). The reactivity of the amine functions in this molecule is modified by a —SCH_2CH_2S— group rather than by two chlorine atoms as in MOCA. The curing characteristics of this curative are somewhat different from that of MOCA. In Figure 3, the cure state (pot life, gel time, and demold time) versus cure time for Cyanaprene A-8HT–Cyanacure and Cyanaprene A-8–MOCA systems are compared [77]. It shows that Cyanacure tends to give a more rapid initial curing, resulting in a faster transformation of the liquid prepolymer to solid, and a shorter pot life. However, the subsequent curing for the Cyanacure system is slower, leading to a longer curing cycle. The processing conditions for these two systems are the same, except the curing temperature, 85°C for Cyanacure and 100°C for MOCA. The reaction exotherm for these TDI–polyester systems is shown in Figure 4. Data indicate that Cyanaprene A-8HT–Cyanacure system compares favorably with the MOCA system. The Cyanaprene A-8–MOCA is known for its low exotherm and is favored for the production of large parts because of this desirable characteristic. In terms of the physical properties of cured elastomers, there is

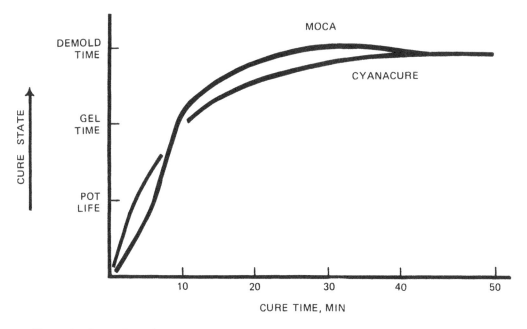

Figure 3 Comparison of curing characteristics. (From Ref. 77.)

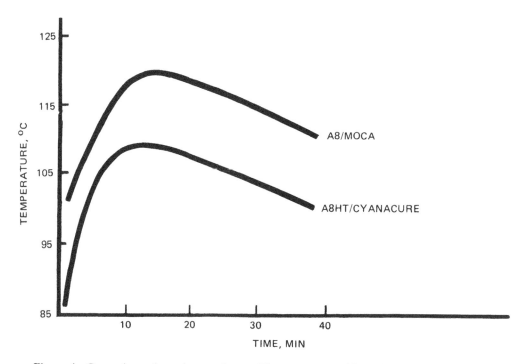

Figure 4 Comparison of reaction exotherms. Elastomer composition: Cyanaprene A-8/MOCA and Cyanaprene A-8HT/Cyanacure. Trademarks: MOCA, Du Pont; Cyanaprene and Cyanacure, American Cyanamid. (From Ref. 77.)

little difference shown by these two systems [77]. However, some differences in the tensile and compression stress–strain curves are noted in other Cyanaprene HT–Cyanacure and the corresponding Cyanaprene–MOCA systems, in which the isocyanate contents of the prepolymers are higher than Cyanaprene A-8HT. In these systems elastomers cured with Cyanacure have a lower or slower rising stress–strain curve than those cured with MOCA.

The difference between Cyanacure and MOCA appears to be much greater when they are compared to elastomers containing polyether soft segments instead of polyester, as in the systems just discussed. For example, the elastomer properties obtained from Cyanacure–Cyanaprene 3090 (%NCO = 6.2 ± 0.2) and MOCA–Adiprene L-100 (%NCO = 4.1 ± 0.2) are comparable [22, 54], but the isocyanate content in the Cyanacure system is considerably higher. This perhaps indicates that a higher hard-segment content is needed for the Cyanacure system to obtain a similar elastomer strength. Typical elastomer properties obtained from two TDI–polyether–Cyanacure systems are given in Tables 10 and 11.

Polacure 740M

The chemical composition of this curative is trimethyleneglycol-di-*p*-aminobenzoate [78]. It is an off-white granular powder and has an equivalent weight of 157 and a melting point range of 125–128°C. The reactivity of the amine functions in this compound are modified

Table 11   Physical Properties of Urethane Elastomers Cured with Non-MOCA Curatives

	Polycure 1000	Polacure 740M	Cyanacure	Caytur 21
Mix ratio, parts by weight:	100:21.7	100:22.7	100:17.9	100:31.7
% theory:	92	95	95	95
Mixing and curing conditions				
RN-1521, temp., °C	80	80	90	90
Curative, temp., °C	135	125	90	55
Mixed viscosity at mix temp., cps	700	1200	2900	1400
Working life, min	11	4	1.5	60
Cure, hr/°C	16/100	16/100	16/100	16/130
Postconditioning (77°F, 50% R.H.), days	7	7	7	7
Physical properties				
Hardness, Shore A/D	97/55	–/60	93/40	95/48
Tensile strength (psi-DIE D)	8000	9588	7140	7488
Modulus, psi				
100%	1700	2000		1450
300%	5463	3763		2600
Tear strength, pli	447	751	472	601
Elongation, %	347	523	450	560
Compression set (16 hr at 100°C plus 3 weeks at R.T.), %	34			

Prepolymer: Conathane RN-1521; % NCO, 6.2–6.4; PTMG polyether.
Trademarks: Conathane, Conap; Polycure, PTM&W; Polacure, Polaroid; Cyanacure, American Cyanamid; Caytur 21, Uniroyal Chemical.
Source: Ref. 18.

by a —OCOCH$_2$CH$_2$CH$_2$OCO— group at the para position so that the pot life obtained from Polacure- and TDI-based prepolymers is suitable for casting applications. Because of certain anomalies in the behavior of Polacure toward 2,4-TDI prepolymers, it is recommended that 80 : 20-TDI or 2,6-TDI prepolymers be used for optimum performance [21, 72]. The mixing and curing temperatures for Polacure are usually at 90 and 120°C, respectively, which are higher than those recommended for MOCA (100°C for both mixing and curing).

Although earlier work shows that there was little difference in either the curing charac- teristics or the physical properties of elastomers when Polacure or MOCA was used to cure prepolymers based on 80 : 20-TDI [14], subsequent work indicates a substantial difference in the reactivity (pot life and demold time) and elastomer properties between the two curatives if the prepolymers used are derived from 2,4-TDI, such as Adiprene L-100 [23]. These differences diminish, however, as the 2,6-TDI isomer contents in the prepolymers are increased from 0 to 20%. With MOCA the effect of varying 2,6-TDI isomer level is an increase in reactivity (shorter pot life) but with little change in physical properties. Consequently, elastomers based on 80 : 20-TDI prepolymers and cured with MOCA or Polacure are generally comparable in physical properties, although in some formulations Polacure tends to give elastomers with higher tensile, modulus, and tear properties (Table 10). Other properties such as hydrolytic stability, high- and low-temperature properties, and electrical properties are also similar [23].

The dependence of Polacure on the TDI isomer content may be attributed to its lower reactivity and lower solubility in prepolymers as compared to MOCA. The higher reac- tivity of 2,6-TDI prepolymers and its ability to crosslink more efficiently than 2,4-TDI prepolymers seem to compensate for the difference. A higher crosslinking is probably necessary to compensate for the less rigid molecules of Polacure to obtain similar proper- ties.

Polacure is used primarily for the production of high-modulus elastomers having the following range of properties: hardness: 90 or 95 Shore A; tensile strength: 7000–8200 psi; tear: 400–620 pli (Die C) or 140–530 pli (split); % elongation: 450–525; compression set: 24–46; and % rebound: 46–54. These ranges of properties are obtained when Pola- thane XPE systems-10, 15, and 20, specifically designed for Polacure, are used [72]. A 90% curative stoichiometry is employed for these systems; some of the properties can be optimized by varying the curative stoichiometry. Elastomers prepared from these systems are suitable for applications as drive belts, timing belts, rollers, and oil field equipment which require superior solvent resistance properties, or as aerospace components includ- ing cast shells, gear, and impellers.

It is interesting to note that among the several alkylene-glycol aminobenzoates evalu- ated, Polacure (trimethylene-glycol di-*p*-aminobenzoate) was identified as the most suit- able compound for application as a curative [78]. The meta-amino derivatives are found to be too reactive and aminobenzoates with even numbers of carbon atoms in the glycol portion have much higher melting points. Furthermore, the para isomer is preferred since there is no possibility of side reactions occurring on curing due to cyclization, as is possible with the ortho derivatives. In addition, the isomeric compounds derived from 1,2-propanediol give elastomers with much weaker physical properties and therefore are not useful as curatives.

Polycure 1000

Methylene-bis(methyl anthranilate) is another non-MOCA diamine available under the trademark of Polycure [79–81]. It is a close analog of MOCA, in which two chlorine atoms are each replaced by a carboxymethyl group, $COOCH_3$. The carboxymethyl group apparently deactivates the amino functions to a greater extent than the chlorine atom; as a result, Polycure is less reactive than MOCA. Polycure is reported to be less sensitive to the 2,4- and 2,6-TDI isomer content than are Polacure and Cyanacure, perhaps because of its closer similarity to MOCA in terms of the position of the substituents. Studies have shown that Polycure imparts better solvent resistance than several other compounds which were being investigated [81]. This could be due to the ability of this curative to form intramolecular hydrogen bonds involving the carboxylate and the —NH— groups in addition to the usual interchain hydrogen bonds. Therefore, Polycure is especially useful for applications where superior solvent resistance is required. It should be pointed out that Polycure gives dark red clear castings rather than translucent amber as in most cast products [71]. The high melting point (140°C) of this curative makes it considerably more difficult to handle, especially in casting machines, compared to MOCA or other non-MOCA diamines. As already discussed, in terms of elastomer physical properties, there is a less difference between Polycure and MOCA as compared to other non-MOCA curatives (Tables 10 and 11).

Ethacure 300

This curative is a mixture of two liquid aromatic diamines, 3,5-dimethylthio-2,4-tolylene-diamine and its 2,6-diamine isomer. It has recently been introduced to the industry as a general-purpose curative for TDI-based prepolymers [82]. This material is the newest member of the non-MOCA curatives. Like other MOCA replacements, Ethacure is shown, based on available toxicity data, to be less toxic than MOCA and does not present carcinogenic or mutagenic risk.

Ethacure 300 is different from other aromatic diamine curatives discussed so far in two main aspects: it is the derivatives of benzene (mononuclear) rather than derived from biphenyl or diphenyl methane (multinuclear) and it is a liquid, which is a distinctive advantage for a curative in castable applications. Ethacure 300 has an equivalent weight of 107, which is relatively low in comparison with other non-MOCA curatives. Ethacure is usually processed at lower temperatures (80–90°C) than MOCA, has the same or slightly shorter pot life, and is equally forgiving.

It is interesting to note that elastomers prepared from Ethacure are similar in physical properties to elastomers cured with either MOCA or other diamines, even though Ethacure has a lower equivalent weight and is a derivative of benzene. These factors, as might be expected, would lead to elastomers with lower hardness content and less rigidity. These factors could result in polymers of lower physical strength [24, 25]. However, as shown in Table 12, the elastomer properties of TDI–polyester prepolymers (Cyanaprene A9QM and D5QM) cured with Ethacure 300, MOCA, and Cyanacure, respectively, are remarkably similar in all physical properties except a slightly higher tensile strength with Ethacure products [82]. Ethacure also behaves similarly to other diamine curatives in TDI–polyether systems, except elastomers prepared from this curative have a slightly lower modulus and tear but are equal in tensile and other properties. Ethacure can be used with a wide range of prepolymers of differing structures and NCO contents. The resulting elastomers, which have a wide range of hardness (86 Shore A to 73 Shore D) and other properties, are presented in Table 13 [73].

Elastomers made from Ethacure 300 have shown interesting thermal aging characteristics which are not commonly observed with other diamine-cured products [73]. Normally, as the elastomers are being aged at elevated temperatures, physical strength in general will decrease. Preliminary results indicate that the tensile strength, tear, elongation, and

Table 12 Physical Properties of Ethacure 300 Curative and Other Curatives in Polyester Prepolymers[a]

Curative: prepolymer, % NCO:	Ethacure 300 4.0	MOCA 4.2	Cyanacure 4.9	Ethacure 300 4.9	MOCA 5.0	Cyanacure 6.0
Processing parameters						
NH$_2$/NCO, %	97	95	85	97	95	85
Mixing temp., °C	90	100	85	90	100	85
Pot life, min	4	7	4	3	5	3
Gel time, min	5	10	10	4	8	7
Postcure, hr/°C	16/85	16/100	16/85	16/100	16/100	16/85
Physical properties						
Hardness (A) D-676	93	91	91			
Hardness (D) D-676				49	51	50
Tensile strength D-412, psi	7710	6600	6000	7400	7000	5800
M-100 D-412, psi	1080	1020	780	1530	1530	1300
M-300 D-412, psi	1700	2030	1330	2580	3000	2150
Elongation D-412, %	600	540	600	500	500	520
Tear D-470, pli	120	110		130	130	
Compression set D-395 (B), %	35	34	34	32	36	30
Resilience D-2632, %	32	32	30	35	33	35

[a]Prepolymer is based on polyethylene adipate and 80 : 20-TDI.

Trademarks: Ethacure, Ethyl Corp.; MOCA, Du Pont; Cyanacure, American Cyanamid.

Source: Ref. 82.

Table 13 Polymer Properties of Prepolymers Reacted with Ethacure 300 Curative

	Adiprene					Vibrathane			
	L-83	L-100	L-167	L-300	L-315	B-600	B-601	B-602	B-839
Prepolymer, pbw	100	100	100	100	100	100	100	100	100
Ethacure 300 curative 195% theory	7.87	9.93	15.3	9.93	22.8	10.05	15.3	7.54	15.5
Mix temp., °F	176	176	176	176	176	176	176	176	176
Pot life, min at °F	6/176	6/176	3/176	3.5/176	1.5/176	6/176	3/176	7/176	1.3/176
Postcure, hr at °F	3/194	2/212	16/221	16/212	3/221	16/212	16/212	16/212	16/212
Physical properties									
Hardness, Durometer A	86	88	95	89		90	96	80	96
Durometer D			44		73		50		49
Modulus, psi									
100%	970	1030	1720	1060		1200	1870	660	1860
300%	1640	1790	3430	2210	4600	2160	4170	1260	3390
Tensile strength, psi	4470	4890	6850	4070	7560	4400	5670	4050	5850
Elongation at break, %	490	500	430	390	260	415	350	485	410
Tear strength									
Trouser tear, pli	100	81	120	65	100	72	95	70	100
Die C tear, pli		340				365	420	250	450
Compression set, %	29	34	46	31		29	36	30	37
Resilience, rebound, %	57	47	44	48	51	48	48	61	49

Trademarks: Adiprene and Vibrathane, Uniroyal Chemical Co.; Cyanaprene, American Cyanamid; Ethacure, Ethyl Corp.
Source: Ref. 82.

compression set improved in elastomers made from Ethacure 300, whereas the reverse is true for MOCA. This is believed to be due to the reorientation of polymeric chains rather than the formation of more secondary bondings under heat, since it would lead to higher crosslinkings, resulting in reduced tear and elongation.

Conacure AH-40

Conacure AH-40 is also a liquid diamine curative for isocyanate-containing prepolymers. The exact composition is not known except that it is based on an aromatic diamine and has the same equivalent weight as MOCA [18]. It is designed specifically to replace MOCA. Since this curative is highly reactive, it is usually processed at room temperature without the need for melting or processing at elevated temperatures. However, because of its relatively short pot life, it is recommended that this curative be processed through automatic mixing and dispensing equipment.

Elastomeric properties obtained from Conacure AH-40 and a large variety of prepolymers, both of polyester and polyether types, are listed in Table 14 [18]. In comparison with MOCA, for example, with Adiprene L-100, Conacure AH-40 gives elastomers with somewhat lower tensile strength and tear but higher compression set properties.

Baytec 1604

Baytec 1604 is another diamine curative containing benzene ring rather than biphenyl or diphenyl methane structures. The chemical composition of this curative is isobutyl 3,5-diamino-4-chlorobenzoate [74]. It has an equivalent weight of 117 ± 4 and a melting range of 80–88°C and is being supplied in the form of dark brown flakes. In many applications it can be used as a replacement for MOCA in combination with TDI–polyether and TDI–polyester prepolymers. For example, when Baytec TE-060, a TDI–polyether prepolymer, is cured with Baytec 1604 or MOCA, the elastomers have the same hardness (95 Shore A), rebound (48%), and 100% modulus (1800 psi). But elastomers made with Baytec 1640 have higher split tear (135 vs. 110), Die C tear (550 vs. 500), % elongation (360 vs. 330), and break set (10 vs. 5). On the other hand, MOCA-cured material has slightly better tensile and compression properties. It is interesting to note that these differences are reversed when a polyester prepolymer, Baytec TS-040, is used; Baytec 1604 curative gives higher tensile strength, 300% modulus, and abrasion resistance [74].

Hydroxy (Diol and Triol) Curatives

Water is the simplest member of this class of curatives. In the early stage of urethane elastomer development, isocyanate prepolymers were cured with water and processed by the conventional rubber technique [4]. Although elastomers with good properties were made, the formation of CO_2 as by-product complicated the processing. The replacement of water by glycols such as ethylene glycol or 1,4-butanediol marked substantial progress and allowed the processing of Vulkollan, an unstable prepolymer based on 1,5-naphthalene diisocyanate, by the liquid casting technique [10, 11]. Some of the better known diol and triol curatives for castable urethanes are 1,4-butanediol [83], Isonal 93 [84], TIPA [8, 85], Pluracol TP440 [86], trimethylolpropane (TMP) [8], 1,4-di(2-hydroxyethyl) hydroquinone (HQEE) [87], and 1,3-di(2-hydroxyethyl) resorcinol (HER) [14, 88]. Others such as ethylene glycol, triethylene glycol, and 1,3-dipropylene glycol, which are important chain extenders, are used occasionally as curatives [4, 47]. Information for most of these hydroxy curatives is presented in Table 2.

Table 14 Urethane Elastomers Cured with Conacure AH-40

	Adiprene L-100	Adiprene L-167	Cyanaprene A-9	Conathane RN-1558	Conathane RN-1559	Conathane RN-1560	Conathane RN-1570
Type	TDI/ether	TDI/ether	TDI/ether	TDI/ether	TDI/ether	TDI/ether	Aliphatic
Polymer, pbw	100.0	100.0	100.0	100.0	100.0	100.0	100.0
Conacure, AH-40, pbw	12.0	18.3	12.1	9.2	12.2	18.5	16.5
% Theory	90	90	90	90	90	90	90
Mixing and curing conditions							
Polymer temp., °C	60	60	60	60	60	60	60
AH-40 temp., °C	25	25	25	25	25	25	25
Initial mixed viscosity at mix temp., cps	5200	8,000	12,800	2,000	4,500	8,000	6,000
Working life, min	3	1	1	4	2	1	10
Cure, hr at 100°C (212°F)	8	8	8	8	8	8	16
Physical properties							
Hardness, Shore A	89	93	92	70	83	90	94
Shore D	39	43	43	20	30	40	45
Tensile strength, psi (Die D)	7,500	6,500	5,600	1,600	5,000	4,000	6,700
Modulus, psi							
100%	1,050	1,250	1,150	400	800	1,150	1,200
200%	1,250	1,800	1,300	400	1,000	1,460	1,800
300%	1,650	3,000	1,600	550	1,200	1,870	3,200
Elongation, %	520	430	570	1,100	700	450	430
Tear strength, pli	400	360	570	260	290	290	360
Compression set (22 hr at 70°C), %	35	72					

Trademarks: Adiprene, Uniroyal Chemical; Cyanaprene, American Cyanamid; Conathane, Conap.
Source: Ref. 18

In general, minor changes in the structure of hydroxy curatives have little impact on elastomer properties [46]. However, gross changes in diol curative such as the introduction of an aromatic ring does have a large effect due to increases in rigidity and aromatic content [24, 25]. Aromatic diols tend to yield elastomers with a greater thermal resistance and a better retention of properties at elevated temperatures. Although diols are commonly used as the sole curative in many formulations, triols are mainly used in conjunction with a diol or diamine to increase crosslinking for better balance of properties. In comparison to diamines, hydroxy compounds have a much smaller share of the total curative market. In the following sections, a few of the commercially important diol curatives are briefly discussed.

1,4-Butanediol

1,4-Butanediol (BDO or 1,4-BD) is a nonviscous clear liquid and like many other hydroxy-containing materials it is highly hydroscopic. The use of "urethane-grade" BDO (moisture content ≤0.2%) is essential for producing elastomers with optimum physical properties [22, 83]. The typical properties of this curative are listed in Table 15.

As already mentioned, 1,4-butanediol is used with either TDI prepolymers to produce elastomers of medium to low hardness or with MDI prepolymers to yield elastomers with high hardness [8, 52, 89–92]. The elastomer properties obtained from these systems were discussed earlier (see Tables 4 and 5).

Studies have shown that in comparison with other C_4-diols, 1,4-butanediol imparts the best combination of physical properties to the resulting elastomers [93]. The presence of a carbon–carbon double bond in *cis*- and *trans*-2-butane-1,4-diol or a triple bond in 2-butyne-1,4-diol is shown to make these compounds not only less reactive and less forgiving but also weakens some elastomer physical properties. As shown in Table 16, fairly consistent elastomer properties are obtained when an MDI–polyester prepolymer (Multrathane F-242, NCO = 6.2%) is cured with 85, 95, or 105% theory of 1,4-butanediol, indicating that 1,4-BDO is quite forgiving. A greater difference, however, is observed when other C_4-diols are used [93]. Data also show that although elastomers having similar tear resistance and compression set properties are obtained for all C_4-diols, the tensile elongation and rebound properties are decreased with unsaturated diols. The presence of carbon–carbon unsaturation, which may require the polymer to adopt a certain

Table 15 Typical Properties of 1,4-Butanediol

Appearance	Clear liquid
Functionality (CH_2OH)	2
Equivalent weight	45.1
Specific gravity at 20°C	1.0154
Refractive index at 20°C	1.4460
Viscosity at 40°C, cP	70–73
Boiling point (°C)	229.5
Flash point (°C)	90–92
Freezing point (°C)	19.3–19.5
Acid value	0
Water content (%)	<0.2
Typical	0.02–0.03

conformation and makes intermolecular hydrogen bonding less effective, is probably the cause of these differences [93]. In addition, the hard segments formed by 1,4-butanediol and MDI can presumably achieve better crystalline order and phase separation than other C_4-diol–based hard segments, thus resulting in a tougher polymer [24, 25].

Further studies show that among the eight C_2–C_6 aliphatic diols tested, 1,4-butanediol has the best overall performance in terms of processing and mechanical and dynamic properties of elastomers [94, 95]. The diols tested include ethylene glycol (EG), propylene glycol (1,3-PD), four of the C_4-diols, pentanediol (1,5-PD), and hexanediol (1,6-HD). Prepolymers used are four MDI–polyether types with % NCO ranging from 6.3 to 10.7. Elastomer properties obtained from two of these prepolymers, Adiprene M483 (%NCO = 6.3) and M415 (%NCO = 10.7), each cured with all eight diols, are presented in Tables 17 and 18, respectively.

Overall results indicate that as the chain length of the diol curative increases, the resulting elastomers become softer and more elastic. The optimum balance of tensile, elongation, and tear property occurs when 1,4-butanediol is used. This conclusion agrees well with an earlier study in which elastomers based on polyester–MDI prepolymers were used [46, 47]. In terms of thermal transition behavior, DSC and TMA data show that all eight diol-cured elastomers have one soft segment T_g in the region of -30 to $-60°C$ depending on the isocyanate content of the prepolymer, while the hard segment has three transitions in the region of 170–220°C. Within each system, T_g data for eight diol-cured elastomers are nearly identical and are almost insensitive to the curative chain length [94]. Results from dynamic mechanical property measurements on elastomers based on Adiprene M483 and cured with eight diols are in good agreement with DSC data, further supporting the conclusion that no major improvement in low-temperature properties is obtained by varying the length of diol curatives from C_2 to C_6.

Table 16 Effect of Diol Level on Properties of Castable Urethane Elastomers

	85% Theory	95% Theory	105% Theory
Multrathane F-242[a]	100	100	100
1,4-Butanediol	5.89	6.46	7.13
% Stoichiometry	85	95	105
Pot life, min	8	7	6
Elastomer properties			
Hardness, Shore A	88	87	86
Shore B	32	32	32
Modulus, psi			
100%	825	780	819
300%	1734	1939	2129
Tensile strength, psi	5006	6273	6701
Elongation, %	596	493	460
Tear strength, grave Die C, pli	556	608	651
Bashore rebound, %	41	44	44
Compression set (method B), %	20	21	31

[a]Polyester–MDI prepolymer, % NCO = 6.3 ± 0.1, is a product of Mobay.
Source: Data from Ref. 93.

Table 17 Properties of Elastomers Prepared from Polyether–MDI Prepolymer (Adiprene M-483)[a] Cured with Diol Chain Extenders (95% Stoichiometry)

Properties	EG	1,3-PD	B₁D	B₂D (cis)	B₂D (trans)	B₃D	1,5-PD	1,6-HD
Hardness, Shore A	84	82	85	84	86	85	80	82
Shore D	35	32	35	34	36	35	30	32
Modulus, psi								
100%	913	900	888	648	748	715	846	792
300%		1822	1943	1378	1610		1440	1383
Tensile strength, psi	1705	2516	4443	2258	2735	1372	3480	3298
Elongation, %	275	402	488	415	390	230	452	528
Tear strength, Die C, pli	409	556	570	562	542	536	447	463
Bashore rebound, %	65	65	63	66	65	65	60	63
Compression set (22 hr/70°C, method B), %	18	23	25	15	19	28	28	26
Pot life, min	6–25	7–15	7–9	8–15	7–14	7–42	7–11	5–7

[a]Adiprene M-483 (6.3% NCO) is a product of Uniroyal Chemical Co., Inc.

Key: EG, ethylene glycol; 1,3-PD, 1,3-propanediol; BD, 1,4-butanediol; B₂D(cis), 2-butene-1,4-diol; B₂D(trans), 2-butene-1,4-diol; B₃D, 2-butyne-1,4-diol; 1,5-PD, 1,5-pentanediol; 1,6-HD, 1,6-hexanediol.

Source: Ref. 94.

Table 18 Properties of Elastomers Prepared from Polyether–MDI Prepolymer (Adiprene M-415) Extended with Diol Chain Extenders (95% Stoichiometry)[a]

Properties	EG	1,3-PD	B₁D	B₂D (cis)	B₂D (trans)	B₃D	1,5-PD	1,6-HD
Hardness, Shore A								
Shore D	58	55	50	50	55	55	55	55
Modulus, psi								
100%	3540	3775	3600	3618	3796	4287	3105	2903
300%								
Tensile strength, psi	4155	5232	6027	4844	5572	5122	5275	4946
Elongation, %	132	178	222	154	194	137	218	192
Tear strength, Die C, pli	615	726	654	630	688	672	680	611
Bashore rebound, %	42	45	51	45	45	41	44	45
Compression set (22 hr/70°C, method B), %	33	31	25	21	25	28	14	42
Pot life, min	6	4	3	6	6	9–16	3	2

[a]Adiprene M-415 (TDl–polyether, 10.7% NCO) is a product of Uniroyal Chemical Co., Inc.

Key: See Table 17.

Source: Ref. 94.

Aromatic Diol Curatives

Two aromatic diols which are commonly used as curatives for castable urethanes are the hydroxyethyl ether of hydroquinone and resorcinol, known as HQEE and HER [87, 88]. These two curatives are used mainly with MDI-based prepolymers of both polyether and polyester types. The main advantages of these curatives are their ability to give elastomers which are higher in strength and can better retain mechanical properties at elevated temperatures as compared to aliphatic hydroxy curatives [47]. In addition, problems with poor green strength associated with MDI–diol systems are less serious with these aromatic diols. These differences may be attributable to the greater rigidity of the aromatic diols as well as their ability to produce hard segments which have higher T_g and are less likely to be affected by temperature. Both HQEE and HER are solids and are not compatible with MDI prepolymers at temperatures below their melting points; consequently, starring or freezing of these curatives, especially with HQEE, is the main disadvantage of using these curatives. The problem of freezing out is much less serious with HER because of its lower melting point (87°C) compared to HQEE (103°C). Using HQEE blended with phenyl ethanolamine (PDEA), trimethylolpropane (TMP), BDO, or Pluracol TP-440 can lower the freezing point of HQEE and thus reduce starring. Data show that similar elastomeric properties are obtained whether HQEE is used alone or blended with these diols or triols. No advantage is evident, however, when HQEE is blended with HER. Use of higher mixing temperatures (120–130°C) and higher curing temperatures (115–130°C) can also effectively eliminate starring for formulations containing HQEE [87].

A direct comparison of HQEE and HER in several MDI prepolymers with % NCO ranging from 3 to 8% is provided in Table 19 [96]. Data show that HER is slower acting, thus has a longer pot life. HER also exhibits a much better processing characteristic than HQEE and no crystallization or starring until temperature is lower than 70°C. HQEE in general gives elastomers which are slightly better in mechanical properties, but these differences diminish as the diisocyanate content of prepolymers increases, except the tear strength; HER tends to give a poorer tear property. With a polyester–MDI prepolymer, both HQEE and HER impart an exceptionally good compression set property (3–26%).

Diol–MDI Systems as MOCA Alternatives

In recent years, many 1,4-BD–MDI–polyether or polyester cast systems have been developed as alternatives to MOCA–TDI systems so that the use of MOCA can be avoided [52, 97]. These alternative systems are primarily for the production of castable elastomers with hardness typically ranging from 85 to 95 Shore A. Although there is little difference in the final elastomer properties (see Table 6), there are some differences in the processing characteristics between diol–MDI and MOCA–TDI systems. For instance, diol–MDI systems tend to gel quickly then pass through a ''green'' or cheesy state resulting in a longer demold time compared to MOCA–TDI systems. For MOCA–TDI systems, both the gel and buildup of strength proceed gradually. In addition, a more frequent check of prepolymer/curative ratio for MDI–diol systems is necessary to ensure uniform product quality since they are less forgiving than MOCA–TDI systems. Furthermore, diol–MDI systems are more moisture-sensitive due to the hydroscopic nature of the hydroxy curatives and the more reactive nature of MDI toward moisture. Therefore, a better preventive measure for excluding moisture is needed during the storage and processing of diol–MDI systems. A shrinkage problem, especially with high-hardness formulations which may cause cracking, is also more likely to happen with diol–MDI alternatives. Use of a

Table 19 Comparison of HQEE- and HER-Cured Castable Urethane Elastomers

	HQEE	HER	HQEE	HER	HQEE	HER
Prepolymer[a]	B605	B605	B635	B635	B6012	B6012
% NCO	3.02	3.02	7.97	7.97	6.33	6.33
NCO/OH	1.051	1.04	1.05	1.043	1.046	1.048
Pot life, min/°C	13/130	20/130	5/130	11/130	6/130	6/130
Properties						
Hardness, Shore A	81	75	94	94	91	90
Modulus, psi						
100%	644	557	1961	2048	1731	1471
300%	1138	959	3315	3194	3225	3021
Tensile strength, psi	4273	4427	4648	4488	4074	3617
Elongation, %	661	700	403	400	404	425
Tear, pli						
Die C	367	323	473	498	480	479
Split, max	120	155	162	179	332	280
Bashore rebound, %	62	64	44	41	38	42
Compression set, %	17	22	24	26	3	7

[a]Prepolymers are Vibrathane series. Vibrathane B605 and B635 are MDI–polyethers. Vibrathane B6012 is MDI–polyester. Vibrathane is a trademark of Uniroyal Chemical Co., Inc.
Source: Ref. 96.

catalyst which can shorten the cheesy stage can often minimize cracking [52]. In addition, using diol–triol curative blends, higher processing temperature, and longer cure time can effectively minimize or reduce these processing difficulties with diol–MDI systems.

The physical properties of elastomers derived from two comparable TDI–polyether–MOCA and MDI–polyether–BDO systems as a function of curative concentrations are given in Table 20. As can be seen, the MDI–diol system is more sensitive to curative stoichiometry than is the TDI–MOCA system. Recommended stoichiometry for MDI–diol systems is 95% of theory while 90% is usually recommended for the best balance of properties in TDI–MOCA systems.

In addition to BDO, aromatic diols such as HQEE or HER can be used in these MOCA alternative systems. As previously mentioned, HQEE in general gives elastomers having a better split tear, a slightly higher tensile strength and rebound, and a lower compression set (see Table 6) [52].

Phosphorus-Containing Diol and Triol Curative

Although many reactive and nonreactive phosphorus-containing compounds are widely used as flame-retardant chemicals in both flexible and rigid urethane foams, these compounds are usually not used in the preparation of castable urethane elastomers [98]. Phosphorus compounds such as tris-(2-hydroxyalkyl) phosphine oxides are disclosed in patents as flame retardants for polyurethanes [91]. Phosphine oxides are known for their superior stability compared to other classes or organophosphorus compounds such as organophosphate esters. Recently, two phosphine oxides, *sec*-butyl bis(3-hydroxypropyl)-phosphine oxide and tris(3-hydroxypropyl)phosphine oxide, designated, respectively, as

Table 20 Comparison of Elastomeric Physical Properties of 90A TDI–Polyether–MOCA and MDI–Polyether–Diol Systems as a Function of Curative Levels

	TDI–MOCA				MDI–BDO			
% Theory	80	90	95	105	80	90	95	105
Shore A	88	90	90	90	86	89	89	90
Modulus, psi								
100%	710	1050	1040	860	1090	1190	1230	1060
300%	2560	2250	1990	2390		3230	2630	2030
Tensile strength, psi	3860	5330	5250	4920	2430	5950	5420	4620
Elongation, %	350	430	450	510	250	390	360	510
Tear, split D470, pli	33	88	98	119	39	66	100	120
Bashore rebound, %	43	44	44	42	36	39	38	39
Compression set, %	27	27	27	38	42	30	31	59
Specific gravity	1.10	1.11	1.11	1.11	1.12	1.12	1.12	1.12

Source: Data from Ref. 52.

C-206 (or C-200) and C-300, have been introduced as developmental products for curing TDI-prepolymers [100–102]. In these compounds, the last two numbers designate the triol content (by wt %). The typical physical properties of C-206 and C-300 are given in Table 21. The diol C-206 is a liquid and the triol C-300 is a low-melting solid, both having a relatively low equivalent weight.

These phosphine oxide curatives are found to behave very differently from the conventional hydroxy-type curatives. They are comparable to MOCA in chemical reactivity and give fast cure to TDI-based prepolymers, as indicated by the remarkably short demold and curing times. In addition, the resulting elastomers have lower hardness and lower moduli than those MOCA-cured products, however; both the tensile and tear strength are remarkably higher for typical diol–triol-cured products. This combination of properties is unusual and is not commonly obtained with either a diamine or diol–triol-cured elastomers. The physical properties of elastomers prepared from a TDI–polyester prepolymer (Vibrathane 8011, 3.2 % NCO) and these phosphine oxide curatives are shown in Table 22. In comparison with a typical triol curative, Isonol 93, the C-200–C-300 blends have a much faster cure, 0.5 hr for the phosphine oxides and 16 hr for Isonal 93. The resulting elastomers from the phosphine oxides are higher in hardness (by 6 Shore A units), tensile strength (by about 1500 psi), and elongation, but are lower in moduli. Compression set properties are what would be expected of hydroxy-cured and improved as the triol C-300 content increases. The enhanced properties are believed to be due to the extra hydrogen-bonding sites provided by the polar phosphorus–oxyl bonds in these curatives and differences in the nature of hardness segments.

The behavior of a C-200–C-300 blend containing 34% triol and designated as C-234 is compared with a 1,4-butanediol–TMP blend of the same triol content in Table 23. It is evident that C-234 is vastly more reactive than the 1,4-BDO–TMP blend, as indicated by the fast curing. The resultant elastomers, however, are similar in hardness and other properties; those cured with C-234 have much higher tensile strength, especially when polyester-based prepolymers are used.

The phosphine oxide curatives can also be used in conjunction with diamines. The presence of a small amount of C-206 in MOCA has an interesting and unusual effect on the elastomer property. As shown in Table 24, elastomers cured with the 90:10 blend of MOCA/C-206 are much softer, but their tensile and compression set are much improved as compared to those cured with MOCA. This seems to indicate that C-206 acted as an internal plasticizer which did not adversely affect elastomer properties as a conventional plasticizer would have. Since the chemical reactivity of these phosphine oxides is compa-

Table 21 Typical Properties of Curatives C-206 and C-300

	C-206	C-300
Appearance	Clear liquid	Solid, mp 109°C
Functionality (CH_2OH)	Two	Three
Equivalent weight	111	75
Specific gravity	1.12	1.24 at 25°C
		1.19 at melt
Viscosity at 77°F	12,000 cps	

Source: Ref. 100.

Table 22 Urethane Elastomers Prepared from Prepolymer Vibrathane 8011 and Curatives C-200 and C-300

Sample No.	1	2	3	4
Vibrathane 8011	100	100	100	100
Curatives, 95% theory				
Parts per 100	8.0	8.21	8.20	Isonol 93[a]
C-200/C-300, % ratio	92/8	67/33	85/16	
Cure time at 212°F				
Tensile sheets, min	20	20	20	16 hr
Compression buttons, min	30	30	30	
Urethane properties				
Hardness, Shore A	63	65	63	57
Modulus, psi				
100%	157	152	156	260
300%	250	305	270	550
500%	470	899	644	
Tensile strength, psi	3677	5513	>5040[b]	3500
Elongation, %	768	682	>742[b]	460
Tear strength				
Die C, pli	90	150	120	190
D470 pli				17
Compression set (22 hr at 158°F, method B), %	41	22	12	3.0
Bashore rebound, %	38	32	38	28

[a]Data for Isonol 93 are taken from Vibrathane 8011 Product Data Sheet, Uniroyal Chemical Co., Inc.
[b]Samples did not break in an Instron tester.
Source: Ref. 100.

Table 23 Elastomers Cured with C-206–C-300 Blends

Prepolymer:[a] curative:	A C-234	A BD/TMP	B C-234	B BD/TMP
Demold time at 212°F				
Tensile sheets, min	20	16 hr	10	16 hr
Compression buttons, min	30	16 hr	15	16 hr
Elastomer properties				
Hardness, Shore A	65	52	61	61
Tensile modulus, psi				
100%	150	160	170	220
300%	300	240	250	340
500%	900	330	500	660
Ultimate tensile, psi	5500	2500	3300	2100
Elongation, %	680	730	700	610
Pot life, min	4–8	>60	4–8	>60

[a]Prepolymer A, TDI–polyester (3.3% NCO); B, TDI–polyether (3.3% NCO).
Source: Ref. 101.

Table 24 Elastomers Prepared from C-206–Diamine Curative Blends[a]

Sample No.	1	2	3
Curative blends[b]			
C-206, %	0	10	30
Diamine, %	100	90	70
Elastomer properties			
Hardness, Shore A	91	83	58
Tensile modulus, psi			
100%	1040	630	240
300%	1610	1210	470
500%	—	—	2020
Ultimate tensile, psi	2180	2690	2440
Elongation, %	440	480	520
Tear, pli			
Die C	290	210	103
D470	110	87	32
Compression set (22 hr at 158°F, method B), %	62	44	30
Bashore rebound, %	31	30	15

[a]Elastomer samples are prepared from C-206 and/or MOCA with a TDI polyether (6.2% NCO).
[b]Curative blends are based on weight percent.
Source: Ref. 101.

rable to that of MOCA, their mixtures should give more uniform curing as compared to mixtures containing MOCA and other less reactive hydroxy curatives.

MANUFACTURING OF CURATIVES

Hydroxy Curatives

In addition to being the most important chain extender and hydroxy curative, 1,4-butanediol has many other industrial applications. Its principal manufacturing process involves ethynization of formaldehyde in an aqueous solution with the presence of a catalyst to produce the intermediate, 2-butyne-1,4-diol (Eq. 6). This intermediate is then hydrogenated to yield 1,4-butanediol (Eq. 7) [103].

$$HC\equiv CH + 2HCHO \xrightarrow{[cat]} HOCH_2-CH=HC-CH_2OH \tag{6}$$

$$HOCH_2-CH=CH-CH_2OH + 2H_2 \xrightarrow{[cat]} HOCH_2CH_2CH_2CH_2OH \tag{7}$$

The diethylol ether of either hydroquinone (HQEE) or resorcinal (HER) can be manufactured by the reactions of ethylene oxide and the corresponding aromatic dihydroxy compounds under conditions similar to those used for the preparation of polypropylene glycols using basic catalysts (Eq. 8) [104].

$$\text{(8)}$$

Amine Curatives

Several patents describe the manufacturing of 4,4'-methylene-bis(2-chloroaniline) (MOCA) from the condensation of chloroaniline and formaldehyde using various acids as catalysts (Eq. 9). The condensation reactions between other anilines and CH_2O are also well known in the preparation of other diamine curatives [105–107].

$$\text{(9)}$$

Aminobenzoate-type curatives can be synthesized in good yields in two steps; reaction of nitrobenzoyl chloride and alcohols or diols followed by the reduction of the nitro groups to the amine [72].

REFERENCES

1. Fundamentals of Plastics and Elastomer, *Handbook of Plastics and Elastomers* (C. A. Harper, ed.), McGraw-Hill, New York, pp. 1–4 (1975).
2. *Kirk-Othmer Encyclopedia of Chemical Technology*, Vol. II, 2nd ed., Wiley, New York (1967).
3. S. Schwartz and S. H. Goodman, *Plastics Materials and Processes*, Van Nostrand Reinhold, New York (1982).
4. J. H. Saunders and K. C. Frisch, *Polyurethane Chemistry and Technology*, Parts I and II, Interscience, New York (1962, 1964).
5. P. Wright and A. P. C. Cumming, *Solid Polyurethane Elastomers*, Gordon and Breach, New York (1964).
6. C. Hepburn, *Polyurethane Elastomers*, Applied Science Publishers, London (1982).
7. *PMA Reference Guide to Polyurethane Processing*, published for Polyurethane Manufacturers Association, Glen Ellyn, Ill. (1983).
8. R. O. Rosenberg, A. Singh, and R. W. Fuest, *Vibrathane^R, Castable Urethane Elastomers for Printing and Coating Rolls*, Uniroyal Bull. ASP 5706.
9. *Caytur^R 21 Urethane Curative*, Uniroyal Chemicals, Bull. ASP 1615 (Apr. 1985).
10. E. Mueller, *Rubber Plast. Age, 39*(3): 195 (1958).
11. E. Mueller, O. Bayer, S. Petersen, H. F. Piepenbrink, F. Schmidt, and E. Weinbrenner, *Angew. Chem., 64*: 523 (1952); *Rubber Chem. Technol., 26*: 493 (1953).
12. R. Wyman, *RPN Technical Notebook* (Dick Walker, ed.) (1979).
13. R. C. Baron, R. E. Brooks, and K. C. Frisch, *Saf. Health Plast., Natl. Tech. Conf., Soc. Plast. Eng.*, pp. 200–209 (1977).
14. K. C. Frisch, *Rubber Chem. Technol., 53*(1): 126 (1980).
15. *Plast. Technol., 1974*: 21 (Oct.); *1975*: 26 (Feb.).
16. 4,4'-Methylene Bis(2-chloroaniline), Notice of Potential Risk, issued by the Office of Pesticides and Toxic Substance (TS-799), U.S. Environmental Protection Agency (June 1985).
17. Palmer Davis Seika, Inc., *Health Hazards and Safe Handling Procedures for Bis-Amine A* (December 1984).
18. Conap, Inc., *Conathane^R Urethane Elastomers*, Bull. R134 and 138.

19. W. Cooper, R. W. Pearson, and S. Darke, *Ind. Chem.*, *36*(421): 121 (1960).
20. R. A. Martin, K. L. Hoy, and R. H. Peters, *Ind. Eng. Chem. Prod. Res. Dev.*, *6*(4): 218 (1967).
21. I. C. Kogan, *J. Org. Chem.*, *23*: 1594 (1958).
22. Uniroyal Chemical Company, Inc., *Adiprene*[R] *L 100 A Liquid Urethane Elastomers* (1976).
23. R. C. Baron and F. O. Shaughenessy, Paper presented to PMA Conference, Point Clear, Alabama (October 1978).
24. D. C. Allport and A. A. Mohajer, in *Block Copolymers* (D. C. Allport and W. H. Janes, eds.), Wiley, New York, pp. 443–487 (1973).
25. J. W. C. Van Bogart, A. Lilaonitkul, and S. L. Cooper, *Multiphase Polymers* (S. L. Cooper and G. M. Estes, eds.), American Chemical Society, Washington (1979).
26. C. G. Seefried, Jr., J. V. Koleskee, and F. E. Critchfield, *J. Appl. Polym. Sci.*, *19*: 3185 (1975).
27. C. E. Wilkes, and C. S. Yusek, *J. Macromol. Sci. Phys. B*, *7*(1): 157 (1973).
28. T. L. Smith, *J. Polym. Sci. Polym. Phys.*, *12*: 1825 (1974).
29. C. G. Seefried, Jr., J. V. Koleskee, and F. E. Critchfield, *J. Appl. Polym. Sci.*, *19*: 2503 (1975).
30. C. P. S. Sung, *Am. Chem. Soc. Polym. Prepr.*, *16*: 1 (1975).
31. R. W. Seymour, G. M. Estes, and S. L. Cooper, *Macromolecules*, *3*(5): 579 (1970).
32. R. J. Zdhahala, R. M. Gerkin, S. L. Hagen, and F. E. Crutchfield, *J. Appl. Polym. Sci.*, *24*: 2041 (1979).
33. C. S. P. Sung, C. B. Hu, and C. S. Wu, *Macromolecules*, *13*: 111 (1980).
34. C. B. Hu, R. S. Ward, Jr., and N. S. Schneider, *J. Appl. Polym. Sci.*, *27*: 2167 (1982).
35. C. C. Lin and B. Z. Yang, *J. Appl. Polym. Sci.*, *25*: 1875 (1980).
36. T. K. Kwei, *J. Appl. Polym. Sci.*, *27*: 2891 (1982).
37. Y. Minoura, S. Yamashita, H. Okamoto, T. Matsuo, M. Izawa, and S. Kohmoto, *J. Appl. Polym. Sci.*, *22*: 1817 (1978).
38. Y. Minoura, S. Yamashita, H. Okamoto, T. Matsuo, M. Izawa, and S. Kohomoto, *J. Appl. Polym. Sci.*, *23*: 1137 (1979).
39. G. B. Guise and G. C. Smith, *J. Appl. Polym. Sci.*, *25*: 149 (1980).
40. H. L. Heiss, *Rubber Age*, *88*: 89 (1960).
41. S. L. Axelrood and K. C. Frisch, *Rubber Age*, *88*(3): 465 (1960).
42. R. J. Ferrari, *Rubber Age*, *53* (1967).
43. J. S. Rugg and C. F. Blaich, Jr., *Ind. Eng. Chem.*, *48*(5): 930 (1956).
44. T. C. Patton, A. Ehrlich, and M. K. Smith, *Rubber Age*, *86*(4): 639 (1960).
45. F. B. Hill, C. A. Young, J. Q. Nelson, and R. G. Arnold, *Ind. Eng. Chem.*, *48*(5): 927 (1956).
46. K. A. Pigott, R. J. Cote, K. Ellegasts, B. F. Frye, E. Mueller, W. Archer, K. R. Allen, and J. H. Saunders, *Rubber Age*, 629 (1962).
47. K. A. Pigott, B. F. Frye, K. R. Allen, S. Steingiser, W. C. Darr, J. H. Saunders, and E. E. Hardy, *J. Chem. Eng. Data*, *5*(3): 391 (1960).
48. U.S. Pat. 2,871,215 (1955); B. F. Goodrich Co.
49. W. J. Pentz, in *PMA Reference Guide to Polyurethane Processing*, Glen Ellyn, Ill., I-15 to I-43 (1983).
50. T. L. Smith and A. D. Magnusson, *J. Appl. Polym. Sci.*, *45*(14): 218 (1961).
51. T. Tanaka and T. Yokayama, *Rubber Chem. Tech.*, *35*(4): 470 (1962).
52. E. L. Hagen, paper presented to PMA, San Francisco, April 1978; *Plast. Technol.*, 95 (September 1978).
53. *Types of Adiprene*[R] *L, M and LW*, Uniroyal Chemical Co. Inc., Tech. Bull. Ap-210.1.
54. *Cyanaprene*[R] *Castable Urethanes*, American Cyanamid Co., Tech. Bull. 2K484.
55. T. M. Vial et al., U.S. Pat. 4,089,822 (1978); American Cyanamid.
56. Jpn. Pat. 52-85297 (1977); Atlantic Richfield.

57. J. Blahak et al., Br. Pat. 1,324,381 (1973); Bayer.
58. W. Mechkel et al., U.S. Pat. 3,794,621 (1974); Bayer.
59. Germ. Pat. 2,623,961 (1977); Bayer.
60. Germ. Pat. 2,635,400 (1978); Bayer.
61. A. Takahashi et al., U.S. Pat. 4,002,584 (1977); Ihara Chemical Kogyo Kabushiki Kaisha.
62. A. J. Sampson and C. F. Blaich, *Rubber Age*, 89: 263 (1961).
63. R. J. Athey, *Rubber Age*, 85(1): 77 (1959).
64. R. J. Athey, *Ind. Eng. Chem.*, 52(7): 611 (1960).
65. *Bis-Amine A*, Palmer Davis Seika, Inc., Tech. Bull.
66. Anderson Development Company, Product Data Curene[R] 442.
67. T. W. Bethea et al., U.S. Pat. 4,071,492 (1978); Firestone Tire & Rubber.
68. I. C. Kogen, U.S. Pat. 3,997,514 (1976); du Pont.
69. *Andur[R] Urethane Products*, Anderson Development Company Bull.; J. F. Thompson and T. H. Porter, Germ. Pat. DE 2241388 (1973); (Anderson Development Co.).
70. Uniroyal Chemical Company, Inc., Preliminary Data Sheet, Vibrathane[R] B835 (March 12, 1980).
71. Uniroyal Chemical Company, Inc., Teletech, Polycure[R] 1000—A MOCA Alternative (January 1983).
72. Polaroid Corporation, *Polacure 740M* (1980); *Polathane XPE System* (1984).
73. Ethyl Corporation, Product Information, Ethacure, Polymer Modifier (1987).
74. Mobay Corporation, Preliminary Product Information, Baytec 1604.
75. P. Conacher, Jr., U.S. Pat. 3,917,702 (1975); du Pont.
76. L. Ahranjian, R. F. Harris, and J. L. Stanton, paper presented at PMA, Buford, Georgia, October 28, 1980.
77. S. R. Harvey, paper presented at the Polyurethane Manufacturers Association, Bloomington, Minnesota, April 29, 1980.
78. R. C. Baron, L. D. Cerankowski, N. Mattucci, and L. D. Taylor, *J. Appl. Polym. Sci.*, 20: 285 (1976).
79. S. W. Wong, A. Damusis, and K. C. Frisch, Elastomerics, 37 (1977).
80. PTM&W Industries, Inc., Product Bulletin, Polycure 1000 Urethane Hardener (May 29, 1982).
81. W. T. Maurice, *Development of Castor Oil Resistant Urethane Sonar Encapsulants*, Naval Research Laboratory, Report No. N00014-81-C-2576 (1983).
82. C. J. Nalepa and A. A. Eisenbraun, paper presented at PMA Conference, Nashville, Tennessee, October 13, 1986; *Proceedings of the SPI 30th Annual Technical/Marketing Conference*, Toronto, Canada (1986).
83. *Polyether Glycol*, E. I. du Pont de Nemours Co., Bull. E50149.
84. *Isonol[R] 93*, Upjohn Polymer Chemicals, Upjohn Co., Bull. PCN No. 10.
85. *Triisopropanolamine (TIPA)*, George Mann Co., Inc., Bull.
86. *Urethane Chemicals, Pluracol[R] TP Polyols*, BASF Wyandotte Corporation, Bull.
87. M. Palmer, J. H. Davis, T. L. Douglas, and M. R. Wilhelm, paper presented at a meeting of the PMA, Chicago, April 17–20, 1977.
88. Anderson Development Co., Preliminary Product Data Sheet, HER[R] Resorcinol Di(beta-hydroxy-ethyl) Ether (November 23, 1982).
89. K. H. Illers et al., U.S. Pat. 4,098,773 (1978); BASF Wyandotte.
90. M. Ulrich et al., U.S. Pat. 3,963,679 (1976); Bayer.
91. A. E. Brachman, Can. Pat. 1,010,186 (1977); GAF; U.S. Pat. 3,838,105 (1974).
92. F. X. O'Shea, U.S. Pat. 3,983,094 (1976); Uniroyal Inc.
93. I. S. Lin, S. J. Gromelski, Jr., and J. D. Pelesko, *Proceedings of the SPI Polyurethane Division's 27th Annual Technical/Marketing Conference*, Bal Harbour, Fla., October 20–22, 1982.

94. I. S. Lin and S. J. Gromelski, Jr., *Proceedings of the SPI Polyurethane Division's 6th International Conference*, San Diego, Calif., November 1983.

95. I. S. Lin, J. Biranowski, and D. H. Lorenz, in *Advances in Urethane Science and Technology*, Vol. 8 (K. C. Frisch and D. Klemner, eds.), Technomic Publishing Co., Westport, Conn. (1981).

96. D. Klemner and K. C. Frisch, eds., *Advances in Urethane Science and Technology*, Vol. 8, Technomic Publishing Co., Westport, Conn. (1981).

97. *Cast Elastomers and RIM Processing: The MDI Bridge*, Upjohn Polymer Chemicals, Upjohn Co., Technical Service Report (April 1983).

98. A. J. Papa, in *Flame Retardancy of Polymeric Materials*, Vol. 3 (W. C. Kuyla and A. J. Papa, eds.), Marcel Dekker, New York (1975).

99. L. Maier, U.S. Pat. 3,666,543 (1972).

100. F. T. Lee, J. Green, and C. Tennesen, *Proceedings of the SPI 6th International Technical/Marketing Conference*, San Diego, Calif., p. 510, November 1983.

101. F. T. Lee, J. Green, and J. A. Vincent, *Proceedings of the SPE 43rd Annual Technical Conference and Exhibition*, Philadelphia, Pa., pp. 1309–1311, May 1985.

102. U.S. Pat. 4,555,562 (1985); F. T. Lee and J. Green (FMC Corp.).

103. *Kirk-Othmer Encyclopedia of Chemical Technology*, Vol. 1, 3rd ed., Wiley, New York, pp. 244–259 (1966).

104. G. J. Dege, R. L. Harris, and J. S. Mackenzie, *J. Am. Chem. Soc., 81*, 3374 (1959).

105. G. K. Hoeschele, Fr. Pat. 2,019,398 (1970); E. I. duPont de Nemours.

106. J. F. Thompson and T. H. Porter, Germ. Pat. 2,241,388 (1973); Anderson Development Co.

107. E. A. T. Foster, A. Odinak, and A. A. R. Sayish, U.S. Pat. 3,358,025 (1967); Upjohn Co.

II

PROPERTIES OF PLASTICS

11

Polyethylene: Synthesis, Properties, and Uses

Dhoaib Al-Sammerrai and Nedhal K. Al-Nidawy
Petroleum Research Center
Jadiriyah, Baghdad, Iraq

INTRODUCTION

Polyethylene, with one of the simplest molecular structures, is the world's largest tonnage thermoplastic [1] and a material about which more has probably been written than any other polymer. It is manufactured by the polymerization of ethylene monomer to essentially $(-CH_2-CH_2-)_n$ of high molecular weights.

Polyethylene was first recognized for its remarkable properties as an electrical insulator [2], and full-scale commercial production began in Great Britain in 1939 and in United States in 1943.

SYNTHESIS OF POLYETHYLENE

There are four distinct routes for the commercial preparation of high polymers from ethylene monomer: High-pressure process, Ziegler process, Phillips process, and Standard Oil (Indiana) process.

The bulk of the ethylene monomer is obtained from petroleum sources [3]. When supplies of natural or petroleum gas are available the monomer is produced in high yield by high-temperature cracking of ethane and propane. Good yields of ethylene may also be obtained if the gasoline fraction from primary distillation of oil is "cracked." The gaseous products of the reaction include a number of lower alkanes and olefins and the mixture may be separated by low-temperature fractional distillation and by selective adsorption.

Since impurities—particularly carbon monoxide, acetylene, oxygen, and water—can affect both the polymerization reaction and the properties of the finished product, they must be completely removed.

High-Pressure Polymerization

High-pressure polymerization processes [4] are based on the use of high pressures from 1.0×10^8 N/m^2 up to 3.0×10^8 N/m^2 and at temperatures ranging up to 350°C. The polymerization reaction is initiated either by oxygen or more commonly by organic peroxides.

The polymerization reaction is typical free-radical polymerization involving free-radical initiation, polymer chain propagation, and radical recombination. Several important side reactions involving chain transfer occur due to the highly reactive nature of the polyethylene free radical and the high temperatures normally employed for manufacture. The molecular weight is usually determined by kinetic chain transfer either by hydrogen abstraction from an added chain transfer agent or by an internal rearrangement forming a double bond and a small radical. Chain transfer agents which are used include alkanes, olefins, ketones, aldehydes, and hydrogen. Radical recombination generally makes only a small contribution to molecular weight control and since chain transfer has a higher activation energy than the propagation reaction, the molecular weight decreases with increasing temperature. Short branches, mainly butyl and ethyl groups [5], are produced by intramolecular chain transfer, probably via six-membered ring transition states [6]. Long-chain branches are produced by intramolecular chain transfer between a dead polymer molecule and a growing polymeric radical [7].

The principal problems of the high-pressure ethylene polymerization process are those concerned with the compression and handling of high-pressure gases and with the control of the highly exothermic reaction. The heat of polymerization of ethylene in the gas phase is about 22–23 kcal/mol or about 800 cal/g. At 1.4×10^8 N/m^2 the specific heat of ethylene in the range 150–300°C is 0.60–0.68 cal/(g)(°C). Unless heat is removed during the reaction the temperature will consequently rise about 12–13°C for each 1% of ethylene converted during polymerization. If the reaction temperature becomes too high, alternative decomposition reactions of ethylene to give a mixture of carbon, methane, and hydrogen may occur. These are also strongly exothermic, for example, the decomposition of ethylene to carbon and methane alone evolves 34 kcal/mol, and once initiated such reactions may cause pressure increases of explosive violence. Thus efficient control of reaction temperature is of paramount importance [4].

Pressure exerts a marked effect on the polymerization reaction rate constant [8] and can be used to control the reaction rate and molecular weight in addition to the more usual variables of initiator concentration and temperature. Since the number of short branches and the molecular weight are determined by chain transfer reactions which are more influenced by temperature and less by pressure than the polymerization reaction, it follows that the molecular weight decreases and the degree of short branching increases with increasing temperature (and vice versa with pressure).

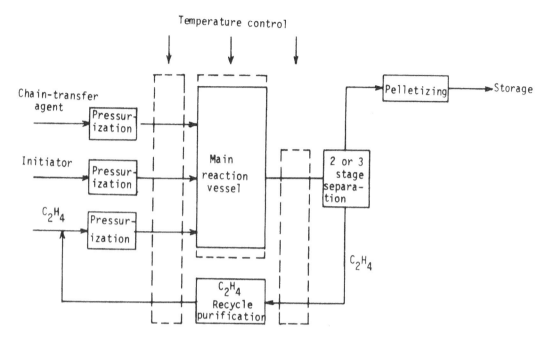

Figure 1 High-pressure polyethylene process. Dashed lines indicate heating control zones in the process.

A typical high-pressure polyethylene synthesis plant is displayed in Figure 1. It produces low-density polyethylenes (LDPE) of density ranging between 0.915 and 0.940 g/cm^3.

Ziegler Process

Ziegler processes are largely known through the work of Ziegler and colleagues [9]. This type of polymerization is sometimes referred to as coordination polymerization since the mechanism involves a catalyst–monomer coordination complex or some other directing force that controls the way in which the monomer approaches the growing chain. The coordination catalysts are generally formed by the interaction of the alkyls of group I–III metals with halides and other derivatives of transition metals in groups IV–VIII of the periodic table. In a typical process the catalyst is prepared from titanium tetrachloride and aluminum triethyl or some related material.

In a typical process, as displayed in Figure 2, ethylene is fed under low pressure into the reactor which contains liquid hydrocarbon to act as diluent. The catalyst complex may be prepared first and fed into the vessel or may be prepared in situ by feeding the components directly into the main reactor. Reaction is carried out at some temperature below 100°C (typically 70°C) in the absence of oxygen and water, both of which reduce the effectiveness of the catalyst through poisoning. The catalyst remains suspended and the polymer, as it is formed, becomes precipitated from the solution. A slurry is formed which progressively thickens as the reaction proceeds. Before the slurry viscosity becomes high enough to interfere seriously with removing the heat of reaction, the reactants

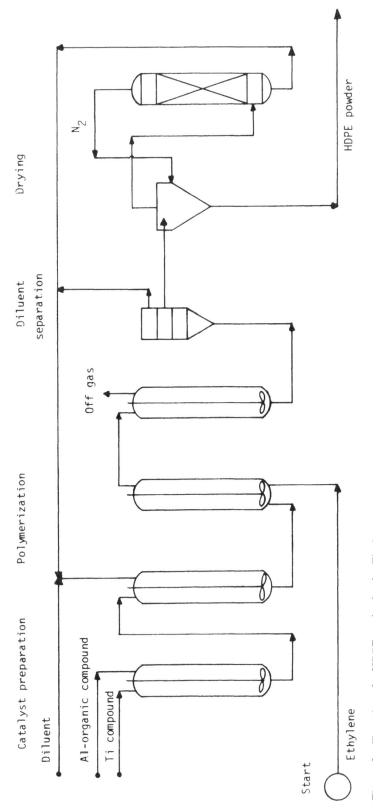

Figure 2 Flow sheet for HDPE production by Ziegler process.

are discharged into a catalyst decomposition vessel. Here the catalyst is destroyed by the action of ethanol, water, or caustic alkali. To reduce the amount of metallic catalyst fragments to the lowest possible value, the processes of catalyst decomposition and subsequent purification are critical, particularly where the polymer is intended for use in high-frequency electrical insulation. A number of variations on this stage of the process have been described [10].

The Ziegler polymers are intermediate in density (about 0.945 g/cm³) between the high-pressure polyethylenes and those produced by the Phillips and Standard Oil (Indiana) processes. A range of molecular weights may be obtained by varying the Al/Ti ratio in the catalyst, by introducing hydrogen as a chain transfer agent, and by varying the reaction temperature [11].

Phillips Process

This method of polymerization is based on the discovery by Phillips Petroleum Company [12] in the postwar period which described the use of heterogeneous catalyst consisting of transition metal compounds and revealed the possibility of carrying out the polymerization at relatively low pressure with these catalysts. Two main variations of this process are known.

Solution Process

In the solution process [13] ethylene dissolved in a liquid hydrocarbon such as cyclohexane is polymerized by a supported metal oxide catalyst at about 130–160°C and 1.4×10^6–3.5×10^6 N/m² pressure. The solvent serves to dissolve the polymer as it is formed and acts as a heat transfer medium but is otherwise inert.

The preferred catalyst is one that contains 5% of chromium oxides, mainly CrO_3, on a finely divided silica–alumina catalyst (75–90% silica) which has been activated by heating to about 250°C. After reaction the mixture is passed to a gas–liquid separator where the ethylene is flashed off; catalyst is then removed from the liquid product of the separator and the polymer is separated from the solvent by either flashing off the solvent or precipitating the polymer by cooling.

Polymers ranging in melt flow index (an inverse measure of molecular weight) from less than 0.1 to greater than 600 can be obtained by this process, but commercial products have a melt flow index of only 0.2–5 and have the highest density of any commercial polyethylenes (~0.96 g/cm³) and are known as high-density polyethylenes (HDPE).

The polymerization mechanism [14] is largely unknown but no doubt occurs at or near the catalyst surface where monomer molecules are both concentrated and specifically oriented so that highly stereospecific polymers are obtained. It is found that the molecular weight of the product is critically dependent on temperature and in a typical process there is a 40-fold increase in melt flow index and a corresponding decrease in molecular weight in raising the polymerization temperature from 140°C to just over 170°C. Above 2.8×10^6 N/m² the reaction pressure has little effect on either molecular weight or polymer yield but at lower pressures there is a marked decrease in yield and a measurable decrease in molecular weight. The catalyst activation temperature also has an effect on both yield and molecular weight. The higher the activation temperature the higher the yield and lower the molecular weight. A number of materials including oxygen, acetylene, nitrogen, and chlorine are catalyst poisons, and very pure reactants must be employed [14].

A typical Phillips solution process for the polymerization of polyethylene is seen in Figure 3.

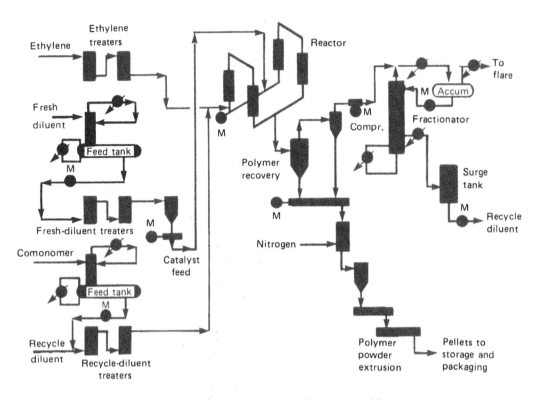

Figure 3 Simplified flow diagram for Phillips polyethylene process. M = meter

Slurry Process

The slurry process [15] for the polymerization of ethylene is carried out at about 90–100°C, which is below the crystalline melting point. At this temperature the polymer has low solubility in the solvent. The polymer is therefore formed and removed as a slurry of granules, each formed around individual catalyst particles. High conversion rates are necessary to reduce the level of contamination of the product with catalyst; in addition, there are problems of polymer accumulation on reactor surfaces. Because of the lower polymerization temperatures polymers of higher molecular mass may be prepared.

A typical Phillips slurry process of polymerization is seen in Figure 4. It can also be used to produce linear low density polyethylene (LLDPE).

Standard Oil Company (Indiana) Process

This process has many similarities to the Phillips process and is based on the use of a supported transition metal oxide in combination with a promoter [16]. Reaction temperatures are of the order of 230–270°C and pressures are 40.5×10^5–81.1×10^6 N/m². Molybdenum oxide is a catalyst that figures in the literature, and promoters include sodium and calcium either as metals or as hydrides. The reaction is carried out in a hydrocarbon solvent.

The process is operated as a continuous system using conventional equipment. A low-cost hydrocarbon solvent is used as the reaction medium, serving both to dissolve the polymer as it is formed and to remove heat from the reaction site. Polymer concentrations

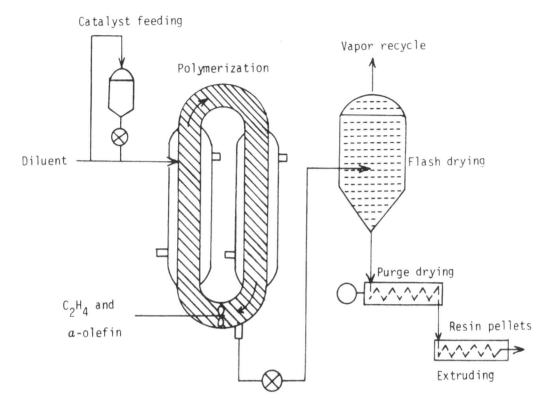

Figure 4 Phillips particle-form process used for producing HDPE and LDPE.

during polymerization and transport are relatively high, permitting the use of minimum-sized equipment.

The products of the process have a density of about 0.96 g/cm³, similar to the Phillips polymers. Another similarity between the processes is the marked effect of temperature on average molecular weight.

Many other methods have been developed in order to produce linear low-density polyethylenes (LLDPE) with short chain branches exclusively [17]. For example, among the earliest of these was a process operated by Du Pont Canada and another developed by Phillips. In 1968 commercial production of LLDPE was pioneered by Phillips Petroleum Company with the introduction of two grades of LLDPE at densities of 0.925 and 0.935 g/cm³.

Recently Union Carbide developed a gas phase process which is similar to that used for producing HDPE (Figure 5) in which gaseous monomers and a catalyst are fed to a fluid bed reactor at pressures of 0.7–2.1 × 10⁵ N/m² and temperatures of 100°C and below. The short branches are produced by including small amounts of higher alkenes such as propene, but-1-ene or hex-1-ene into the monomer feed. Somewhat similar products are produced by Dow using a liquid phase process, though it is based on a Ziegler-type catalyst system and again uses higher alkenes to introduce branching.

LLDPE is growing rapidly in importance [18] owing to the excellent physical properties that it displays in a number of applications and because of economic factors that favor low-pressure technology.

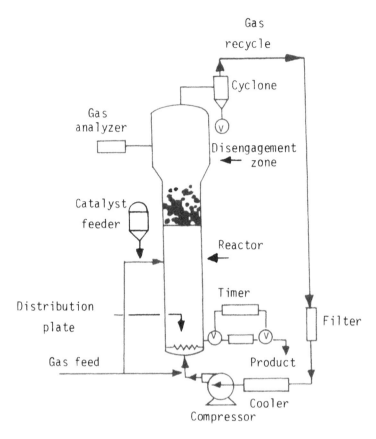

Figure 5 A gas phase process used for producing LLDPE and HDPE.

PROPERTIES OF POLYETHYLENE

Polyethylene is a waxlike thermoplastic with a softening range of 80–130°C and density less than that of water. The ASTM [19] classifies polyethylene into three main classes as shown in Table 1.

Polyethylene is tough but has a moderate tensile strength, very good chemical resistance, and excellent electrical insulation properties. In the mass it is translucent or opaque but thin films may be transparent.

Table 1 Three Main Classes of Polyethylene

Type	Density (g/cm³)
Low-density polyethylene (LDPE)	0.910–0.925
Medium-density polyethylene (MDPE)	0.926–0.940
High-density polyethylene (HDPE)	0.941–0.965

Structure of Polyethylene

The polymer is essentially a long-chain aliphatic hydrocarbon of the following structure:

$$-CH_2-CH_2-CH_2-CH_2-$$

The flexibility of C—C bonds would be expected to lead to low values for glass transition temperatures (T_g). The T_g, however, is associated with the motion of comparatively long segments in amorphous matter and since in a crystalline polymer there is only a small number of such segments the T_g has little physical significance. The values of T_g of polyethylene quoted in the literature [20] range between -60 and $130°C$. However, the most likely value for the T_g is considered to be $-20°C$.

The T_g value has little technological significance when compared to the far more important crystalline melting point (T_m), which is in the range $108–132°C$, the exact value depending on the detailed molecular structure of a particular commercial polyethylene. Low T_m values are to be expected of a structure with flexible backbone and no strong intermolecular forces.

The polymer has a low cohesive energy density; the solubility parameter is about 16.1×10^3 $(N/m^2)^{1/2}$ and would be expected to be resistant to solvents of solubility parameters greater than 18.5×10^3 $(N/m^2)^{1/2}$. Because it is a crystalline material and does not enter into specific interaction with any liquids, there is no solvent at room temperature. However, at elevated temperatures the thermodynamics are more favorable to solution and the polymer dissolves in a number of hydrocarbons of similar solubility parameter.

There are available many hundreds of grades of polyethylene which differ in their properties in one way or another.

The differences arise from the following variables in the polymer structure: degree of short-chain branching; degree of long-chain branching; average molecular weight; and molecular weight distribution. In addition, the presence of small amounts of comonomer residues or impurities or compounding and crosslinking may cause differences.

Possibilities and types of branching in polyethylenes are usually investigated by infrared and nuclear magnetic spectroscopy [21, 22].

In high-density polyethylenes, there are about 20–30 methyl groups per 1000 carbon atoms. Therefore, in a polymer molecule having molecular weight of 26,000 there will be present about 40–60 methyl groups, which is far in excess of the one or two methyl groups expected from normal chain ends. The methyl groups are probably part of ethyl and butyl groups. The most reasonable explanation is that these groups arise owing to a backbiting mechanism during polymerization:

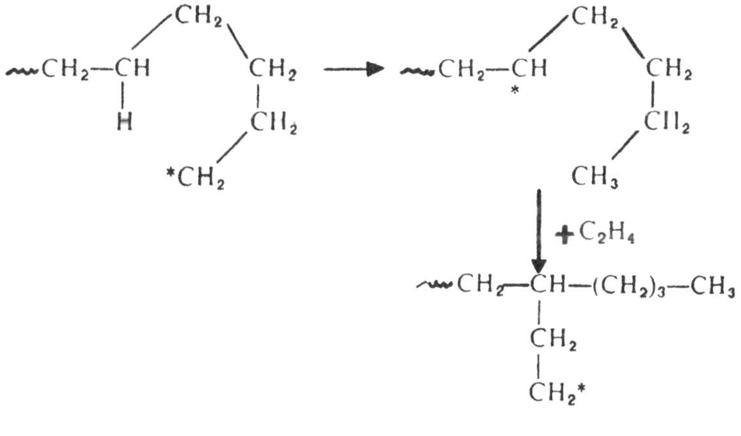

Polymerization could proceed from the radical in the normal way or alternatively chain transfer may occur by a second backbiting stage either to the main chain

$$
\begin{array}{ccc}
& \overset{\displaystyle Bu}{\underset{\displaystyle |}{\diagup}} & & & \overset{\displaystyle Bu}{\underset{\displaystyle |}{\diagup}} \\
& CH_2{-}CH & & & CH_2{-}CH \\
\diagup & \diagdown & & \diagup & \diagdown \\
{\sim\!\sim} CH & CH_2 & \longrightarrow & {\sim\!\sim} CH & CH_2{-}CH_3 \\
\diagdown & \diagup & & | & \\
H & {}^*CH_2 & & {}^* &
\end{array}
$$

or to the butyl group.

$$
\begin{array}{ccc}
& \overset{\displaystyle CH_2{-}CH_3}{\underset{\displaystyle /}{}} & & & CH_2{-}CH_3 \\
CH_2{-}CH & & & & CH_2{-}CH_* \\
\diagup \quad | & & & \diagup & \\
{\sim\!\sim} CH \quad H & & \longrightarrow & {\sim\!\sim} CH & \\
\diagdown & & & \diagdown & \\
CH_2{-}CH_2{}^* & & & CH_2{-}CH_3 &
\end{array}
$$

According to the preceding schemes a third backbite is also possible:

$$
\begin{array}{ccc}
& \overset{\displaystyle Et}{\underset{\displaystyle /}{}} & & \overset{\displaystyle Et}{\underset{\displaystyle /}{}} \\
CH_2{-}CH & & CH_2{-}CH & \\
\diagup \quad \diagdown & & \diagup \quad \diagdown & \\
{\sim\!\sim} C{-}Et \quad CH_2 & \longrightarrow & {\sim\!\sim}C{-}Et \quad CH_2 & \\
\diagdown \quad \diagup & & {}^* \quad \diagup & \\
H \quad {}^*CH_2 & & CH_3 &
\end{array}
$$

$$
{\sim\!\sim}CH_2{}^* + CH_2{=}C\overset{CH_2{\sim\!\sim}}{\underset{|}{\diagup}}\underset{Et}{}
\qquad
{\sim\!\sim}C{=}CH_2 + {}^*CH
$$

with Et substituents as shown.

Here a tertiary radical is formed which could then depolymerize by β-scission. This will generate vinylidene groups, which have been observed and found to provide about 50% unsaturation in high-pressure polymers, the rest being about evenly divided by vinyl and in-chains trans double bonds.

Short chain branching is negligible with Ziegler and Phillips homopolymers but they could be deliberately introduced by up to about seven ethyl side chains per 1000 carbon atoms in the Ziegler polymers. The presence of these branch points is bound to cause interference with the ease of crystallization, and this is shown clearly in the differences of the polymers properties.

The branched high-pressure polymers have the lowest density, least opacity, lower melting point, surface hardness, yield point, and Young's modulus in tension, which are all related to the lower degree in crystallinity. In addition, the lower crystallinity and the more branching will lead to greater permeability to gases and vapors. For general technological purposes the density of the polyethylene (as prepared from the melt under standard conditions) is taken as a measure of short chain branching [23].

In addition to short chain branches there is evidence in high-pressure polyethylenes for the presence per chain of a few long chain branches, which could be several tens of carbon atoms long. These could arise from the transfer mechanism during polymerization:

$$\text{ᴍᴍCH}_2\text{—CH}_2^* + \text{ H—}\{ \rightarrow \text{ᴍᴍCH}_2\text{—CH}_3 + \{\text{—}^*$$

Growing	"Dead"	"Dead"	Radical
Radical	Polymer	Polymer	

$$\{\text{—}^* + \text{CH}_2\text{=CH}_2 \rightarrow \{\text{—CH}_2\text{—CH}_2^* \xrightarrow{\text{CH}_2\text{=CH}_2} \text{etc. etc.}$$

Such side chains may be as long as the original main chain and like the original main chain will produce a wide distribution of lengths. It is therefore possible to obtain fairly short side chains grafted on to short main chains, long side chains on to long main chains, and a vast range of intermediate cases. In addition, subsequent chain transfer reactions may occur on side chains and the larger the resulting polymer the more likely it will be attacked. These features and variations tend to cause a wider molecular weight distribution for these materials and it is occasionally difficult to check whether an effect is due inherently to a wide molecular weight distribution or simply to long chain branching.

Another effect of long chain branches is on flow properties. Unbranched polymers have higher melt viscosities compared to that of long-branched polymers of similar weight average molecular weight. This is expected since long-branched molecules would be more compact and tend to entangle less with other molecules [24].

Linear low-density polyethylenes are virtually free of long chain branches but do contain short side chains as a result of copolymerizing ethylene with a small amount of higher alkenes such as hex-1-ene. Such branching interferes with the ability of the polymer to crystallize, as with the older low-density polymers, and like them possess low densities. The word linear is understood to imply the absence of long chain branches.

The polymer produced from diazomethane, which is free from both long and short branches and consists only of methylene groups (apart from the end groups), is particularly useful for refereeing purposes and is generally known as polymethylene [25]. This name is recommended by IUPAC to describe polyethylenes in general. The diazomethane-derived polymer has the highest density of this family of materials, about 0.98 g/cm^3.

Variation in molecular weight will also lead to differences in physical properties, i.e., the higher the molecular weight the greater the number of points of attraction and entanglement between molecules, whereas differences in short chain branching and hence degree of crystallinity will largely influence properties characterized by small solid displacement. Meanwhile, molecular weight differences will also affect properties related to large deformations, e.g., ultimate tensile strength, elongation at break, low temperature brittle point, and most importantly melt viscosities [24].

It was common practice [26] to characterize the molecular weight for technological purposes by the melt flow index (MFI), which is defined as the weight in grams extruded under a standard load in a standard plastomer at 190°C in 10 min. This test also proved useful for quality control and as a very rough guide to processability. Various workers [27] have calculated the apparent viscosity of the polymer from MFI measurements and correlated the values with both weight average and number average molecular weight, which indicated that the higher the MFI the lower is the molecular weight. Such measurements are considered to be unreliable due to the experimental errors involved. However,

Table 2 Effect of Molecular Weight and Density (Branching) on Some Mechanical and Thermal Properties of Polyethylene

Property	Test	Density 0.92 g/cm³ (high-pressure polymers)					Density = 0.94 g/cm³ (high-pressure polymers)	Density = 0.95 g/cm³ (Ziegler-type polymers)			Density = 0.96 g/cm³ (Phillips-type polymers)	Density = 0.98 g/cm³ (Polymethylene)
Melt flow index	BS2782	0.3	2	7	20	70	0.7	0.02	0.2	2.0	1.5	
Tensile strength × 10⁻⁶ N/m²	BS903	15.3	12.5	10.2	8.9		20.7	22.0	23.0	23.0	~27.5	~34.5
Elongation at break, %	BS903	620	600	500	300	150		>800	380	20	500	~500
Izod impact strength, J	BS2782	~13.5	~13.5	~13.5	~13.5	13.5		4.3	2.7	2.0	6.8	
Vicat softening point, °C	BS2782	98	90	85	81	77	116	124	122	121	122	
Softening temp., °C	BS1493							110	110	106		
Crystalline melting point, °C		~108	~108	~108	~108	~108	125	~130	~130	~130	~133	136
Number average molecular weight		48,000	32,000	28,000	24,000	20,000						
CH₃ groups per 1000 C atoms		20	23	28	31	33		5–7	5–7	5–7	<1.5	Unbranched

the value of MFI as a measure for molecular weight diminished with the availability of higher density polymers. Thus, for example, two polymers of the same average molecular weight of 4.2×10^5 and of densities of 0.92 g/cm³ for the branched polymer and 0.96 g/cm³ for the unbranched polymer possessed viscosities at a ratio of 1 : 50. This could be attributed to long chain branches, as explained previously. Commercial polyethylenes also vary in their molecular weight distribution.

The ratio of weight average molecular weight to number average molecular weight ($\overline{M_w}/\overline{M_n}$) usually is a useful parameter, although in some cases full description of the distribution is required. Its main deficiency is that it provides no information about unusual high or low molecular weight tail, which might have profound significance. For polymethylenes the ratio is about 2 whereas with low-density polymers values varying from 1.9 to 100 have been recorded [28, 29], with values of 20–50 being very common. High-density polyethylenes have values ranging between 4 and 15.

The very high values for low-density materials are in part a result of long chain branching and, as has already been stated, it is sometimes not clear if an effect is due to branching or to molecular weight distribution. It is generally considered, however, that with other structural factors constant, a decrease in $\overline{M_w}/\overline{M_n}$ will lead to an increase in impact and tensile strength, toughness, softening point, and resistance to environmental stress cracking. There is also a pronounced influence on melt flow properties, the narrower distribution materials being less sensitive to shear rate but more susceptible to sharkskin effects [30].

Mechanical and Thermal Properties

The molecular weight and degree of branching in polyethylene usually affect its mechanical properties, which are also dependent on the rate of testing, the temperature of test, specimen preparation, and to a smaller extent the conditioning of samples before testing. The data in Table 2, which are obtained from different sources under same test conditions, show clearly the effect of branching (density) and molecular weight on some of the

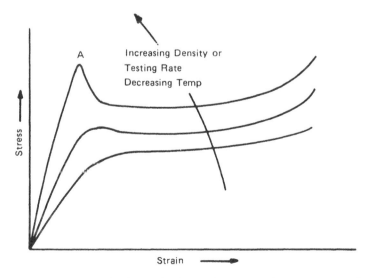

Figure 6 Effect of polymer density, testing rate, and temperature on the stress–strain curve for polyethylene.

polymer properties. It should be remembered that polyethylenes of different density but with same MFI do not have same molecular weight.

The general effects of changing rate of testing, temperature, and testing on the tensile stress–strain curves are shown in Figure 6. It is obvious that as the test temperature is decreased or the testing rate increased a pronounced hump in the curve becomes apparent, the peak of the hump being the yield point. Up to the yield point deformations are recoverable and the polymer is almost Hookean in its behavior. The working of the sample, however, causes strain softening by, for example, spherulite breakdown or in some cases by crystal melting so that the polymer extends at constant stress.

Molecular orientation and induced crystallization are created due to this cold-drawing

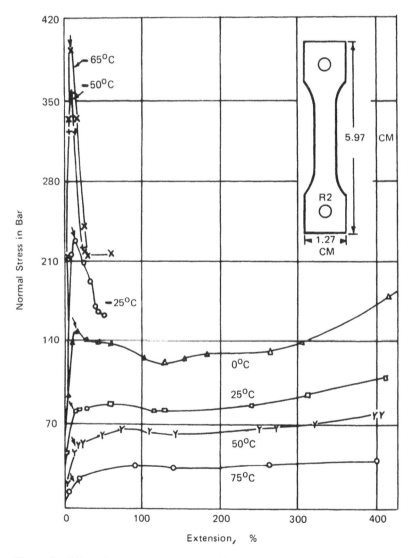

Figure 7 Effect of temperature on the tensile strain curve of polyethylene. LDPE $\sim$ 0.92 g/cm^3; MFI = 2. Rate of extension 190%/min.

behavior causing stiffness in the sample and an upward sweep of the stress–strain curve. The effect of temperature on a sample of low-density polyethylene having an MFI of 2 is shown in Figure 7. The varying influence of rate of strain on test results is clearly displayed in Table 3 as figures obtained with two commercial polyethylene samples. It is obvious that in one case an increase in rate of strain is accompanied by increase in tensile strength and a reduction in the other case.

The elongation at break of polyethylene is strongly dependent on density. As shown in Figure 8, the more highly crystalline high density polymer, the less ductility. This lack of ductility results in high-density polyethylene being brittle, mainly with low molecular weight materials. Figure 9 shows the tough–brittle dependence on MFI and density.

A knowledge of creep behavior (continuous deformation under load) of polyethylene is of importance in load-bearing applications such as water piping. In general there will be an increase in creep in polyethylene with increased load, increased temperature, and decreased density [31].

As mentioned earlier, the reported T_g of polyethylene varies between -60 and $120°C$. However, the truest value of that transition is considered to be $-20°C$. The polymer usually becomes brittle on cooling and some samples do not become brittle until about $-70°C$. The general tendency is for the brittle point to decrease as branching and molecular weight increase. The brittle point also depends on method of sample preparation, thus indicating that the polymer is sensitive to surface imperfections.

The specific heat (Cp) of polyethylene, which is higher than that of most thermoplastics, is very much dependent on temperature. Low-density polyethylene has a Cp value of about 2.3 J/g at room temperature while at $120–140°C$ the value is 2.9 J/g. This relationship is seen in Figure 10.

The melting points (T_m) for polyethylenes of different grades are displayed in Table 2. It is clear that T_m varies with density.

Table 3 Effect of Straining Rate on the Measured Tensile Strength and Elongation at Break of Two Samples of Polyethylene

Rate of strain (cm/min)	Polymer A	Polymer B
	Tensile strength $\times 10^{-6}$ N/m^2	
15.24	18.48	11.03
30.48	18.96	10.90
45.72	20.00	10.34
76.20	22.07	9.66
	Elongation at break (%)	
15.24	380	450
30.48	300	490
45.72	200	490
76.20	180	500

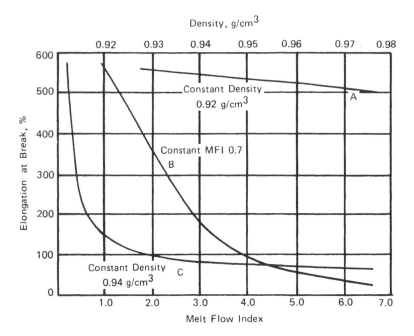

Figure 8 Effect of density and melt flow index on elongation at break (separation rate 45 cm/min on specimen of 2.54 cm gage length). A, constant density (0.92 g/cm³); B, constant MFI (0.7); C, constant density (0.94 g/cm³).

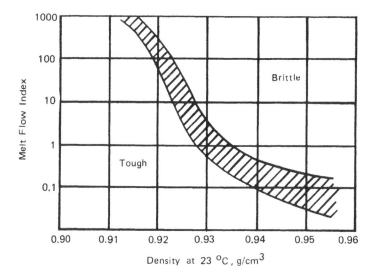

Figure 9 Effect of melt flow index and density on the room temperature tough–brittle transition of polyethylene.

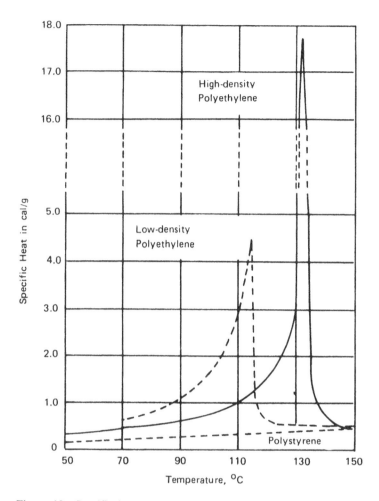

Figure 10 Specific heat–temperature relationships for low-density polyethylene, high-density polycthylene, and polystyrene.

The melts of polyethylenes are pseudo-plastic, in common with most thermoplastics. The zero rate apparent viscosity of linear polyethylenes is related to the weight average molecular weight by the relationship

$$\log (\eta_{a,0}) = K + 3.4 \log \overline{M_w}$$

for polymers of molecular weight exceeding 5000. However, for polyethylenes with different degrees of long branching a different equation is used which gives a good fit to the data:

$$\log (\eta_{a,0}) = A + B\overline{M_n}^{1/2}$$

It is interesting to note that linear low-density polyethylenes are less pseudo-plastic than conventional low-density polyethylenes. Thus on comparing the two materials at the same MFI, the linear polymer is more viscous at the higher shear rates usually encountered during processing.

As usual an increase in temperature reduces the melt viscosity of polyethylene. Melt processing is usually carried out in the range of 150–210°C but temperatures as high as 300°C may be encountered in some applications [32]. However, such applications must be performed in an inert atmosphere to avoid oxidative degradation.

Elastic melt effects are of considerable importance in understanding much of the behavior of polyethylene when processing by film extrusion techniques and when blow molding is commonly encountered. The phenomenon of elastic turbulence is also observed in low-temperature processes such as bottle blowing and when extruding at very high rates such as in wire drawing. This behavior is generally enhanced by high molecular weights and low temperature but reduced by long chain branching and increase in the molecular weight distribution [32].

In addition to elastic turbulence, another phenomenon known as sharkskin may be observed [30]. This is the appearance of a number of ridges transverse to the extrusion direction which are often just barely discernible to the naked eye. These often appear at lower shear rates than the critical shear rate for elastic turbulence and seem more related to the linear extrudate output rate, suggesting that the phenomenon may be due to some form of slip-stick at the die. It appears to be temperature-dependent in a complex manner and occurs with polymers of narrow molecular weight distribution.

Chemical Properties

The chemical behavior of polyethylene is similar to that of paraffins. It is not chemically attacked by nonoxidizing acids, alkalis, and aqueous solutions [33].

Oxidizing acids, such as nitric acid, oxidize the polymer, leading to a rise in the power factor and deterioration in mechanical properties. Halogens combine with polyethylene by means of substitution reactions, and when chlorine is used in the presence of sulfur dioxide, sulfonyl chloride as well as chlorine may be incorporated into the polymer, leading to the production of a useful polymer known as Hypalon [34].

Oxidation of polyethylene, which leads to structural and chemical changes, can occur to a measurable extent at temperatures as low as 50°C, and under the influence of UV light the reaction may occur at room temperature. Thermoanalytical techniques such as differential thermal analysis (DTA), differential scanning calorimetry (DSC), and thermogravimetery (TG) to measure the oxidation stability of polyethylene have been used extensively as alternatives to the standard oxygen adsorption method [35, 36].

The DTA method for the measurement of oxidation stability usually involves heating the test sample at a constant elevated temperature (~200°C) followed by determination of oxidative induction time [37]. Figure 11 shows a schematic DTA and temperature curve for the determination of the oxidative induction time of polyethylene while Table 4 shows variation in oxidative induction time with sample thickness.

Oxidation of polyethylene, which is usually reduced by the incorporation of antioxidants, leads to discoloration, degradation, and serious deterioration in power factor and electrical properties.

The chlorination of polyethylene was first reported by ICI in 1938 [38]. The introduction of chlorine atoms in the polyethylene backbone reduces the ability of the polymer to crystallize and the material becomes rubbery at a chlorine level of 20%, providing the distribution of chlorine atoms is random. An increase in the chlorine level beyond this point, and indeed from zero chlorination, causes an increase in the T_g so that at a chlorine level of about 45% the polymer becomes stiff at room temperature and with further increase the polymer becomes brittle.

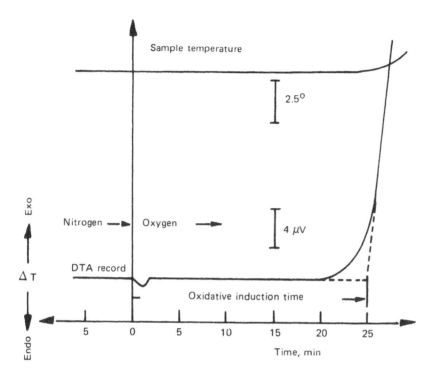

Figure 11 Schematic DTA and temperature curve for the determination of the oxidative induction time for polyethylene.

Table 4 Variation in Oxidative Induction Time with Sample Thickness

Sample thickness (mm)	Sample weight (mg)	Sample temperature (°C)	Sample induction time (min)
0.1	1.65	199.3	13.5 ± 2.6
0.3	4.91	199.8	23.2 ± 0.6
0.45	7.87	200.0	22.7 ± 1.0
0.75	12.93	199.8	24.6 ± 1.3
1.00	17.53	199.8	26.0 ± 0.9

Chlorination may be carried out with both high-density and low-density polyethylene. In solution the chlorination is random but when carried out with the polymer in the form of slurry the chlorination is uneven. Due to residual crystalline zones of unchlorinated polyethylene the material remains a thermoplastic.

Crosslinking of polyethylene usually interferes with its molecular packing and reduces the level of crystallization. Consequently, the polymer has a lower modulus, hardness, and yield strength than the corresponding uncrosslinked material. More important, because the network structure still exists above the crystalline melting point the material retains a measure of strength typical of a rubber material.

The main approaches used for crosslinking polyethylene are through radiation, use of peroxide, and vinyl silane crosslinking [39, 40]. Radiation crosslinking requires expensive equipment and extensive protective measures. The technique is used commercially and is most suitable with thin sections. Equipment required for peroxide crosslinking is cheaper but the method requires close control. Dicumyl peroxide is frequently used for low-density polyethylene but more stable peroxides are necessary for higher density materials. Vinyl silane crosslinking involves the grafting of an easily hydrolyzable trialkoxyvinyl silane onto the polyethylene chain activated by a small amount of peroxide followed by treatment with hot water; this hydrolyzes the alkoxy group, which then condenses to form a siloxane crosslink.

Since polyethylene is a crystalline hydrocarbon polymer incapable of specific interaction and with a melting point of above 100°C, there are no solvents at room temperature. Low-density polymers will dissolve in benzene at about 60°C but the more crystalline high-density polymers only dissolve at about 80–90°C. Materials of similar solubility parameter and low molecular weight, however, will cause swelling, the more so in low-density polymers.

Low-density polyethylene has a gas permeability in the range expected with rubbery materials, because in the amorphous zones the free volume and segmental movements facilitate the passage of small molecules. Polymers of density 0.96 g/cm^3 have a gas permeability of about 1/50 that of polymer with density of 0.92 g/cm^3.

Although polyethylene is often used without additives, different types of additives may be incorporated into the polymer in order to impart various properties [41]. When polyethylene is to be used in long-term application where a low power factor is to be maintained or to provide protection against thermal degradation, antioxidants mainly of the phenol type derivatives are usually incorporated in low concentrations. Antistatic additives of the glycol alkyl esters type are widely used in polyethylene to reduce dust attraction and also in films to improve handling behavior on certain types of bag making and packing equipment.

Polyethylenes burn readily and a number of materials have been used as flame retarders. These include antimony trioxide and a number of halogenated materials.

Layers of low-density polyethylene film often show high cohesion or blocking, a feature that is often a nuisance in both processing and use. One way of overcoming this defect is to incorporate antiblocking agents such as fine silicas. In addition, slip agents may be added to reduce the friction between layers of films. Fatty acid amides such as oleamide and, more importantly, erucamide are widely used for this purpose. Polymers with densities over 0.935 g/cm^3 show good slip properties and slip agents are not incorporated.

Products with very low dielectric constant ($\sim$1.45) can be obtained by the use of cellular polymers. Blowing agents such as 4,4-oxybisbenzenesulfonohydrazide and azocarbonamide are incorporated in the polymer. On extrusion, the blowing agent decomposes with evolution of gas and gives rise to cellular extrudate.

Among other additives incorporated in polyethylene are fillers such as carbon black and still better silane and titanate coupling agents. Pigments based on cobalt, cadmium, manganese, iron, and chromium compounds have been incorporated in polyethylene. However, care must be taken when choosing such compounds since they can catalyze oxidation and can cause a rapid rise in the power factor.

USES OF POLYETHYLENE

Polyethylene, which was introduced as a special-purpose dielectric material of particular value for high-frequency insulation [42], is used in a wide variety of applications where its general toughness and resistance to moisture as well as its chemical inertness make it desirable. Since World War II there has been a considerable and continuing expansion in polyethylene production and this, together with increasing competition between manufacturers, has resulted in the material becoming widely available in a range of grades.

The characteristics of polyethylene which lead to its vast use may be summarized as follows:

1. Low cost.
2. Easy processability.
3. Excellent electrical insulation properties.
4. High chemical inertness.
5. Toughness and flexibility.
6. Availability of information regarding its processing and properties.
7. Nontoxicity and odorlessness.

There are, however, the following limitations of the polymer:

1. Low softening point.
2. Susceptibility to oxidation.
3. Opacity of the material in bulk.
4. Poor scratch resistance.
5. Lack of rigidity.
6. Low tensile strength.
7. High gas permeability.

Many of these limitations can be overcome by the correct choice of polymer, additive incorporation, processing conditions, and aftertreatment.

There are many different methods for the processing of polyethylenes [43]. However, polyethylene films are the most widely processed form. Figure 12 shows the methods employed in producing polyethylene films.

Polyethylene film is the most important application of LDPE [44]. It is used for making heavy-duty sacks, refuse bags, carrier bags, general packaging, and in the building and construction industry. Many of these uses are now considered to be fully mature. Areas capable of further development are shrink film for food wrapping and film for agricultural uses.

LDPE and blends of LDPE and polypropylene have also been used as gellants in mineral oils for producing lubricating greases of improved thermal stability [45]. In addition to their use as sole gellants, polyethylenes are often incorporated as supplements to other types of thickening agents in order to improve the adhesion and cohesion characteristics of lubricating greases [46]. Table 5 shows properties of polyethylene–polypropylene-based grease and of a soap-based grease.

LLDPE has caused some threat to the LDPE market since polymerization plants of LLDPE are cheaper to build and easier to operate and to maintain than are high-pressure plants [18]. There are also some technical advantages to the user since films of LLDPE possess higher impact strength, tensile strength, and extensibility. Such properties will yield film of lower gage but of the same mechanical performance. However, LLDPEs

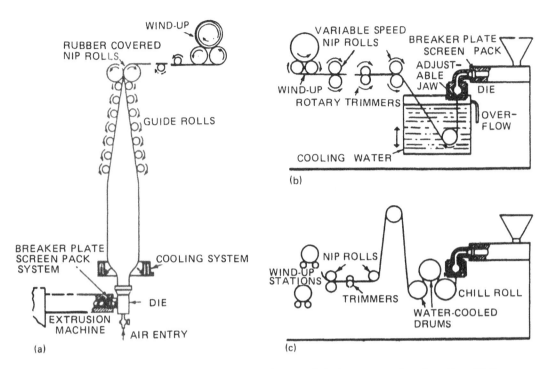

Figure 12 Methods of producing polyethylene film. (a) Tubular process using air cooling; (b) flat film process using water bath cooling; (c) flat film using chill roll cooling.

exhibit some undesirable properties such as lower gloss, greater haze, and narrower heat-sealing range.

HDPE film is now increasingly replacing paper for carrier bags and as a wrapping material for food products because of its crisp feel and greaseproof nature.

HDPE has also been used in some pseudo-fiber applications and in a process developed by Smith and Nephew in England is now used to make nets varying in appearance from a fine web to coarse open structures.

Polyethylene is an important injection molding material [47]. Although the percentage of LDPE used for injection molding is small compared to HDPE and polypropylene, the actual tonnage is substantial. In addition, there is considerable [48] use of blends of LDPE and HDPE and a very wide range of products is manufactured, including toys, chemical plants, electrical fittings, seals, bushes, and many other items which were made at one time from rubber. Injection molding is somewhat more important with HDPE both in percentage and absolute terms. Uses include industrial containers, crates, cases, pails, food tubs, closures, caps, housewares, and toys.

Blow molding is widely used [47] for HDPE for the production of bottles for milk and other foodstuffs, household detergents and chemicals, personal toiletries and drug packaging; these constitute the majority of the blow molding market. In some cases the selection between blow molding and rotational molding requires careful consideration. However, the tonnage rotationally molded is small compared to that blow molded.

Blow molding of LDPE is mainly limited to squeeze bottles, particularly for detergents, and in this case is a less important process than rotational molding.

Table 5 Specifications of Prepared Lubricating Greases

Specification	Polymer-based	Soap-based
Color	Brown	Brown
Texture	Smooth	Smooth
Dropping point of lubricating grease, ASTM[a] D2265, °C	149	197
Cone penetration of lubricating grease (0.1 mm.), IP[b] 310	305	310
Grade number	1	1
Acidity of grease of oleic acid, IP 37, wt %		0.27
Ash determination of lubricating grease, ASTM D128, wt %		0.64
Wear preventive characteristic of lubricating grease (four-ball method), scar diameter, IP 239, mm	0.75	0.6
Measurement of extreme-pressure properties of lubricating grease (four-ball method), ASTM D 2596		
Mean load, Hz	39.2	21.2
Welding point, kg	200	126
Oxidation stability of lubricating grease by the oxygen bomb method, ASTM D942	552 h	300 h

[a]ASTM: standard methods of the American Society of Testing and Materials.
[b]IP: standard methods of the Institute of Petroleum.

Table 6 Electrical Properties of Polyethylene

Volume resistivity	$>10^{22}$ Ω/m
Dielectric strength	700 kV/mm
Dielectric constant	
Density = 0.92 g/cm³	2.28
Density = 0.96 g/cm³	2.35
Power factor	$\sim 1{-}2 \times 10^{-4}$

The excellent electrical insulation properties of polyethylene, which are listed in Table 6, and its moisture resistant make it ideal for wire and cable coatings. Uses include power, communications, and control applications. One particular trend is the increasing use of the crosslinked polyethylene for this area of use. Such materials have improved resistance toward heat and stress cracking [49].

Another substantial area for polyethylene use is in making pipes such as domestic water and gas piping, agricultural piping, and ink tubes for ballpoint pens. Other end uses for polyethylene include filament for ropes, fishing nets, and fabrics, which are an important outlet for HDPE. Polyethylene in the powder form is used for dip coating and in flame spraying.

Treatment of solutions of HDPE has led to formation of products with a celluloselike morphology, known as fibrides or synthetic wood pulp [50]. They are used for finishing paper and special boards to impart such features as sealability and improved wet strength. Other specialized uses are as tile adhesives, thixotropic agents, and battery separators.

Controlled oxidation of low molecular weight polyethylenes ($\overline{M_v} = 1000$) leads to the formation of an interesting class of products [51] referred to as oxidized polyethylene waxes, which have found wide application in the petroleum industry. They usually impart good anticorrosion properties in lubricating oils and greases and have also been incorporated in petroleum-derived and naturally occurring waxes to improve their thermal properties.

There are many other specialized uses of polyethylenes including medical and biological applications. Most important among them are the millions of disposable syringes that are manufactured every year and the use of the polyethylene bags and tapes in the culture and study of organisms and fungi [52, 53].

REFERENCES

1. R. Martino, ed., *Mod. Plast. Int., 17*: 20 (1987).
2. E. W. Fawcett, R. O. Gibson, M. H. Perrin, J. G. Patton, and E. G. Williams, Br. Pat. 471,590 (1937); ICI.
3. J. Haley and L. Turner, in *Modern Petroleum Technology* (G. D. Hobson, ed.), Applied Science Publishers, London, p. 440 (1975).
4. H. Mark and R. Raff, in *High Polymers*, Vol. 3, Interscience, New York, p. 233 (1941).
5. A. H. Willbourn, *J. Polym. Sci., 34*: 569 (1959).
6. M. J. Roedel, *J. Am. Chem. Soc., 75*: 6110 (1953).
7. G. A. Tirpak, *Polym. Lett., 4*(2): 111 (1966).
8. R. O. Symcox and P. Ehrlich, *J. Am. Chem. Soc., 84*: 531 (1962).
9. K. Ziegler, E. Holzkamp, H. Breil, and H. Martin, *Angew. Chem., 67*: 541 (1955).
10. W. Grundmann, H. Bestian, and S. Sommer, U.S. Pat. 3,066,130 (1962); Hercules Inc.
11. K. Ziegler, Belg. Pat. 540,459 (1956).
12. A. Clark, J. P. Hogan, R. L. Banks, and W. C. Lanning, *Ind. Eng. Chem., 48*: 1152 (1956).
13. J. P. Hogan and R. L. Banks, U.S. Pat. 2,825,721 (1958); Phillips Petroleum Company.
14. A. Standen, ed., *Encyclopedia of Chemical Technology*, Vol. 14, 2nd ed., Interscience, New York, p. 251 (1968).
15. G. T. Leatherman, Br. Pat. 853,414 (1957); Phillips Company.
16. E. L. d'Ouville, in *Polyethylene*, 2nd ed. (A. Renfrew and P. Morgan, eds.), Interscience, New York, p. 35 (1960).
17. M. Grayson, ed., *Encyclopedia of Chemical Technology*, Vol. 16, 3rd ed., Wiley-Interscience, New York, p. 386 (1978).
18. C. T. Levett, J. E. Pritchard, and R. J. Martinovich, *SPE J., 26*(6): 40 (1970).
19. Standard Specification ANSI/ASTM D1248, Philadelphia (1974).
20. A. D. Jenkins, *Polymer Science*, North-Holland, Amsterdam (1972).
21. C. Baker and W. F. Maddams, *Makromol. Chem., 177*: 437 (1976).
22. D. E. Axelson, G. C. Levy, and L. Mandelkern, *Macromolecules, 12*: 41 (1979).
23. M. Grayson, ed., *Encyclopedia of Chemical Technology*, Vol. 16, 3rd ed., Wiley-Interscience, New York, p. 387 (1978).
24. D. Romanini, *Polym. Plast. Technol. Eng., 19*(2): 201 (1982).
25. J. A. Brydson, in *Plastics Materials*, 4th ed., Butterworth, London, p. 187 (1982).
26. Standard Specification ANSI/ASTM D1238, Philadelphia (1974).
27. C. J. Stacey and R. L. Arnett, *J. Polym. Sci., A2*(2): 167 (1964).

28. J. C. Moore, *J. Polym. Sci.*, *A2*(2): 835 (1964).
29. J. C. Moore and J. G. Hendrickson, *J. Polym. Sci.*, *C*(8): 233 (1965).
30. P. L. Clegg, *Plast. Inst. (London) Trans. J.*, *26*: 151 (1958).
31. S. Turner, *Br. Plast.*, *38*(1): 44 (1965).
32. J. A. Brydson, *Flow Properties of Polymer Melts*, 2nd ed., Godwins, London (1981).
33. E. L. d'Ouville, in *Polyethylene*, 2nd ed. (A. Renfrew and P. Morgan, eds.), Interscience, New York (1960).
34. J. L. Lann and M. Hunt, U.S. Pat. 2,556,879 (1951); Du Pont.
35. I. Gömory, *J. Therm. Anal.*, *11*: 327 (1977).
36. J. Chiu, *Polym. Prep. Am. Chem. Soc. Div. Polym. Chem.*, *14*: 846 (1973).
37. E. L. Charsley and J. G. Dunn, *J. Therm. Anal.*, *17*(2): 537 (1979).
38. E. W. Fawcett, Br. Pat. 481,515 (1938); ICI.
39. R. W. Ivett, U.S. Pat. 2,826,570 (1958); Hercules Incorporated.
40. A. Charlesby, *Proc. R. Soc. (London)*, *215A*: 187 (1952).
41. E. L. Hawkins, ed., *Polymer Stabilization*, Wiley-Interscience, New York (1972).
42. R. Raff and K. W. Doak, eds., *High Polymers Series*, Part 2, Wiley, New York, Chap. 3 (1964).
43. D. Renfrew and P. Morgan, eds., *Polyethylene*, 2nd ed., Illife, London, Chap. 8 (1960).
44. M. Grayson, ed., *Encyclopedia of Chemical Technology*, 3rd ed., Wiley-Interscience, New York, Vol. 10, p. 219, and Vol. 11, p. 119 (1978).
45. D. Al-Sammerrai, *J. Appl. Polym. Sci.*, *31*: 1 (1986).
46. B. Mitacek, U.S. Pat. 3,112,270 (1963).
47. M. Sitting, *Polyolefins Resin Processes*, Gulf, Houston (1961).
48. H. F. Mark, ed., *Encyclopedia of Polymer Science and Technology*, Vol. 6, Interscience, New York, p. 316 (1967).
49. *Chem. Eng. News*, *44*(17): 28 (1966).
50. H. F. Mark, ed., *Encyclopedia of Polymer Science and Technology*, Vol. 6, Interscience, New York, p. 324 (1967).
51. D. Al-Sammerrai and W. Selim, *Polym. Degradation Stab.*, *15*: 183 (1986).
52. AY Malkina, *Mikol. Fitopatol.*, *16*(6): 543 (1983).
53. C. Barker and N. Dixon, *Ann. Ap. Biol.*, *103*(3): 485 (1983).

12

Crystal Growth in Polyethylene

M. Dosière
University of Mons
Mons, Belgium

INTRODUCTION

Among the large variety of synthetic polymers, polyethylene (PE) possesses the simplest monomeric unit. Polyethylene is the most studied of all organic polymers and its morphological behavior has been taken to be representative of almost all synthetic polymers, sometimes unfortunately. Its use in the industrial field is also one of the most important.

Low-density polyethylene (LDPE) is produced under high pressure with free radical initiators [1] and contains some long branches which are formed as a result of chain transfer of an active macroradical onto an already terminated polymer chain. This phenomenon has an important influence on the rheology and processing of this material. Another structural modification occurring in radical polymerization of LDPE is the formation of some short chain branches via a ''backbiting'' mechanism originally suggested by Roedel [2]. These short chain branches are very important in disrupting the chain packing in LDPE and are principally responsible for lowering the melting temperature (by about 25°C), the crystal density [3], and the degree of crystallinity.

The advent of Ziegler–Natta catalysis [4–7] amplified and greatly expanded the idea that ionic polymerizations can be stereochemically controlled via coordination of the growing chain with its monomer and counterion. The industrial importance of these reactions may be noted by stating that coordination polymerization is used to produce linear high-density polyethylene (HDPE) and linear low-density polyethylene (LLDPE).

The orthorhombic unit cell [8] of polyethylene is shown in Figure 1a. The value of the crystallographic parameters are $a = 0.736$ nm, $b = 0.492$ nm, and $c = 0.254$ nm. The space group is P_{nam}. The chains have a planar zigzag conformation. Two chains pass

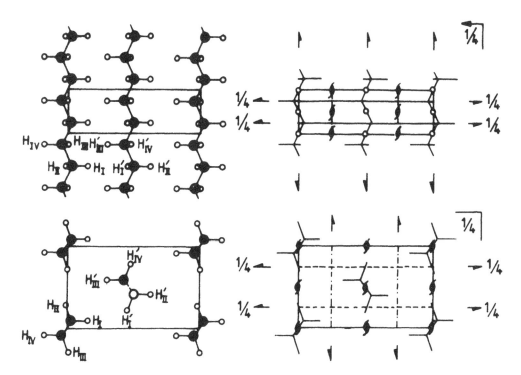

Figure 1 Orthorhombic unit cell of polyethylene ($P_{nam}-D_{2h}$). The diagonal glide planes (········) are normal to the a axis and translate by $(b + c)/2$. The a glide planes (---- or ➔) and the mirror planes (--- or ⌐) are normal to the b and c axes, respectively. The molecular axis of polyethylene coincides with the twofold screw axis. § denotes the 2_1 screw axis perpendicular to the page and ¼ denotes the glide plane parallel to the page with the glide translation along the axis indicated by the arrow. The center of symmetry is denoted by a single-edged arrow (→) and dashed lines. (From Ref. 8.)

through the unit cell, which therefore contains four —CH$_2$ groups. The temperature coefficients of lattice parameters have been determined by Swan over a wide temperature range [9]. Essentially the a and b parameters relating to interchain separation are markedly dependent on temperature. In the temperature range around 30°C, Swan reported $\alpha_a = 22 \times 10^{-5}$ K^{-1} for a and $\alpha_b = 3.8 \times 10^{-5}$ K^{-1} for b. Thermal shrinkage along the chain direction has been observed from a detailed precision x-ray investigation of the (002) reflection of polyethylene single crystals as precipitated from dilute solution [10]. The thermal change is more pronounced in the upper temperature range. Experimental values of $\alpha_c = -12 \times 10^{-6}$ K^{-1} between 20 and 65°C and $\alpha_c = -21 \times 10^{-6}$ K^{-1} between 65 and 120°C have been obtained. Therefore, a mean value of the expansion coefficient of $\alpha_c = -18 \times 10^{-6}$ K^{-1} can be proposed. Nakafuku reported unit cell variations of PE crystal with temperature and pressure [11]. For the a axis direction, a drastic increase of the compressibility (30%) was observed above around 90°C, but for the b and c axis directions, the compressibilities remain constant for all the temperature range of the measurement. The equation for the volume compression of the unit cell is

$$\frac{V_0 - V_p}{V_0} = Ap - Bp^2 + Cp^3$$

At 30°C, these coefficients are respectively the following: $A = 16 \times 10^{-6}$ cm^2/kg, $B = 0.80 \times 10^{-9}$ cm^4/kg^2, and $C = 0.020 \times 10^{-12}$ cm^6/kg^3. Extra diffractions have been observed in the x-ray patterns of highly mechanically deformed samples [12, 13] and have been indexed on the basis of a monoclinic lattice [14]. The crystallographic parameters are $a = 0.809$ nm, $b = 0.253$ nm, $c = 0.479$ nm, $\beta = 107.9°$ (b is the chain axis). The space group is C_2/m. The unit cell contains two monomer units and its projection is seen in Figure 2. Turner-Jones [15] gave a unit cell which corresponds to the triclinic form of

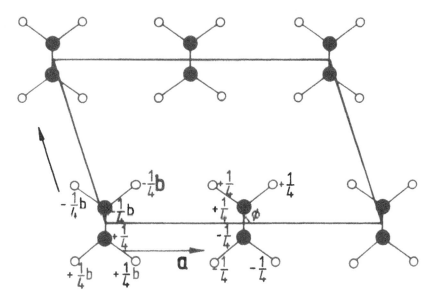

Figure 2 Crystal structure of the monoclinic form of polyethylene viewed in b projection (○, H atom; ●, C atom). (From Ref. 14.)

paraffins in which the projections along the axis of the molecular chain are in conformity with the monoclinic structure [13]. A pseudo-hexagonal structure with positional disorder due to a random arrangement of skeletal zigzag planes about the chain axis was suggested by McCall and Slichter [16].

MORPHOLOGY OF POLYETHYLENE CRYSTALS

Closely related to crystalline growth is the subject of polyethylene morphology. Crystalline polyethylene in the solid state has very complex morphologies. This is due to the chainlike nature of its constituent molecules, which leads to crystallization behavior and morphologies that are in most aspects only rarely encountered in traditional solids made of "small molecules" and in other aspects are unique to synthetic organic polymers. Polyethylene crystals exhibit a large variety of habits depending on their crystallization conditions. Since normal alkanes are used as a model for polyethylene, a short summary of the oligomers of polyethylene is given first.

Normal Alkanes as a Model for Polyethylene

Normal alkanes are aliphatic hydrocarbons of general formula C_nH_{2n+2} designated hereafter C_n for simplicity. At room temperature, these alkanes are gaseous from C_1 to C_4, liquid from C_5 to C_{17}, and solid above C_{18}. Since 1925, the crystal structures of various n-alkanes, the oligomers of PE, have been fully investigated [17]. In n-alkane crystals the chains are extended, forming layers with the methyl end groups at the layer surface. The molecules have a zigzag arrangement of the carbon atoms in the transconformation. All carbon atoms are coplanar while the hydrogen atoms lie in planes normal to the chain axis and going through the carbon atoms. The C—C and C—H distances are close to 0.153 and 0.105 nm; the CCC and HCH angles are around 112 and 108° respectively, which results in a chain length for the extended configuration of the molecule C_n of 0.127 ($n -$ 1) nm.

The layers of extended chains are stacked on top of each other; various stacking modes corresponding to slight differences in methyl group packing give rise to different polymorphic forms. In the vertical structures (orthorhombic and hexagonal), the chains are parallel to the c axis, being therefore perpendicular to the molecular layers, i.e., to (001) crystallographic planes. In the oblique structures (triclinic and monoclinic) the chains are nearly parallel to the c axis and consequently tilted with respect to the molecular layers. The orthorhombic crystalline form is the stable form at low temperature for odd $n \geq 9$ and for even $n \geq 44$. Transition from the orthorhombic to the hexagonal form precedes melting for odd n in the range $11 \leq n \leq 43$ and for even n, $34 \leq n \leq 42$. For even $n \leq 32$ a stable orthorhombic form does not seem to occur at any temperature. For all alkanes with $n \geq 44$ the orthorhombic form melts without polymorphic transition. Uniform preparations of n-alkanes of several hundred carbon atoms have become available as a result of a new synthesis [18, 19]. Although n-alkane C_{102} crystallizes in an extended chain form, the long paraffins C_{150}, C_{198}, C_{246}, and C_{390} can crystallize in a chain-folded manner [20]. The fold length is a function of crystallization temperature T_c or more precisely of undercooling ΔT, higher crystallization temperatures yielding higher lamellar thicknesses. The fold length is quantized: the chains complete almost exactly one, two, three, and four folds up the largest chain examined, C_{390} [20]. Since the fold lengths are integral reciprocals of the full chain length, the chain ends must lie at the layer surface. As the folds contain only a few ethylene units, they must be sharp with adjacent reentry.

Time-resolved small-angle x-ray scattering on C_{246} allows the identification of transient fold lengths, which were noninteger fractions of the chain length, during the initial stages of crystallization in the melt [21]. This intermediate structure transforms subsequently into forms with integer fraction fold lengths; for C_{246} the noninteger fraction of the chain length has a fold length between extended chain and once-folded configurations. This transformation into forms with integer fraction fold lengths occurs either by lamellar thickening or thinning or by both processes. The noninteger fold length state has a more disordered layer surface compared to the final structure with integer fold length. As emphasized by Keller, "the existence of an initial structure with noninteger fold length focusses attention to the importance of the fastest kinetic pathway as the determining factor of chain-folded crystal growth" [22].

To obtain the equilibrium thermal properties of PE, extrapolations of melting temperatures of *n*-alkanes to longer chain length were done by several authors [23–28] with results ranging from 385 to 420 K. Experimental results obtained on very long paraffins [20] have been included in data collected by Wunderlich and Czornyj [28] and a melting temperature of 413.2 K was obtained by a least square fitting of T_m versus n^{-1}. A value of 414.6 K was reported [28].

Crystallization from Solution

Single Crystals

The modern study of polymer morphology began with the discovery that single crystals of polyethylene could be grown from dilute solution [29–34]. These single crystals provided the basis for an academic study of the nature and process of polymer crystallization. A typical electron micrograph and electron diffraction pattern of a monolayered single crystal of LDPE having a lozengelike lateral shape are seen in Figure 3. The lateral dimensions of PE single crystals are around 10 to some tens of microns. The lamellar thickness *l* experimentally determined by electron microscopy and small-angle x-ray scattering (Figure 4) takes values in the range of 10 to some tens of nanometers. The necessity to reconcile the far excess of the mean length of the polyethylene macromolecule over the small thickness *l* of the monolayered crystal with the fact that the *c*-crystallographic axis is tilted by at most 45° to the normal to the surface of the lamella led Keller [32] in 1957 to propose that the macromolecules in a polyethylene lamella are folded back and forth between the upper and lower surfaces, as shown schematically in Figure 5. The loops resulting from these folds form the two surfaces of the lamellar crystals, also termed fold surfaces. The lamellar character is the fundamental feature of the habits of polyethylene crystals (and polymer crystals in general) and is the consequence of the fact that the macromolecules adopt a folded conformation during the process of crystallization [35–37].

The lateral growth faces are parallel to [110] crystallographic planes [32]. The different sectors in PE single crystals are often referred to as fold domains, a term originally coined by Reneker and Geil [34]. The central pleat in the PE single crystal shown in Figure 3 proves that the crystal did not grow flat in solution but collapsed on sedimentation [38]. Collapse can occur by shear of molecular chains parallel to the *c* axis or by tilting each sector so that it becomes flat on the substrate or by a combination of both these mechanisms. Sectorization allows definition of the fold plane direction and provides a simultaneous record of the present state of a crystal and its growth history: the position of the growth front and therefore the shape of the crystal at any stage of its development can be

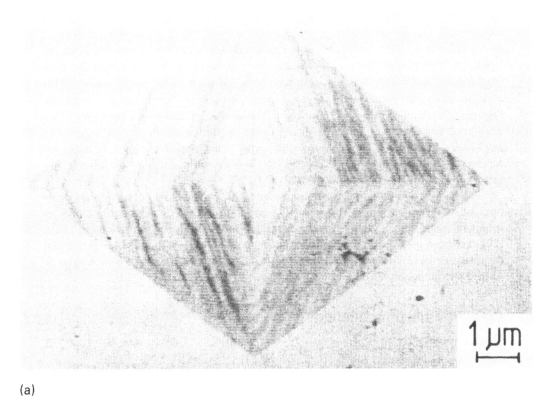

(a)

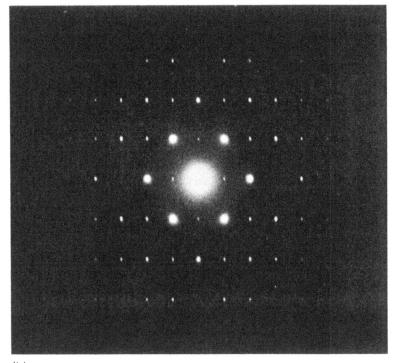

(b)

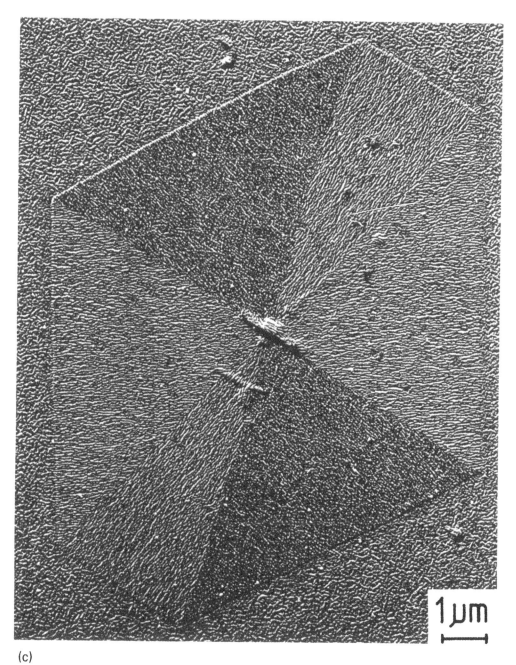

(c)

Figure 3 (*a*) Bright-field image exhibiting diffraction contrast of a single crystal of linear polyethylene grown from dilute solution. (*b*) Associated electron diffraction pattern (courtesy of J. C. Wittmann). (*c*) Six-sectored single crystal of linear polyethylene ($M_w = 17,000$, $M_w/M_n = 1.1$) decorated with vaporized polyethylene. (M. Dosière, unpublished results.) (*d*) Bright-field pattern of a single crystal of linear polyethylene ($M_w = 17,000$, $M_w/M_n = 1.1$) grown from *p*-xylene solution ($C = 2.18 \times 10^{-3}$ wt %) and decorated following the isochronous decoration method ($T_{c1} = 88°C$, $T_{c2} = 81°C$). (From Ref. 285.)

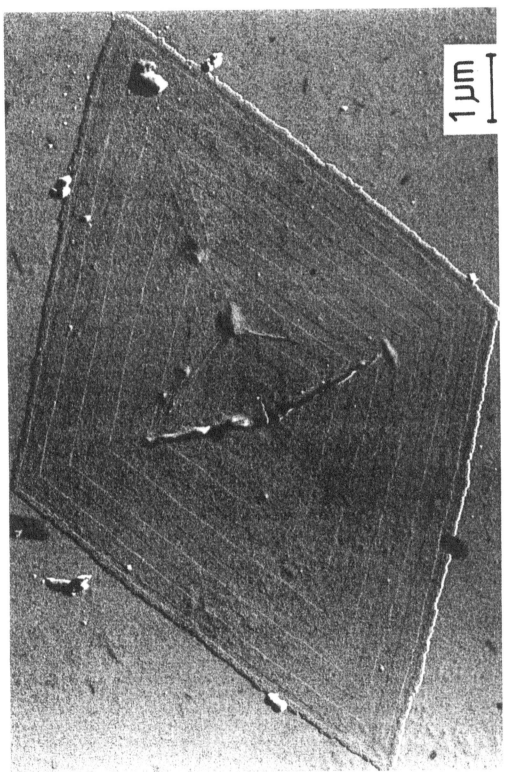

1 μm

(d)   **Figure 3** (*continued*)

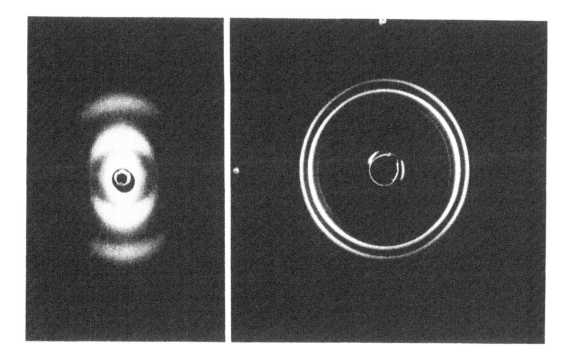

Figure 4 Small- and wide-angle x-ray patterns of a mat of polyethylene single crystal grown from xylene at 80°C. The plane of the mat is horizontal. Four orders can be observed in the original small-angle x-ray pattern. (M. Dosière, unpublished results.)

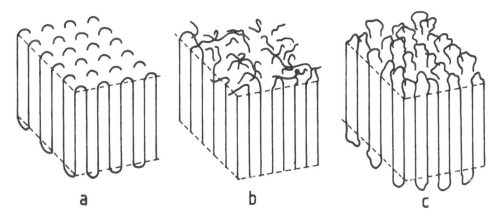

Figure 5 Schematic drawing of a lamellar crystal: (*a*) the sharp fold model; (*b*) the switchboard model; (*c*) the loose loop model. The sharp fold model and the switchboard model are the two extreme models proposed for describing lamellar crystals. The sharp fold model is characterized by a regular folding of the chains with adjacent reentry; the lamellar crystals can be considered as independent entities. In the switchboard model, the lamellar crystals are extensively intertwined; the adjacent reentry of a macromolecule in a given lamellar is of low frequent occurrence.

detailed [39]. Dark-field electron microscopy allows a demonstration of sectorization. The gold-decoration technique proposed by Spit [40] has been used for the electron microscopic study of the nature of the fold surface of PE single crystals [41, 42]. The sensitive decoration technique involving the vacuum deposition of thermally degraded polyethylene onto the crystal surface now allows one to obtain additional and very accurate information on the nature of sectorization [43–45].

Cracks traversing the [110] fold planes contain fibrils, although no fibrils are observed in cracks parallel to [110] folds [34]. Crystals bounded by four [110] faces with truncating [100] faces are obtained as the concentration or crystallization temperature is raised [46–49]. The [100] faces become longer than the [110] faces in crystals grown at high enough temperature. The upper limit of the crystallization temperature range can be extended by using poorer solvents such as alcohols, esters, and alkanes. From crystals grown from n-alkane, mainly n-$C_{32}H_{66}$, Keith [50] observed that the axial ratio of the crystals increased with crystallization temperature and with concentration, crystals becoming more elongated in the b direction. Crystals grown from low molecular weight PE have axial ratios higher than those grown from high molecular weight PE under the same conditions. Keith also obtained radially arranged aggregates of crystals whose structure can be compared to the radial crystalline units in spherulites grown from the melt. The findings of Keith with respect to crystallization temperature, concentration, and molecular weight were confirmed by Khoury and Boltz [51], who used mainly dodecanol and heptyl acetate as solvents to grow PE crystals at temperatures up to 120°C. Khoury pointed out the increasing curvature of the [100] faces as the axial ratio increases. Higher axial ratios are obtained with poorer solvents (heptyl acetate, for example) for a given crystallization temperature and molecular weight. An extensive study was recently carried out by Organ and colleagues [52–57] on PE crystals grown from poor solvents at temperatures up to 120°C, achieving an overlap with conditions used for melt growth (supercooling $\Delta T = T_d^0 - T_c$). The long spacing l of PE single crystals is given versus the crystallization temperature for various solvents in Figure 6. The conclusions of Organ and colleagues are the following:

1. The axial ratio increases with increasing crystallization temperatures for crystals grown from a given solvent.
2. The axial ratio behavior is roughly related to the crystallization temperature when crystals grown from different solvents are compared.
3. [100] crystal faces are curved and the degree of curvature increases with increasing crystallization temperature.
4. [100] crystal faces are straight at low crystallization temperatures but show an increasing slight curvature with increasing crystallization temperature.
5. At high temperatures, the crystals tend to grow in clusters, with many crystals growing from a common center in all directions.

The preferential growth of crystals in the b-crystallographic direction and the cluster morphology occurring at high crystallization temperatures present some similarities with spherulitic growth; however, even the most elongated single crystals appear to grow with constant shape, whereas the lamellae in spherulites grow radially out.

The determination of the chain conformation of polyethylene in single crystals is one of the most important problem in polymer physics and its knowledge is fundamental to obtain information about the conformation of a single macromolecule in the crystalline state. Several models have been proposed to describe the conformation of the chains in the

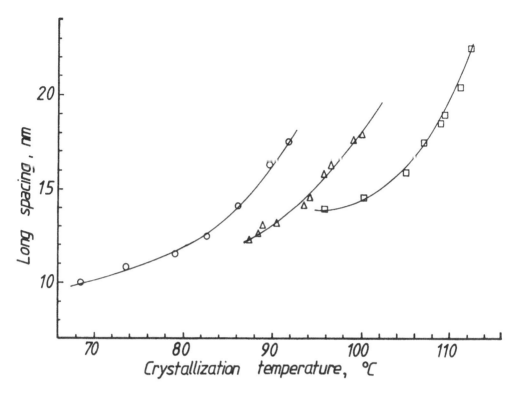

Figure 6 Long spacing *l* of mats of polyethylene single crystals grown from various solvents versus crystallization temperature T_c (○, *p*-xylene; △, octane; □, hexylacetate).

fold regions. The three main models are given in Figure 5*a*, the sharp fold model [32]; Figure 5*b*, the switchboard model [58]; and Figure 5*c*, the loose loop model [59]. From high-resolution ^{13}C nuclear magnetic resonance spectra of PE single crystals measured by the cross-polarization/magic angle spinning technique, Ando et al. [60] concluded that PE single crystals mainly contain sharply folded structures. Petraccone and colleagues [61, 62] performed calculations of potential energy minima for the possible conformations of folds and kinks in a polyethylene crystal. The conformations of minimum energy for tight folds in the (110) and (200) planes of polyethylene have energies of the order of 16.7 and 12.5 kJ/mol.

Twinned Crystals

Twinned polyethylene crystals having a lath shape were first observed by Keller and O'Connor [63] and Sella and Trillat [64]. A twinned crystal is a single entity composed of two (or more, in which case the edifice is called a multiple twin) identical single crystals joined macroscopically in a symmetrical fashion (Figure 7). The twin contains an added element of macroscopic symmetry with respect to whose component each possesses. Khoury and Padden [65] pointed out, with reference to the Dawson's data [66] on (110) twins of *n*-hectane ($C_{100}H_{202}$), two important differences between the (110) twins of paraffins and polyethylene. First, whereas in *n*-hectane the molecules are fully extended in the *c* axis of the lath, in PE twinned crystals the molecular chains are folded. Second,

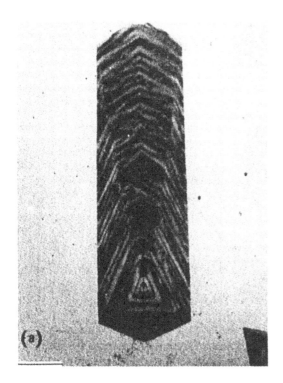

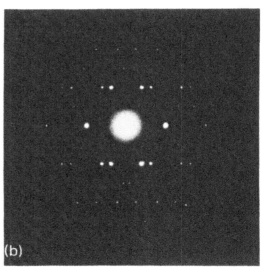

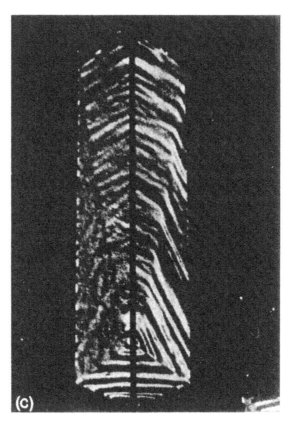

Figure 7A Typical solution (110) twinned crystal of polyethylene; the twin boundary is vertical. (*a*) Bright field electron micrograph (scale bar, 2 μm). (*b*) Associated electron diffraction pattern in corresponding orientation. (*c*) Dark field electron micrograph. 110 and 200 diffracted beams selected by aperture displacement. (Courtesy of J. C. Wittmann.)

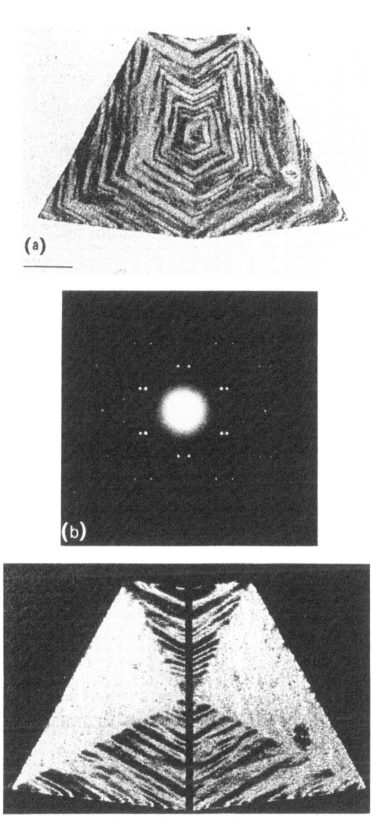

Figure 7B (*continued*) (310) twinned crystal of polyethylene; the twin boundary is vertical. (*a*) Bright field electron micrograph (scale bar 2 μm). (*b*) Corresponding electron diffraction pattern. (*c*) 110–110 twin-beam dark field electron micrograph. (Courtesy of J. C. Wittmann.)

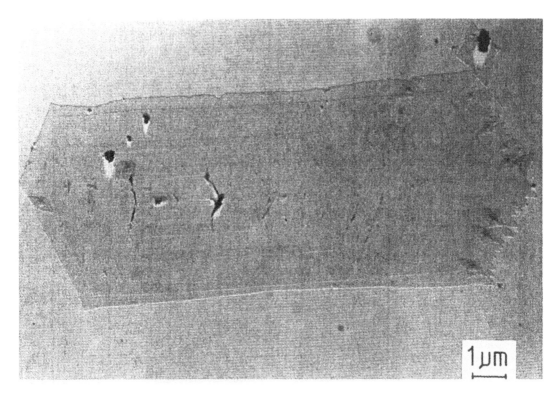

Figure 7C (*continued*) (110) twinned crystal of linear polyethylene (M_w = 17,000, M_w/M_n = crystallized from 1.287×10^{-3} wt % *p*-xylene and decorated following the isochronous method (T_{c1} = 89.0°C; T_{c2} = 83.2°C). (From Ref. 74.)

both *n*-hectane and PE laths exhibit a single direction of rapid growth, although no distinct reentrance at the rapid growing edge is obvious in PE. The justification of the overall lath-shaped appearance of the (110) twins of paraffins is known since Stranski [67] and Frank [68] made the observation that enhanced growth at a twin reentrant corner demonstrates the existence of a regime of nucleation control of growth: the nucleation barrier to growth on the crystal face is absent because a step always exists at the corner. An exhaustive description of the structure of PE twins can be found in the comprehensive studies of Wittmann [69] and Wittmann and Kovacs [70] and in the review of Khoury and Passaglia [71]. In his careful study of low molecular weight PE twins, Wittmann elucidates not only the structure of (110) reflection twins, but also that of numerous other twins made up of two or more crystals. The various filling sectors were clearly identified using dark- and bright-field electron microscopy. Boistelle and Aquilano [72] quantified the free energy gain associated with the nucleation in a reentrant corner by detailed calculations. Blundell [73] described (110) PE twins as objects growing with a constant shape, but other authors allude to progressive filling of the reentrant angle or refer to twins as objects with moving shape during their growth process. Recently Colet et al. [74] reported new morphological observations about (110) PE twinned crystals: the fast-growing tip of the laths presents, in addition to the two (100) facets usually observed, a small stable reentrant (110) corner.

This is a situation intermediate between the twin facies described by Dawson and Keller. The slow tip of the laths presents various degrees of asymmetry with respect to the junction plane. Sadler et al. [75, 76] recognized the importance of twin morphology for the kinetic theory of crystallization.

Crystal Growth from Concentrated Solutions: Hedrites and Axialites

Multilayer PE crystals can be found in very dilute solutions where monolayer crystals predominate. Multilayer crystals are often found in a splayed configuration where consecutive chain-folded layers are not in contact except in a small region [77]. The splaying multilayer character of the crystals increases continuously with xylene concentration up to about 0.3% and also beyond, for crystallization carried out below 75°C, without showing any fundamental change. At concentrations exceeding 0.3% and for crystallization temperatures higher than 75°C, a new habit with a much greater compactness, termed axialite, arises [78–81]. Transitional objects between single crystals and spherulites, called hedrites, are flat bodies. They can be polygonal, oval, or sheaflike, approximating two-dimensional spherulites. Hedrites might correspond to axialites, the development of which is restricted by the substrate. If hedrites grow with molecules normal to the substrate, polygonal forms would result as in crystals; if nucleation and growth are initiated by molecules lying along the substrate, the sheaf form would dominate. Dendritic growth in PE crystals [80, 81] occurs only when the crystal is growing into a highly supercooled melt or solution.

Hydrodynamically Induced Crystallization from Solution

In most stirred solutions, spontaneous nucleation takes place at a lower degree of supercooling than in the undisturbed solutions. The rate of crystallization of polymer molecules in solution highly increases upon introduction of a velocity gradient. Crystallization of linear PE carried out in a hydrodynamic field results in formation of fibrillar polymer crystals [82–87]. The very high molecular weight components of the polydisperse polymer form thin central threads of extended chain type which act as nuclei for transverse growth of folded chain lamellae. This fibrillar morphology, called a shish kebab, is a general polymer crystal structure encountered in oriented crystallization of polymers from solution as well as from the melt. Smooth ribbonlike PE crystals display striations arranged perpendicular to the fiber axis and spaced 50–100 nm apart. The thickness, diameter, and shape of the lamellar overgrowth depend on the crystallization condition. Small-angle x-ray scattering has been used to characterize various PE fibers of the shish kebab type [88]. The melting of such fibrillar PE crystals has been investigated extensively [89–95]. These shish kebabs show considerable shrinkage during melting as well as persistent birefringence and infrared dichroism [96] at temperatures far above the equilibrium melting temperature (145°C). The fibrous backbone and the lamellar overgrowth of high molecular weight PE are made of highly superheatable material. The backbone melting temperature at zero scan speed is found to be 148°C [95]. The stress–strain behavior of fibrillar crystals precipitated above 100°C from xylene is found to be essentially Hookean, and the maximum values determined for Young's modulus and tensile strength at break amounted to 27×10^4 kg/cm^2 and 10^4 kg/cm^2 respectively [96].

Ultra-High-Modulus PE Structure

It is recognized that the ultimate mechanical properties of PE, like those of polymers in general, depend on the molecular weight and its distribution and on the alignment and extension of the macromolecules. One of the primary objectives of studies devoted to the drawing behavior of melt-crystallized linear PE at elevated temperatures was to produce ultra-high-modulus material. Melt-spun fibers and solid-state extrusion have been used to orient high molecular weight PE but with a nonadvantageous manner due to the high viscosity of these high molecular weight materials, which causes fracture upon drawing [97–99]. Ultradrawing (i.e., drawing to ratios of 20–40) of PE has been used successfully to produce melt-spun fibers and solid-state extruded structures with Young's modulus up to 70 GPa [100–102], which compares with modulus of glass and aluminum. However, melt-crystallized polyethylene with a molecular weight exceeding 10^6 cannot be drawn effectively to ratios higher than about 5–10 [97–102], and it was of significance to orient very high molecular weight PE by elongation to similar high draw ratios to improve creep properties.

Smith and Lemstra reported a drastically enhanced effective drawability of high molecular weight that was spun [104, 105] or cast [106] from semidilute solutions to form macroscopic gels. PE gel fibers, produced by spinning of a 2% w/w solution of PE ($\bar{M}_w = 1.5 \times 10^6$) in decalin, can be drawn to a draw ratio of 30 at a temperature of 120°C and a strain rate of 1 sec^{-1}. As a comparison, melt-crystallized fibers of the same polymer sample cannot be elongated more than fivefold under the same drawing conditions [104]. Films "gelled" from a 1% w/w solution of high molecular weight PE in decalin, which were subsequently free from solvent, can readily be drawn to a draw ratio of 70 at 120°C; such highly drawn films show excellent axial molecular orientation and have a very high room temperature Young's modulus of 120 GPa. The effectiveness of hot drawing of high molecular weight PE gels is not influenced by the presence of solvent in the gel fibers [107]. The maximum draw ratio–drawing temperature curve exhibits a maximum for wet gels: at temperatures higher than 120°C, the maximum draw ratio attainable for wet gel fibers rapidly drops due to the lack of strength of the PE gels, which are dissolved at such high temperatures [107]. The influence of the initial PE volume fraction ϕ on the maximum draw ratio λ_{max} of the dried films has been investigated in the temperature range 90–130°C; the experimental results are well described by the relation

$$k_{max} = \frac{\lambda'_{max}}{\sqrt{\phi}}$$

where λ'_{max} is the temperature-dependent maximum value of the draw ratio of the melt-crystallized film, which takes the value 9.5 at 130°C [108]. Using a volume fraction $\phi = 0.006$, calculated maximum draw ratio of 123 for solution-cast film at 130°C is obtained, nearly equal to the experimental value of 130.

Very highly extended drawn films ($\lambda = 130$) evidently exhibit a near-perfect c axis orientation in the draw direction but also a strong orientation of the b axis of the PE unit cell in the plane of the film and therefore an a axis orientation perpendicular to the film [109]. No double orientation of the crystalline lattice is observed in these 130 times drawn films of high molecular weight PE [110].

Crystallization from the Melt

Crystallization under Atmospheric Pressure

Crystallization of polyethylene from the melt can give many complex morphologies but most commonly it gives spherulites, i.e., spherical aggregates of crystals [111, 112]. The investigation of the morphology of PE crystallized from the melt essentially concerns the determination of the structure of spherulites and the mechanism of their growth. The size of

PE spherulites depends on the magnitude of the undercooling and is around 100 μm. Spherulites observed in a thin film of PE crystallized from the melt under the optical microscope with crossed polarizers are shown in Figure 8; in fact, such a preparation is a diametral section of three-dimensional spherulites. The shape of the boundaries between spherulites allows the distinction as to whether nucleation has occurred at different times or at the same time.

Polyethylene is a biaxial material having the following indices: $n_p = 1.51$, $n_m = 1.52$, $n_g = 1.57$. The small n_p, medium n_m, and large n_g indices are parallel to the crystallographic axes *a*, *b*, and *c* of the unit cell respectively. As twisting occurs along the *b* crystallographic axis, i.e., also n_m, it is easily shown that the angle between the optic axes is 18°. Rings with a double periodicity occur therefore in PE spherulites.

Microbeam x-ray diffraction studies [113] have been carried out on selected regions of large PE spherulites in order to obtain the molecular orientation with respect to the radial direction. In polyethylene spherulites where the radial direction is the *b* crystallographic axis, the molecular chains are normal to the radius. In morphological studies carried out by optical microscopy, the quantity of interest is the birefringence, defined as the difference between the refractive index for light polarized with its electric vector along the radial direction of the spherulite and the refractive index for light polarized with its electric vector normal to the radial direction. Bunn and Alcock [114] determined that PE spherulites were negatively birefringent. Each spherulite exhibits an extinction Maltese cross centered at its center with the arms of the Maltese cross oriented along the vibration directions of the polarizer and analyzer of the polarizing microscope. Concentric extinction rings with a regular double periodicity are also observed [115]. Morphological data obtained by x-ray diffraction and optical microscopy show quite unambiguously that the crystal orientation rotates around the radius *b* in progressing out from the center of the spherulite.

Fischer and colleagues [116] showed from electron micrographs that the optical banding corresponds to an alternation of regions with radial lamellae lying successively in and perpendicular to the plane of the diametral section of the spherulite. From their study of banded polyethylene spherulites, Keller and Sawada [117] concluded that the twisting airscrew units were built of "crystals of length comparable with one or half a turn, stacked in a roof tile fashion along the spherulite radius." Helicoidal twisting, however, has been questioned [118, 119]. The helicoidal twisting of lamellar crystals is achieved by interleaving sequences of curved lamellar units, with slightly different azimuthal orientation about the spherulitic radius: the length of these untwisted sequences is about one-fourth to one-third of the band period [120]. The ring spacing, i.e., the period of the twist, depends on the crystallization temperature being higher the higher is the temperature [111, 112, 121]: ring spacing varies from about 1 to some tens of micrometers for banded PE spherulites. Several authors [121–127] have proposed various mechanisms to explain twist but none of them can account convincingly for experimental data. Additional observations on more complex extinction pattern were carried out by Keith and Padden [128, 129], Price [130], and Keller [131] simultaneously. A phenomenological theory for the spherulitic crystallization has been developed by Keith and Padden [132–135]. These authors assume that a nucleus develops into a spherically symmetric structure with a fibrillar habit. As the fibrils must fill space to confer a uniform density, they must "branch." Extending arguments advanced for metallic alloys [137, 138], Keith and Padden suggest that fibrillation results from cellulation due to the segregation of "impurities" during the spherulitic growth; impurities should be low molecular weight PE, i.e., slow or noncrystallizable species. The presence at the solid–melt interface of an impurity layer of width δ causes cellulation. This characteristic

Figure 8 Optical micrograph of banded spherulites of polethylene photographed between crossed polarizers crystallized in nonisothermal conditions. (M. Dosière, unpublished results.)

length δ is equal to D/G where D is the diffusion coefficient of the segregating species and G the radial growth rate. Progress in the investigation of melt-crystallized morphologies has been severely limited by the inability to observe representative structures by electron microscopy. Recently a chlorosulfonation staining technique [140] and a permanganic etching technique [141, 142] have allowed morphological details in melt-crystallized polyolefins to be examined with much better resolution than previously attainable. Using their permanganic etching, Bassett et al. [142–147] showed that the width of lamellar crystals does not agree with the characteristic length δ originally suggested by Keith and Padden [132–136]. The spherulites form by the outward growth of "dominant lamellae" along the crystallographic b axis, the intervening spaces being subsequently filled by subsidiary lamellae. The phenomenological theory developed by Keith and Padden is not general and applies only to a restricted range of spherulitic morphologies. There is agreement only on the following facts for the spherulitic crystallizations [148, 149]: occurrance of segregation, ubiquitous branching and splaying of dominant lamellae to form the framework of the spherulites, existence of a characteristic length which is the separation of adjacent dominant lamellae and not their width (Figure 9).

Crystallization from the melt in a gradient temperature [150] or in contact with a surface acting as a nucleating substrate [151, 152] may result in a transcrystalline structure [153] with the growth axis of the crystals, the b axis in the case of polyethylene [154], parallel to the gradient or perpendicular to the substrate. The orientation of the other two crystallographic axes is at random in the plane perpendicular to the growth direction. A basic requirement for the formation of such a structure is the absence of heterogeneous nuclei, which would initiate a random nucleation in the supercooled melt and therefore interfere with the well-aligned crystal growth initiated by homogeneous nucleation in the cooler section of the melt or on the nucleating substrate. The high concentration of primary nuclei in the growth plane very soon limits the lateral growth of spherulites so that the subsequent growth is confined to the direction perpendicular to the plane of primary nuclei.

The crystallization from strained melt, for instance, in a blown film or in the jet during fiber spinning, produces a row-nucleated structure [155]. Linear nuclei containing more or less extended polyethylene chains are formed parallel to the strain direction. Secondary epitaxial nucleation on the surface of such linear row nuclei produces folded-chain lamellae oriented perpendicular to the strain direction. In such a case the uniaxial orientation of chain axes is in the strain direction with, in the perpendicular plane, a random orientation of the a and b axes. If the growing lamellae exhibit a helical twist, the chain orientation in the strain direction is very often replaced by the orientation of the b crystallographic axis, i.e., the axis of maximum growth rate perpendicular to the strain direction, and a more random orientation of the remaining two a and c axes with a maximum in the strain direction. Such a row-nucleated structure has parallel cylindrical spherulites as its basic supercrystalline element. The row-nucleated structure contains two types of crystals: a small fraction of fibrillar crystals with partially or fully extended chains and the normal type of folded-chain lamellae.

Crystallization under High Pressure

Since pressure increases the melting point of polyethylene and synthetic polymers generally by about 0.2 K/b, crystallization under pressure can be carried out at much higher crystallization temperatures than achieved at atmospheric pressure. The influence of hydrostatic pressure on the crystallization of polyethylene has been an area of interest for

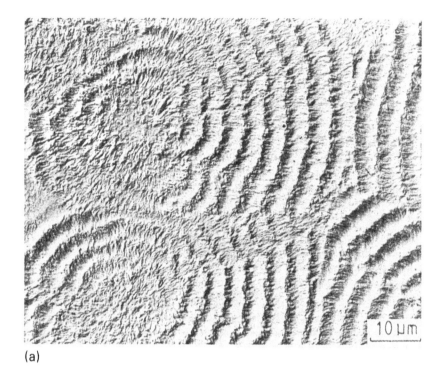

(a)

(b)

Figure 9 (*a*) Surface replica of banded spherulites revealed in a cut surface of linear polyethylene following permanganic etching. (*b*) Detail of (*a*). (Courtesy of D. C. Bassett.)

polymer scientists and engineers. In 1963 Wunderlich [156] used interference microscopy to characterize linear PE (Marlex 50) isothermally crystallized from 0.05% toluene solution under pressures ranging from atmospheric pressure to 6 kb: single crystals, twinned-crystals, dendrites, and high-temperature crystals are obtained. However, crystallization at such elevated pressures in a range of crystallization temperatures of 70–170°C does not markedly increase the crystal thickness.

A new type of morphology [157] was observed by electron microscopy of fracture surfaces of Marlex 50 crystallized from the melt under pressure higher than 2 kb: the chain axes are oriented along the fine striations shown in Figure 10; fractured lamellae shown in the micrograph are lamellae seen edge-on. The thickness of these lamellae is much larger than the thickness of chain-folded crystals grown at atmospheric pressure: thicknesses of 1 μm are frequently observed. These novel types of crystals have been termed extended-chain crystals [158]. Wunderlich and Davidson [159] defined an extended-chain (EC) crystal as "any crystal of a flexible high polymer having a chain length of over 200 nm in the molecular direction." These authors mentioned that such a length corresponds to molecular weights of over 20,000, which is of the order of the accepted lower limit of a high polymer. The term fully extended-chain (FEC) crystal was reserved for a high-polymer crystal whose length is equal to or greater than the length of the polymer chain. Table 1 lists crystallization conditions and some properties of high-pressure PE samples crystallized from the melt used by Geil et al. [157]. A sharp increase was observed in density at about 2.3 kb [160]; above this pressure, melt-crystallized linear PE is predominantly made of EC crystals. From melting and dissolution data on high-pressure crystallized PE, Mandelkern et al. [161] obtained a value of the interfacial free energy about two orders of magnitude higher than the value obtained for the same material crystallized at atmospheric pressure. The size of these extended crystals was questioned and the orientation of molecular chains normal rather than parallel to the striations was suggested by Mandelkern et al. [161]. Wunderlich [162] confirmed Geil's results and reaffirmed that PE crystallized under high pressure from the melt contains crystals whose thickness exceeds 1 μm; these EC crystals result from isothermal thickening of the preformed folded-chain crystals [162–164].

The thickness of the lamellae increases with crystallization temperature at a given pressure, with crystallization pressure at constant supercooling and with molecular weight at least in the range 20,000–50,000 for crystallization at the same pressure and temperature [166]. Weak reflections corresponding to the triclinic unit cell [15] have been observed in the wide angle x-ray patterns of linear PE crystallized at 4.36 and 5.07 kb [167, 168]. Jackson et al. [169] observed the process of crystallization of linear PE under pressures up to 10 kb by the use of optical microscopy: at 5–6 kb, PE crystallizes into shaped rods having a length of 10–20 μm and a diameter of 0.5–1 μm. Differential scanning calorimetry carried out on these rods showed that they were not EC crystals [170]. Several different mechanisms have been proposed for the formation of EC crystals from the melt under elevated pressure. Calvert and Uhlmann [171] observed extended-chain crystal PE to form directly from the melt under pressures up to about 5 kb both isothermally and in a temperature gradient. Following the first mechanism, crystals directly nucleate in the EC conformation and subsequently grow to larger thicknesses. In the second mechanism, the crystals could nucleate first to a folded-chain conformation and then thicken isothermally [160, 162–165]. Bassett and Turner indicated from studies of melting and crystallization of PE under elevated pressure that both processes occurred in two stages [172, 173]. It was extremely attractive to assume that chain-extended PE

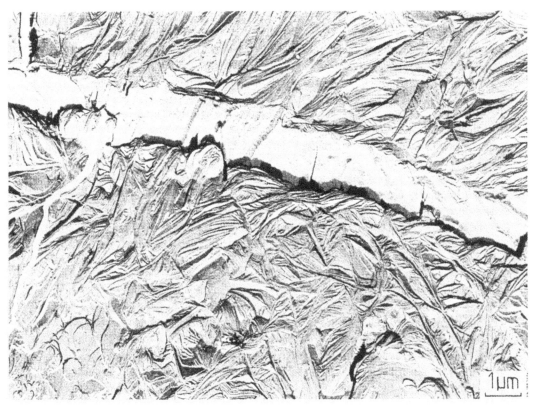

(a)

Figure 10 (*a*) Isolated thick lamella in a matrix of thin lamellae following crystallization of linear polyethylene from the melt at 2.6 kb. The former crystallized as the high-pressure (anabaric) phase, the matrix directly as the orthorhombic structure. Replica of a permanganic etched cut surface. (*b*) Fracture surface replica of a linear polyethylene crystallized from the melt at 5 kb. (Courtesy of D. C. Bassett.)

Table 1 Some Physical Properties of LPE Samples Crystallized under High Pressure

Crystallization conditions			Density measurements		X-ray measurements		Extrapolated melting point (DTA)/°C
Temp. (°C)	Pressure (atm)	Time (hr)	Density (25°C)	Crystallinity (%)	Crystallinity (%)	Long period (nm)	
130	1	8	0.9799	89	89	40	
130	1	8	0.9785	88			134.8
170.1	2000	8.5	0.9826	90	91	60	136.6
186	2760	8	0.9860	92	91	40	133.9
226	4800	8	0.9938	97	97	No discrete	140.1
236	5300	49	0.9921	96	93	Or diffuse scatter	139.9

Source: Ref. 157.

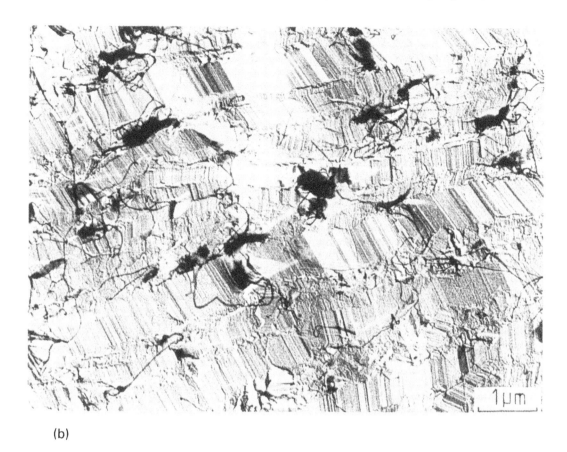

(b)

resulted from crystallization via an intermediate phase, in contrast to conventional chain-folded crystals, which formed directly from the melt. The existence of a new phase in PE at elevated temperature and pressure was experimentally shown some time later [174]; this hexagonal phase coexists with the conventional orthorhombic crystalline phase and liquid PE at approximately 220°C and 3.5 kb. At higher pressures, the orthorhombic phase transforms into the hexagonal phase before melting. Inversely, crystallization into the hexagonal phase occurs above 3 kb and at low undercooling; transformation to the orthorhombic phase takes place when the temperature is lowered. The intervention of the new hexagonal phase in extended-chain crystallization is now commonly accepted. Two processes must be distinguished for crystallization from the melt: chain-folded crystallization of orthorhombic phase, which predominates around atmospheric pressure, and chain-extended crystallization of the hexagonal phase.

Crystallization under elevated pressures in an Instron capillary [175–178] combines influence of hydrostatic pressure and shear and yields a high-modulus and highly transparent sample with unusually high orientation of the molecular axis along the flow axis. The Young's modulus of such a sample is given in Table 2 with that of other materials for comparison.

Large polyethylene crystals having a thickness of about 3–30 μm and width of about 100–600 μm have been obtained with unfractionated and fractionated high-density PE

Table 2 Young's Modulus of Some Materials

Material	Young's modulus $(dyn/cm^2 \times 10^{-10})$
Polyethylene (Porter et al.)	70
Aluminum alloys	70
Glass fibers	72
Asbestos fibers	160
Steel	200

Source: Ref. 178.

$(M_n = 50,000$ and $M_w/M_n = 2.7$ and 1.2 respectively) [179]. As the crystal thickness largely exceeds the average chain length (0.48 μm), the large crystals do not consist of lamellae of long and continuous chain molecules but rather one lamella of the large crystal consists of several straight-chain molecules stacked end to end along the thickness of the crystal [180]. Etching of extended-chain PE crystals allows separation of the polycrystalline aggregates into single crystals [181]. The effects of hydrostatic pressure on the mechanical behavior, compressibility, crystallization, and melting of polyethylene have been extensively reviewed by Pae and Bhateja [182, 183].

Miyashita et al. [184] isolated for the first time a single crystal of PE about 2 μm thick crystallized under high pressure. A bulk specimen composed of several bands stacked approximately parallel gives hexagonal symmetry in an x-ray oscillation photograph when the oscillation axis is along the end surface of the bands. Therefore, a thick single crystal in the hexagonal phase splits into several bands on the transition from the hexagonal phase to the orthorhombic phase.

Crosslinked Polyethylenes

Low- and high-density crosslinked polyethylenes are now commonly used in industrial fields. Indeed the irradiation processing of manufactured devices for dimensional stability as well as the interest in aging effects on polyethylene electrical insulation are of industrial importance. The effect of a crosslinked network on the crystallization process is of particular interest. Generally irradiation, often carried out in the solid state, has been used to produce crosslinks. High-energy radiation produces crosslinks essentially in the amorphous phase and in the fold surfaces [185], so that unhomogeneously crosslinked networks are produced. Charlesby et al. [186] and Mandelkern et al. [187] have investigated PE crosslinked by irradiation. Although details of the crosslinking process remain controversial, the overall effects on the morphology are well known.

Recent work on crosslinking by irradiation of linear PE has shown large differences between solid-state and melt irradiation [188]. Much of the literature on the influence of crosslinking on crystallization has been aimed at crystallization under stress on linear PE [189, 190].

Chemical crosslinking produces a lower chain scission than irradiation crosslinking. A few studies have been devoted to chemically crosslinking with ditertiary butylperoxide on linear PE [191–193]. The presence of crosslinks distorts the crystal structure of PE: small [188–194] to considerable [195] variations in unit cell parameters have been reported for crosslinked PE. Orth and Fischer [196] and Ungar and Keller [197] have observed a hexagonal phase in unextracted samples of high-density PE. An intensive morphological and crystallization kinetics investigation of chemically crosslinked low-density PE was carried out by Phillips and colleagues [198–200]. The boiling xylene method [195] appears the most effective among several methods of solvent extraction used for crosslinked PE. The sol fraction (extractable fraction) and its molecular weight, the molecular weight between crosslinks for the gel (estimated from swelling ratio in xylene at 100°C), decrease with increasing percentage of peroxide.

Values of $\sigma\sigma_e$ (1426–2295 erg²/cm⁴) comparable with those obtained for linear PE are obtained for crosslinked low-density PE and their gel fractions [198] when crystallization kinetics data are analyzed using regime III kinetics. Data for noncrosslinked low-density polyethylene and sol fractions give values of $\sigma\sigma_e$ in the range 1900–2040 erg²/cm⁴ when regime II is assumed. For low levels of crosslinking, Avrami exponents (see "The Overall Crystallization" section) indicate a sheaflike morphology which changes to fibrillar morphology as crosslink density is increased.

CRYSTALLIZATION THEORIES

Experimental Data About Polyethylene to Be Theoretically Explained

The prevailing theories of polymer crystal growth were developed to account for chain folding, variations of the lamellar thickness, and linear growth rates. If extensive experimental results are now available about crystallization of polyethylene, all data which are necessary to fully check theories of crystallization have unfortunately not always been collected. The experimental results can be summarized as follows:

1. All kinds of polyethylene crystallize from solution and from the melt in the form of lamella in which the macromolecules are folded.
2. The lamellar thickness is on the order of 10 nm. The lamellar thickness is experimentally determined by electron microscopy and small-angle x-ray scattering. The thickness of the crystalline core can be obtained from low acoustic mode vibration measurements by Raman spectroscopy. The lamellar thickness depends on the nature of the solvent and on the crystallization temperature T_c increasing with increasing temperature.
3. At a given crystallization temperature T_c, the number of orders observed in small-angle x-ray diffraction patterns indicate that the lamellar thickness of solution-grown crystals differs by less than 10% from crystal to crystal.
4. The development of a hollow pyramidal crystal structure during crystallization from dilute solution suggests that there is a sharply defined fold period in a given single crystal.
5. Discrete changes either up or down in crystallization temperature result in corresponding substeps in the growing lamella. This observation is supported by the isochronous decoration method recently developed for measuring linear rates in one single crystal during its growth.
6. The lamellar thickness increases if a lamella crystallized from solution or the melt is heated at temperatures above the original crystallization temperature.
7. The concentration dependence of the linear growth of a PE crystal may be described by the relation $G \sim C^\alpha$ where G is the linear growth rate measured along the normal to the growing faces (110) or (200), C is the concentration, and the concentration exponent α falls in the range 0.3–2.0.

The following questions relative to various aspects of chain folding have been discussed in the past but remain fundamental questions. Are the folds tight or loose in a lamella? To what extent does nonadjacent reentry manifest itself, if it does manifest at all? Is the fold period (the lamellar thickness) regular or not at a given crystallization temperature? Are the chain ends incorporated in the crystal or are they excluded, as cilia are at the level of the fold regions of the lamellae?

The Overall Crystallization

The overall crystallization rate can be computed from the knowledge of the linear growth rate and the primary nucleation rate. The expected number E of spherulites which start growing at time zero and reach a given point P at time t is analogous to Poisson's raindrop problem [201]. Kolmogoroff [202], Johnson and Mehl [203], Avrami [204, 205], and Evans [206] describe quantitatively macroscopic development of crystallinity with respect to nucleation and linear crystal growth. The classical equation describing the time dependence of the melt to solid transformation is derived from

$$1 - v_c = \exp(-Kt^n) = e^{-E}$$

where v_c is the volume fraction of the material crystallized by time t. The exponent n should be a whole number and contain contributions related to the crystal growth geometry and the time dependence of the nucleation rate, and K is the overall crystallization rate constant involving contributions from crystal growth and nucleation.

The Avrami equation has been applied to the spherulitic crystal growth in two and three dimensions for athermal and thermal nucleations (when nuclei are created sporadically in time and space, nucleation is called thermal nucleation).

Several simplifications have been made in the derivation of the Avrami equation.

1. The volume is assumed to stay constant on crystallization. Price [207] has discussed the effect of volume change on crystallization.
2. The linear growth rate does not keep a constant value with time in certain cases and the E value is therefore affected. This is more particularly the case of crystallization governed by transport processes.
3. The number of nuclei may not increase continuously.
4. The morphology of the crystals does not correspond to circular or spherical habit for the two- or three-dimensional approximation respectively. As is well known, semicrystalline PE samples crystallize with a lamellar or fibrillar morphology. The case of the fibrillar growth has been discussed by Morgan [208].
5. A two-stage crystallization can occur. This complication of the overall crystallization rate has been examined by Price [209] and Hillier [210].
6. Modification after crystallization can lead to more perfect crystals. Detailed investigation of secondary crystallization were realized by Majer [211] for low-pressure PE. As used by Wunderlich [201], the term secondary crystallization is often applied to all effects increasing the crystallinity after the primary crystallization described by an Avrami equation. The first, called secondary nucleation, involves further crystallization; the second, called crystal perfection, involves further perfection of initially poorly crystallized macromolecules. Several authors [212, 213] have noted that low-density PE shows crystallization isotherms which cannot be explained on the basis of a simple Avrami equation. Strobl et al. [213] analyze data of isothermal crystallization of branched PE starting from the Avrami equa-

Table 3 Avrami Equation

	Number of spherulites E	
Crystallization	Athermal nucleation	Thermal nucleation
Bidimensional	$\pi N G^2 t^2$	$\pi I^* G^2 t^3/3$
Tridimensional	$\frac{4}{3}\pi N G^3 t^3$	$\pi I^* G^3 t^4/9$

Key: I^* is the nucleation rate; N is the number of nuclei per unit area; G is the linear rate of crystal growth.
Source: Ref. 201.

Table 4 Exponents of Time in the Avrami Equation

Type of crystallization	Nucleation	n
Linear problem		
Line	Athermal	1
Line	Thermal	2
Two-dimensional problem		
Ribbon	Athermal	≤ 1
Ribbon	Thermal	≤ 2
Circular	Athermal	2
Circular	Thermal	3
Circular, diffusion control	Athermal	1
Circular, diffusion control	Thermal	2
Circular	Thermal, exhaustion	$3 \rightarrow 2$
Three-dimension problem		
Fibrillar	Athermal	≤ 1
Fibrillar	Thermal	≤ 2
Circular lamellar	Athermal	≤ 2
Circular lamellar	Thermal	≤ 3
Spherical	Athermal	3
Spherical	Thermal	4
Spherical, diffusion control	Athermal	3/2
Spherical, diffusion control	Thermal	5/2
Spherical	Thermal, exhaustion	$4 \rightarrow 3$
Two-stage	Athermal/thermal	Fractional
Branching fibrillar	Athermal/thermal	$1,2 \rightarrow$ large
Solid sheaf	Athermal	≥ 5
Solid sheaf	Thermal	≥ 6
Truncated sphere	Athermal	2–3
Truncated sphere	Thermal	3–4
Volume decrease on crystallization	Athermal/thermal	Fractional increase
Perfection after initial crystallization	Athermal/thermal	Decrease

Source: Ref. 201.

tion. Although a strict proportionality between small-angle scattering intensities and density changes was observed for all times, indicating a complete absence of crystal thickening and perfectioning processes, Strobl et al. found values for the Avrami coefficient which change with time and lie in the range between 1 and 2.5 rather than taking on the integral value $n = 3$ or 4 as would be expected for the growth of spherulites after athermal and thermal nucleation (Tables 3 and 4).

In conclusion, the Avrami equation is a convenient means to collect data of crystallization only when the microscopic mechanism of crystallization is unknown.

Thermodynamics Theories

In the few years following to the discovery of folded-chain polymer crystals, essentially two theories have been proposed. One, developed by Fischer, Peterlin, and Reinhold [214–217], suggested that the lamellar thickness was determined thermodynamically, corresponding to a minimum in the free enthalpy of the crystal at the crystallization temperature. They point out that a polymer lattice is different from the lattice of most low molecular weight organic materials and of metals, in which the binding forces are highly anisotropic. In a polyethylene crystal, the binding forces that oppose axial translation or rotation of one segment with regard to its neighbors are much less than the forces in the chain, due to primary valence bonds, which oppose bending, stretching, or partial lateral translation of the chain. Peterlin and colleagues examined the effect on the free energy density of a macromolecule (free energy per backbone atom) and its neighbors of longitudinal vibrations along the axis of a given chain and torsional oscillations. A minimum in the free energy density was found for some finite N^* value of the number of backbone atoms N in the segment. This N^* value, which depends on temperature, is thermodynamically stable and the length of the crystal will be restricted to it. This theory was not specific to chain folding, the most likely means of length restriction, in view of the known morphology of polyethylene. Reversible changes of the long spacing in bulk polyethylene as a function of temperature have been reported by O'Leary and Geil [218] from this point of view. Nevertheless, equilibrium theories were quickly abandoned because of the advantage of kinetic theories.

Homogeneous Nucleation

Nucleation theory frequently leads to an expression of the form [219]

$$\frac{\Delta G_i}{kT} = Ai^{2/3} - Bi \tag{1}$$

for the free enthalpy associated with the formation of a region of a new phase β in a parent phase α. In this equation i is the number of atoms or molecules in the transformed region; A is proportional to the interfacial free enthalpy per unit area of α–β interface, and B is proportional to the bulk free enthalpy difference between β and α in the absence of surfaces. The curve of $\Delta G_i/kT$ versus i passes through a maximum $\Delta G_i^*/kT = 4A^3/27B^2$ at $i^* = 2A/3B$ then decreases without limit [220]. Subcritical nuclei containing fewer i^* atoms require free enthalpy for further growth, while those containing more than i^* grow freely with decreasing free enthalpy.

On the basis of this nucleation theory and the theory of absolute reaction rates, Turnbull and Fisher [221] derived the following relation for the absolute rate of nucleation in condensed systems:

$$I^* = \left(\frac{NkT}{h}\right)\exp\left[-\frac{(\Delta G^* + \Delta G_\eta)}{kT}\right] \tag{2}$$

The rate of nucleation I^* is given in nuclei per second and refers to the number of uncrystallized elements N able to participate in nucleation by a single step. ΔG_η is the free enthalpy of activation for the short-range diffusion of molecules moving across the interface to join the lattice. ΔG^* stands for the free enthalpy of crystallization of a nucleus of critical size.

Equation 2 was obtained with the following assumptions:

1. A subcritical nucleus β_m is formed from m units α_1

$$m\alpha_1 \rightleftharpoons \beta_m$$

and the number of subcritical nuclei n_m is given by the Boltzmann distribution

$$n_m = N \exp\left(-\frac{\Delta G_m}{kT}\right) \tag{3}$$

2. Each nucleus grows by successive additions of units α_1 as shown by the generalized equation

$$\beta_i + \alpha_1 \rightleftharpoons \beta_{i+1} \qquad \text{where } i \geq m$$

The forward and reverse rates at each step are given by the following equations respectively according to the theory of absolute reaction rates:

$$I^+ = n_i ai^{2/3}\left(\frac{kT}{h}\right)\exp\left[\frac{(\Delta G_i - \Delta G_{i+1})/2 - \Delta G_\eta}{kT}\right] \tag{4}$$

$$I^- = n_{i+1} ai^{2/3}\left(\frac{kT}{h}\right)\exp\left[\frac{(\Delta G_i - \Delta G_{i+1})/2 - \Delta G_\eta}{kT}\right] \tag{5}$$

where $ai^{2/3}$ is the number of α elements in contact with β_i or β_{i+1} nuclei; n_i and n_{i+1} are the steady numbers of β_i and β_{i+1} nuclei. A β_{i+1} nucleus in a first approximation has the same surface as a β_i nucleus.

3. The difference between the free enthalpies of crystallization $(\Delta G_{i+1} - \Delta G_i)$ of nuclei made of $(i + 1)$ and i elements is very small compared to kT.

Let us consider y_i defined as follows:

$$y_i = n_i \cdot ai^{2/3}\left(\frac{kT}{h}\right)\exp\left[\frac{\Delta G_i}{kT} - \frac{\Delta G_\eta}{kT}\right] \tag{6}$$

as variables instead of n_i. The forward and backward rates can be written as

$$I^+ = y_i \exp\left[-\frac{(\Delta G_i + \Delta G_{i+1})}{2kT}\right] \tag{7}$$

$$I^- = y_{i+1} \exp\left[-\frac{(\Delta G_i + \Delta G_{i+1})}{2kT}\right] \tag{8}$$

Assuming steady state at every step of the growth, the net forward rate of nucleation I^* is, according to Turnbull and Fisher [221], given by

$$I^* = I^+ - I^-$$

Therefore

$$y_i = y_{i+1} + I^* \exp \frac{\Delta G_i + \Delta G_{i+1}}{2kT} \tag{9}$$

and

$$y_m = \lim_{i \to \infty} y_i + I^* \sum_{i=m} \frac{\Delta G_i + \Delta G_{i+1}}{2kT} \tag{10}$$

which leads to

$$I^* = \frac{n_m a m^{2/3} (kT/h) \exp[(\Delta G_m - \Delta G_\eta/kT)]}{\sum\limits_{i=m}^{\infty} \exp[(\Delta G_i + \Delta G_{i+1})/2kT]} \tag{11}$$

Using Eq. 3 giving the number of subcritical nuclei n_m, the net forward rate of nucleation I^* is

$$I^* = N \left(\frac{kT}{h}\right) a \left(\frac{A}{9\pi}\right)^{1/2} \left(\frac{m}{i^*}\right)^{2/3} \exp\left(-\frac{\Delta G^* + \Delta G_\eta}{kT}\right) \tag{12}$$

where the number of elements i^* in the critical nucleus is given by $2A/3B$, as previously defined. The rate of nucleation I^* depends on the affinity of crystallization of the critical nucleus ΔG^* on the temperature and on the dimensions of the subcritical and critical nuclei.

The enthalpy of formation of a nucleus made of i elements of quadratic cross section a^2 and height l is

$$\Delta G = 4al\sigma \sqrt{i} + 2ia^2\sigma_e - ia^2l \, \Delta g_m \tag{13}$$

The parameters of the critical nucleus are obtained by differentiation with respect to the dimensions a and l:

$$l^* = \frac{4\sigma_e}{\Delta g_m} \tag{14}$$

$$i^* = \left(\frac{4\sigma}{a\Delta g_m}\right)^2 \tag{15}$$

$$\Delta G^* = \frac{32\sigma^2\sigma_e}{\Delta g_m^2} = \frac{32\sigma^2\sigma_e T_m^{\circ \, 2}}{(\Delta H_m \Delta T)^2} \tag{16}$$

General homogeneous nucleation experiments have been carried out on polyethylene [222, 227] in order to obtain $\sigma^2\sigma_e$.

The fraction of droplets nucleated after time t in an isothermal experiment is given by

$$\frac{N}{N_0} = 1 - \exp(-I^*vt) \tag{17}$$

where v is the volume of the droplet ($v \simeq 10^{-11}$ cm^3) and I^* the nucleation rate given by Eq. 12. The half-life $t_{1/2}$ which is necessary to crystallize half of the droplets is given by

$$t_{1/2} = \frac{0.693}{I^*v} \tag{18}$$

By replacing I^* by its value given by Eq. 18, one obtains

$$\ln t_{1/2} = C + \frac{32\sigma^2\sigma_e T_m^{\circ 2}}{\Delta H_m^2 kT(\Delta T)^2} \tag{19}$$

The slope of the straight line obtained in a plot of $\ln t_{1/2}$ versus $1/[T(\Delta T)^2]$ allows one to obtain $\sigma^2\sigma_e$. Some experimental results are given in Table 5.

Values of $\sigma^2\sigma_e$ obtained from nucleation data show a wide spread and must be considered estimates, as indicated by Wunderlich [201]. Indeed, it is not clear that really homogeneous nucleation occurs during these droplet experiments, and according to Binsbergen [228–232] the shape of the nucleus assumed is too simple.

Polyethylene melts cannot be crystallized in an isothermally controlled manner below 110°C because crystallization is too fast owing to the presence of preexisting nuclei; thus Barham et al. [225–227] proposed a simple method for producing droplets enabling crystallization to be conducted isothermally down to 75°C. Small droplets of a solution of linear PE Sclair 2907 ($M_w = 53,600$, $M_n = 9300$) are sprayed onto warmed glass slides. Isolated patches of polyethylene obtained after evaporation of the solvent melt on heating to 160°C. The preparation then is cooled at constant rate until all droplets are fully crystallized. Barham et al. observed that 10–20% of droplets (fraction 1) crystallized around 125°C although the remainder of the preparation could be cooled to 70°C prior to crystallization. The first population of droplets crystallized at ~125°C because impurities remain even after cleansing of the polymer. These authors suggest that nucleation of the second population occurs preferentially at the polymer–substrate interface and is strongly affected by the nature of the interface in an unknown way. They define the nucleation temperature T_n as the temperature at which half of the droplets that have not crystallized at ~120°C have crystallized. Since the lowest values of T_n obtained by Barham (74.9°C) are considerably lower than T_n values taken from previous works (85°C), the experiments being carried out on the same substrate, these authors conclude that homogeneous nucleation has not been observed in polyethylene. At 80°C, the value of the growth rate reported by Barham et al. is around 2 m/s. Dupire [233] underlined that the two following main points must be taken into account during isothermal crystallization of polymers. First, coming from the seeding temperature T_s, the solution or the melt must reach the crystallization temperature without extensive nucleation. Second, the crystallization heat must flow at the interface between the just crystallized material and the melt or solution. As the coefficients of thermal conductivity of organic solvents and polyethylene are low, temperature at the liquid–solid interface can be very much higher than the crystallization

Table 5 Homogeneous Nucleation Data

Sample	T (°C)	ΔT (°C)	Slope	$\sigma_e\sigma^2$ (erg/cm^2)^2	Ref.
Linear PE	86	56		15,500	223
Linear PE	87	55	$5.24 10^7$	11,000	224
Linear PE	88	54		9,260	222
Branched PE	84	58		9,260	222

Source: Ref. 201.

temperature when the growth rate becomes larger than a critical value depending of the experimental conditions. The calculated distribution of temperatures near a PE crystal having a thickness of 10 nm growing from the melt when no heat dissipation occurs shows that the temperature rise reaches 8°C at 4 nm and 2°C at 60 nm from the crystallization interface. When heat dissipation at the interface is concerned and more particularly if the hypothesis of ideal conditions of thermostatization on one face of the crystal is made, the distribution of temperatures near the crystallization front is greatly reduced.

Kinetic Theories

Kinetic theories assume that the lamellar thickness of a crystal is that which corresponds to the fastest growth, although the crystal so obtained does not necessarily correspond to the most stable crystal that could have been grown. This review on polyethylene growth is devoted almost exclusively to a critical and historical examination of the theoretical work in this field almost completely due to Hoffman and Lauritzen's team. Hoffman and Lauritzen (H–L) have modified their first proposition several times [234, 235] to take account of new experimental data.

First Version of the Hoffman–Lauritzen Theory

The schematic model for surface nucleation and growth of chain-folded crystal is shown in Figure 11. As the crystallizing molecule puts down step elements corresponding to the first step, new surfaces are created: two new lateral surfaces and two folds having as area $2b_0l$ and $2a_0b_0$ respectively.* No new lateral surface is created as the crystallizing macromolecule puts down step elements corresponding to the adjacent stem.

The free enthalpies ΔG_1 and ΔG_v corresponding to the crystallization in *one step* of the first stem having a length l and an adjacent stem v respectively are given by

$$\Delta G_1 = 2b_0l\sigma + 2\epsilon a_0b_0\sigma_e - a_0b_0l\,\Delta g \tag{20}$$

and

$$\Delta G_v = 2a_0b_0\sigma_e - a_0b_0l\,\Delta g \tag{21}$$

where Δg is the free enthalpy of melting by unit volume of a crystal having infinite dimensions.

Until 1966, Hoffman and Lauritzen used for the variation of the free enthalpy corresponding to the crystallization of the first stem of length l the following relation obtained by setting $\epsilon = 1$ in Eq. 20:

$$\Delta G_1 = 2b_0l\sigma + 2a_0b_0\sigma_e - a_0b_0l\,\Delta g \tag{22}$$

The term $2b_0l\sigma - a_0b_0l\,\Delta g$ is the network for putting the first stem of the crystallizing macromolecule on the substrate. After a stem has attached itself as an initial nucleus on the substrate at a rate denoted i which has the units of stems in inverse seconds per centimeter, the nucleus then pulls other stems that attach to the substrate in an adjacent fashion, forming a surface patch with a spreading rate g cm/sec. The overall process

---

*a_0 and b_0 are used for the thickness and width of a stem respectively and must not be confused with the parameters of the orthorhombic unit cell. Moreover, growth planes in PE single crystals are parallel to (110) and (200) crystallographic planes, and therefore the growth rate vector G is along the normal to (110) or (200) planes.

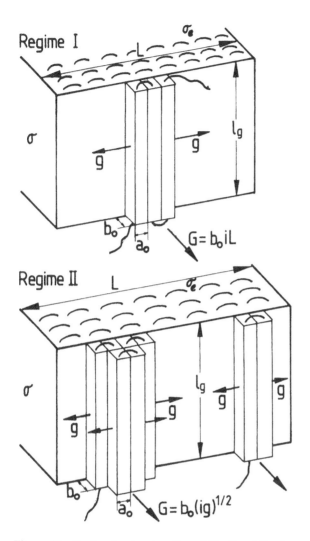

Figure 11 Regime I and regime II crystallization following surface nucleation on substrate having a length L. a_0 is the molecular width; b_0 is the layer thickness; σ is the lateral surface free enthalpy; σ_e is the fold surface free enthalpy. Each stem attaches at a rate denoted i. The nucleus pulls in other segments that attach to the substrate in an adjacent manner, forming a surface patch with a spreading rate g. The overall process causes the crystal to grow with a rate G. The mean lamellar thickness of the crystal is $l = l_g$.

causes the crystal to grow with a rate G cm/sec along the normal to the substrate (Figure 11). The free enthalpy of formation of such a surface patch of ν stems [and $(\nu - 1)$ folds] is

$$\Delta G_\nu = 2b_0 l \sigma - a_0 b_0 l \, \Delta g + (\nu - 1)a_0 b_0 (2\sigma_e - l \, \Delta g) \tag{23}$$

Equation 23 describes the barrier system seen in Figure 12, where the A's and B's represent the rate constants of the forward and backward reactions noted.

Three suggestions can be made to support the existence of $2a_0 b_0 \sigma_e$ in Eq. 22: (1) it

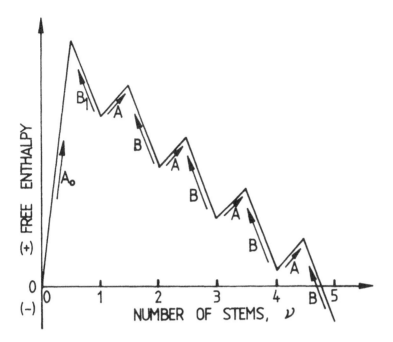

Figure 12 Barrier system showing rate constants of surface completion process. A_0 and B_1 are the forward and backward reaction constants of the first stem respectively ($\nu = 1$). A and B are the forward and backward reactions of the next stems ($\nu > 1$). ν is the number of stems.

accounts for the building of the fold during the deposit of the first stem; (2) it results from the existence of an uncrystallized chain end at one end of the first stem; and (3) it results from the necessity to fasten onto the substrate a molecule previously free to move in the solution or the melt. So A_0 and B_1 are the rate constants relative to the forward and backward reactions of spreading of the first step element on the substrate:

$$A_0 = \frac{kT}{h} \exp\left[-\frac{(\Delta G_\eta + \frac{1}{2}\Delta G_1)}{kT} \right] \tag{24a}$$

$$B_1 = \frac{kT}{h} \exp\left[-\frac{(\Delta G_\eta - \frac{1}{2}\Delta G_1)}{kT} \right] \tag{24b}$$

$$A = \frac{kT}{h} \exp\left[-\frac{(\Delta G_\eta + \frac{1}{2}\Delta G_\nu)}{kT} \right] \tag{24c}$$

$$B = \frac{kT}{h} \exp\left[-\frac{(\Delta G_\eta - \frac{1}{2}\Delta G_\nu)}{kT} \right] \tag{24d}$$

From the theory of absolute rates of Eyring, the ratio of these rate constants is

$$\frac{A_0}{B_1} = \exp\left(\frac{-\Delta G_1}{kT} \right) \tag{25}$$

where k is the Boltzmann constant.

The rate constants relative to the forward and backward reactions of spreading of an adjacent stem ν ($\nu > 1$) is similarly given by

$$\frac{A}{B} = \exp\left(\frac{-\Delta G_\nu}{kT}\right) \tag{26}$$

In the steady-state conditions, the net rate of formation of nuclei of length l is given by

$$S(l,T) = \frac{N_0 A_0 (A - B)}{A - B + B_1} \tag{27}$$

for all values of thickness l larger than a critical thickness l_c obtained by the condition that the free enthalpy is a maximum:

$$l_c = \frac{2\sigma_e}{\Delta g}$$

N_0 is the number of nuclei of one stem appearing in time unit on the substrate.

Assuming that $B_1 = B$, the flux can be expressed as

$$S(l,T) = N_0 A_0 \left(1 - \frac{B}{A}\right) \qquad \text{for } l > l_c \tag{29a}$$

and

$$S(l,T) = 0 \qquad \text{for } l < l_c \tag{29b}$$

The third term on the right-hand side in Eq. 22 increases with l. The limitation of the length of the stem is therefore obtained because folded-chain nuclei of very large thickness have a high lateral growth rate g but appear with a low frequency. The relation between the lamellar thickness versus the crystallization temperature was, for low degrees of undercooling,

$$l_g = \frac{2\sigma_e}{\Delta g} + \frac{kT}{b_0 \sigma} = \frac{2\sigma_e}{\Delta g} + \delta l \tag{30}$$

First Modifications of the Kinetic Constants

In 1966 Gornick and Hoffman [236] used the equations of the kinetic constants proposed earlier by Frank and Tosi [237]:

$$A_0 = \frac{kT}{h} \exp\left[-\frac{(\Delta G_\eta + \Delta G_1)}{kT}\right] \tag{31a}$$

$$B = B_1 = \frac{kT}{h} \exp\left(-\frac{\Delta G_\eta}{kT}\right) \tag{31b}$$

$$A = \frac{kT}{h} \exp\left[-\frac{\Delta G_\eta + \Delta G_\nu)}{kT}\right] \tag{31c}$$

and the lamellar thickness was given by

$$l_g = \frac{2\sigma_e}{\Delta g} + \frac{kT}{2b_0\sigma}\left(\frac{4\sigma/a_0 - \Delta g}{2\sigma/a_0 - \Delta g}\right) = \frac{2\sigma_e}{\Delta g} + \delta l \tag{32}$$

where l_g is the initial thickness of the crystal. As shown in Appendix 1, the variation of the free enthalpy of melting Δg can be approximated as

$$\Delta g \simeq \Delta h_m \frac{\Delta T}{T_m^\circ} f \tag{33}$$

where Δh_m is the melting heat by volume unit; T°_m is the equilibrium melting temperature, i.e., the melting temperature of a crystal having a very large thickness l; and $\Delta T = (T^\circ_m - T)$ is the degree of undercooling. For crystallization from dilute solution, the equilibrium dissolution temperature T°_d is substituted to T°_m. The f factor takes into account the variation of the melting heat Δh_m with temperature $[f = 2T/(T + T^\circ_m)]$. The "δl catastrophe" is obtained when $\Delta g = 2\sigma/a$. For polyethylene the value of the critical supercooling ΔT_c was 70 K using the values of the parameters given in the list of symbols.

In 1970 observable crystallization rates occurred between 120 and 131°C for LPE. The calculated value for the critical undercooling ΔT_c was therefore larger than the usual experimental undercooling (28–12°C). A hypothetical increase of the lamellar thickness for very large supercoolings always was plausible. As experimental and calculated data from Eq. 32 were in agreement, the H–L kinetics model for describing crystallization of linear polyethylene, and semicrystalline polymers more generally, seemed adequate.

This δl catastrophe, as it was referred to by Hoffman, gives the result that, for supercoolings higher than ΔT_c, ΔG_1 and therefore $[\partial(\Delta G_1)/\partial l]$ take negative values. As the rate constants of each step deposit $(A_0, A_1, \ldots)$ are increasing with l, there is no more limitation to the length of the stem.

On a molecular level, Gornick and Hoffman [236] considered the linear growth rate along the normal to (110) and (200) growing face as

$$G = b_0 iL \tag{34}$$

where L is the length of the substrate and i is the number of nucleation events per unit length and time on a growing face. They indicated also that

$$i = S(l_g, T) \tag{35}$$

The mean rate of formation of nuclei (expressed as number of nuclei per unit time) is obtained by integration of the flux $S(l, T)$ on all available l values:

$$S_T = \int_{l_c}^{\infty} S(l, T)\, dl \tag{36}$$

This allows one to obtain the linear growth rate G along the normal to the nucleation face:

$$G = \frac{kT}{h}\left\{\left[1 - \exp\left(\frac{a_0\Delta g}{\sigma}\right)\right]\exp\left(\frac{a_0\Delta g}{\sigma}\right)\right\}\exp\left(-\frac{\Delta G_\eta}{kT}\right)\exp\left(-\frac{4b_0\sigma\sigma_e T^\circ_m}{\Delta h_m \Delta T kT}\right) \tag{37}$$

As the factor between brackets was around unity, Gornick and Hoffman [236] proposed the following relation:

$$G = G_0 \exp\left(-\frac{\Delta G_\eta}{kT}\right)\exp\left(-\frac{4b_0\sigma\sigma_e T^\circ_m}{\Delta h_m kT \Delta T}\right) \tag{38}$$

or

$$G = G_0 \exp\left(-\frac{\Delta G_\eta}{kT}\right)\exp\left(-\frac{jb_0\sigma\sigma_e}{\Delta g kT}\right) \tag{39}$$

where G_0 includes all terms having a slow dependence versus temperature, j takes values in the range 2–4, depending on the type of growth regime, and

$$\Delta G_\eta = kT\frac{U^*}{R(T - T_\infty)} \tag{40}$$

where U^* is an energy taking into account the thermal variations on the rate of transport at the liquid–crystal interface; T_∞ is the temperature where the viscosity of the liquid becomes infinite; and R is the constant of perfect gas.

The values of the free enthalpy of lateral faces and fold ends obtained from experimental data available in 1966 for polyethylene were independent of the crystallization temperature and respectively equal to 10 and 60 ergs/cm².

The Adsorption Factor

To avoid the δl catastrophe, Lauritzen and colleagues [238, 239] suggested the following hypothesis. The free enthalpy balances relative to the deposit of the first stem on the one hand and the following stem on the other hand were in deficit and in excess respectively for the range of crystallization temperatures. This condition was filled by a composition of the crystallization free enthalpy by means of the ψ parameter, which value falls between 0 and 1. The ψ parameter took partially into account the adsorption of chains on the substrate.

The kinetics constants relative to the deposit and to the removal of the first stem were therefore the following:

$$A_0 = \beta \exp\left(\frac{-2b_0 l\sigma + \psi a_0 b_0 l\, \Delta g}{kT}\right) \tag{41}$$

$$B = B_1 = \beta \exp\left[\frac{(\psi - 1)a_0 b_0 l\, \Delta g}{kT}\right] \tag{42}$$

$$A = \beta \exp\left(\frac{-2a_0 b_0 \sigma_e + \psi a_0 b_0 l\, \Delta g}{kT}\right) \tag{43}$$

with

$$\beta = \left(\frac{kT}{h}\right) J_1 \exp\left[-\frac{U^*}{R(T - T_\infty)}\right] \tag{44a}$$

where the parameter J_1 takes into account of potential barriers not explicitly contained in $\exp[-U^*/R(T - T_\infty)]$. For dilute PE solutions, the following relation was used [238]:

$$\beta = C'\left(\frac{kT}{h}\right) J_1 \exp\left(-\frac{\Delta H^*}{RT} + \frac{\Delta S^*}{R}\right) \tag{44b}$$

where C' was a function of concentration whose analytical form was not given.

The flux $S(l,T)$ versus $\delta l = l - l_c$ is zero for $l = l_c$ and increases quickly with δl in order to reach a maximum value when

$$l_{max} = l_c + \frac{kT}{a_0 b_0 \Delta g} \ln\left[\frac{(\psi - 1)a_0 b_0 \Delta g - 2b_0 \sigma}{\psi a_0 b_0 \Delta g - 2b_0 \sigma}\right] \tag{45}$$

It then decreases to zero when δl is very large. The second term on the right side of Eq. 45 is relatively independent of T and is approximately equal to 1 nm.

The mean thickness of the lamellar crystal is

$$l = \langle l \rangle = l_c + \langle \delta l \rangle = \frac{2\sigma_e T_m^\circ}{\Delta h_m \Delta T} + \langle \delta l \rangle \tag{46}$$

where

$$\langle \delta l \rangle = \frac{\sum_{l=l_c}^{\infty} (l - l_c) S(l,T)}{\sum_{l=l_c}^{\infty} S(l,T)} \tag{47}$$

If l_u is the length of the monomer unit, $\langle \delta l \rangle$ is given by

$$\langle \delta l \rangle = \frac{1}{l_u} \frac{\int_{l_c}^{\infty} (l - l_c) S(l,T) \, dl}{\int_{l_c}^{\infty} S(l,T) \, dl} \tag{48}$$

In 1976 [227] the value of the mean lamellar thickness $\langle l \rangle$ was given by

$$\langle l \rangle = l_g = \frac{2\sigma_e}{\Delta g} + \frac{kT}{2b_0 \sigma} \frac{2 + (1 - 2\psi)a_0 \Delta g/2\sigma}{(1 - a_0 \psi \Delta g/2\sigma)[1 + a_0(1 - \psi)\Delta g/2\sigma]} \tag{49}$$

$\langle \delta l \rangle$ approximately equals $kT/b_0 \sigma$ and is not very sensitive to temperature. Therefore, Eq. 49 can be written in the following simplified form:

$$l_g = \frac{C}{\Delta T} + D \tag{50}$$

Both $\langle \delta l \rangle$ and l_g become infinite at the critical undercooling

$$\Delta T_c = \frac{2\sigma T_m^\circ}{a_0 \psi \Delta h_m} \tag{51}$$

The presence of the parameter ψ, of which the numerical value falls in the range 0–1, does not change the order of magnitude of $\langle \delta l \rangle$. The parameter ψ in the denominator of the relation giving ΔT_c allows one to take into account the extent of the experimental range of crystallization temperatures. To obtained calculated values of the lamellar thickness in perfect agreement with experimental data, the parameter ψ takes the well-defined value of 0.7 for a range of crystallization temperatures of 90 K.

Thermal Dependence of the Free Enthalpy of Fold Ends σ_e

The experimental range of crystallization temperatures becoming larger, it quickly appeared that the new experimental data obtained at higher undercoolings [227] could not be taken into account using Eq. 32. Indeed, the first term in Eq. 45 continuously decreases with increasing ΔT values, and such a decrease is not balanced by an increase of $\langle \delta l \rangle$ because $\langle \delta l \rangle$ is virtually independent of ΔT except in the neighborhood of the critical degree of supercooling ΔT_c. From lamellar thickness of single crystals grown from p-xylene solution, use of Eq. 50 gives $D = 5.5$ nm. However, a value of $\langle \delta l \rangle = 1$ nm is obtained from Eq. 49 using the usual values of various parameters for PE. To resolve this discrepancy Hoffman et al. [238] suggested a thermal dependence for the fold surface free energy σ_e:

$$\sigma_e = \sigma_{e0}(1 + y \, \Delta T) \tag{52}$$

In this case, Eq. 41 gave

$$D = \langle \delta l \rangle + \frac{2T_m^\circ y}{\Delta h_m} \sigma_{e0} \tag{53}$$

Following Hoffman et al. [238], Eq. 52 takes into account the roughness of the fold surface with respect to the degree of supercooling.

Regimes of Crystalline Growth

Regimes I and II. We first consider regime I and regime II or growth rate for the case where the formation of a surface nucleus is followed by rapid completion of the substrate (regime I) and where numerous nuclei form on substrate and spread slowly (regime II).

The plots of the growth rate of PE fractions having a mean molecular weight between 18,000 and 115,000 versus crystallization temperature show a slope discontinuity for an undercooling degree $\Delta T_b = 17.5 \pm 1$ K (Figure 13). The experimental data in a plot of $(\ln G - \ln \beta)$ versus $1/(T\Delta T)$ (Figure 14) are distributed on two lines whose slope values K_g are in the ratio $1:2$ [240]. Since such results cannot be explained from Eq. 38, Hoffman introduced two types of crystalline growth, regime I and regime II respectively.

For degrees of undercooling lower than ΔT_b, the main stage in the crystallization of a monomolecular row is determined by the deposit of a single secondary nucleus with a rate i; this stage is followed by the quick deposit of the macromolecule with a lateral growth rate g before a new nucleus can be deposited on the substrate. Crystalline growth follows a mononucleus regime called regime I. It clearly appears that the lateral growth rate g is much larger than i. It must be noted that i and g cannot be directly compared because i and g are a number of events by time and length unit and a number of events by time unit respectively. For regime I, the growth rate along the normal to the crystalline growth face is

$$G = b_0 i L \tag{54}$$

$b_0 i$ being the number of monomolecular layers appearing by time unit on the length L of the substrate. If in Eq. 54 L was the length of the growing face, the growth rate G should increase with the crystal size, a fact which is not supported by experiments. Hoffman and Lauritzen [239] proposed a model where the growing crystalline phase should be made of

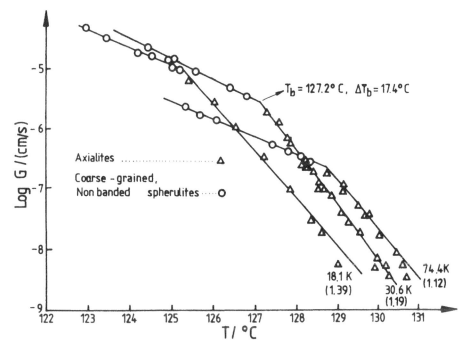

Figure 13 Log G (cm/sec) versus the temperature T_c for crystallization from the melt of intermediate molecular weight linear polyethylenes (18,100–115,000) having a sharp polydispersity. (From Ref. 240.)

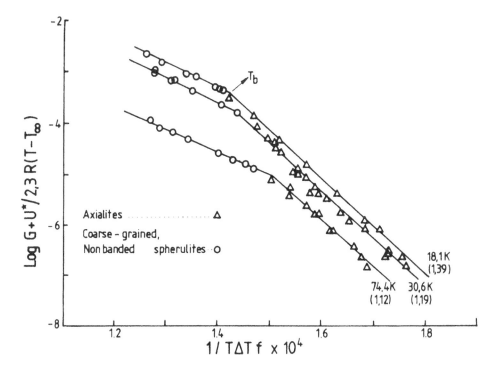

Figure 14 Log $G + U^*/2.3\,R(T - T_\infty)$ versus $(T\Delta T)^{-1}$ for linear polyethylenes of intermediate molecular weight having a sharp polydispersity (18,100–115,000). (From Ref. 240.)

small monolayered entities having a constant length L_p, called the persistence length. Crystal defects limit these sections: the value of L_p is around 1 μm. The growth rate therefore becomes

$$G = b_0 i L_p \tag{55}$$

and is independent of the size of the crystal.

The secondary nucleation rate i is now given by

$$i = \frac{S_T}{a_0 N_0} \tag{56}$$

and the resulting growth rate G is

$$G_{(1)} = G_{0(1)} \exp\left[-\frac{U^*}{R(T - T_\infty)}\right]\exp\left(-\frac{4b_0\sigma\sigma_e}{\Delta g kT}\right) \tag{57}$$

where $G_{0(1)}$ contains all terms having a very slow dependence versus crystallization temperature:

$$G_{0(1)} = b\left(\frac{kT}{h}\right) n_s \frac{J_1}{l_u}\left[\frac{kT}{2b_0 - a_0\beta_0\psi\,\Delta g}\right.$$

$$\left. - \frac{kT}{2b_0\sigma + (1 - \psi)a_0b_0\,\Delta g}\right]\exp\left(\frac{2a_0b_0\psi\sigma_e}{kT}\right) \tag{58}$$

where $n_s = L/a_0$.

For degrees of supercooling higher than ΔT_b, Hoffman et al. [240] consider that secondary nuclei appear with a high frequency i by length unit, when the lateral growth rate g (expressed as length unit versus time unit) of deposit of new layers on the substrate is low. This other regime, called regime II, is a multinuclei regime. More irregular crystalline faces must result from multinuclei growth following regime II.

The growth rate is given as

$$G = b_0 \sqrt{ig} \tag{59}$$

For a nucleus having a thickness l, the lateral growth rate is

$$g = a_0(A - B) \tag{60}$$

Substituting for l its mean value $l_g = l_c + \langle \delta l \rangle$, the growth rate G is

$$G_{(II)} = G_{0(II)} \exp\left[-\frac{U^*}{R(T - T_\infty)}\right] \exp\left[-\frac{2b_0\sigma\sigma_e}{\Delta g k T}\right] \tag{61}$$

where

$$G_{0(II)} = b_0 \left(\frac{kT}{h}\right) J_1 \exp\left[\frac{a_0 b_0 \sigma_e(2\psi - 1)}{kT}\right] \tag{62}$$

The estimate of Z, a dimensionless parameter defined by Lauritzen and co-workers [239, 241, 242] as

$$Z = \frac{iL^2}{4g} \tag{63}$$

allows a choice between both regimes: the regime is mononucleus if $Z < 0.1$ (regime I) and polynucleus if $Z \gg 1$ (regime II).

Regime III. Not all experimental data can be explained in terms of regime I or regime II. More particularly, crystalline growth of high molecular weight fractions of polyethylene operates according to a regime lying between regimes I and II and called regime III.

Hoffman [243] recently showed the existence of a second abrupt change of slope in the growth rate versus crystallization temperature for crystallization of polyethylene and polyoxymethylene from the melt at relatively low crystallization temperature. When the slope of the growth rate falls off with lowering temperature at the I $\rightarrow$ II transition, the slope increases at the II $\rightarrow$ III transition. The regime II $\rightarrow$ regime III transition is predicted to occur at an undercooling degree $\Delta T \sim 23$ K. In regime II, multiple surface nuclei occur on the substrate because the nucleation rate i quickly increases with larger undercooling. The mean separation of these surface nuclei, called the niche separation, is given with suitable accuracy by the nucleation theory. Transition from regime II to regime III occurs when the niche separation in regime II approaches the width of a stem a_0. In regime III, the crystal growth rate G is controlled by the rate i of deposition of nuclei on the substrate rather than $\sqrt{i}$ as in regime II. Each nucleus depositing on the substrate possesses two niches that promote rapid completion of the substrate at a rate g. The resultant observable growth rate is

$$G_{III} = b_0 i L' = \left(\frac{C_{III}}{n}\right) \exp\left(\frac{-Q_D^*}{RT}\right) \exp\left[\frac{-K_g(III)}{T \Delta T}\right] \tag{64}$$

where b_0 is the layer thickness, i is the nucleation rate in nuclei per second per centimeter, L' is an "effective" substrate length $= n'_s \cdot a_0$ and

$$K_{g(III)} = \frac{4b_0\sigma\sigma_e T_m^\circ}{(\Delta h_m)k} \qquad (65)$$

$$C_{III} = K\rho_i(C_nP_0)\left(\frac{kT}{h}\right)b_0\tilde{Z}n'_s \qquad (66)$$

n'_s is the mean number of stems laid down in the niche adjacent to the newly nucleated stem and Q_D^* is the activation energy of the steady-state reptation process in the melt ($Q_D^* = 7000$ cal/mol for PE).

Regime III and regime I have identical nucleation exponents but the preexponential factors of both regimes are largely different. For PE, Hoffman suggests for the ratio of the preexponential factors $G_{0,I}/G_{0,III} \approx 80$ [243].

Some Criticisms of the Hoffman–Lauritzen Theory

General Survey

The mechanism of limitation of the length of the stems clearly results from the competition between the free enthalpies of edge and fold surfaces. This theory contains all that is necessary to relate quantitatively all the experimental data. So, for example, σ_e is the relation between l and crystallization temperature and appears to justify the quick decrease of the lamellar thickness versus the undercooling ΔT for crystallization temperatures near the melting or dissolution temperature. ψ defines the temperature range where experimental data have been collected.

Many restricted experimental data about the lamellar thickness or/and the growth rate versus crystallization temperature have been interpreted using the Hoffman–Lauritzen theory; the numerical values of the free enthalpy of end folds σ_e so obtained generally agree with expected values. A posteriori, this agreement is used as a proof of the solid foundation of the theory.

The classical model of Hoffman–Lauritzen implies the adjacent reentry of the crystallized chains, the mode of deposit of a stem, the dependence of σ_e versus temperature and molecular weight, the value of the theoretical crystallinity, the roughness of the lamellar surface, the meaning of growth regimes, the adsorption factor ψ, the uniformity of the length of each stem. A few of these hypotheses being interconnected, they are examined together.

Adjacent Reentry and Its Experimental Proof

The simplification of the mathematical formulation is generally invoked to justify the strictly adjacent reentry of the different stems for a given macromolecule. But in fact the adjacent reentry is a necessary condition for the coherence of the kinetic theory of Hoffman and Lauritzen. As shown here, this hypothesis is constraining. During the first stages of growth of the secondary nucleus, the free enthalpy of formation ΔG_ν reaches a maximum corresponding to the deposit of the first stem and to the building of new edge surfaces. The nucleus then grows by successive deposits of adjacent stems resulting in the folding of the macromolecule, and its free enthalpy progressively decreases, becoming zero for a ν^* value of the number of stems. For higher values of ν^*, the free enthalpy of formation of the nucleus takes negative values and the nucleus is finally stable (Figure 12).

The free enthalpy of formation ΔG_ν of a nucleus made of ν stems with strictly adjacent reentry and length l is given by

$$\Delta G_\nu = \nu(2a_0b_0\sigma_e - a_0b_0l\,\Delta g) + 2b_0l\sigma \tag{67}$$

The domain of stability is obtained when

$$\nu > \nu^* = \frac{2l\sigma}{a_0l\,\Delta g - 2a_0\sigma_e} \tag{68}$$

For a nucleus made of ν stems but with j nonadjacent reentries, the free enthalpy of formation $\Delta G_\nu'$ contains the supplementary positive part $j(2b_0l\sigma + 2a_0b_0\sigma_e')$ where σ_e' is the free enthalpy associated with the formation of a loose fold:

$$\Delta G_\nu' = \nu(2a_0b_0\sigma_e - a_0b_0l\,\Delta g) + 2(j + 1)b_0l\sigma + 2a_0b_0j\sigma_e' \tag{69}$$

The stability condition becomes

$$\nu'^* = \frac{(2j + 1)l\sigma + 2ja_0\sigma_e'}{a_0l\,\Delta g - 2a_0\sigma_e}j\nu^* \tag{70}$$

Equation 70 shows that the region of stability can never be reached if the number of nonadjacent reentries j is large. Figure 15 shows ν^* values versus the supercooling ΔT for PE crystallized from dilute solution of xylene (a) and from the melt (b). For low or moderate supercooling ΔT, the theory is adequate only if the adjacent reentry condition is fulfilled on large lengths $L^* = a\nu^*$. For example, for single crystals grown from xylene, curve a in Figure 15 gives $\nu^* = 190$. Before reaching a domain of thermodynamic

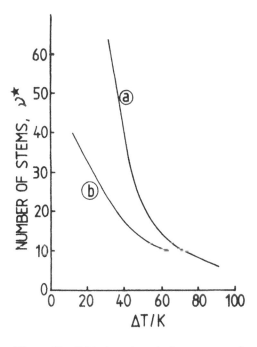

Figure 15 Critical number ν^* of stems versus the undercooling ΔT for linear polyethylene crystallized from xylene solution (curve a) and from the melt (curve b).

stability, the nucleus passes through a very large number of unstable states during which the probability of reorganization is high. The "normal" growth of such a crystal is uncertain and the hypothesis of persistence of the fold length seems not very likely.

Adjacent reentry has been discussed from experimental data obtained by neutron scattering and infrared spectroscopy obtained on mixtures of hydrogenated and deuterated polyethylene.

Krimm and co-workers [244–246] exploited the isotope effect to label one PE chain. As PE has two chains per unit cell, the CH_2 bending and rocking vibrations active in infrared are split by about 10 cm^{-1}. The deuterated PE (DPE) has a vibration frequency about $\sqrt{2}$ lower than hydrogenated PE (HPE). In an isotopic blend, only stems with the same isotopes and separated by $\langle \frac{1}{2}, \frac{1}{2}, 0 \rangle$ planes will interact so as to contribute to a splitting. A doublet is observed for solution-grown crystals, indicating that the stems are arranged with adjacent reentry along the crystallographic direction (110).

Krimm and Cheam [246] showed by infrared spectroscopy on single crystals grown from xylene around 70°C that the probability of adjacent reentry is nearly unity. However, the results of neutron scattering carried out on crystals grown from solution at 70°C seems to show that large sequences of adjacent reentry corresponding to $v > v^* = 30$ as implied by the classical kinetic theory are not of frequent occurrence [247, 248].

It must be emphasized that neutron scattering data do not give direct information about reentry. The basic challenge in dealing with a model that is used to predict a neutron scattering curve lies in simultaneously meeting the following criteria: (1) the observed radius of gyration from small-angle neutron scattering data should be predicted correctly; (2) the shape and absolute value of the observed scattering function $F_n(\mu)$ at moderate and high angles should be matched within experimental error; (3) the degree of crystallinity consistent with the model should match the experimental value; (4) the proposed model must be realistic, i.e., not show large density anomalies (the computed densities of the crystalline and amorphous regions must be as near as possible to the experimental estimates for densities ρ_a and ρ_c).

The interpretation of such neutron scattering data, however, is not unambiguous. Data of Schelten et al. [249] on a specimen where the deuterated species correspond to $n = 3750$ and crystallized at $\Delta T = 27.5 \pm 3.5$°C have been analyzed respectively by Flory [250, 251] and Hoffman [252] teams. Flory stated that adjacent reentry in melt-crystallized polyethylene is an improbable event. However, in the Monte Carlo simulation of Flory, a chain was allowed to emanate from the crystal into the interfacial layer, but an immediate return in a crystallographic adjacent position was not allowed. When a chain in random flight reached the boundary between the interfacial and amorphous zone, it was allowed to escape into the amorphous zone with a probability of 0.3 and was turned back toward the lamella of origin with a probability of 0.7. Using the "central core" model suggested by Roe [253], Guttman et al. [254], allowing adjacent and nonadjacent reentry, obtain a probability of adjacent reentry of 0.65.

Isotopic segregation occurring during crystallization with a low degree of supercooling does not allow the collection of correct data. However, permanganic etching on polyethylene crystallized from the melt at a small degree of supercooling permitted the Bassett group [255] to suggest that a regular folding of chain with adjacent reentry occurs in highly regular lamellar crystals. According to Hoffman [242], crystallization of PE fractions from the melt at temperatures near the melting temperature is described by a mononucleus growth regime. Crystallization with adjacent reentry leads to regular growth faces parallel to crystallographic planes with low Miller indexes. Conversely, crystallization from the melt at a low degree of supercooling, as carried by Labaig, gives crystals with curve-limiting faces [256].

Thermal Dependence of the End Fold Free Enthalpy σ_e

The melting temperature T_m (the dissolution temperature T_d) of crystals is a few degrees Centigrade higher than their crystallization temperature T_c. Such an experimental result implies that $\langle \delta l \rangle$, the difference between the thickness of a crystal built at the crystallization temperature T_c and the thickness l_c of a crystal of which the melting temperature is T_c, must take a large value. According to Mandelkern, it is impossible that a regularly folded chain nucleus forms and subsequently develops into a mature crystal with the same structure [257]. The structure of the crystallite is modified either to decrease σ_e or to increase its thickness l. Following the first hypothesis, the free enthalpy of the fold during the crystallization process (σ_e^*) is an unknown parameter which is larger than the fold free enthalpy in the mature crystal σ_e. Hoffman et al. estimated δl to be about 0.9 nm [242], which is less than 10% of the crystal thickness. Such a δl is much lower than the computed value from the Tamann equation

$$T_m = T_m^\circ \left(1 - \frac{2\sigma_e}{l\,\Delta h_m} \right) \tag{71}$$

written for a crystal of thickness $\langle l \rangle = l_c + \langle \delta l \rangle$ having grown at a degree of undercooling $\Delta T = T_m^\circ - T_c$

$$T_m = T_c + \Delta T \frac{\delta l}{l_c + \delta l} = T_c + \Delta T' \tag{72}$$

From data of Organ and Keller [57] relative to lamellar thickness and dissolution temperature of PE single crystals grown from xylene at 90°C, a value of 3.9 nm for $\langle \delta l \rangle$ is computed from the following measured data:

$l_{exp} = 16.4$ nm
$\Delta T'_{exp} = 4.7$ K, $\quad \langle \delta l \rangle = 3.9$ nm
$\Delta T'_{H\text{-}L} = 1.2$ K, $\quad \langle \delta l \rangle = 0.9$ nm

The stability of the crystal is therefore only theoretically predicted for temperatures near the crystallization temperature. Moreover, the empirical relation $l = C/\Delta T + D$ [50] gives D values much larger than 0.9 nm.

To outline these difficulties, Hoffman and Lauritzen suggested a thermal dependence for the fold surface interfacial free enthalpy:

$$\sigma_e = \sigma_{e0}(1 + y\Delta T) \tag{52}$$

where $y = 0.014$ [238].

Figure 16 shows the computed lamellar thickness l_g assuming the thermal dependence for σ_e given by Eq. 52. The variation of σ_e with temperature is almost completely responsible for the nondecrease of the lamellar thickness for a high degree of supercooling. The parameter y in Eq. 52 is not an elementary corrective parameter.

Equation 52 giving the thermal dependence of σ_e can be written as

$$\sigma_e = \sigma_{e0}(1 + 0.014) = \Delta H_e - T\,\Delta S_e \tag{73}$$

to show the enthalpic and entropic components respectively. Two comments can be made about Eq. 73. First, let us assume that the entropic term can be disregarded and the fold

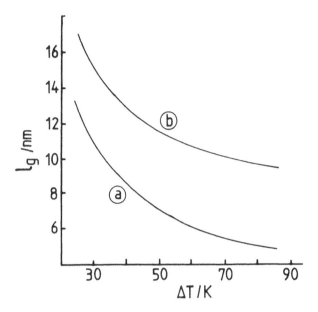

Figure 16 Computed lamellar thickness l assuming a thermal dependence for the fold surface free enthalpy σ_e. $y = 0$ (a); $y = 0.074$ (b).

enthalpy ΔH_e does not depend on the lamellar structure. Mandelkern et al. [258] derived the following general expression to describe the difference of enthalpy between the crystalline polymer and the melt for various types of morphology and interfacial structures that can evolve:

$$\Delta H^* = w_c \left(\Delta H_u - \frac{2\Delta H_e}{l} + a \, \Delta H_d \right) \tag{74}$$

where ΔH^* is the measured enthalpy of fusion per repeating unit; ΔH_u is the corresponding quantity for melting of a fully crystalline PE; ΔH_e is the enthalpy deficiency per sequence at each end of crystalline sequence, i.e., the interfacial enthalpy; ΔH_d is the enthalpic contribution from any defected structure within the interior of the crystal having a thickness l.

For a regularly structured interface, exemplified by regularly folded chains, Eq. 74 can therefore be rewritten as

$$\Delta H^* = \Delta H_u - \frac{2\Delta H_e}{l} \tag{75}$$

Figure 17 is a plot of ΔH^* versus $(1 + y\,\Delta T)/l$ for two values of the y parameter (0.014 and 0) for PE single crystals grown from xylene. For $y = 0$ (no thermal dependence of σ_e) the intercept, corresponding to $1/l = 0$, is 299 J/g, an extrapolation slightly higher than the generally accepted value of 292.8 J/g for ΔH_m° [250]. From the slope of the straight line ΔH_e is found to be 6550 J/mol of sequence, a value in agreement with the estimated ΔH_e values of 6300 J/mol of sequence for an interface made of regularly folded chains [251]. The straight line obtained from data relative to $y = 0.014$ (thermal dependence for σ_e) has the following equation:

$$\Delta H^* \text{ (J/g)} = 272 - \frac{402}{l(\text{nm})} \tag{75}$$

From the poor values of ΔH_m° (272 J/g) and ΔH_e (2800 J/mol CH$_2$) obtained by the least square linear fit, it seems justified to consider that ΔH_e does not depend on temperature.

Adsorption Factor

Although ψ is present in the expression of the crystal thickness l_g (Eq. 49), the "adsorption factor" has no real influence about the value of $\langle\delta l\rangle$. However, the presence of ψ in the denominator of the relation giving the critical degree of supercooling Δl_c (Eq. 51) defines the experimental range of crystallization temperatures. A value of 0.7 for PE gives a $\Delta T_{max} = 85°C$. Moreover, if one assumes that only a fraction ψ of the macromolecule is adsorbed on the substrate, the kinetic constant A_0 must not contain the totality of the term $2b_0 l/\sigma$. More generally, it is surprising that this adsorption factor keeps a constant value over the crystallization temperature range.

It seems that there is no agreement on the writing of the variation of free enthalpy occurring during the deposit of the first stem. Putting $\nu = 1$ in Eq. 67 leads to

$$\Delta G_1 = 2a_0 b_0 \sigma_e + 2b_0 l\sigma - a_0 b_0 l \,\Delta g \tag{22}$$

which was proposed by Gornick and Hoffman [236] and Frank and Tosi [237].

In 1969 Hoffman et al. [238] used the relation

$$\Delta G_1' = 2b_0 l\sigma - a_0 b_0 l \,\Delta g \tag{76}$$

which indicated that the building of the first fold occurs simultaneously with the deposit of the second stem. However, the term $2a_0 b_0 \sigma_e$ in Eq. 22 can take into account the free

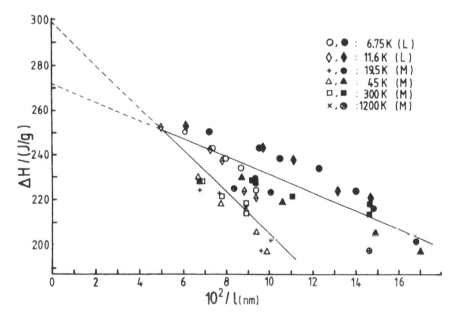

Figure 17 Melting enthalpy ΔH versus $(1 + y\Delta T)/l$ for single crystals of linear polyethylene grown from *p*-xylene. (Data from Refs. 258, 259, 272–274.)

enthalpy balance due to the setting of an uncrystallized part of the macromolecule (cilia); Eq. 76 should indicate that one end of the macromolecule can deposit on the substrate without creating a cilia. Let us examine the consequence of using Eq. 22 or 76 on the growth rate for regime II. From the conventional equations giving the linear growth rate G

$$G \div (S_T \cdot g)^{1/2} \tag{77}$$

and the spreading rate g, one obtains

$$g = a_0(A - B) = a_0\beta \exp\left[-\frac{2a_0b_0\sigma_e(1 - \psi)}{kT}\right]\left[\exp\frac{\psi a_0b_0\langle\delta l\rangle\,\Delta g}{kT}\right.$$
$$\left. - \exp\frac{(1 - \psi)a_0b_0\langle\delta l\rangle\,\Delta g}{kT}\right] \tag{78}$$

and using Eqs. 22 and 76, the following relations are respectively obtained:

$$\frac{G_1}{\beta} \div \left[\frac{kT}{2b_0\sigma - a_0b_0\psi\,\Delta g} - \frac{kT}{2b_0\sigma + (1 - \psi)a_0b_0\,\Delta g}\right]^{1/2} \exp\left[\frac{2a_0b_0\sigma_e(\psi - 1)}{kT}\right]$$
$$\exp\left(-\frac{2b_0\sigma\sigma_e}{\Delta g \cdot kT}\right)\left[\exp\left(\frac{\psi a_0b_0\langle\delta l\rangle\,\Delta g}{kT}\right) - \exp\frac{(1 - \psi)a_0b_0\langle\delta l\rangle\,\Delta g}{kT}\right]^{1/2} \tag{79}$$

$$\frac{G_1'}{\beta} \div \left[\frac{kT}{2a_0b_0\sigma - a_0b_0\psi\,\Delta g} - \frac{kT}{2b_0\sigma + (1 - \psi)a_0b_0\,\Delta g}\right]^{1/2} \exp\left[\frac{a_0b_0\sigma_e(2\psi - 1)}{kT}\right]$$
$$\exp\left(-\frac{2b_0\sigma\sigma_e}{\Delta g \cdot kT}\right)\left[\exp\frac{\psi ab\langle\delta l\rangle\,\Delta g}{kT} - \exp\frac{(1 - \psi)ab\langle\delta l\rangle\,\Delta g}{kT}\right]^{1/2} \tag{80}$$

Experimental data on the crystallization of linear polyethylene are available only in a small range of crystallization temperatures so that no deviation with respect to linearity in a plot of log (G/β) versus $(T\,\Delta T)^{-1}$ can be found for values of undercooling agreeing with regime II. Experimental kinetic data on the growth of polystyrene from the melt are available on a wider range of crystallization temperatures. Dupire [233] showed that for undercooling degrees larger than 30 K, i.e., in the range where crystallizations are experimentally performed for polystyrene, log (G/β) does not show a linear behavior versus $(T\,\Delta T)^{-1}$.

Depending on the temperature and on the "choice" of the analytic relation for A_0, G_0 cannot be considered a constant. As a consequence, an "objective estimation" of β and finally of $(\nu\nu_e)$ seems questionable. Let us finally remark that $G \exp [U*/R(T - T_\infty)]$ is obtained from theoretical considerations when only G values are obtained from experimental works. The numerical amplitude of the β semiempirical parameter seems to be the only way to obtain a linear behavior of (log G) versus $(T\Delta T)^{-1}$ in a wide range of crystallization temperatures.

Fluctuation of the Fold Period

One of the main criticisms of the kinetic model of Hoffman and Lauritzen is the uniformity of the length of each stem. Lauritzen and Passaglia [263] considered the deposit of a molecule on a smooth substrate by successive additions of stems of nonuniform lengths and formation of adjacent folds. Computed values of growth rates and mean lamellar thickness obtained with this assumption are similar to computed values obtained with the classical model characterized with stems of uniform length if the fold end free energy σ_e is replaced by σ_{eff}, which is an increasing function of the degree of undercooling. Such substitution explains the increase of roughness of the lamellar surfaces with the undercooling ΔT. The model of Lauritzen and Passaglia can take into account a slight fluctuation of the fold length. At the melting temperature (or the dissolution temperature), the standard deviation on the lamellar thickness is 0.8 nm for $\sigma = 10$ ergs/cm^2. The standard deviation is 3.5 nm for $\Delta T = 35$ K; it increases with the undercooling degree ΔT. Such a model, however, is not fully satisfactory because the limit value of the lamellar thickness is infinite when $\Delta T = \Delta T_c = \sigma T_m^\circ / a_0 \Delta H_m$. For this temperature, the kinetic constant relative to the addition of a stem of length l_j on the previous stem of length l_i is independent of l_j. The important fluctuations of thickness are as probable as the small ones and the lamellar thickness goes therefore to infinity.

Sequential Processes During Crystallization

Kinetic problems are concerned with the dynamics of the pathways between end points and are therefore orders of magnitude more complex than the corresponding equilibrium problems. To assign path probabilities and then to sum over the large number of paths is a very difficult and perhaps impossible task since the number of kinetics paths is infinite. Well-founded experimental results allow simplification of the theoretical work. Thus crystallization occurring in regime I, i.e., at small undercooling ΔT, is essentially two dimensional as the crystal grows as a thin lamella. The interface between the young crystal and liquid is sharp. Moreover, as the lamella grows by increase of its outer perimeter, the problem becomes essentially one dimensional. The macromolecules add on one at a time to the perimeter of the crystal. Even with such simplifying hypotheses, models for the attachment of a long stem of flexible molecular chain onto the crystallographic face of a growing PE crystal remain a complex process. The process was first noted by Frank and Tosi [237] in the early days of the development of kinetic theory. They made a clear distinction between two extreme academic situations. First, a range of configurations corresponding to the partial attachment of a stem is considered as belonging to a single stage. In the second situation, a stem attaches in successive segments called "flexibility units." The comparison of the height and the shape of the free enthalpy barriers was the main point of difference between these two situations. Whereas the successive fold segments are of the same length as the first on the growing crystal face, Frank and Tosi [237] show that there must be a large fluctuation in length from segment to segment around a characteristic segment length l^* or a stable mean strip width into which the macromolecule folds.

Other authors [264–267] have treated the problem of fluctuation of fold period. We emphasize the treatment of Point [265, 266], where the main idea of attachment of segments shorter than the full fold stem but longer than a monomer unit is kept. For Frank and Tosi [237], however, the objective of Point is not to consider the shape and the height of the free enthalpy barrier but to examine the opportunity for the macromolecule to fold back after each stage of deposition. Therefore, when the first stem of the secondary nucleus deposits on the substrate, the macromolecule does not anticipate the end length l of the full fold stem. The building of the secondary nucleus starts when a segment of length Δl comes into crystallographic register with the substrate.

The segment of the macromolecule folds back and starts to crystallize in a position adjacent to the first stem, or the crystallization of a further segment of length Δl without folding and along the direction of the preceding one occurs. Only the first stem has a choice of folding at every segment; subsequent stems are forced to fold with the same constant length l. As quoted by di Marzio and Guttman [267], such a process results in a cluster of stems in which each stem in a given cluster has the same length but different clusters have stems of different length. Numerical simulations have been carried out to model the case where stems in a given cluster have varied length [266]. Two main advantages result from the zippering process suggested by Point: the δl catastrophe may be avoided and the value of the variation of crystallization free enthalpy of the first stem is considerably reduced with respect to kT.

The treatment of Point has been analytically augmented by allowing each of the subsequent stems to fold at any stage in the zippering process rather than only during the building of the first stem. As quoted by di Marzio [267], such a problem is isomorphic to the mathematical problem of the growth of a Cayley tree with infinite branching. Two variations have been treated: (1) all subsequent stems have the same length, different from the length of the first stem; and (2) folding back of any stem can occur at any stage of zippering. The net result of these theoretical investigations is that classical nucleation theory remains valid at low and moderate undercoolings.

A Fundamental and Accurate Criticism of the Kinetic Theory of Hoffman et al.

As already emphasized, three extensive sets of experimental data are available on the crystallization of various sharp molecular weight fractions of poly(ethylene oxide) from the melt [268], linear polyethylene from dilute solution [269–276] and from the melt [240, 256]. In the first two systems, adjacent reentry can be reasonably assumed; precise values of the crystal thickness and extensive data on the growth rate of single crystals are available. Analyzing data of Kovacs relative to the growth rate of extended-chain poly-(ethylene oxide) crystals, Hoffman [277] makes the suggestion of the "presence in poly(ethylene oxide) of a mode of growth very different from that which occurs in the corresponding hydrocarbons." Analyzing data of Leung et al. [274], Hoffman [277] reached quite different conclusions despite the similarities of both sets of experimental data. Hoffman found a numerical agreement between experimental and predicted values of the linear growth rate G when extended-chain growth rate data are considered. Data of Leung et al. [274] relative to the polyethylene fraction denoted 3100 (M_n = 2900 and M_w/M_n = 1.07) have been also analyzed by Point [278]. For extended-chain growth rate data, Point calculates not only the linear growth rate G, but the initiation rate i, the rate of lateral spreading g, the kinetic length L_k, and the area A_n of the nucleation site for a full stem. These results are given here, being particularly critical for the Hoffman–Lauritzen theory:

$$L_k = 0.55 \text{ cm}; \quad i = 3.10^4 \text{ cm}^{-1}/\text{sec}; \quad g = 45.8 \text{ m/sec}; \quad A_n = 1.2 \ 10^{-7} \text{ cm}^2$$

As Leung et al. [274] found a decrease of the growth rate for crystals smaller than 1 μm, Point [278] concluded that either $L_k \ll 1$ μm and/or $L_p \ll 1$ μm. Indeed, it is admitted that the growth rate of crystals whose size is smaller than both the persistence length L_p and the kinetic length L_k cannot be a constant. The values of A_n and g are too small and too large respectively to be credible. The origin and these bad predictions result from an overestimation of the kinetic length $L_k = (2g/i)^{1/2}$ due to a factor representing the effect of the large free enthalpy barrier associated with the deposition of the first stem in a single step. The assumption that the secondary nucleus builds up following a zippering process [265–267] of macromolecular segments shorter than the full fold stem avoids accepting incredible numerical values for important theoretical parameters having a well-defined physical meaning.

The Concentration Dependence of the Linear Growth Rate of LPE Single Crystals from Solution

The dependence of the linear growth rate G on concentration in dilute solutions shows interesting and unexpected behavior; it may be described by the relation $G \sim C^\alpha$, where G is the linear growth rate, C is the concentration, and α is the concentration exponent. The principal finding of the first study of Blundell et al. [279, 280] was a surprisingly weak concentration dependence of the growth rate: the rate increases only with the cube root of the concentration for Marlex 6009. Following these authors "the $C^{1/3}$ factor has a special geometrical significance as it is inversely proportional to the average separation of molecular centers of macromolecules, hence will also be a measure of the number of macromolecules in contact with, or within a certain reach of the crystal face." Using a commercial PE with an average molecular weight about 58,000 (Shohex 6050), Seto and Mori [281] observed that, over a wide range of concentration C ($1.6 \times 10^{-2} < C < 5.00 \times 10^{-4}$ wt%), the growth rate is approximately proportional to the square root of the concentration. In the range of very low concentrations (6.25×10^{-5} to 5×10^{-4}), the growth rate becomes proportional to the concentration and the temperature where this change occurs gradually shifts to lower concentration as the crystallization temperature increases.

Keller and Pedemonte [282] followed up the preliminary work of Blundell et al. [280] using different molecular weight PE and two solvents (xylene and octane). The most extensive results [269–274] have been reported on very sharp fractions of LPE covering the molecular weight range from 3100 to 451,000: the polydispersity index of all these fractions is around 1.10 except for the highest molecular weight fraction where it is 1.18. The growth rate is found to be proportional to the concentration raised to a power less than unity in Manley's data; Keller and Pedemonte [282] have reported values as high as 2. The magnitude of the concentration exponent at a given crystallization temperature decreases as the molecular weight increases. At a given molecular weight, the concentration exponent increases as the crystallization temperature increases. These results can be satisfactorily accounted for by the Sanchez–di Marzio theory of self-nucleation [283, 284]. These authors have studied from a theoretical point of view the relation between the linear growth rate G and the concentration C. They introduce a self-nucleating mechanism where cilia, i.e., the portion of the chains which are not incorporated into the crystal and dangling in the solution, can nucleate on the next layer. When a macromolecule of finite length attaches to a next step from a solution, crystallization may start from anywhere in the chain and particularly from its middle. In this case the longer portion of the chain folds back easily and crystallizes. The shorter portion cannot crystallize on this layer and stays dangling in the solution, forming what is termed a primary cilia. Sanchez et al. showed that the concentration exponent α for crystalline growth governed by this molecular mechanism increases from 0 to 1 with decreasing molecular weight.

Unexpected interesting results concern the temperature variation of the linear growth rates for fractions in the molecular weight range of 3000–4000: the growth curves show a notched appearance in which two branches can be distinguished. These branches intersect at a transition temperature where the temperature coefficient discontinuously changes. It has been suggested that above and below these transition temperatures, the crystals are composed of n times and $(n + 1)$ times folded chains, respectively. However, the lamellar thickness shows a continuous variation with respect to crystallization temperature for both these fractions. For all fractions of low molecular weight (3100–11,000), a minimum appears in the plot of the linear growth rate versus molecular weight. No detailed interpretation of this phenomenon is yet available [274].

The isochronous decoration method (see Appendix 2) has been used to show quantitatively that the linear growth rate does not remain constant during the overall growth of a single crystal [285]. As the mean values of the lamellar thicknesses versus crystallization temperature are known, the concentration dependence of the linear growth rate G_{110} has been measured in one LPE single crystal. Two sharp fractions having a very low polydispersity index have been investigated ($M_w = 17,000$, $p = 1.11$; $M_w = 115,000$, $p = 1.15$). The values of the initial G linear growth rate agree with values reported by Cooper and Manley on similar samples [270]. However, the values of the instantaneous linear growth rate measured throughout the growth of the crystal show appreciable deviations with values expected from the concentration dependence reported by Cooper and Manley [269]: values of the concentration exponent ranging from 0.46 to 2.35 have been obtained [285]. The decrease of the growth rate with the concentration is much sharper for the LPE 17 K fraction than for the LPE 115 K. Although these LPE samples have a sharp molecular weight distribution, a segregation of macromolecules according to their molecular weight during crystallization is involved. Such a segregation has been reported for LPE 17 K single crystals grown at high crystallization temperatures [286].

Molecular Fractionation During Crystallization

Experimental evidence has been accumulated to indicate that the crystallization from solution [287–292] and from the melt [293–296] of polyethylene which is always polydispersed in molecular weight is accompanied by a molecular weight fractionation from the crystal phase of a part of macromolecules which are not able to crystallize under the given experimental conditions. From an experimental point of view, separation of the fractionated macromolecules from the crystals is easily effected by filtering at the crystallization temperature. Once this separation has been achieved, measurement of the molecular weights in the filtrate and crystalline phases allows the determination of the critical molecular length (Table 6).

Table 6 Variation of Critical Molecular Weight with
Temperature of Crystallization[a]

Crystallization temperature T_c (°C)	Critical molecular weight	Critical molecular length (nm)
50	1050	9.5
65	1300	11.8
70	1700	15.4
85	3800	34.5
90	7000	63.5
92.8	15000	136.1

[a]Data obtained on linear PE crystallized from xylene solution (0.5–0.7 wt %).
Source: Ref. 296.

Taking into account the experimental difficulties, fractionation from the melt has not been analyzed to any large degree; Mehta and Wunderlich [296] used extraction by dissolution in *p*-xylene to selectively remove the polymer fractionated during crystallization from the melt at atmospheric pressure and under elevated pressure (5 kb). Details of the fractionation behavior during crystallization from the melt were obtained from a quantitative analysis of the low-temperature melting peak from the differential scanning calorimetry (DSC) curves of a crystallized PE sample and from measurements of the viscosity average molecular weight of the fractionated samples by selective extraction. Table 7 presents experimental and calculated data for a linear PE sample ($\bar{M}_w = 153,000$, $\bar{M}_n = 8530$, $p = 18$).

The model used for equilibrium melting calculations is that of a homopolymer system with a broad molecular weight distribution showing complete segregation into crystals of different molecular weight [294]. In such equilibrium crystallizations, the lamellar thickness would be the molecular length of the crystallizable species; macromolecules shorter or longer than a given lamellar thickness would not be included on the crystal since they would introduce defects. The effect of the different species in the melt has been described using the following equation, a modification of the original Flory–Huggins equation [297]:

$$\frac{1}{T_m} - \frac{1}{T_m^\circ} = \frac{r}{\Delta H_p}[-\ln v_p + (\bar{x} - 1)(1 - v_p) - \bar{\chi} \times (1 - v_p)^2] \tag{81}$$

Table 7 Variation of Critical Molecular Weight with Crystallization Temperature for Crystallization from the Melt

Crystallization temperature T_c (°C)	Critical molecular weight			Critical molecular length (nm)		
	—[a]	—[b]	—[c]	—[a]	—[b]	—[c]
90			940			8.5
100	600		1,180	5.4		10.7
110	1,100		1,800	10.0		16.3
115	3,100		2,200	28.1		20.0
120	3,700		2,920	33.6		26.5
125	5,700	4,400	3,860	51.7	39.9	35.0
127	6,300	7,300	4,520	57.1	66.2	41.0
129	7,700	22,300[d]	5,240	69.8	202.3	47.5
238[e]		30,300			274.9	

[a]Data obtained from analysis of DSC curves.

[b]Data obtained from molecular weight analysis of extracted polymer.

[c]Data calculated from equation.

[d]The discrepancy between both estimations is due following Wunderlich to the good crystallization conditions (T_c = 129°C, t = 21 days) which increase the area of the main melting peak.

[e]Sample crystallized at 5.5 kb followed by slow cooling.

Source: Ref. 296.

where T_m is the temperature of the melting–crystallization equilibrium under consideration; T_m° is the equilibrium melting temperature of the considered molecular weight species; v_p and ΔH_p are the volume fraction and the molar melting heat respectively of the considered molecular weight species; $\bar{\chi}$ is the well-known interaction parameter; and $\bar{x}$ is an average molecular volume ratio of the crystallizing species to all other species, where

$$\bar{x} = \Sigma \, v_i x_i / \Sigma \, v_i \tag{82}$$

Comparison of solution and melt crystallization data shows that a shift of 36°C upwards of the solution crystallization data bring them into almost exact coincidence with the curve of melt crystallization. The experimental critical molecular lengths show a $1/\Delta T^2$ dependence on supercooling contrasting with the fold length dependence of solution-grown lamellar crystals. From Tables 6 and 7 it clearly appears that the lengths of the rejected species far exceed the initial fold period of the lamellae. According to Wunderlich [298], "A single traverse of which the length is the fold period is by far not a sufficient molecular length for a molecule to be accepted into the crystal. A multiple traverse length requirement exists." As molecular weight species are rejected from the substrate at temperatures below their respective equilibrium melting temperatures, Wunderlich suggests that each macromolecule has to go through a nucleation stage before being accepted into the growing crystal, a process termed molecular nucleation [298].

Model of Chain Folding

Model with Smooth Faces

The statistics of a random walk between two absorbing parallel planes allows one possible theoretical treatment of the amorphous regions between crystal cores in lamellar crystals of semicrystalline polymers. From the so-called gambler's ruin model and using random walks on a simple cubic lattice, Guttman et al. [299, 300] obtained the two following conclusions: many walks, initiated simultaneously at both faces, fill the amorphous domain of thickness l_a and the average lengths of such walks is $3l_a$. To keep an agreement between the experimental and calculated values of the amorphous density, the gambler's ruin model requires that only one-third of the stems originating from the crystal walk into the amorphous regions and the remaining two-thirds fold back sharply with strictly and nearly adjacent reentry [301]. Since Vonk [302] had argued that when walks are performed off-lattice, i.e., on the freely jointed chain, they have an average length equal to $2l_a$ rather than $3l_a$, the preceding results [299, 300] were confirmed by Mansfield et al. [303, 304].

Model with Rough Growth Faces

Curved as well as faceted PE crystals are currently observed [50–52], as reported earlier. As the crystallization kinetic theory and more particularly the secondary nucleation theory was conceived when all morphological observations showed facets on lamellar crystals, it is clearly necessary, according to Sadler and Gilmer [305, 306], to develop a more general approach which does not rely on facets and does not invoke surface nucleation. The Sadler model satisfies the following principal requirements: (1) applicability for both faceted and rough crystalline growth faces; (2) extrapolation of the lamellar thickness obeying a relation such as $l = C/\Delta T + \delta l$ (Eq. 50); (3) explanation of the acceleration speed of the crystalline growth front G such as $G \div \exp(-K_g/T_c \, \Delta T)$; (4) generality of the model. The Sadler model depends explicitly on the coiled conformations existing before crystallization in the uncrystallized material. A large number of processes and possible configurations are initially included and the use of Monte Carlo simulation techniques helps to identify the most frequent processes. The crystal is set up as a simple rectangular lattice and the energy of the crystal is calculated according to nearest neighbor interaction. The probabilities of addition of flexible units made of several monomers are independent of the nature of the site and of the temperature. The probability of removal is given by a Boltzmann weighting factor in which the energy is simply given by the sum of bond energies with nearest neighbors. Only a poor qualitative agreement is obtained between calculated values and experimental values of the following parameters: the growth rate G of the crystalline growth interface, the spreading rate g, and the lamellar thickness l.

Hoffman [307] proposed a model for the physical origin of the substrate length L that appears in the treatment of the regime I $\rightarrow$ II growth rate transition during crystallization from the melt. L is treated as a "persistence length" between defects that have the ability to inhibit substrate completion. These defects are cilia, loose loops, interlamellar links, and so on.

Unidimensional Kinetic Model for Crystalline Growth

As detailed versions of Frank kinetics have been constructed by Toda et al. [309, 310] and Point et al. [286], it appears necessary to review the kinetic equations describing traveling steps during crystallization on the substrate due to Frank [308] and Seto and Mori [281]:

$$\frac{\partial l}{\partial t} + \frac{\partial}{\partial x}(-gl) = i - 2glr \tag{83}$$

and

$$\frac{\partial r}{\partial t} + \frac{\partial}{\partial x}(+gr) = i - 2glr \tag{84}$$

where $l(x)$ and $r(x)$ are the density of steps at x which travel to the negative x direction (left) and the positive x direction (right) on the substrate of thickness L; i is the secondary nucleation per unit length and g is the spreading velocity assumed constant (crystallization of macromolecules of infinite chain length). The terms $-gl$ and gr are the flux of spreading steps to the left and right respectively. The term i represents generation of steps by secondary nucleation and $-2glr$ annihilation by running into each other. In the steady-state approximation, Eqs. 83 and 84 become

$$-\frac{dl}{dx} = \frac{dr}{dx} = \frac{i}{g} - 2lr \tag{85}$$

The linear growth rate G is given by

$$G = b_0(l + r)g \tag{86}$$

where b_0 is the thickness of a stem. One had $(l + r) \equiv 2c$ where c is a constant determined by Frank by solving Eq. 85 with given boundary conditions

$$c = x\, tgLx \tag{87}$$

where

$$x = \left(\frac{i}{2g} - c^2\right)^{1/2} \tag{88}$$

Frank estimates the growth rate at two extreme cases:

Regime I:

$$\frac{iL^2}{2g} \ll 1, \qquad G = biL\left(1 - \frac{iL^2}{3g} + \ldots\right) \tag{89}$$

Regime II:

$$\frac{iL^2}{2g} \gg 1, \qquad G = \sqrt{2ig}\left(1 - \frac{\pi^2 g}{7iL^2} + \ldots\right) \tag{90}$$

Toda et al. [309] resolve the case of finite molecular chain length, considering that a traveling step stops when the macromolecule terminates its crystallization and begins to start again when another macromolecule coming from the solution attaches to the step. Toda et al. suggest the following mechanism of cilia formation, which is different from the Sanchez–di Marzio model [311]. When a traveling step runs into another step which is traveling in the opposite direction, a pair of cilia is generated. Such a mechanism is valid for finite or infinite chain length. The linear growth rate G is derived as a function of the nucleation rate i, the velocity of steps, and the finite length of the substrate. The dependence of these microscopic rates as a function of concentration and supercooling is obtained later. In the case of finite molecular chain length, the asymptotic growth modes are expressed in regime I as

$$G \sim C \exp\left(\frac{-K}{T \, \Delta T}\right) \tag{91}$$

and in regime II as

$$G \sim C^{1/2} \exp\left(\frac{-K}{2T \, \Delta T}\right) \qquad \text{when} \quad \frac{hi'a^2}{g} \gg 1 \quad \text{or} \quad \frac{i(ha)^2}{2g} \gg 1 \tag{92}$$

$$G \sim C \exp\left(\frac{-K}{2T \, \Delta T}\right) \qquad \text{when} \quad \frac{hi'a^2}{g} \ll 1 \tag{93}$$

$$G \sim C \exp\left(\frac{-K}{T \, \Delta T}\right) \qquad \text{when} \quad \frac{i(ha)^2}{2g} \ll 1 \tag{94}$$

where $i'a$ is the rate of the attachment of macromolecules to an immobile step, a_0 is the width of a stem, and h is the ratio of the molecular chain length to the length of a stem.

One has also the following relations:

$$\frac{i'a_0^2}{g} \sim C \qquad \text{and} \qquad \frac{i}{2i'} \sim \exp\left(\frac{-K}{T \, \Delta T}\right) \tag{95}$$

Toda et al. [309] obtain the following relations for the case of cilia nucleation:

$$G \sim C \exp\left(\frac{-K}{T \, \Delta T}\right) \qquad \text{when} \quad \frac{iL^2}{2g}\left(1 + \frac{i''a_0}{iL}\right) \ll 1 \tag{96}$$

$$G \sim C^{1/2} \exp\left(\frac{-K}{2T \, \Delta T}\right) \qquad \text{when} \quad \frac{8ig}{27}(i''a_0)^2 \gg 1 \quad \text{and} \quad \frac{iL^2}{2g} \gg 1 \tag{97}$$

$$G \div C^{1/3} \exp\left(\frac{-2K}{3T \, \Delta T}\right) \qquad \text{when} \quad \frac{8ig}{27}(i''a_0)^2 \ll 1 \quad \text{and} \quad \frac{i^2L^3}{2gi''a} \gg 1 \tag{98}$$

where $i''a$ is the secondary nucleation rate of a pair of cilia and

$$K = \frac{4b_0 \sigma\sigma_e T_m^\circ}{k\Delta h_m} \tag{99}$$

The rate of attachment of a macromolecule to an immobile step i' is proportional to the concentration C as

$$i' = \mu' \exp\left(-\frac{G^*}{kT}\right) \tag{100}$$

where G^* is an activation free enthalpy. The secondary nucleation rate of cilia i'' depends on supercooling only:

$$i'' \sim \exp\left(\frac{-K}{T\,\Delta T}\right) \tag{101}$$

One has the following relations:

$$\frac{i''a^2}{2g} \sim \exp\left(\frac{-K}{T\,\Delta T}\right) \quad \text{and} \quad \frac{i}{i''} \sim C \tag{102}$$

Analyzing the extensive set of experimental data of Cooper and Manley [270, 271], Toda et al. [309] found a value about 0.5 for the concentration exponent α. For low molecular weight ($\simeq 10^4$), nucleation of cilia is neglected because each macromolecule only folds around 10 times and $\alpha = \frac{1}{2}$ (regime II). Data on relatively high molecular weight materials show a value of $\alpha = \frac{1}{3}$, which can be explained assuming only an effect of cilia, the effect of finiteness been ignored. Values of α calculated by Toda et al. from data of Cooper and Manley are listed in Table 8.

Recent Work on Crystallization of Polyethylene

Experimental Criterion for the Crystallization Regime in PE Single Crystals

As has been shown, the crystallization regime depends on three characteristic lengths [308, 312]: the persistence length L_p; the kinetic length $L_k = (2g/i)^{1/2}$; and a_0, the distance between adjacent stems crystallizing on the substrate. Although the currently

Table 8 Average Value of the Concentration Coefficient α and Values of $\sigma\sigma_e$ Calculated from Experimental Data of Cooper and Manley[a]

Molecular weight $\langle M_w \rangle$	15,700	25,200	61,600	83,900	195,000
$\sigma\sigma_e$ (1)	1335	1480	1590	1715	1765
$\sigma\sigma_e$ (2)	1001	1110	1192	1286	1323
α	0.52	0.42	0.38	0.36	0.24

[a]The value $\sigma\sigma_e$ (1) is obtained from Eq. 97 and $\sigma\sigma_e$ (2) from Eq. 98.
Source: Ref. 309.

accepted value of the persistence length L_p is around 1 μm, no pertinent proof of the validity of such an estimate was available until the recent proposition made by Hoffman [307]. For PE, numerical estimates of the kinetic length lead to values between 1 nm and 1 μm at supercooling between 17 and 30 K. As noted by Sanchez and di Marzio [311], the fundamental rate constants appearing in the kinetic equations are not known with sufficient precision, with regard to their functional forms or the values of the basic thermodynamic parameters. Since the kinetic length L_k plays a vital role in the description of the crystallization mechanism, it is critical to remove these uncertainties. Indeed, if L_p is larger than L_k, L_k is nearly equal to the average distance between two nuclei on the substrate ($L_k = \sqrt{g/i}$). Knowledge of G and L_k provides the value of g. Knowledge of g should make it possible to appreciate to the extent to which the folded structures are compatible with current ideas of mobility.

The isochronous decoration method [313] allows measurement of the quasi-instantaneous growth rate when the single crystals have very small dimensions. Point et al. [286] observed that the linear growth rate G did not increase with the size of the crystals for a sharp fraction of moderate molecular weight PE ($M_w = 17,000$, polydispersity $p = 1.11$). The kinetic theory predicts that the growth rate increases with crystal size as long as this latter dimension is smaller than the persistence length and/or the kinetic length. Therefore, both these characteristic lengths (L_p and L_k) are beyond the resolution limit of electron microscopy, and crystallization occurs in the polynucleated regime (in regime II, $G = b_0 \sqrt{2i/g}$). As experimental values of G range from 0.01 to 5×10^{-9} nm/sec and as a value lower than 20 nm for L_k ($= \sqrt{2g/i}$) very probably holds, an upper limit g_{max} is derived for the lateral spreading rate of a nucleus:

$$g = \frac{1}{2} \frac{GL_k}{b_0} < g_{max}$$

with g_{max} ranging from 0.2 nm/sec to 0.1 μm/sec. These values are much lower than previous estimates. This new estimate of g shows that the crystallizing macromolecules have adequate time to fold in a regular fashion during their growth.

To determine L_k and L_p, Point et al. [286], using the isochronous decoration method, observed that the growth rate of the crystals decreases as crystallization proceeds. Experimental measurements of the rate G during all the growth of the crystal cannot be explained quantitatively by the progressive exhaustion of the solution, a currently accepted qualitative explanation for the decrease of linear growth rate G. The fall in the growth rate is essentially due to molecular weight fractionation occurring to an unexpected extent even in sharp PE fraction.

New Morphological Observations on (110) Twinned PE Crystals

The importance of twin morphology for theories of crystallization of PE and polymers in general is well recognized [73]. Sadler et al. [75, 76] recently discussed the enhancement in growth due to reentrant corners.

New morphological observations about the growth of (110) twinned crystals of a sharp fraction of LPE ($M_w/M_n = 1.1$) of moderate molecular weight ($M_w = 17,000$) has been reported by Colet et al. [74]. The fast-growing tip of the laths presents, in addition to the two (100) facets usually observed, a small reentrant (110) corner. This is a situation intermediate between the facies described by Dawson [66] and Blundell and Keller [73]. The slow tip of these laths presents various degrees of asymmetry with respect to the junction plane. These morphological observations on (110) twinned crystals grown from dilute xylene solution can be explained by the introduction of a new characteristic length $L_n = j/i$ where j is the nucleation rate at a reentrant corner. Three linear growth rates G_{hkl} are calculated as a function of the length L of the growing crystallographic face (hkl): G_{hkl} and G_{hkl} (or G'_{hkl}) are the growth rates of a face bordered by two salient corners and by a reentrant corner respectively. A distinction between G_{hkl} and G'_{hkl} is introduced to take into account the relative sizes of the two faces of the reentrant dihedral corner. These authors also discussed the stability of the small (110) reentrant corner of the fast tip of the lath, the effects of dislocations incorporated in the fast edge of the laths, and the various asymmetries observed in the slow tip of the laths.

Evidence for Nucleation and Growth Processes in the Crystallization of PE Twins

The attempt made by Colet et al. [74] to obtain a value of the spreading rate g led to a precise upper limit for this parameter. In polynucleated growth, the linear growth rate G is given by $G = b_0\sqrt{2ig}$. The investigation of the dependence of growth rate of moderate molecular weight PE fractions with respect to the concentration of the dilute solution gave a concentration exponent α equal to 0.5. Toda et al. [309, 310] explained this result by assuming—without experimental proof, however—that the nucleation rate i is proportional to the concentration C and the spreading rate g is to a first approximation independent of it over the usual concentration range. As already emphasized, it is difficult to separate i and g from measurements of the linear growth rate. Toda and Kiho [314] have been successful by utilizing the enhanced growth on the (110) PE twins having reentrant corners when crystallizing from tetrachloroethylene, a better solvent than xylene for sharp fraction of moderate molecular weight. Since at the reentrant corner steps are generated without nucleation, the enhancement of growth rate on the surface whose end is a reentrant corner suggests that on the normal surface, the growth is controlled by nucleation. The net result is that the surface is inclined to the (110) face at an angle θ. The density of steps originating from the reentrant corner per unit length (c_t) is given by

$$c_t = \frac{\tan\theta}{b_0} = \frac{i_t \cdot a_0}{g} \tag{103}$$

$i_t \cdot a_0$ being the rate of step generation at the reentrant corner. The macroscopic linear growth rate on the reentrant corner is given by

$$G_t = b_0\sqrt{(i_t \cdot a_0)^2 + 2ig} \tag{104}$$

The supercooling dependence of the rate i, i_t, and g led to the conclusions that the slope of i is twice that of G and g is almost constant. The rate i_t slightly decreases as the supercooling is lowered. From Eq. 103, expected to give the supercooling dependence of the nucleation rate on a smooth surface,

$$i \sim \exp\left(\frac{-K}{T\,\Delta T}\right) \tag{105}$$

with

$$K = \frac{4b_0\sigma\sigma_e T_d^\circ}{k\,\Delta h_m} \tag{106}$$

a value of 1500 erg²/cm⁴ is obtained for the product $\sigma\sigma_e$.

The experimentally observed concentration dependence of the rates i and g is the opposite of what was expected [309]: the spreading rate g is proportional to concentration

C over a 10^{-5} to 5×10^{-2} wt % range of concentration and the nucleation rate i is constant over the usual concentration range but becomes proportional to concentration for concentrations lower than 10^{-4} wt %. Such a dependence of the traveling of steps g on the substrate seems therefore controlled by the volume diffusion in the solution. The dependence of the nucleation rate i with respect to the concentration suggests on the other hand that the surface of the growing interface is saturated with adsorbed molecules over the usual range of concentrations.

It is now generally accepted that the linear growth rate G increases with a fractional power of concentration C: $G \sim C^{1/2}$ for low molecular weight PE fractions and $G \sim C^{1/3}$ for higher ones, and molecular models involving ciliation have been invoked to explain the experimental data. A linear dependence of the growth rate with respect to the concentration should be explained in a simpler manner. Toda et al. [310] observed the crossover of concentration exponent α from $\frac{1}{2}$ in the usual concentration range (10^{-1} to 10^{-3} wt %) to 1 with decreasing concentration (10^{-3} to 10^{-6} wt %). Such behavior for the very dilute solution had not been reported because most of the previous experiments had been performed in the concentration range 10^{-1} to 10^{-3} wt %.

APPENDIX I. UPPER AND LOWER LIMITS FOR THE FREE ENTHALPY OF MELTING ΔG^*

The differentials of thermodynamic functions such as enthalpy H, entropy S, and free enthalpy G are given respectively by

$$dH = C_p \, dT + \left[V - T\left(\frac{\partial V}{\partial T}\right)_p \right] dp \tag{1}$$

$$dS = \frac{C_p}{T} \, dT - \left(\frac{\partial V}{\partial T}\right)_p dp \tag{2}$$

$$dG = -S \, dT + \frac{V}{dp} \tag{3}$$

At constant pressure, the difference between the melting heat at temperature T_m° and T, obtained from Eq. 1, is given by

$$\Delta H(T_m^\circ) - \Delta H(T) = \int_T^{T_m^\circ} \Delta C_p \, dT \tag{4}$$

where

$$\Delta C_p = C_{p,\text{liq}} - C_{p,\text{sol}} \tag{5}$$

$C_{p,\text{liq}}$ and $C_{p,\text{sol}}$ are the specific heat at constant pressure for the liquid and solid phases respectively.

In the same way, the difference between the entropy at temperatures T_m° and T, obtained from Eq. 2, is given by

$$\Delta S(T_m^\circ) - \Delta S(T) = \int_T^{T_m^\circ} \frac{\Delta C_p}{T} \, dt \tag{6}$$

---

*The discussion in this appendix follows Sanchez and di Marzio [284].

For $T < T_m^\circ$

$$T[\Delta S(T_m^\circ) - \Delta S(T)] \leq \int_T^{T_m^\circ} \Delta C_p \, dT = \Delta H(T_m^\circ) - \Delta H(T) \tag{7}$$

or

$$\Delta H(T) - T \, \Delta S(T) \leq \Delta H(T_m^\circ) - T \, \Delta S(T_m^\circ) \tag{8}$$

As

$$\Delta S(T_m^\circ) = \frac{\Delta H(T_m^\circ)}{T_m^\circ} \tag{9}$$

Eq. 8 becomes

$$\Delta H(T) - T \, \Delta S(T) \leq \Delta H(T_m^\circ) \times \left(1 - \frac{T}{T_m^\circ}\right) \tag{10}$$

i.e.,

$$\Delta G(T) \leq \Delta H(T_m^\circ) \frac{\Delta T}{T_m^\circ}$$

where

$$\Delta T = T_m^\circ - T$$

The lower limit for $\Delta G(T)$ is obtained from the inequality

$$\int_T^{T_m^\circ} \frac{T_m^\circ}{T} \Delta C_p \, dT \geq \int_T^{T_m^\circ} \Delta C_p \, dT \tag{11}$$

or

$$T_m^\circ [\Delta S(T_m^\circ) - \Delta S(T)] \geq \Delta H(T_m^\circ) - \Delta H(T) \tag{12}$$

which implies that

$$\Delta S(T) \leq \frac{\Delta H(T)}{T_m^\circ} \tag{13}$$

and therefore

$$\Delta H(T) - T \, \Delta S(T) \geq \Delta H(T) - \frac{T}{T_m^\circ} \Delta H(T) \tag{14}$$

and

$$\Delta G(T) \geq \Delta H(T) \frac{\Delta T}{T_m^\circ} \tag{15}$$

Equations 10 and 15 can be collected as

$$\Delta H(T) \frac{\Delta T}{T_m^\circ} \leq \Delta G \leq \Delta H(T_m^\circ) \frac{\Delta T}{T_m^\circ} \tag{16}$$

and following Sanchez and di Marzio [284] the error inherent in this approximation is given by

$$\frac{[\Delta H(T_m^\circ) - \Delta H(T)]\, \Delta T/T_m^\circ}{\Delta H(T_m^\circ)\, \Delta T/T_m^\circ} = 1 - \frac{\Delta H(T)}{\Delta H(T_m^\circ)} = \frac{\Delta C_p \cdot \Delta T}{\Delta H(T_m^\circ)} \tag{17}$$

Using values for ΔC_p and $\Delta H(T_m^\circ)$ for PE, the maximum error is given by $\Delta T/1000$.

APPENDIX 2. EXPERIMENTAL METHODS

Self-Seeding Technique

In the self-seeding method [263, 303], a dilute suspension of polyethylene crystals in a solvent is dissolved slowly to a temperature T_s a few degrees below the dissolution temperature in order to obtain a more or less important number of submicroscopic seeds. On cooling to the crystallization temperature T_c, these submicroscopic seeds act as primary nuclei for crystal growth. The growth rate of the crystals is uniform from the beginning of the growth and throughout the preparation so that at any time all the crystals have the same size. Each crystal is therefore characteristic of the whole preparation. As the seeding temperature increases, the number of seeds decreases and the ultimate crystal size increases [315]. The sizes of self-seeded PE crystals (0.01 % solution of Marlex 6009 in xylene) at 97 and 105°C after having grown at 80°C are 1 and 50 μm respectively [315]. After regular time intervals, a few drops of the dilute solution are quenched into a tube containing pure solvent at a temperature ~5°C lower than the crystallization temperature. The size of the crystal at the time of sampling is therefore labeled by the growth step resulting from the change in the lamellar thickness. This step is easily observed after the crystals have been metal-shadowed.

Isochronous Decoration Method

The published curves giving the size of different single crystals as a function of time and our own experience show that the derivative of this curve cannot be obtained with sufficient accuracy. Recently an isochronous decoration method to determine both the shape and the dimension of a single crystal or twinned crystals of polyethylene as a function of the time of crystallization was developed [285]. The basic features of the new experimental technique are found in the pioneering work of Bassett and Keller [35]: thicker or thinner borders resulted in single crystals of LPE whenever the temperature rose or fell during crystallization. We have used the property that the lamellar thickness is increased (or decreased) when the crystallization temperature is increased (or decreased). For crystallization temperatures higher than 80°C in *p*-xylene, a variation of 5 K in the crystallization temperature results in a modification of 2 nm in the lamellar thickness. Such a step, if it is sharp, can be observed by electron microscopy after conventional shadowing. Elementary calculations of heat transfer show that it is possible to change the temperature of a dilute polymer solution in a thin-walled glass tube having a radius of 1 mm in a few seconds. The crystallization device is fully described elsewhere [313]. A typical "isochronous decorated" LPE single crystal is seen in Figure 3*d*. The steps observed on the micrograph prove that the lamellar thickness is increased or decreased when the crystallization temperature is increased or decreased respectively. With this isochronous decoration method, it is not necessary to sample at given time intervals with a previously warmed teat pipette. All measurements can be made on the same single crystal and the quasi-instantaneous linear growth rate can be estimated during the crystal growth [313]. The exhausting of the mother solution can be quantitatively obtained.

Objections that can be made to the validity of this new method of measurement of the size of a single crystal during its growth include partial dissolution, reorganization, and nucleation on a substrate having a finite height. These have been discussed elsewhere [313].

ACKNOWLEDGMENTS

I wish to thank Dr. R. Andrew for assistance in revision of this manuscript. For the task of typing the manuscript and making the revisions and corrections, I would like to thank Mrs. Ch. Lemaire. Illustrations for this review were through the generous contribution of Prof. D. C. Bassett and Dr. J. C. Wittmann. I would like to thank Prof. J. D. Hoffman and Dr. H. D. Keith for their numerous and extensive comments. In addition, permission was generously granted by the copyright holders.

SYMBOLS, ABBREVIATIONS, AND PHYSICAL CONSTANTS

a, b, c	parameters of the unit cell
β	angle between a and c parameters of the unit cell
a_0	width of a stem (0.455 nm)
b_0	thickness of a stem (0.415 nm)
C	concentration
d_{hkl}	Bragg spacing between (hkl) crystallographic planes
G	free enthalpy or Gibbs energy
G_{hkl}	linear growth rate along the normal to an (hkl) crystallographic face
g	spreading rate of a new layer on the substrate
(hkl)	Miller indices of a crystallographic plane
H	enthalpy
Δh_m	melting enthalpy (2.80×10^9 ergs/cm^3 or 2.80×10^8 J/m^3)
i	nucleation rate on a smooth surface, i.e., the number of nucleation events occurring on the edge of a lamellar crystal per time unit and length unit
j	nucleation rate at a reentrant corner
J	Joule
k	Boltzmann constant (1.3805×10^{-23} J/K)
l	lamellar thickness; $l = l_c + l_a$
l_c	thickness of the crystalline core
l_a	thickness of the amorphous region
L	length of the growing crystallographic face (substrate)
L_k	kinetic length $(2g/i)^{1/2}$
L_p	persistence length, i.e., hypothetical length delimited by obstacles to growth such as boundaries of imperfect lattice coherence
l_g	initial thickness
p	pressure (1 standard atmosphere = 101325 Pa = 760 torr = 1.013 bar; 1 Gbar = 10^9 bar)
Pa	Pascal
Q_D^*	activation energy for the steady-state reptation process in the melt ($Q_D^* = 29.3$ KJ/mol)
R	gas constant (8.3143 J/K/mol)
S	entropy
σ	surface free energy of the edge surface [14.1 ± 1.7 ergs/cm^2 or $(14.1 \pm 1.7) 10^{-3}$ J/m^2]
σ_e	surface free energy of the folding surface [93 ± 8 ergs/cm^2 or $(93 \pm 8) 10^{-3}$ J/m^2]
ψ	adsorption factor (0.7)
T	absolute temperature

ΔT	undercooling degree, $\Delta T = T_m^\circ - T$ or $T_d^\circ - T$
T_m°	equilibrium melting temperature (416 K)
T_d°	equilibrium dissolution temperature
T_∞	temperature where the viscosity of the liquid becomes infinite
U^*	energy taking into account of the thermal variations on the rate of transport at the liquid–crystal interface
q	work of chain folding ($= 17.7$–21.1 kJ/mol)
y	thermal dependence parameter of the fold surface free enthalpy σ_e (0.014 K^{-1})

REFERENCES

1. R. A. V. Raff and K. W. Doak, *Crystalline Olefin Polymers*, Wiley, New York (1965).
2. W. J. Roedel, *J. Am. Chem. Soc.*, *75*: 6110 (1953).
3. P. R. Swan, *J. Polym. Sci.*, *56*: 409 (1962).
4. G. A. Natta and F. Danusso, *Stereoregular Polymers and Stereospecific Polymerizations*, Vols. 1 and 2, Pergamon Press, New York (1967).
5. J. Boor, *Ziegler–Natta Catalysts and Polymerizations*, Academic, New York (1979).
6. P. Pino and R. Mulhaup, *Angew. Chem.*, *19*: 857 (1980).
7. J. E. McGrath, *J. Chem. Educ.*, *58*: 844 (1981).
8. C. W. Bunn, *Trans. Faraday Soc.*, *39*: 482 (1939).
9. P. R. Swan, *J. Polym. Sci.*, *56*: 403 (1962).
10. Y. Kobayashi and A. Keller, *Polymer*, *11*: 114 (1970).
11. C. Nakafuku, *Polymer*, *19*: 149 (1978).
12. F. C. Frank, A. Keller, and A. O'Connor, *Phil. Mag.*, *8*: 64 (1958).
13. K. Tanaka, T. Seto, and T. Hara, *J. Phys. Soc. Jpn.*, *17*: 873 (1962).
14. T. Seto, T. Hara, and K. Tanaka, *Jpn. J. Appl. Phys.*, *7*: 31 (1968).
15. A. Turner-Jones, *J. Polym. Sci.*, *62*: 559 (1962).
16. D. W. McCall and W. P. Slichter, *J. Polym. Sci.*, *26*: 171 (1957).
17. R. Boistelle, in *Current Topics in Materials Science*, Vol. 4 (E. Kaldis, ed.), North Holland, Amsterdam, p. 413 (1980).
18. O. I. Paynter, D. J. Simmonds, and M. C. Whiting, *Chem. Commun.*, *1982*: 1165.
19. J. Bidd and M. C. Whiting, *Chem. Commun.*, *1985*: 543.
20. G. Ungar, J. Stejny, A. Keller, I. Bidd, and M. C. Whiting, *Science*, *229*: 386 (1985).
21. G. Ungar and A. Keller, *Polymer*, *27*: 1835 (1986).
22. A. Keller, in *Morphology of Polymers* (B. Sedlacek, ed.), Walter de Gruyter, Berlin, New York, pp. 1–26 (1986).
23. M. L. Huggins, *J. Phys. Chem.*, *46*: 151 (1942).
24. M. G. Broadhurst, *J. Res. Nat. Bur. Stand.*, *67A*: 233 (1963).
25. M. G. Broadhurst, *J. Chem. Phys.*, *36*: 2578 (1962).
26. P. J. Flory and A. Vrij, *J. Am. Chem. Soc.*, *85*: 3548 (1963).
27. M. G. Broadhurst, *J. Res. Nat. Bur. Stand.*, *70A*: 481 (1966).
28. B. Wunderlich and G. Czornyj, *Macromolecules*, *10*: 906 (1977).
29. K. H. Stoks, *J. Am. Chem. Soc.*, *60*: 1753 (1938).
30. O. V. R. Jacodine, *Nature*, *176*: 305 (1955).
31. P. H. Till, Jr., *J. Polym. Sci.*, *204*: 301 (1957).
32. A. Keller, *Phil. Mag.*, *2*: 1171 (1957).
33. E. W. Fischer, *Z. Naturforsch.*, *12a*: 753 (1957).
34. D. H. Reneker and P. H. Geil, *J. Appl. Phys.*, *31*: 1916 (1960).
35. D. C. Bassett and A. Keller, *Phil. Mag.*, *7*: 1553 (1962).
36. D. C. Bassett, F. C. Frank, and A. Keller, *Phil. Mag.*, *8*: 1753 (1963).
37. A. Keller and A. O'Connor, *Faraday Soc. Disc.*, *25*: 114 (1958).
38. D. C. Bassett, F. C. Frank, and A. Keller, *Phil. Mag.*, *8*: 1739 (1963).

39. D. C. Bassett, F. C. Frank, and A. Keller, *Nature, 184*: 810 (1959).
40. B. J. Spit, *J. Macromol. Sci., B2*: 45 (1968).
41. G. A. Bassett, D. J. Blundell, and A. Keller, *J. Macromol. Sci., B1*: 161 (1967).
42. D. J. Blundell and A. Keller, *J. Macromol. Sci., B7*: 253 (1973).
43. J. C. Wittmann and B. Lotz, *Makromol. Chem. Rapid. Commun., 3*: 733 (1982).
44. J. C. Wittmann and B. Lotz, *J. Polym. Sci. Phys. Ed., 23*: 205 (1985).
45. B. Lotz and J. C. Wittmann, *J. Microsc. Spectrosc. Electron., 10*: 209 (1985).
46. V. F. Holland and P. H. Lindenmeyer, *J. Polym. Sci., 57*: 589 (1962).
47. P. H. Lindenmeyer, *J. Polym. Sci., C1*: 5 (1963).
48. T. Kawai and A. Keller, *Phil. Mag., 11*: 1165 (1965).
49. B. Valenti and E. Pedemonte, *Makromol. Chem., 175*: 1917 (1974).
50. H. D. Keith, *J. Appl. Phys., 35*: 3115 (1964).
51. F. Khoury and L. H. Boltz, "Aspects and Implications of the Diversity in the Habits Exhibited by Polymer Crystals Grown from Solution," Proceedings of the 26th International Symposium on Macromolecules, IUPAC, Mainz, p. 1302 (1979).
52. S. J. Organ and A. Keller, *J. Mater. Sci., 20*: 1571 (1985).
53. S. J. Organ and A. Keller, *J. Mater. Sci., 20*: 1586 (1985).
54. S. J. Organ and A. Keller, *J. Mater. Sci., 20*: 1602 (1985).
55. S. J. Organ and A. Keller, *J. Polym. Sci. Polym. Phys. Ed., 24*: 2319 (1986).
56. P. J. Barham, R. A. Chivers, A. Keller, J. Martinez-Salazar, and S. J. Organ, *J. Mater. Sci., 20*: 1625 (1985).
57. S. J. Organ and A. Keller, *J. Polym. Sci. Polym. Let. Ed., 25*: 67 (1987).
58. P. J. Flory, *J. Am. Chem. Soc., 84*: 2857 (1962).
59. E. W. Fischer and R. Lorentz, *Kolloid Z. Z. Polym., 189*: 97 (1963).
60. I. Ando, T. Sorita, T. Yamanobe, T. Komoto, H. Sato, K. Deguchi, and M. Imanari, *Polymer, 26*: 1864 (1985).
61. P. Corradine, V. Petraccone, and G. Allegra, *Macromolecules, 4*: 770 (1971).
62. V. Petraccone, G. Allegra, and P. Corradini, *J. Polym. Sci., C38*: 419 (1972).
63. A. Keller and A. O'Connor, *Disc. Faraday Soc., 25*: 114 (1958).
64. C. Sella and J. J. Trillat, *Compt. Rend., 248*: 410 (1959).
65. F. Khoury and F. J. Padden, Jr., *J. Polym. Sci., 47*: 455 (1960).
66. I. M. Dawson, *Proc. Roy. Soc. London, A214*: 72 (1952).
67. I. Stranski, *Disc. Faraday Soc., 5*: 69 (1949).
68. F. C. Frank, *Disc. Faraday Soc., 5*: 186 (1949).
69. J. C. Wittmann, Thesis, University Louis Pasteur, Strasbourg (1971).
70. J. C. Wittmann and A. J. Kovacs, *Ber. Buns. Phys. Chem., 74*: 901 (1970).
71. F. Khoury and E. Passaglia, in *Treatise on Solid State Chemistry* (N. H. Hannay, ed.), Plenum, New York, p. 335 (1976).
72. R. Boistelle and D. Aquilano, *Acta Cryst., A34*: 406 (1978).
73. D. J. Blundell and A. Keller, *J. Macromol. Sci., B2*: 337 (1968).
74. M. C. Colet, J. J. Point, and M. Dosière, *J. Polym. Sci. Polym. Phys. Ed., 24*: 1183 (1986).
75. D. M. Sadler, *Polym. Commun., 25*: 196 (1984).
76. D. M. Sadler, M. Barber, G. Lark, and M. J. Hill, *Polymer, 27*: 25 (1986).
77. S. Mitsuhashi and A. Keller, *Polymer, 2*: 109 (1961).
78. D. C. Bassett, A. Keller, and S. Mitsuhashi, *J. Polym. Sci., A1*: 763 (1963).
79. P. H. Geil, in *Polymer Single Crystals*, Interscience, New York, p. 189 (1963).
80. P. H. Geil and D. H. Renecker, *J. Polym. Sci., 51*: 569 (1961).
81. B. Wunderlich and A. Mehta, *J. Mater. Sci., 5*: 248 (1970).
82. A. J. Pennings, *J. Polym. Sci., C16*: 1799 (1967).
83. A. J. Pennings and A. M. Kiel, *Kolloid Z. Z. Polym., 205*: 160 (1965).
84. A. J. Pennings, "Crystal Growth," Proceedings of the International Conference on Crystal Growth, Boston, p. 389 (1966).

85. A. J. Pennings, J. M. A. A. Van der Mark, and H. C. Booy, *Kolloid Z. Z. Polym.*, *236*: 99 (1970).
86. A. J. Pennings, J. M. A. A. Van der Mark, and A. M. Kiel, *Kolloid Z. Z. Polym.*, *237*: 336 (1970).
87. A. M. Rijke, J. T. Hunter, and R. D. Flanagan, *J. Polym. Sci.*, *A2, 9*: 531 (1971).
88. P. F. Van Hutten and A. J. Pennings, *J. Polym. Sci. Polym. Phys. Ed.*, *18*: 927 (1980).
89. B. Wunderlich, C. M. Cormier, A. Keller, and M. J. Machin, *J. Macromol. Sci.*, *B1*: 93 (1967).
90. T. W. Huseby and H. E. Bair, *J. Polym. Sci. Polym. Lett.*, *5*: 2651 (1967).
91. A. G. Wikjord and R. St. J. Manley, *J. Macromol. Sci.*, *B2*: 501 (1968).
92. J. Kavesh and J. M. Schultz, *J. Polym. Sci.*, *A2*: 255 (1970).
93. A. Peterlin and G. Meinel, *J. Appl. Phys.*, *36*: 3028 (1965).
94. E. W. Fischer, *Kolloid Z. Z. Polym.*, *231*: 458 (1969).
95. A. J. Pennings and J. M. A. A. Van der Mark, *Rheol. Acta, 10*: 174 (1971).
96. A. J. Pennings, C. J. H. Schouteten, and A. M. Kiel, *J. Polym. Sci.*, *C38*: 167 (1972).
97. C. Cappacio, T. A. Crompton, and I. M. Ward, *Polymer, 17*: 644 (1976).
98. S. B. Warner, *J. Polym. Sci. Polym. Phys. Ed.*, *16*: 2139 (1978).
99. A. E. Zachariades, M. P. C. Watts, T. Kanamoto, and R. S. Porter, *J. Polym. Sci. Polym. Let. Ed.*, *17*: 485 (1979).
100. G. Cappacio and I. M. Ward, *Polymer, 15*: 233 (1974).
101. P. J. Barham and A. Keller, *J. Mater. Sci.*, *11*: 27 (1976).
102. C. Jarecki and D. J. Meier, *Polymer, 20*: 1078 (1979).
103. G. Cappacio, A. G. Gibson, and I. M. Ward, in *Ultra-High Modulus Polymers* (A. Ciferri and I. M. Ward, ed.), Applied Science Publishers, London (1979).
104. P. Smith and P. J. Lemstra, *Makromol. Chem.*, *180*: 2986 (1979).
105. P. Smith and P. J. Lemstra, *J. Mater. Sci.*, *15*: 5050 (1980).
106. P. Smith and P. J. Lemstra, *Colloid Polym. Sci.*, *258*: 891 (1980).
107. P. Smith and P. J. Lemstra, *Polymer, 21*: 1341 (1980).
108. P. Smith, P. J. Lemstra, and H. C. Booij, *J. Polym. Sci. Polym. Phys. Ed.*, *19*: 877 (1981).
109. P. Smith, P. J. Lemstra, J. P. L. Pijpers, and A. M. Kich, *Colloid Polym. Sci.*, *259*: 1070 (1981).
110. Y. Ohde, H. Miyagi, and K. Asai, *Jpn. J. Appl. Phys.*, *10*: 171 (1971).
111. J. J. Point, *Bull. Acad. Roy. Belg.*, *41*: 982 (1955).
112. A. Keller, *J. Polym. Sci.*, *17*: 351 (1955).
113. Y. Fujiwara, *J. Appl. Polym. Sci.*, *4*: 10 (1960).
114. C. W. Bunn and T. C. Alcock, *Trans. Faraday Soc.*, *41*: 317 (1945).
115. A. Keller, *Nature, 169*: 913 (1952).
116. R. Eppe, E. W. Fischer, and M. A. Stuart, *J. Polym. Sci.*, *34*: 721 (1959).
117. A. Keller and S. Sawada, *Makromol. Chem.*, *74*: 190 (1964).
118. J. E. Breedon, J. F. Jackson, M. J. Marcinkowski, and M. E. Taylor, *J. Mater. Sci.*, *8*: 1071 (1973).
119. G. Kanig, *Kolloid Z.*, *251*: 782 (1973).
120. D. C. Bassett and A. M. Hodge, *Polymer, 19*: 469 (1978).
121. P. M. Lindenmeyer and V. F. Holland, *J. Appl. Phys.*, *35*: 55 (1964).
122. M. D. Keith and F. J. Padden, *J. Polym. Sci.*, *51*: S4 (1961).
123. J. D. Hoffman and J. I. Lauritzen, *J. Res. Nat. Bur. Stand.*, *65A*: 297 (1961).
124. J. R. Burns, *J. Polym. Sci.*, *A2, 7*: 593 (1969).
125. S. Sato and T. Seto, *Rep. Prog. Polym. Phys. Jpn.*, *12*: 161 (1969).
126. J. M. Schultz and D. R. Kinloch, *Polymer, 10*: 271 (1969).
127. P. D. Calvert and D. R. Ulhmann, *J. Polym. Sci. Polym. Phys. Ed.*, *11*: 457 (1973).
128. H. D. Keith and F. J. Padden, Jr., *J. Polym. Sci.*, *39*: 101 (1959).
129. H. D. Keith and F. J. Padden, Jr., *J. Polym. Sci.*, *39*: 123 (1959).

130. F. P. Price, *J. Polym. Sci.*, *39*: 139 (1959).
131. A. Keller, *J. Polym. Sci.*, *39*: 151 (1959).
132. H. D. Keith and F. J. Padden, Jr., *J. Appl. Phys.*, *34*: 2409 (1963).
133. H. D. Keith and F. J. Padden, Jr., *J. Appl. Phys.*, *35*: 1270 (1964).
134. H. D. Keith and F. J. Padden, Jr., *J. Appl. Phys.*, *35*: 1284 (1964).
135. H. D. Keith, *J. Polym. Sci.*, *2*: 4339 (1964).
136. H. D. Keith, *J. Appl. Phys.*, *35*: 3115 (1964).
137. W. A. Tiller, K. A. Jackson, J. W. Rutter, and B. Chalmers, *Acta. Met.*, *1*: 428 (1953).
138. J. Rutter and B. Chalmers, *Can. J. Phys.*, *31*: 15 (1953).
139. R. L. Parker, in *Solid State Physics,* Vol. 25 (D. Turnbull and F. Seitz, eds.), Academic, New York p. 260 (1970).
140. G. Kanig, *Colloid Polym. Sci.*, *261*: 373 (1983).
141. R.H. Olley, A. M. Hodge, and D. C. Bassett, *J. Polym. Sci. Polym. Phys. Ed.*, *17*: 627 (1979).
142. R. H. Olley and D. C. Bassett, *Polymer, 23*: 1707 (1982).
143. D. C. Bassett and A. M. Hodge, *Proc. Roy. Soc. London, A359*: 121 (1978).
144. D. C. Bassett, A. M. Hodge, and R. H. Olley, *Faraday Disc. Chem. Soc.*, *68*: 218 (1979).
145. D. C. Bassett and A. M. Hodge, *Proc. Roy. Soc. London, A377*: 25 (1981).
146. D. C. Bassett and A. M. Hodge, *Proc. Roy. Soc. London, A377*: 39 (1981).
147. D. C. Bassett and A. M. Hodge, *Proc. Roy. Soc. London, A377*: 61 (1981).
148. H. D. Keith and F. J. Padden, Jr., *Polymer, 27*: 1463 (1986).
149. D. C. Bassett and A. S. Vaughan, *Polymer, 27*: 1472 (1986).
150. R. K. Eby, *J. Appl. Phys.*, *35*: 2720 (1964).
151. H. Schonhorn and F. W. Ryan, *J. Polym. Sci.*, *A2, 6*: 231 (1968).
152. D. R. Fitchmun and S. Newman, *J. Polym. Sci.*, *A2, 8*: 1545 (1970).
153. E. Jenckel, E. Teege, and W. Inrichs, *Kolloid Z.*, *129*: 19 (1952).
154. J. D. Hoffman, *S.P.E. Trans.*, *4*: 315 (1964).
155. A. Keller and M. J. Machin, *J. Macromol. Sci.*, *B1*: 41 (1967).
156. B. Wunderlich, *J. Polym. Sci.*, *A1*: 1245 (1963).
157. P. H. Geil, F. R. Anderson, B. Wunderlich, and T. Arakawa, *J. Polym. Sci.*, *A2*: 3707 (1964).
158. B. Wunderlich and T. Arakawa, *J. Polym. Sci.*, *A2*: 3694 (1964).
159. B. Wunderlich and T. Davidson, *J. Polym. Sci.*, *A2, 7*: 2043 (1969).
160. T. Arakawa and B. Wunderlich, *J. Polym. Sci.*, *C16*: 653 (1967).
161. L. Mandelkern, M. R. Gopalan, and J. F. Jackson, *J. Polym. Sci.*, *B5*: 1 (1966).
162. B. Wunderlich, *J. Polym. Sci.*, *B5*: 7 (1966).
163. A. Peterlin, *Polymer, 6*: 25 (1965).
164. B. Wunderlich and L. Melillo, *Makromol. Chem.*, *118*: 250 (1968).
165. E. W. Fischer and H. Puderbach, *Kolloid Z. Z. Polym.*, *235*: 1260 (1969).
166. D. C. Bassett and B. Turner, *Phil. Mag.*, *29*: 285 (1974).
167. P. J. Holdsworth and A. Keller, *J. Macromol. Sci.*, *B1*: 595 (1967).
168. A. Van Valkenburg and J. Powers, *J. Appl. Phys.*, *34*: 2433 (1963).
169. J. F. Jackson, T. S. Hau, and J. W. Brasch, *J. Polym. Sci. Polym. Lett. Ed.*, *10*: 207 (1972).
170. D. C. Bassett, S. Block, and G. J. Piermarini, *J. Appl. Phys.*, *45*: 4146 (1974).
171. P. D. Calvert and D. R. Uhlmann, *J. Polym. Sci.*, *B8*: 165 (1970).
172. D. C. Bassett and B. Turner, *Nature, 240*: 146 (1972).
173. D. C. Bassett and B. Turner, *Phil. Mag.*, *29*: 925 (1974).
174. D. C. Bassett, S. Block, and G. J. Piermarini, *J. Appl. Phys.*, *45*: 4146 (1974).
175. A. E. Zachariades, W. T. Mead, and R. S. Porter, *Chem. Rev.*, *80*: 351 (1980).
176. J. H. Southern and R. S. Porter, *J. Polym. Sci.*, *A2, 10*: 1135 (1972).
177. G. R. Desper, J. H. Southern, R. D. Ulrich, and R. S. Porter, *J. Appl. Phys.*, *41*: 4284 (1970).

178. N. E. Weeks and R. S. Porter, *J. Polym. Sci. Polym. Phys. Ed., 12*: 635 (1974).

179. K. Nakayama and H. Kanetsuna, *J. Mater. Sci., 12*: 1477 (1977).

180. T. Hatakeyma, M. Kanetsuna, and T. Hashimoto, *J. Macromol. Sci., B7*: 411 (1973).

181. C. Czornyj and B. Wunderlich, *J. Polym. Sci. Polym. Phys. Ed., 15*: 1905 (1977).

182. K. D. Pae and S. K. Bhateja, *J. Macromol. Sci. Rev. Macromol. Chem., C13*: 1 (1975).

183. S. K. Bhateja and K. D. Pae, *J. Macromol. Sci. Rev. Macromol. Chem., C13*: 77 (1975).

184. S. Miyashita, T. Asahi, H. Miyaji, and K. Asai, *Polymer, 26*: 1791 (1985).

185. G. Ungar, *J. Mater. Sci., 16*: 2635 (1981).

186. A. Charlesby, P. Kafer, and R. Folland, *Rad. Phys. Chem., 11*: 83 (1978).

187. L. Mandelkern, D. E. Roberts, J. C. Halpin, and F. P. Price, *J. Am. Chem. Soc., 82*: 46 (1960).

188. A. Jurkiewicz, J. Tritt-Goc, N. Pislewski, and K. A. Kunert, *Polymer, 26*: 557 (1985).

189. M. J. Hill and A. Keller, *J. Macromol. Sci., B3*: 153 (1969).

190. R. Kitamaru and S. H. Hyon, *J. Polym. Sci. Macromol. Rev., 14*: 207 (1979).

191. T. R. Manley and M. Qayyum, *Polymer, 12*: 176 (1971).

192. D. E. Roberts and L. Mandelkern, *J. Am. Chem. Soc., 82*: 1091 (1960).

193. A. Posthume de Boeck, Thesis, Ryksuniversiteit te Groningen (1980).

194. Y. H. Kao and P. J. Phillips, *Polymer, 27*: 1669 (1986).

195. K. A. Kunert, H. Soszinska, and N. Pislewski, *Polymer, 22*: 1355 (1981).

196. H. Orth and E. W. Fischer, *Makromol. Chem., 88*: 188 (1965).

197. G. Ungar and A. Keller, *Polymer, 21*: 1273 (1980).

198. P. J. Phillips and Y. H. Rao, *Polymer, 27*: 1679 (1986).

199. R. M. Gohil and P. J. Phillips, *Polymer, 27*: 1687 (1986).

200. R. M. Gohil and P. J. Phillips, *Polymer, 27*: 1696 (1986).

201. B. Wunderlich, *Macromolecular Physics*, Vol. 2, Academic, New York, Chap. VI (1976).

202. A. N. Kolmogoroff, *Isvest. Akad. Nauk. S.S.R. Ser. Math., 1*: 335 (1937).

203. W. A. Johnson and R. T. Mehl, *Trans. A.I.M.E., 135*: 416 (1939).

204. M. Avrami, *J. Chem. Phys., 7*: 1103 (1939).

205. M. Avrami, *J. Chem. Phys., 8*: 212 (1940).

206. U. R. Evans, *Trans. Faraday Soc., 41*: 365 (1945).

207. F. Price, *J. Appl. Phys., 36*: 3014 (1965).

208. L. B. Morgan, *Phil. Trans. Roy. Soc. London, 247*: 13 (1954).

209. F. P. Price, *J. Polym. Sci., A3*: 3079 (1965).

210. I. L. Hillier, *J. Polym. Sci., A3*: 3067 (1967).

211. J. Majer, *Collect. Czech. Commun., 25*: 2454 (1960).

212. A. J. Kovacs, *Ric. Sci., 251*: 669 (1955).

213. G. R. Strobl, T. Engelke, E. Madereck, and G. Urban, *Polymer, 24*: 1585 (1983).

214. E. W. Fischer, *Z. Naturforsch., 14a*: 584 (1959).

215. A. Peterlin, *J. Appl. Phys., 31*: 1934 (1960).

216. A. Peterlin and E. W. Fischer, *Z. Phys., 159*: 272 (1960).

217. A. Peterlin, E. W. Fischer, and C. Reinhold, *J. Chem. Phys., 37*: 1403 (1962).

218. K. O'Leary and P. H. Geil, *J. Macromol. Sci., B1*: 147 (1967).

219. R. Becker, *Ann. Phys., 32*: 128 (1938).

220. R. Becker and W. Döring, *Ann. Phys., 24*: 719 (1935).

221. D. Turnbull and J. C. Fisher, *J. Chem. Phys., 17*: 71 (1949).

222. J. A. Koutsky, A. G. Walton, and E. Baer, *J. Appl. Phys., 38*: 1832 (1967).

223. R. L. Cormia, F. P. Price, and D. Turnbull, *J. Chem. Phys., 37*: 1333 (1962).

224. F. Gormick, G. S. Ross, and L. J. Frolen, *J. Polym. Sci., C18*: 79 (1967).

225. P. J. Barham, R. A. Chivers, D. A. Jarvis, J. Martinez-Salazar, and A. Keller, *J. Polym. Sci. Polym. Lett. Ed., 19*: 539 (1981).

226. R. A. Chivers, P. J. Barham, J. Martinez Salazar, and A. Keller, *J. Polym. Sci. Polym. Phys. Ed., 20*: 1717 (1982).

227. P. J. Barham, D. A. Jarvis, and A. Keller, *J. Polym. Sci. Polym. Phys. Ed.*, *20*: 1733 (1982).
228. F. L. Binsbergen, *Kolloid Z. Z. Polym.*, *237*: 289 (1970).
229. F. L. Binsbergen, *Polymer, 11*: 253 (1970).
230. F. L. Binsbergen, *Kolloid Z. Z. Polym.*, *238*: 389 (1970).
231. F. L. Binsbergen, *J. Cryst. Growth, 16*: 249 (1972).
232. F. L. Binsbergen, *J. Polym. Sci. Polym. Phys. Ed., 11*: 117 (1973).
233. M. Dupire, Thesis, Université de l'Etat à Mons (1984).
234. J. I. Lauritzen, Jr., and J. D. Hoffman, *J. Res. Nat. Bur. Stand., 64A*: 73 (1960).
235. J. D. Hoffman and J. I. Lauritzen, Jr, *J. Res. Nat. Bur. Stand., 65A*. 297 (1961).
236. F. Gornick and J. D. Hoffman, *Ind. Eng. Chem., 58*: 41 (1966).
237. F. C. Frank and M. Tosi, *Proc. Roy. Soc. London, A263*: 323 (1961).
238. J. D. Hoffman, J. I. Lauritzen, Jr., E. Passaglia, G. S. Ross, L. J. Frolen, and J. J. Weeks, *Kolloid Z. Z. Polym., 231*: 564 (1969).
239. J. I. Lauritzen and J. D. Hoffman, *J. Appl. Phys., 44*: 4340 (1973).
240. J. D. Hoffman, L. J. Frolen, G. S. Ross, and J. I. Lauritzen, Jr., *J. Res. Nat. Bur. Stand., 79A*: 671 (1975).
241. J. I. Lauritzen, Jr., *J. Appl. Phys., 44*: 4353 (1973).
242. J. D. Hoffman, G. T. Davis, and J. I. Lauritzen, Jr., in *Treatise on Solid State Chemistry*, Vol. 3 (N. B. Hannay, ed.), Plenum, New York, p. 497 (1976).
243. J. D. Hoffman, *Polymer, 24*: 3 (1983).
244. I. M. Bank and S. Krimm, *J. Polym. Sci., A2, 7*: 1785 (1969).
245. I. M. Bank and S. Krimm, *J. Polym. Sci., B8*: 143 (1970).
246. S. Krimm and T. C. Cheam, *Faraday Disc. Chem. Soc., 68*: 218 (1979).
247. D. M. Sadler and A. Keller, *Polymer, 17*: 37 (1976).
248. D. M. Sadler and A. Keller, *Macromolecules, 10*: 1128 (1977).
249. J. Schelten, D. G. H. Ballard, G. D. Wigmall, G. Longman, and W. Schmatz, *Polymer, 17*: 751 (1976).
250. D. Y. Yoon and P. J. Flory, *Polymer, 18*: 509 (1977).
251. P. J. Flory and D. Y. Yoon, *Nature, 272*: 226 (1978).
252. J. D. Hoffman, C. M. Guttman, and E. A. di Marzio, *Faraday Disc. Chem. Soc., 68*: 177 (1979).
253. R. J. Roe, *J. Chem. Phys., 53*: 3026 (1970).
254. C. M. Guttman, J. D. Hoffman, and E. A. di Marzio, *Faraday Disc. Chem. Soc., 68*: 297 (1979).
255. A. M. Freedman, D. C. Bassett, A. S. Vaughan, and R. H. Olley, *Polymer, 27*: 1163 (1986).
256. J. J. Labaig, Thesis, Université Louis Pasteur, Strasbourg (1978).
257. L. Mandelkern, *Faraday Disc. Chem. Soc., 68*: 375 (1979).
258. L. Mandelkern, A. L. Allou, Jr., and M. R. Gopalan, *J. Phys. Chem., 72*: 309 (1968).
259. J. F. Jackson and L. Mandelkern, *Macromolecules, 1*: 546 (1968).
260. I. R. Harrison and J. Runt, *J. Macromol. Sci., B17*: 83 (1980).
261. J. Runt, I. R. Harrison, and S. Dobson, *J. Macromol. Sci., B17*: 99 (1980).
262. J. D. Hoffman, *Soc. Plast. Eng., 4*: 315 (1964).
263. J. I. Lauritzen, Jr. and E. Passaglia, *J. Res. Nat. Bur. Stand., 71A*: 261 (1967).
264. F. P. Price, *J. Chem. Phys., 35*: 1884 (1961).
265. J. J. Point, *Macromolecules, 12*: 770 (1979).
266. J. J. Point, *Faraday Disc. Chem. Soc., 68*: 167, 365 (1979).
267. E. A. di Marzio and C. M. Guttman, *J. Appl. Phys., 53*: 6581 (1982).
268. A. J. Kovacs, G. Straupe, and A. Gonthier, *J. Polym. Sci. Polym. Symp., 59*: 31 (1977).
269. M. Cooper and R. St. J. Manley, *J. Polym. Sci. Polym. Let. Ed., 11*: 363 (1973).
270. M. Cooper and R. St. J. Manley, *Macromolecules, 8*: 219 (1975).

271. M. Cooper and R. St. J. Manley, *Colloid Polym. Sci., 254*: 542 (1976).
272. W. M. Leung, R. St. J. Manley, and A. R. Panaras, *Macromolecules, 18*: 746 (1985).
273. W. M. Leung, R. St. J. Manley, and A. R. Panaras, *Macromolecules, 18*: 753 (1985).
274. W. M. Leung, R. St. J. Manley, and A. R. Panaras, *Macromolecules, 18*: 760 (1985).
275. L. Mandelkern, in *Progress in Polymer Science*, Vol. 2 (A. D. Jenkins, ed.), Pergamon Press, Oxford and New York, p. 165 (1970).
276. R. L. Miller, *Kolloid Z. Z. Polym., 225*: 62 (1968).
277. J. D. Hoffman, *Macromolecules, 18*: 772 (1985).
278. J. J. Point, *Macromolecules, 19*: 929 (1986).
279. D. J. Blundell, A. Keller, and A. J. Kovacs, *J. Polym. Sci., B4*: 481 (1966).
280. D. J. Blundell and A. Keller, *J. Polym. Sci. Polym. Let., 6*: 433 (1968).
281. T. Seto and N. Mori, *Rep. Prog. Polym. Phys. Jpn., 12*: 157 (1969).
282. A. Keller and E. Pedemonte, *J. Cryst. Growth, 18*: 113 (1973).
283. I. C. Sanchez and E. A. di Marzio, *J. Chem. Phys., 55*: 893 (1971).
284. I. C. Sanchez and E. A. di Marzio, *Macromolecules, 4*: 677 (1971).
285. M. Dosière, M. C. Colet, and J. J. Point, in *Morphology of Polymers* (B. Sedlacek, ed.), Walter de Gruyter, Berlin and New York, p. 171 (1986).
286. J. J. Point, M. C. Colet, and M. Dosière, *J. Polym. Sci. Polym. Phys. Ed., 24*: 357 (1986).
287. R. Koningsveld and A. H. Pennings, *Rec. Trav. Chim., 83*: 552 (1964).
288. T. Kawai and A. Keller, *J. Polym. Sci., B2*: 333 (1964).
289. A. Peterlin and G. Meinel, *J. Appl. Phys., 35*: 3221 (1964).
290. A. Peterlin and G. Meinel, *J. Polym. Sci., B2*: 751 (1964).
291. R. B. Prime and B. Wunderlich, *J. Polym. Sci., A2, 7*: 2061 (1969).
292. D. M. Sadler, *J. Polym. Sci., A2, 9*: 779 (1971).
293. M. D. Keith, F. J. Padden, Jr., and R. Vandimsky, *J. Polym. Sci., A2, 4*: 267 (1966).
294. R. B. Prime, B. Wunderlich, and L. Melilli, *J. Polym. Sci., A2, 7*: 2091 (1969).
295. C. L. Gruner, B. Wunderlich, and R. C. Bopp, *J. Polym. Sci., A2, 7*: 2099 (1969).
296. A. Mehta and B. Wunderlich, *Colloid Polym. Sci., 253*: 193 (1975).
297. P. J. Flory, *Principles of Polymer Chemistry*, Cornell University Press, Ithaca, N.Y. (1953).
298. B. Wunderlich, *Faraday Disc. Chem. Soc., 68*: 239 (1979).
299. C. M. Guttman, E. A. di Marzio, and J. D. Hoffman, *Polymer, 21*: 597 (1981).
300. C. M. Guttman and E. A. di Marzio, *Macromolecules, 15*: 525 (1982).
301. F. A. M. Leermakers, J. M. H. M. Scheutzen, and R. J. Gaylord, *Polymer, 25*: 1577 (1984).
302. C. G. Vonk (referenced as preprint in Ref. 303).
303. M. L. Mansfield, C. M. Guttman, and E. A. di Marzio, *J. Polym. Sci. Polym. Lett. Ed., 24*: 565 (1986).
304. M. L. Mansfield, *Macromolecules, 16*: 914 (1983).
305. D. M. Sadler, *Polymer, 24*: 1401 (1983).
306. D. M. Sadler and G. H. Gilmer, *Polymer, 25*: 1446 (1984).
307. J. D. Hoffman, *Polymer, 26* : 1763 (1985).
308. F. C. Frank, *J. Cryst. Growth, 22*: 233 (1974).
309. A. Toda, H. Kiho, H. Miyagi, and K. Asai, *J. Phys. Soc. Jpn., 54*: 1411 (1985).
310. A. Toda, H. Miyagi, and H. Kiho, *Polymer, 27*: 1505 (1986).
311. J. C. Sanchez and E. A. di Marzio, *J. Res. Nat. Bur. Stand., A76*: 213 (1972).
312. J. I. Lauritzen, Jr., *J. Appl. Phys., 44*: 4353 (1973).
313. M. Dosière, M. C. Colet, and J. J. Point, *J. Polym. Sci. Polym. Phys. Ed., 24*: 345 (1986).
314. A. Toda and H. Kiho, *J. Phys. Soc. Jpn., 56*: 1631 (1987).
315. D. J. Blundell and A. Keller, *J. Macromol. Sci., B2*: 301 (1968).

13

Branching Analysis for Polyethylenes

Takao Usami
Mitsubishi Petrochemical Company, Ltd.
Yokkaichi, Japan

INTRODUCTION

Polyethylene (PE) has been the simplest polymer molecule examined for both its dynamic and structural characteristics. At first sight, a polymer molecule which is essentially a "polymethylene" would not seem to possess enough different structural characteristics to warrant more than a casual investigation. However, numerous PEs are available commercially that have substantially different physical properties and subsequently different end-use applications.

For PEs, it is branching which is of particular interest from the point of view of resin properties and practical use. Branching, which may be characterized as long chain or short chain, can arise through chain transfer reactions during free radical polymerization at high pressure or by copolymerization with 1-olefins. Short-chain branches (SCBs) are particularly critical in their effects on the morphology and solid-state properties [1–4] because the branched structure of some parts of the molecules prevents a completely ordered arrangement of the chains, while long-chain branches (LCBs) comparably have a profound effect on solution viscosity and melt rheology because of molecular size reduction and entanglements [5]. Commercial PEs have a wide range of densities (from 0.915 to 0.965 g/cm^3) depending on their branching levels.

High-pressure low-density, or branched, polyethylenes (HPLDPEs or BLDPEs) with densities around 0.92 g/cm^3 prepared conventionally by free radical polymerization at high temperature (100–300°C) and high pressure (1000–3000 atm) contain many branches of various lengths. SCBs are mainly formed by intramolecular hydrogen atom abstraction, called the "backbiting" mechanism in a cyclic intermediate [1]. In addition to the SCBs originating by the backbiting mechanism, SCBs sometimes can be incorporated into the chain by the copolymerizations with 1-olefins which are used for controlling the molecular weight of HPLDPEs. On the other hand, LCBs are known to be formed via an intermolecular hydrogen transfer [6, 7] and have a range of lengths, with an upper limit approaching the length of the main chain.

High-density polyethylenes with densities around 0.96 g/cm^3 (HDPE) produced using a Ziegler catalyst at low temperature (50–100°C) and low pressure (5–50 atm) are substantially linear, although the physical and rheological properties of some HDPEs have suggested the presence of small amounts of SCBs and LCBs [8].

In linear (no LCB) low-density polyethylenes (LLDPEs) with densities around 0.92 g/cm^3, SCBs can be introduced deliberately in a controlled manner into the main chain by copolymerizing ethylene with 1-olefins via almost the same polymerization processes as those of HDPEs. LLDPEs show some specific characteristics not only in viscoelastic behavior, due to no LCB, but also in melting and impact strength behavior, due to the type [4, 9] and the heterogeneity of SCBs [10].

It is therefore important to have as much information as possible concerning the type, distribution, and concentration of the branches by branching analysis for various LDPEs.

IDENTIFICATION OF SHORT-CHAIN BRANCHES

The presence of branches in PEs was primarily deduced from the intensity of the methyl vibration at 2960 cm^{-1} (3.38 μm) in the infrared (IR) spectrum by Fox and Martin [11]. Cross et al. [12] and Bryant and Voter [13] identified the branches using the 1378 cm^{-1} (7.26 μm) methyl absorption with suitable correction for overlapping methylene absorptions. The branches were postulated to be n-butyl groups by Roedel [1], who proposed an intramolecular hydrogen atom abstraction reaction (backbiting) involving a 6-membered cyclic transition state shown in Figure 1. Willbourn [14] subsequently resolved two small peaks at 745 and 770 cm^{-1} in the IR spectrum and attributed them to butyl and ethyl branches respectively. He deduced that the ethyl and butyl branches were 67 and 33% respectively and proposed a secondary intramolecular hydrogen atom transfer reaction to form ethyl branches as shown in Figure 1. Willbourn's IR analysis was supported by the measurements of the composition of the volatile products from the radiolysis of PEs [15–18].

Recently carbon-13 (^{13}C) nuclear magnetic resonance (NMR), with its greater sen-

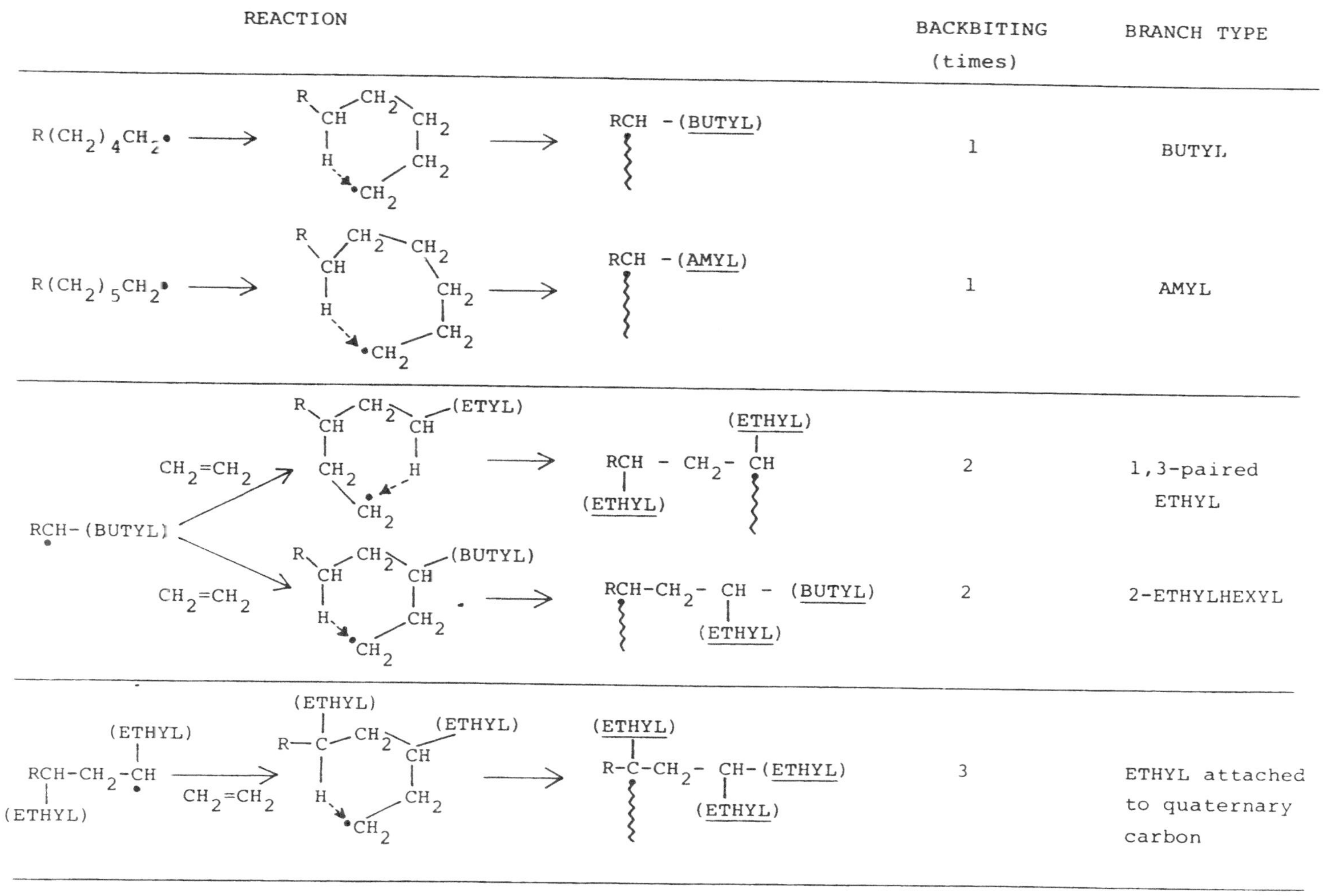

Figure 1 SCB formation mechanism by the backbiting reaction in HPLDPE.

sitivity to subtle details in molecular structure, offers a new and promising tool for branching analysis for PEs [19–24]. In a study of a series of ethylene–1-olefin copolymers ranging from propylene to 1-octene, Randall [20] reported that an array of resonances characteristic of the type of branch were observed for linear C_1 through C_5 branches but the ^{13}C chemical shifts became insensitive to branch length with hexyl and longer branches.

Pyrolysis–hydrogenation gas chromatography (PyHGC) methods have also been utilized for the study of SCB [25–37]. The minor peaks between the serial main peaks of *n*-alkanes are mostly attributed to isoalkanes, which are closely associated with the branching structures in polymer backbones.

Carbon-13 NMR Method

Randall [20] demonstrated that ^{13}C NMR spectra are sufficiently characteristic to distinguish LLDPEs with various types of branches. Thus ^{13}C NMR studies have a great potential to specifically identify a particular branch group and location relative to other branches and, if properly employed, determine its concentration. A typical ^{13}C NMR spectrum of an HPLDPE is shown in Figure 2 and the peak assignments are given in Table 1. A series of ^{13}C NMR studies [19–24] of SCBs in HPLDPEs was carried out and led to the general conclusions that (1) butyl branches were usually found in substantially higher concentrations than the other branches, (2) ethyl and amyl branches were present and the

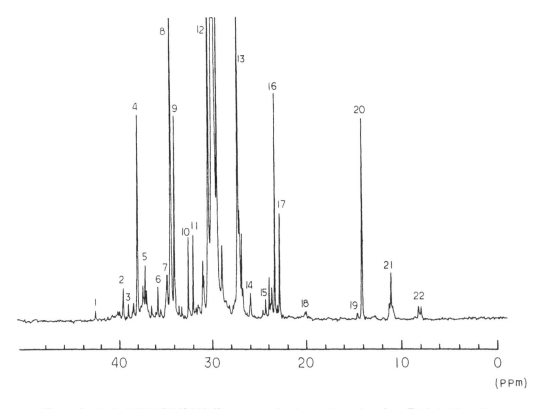

Figure 2 Typical HPLDPE ^{13}C NMR spectrum showing peak numbers from Table 1. (From Ref. 42.)

Table 1 Chemical Shifts and Assignments for the Numbered Peaks in Figure 2

Peak no.	Chemical shift	Assignment[a]
1	42.86	α-CH_2 to C=O
2	39.75	α-CH_2 to quaternary carbon attached to ethyl branch
3	39.19	α,α'-CH_2 of 1,3-paired ethyl branches
4	38.23	CH of butyl and longer-than-butyl branches
5	37.38	CH of 1,3-paired ethyl branches
6	35.99	CH of 1,3-paired ethyl branch to the ethyl branch attached to quaternary carbon
7	35.00	α,γ-CH_2 of 1,3-paired ethyl branches
8	34.61	α-CH_2 of butyl and longer-than-butyl branches
9	34.22	Fourth carbon from the branch end of butyl branch
10	32.70	Third carbon from the branch end of amyl branch
11	32.18	Third carbon from the branch end of longer-than-amyl branch
12	30.00	Backbone CH_2
13	27.33	β-CH_2 of butyl and longer-than-butyl branch
14	25.99	β-CH_2 to quaternary carbon attached to ethyl branch
15	24.36	β-CH_2 to C=O
16	23.36	Second carbon from the branch end of butyl branch
17	22.86	Second carbon from the branch end of longer-than-butyl branches
18	20.04	CH_3 of methyl branch
19	14.59	CH_3 of propyl branch
20	14.08	CH_3 of butyl and longer-than-butyl branches
21	11.01	CH_3 of 1,3-paired ethyl branches
22	8.15, 7.87	CH_3 of ethyl branch attached to quaternary carbon

[a]Greek letters denote the position to the key carbons along the main chain. The key carbons are specified in the table or are the methin carbons attached to the branches if there is no specification.

ethyl branches seemed to be isolated, (3) methyl and propyl branches were present very little, if at all, and (4) hexyl and longer-than-hexyl branches were present, but they could not be distinguished. However, those results were still ambiguous in some respects. The problems were (1) a significant proportion of the ethyl branches should be in 1,3-paired or possibly 2-ethylhexyl structures as required by the double backbiting mechanism, and (2) the occurrence of hexyl, heptyl, and octyl branches through the 8-, 9-, 10-membered cyclic intermediates are also expected because amyl branches through the 7-membered intermediate are present.

Some recent progress in these fields is mentioned later.

Ethyl Branches

Although ethyl groups are expected to occur mainly as 1,3-paired ethyls and as constituents of 2-ethylhexyl branches by the backbiting mechanism, there had been very few ^{13}C NMR studies [22, 38–42] on the types of ethyl branches. Cudby and Bunn [22] reported that the observed broadening of the methyl resonance around 11 ppm arising from ethyl branches would indicate the variations of ethyl branches. Axelson et al. [38] suggested the presence of nonisolated ethyl branches in structures such as 1,3-paired ethyl branches. They pointed out that the 39.86 ppm resonance, representing the methin carbon of an isolated ethyl branch, could not be observed, whereas there was a significant 11 ppm CH$_3$ peak present and there was an appreciable amount of resonance around 37.5 ppm, which can be assigned to the methin signal in 1,3-paired ethyl branches.

Nishioka et al. [39] tried to assign the characteristic signals to nonisolated ethyl branches by the empirical chemical shift calculations such as that of Lindeman and Adams [43]. However, there had been some ambiguities concerning the type of ethyl branches. Grenier-Loustalot [41] pointed out that there was about 0.2 ppm chemical shift difference

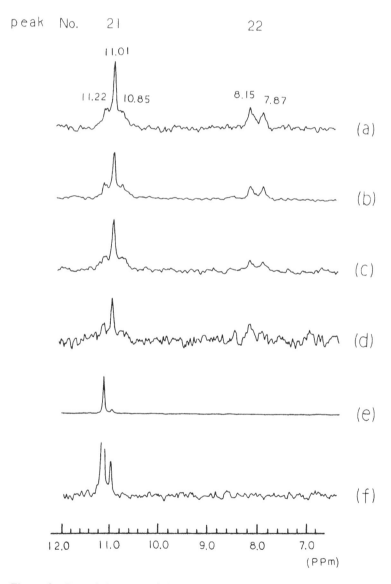

Figure 3 Expanded spectra of the methyl resonance region for ethyl branches: (a) HPLDPE-A; (b) HPLDPE-B; (c) HPLDPE-C; (d) HPLDPE-D; (e) ethylene–1-butene (E–B) copolymer (showing the isolated ethyl branch); (f) E–B copolymer (vertically expanded spectrum showing the 1,3-paired ethyl group). (From Ref. 42.)

between the methyl signals of the 1,3-paired and the isolated ethyl branches and these two signals can be distinguished in a series of branched alkanes as model compounds. Recently Usami and Takayama [42] demonstrated that the ethyl CH_3 signal around 11 ppm showed the multiplicity due to the diversity of the ethyl branch structure, and the predominant component of ethyl branches was 1,3-paired ethyl groups and the peak at 37.38 ppm should be assigned to the methin in the 1,3-paired ethyl branches by comparing the ^{13}C NMR spectra of HPLDPE with those of ethylene–1-butene copolymer as shown in Figures 3 and

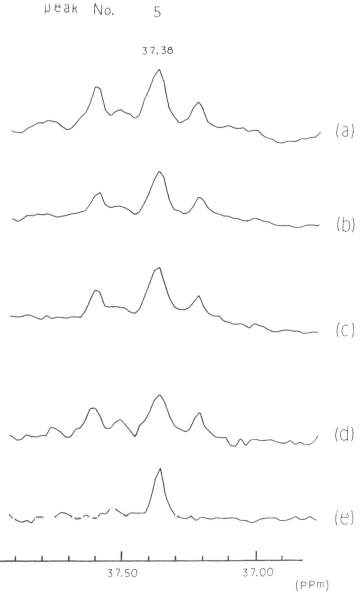

Figure 4 Expanded spectra of the methin resonance region for the 1,3-paired ethyl group: (a) HPLDPE-A; (b) HPLDPE-B; (c) HPLDPE-C; (d) HPLDPE-D; (e) E–B copolymer (showing the methin carbon signal of the 1,3-paired ethyl branches). (From Ref. 42.)

4. The validity of these assignments was assured by the ^{13}C NMR spectrum [44] shown in Figure 5, obtained by the insensitive nuclei enhanced by polarization transfer (INEPT) method where CH and CH$_3$ signals turn down and CH$_2$ signals turn up. The contribution of 2-ethylhexyl branches seems to be minor because any unique signal assigned to this branch type has not been clearly observed yet [38, 42]. There exists an appreciable amount of high-field resonance around 8 ppm and this resonance also shows the multiplicity due to the varieties of the structure. As pointed out by Axelson et al. [38], this resonance is compatible with the predicted shifts of the CH$_3$ carbon of ethyl groups attached to quaternary backbone carbons. The presence of this branch type indicated the possibility of a triple backbiting reaction as shown in Figure 1.

Dechter and Mandelkern [40] reported that 1,2-diethyl branches could be present and the CH$_3$ signal would appear around 11.8 ppm by Lindeman–Adams calculations. However, Grenier-Loustalot [41] showed that the chemical shifts of the CH$_3$ resonance should be 12.8 ppm from the observation of a model compound. Although they did not clearly observe the 1,2-ethyl pair for an HPLDPE, the hexane-extractable component of HPLDPE showed distinct signals around 12.8 ppm, as seen in Figure 6 [45].

Thus ethyl branches have a variety of structures arising from the extensions of the backbiting mechanism and give multiple peaks ranging from 13 to 8 ppm.

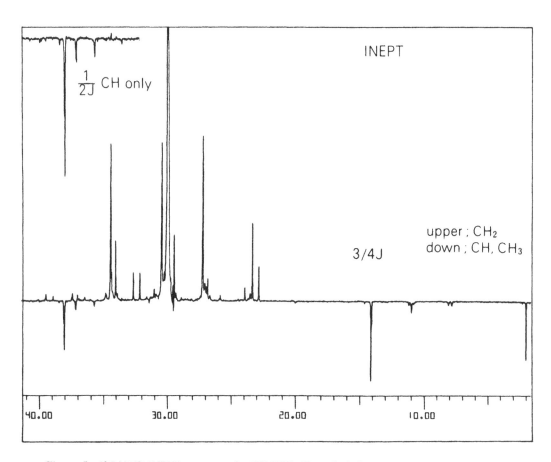

Figure 5 ^{13}C NMR INEPT spectrum of a HPLDPE. (From Ref. 44.)

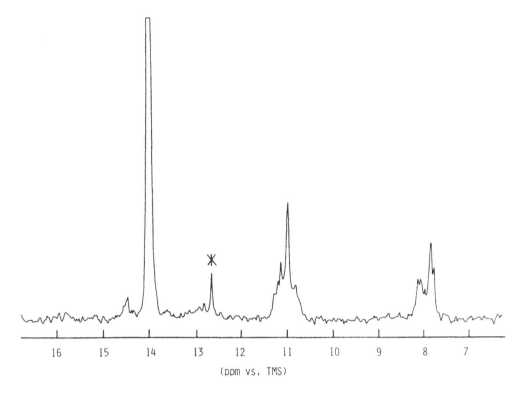

Figure 6 ^{13}C NMR CH_3 region spectrum of *n*-hexane-extractable component of HPLDPE (asterisk indicates the signal of 1,2-diethyl branches).

Hexyl Branches

The presence of amyl branches [46] by the backbiting reaction through a 7-membered ring intermediate suggested the analogous reaction through an 8-membered ring intermediate, which generates a hexyl branch. Mattice and Stehling [47] reported that calculations using a rotational isomeric state model for the chain statistics predicted that the concentrations of intermediate-length branches such as hexyl, heptyl, and octyl branches would not be negligibly small. Hexyl and longer-than-hexyl branches had not been able to be distinguished, but Cavagna [48] pointed out the possibility of their discrimination. Freche and Grenier-Loustalot [49] showed that there was a small chemical shift difference (0.02 ppm) on the third carbon signal from the branch end. Recently Usami and Takayama [42] demonstrated that hexyl branches were distinguished from LCBs and there was no appreciable amount of hexyl branches in HPLDPEs by comparing the ^{13}C NMR spectra of *n*-hexatriacontane (n-$C_{36}H_{74}$), ethylene 1 octene (E–O) copolymer (hexyl branches only), HPLDPEs, and their mixtures as shown in Figure 7.

Quantitative Analysis

For quantitative analyses by ^{13}C NMR, the use of proper experimental parameters for data acquisition is often critical. Many problems arise when nonequilibrium magnetization data are acquired. For instance, short-pulse repetition rates selectively saturate carbons having the longest relaxation times to lead to the conclusion that branch types associated with such

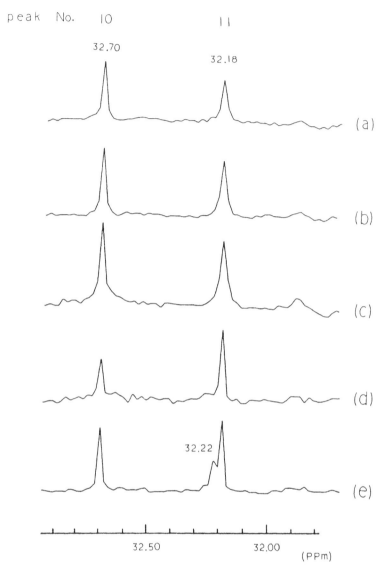

Figure 7 Expanded spectra of the signals of the third carbon from the longer-than-butyl chain end: (a) HPLDPE-A; (b) HPLDPE-B; (c) HPLDPE-C; (d) HPLDPE-D; (e) mixture of HPLDPE-B and ethylene–1-octene (hexyl branch) copolymer. (From Ref. 42.)

carbons might have existed less than they actually do. In an ideal experiment, each pulse should sample the equilibrium magnetization of all carbons. This would require a pulse interval at least five times that of the longest carbon spin-lattice relaxation time, T_1. Until recently no measurements had been made of T_1 for the major carbons of interest in LDPEs. Thus the optimum delay times were essentially unknown. Axelson et al. [50] showed that there was a large gradation in relaxation times in the branch carbons as shown in Figure 8. These values range from approximately 1.1 to 7 sec at 118°C in solution. With this wide variation in T_1, delay time of about 40 sec should be required for completely quantitative analysis.

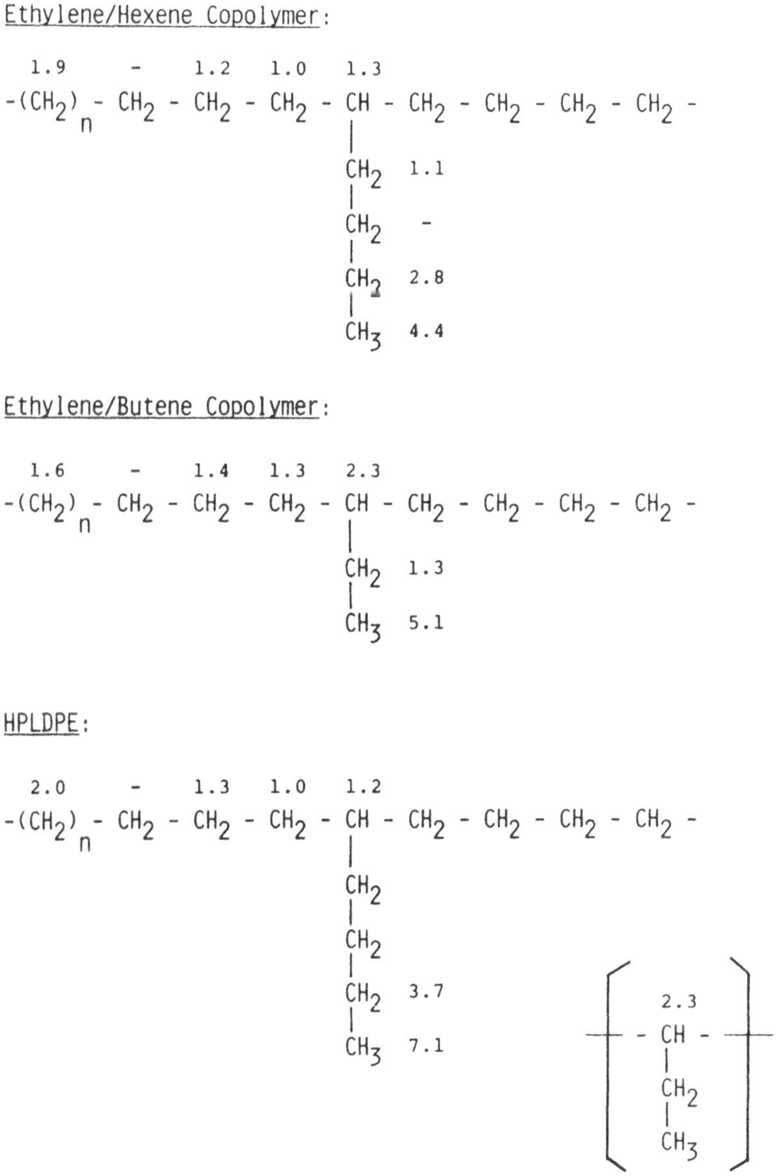

Ethylene/Hexene Copolymer:

```
    1.9         -    1.2   1.0   1.3
 -(CH2) - CH2 - CH2 - CH2 - CH - CH2 - CH2 - CH2 - CH2 -
       n                        |
                                CH2   1.1
                                |
                                CH2    -
                                |
                                CH2   2.8
                                |
                                CH3   4.4
```

Ethylene/Butene Copolymer:

```
    1.6         -    1.4   1.3   2.3
 -(CH2) - CH2 - CH2 - CH2 - CH - CH2 - CH2 - CH2 - CH2 -
       n                        |
                                CH2   1.3
                                |
                                CH3   5.1
```

HPLDPE:

```
    2.0         -    1.3   1.0   1.2
 -(CH2) - CH2 - CH2 - CH2 - CH - CH2 - CH2 - CH2 - CH2 -
       n                        |
                                CH2
                                |
                                CH2
                                |
                                CH2   3.7              2.3
                                |                   - CH -
                                CH3   7.1              |
                                                      CH2
                                                      |
                                                      CH3
```

Figure 8 T_1 map of ethylene copolymers and HPLDPE in 40% w/v trichlorobenzene. Data obtained at 118°C and 67.905 MHz. (From Ref. 50.)

The nuclear Overhauser enhancement factors (NOEFs) must also be evaluated for the pertinent carbons. HPLDPEs and a related ethylene–1-olefin copolymer were confirmed to exhibit a maximum NOEF of 2.0 [38, 50]. To keep the total accumulation time reasonable, a compromise can be made between the delay time and minimum number of pulses necessary for adequate signal to noise ratio to allow identification and integration of branching at very low concentrations because various types of ethyl branches are expected to give weak and broad signals. Usami and Takayama [42] employed a calibration method.

Table 2 Calibration Factors for Various Carbons Compared to the β-CH_2 in Ethylene–1-Olefin Copolymers Determined with 2-sec Pulse Delays

Branch type	Integral intensity ratios to β-CH_2		
	Third[a] carbon	Second[a] carbon	CH_3
$n = 1$			0.90
$n = 2$		0.95	0.84
$n = 3$			0.83[b]
$n = 4$		0.90	0.81
$n = 5$	0.90[b]	0.81[b]	0.72[b]
$n = n$	0.80	0.72	0.67

[a]From the chain end.
[b]Tentatively determined from other results but would be considered as reasonable values and not bring intolerable errors.
Source: Ref. 42.

Table 3 Branch Concentrations and Distributions Determined by the ^{13}C NMR Calibration Method

Sample	Total CH_3/1000 C	Methyl	Ethyl[a]	Ethyl[b]	Propyl	Butyl	Amyl	Longer[c]
HPLDPE-A	19.9	0.4	4.4	2.0	0.3	7.4	2.6	2.8
(rel. conc., %)		(2)	(22)	(10)	(2)	(37)	(13)	(14)
			(32)					
HPLDPE-B	24.9	0.5	6.2	3.0	0.5	8.5	2.7	3.5
(rel. conc., %)		(2)	(25)	(12)	(2)	(34)	(11)	(14)
			(37)					
HPLDPE-C	17.3	0.5	3.8	1.6	0.4	6.4	2.2	2.4
(rel. conc., %)		(3)	(22)	(9)	(2)	(37)	(13)	(14)
			(31)					
HPLDPE-D	6.7	0.1	1.6	0.7	0.2	2.4	0.7	1.0
(rel. conc., %)		(2)	(23)	(11)	(3)	(35)	(11)	(15)
			(34)					

[a]Ethyl branches around 11 ppm including 1,2- and 1,3-paired ethyl branches.
[b]Ethyl branches around 8 ppm including ethyl branches attached to quaternary carbons.
[c]Branches longer than amyl including linear terminal methyls.
Source: Ref. 42.

In this method, the calibration factors for the signals of various kinds of carbon to the main-chain methylene signal were determined experimentally with 2-sec pulse delays as shown in Table 2. Using these calibration factors, the corrected SCB concentrations of HPLDPEs can be determined from the 2-sec pulse delay measurements. Table 3 lists the SCB concentrations of some HPLDPEs determined by the calibration method. In conventional polymerization the differences in the conditions, which give total branch concentrations between 6.7 and 24.9 (per 1000 carbon atoms), do not have any appreciable effect on the SCB-type distribution and the ethyl branch concentrations are found to be almost comparable to those of butyl branches by summarizing the weak and broad signals between 13 and 8 ppm.

Infrared Spectroscopic Method

Among all the infrared active vibrational modes in the branches of PEs, the methyl symmetrical deformation band near 1378 cm^{-1} [4, 9, 12–14, 51–63], the methyl rocking band [12] ranging from 880 to 940 cm^{-1}, and the methylene rocking band [14, 64, 65] ranging from 720 to 770 cm^{-1} are of interest for qualitative and quantitative analysis of SCBs.

Methyl Symmetrical Deformation Band (1378 cm^{-1})

The 1378 cm^{-1} band is strongly overlapped by the doublet at 1367 and 1350 cm^{-1} due to the wagging modes of short sequences of methylene in nonplanar conformation [66, 67], as seen in Figure 9. This overlapping makes the identification of the branches difficult and often inaccurate. Earlier investigations used graphic band decomposition procedures [13] for the analysis of the band. Willbourn [14] later resolved the problem of the interference of the doublet by using a compensation method for the investigation of model compounds such as branched PE. This method is simply based on a polymethylene wedge placed in the reference beam which compensates the interference bands at 1367 and 1350 cm^{-1}. The testing method for the branching of PEs by ASTM employs a similar wedge technique [68].

Recently Rueda et al. [62] and Müller et al. [63] demonstrated that a systematic and reproducible decomposition of bands in the 1320–1400 cm^{-1} region was accomplished by means of a computer simulation method, as shown in Figure 10. Although the computer simulation method greatly improved the branching analysis using the 1378 cm^{-1} band, Reding and Lovell [9] and Teranishi and Sugawara [58] pointed out the problem that the absorption coefficients of the band change depended on the branch length such as methyl, ethyl, butyl. Therefore, it was necessary to identify the branch type and determine the absorption coefficient for each branch. Shirayama et al. [57] determined the absorption coefficients for various branches shown in Table 4 with a compensation method using *n*-octadecane. However, there were problems in the identification of the branch types and ambiguity in the compensation of the overlapping peaks. Müller et al. [63] found that the peak positions of the 1378 cm^{-1} methyl band also changed depending on the branch length for some copolymers of 1-olefins by a peak deconvolution method using computer simulation as shown in Table 5.

Recently Usami and Takayama [69] carried out detailed analysis of the methyl band by a Fourier transform infrared spectroscopic (FTIR) method, which has an excellent reproducibility for peak position (<0.1 cm^{-1}) and is easily applied for obtaining an accurate difference spectrum. The spectra before and after the subtractive operation for an HPLDPE sample are shown in Figure 9 along with the spectrum of HDPE used as a

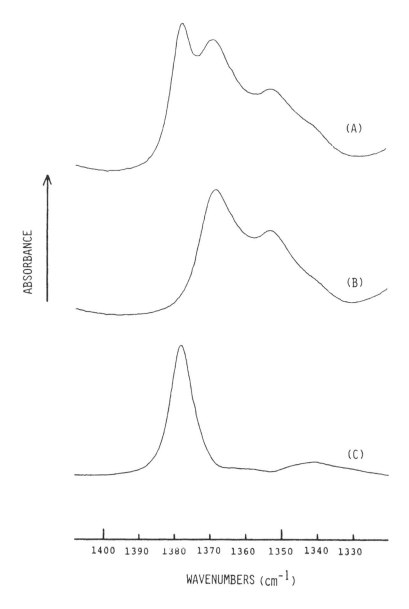

Figure 9 Difference spectrum of the methyl symmetrical deformation band for a HPLDPE. Curve A, HPLDPE; curve B, high-density PE; curve C, A–B spectrum. (From Ref. 69.)

subtraction reference. In the same way, differences of spectra were obtained for LDPEs having various types of branches. The peak positions of those branches were determined as shown in Table 5. The absorption coefficients for various types of branches were also determined by applying the difference spectrum method to LLDPEs with certain kinds of branches and known branch concentrations determined by ^{13}C NMR [69]. The absorption coefficient for HPLDPE was estimated by averaging the absorption coefficients of various branches by considering the branch type distribution determined by ^{13}C NMR. The coefficient values are listed in Table 4.

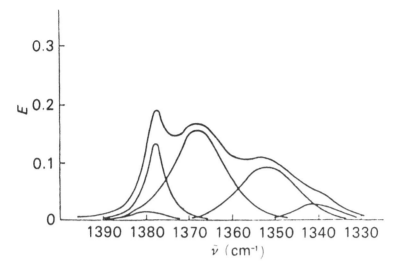

Figure 10 Decomposition of the bands in 1330–1400 cm^{-1} region for an HPLDPE by a computer simulation method. (From Ref. 63.)

Table 4 The Absorption Coefficients[a] of the Methyl Symmetrical Deformation Bands for Various Branches

Branches	Absorption coefficient (K)	
	Usami and Takayama [69]	Shirayama et al. [57]
Methyl	0.039	0.053
Ethyl	0.059	0.086
n-Butyl	0.070	0.092
Isobutyl	0.042	0.114
n-Hexyl	0.076	
n-Decyl		0.097
HPLDPE	0.067[b]	

[a]The branch concentrations are calculated from the equation $N = K \times (A/T)/D$, where N is the branch concentration (per 1000 C), A is the absorbance, T is the film thickness in centimeters, and D is the film density in grams per cubic centimeter.

[b]The value is determined by averaging the K values of the branches in HPLDPE from the results by ^{13}C NMR.

Table 5 Wavenumbers of the Methyl Symmetrical Deformation Bands and the Methyl Rocking Bands for Various Branch Types

Branch type	Methyl	Ethyl	*n*-Butyl	*n*-Hexyl	Isobutyl	HPLDPE
Deformation band						
Wavenumber of the peak (cm^{-1})	1377.3	1379.4	1378.1	1377.9	1383.6 1365.6	1378.3
Müller's value [63]	1378.0	1379.4	1378.3	1378.1		1378.5
Rocking band						
Wavenumber of the peak (cm^{-1})	936.7	Not detectable	894.3	889.8	952.5 919.5	Broad peak at 893.7

Source: Ref. 69.

Methyl Rocking Band (880–940 cm^{-1})

Methyl rocking bands are also used to identify branches [12]. However, these bands are overlapped by the double bond absorptions which appear in the 800–1000 cm^{-1} region. In this region the absorption bands due to unsaturation are respectively located at 888 cm^{-1} for the vinylidene, at 903 and 984 cm^{-1} for the vinyl, and at 960 cm^{-1} for the vinylene. The elimination of the interference of these bands can be attained by bromination of the double bonds. Usami and Takayama [69] determined the positions of the methyl rocking bands after bromination of LLDPEs with various types of branches and the specific values for the methyl rocking bands are listed in Table 5.

Methylene Rocking Band (720–770 cm^{-1})

Willbourn [14] demonstrated that the ratio of ethyl to butyl branches could be determined from the methylene rocking bands at 772 and 745 cm^{-1} by the compensation method. Although the absorption due to the ethyl branches was clearly observed at 772.2 cm^{-1}, it was pointed out [69] that the butyl branch absorption seemed to be located very close to 730 cm^{-1} rather than at 745 cm^{-1} and was very difficult to be found because of the large peak at 730 cm^{-1}, due to the methylene in the crystalline phase.

Pyrolysis–Hydrogenation Gas Chromatography Method

Earlier work [26–30] with a packed separation column showed that PyHGC was a very promising technique for studying the branch structure of PEs. Michailov et al. [26] identified some of the isoalkane peaks on the hydrogenated pyrograms of HPLDPEs and pointed out that ethyl and butyl branches were predominant. A firm relationship between chain branching in the polymer backbone and the degradation products appearing on pyrograms of PE was established by Seeger et al. [27–29]. Recently the resolution of pyrograms has been drastically improved using the fused-silica capillary column [34, 35, 37]. Figures 11 and 12 show a typical high-resolution hydrogenated product pyrogram obtained for an HPLDPE and the expanded portions of the pyrograms for reference ethylene–1-olefin copolymers (LLDPEs) along with an HPLDPE between the *n*-C$_{10}$ and *n*-C$_{11}$ regions, respectively. The minor peaks between the serial main peaks of *n*-alkanes are mostly isoalkanes, which are closely related to the SCB in the polymer chain.

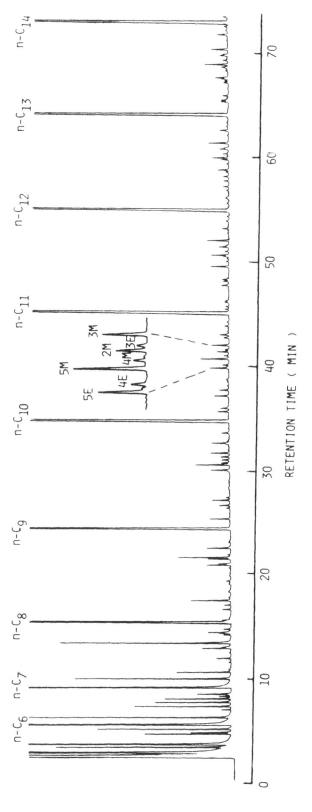

Figure 11 A high-resolution program of HPLDPE at 650°C. 2M. 2-methyldecane; 3M. 3-methyldecane; 4M. 4-methyldecane; 5 M. 5-methyldecane; 3E. 3-ethylnonane; 4E. 4-ethylnonane; 5E. 5-ethylnonane. (From Ref. 37.)

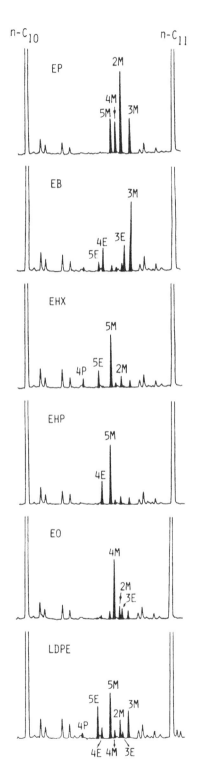

Figure 12 Expanded programs of reference copolymers and HPLDPE in the C_{11} region. 2M, 3M, 4M, 5M, 3E, 4E, and 5E are defined in Figure 11; 4P is 4-propyloctane. (From Ref. 37.)

Table 6 Possible Fragmentation Products of C_{11} from the Site of Branching in PE

scission point	branch type α'	β'	γ'	δ'	ε'	ξ'	η'	θ'	ι'	κ'
α	n									
β	n	2M								
γ	n	3M	3M							
δ	n	4M	3E	4M						
ε	n	5M	4E	3E	5M					
ξ	n	5M	5E	4P	4E	5M				
η	n	4M	4E	4E	4P	5E	4M			
θ	n	3M	3E	5E	5E	4P	3E	3M		
ι	n	2M	3M	4M	5M	4E	4E	3E	2M	
κ	n									n

(The carbon positions of the drawn chain are labelled C_{10}, C_9, C_8, C_7, C_6, C_5, C_4, C_3, C_2, C_1, corresponding to branch types α', β', γ', δ', ε', ξ', η', θ', ι', κ'. R : short branches)

Legend:

n : n-undecane
2M : 2-methyldecane
3M : 3-methyldecane
4M : 4-methyldecane
5M : 5-methyldecane
3E : 3-ethylnonane
4E : 4-ethylnonane
5E : 5-ethylnonane
4P : 4-propyloctane

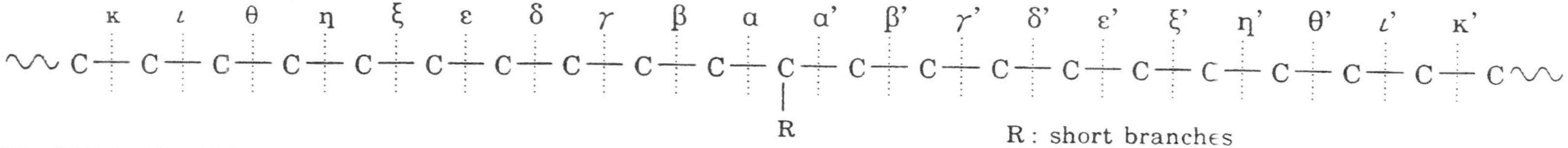

Source: Ref. 37.

Table 6 summarizes the possible fragmentation products of C_{11} formed from the SCB structures provided that any side-chain cleavages do not occur. Each compartment divided by diagonal dashed lines depicts the fragments corresponding to each type of branching from C_1 to C_{10}. Relating the relative peak intensities of characteristic isoalkanes to those of reference model ethylene–1-olefin copolymers, Sugimura and Tsuge [32] determined methyl, ethyl, and butyl branch contents of HPLDPEs. Liebman et al. [35] reported a comparable study on the SCB of PEs by fused-silica capillary PyHGC and ^{13}C NMR and suggested that PyHGC can estimate concentrations of SCB as low as one branch per 10,000 CH_2. Ohtani et al. [37] determined the branch concentration of HPLDPEs from the relative peak intensities of the characteristic key isoalkanes in the C_{11} region by comparing the data of reference model ethylene–1-olefin copolymers (LLDPEs) with known concentrations of methyl, ethyl, butyl, amyl, and hexyl branches, respectively. The results listed in Table 7 were in good agreement with those found by ^{13}C NMR. They also suggested that paired and/or branched branches of ethyl branch were present in HPLDPEs.

HETEROGENEITY OF SHORT-CHAIN BRANCHES

Most of the SCB studies by ^{13}C NMR, IR, and PyHGC just mentioned were carried out for a whole polymer. Therefore, the results such as the concentrations of various kinds of branches represent only the average figure of a PE sample. However, polymers are basically heterogeneous materials. Heterogeneity may be exhibited in a number of ways: differences in chain length and chemical composition between chains, for example. Each form of heterogeneity and possible combinations are expected to exert an influence on the properties of the polymer.

Analyses can be accomplished only by fractionation to separate the molecules accord-

Table 7 Estimated SCB Content in HPLDPEs by PyHGC and by ^{13}C NMR Spectroscopy

Sample	Estimated SCB content per 1000 carbon atoms[a]						
	Methyl	Ethyl	Butyl	Amyl	Hexyl	Longer[b]	Total[c]
HPLDPE-A	1.3	4.9	8.3	2.3	0.2		17.0
	(0.4)	(6.4)	(7.4)	(2.6)		(2.8)	(19.9)
HPLDPE-B	1.5	7.2	11.2	2.2	0.3		22.4
	(0.5)	(9.2)	(8.5)	(2.7)		(3.5)	(24.9)
HPLDPE-C	1.3	4.8	8.6	1.9	0.4		17.0
	(0.5)	(5.4)	(6.4)	(2.2)		(2.4)	(17.3)
HPLDPE-D	1.0	2.1	4.5	0.6	0.3		8.5
	(0.1)	(2.3)	(2.4)	(0.7)		(1.0)	(6.7)

[a]Estimated SCB content by PyHGC is given first, and estimated SCB content by ^{13}C NMR spectroscopy is given in parentheses.
[b]The longer chain branches are not taken into consideration in the case of PyHGC.
[c]Content by NMR also involves propyl branch content between 0.2 and 0.5.
Source: Ref. 37.

ing to their differing structures. The most popular technique is size exclusion chromatography (SEC), which, through a size separation, provides a measure of the chain length or molecular weight distribution (MWD). Further analysis of species separated by molecular weight (MW) usually reveals that they differ in terms of their chemical composition [4, 57]. However, this procedure is not effective in determining compositional distributions within polymers [70]. A more effective way seems to be first to separate molecules based on compositional difference and next to separate by molecular size by SEC [71, 72]. This procedure is called cross-fractionation.

Concerning SCBs of LDPE, the distribution of molecules having different degrees of SCB (SCB distribution, SCBD) is an important heterogeneous character [57, 73, 74]. Considering the variety of processes used in LDPE manufacture, extending from free radical polymerization to copolymerization by Ziegler-type catalyst, a variety of SCBD would be expected to be formed.

Evidence for differences in SCB concentrations between molecules in HPLDPEs were reported [75, 76]. However, these studies considered branching only in terms of its molecular weight dependence and provided somewhat incomplete indications of the actual SCBD. As suggested by Riess and Callot [77], what is needed is an additional and complementary fractionation technique that will provide a separation based on composition variation, independent of molecular weight. From an experimental standpoint, a stepwise temperature-rising elution fractionation (TREF) method similar to the technique described by Desreux and Spiegels [78] should be used to achieve a separation based on only the degrees of branching.

Fractionation based on crystallizability for the purpose of achieving compositional separation of molecular species has been performed by isothermal crystallization at successively lower temperatures or, conversely, by isothermal dissolution at a series of rising temperatures [57, 71, 73, 74]. The work of Shirayama et al. [57], in particular, clearly demonstrated that a TREF technique provided a means of determining SCBD in LDPEs. This is because crystallinity of PEs is directly related to the degree of branching. Although TREF is cumbersome and time-consuming, it does suggest an approach to determining SCBD.

Distribution of SCBs of LDPEs

Shirayama et al. [57] and Bergström and Avela [73] applied TREF to some HPLDPEs and indicated that the difference of SCBD was not great in the different samples in spite of their drastically different polymerization conditions. These results implied that HPLDPE would have a rather uniform character in SCBD. On the other hand, SCBD is a remarkable phenomenon for LLDPEs [4, 79]. It can be observed in the melting behavior as the appearance of several peaks in the data of differential scanning calorimetry (DSC), whereas HPLDPE shows only one melting peak, as seen in Figure 13 [80]. These DSC thermograms suggest that LLDPEs are not uniform in composition and HPLDPE is much more uniform, as expected from the TREF results of Shirayama et al. [57]. The cause of this nonuniformity in melting behavior of LLDPE would be attributed to a heterogeneous SCBD which gives a multipeak ethylene sequence length distribution. A question that comes up is whether the observed heterogeneity exists within individual molecules (intramolecules) or between different molecules (intermolecules) [81]. Unequivocal answers to these problems can be provided by TREF.

Wild et al. [74] carried out the investigation on SCBD of some HPLDPEs and LLDPEs

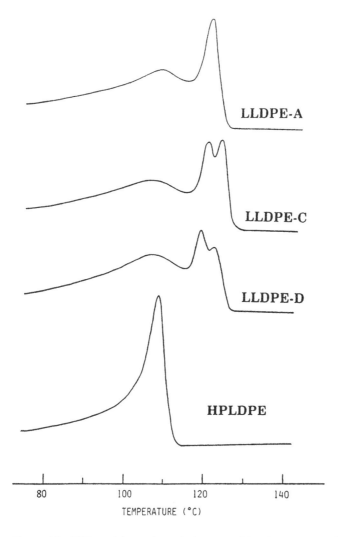

Figure 13 Differential scanning calorimetry melting thermograms of various LDPEs (cooling and heating rates are both 10°C/min).

by their analytical-scale TREF system and got the results shown in Figures 14–16. They found that (1) LLDPEs exhibited extremely wide SCBD with a very low SCB concentration peak and the SCBD of LLDPEs was much wider than that of HPLDPEs, (2) HPLDPE by a tubular reactor exhibited broader SCBD than that by an autoclave reactor, (3) changes in polymerization conditions can influence the SCBD of HPLDPEs produced in the same reactor. The first result has a good correlation with the DSC melting behaviors.

Recently Nakano and Goto [71] improved the TREF technique to provide a more effective and efficient means of determining SCBD in LDPEs by constructing an automated analytical TREF–SEC online system. The procedure of analysis and the design of their automated TREF–SEC system are shown in Figures 17 and 18, respectively. TREF results shown in Figure 19 are obtained at the first part of the system [82].

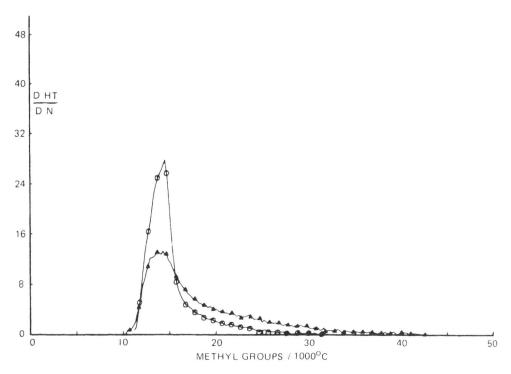

Figure 14 Comparison of the SCB distribution of an autoclave reactor HPLDPE. ○, MI 3.0, density 0.924, with that for a typical HPLDPE from a tubular reactor; ▲, MI 2.2, density 0.921. (From Ref. 74.)

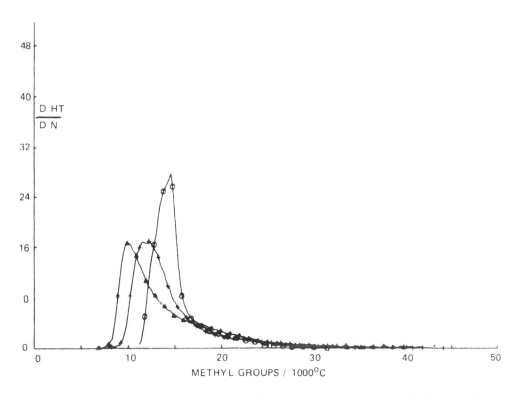

Figure 15 SCB distributions for HPLDPEs made in an autoclave reactor under differing operating conditions. ○, MI 3.0, density 0.924; ▲, MI 1.5, density 0.923; +, MI 1.9, density 0.922. (From Ref. 74.)

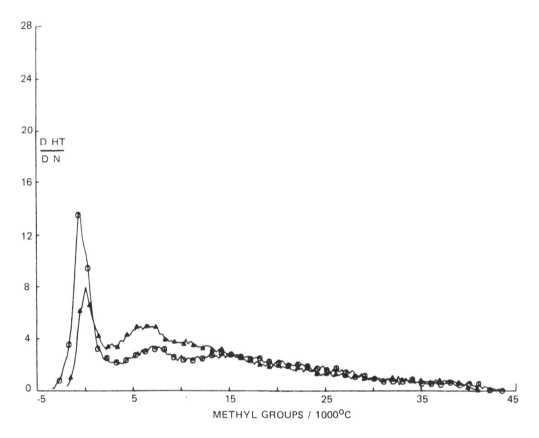

Figure 16 SCB distributions for two commercial LLDPE resins made by different processes. ○, MI 1.2, density 0.919; ▲, MI 1.0, density 0.919. (From Ref. 74.)

Usami et al. [82] have investigated an HPLDPE and LLDPEs manufactured by different processes (gas, slurry, bulk, and solution) by using the automated TREF–SEC system developed by Nakano and Goto [71]. Their TREF results are shown in Figure 19. All LLDPEs show bimodal TREF elution curves. Since there was a linear relationship between the SCB concentration and the elution temperature [74, 82], these TREF curves indicate that LLDPEs manufactured under quite different conditions of temperature (85–230°C), pressure (21–1200 atm), and solvent have in common a characteristic bimodal SCBD, whereas HPLDPE exhibits a single-peak SCBD. A possible explanation for the characteristic bimodal SCBD of LLDPEs is the existence of two groups of active sites in Ti-based heterogeneous Ziegler-type catalysts. Identification of the two groups of active sites has been carried out by ^{13}C NMR sequential analysis for some TREF fractions fractionated from a given LLDPE sample [82]. The reactivity ratio product ($r_1 r_2$) of copolymerization has been determined for each fraction [83] by the equation

$$r_1 r_2 = 1 + f(\chi + 1) - (f + 1)(\chi + 1)^{1/2}$$

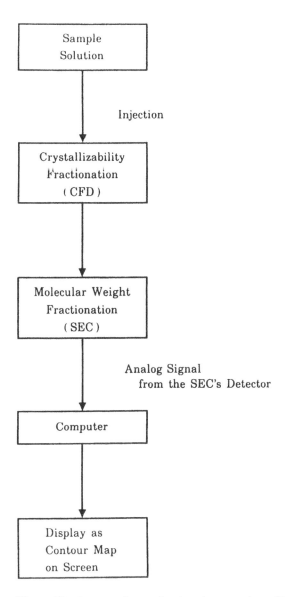

Figure 17 Automated cross-fractionation procedure. (From. Ref. 71.)

where f is the mole ratio of ethylene to 1-olefin in the LLDPE and χ is the mole ratio of 1-olefin in sequences of two or more to isolated 1-olefin in the LLDPE. The results shown in Figure 20 and Table 8 demonstrate that the values are completely different between the two peaks of the bimodal SCBD. Therefore, the two groups of active sites are different in copolymerization character. A discontinuous change in vinyl concentrations of the fractions between the two peaks shown in Figure 21 is further evidence for the two groups of active sites.

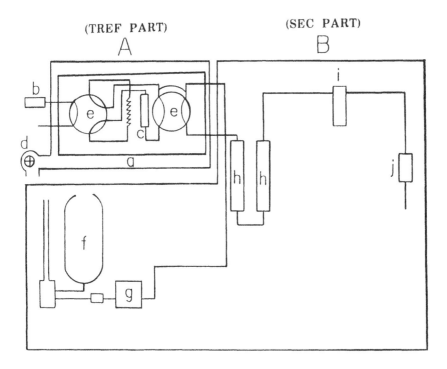

Figure 18 Schematic diagram of the automatic analytical-scale TREF–SEC system. (A) TREF part: (a) block heater; (b) syringe; (c) column; (d) fan cooler; (e) valve. (B) SEC part: (f) solvent tank; (g) pump; (h) column; (i) detector; (j) siphon. (From Ref. 82.)

Short-Chain Branch Concentration as a Function of Molecular Weight

The heterogeneous character of SCBs appears in various forms. Mechanical properties of polymers are markedly affected not only by average molecular weight (MW) and molecular weight distribution (MWD) but also by SCBs. Although an exact characterization of SCB heterogeneity can be accomplished only by cross-fractionation, the type and concentration of SCBs as a function of MW is one of the important forms [84].

Shirayama et al. [57] fractionated commercial HPLDPEs according to MW and determined the SCB concentrations of the fractions by an IR method for six samples designated A to F in Figure 22. They found that some samples had broad distribution of SCB while others had narrow distribution; the degree of SCB generally increased with decreasing MW. However, in some instances the degree of SCB increased with MW at high MW region. On the other hand, Otocka et al. [75] found that SCB concentration was proportional to MW in some cases for HPLDPE.

Recently Usami et al. [85] determined the type and concentration of SCBs as a function of MW for fractions of an HPLDPE by ^{13}C NMR and PyHGC. Their results are seen in Figure 23 and Tables 9 and 10. They found that (1) total SCB and butyl and ethyl branch concentrations increase steeply with decreasing MW at the low MW region while the tendency is more gradual at medium and high MW regions, (2) the change of ethyl branch concentration as a function of MW was greater than that for butyl branch, (3) the type of ethyl branches were mainly 1,3-paired ethyl and/or 2-ethylhexyl branches and ethyl

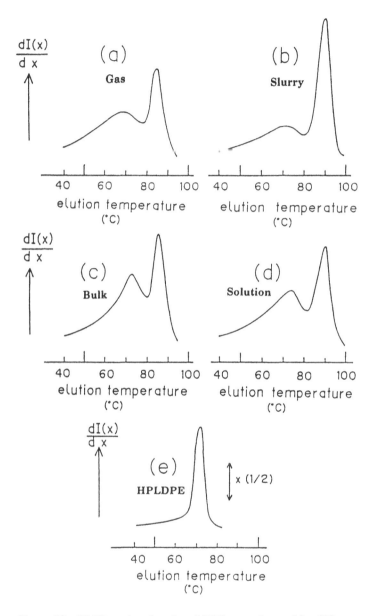

Figure 19 TREF results of various LDPEs manufactured by different processes. (a) LLDPE-A (gas); (b) LLDPE-B (slurry); (c) LLDPE-C (bulk); (d) LLDPE-D (solution); (e) HPLDPE. (From Ref. 82.)

branches attached to quaternary carbon atom at all MW regions, and (4) the concentration of methyl branch incorporated by copolymerization with propylene as a MW modifier was almost constant over all MW regions. These results seem to support the findings of Shirayama et al. [57]. PyHGC may have the advantages of minimal sample size and short analysis time in application to fractionated samples and SEC fractions.

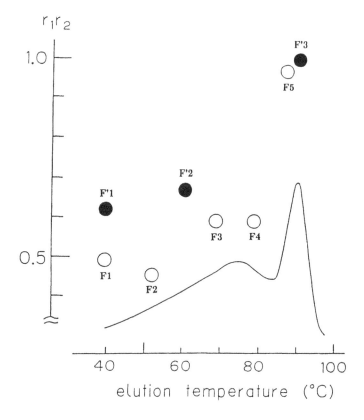

Figure 20 Plots of r_1r_2 values for the TREF fractions showing the TREF diagram. (From Ref. 82.)

Table 8 Values of f, χ, and r_1r_2 for the Fractions by TREF

Fraction	f	χ	r_1r_2
F1	6.51	0.170	0.49
F2	14.0	0.0686	0.45
F3	22.9	0.0530	0.59
F4	34.7	0.0347	0.59
F5[a]	62.4	0.0313	0.97
F'1	6.45	0.217	0.63
F'2	14.7	0.0956	0.67
F'3[a]	32.0	0.0643	1.00

[a]F5 and F'3 belong to the lower SCB concentration peak of the bimodal SCB distribution.

Source: Ref. 82.

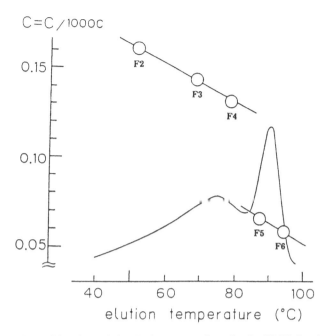

Figure 21 Plots of the vinyl concentrations for the TREF fractions showing the TREF diagram. (From Ref. 82.)

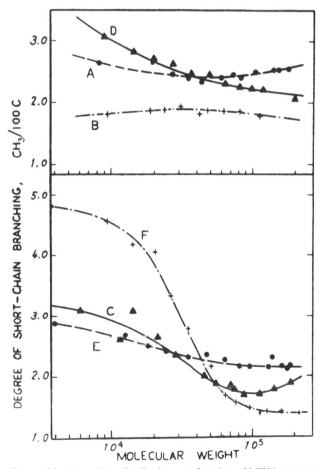

Figure 22 Branching distribution as a function of MW in commercial HPLDPEs. (From Ref. 57.)

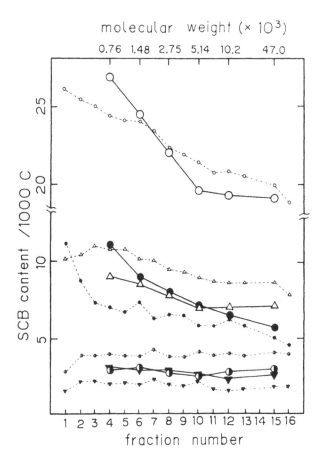

Figure 23 SCB concentrations as a function of MW of an HPLDPE. Solid curve, ^{13}C NMR; dashed curve, PyHGC. (From Ref. 85.)

Table 9 Branch Distributions Determined by ^{13}C NMR for the HPLDPE Fractions

Fraction no.	Total	$-CH_3$	$-C_2H_5$	$-\overset{\displaystyle \mid}{\underset{\displaystyle \mid}{C}}-$ C_2H_5	$-C_3H_7$	$-C_4H_9$	$-C_5H_{11}$	Longer[a]
4	31.3	2.9	7.7	3.3	0.9	9.0	3.1	4.4
6	28.1	3.1	6.0	3.0	0.9	8.5	3.0	3.6
8	25.4	2.8	5.9	2.1	0.6	7.8	2.9	3.3
10	22.7	2.5	5.2	1.8	0.4	7.1	2.6	3.1
12	22.2	2.8	4.8	1.7	0.6	7.0	2.4	2.9
15	22.2	3.0	4.4	1.3	0.7	7.1	2.6	3.1

[a]Branches longer than amyl and linear terminal methyls.

Source: Ref. 85.

Table 10 Branch Distributions Determined by PyHGC for the HPLDPE Fractions

Fraction no.	Branch content per 1000 carbon atoms					
	Total	$-CH_3$	$-C_2H_5$	$-C_4H_9$	$-C_5H_{11}$	$-C_6H_{13}$
1	26.2	2.9	11.2	10.2	1.7	0.2
2	25.6	3.9	8.7	10.4	2.3	0.3
3	25.0	4.0	7.3	11.0	2.3	0.4
4	24.4	4.1	7.0	10.8	2.1	0.4
5	24.1	3.9	6.7	10.8	2.2	0.5
6	24.1	3.9	7.3	10.2	2.1	0.6
7	23.5	4.3	6.3	10.1	2.4	0.4
8	22.3	3.8	6.5	9.5	2.1	0.4
9	21.9	3.8	6.5	9.3	2.0	0.3
10	21.4	4.1	5.8	8.9	2.3	0.3
11	20.7	3.9	5.9	8.7	1.8	0.4
12	20.8	4.0	6.2	8.6	1.7	0.3
13	20.5	3.9	5.8	8.6	1.8	0.4
15	19.9	4.1	5.0	8.6	1.9	0.3
16	18.8	4.0	4.6	7.8	2.0	0.4

Source: Ref. 85.

The study of SCBs as a function of MW has not received much attention so far. This is due primarily to the difficulty in fractionating PE, accurately determining the MW of the fractions, and recovering enough sample from each fraction to determine the SCB concentration. Recent advances in analytical techniques have improved this situation. FTIR offers many advantages as a valuable online detector for SEC [86–90]. Briefly, the advantages are (1) all wavelengths of the IR spectrum can be measured simultaneously, (2) the lack of slits and dispersive optics allows more energy to reach the detectors, thereby resulting in higher sensitivity, and (3) highly accurate wavelength calibration enables accurate postrun spectral manipulation such as subtraction/ratio techniques. By constructing the online SEC–FTIR system [91] shown in Figure 24, the methyl concentration at a

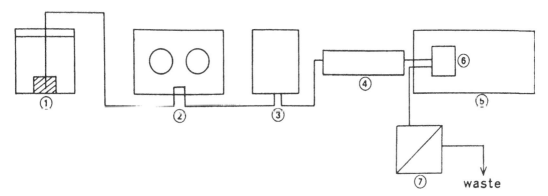

Figure 24 Schematic diagram of the online SEC–FTIR system for the determination of branch concentrations as a function of MW. (1) Solvent (ODCB); (2) HPLC pump; (3) injection; (4) GPC column; (5) FTIR; (6) IR cell for high temperature; (7) RI detector. (From Ref. 91.)

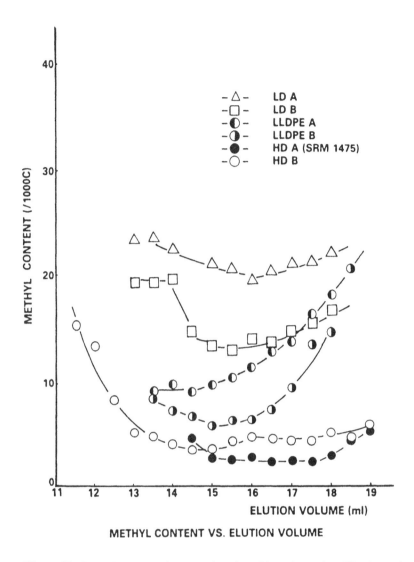

Figure 25 Methyl concentrations as a function of MW for various PEs determined by the online SEC–FTIR system. (From Ref. 91.)

desired elution time can be determined from the ratio of the methyl stretching absorption around 2955 cm^{-1} to the methylene absorption around 2828 cm^{-1}. Thus the methyl concentrations as a function of MW shown in Figure 25 were obtained for various PEs [91].

Cross-Fractionation

Exact analysis for SCB heterogeneity can be done by cross-fractionation. The cross-fractionation of LDPEs includes both TREF and MWD determination in the procedure because a polymer must be separated in two variables, elution temperature (SCB concentration) and MW.

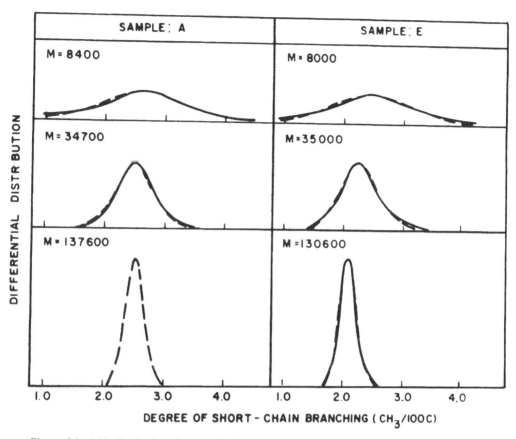

Figure 26 SCB distributions in some fractions of HPLDPE-A and E. (From Ref. 57.)

Shirayama et al. [57] made the initial attempt at cross-fractionation of HPLDPEs. They first performed the MW fractionation and next TREF for the fractions. The results are seen in Figure 26. The SCBD changed markedly depending on MW, but the difference between HPLDPE samples was not significant. Nakano and Goto [71] developed an automatic cross-fractionation system as shown in Figure 17. They placed the automatically eluted TREF column before SEC and the analysis was carried out based on the schematic diagram seen in Figure 18. Figure 27 is a three-dimensional view of their results.

Usami et al. [82] determined the MW of the polymers polymerized at the two groups of active sites which gave the bimodal SCBD of LLDPE using the automated cross-fractionation system. They found that the active sites with higher SCB concentration peak of the bimodal SCBD gave lower MW polymer, as shown in Figure 28. Wild et al. [72] also performed off-line cross-fractionation for some LDPEs by SEC analysis after preparative scale TREF and found that there was a significant trend toward the lower MW species to be more branched, but the strength of the trend varied considerably.

From these cross-fractionation data, it seems to be concluded that in LDPEs the lower MW species generally have more SCBs, except for a very high MW component of HPLDPE.

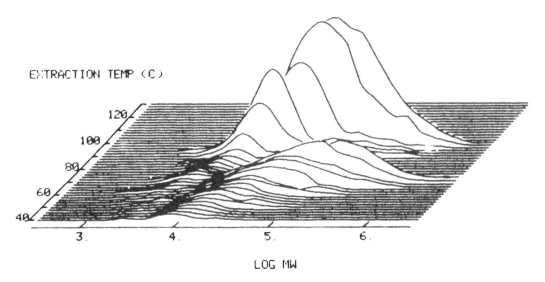

Figure 27 A three-dimensional view of the cross-fractionation result for a PE mixture. (From Ref. 71.)

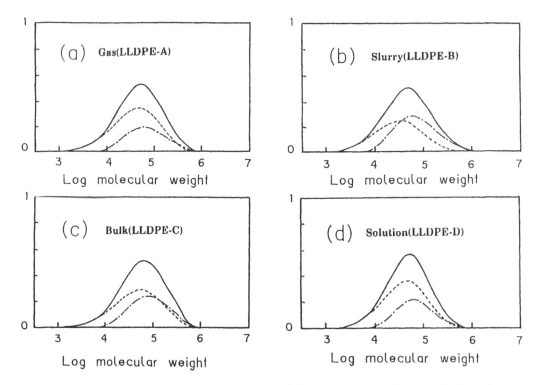

Figure 28 Molecular weight distribution curves of the polymers eluted below and above the boundary temperature (T_b) of the bimodal SCB distribution. (a) LLDPE-A; (b) LLDPE-B; (c) LLDPE-C; (d) LLDPE-D. Solid curve, whole polymer; dashed curve, polymer eluted below T_b; dot–dash curve, polymer eluted above T_b. (From Ref. 82.)

LONG-CHAIN BRANCHING

It was not until the appearance of the important papers in 1953 [1, 5, 92, 93] that any satisfactory evidence became available concerning the nature of the branches in HPLDPEs. Roedel [1] showed that the free radical polymerization mechanism could be expected to lead not only to short- but also to long-chain branches (LCBs) by a mechanism [6] involving the attack of growing polymer radicals on dead polymer chains and subsequent propagation of the branches comparable in length with the main chain from the active centers.

LCB strongly influences the dimensions of the molecules, and hence the viscosity in solution. The synthesis of well-defined branched polymers and the correlation of their structure with their properties have brought an important advance in the study of LCB. Small [94] and Scholte [95] gave extensive reviews of the characterization of LCB in polymers. However, a complete characterization of LCB is still impossible. This is mainly due to the fact that the relation between the solution properties determined on the polymer and the degree of LCB is not fully known. In such cases one has to make use of experimentally defined branching parameters, which bear a not yet fully known relation to the actual chain-branching structure.

It is well accepted that long-chain branched molecules in solution are more compact than their linear analogs. A dimensional parameter that is often used is the so-called mean-square radius of gyration, $\langle s^2 \rangle$. A factor g is used to indicate the ratio of the $\langle s^2 \rangle$ of a branched polymer to the $\langle s^2 \rangle$ of a linear polymer at same MW:

$$g = \frac{\langle s^2 \rangle_{br}}{\langle s^2 \rangle_{lin}} \tag{1}$$

The value of g is a function of the number and type of LCB points in a molecule. Zimm and Stockmayer [96] have derived equations relating g and the number of branch points for randomly branched polymer having trifunctional branch points for both monodisperse and polydisperse systems:

$$g_{LCB} = \left[\left(1 + \frac{\lambda M}{7} \right)^{1/2} + \frac{4}{9\pi} \lambda M \right]^{-1/2} \tag{2}$$

and

$$g_{LCB} = \frac{6}{\lambda \bar{M}_w} \left[\frac{1}{2} \frac{(2 + \lambda \bar{M}_w)^{1/2}}{(\lambda \bar{M}_w)^{1/2}} \ln \frac{(2 + \lambda \bar{M}_w)^{1/2} + (\lambda \bar{M}_w)^{1/2}}{(2 + \lambda \bar{M}_w)^{1/2} - (\lambda \bar{M}_w)^{1/2}} - 1 \right] \tag{3}$$

where Eqs. 2 and 3 refer to the monodisperse and polydisperse systems, respectively. Here λ is the number of branches per unit MW, M is the MW, and $\bar{M}_w$ is weight average MW.

Most of the methods that are used to determine the degree of LCB in polymers in essence comprise the determination of the molecular dimensions in a solution. The intrinsic viscosity [η] of a polymer in solution is a parameter that is directly dependent on molecular dimensions [97]. Also, the coefficient of friction of a dissolved polymer is strongly dependent on molecular dimensions and can be determined by sedimentation in ultracentrifuge [98]. SEC, the method used to determine molecular size [99, 100], is quite suitable in combination with other methods [101–107] which determine absolute MW for obtaining information on the reduction in molecular size as a result of branching. Since the use of a combination of light scattering and viscometry [5], various combinations of solution methods have been developed [101–110]. Carbon-13 NMR, developed in recent years, can give information on the number of side chains up to a certain length [23, 38,

111–113]. This method gives a number average estimation of the end groups or branch points, and correspondingly the number average MW must be known before the degree of branching can be found. Carbon-13 NMR has the considerable advantages of directness and freedom from the difficulties of interpretation inherent in the methods previously described.

Intrinsic Viscosity Method

The ratio of the hydrodynamic volume or intrinsic viscosity of a branched molecule to that of a linear polymer having the same molecular weight has been used to obtain g [96, 114–118]:

$$g^\epsilon = \frac{[\eta]_{br}}{[\eta]_{lin}} \tag{4}$$

The value of the exponent ϵ depends on the type of branching. For the star-shaped branched polymer whose functionality is not too high, ϵ has the value of $\frac{1}{2}$ [115]. In the case of polymers showing a comb-shaped structure, with side chains that are short compared with the backbone, the configuration in solution is determined mainly by the linear backbone, which means that there is the same ratio of the hydrodynamic radius to the radius of gyration as for linear molecules, ϵ has the value of $\frac{3}{2}$ [119, 120]. Therefore, for comb-shaped branching with longer side chains and for random branching, an exponent value between $\frac{1}{2}$ and $\frac{3}{2}$ may be expected.

SCB will also contribute to a reduction in viscosity [121–124], but the effect is generally small and constant. The corrected equation was developed by Billmeyer [5]:

$$g_{SCB} = \left[\frac{1}{(S + 1)} \right] [1 + S(1 - 2f + 2f^2 - 2f^3) + S^2(-f + 4f^2 - f^3)] \tag{5}$$

In Eq. 5 S is the number of SCBs per molecule and f is the fractional branch length (number of carbons in the branch divided by the number of carbons in the backbone). For molecules with both LCB and SCB, the effects on the molecular dimensions are cumulative, so g can be written as

$$g = g_{LCB} \times g_{SCB} \tag{6}$$

$[\eta]_{lin}$ can be calculated from the molecular weight M, using the Mark–Houwink equation:

$$[\eta]_{lin} = KM^a \tag{7}$$

By combining Eqs. 4 and 7, we obtain

$$[\eta]_{br} = KM^a g^\epsilon \tag{8}$$

The substitution for g in Eq. 2 or 3 and the correction for SCB are made, and the intrinsic viscosity of branched polymer as function of λ, M, S, and f is given by

$$[\eta]_{br} = KM^a [g(M,\lambda,S,f)]^\epsilon \tag{9}$$

Hence measurements of $[\eta]_{br}$ and M, knowledge of constant K and a, assumption of the value of ϵ, and a short computer program by varying values assigned to λ are required for determination of λ. M can be determined by, for example, light scattering.

Sedimentation Method

The sedimentation constant of a polymer at zero concentration, S_0, can be written as

$$S_0 = \frac{kM}{R} \tag{10}$$

where R is the hydrodynamic radius, M is the MW, and k is a constant containing the partial specific volume of the polymer and the density and volume of the solvent; it is independent of branching. By introducing the exponent ϵ, the equation

$$\frac{[\eta]}{S_0} = k_3 \alpha^4 g^{(4/3)\epsilon} \tag{11}$$

can be derived. Here k_3 is a constant independent of MW or branching and α is the Flory's coil expansion factor. Since this equation does not contain M, it is not necessary to determine the MW of the polymer. If $[\eta]$ and S_0 are determined under conditions where $\alpha = 1$ and k_3 is determined from measurements on linear polymer, g can be found [108].

Size Exclusion Chromatography Method

In SEC it is primarily the molecular dimensions that determine the elution volume V_E in the separation process. The hydrodynamic radius is a unique function of the product $[\eta]M$ [99, 100]. The universal calibration procedure [100] relates V_E to $[\eta]M$. SEC is calibrated with monodisperse linear polymers and gives the correct MWD for a polydisperse linear polymer. With long-chain branched polymers the data can be treated as if the polymer were linear, so that an apparent MW, M_{lin}, and an apparent MWD will be obtained. M_{lin} and $[\eta]_{lin}$ are the MW and the intrinsic viscosity of linear molecules leaving the columns at the same elution volume as the branched molecules with M_{br} and $[\eta]_{br}$. For branched molecules, the apparent MW, M_{lin}, will be smaller than the real MW, M_{br}. The ratio M_{lin}/M_{br} therefore is an index of LCB. For a monodisperse polymer, the following relations hold:

$$[\eta]_{lin}M_{lin} = K(M_{lin})^{a+1} \tag{12}$$

and

$$[\eta]_{br}M_{br} = g^\epsilon [\eta]_{lin}M_{br} = g^\epsilon K M_{br}^{a+1} \tag{13}$$

Application of the universal calibration principle

$$[\eta]_{lin}M_{lin} = [\eta]_{br}M_{br} \tag{14}$$

to Eqs. 12 and 13 [125] leads to

$$g^\epsilon = \left(\frac{M_{lin}}{M_{br}}\right)^{a+1} \tag{15}$$

This relation between M_{lin}/M_{br} and g is valid only for monodisperse polymers. For polydisperse samples the relation is more complicated. The SEC method has been used in combination with viscometry [75, 76, 102, 125–128], sedimentation [129], and light scattering [104–107, 130] for the determination of LCB.

Carbon-13 NMR Method

In the ^{13}C NMR spectrum of HPLDPE, a resonance at 32.16 ppm (from TMS), corresponding to the third carbon (C-3) from the branch end, provides a measure of branches longer than *n*-amyl but does not distinguish long SCBs [47], presumably formed by intramolecular backbiting reaction, from the true LCBs, which contain possibly many tens or hundreds of carbons and are formed by intermolecular chain transfer to other polymer chains. However, the evidence of no appreciable amount of hexyl branch [42] implies that in HPLDPE short branches longer than *n*-amyl are of negligible probability and the resonance provides in fact a direct measure of the LCB concentration.

LONG-CHAIN BRANCH STUDIES FOR HPLDPEs

Billmeyer [5] first reported that the intrinsic viscosities of HPLDPE samples were lower than would be expected for linear polymers with the same MW and LCB. Billmeyer prepared three fractions for each of three different HPLDPEs and found in each case that the degree of LCB increased with increasing MW. The results indicated that the larger molecule had the higher probability of the chain branching reaction by a transfer mechanism [92]. The study was based on solution viscometry and light scattering measurements. A correction was made for SCB by means of a theoretical technique. Since then, many studies have been carried out, as seen in Table 11.

Size Exclusion Chromatography–Viscometry Combination Methods

After the appearance of SEC, it was soon used in combination with viscometry [75, 76, 102, 103, 125–128, 139]. The combination was first developed by Drott and Mendelson [102] and became very popular. In their method with an iterative computer program, the branching index λ is varied until the solution viscosity calculated using SEC data matches the experimentally measured viscosity of the whole polymer. Using this value of λ, the raw SEC curve is interpreted to calculate the MWD for the whole polymer. The procedure is as follows:

1. $[\eta]_{\text{lin}}M_{\text{lin}} = [\eta]_{\text{br}}M_{\text{br}}$.
2. $g^{\epsilon} = [\eta]_{\text{br}}/[\eta]_{\text{lin}}$, where g is defined by Eq. 3 for a polydispersed polymer.
3. $[\eta]_{\text{br}} = \Sigma W_i[\eta]_{\text{br}_i} = K\Sigma W_i M_i^a[g(\lambda M_i)]^{\epsilon}$.
4. λ is assumed to be constant for the whole spectrum of MWs that make up the whole polymer.
5. In the computer program, λ is varied until calculated and measured viscosities are in good agreement.
6. This is achieved through the use of the two functions, $[\eta]_{\text{br}_i} = f(\lambda M_i)$ and $[\eta]_{\text{br}_i}M_i = f(V_E)$.
7. This leads to the branched calibration curve, $V_E = f(M)$, and gives the corrected MWD of the whole polymer.

Drott and Mendelson assumed that the frequency of LCB was constant as MW varied because they found that it remained constant within experimental error for three fractionated HPLDPEs.

Ram and Miltz [103] developed a comparable method. The procedure is as follows:

1. A polynomial expression of a Mark–Houwink-type relationship is assumed for a branched polymer: $\ln[\eta] = \ln K + a \ln M$ for $M \leqq M_0$; $\ln[\eta] = \ln K + a \ln M +$

Table 11 LCB Studies on HPLDPE

Method[a]	ϵ[b]	LCB (m/1000 C)[c]	Ref.
[η], LS	½	0.12–1.02	5
[η], LS	½	0.35–0.74	134
[η], LS	½	0.07–11.1	122
[η], LS		1.4	133
[η], LS, S	½		109
[η], LS, S			135
[η], OS			132
[η], LS, S	½	0.41–1.02	123
[η], SEC, LS, OS	½		76
[η], SEC, LS, OS			131
[η], SEC	½	0.35–1.60	139
[η], OS	½	2.94–5.46	118
[η], SEC	½	1.96–7.7	102
SEC, S			129
[η], S	½		136
[η], LS	1.3	0.07–0.15	121
[η], SEC	⅓	0.84–4.48	126
[η], SEC	½, ⅓	0.28–12.9	75
[η], SEC	½	4.48–7.56	127
[η], LS, OS	1.0 ± 0.3	0.36–0.94	137
[η], SEC			125
[η], SEC	½	0–56	128
[η], SEC, ¹³C NMR	½	0.6–0.8	23
[η], SEC, LS, OS	½	0–1.23	138
[η], SEC, LS, ¹³C NMR	¾	1.2–4.9	111, 140
[η], SEC, ¹³C NMR	½	0.5–26	38
SEC, LALLS	½	0–1.5	130
¹³C NMR		1.96–2.94	113
[η], LS, OS, ¹³C NMR	½, ¼, 1, ½	0.45–9.5	85

[a][η], Viscosity; LALLS, low-angle laser light scattering; SEC, size exclusion chromatography; OS, OSMO; S, sedimentation.

[b]$g^\epsilon = [\eta]_{br}/[\eta]_{lin}$.

[c]This column shows the range of values of LCB per 1000 carbon atoms stated by the authors or calculated from the data presented.

Source: Ref. 94.

    $b(\ln M)^2 + c(\ln M)^3$ for $M > M_0$. Where $c - b/\ln M_0$ and M_0 represents the MW at which the LCB begins to have an effect (5×10^3 to 1×10^4).

2. $[\eta]_{total} = \Sigma W_i [\eta]_i$.

3. $[\eta]_i M_i = K M_i^{(1+a+b\ln M_i + c\ln^2 M_i)}$. Applying universal calibration, it is possible to use the SEC curve and the solution viscosity of the whole polymer to obtain the best values of b and c by trial and error.

Some important studies using these combination techniques were carried out. Westerman and Clark [128] concluded that λ could be a constant for HPLDPE. The error involved in this assumption was shown to be no greater than the precision with which molecular weight averages could be evaluated by SEC. However, they demonstrated that λ varied somewhat with MW. Essentially the results suggested that λ varied appreciably with MW at lower MW region while λ approached a constant value as MW increased.

Long-Chain Branch Concentration as a Function of Molecular Weight

Just as is the case with SCB, LCB heterogeneity also seems to be likely. However, an exact analysis for LCB heterogeneity cannot be done because a method like TREF for SCB, which determines the intermolecular LCB distribution, does not exist. Analysis for LCB heterogeneity has been carried out only for the LCB concentration as a function of MW.

Guillet [123] found that the frequency of LCB in fractions of a typical commercial HPLDPE remained relatively constant with increasing MW of the fractions. The data were obtained from chromatographic fractionation of the polymer and subsequent solution viscometry, light scattering, and sedimentation measurements. A correction was made for SCB by means of a theoretical technique [5].

Wild and Guliana [131] found that the frequency of LCB remained constant as MW varied. They studied fractionated HPLDPE, employing solution viscosity, light scattering, and osmometry measurements without making any correction for SCB.

Mendelson and Drott [118] determined that the LCB frequency remained constant as MW increased within broad limits of experimental error. The data were obtained on fractionated HPLDPE with subsequent solution viscometry and osmometry measurements on the fractions without making any correction for SCB.

Moore [133] prepared 10 fractions of an HPLDPE and found that the number of LCBs per molecule increased as the MW increased. The study was based on sedimentation, solution viscometry, and light scattering measurements without making any correction for SCB.

Mirabella and Johnson [84] gave an extensive review of the LCB concentration as a function of MW in various polymers including HPLDPE.

Problems in the Determination of LCB

Otocka et al. [75] indicated that the assumption of λ being a constant independent of MW within a given sample is not generally valid. They found that the LCB coefficient λ varied with MW and could not be assumed to be a constant with all types of HPLDPEs but rather that the LCB distribution can vary appreciably with the polymerization conditions employed. The data were obtained by fractionation of the HPLDPEs with subsequent SEC and solution viscosity measurements on the fractions and the whole polymer. Therefore, there still exist two important problems. One is that λ cannot be assumed to be a constant independent of MW within a given sample and the other one is that there is no agreement on the assumed value of the exponent ϵ.

Solutions to the Problems

Recent advances in low-angle laser light scattering (LALLS) make it practical to combine SEC with LALLS to provide a real time analysis of the SEC effluent. Axelson and Knapp [130] used LALLS not only as a continuous monitor of MW but also as a continuous monitor of LCB. The procedure is as follows:

1. At a given elution volume, the mean square radii of gyration for branched and linear molecules are equal: $M_{\mathrm{br}}^{a+1}(V_E)g^\epsilon\{\lambda M_{\mathrm{br}}(V_E)\} = M_{\mathrm{lin}}^{a+1}(V_E)$, where $M_{\mathrm{br}}(V_E)$

is the MW of the branched PE obtained in the SEC separation at elution volume V_E and $M_{lin}(V_E)$ is the MW of the linear PE having the same elution volume as the branched PE with weight $M_{br}(V_E)$.

2. LALLS provides an absolute measure of $M_{br}(V_E)$.
3. $M_{lin}(V_E)$ is calculated from calibration constants determined from analysis of a sample of known MW.

Axelson and Knapp showed that the results of SEC–LALLS for NBS-1476 are in good agreement with the results by Wagner and McCrackin [138] and Bovey et al. [23] for the fractions of NBS 1476. Therefore, SEC LALLS can be the technique of fast and accurate continuous determination of LCB as a function of MW. Figure 29 presents the results of the studies discussed.

The most promising technique appears to be ^{13}C NMR because this method has advantages in directness and is free from the assumptions of the solution methods. Bovey et al. [23] found that the agreement between the viscosity and ^{13}C NMR methods is within the probable experimental error for subfractions of NBS-1476. Axelson and Knapp [38] compared ^{13}C NMR results with the results of solution methods for 15 HPLDPEs, but there was a considerable discrepancy between these methods for parts of the sample. Usami et al. [85] determined the LCB concentration as a function of MW by ^{13}C NMR for fractions of an HPLDPE prepared by a tubular-type reactor. They found a gradual increase of λ with MW, but the values did not agree with those obtained by the viscosity and light-scattering methods for various ϵ values as listed in Table 12.

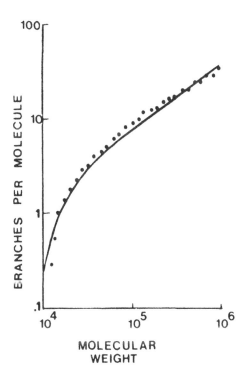

Figure 29 Branch points per molecule as a function of MW for NBS-1476. Solid curve represents data from Ref. 23 and Ref. 138; closed circles are data from the SEC–LALLS experiment [130].

Table 12 LCB Content Determined by ^{13}C NMR and Viscosity Methods for the HPLDPE Fractions

Fraction no.	LCB content per 1000 carbon atoms (by ^{13}C NMR)	LCB content per 1000 carbon atoms by the viscosity method			
		$\epsilon = 0.5$	$\epsilon = 0.75$	$\epsilon = 1.0$	$\epsilon = 1.5$
8	2.8				
10	2.9				
12	3.1	3.1	1.4	0.85	0.45
15	3.4	9.5	2.4	1.1	0.45

Source: Ref. 85.

For the other problem, the value of ϵ, there is no agreement, but most LCB studies by solution methods employed the value of $\frac{1}{2}$ or $\frac{3}{2}$, except for Völker and Luig [121] ($\epsilon = 1.3$), Cote and Shida [126] ($\epsilon = \frac{4}{3}$), and Hama et al. [137] ($\epsilon = 1.0–1.3$).

Foster et al. [140] developed a method called the molecular weight and branching distribution (MWBD) method. This method requires intrinsic viscosity [η] and SEC data on the polymer, and a polystyrene calibration determined under the same solvent–temperature conditions (a universal calibration curve). The MWBD method expresses the size separation with elution volume V as

$$\ln([\eta]_i M_{N^i}) = f(V_i)$$

and assumes the intrinsic viscosity described by a polynomial of the form

$$-\ln[\eta]_i = \ln K + a \ln M_{N^i} + b(\ln M_{N^i})^2 + c(\ln M_{N^i})^3$$

The number average molecular weight and the number of branch points on a molecular basis are obtained as a function of elution volume. From this, integration across the chromatogram provides the number average molecular weight $\bar{M}_N$ and the number average number of branch points per 1000 carbon atoms (λ_N). Their results on the degree of LCB as a function of MW are shown in Figure 30. By comparing the results of the MWBD method with those of the ^{13}C NMR method, they examined the ϵ value. The results are shown in Figure 31 and Table 13 [111]. They found that (1) the use of $\epsilon = 0.75$ gave excellent agreement for the branching frequencies when compared with ^{13}C NMR results, (2) the LCB–MW distribution for NBS-1476 was in good agreement with the data for fractions by Wagner and McCrackin [138], and (3) the LCB–MW distribution changed drastically from sample to sample depending on polymerization conditions.

ACKNOWLEDGMENT

The author thanks Prof. Shin Tsuge, Shigeru Takayama, Yukitaka Gotoh, Hironari Sano, and John Summers for their helpful discussion and assistance in preparing the manuscript.

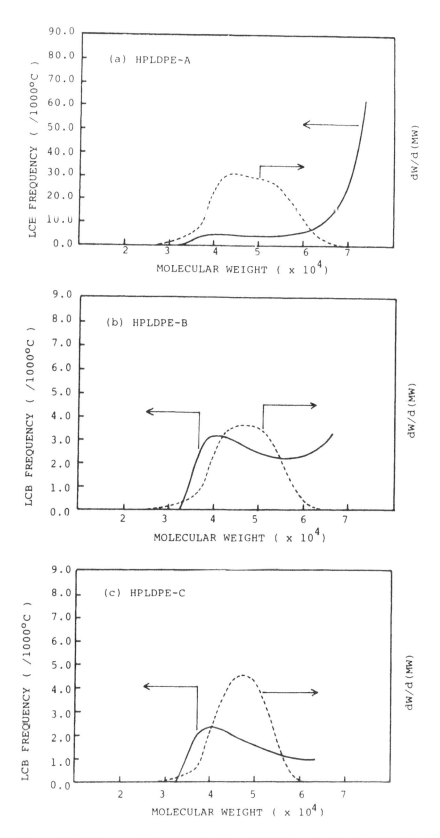

Figure 30 LCB concentrations as a function of MW determined by the MWBD method [140]. Dashed curves, molecular weight distributions; solid curves, LCB concentrations.

Table 13 Comparison of the LCB Concentration per 1000 Carbon Atoms Determined by ^{13}C NMR, [η]-Light Scattering, and MWBD Methods

HPLDPE	^{13}C NMR	[η]-Light scattering (ε = 0.75)	MWBD method (ε = 0.75)
Tubular reactor			
A	4.5	4.9	4.5
D	2.2	2.4	2.6
E	1.8	1.6	1.7
F	3.8	3.8	3.4
Autoclave reactor			
H	2.6	2.5	2.9
I	3.8	4.2	3.7
NBS SRM 1476	1.2	1.4	1.3

Source: Ref. 111.

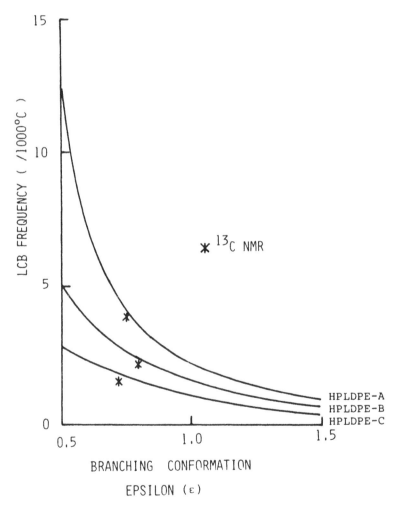

Figure 31 Effect of epsilon (ε) on LCB frequency for HPLDPE-A, B, and C (MWBD method). (From Ref. 140.)

REFERENCES

1. M. J. Roedel, *J. Am. Chem. Soc.*, *75*: 6110 (1953).
2. M. J. Richardson, P. J. Flory, and J. B. Jackson, *Polymer*, *4*: 221 (1963).
3. F. J. Balta-Calleja and D. R. Rueda, *Polym. J. (Jpn.)*, *6*: 216 (1974).
4. K. Shirayama, S. Kita, and H. Watabe, *Makromol. Chem.*, *151*: 97 (1972).
5. F. W. Billmeyer, Jr., *J. Am. Chem. Soc.*, *75*: 6118 (1953).
6. P. J. Flory, *J. Am. Chem. Soc.*, *59*: 241 (1937).
7. P. J. Flory, *J. Am. Chem. Soc.*, *69*: 2893 (1947).
8. J. P. Hogan, C. T. Levett, and R. T. Werkman, *SPE J.*, *23*: 87 (1967).
9. F. P. Reding and C. M. Lovell, *J. Polym. Sci.*, *21*: 157 (1956).
10. L. D. Cady, *Broadening Horiz. Linear Low Tech.*, *1985*: 107.
11. J. J. Fox and A. E. Martin, *Proc. Roy. Soc. (London) A*, *175*: 208 (1940).
12. L. M. Cross, R. B. Richards, and M. A. Willis, *Disc. Faraday Soc.*, *9*: 235 (1950).
13. W. M. D. Bryant and R. C. Voter, *J. Am. Chem. Soc.*, *75*: 6113 (1953).
14. A. H. Willbourn, *J. Polym. Sci.*, *34*: 569 (1959).
15. D. A. Boyle, W. Simpson, and J. D. Waldron, *Polymer*, *2*: 323, 335 (1961).
16. R. Salovey and J. V. Pascale, *J. Polym. Sci. A*, *2*: 2041 (1964).
17. P. M. Kamath and A. Barlow, *J. Polym. Sci. A1*, *5*: 2023 (1967).
18. T. N. Bowmer and J. H. O'Donnell, *Polymer*, *18*: 1032 (1977).
19. D. E. Dorman, E. P. Otocka, and F. A. Bovey, *Macromolecules*, *5*: 574 (1972).
20. J. C. Randall, *J. Polym. Sci. Polym. Phys. Ed.*, *11*: 275 (1973).
21. T. Hama, T. Suzuki, and K. Kosaka, *Kobunshi Ronbunshu*, *32*: 91 (1975).
22. M. E. A. Cudby and A. Bunn, *Polymer*, *17*: 345 (1976).
23. F. A. Bovey, F. C. Schilling, F. L. McCrackin, and H. L. Wagner, *Macromolecules*, *9*: 76 (1976).
24. H. N. Cheng, F. C. Schilling, and F. A. Bovey, *Macromolecules*, *9*: 363 (1976).
25. J. Van Schooten and J. K. Evenhuis, *Polymer*, *6*: 343 (1965).
26. L. Michailov, P. Zugenmaier, and H.-J. Cantow, *Polymer*, *9*: 325 (1968).
27. M. Seeger and E. M. Barrall, II, *J. Polym. Sci. Polym. Chem. Ed.*, *13*: 1515 (1975).
28. M. Seeger, E. M. Barrall, II, and M. Shen, *J. Polym. Sci. Polym. Chem. Ed.*, *13*: 1541 (1975).
29. M. Seeger and R. J. Gritter, *J. Polym. Sci. Polym. Chem. Ed.*, *15*: 1393 (1977).
30. D. H. Ahlstrom and S. A. Liebman, *J. Polym. Sci. Polym. Chem. Ed.*, *14*: 2479 (1976).
31. O. Mlejnek, *J. Chromatog.*, *191*: 181 (1980).
32. Y. Sugimura and S. Tsuge, *Macromolecules*, *12*: 512 (1979).
33. S. Tsuge, Y. Sugimura, and T. Nagaya, *J. Anal. Appl. Pyrolysis*, *1*: 221 (1980).
34. Y. Sugimura, T. Usami, T. Nagaya, and S. Tsuge, *Macromolecules*, *14*: 1787 (1981).
35. S. A. Liebman, D. H. Ahlstrom, W. H. Starnes, Jr., and F. C. Schilling, *J. Macromol. Sci. Chem.*, *A17*: 935 (1982).
36. M. A. Haney, D. W. Johnston, and B. H. Clampitt, *Macromolecules*, *16*: 1775 (1983).
37. H. Ohtani, S. Tsuge, and T. Usami, *Macromolecules*, *17*: 2557 (1984).
38. D. E. Axelson, G. C. Levy, and L. Mandelkern, *Macromolecules*, *12*: 41 (1979).
39. A. Nishioka, Y. Mukai, M. Ohuchi, and T. Imanari, *Bunseki Kagaku*, *29*: 774 (1980).
40. J. J. Dechter and L. Mandelkern, *J. Polym. Sci. Polym. Phys. Ed.*, *18*: 1955 (1980).
41. M. Grenier-Loustalot, *J. Polym. Sci. Polym. Chem. Ed.*, *21*: 2683 (1983).
42. T. Usami and S. Takayama, *Macromolecules*, *17*: 1756 (1984).
43. L. P. Lindeman and J. Q. Adams, *Anal. Chem.*, *43*: 1245 (1971).
44. M. Ohuchi and A. Nishioka, *GX Series SPECTRA No. 1*, Spectra Library, JEOL: 12.
45. T. Usami, to be published.
46. J. C. Randall, *J. Appl. Polym. Sci.*, *22*: 585 (1978).
47. W. L. Mattice and F. C. Stehling, *Macromolecules*, *14*: 1479 (1981).
48. F. Cavagna, *Macromolecules*, *14*: 215 (1981).

49. P. Freche and M. Grenier-Loustalot, *J. Polym. Sci. Polym. Chem. Ed., 21*: 2755 (1983).
50. D. E. Axelson, L. Mandelkern, and G. C. Levy, *Macromolecules, 10*: 557 (1977).
51. E. J. Slowinski, Jr., H. Walter, and R. L. Miller, *J. Polym. Sci., 19*: 353 (1956).
52. G. L. Collier and A. C. M. Panting, *Spectrochim. Acta, 14*: 104 (1959).
53. M. C. Harrey and L. L. Peters, *Anal. Chem., 32*: 1725 (1960).
54. H. Günzler, *Z. Anal. Chem., 170*: 152 (1959).
55. M. Rohmer, *Z. Anal. Chem., 170*: 147 (1959).
56. J. P. Luongo, *Mod. Plastics, 1961*: 132.
57. K. Shirayama, T. Okada, and S. Kita, *J. Polym. Sci. A2, 3*: 907 (1965).
58. K. Teranishi and K. Sugawara, *Kobunshi Kagaku, 23*: 512 (1966).
59. S. Badilescu, M. Toader, H. Oprea, and I. Badilescu, *Mater. Plastics, 7*: 468 (1970).
60. C. Baker and W. F. Maddams, *Makromol. Chem., 177*: 437 (1976).
61. M. A. McRae and W. F. Maddams, *Makromol. Chem., 177*: 449 (1976).
62. D. R. Rueda, F. J. Balta-Calleja, and A. Hidalgo, *Spectrochim. Acta, 35A*: 847 (1979).
63. G. Müller, E. Schröder, and J. Osterode, *Acta Polym., 32*: 694 (1981).
64. H. L. McMurry and V. Thornton, *Anal. Chem., 24*: 318 (1952).
65. M. C. Harvey and A. D. Ketley, *J. Appl. Polym. Sci., 5*: 247 (1961).
66. R. G. Snyder, *J. Chem. Phys., 47*: 1316 (1967).
67. D. R. Rueda, A. Hidalgo, and F. J. Balta-Calleja, *Spectrochim. Acta, 34A*: 475 (1978).
68. The American Society of Testing and Materials, *Spec. Tech. Publ.*, D2238-64 (1964).
69. T. Usami and S. Takayama, *Polym. J. (Jpn.), 16*: 731 (1984).
70. T. Ogawa, *J. Appl. Polym. Sci., 23*: 2315 (1979).
71. S. Nakano and Y. Goto, *J. Appl. Polym. Sci., 26*: 4217 (1981).
72. L. Wild, T. R. Ryle, and D. C. Knobeloch, *Polym. Prepr. (Am. Chem. Soc., Div. Polym. Chem.), 23*: 133 (1982).
73. C. Bergström and E. Avela, *J. Appl. Polym. Sci., 23*: 163 (1979).
74. L. Wild, T. R. Ryle, D. C. Knobeloch, and I. R. Peat, *J. Polym. Sci. Polym. Phys. Ed., 20*: 441 (1982).
75. E. P. Otocka, R. J. Roe, M. Y. Hellman, and P. M. Muglia, *Macromolecules, 4*: 507 (1971).
76. L. Wild, R. Ranganath, and T. Ryle, *J. Polym. Sci. A2, 9*: 2137 (1971).
77. G. Riess and P. Callot, in *Fractionation of Synthetic Polymers* (L. H. Tung, ed.), Marcel Dekker, New York, p. 445 (1977).
78. V. Desreux and M. C. Spiegels, *Bull. Soc. Chim. Belg., 59*: 476 (1950).
79. R. A. Bubeck and H. M. Baker, *Polymer, 23*: 1680 (1982).
80. T. Usami, Y. Gotoh, and S. Takayama, *Polym. Prep. (Am. Chem. Soc. Div. Polym. Chem.), 27*: 110 (1986).
81. D. E. Axelson, *J. Polym. Sci. Polym. Phys. Ed., 20*: 1427 (1982).
82. T. Usami, Y. Gotoh, and S. Takayama, *Macromolecules, 19*: 2722 (1986).
83. C. Wilkes, C. Carman, and R. Harrington, *J. Polym. Sci. Polym. Symp., 43*: 237 (1973).
84. F. M. Mirabella, Jr., and J. F. Johnson, *J. Macromol. Sci. Rev. Macromol. Chem., C12*: 81 (1975).
85. T. Usami, Y. Gotoh, S. Takayama, H. Ohtani, and S. Tsuge, *Macromolecules, 20*:1557 (1987).
86. K. L. Kiser and L. C. Bonar, *Am. Lab., 7*: 85 (1975).
87. D. W. Vidrine, *J. Chromatogr. Sci., 17*: 477 (1979).
88. D. W. Vidrine and D. R. Mattson, *Appl. Spectrosc., 32*: 502 (1978).
89. R. S. Brown, D. W. Hausler, L. T. Taylor, and R. O. Carter, *Anal. Chem., 53*: 197 (1981).
90. L. T. Taylor, *J. Chromatogr. Sci., 23*: 265 (1985).
91. K. Nishikida and M. Morimoto, *FTIR FORUM '86* (by *Perkin-Elmer JAPAN*), 115, 121 (1986).
92. J. K. Beasley, *J. Am. Chem. Soc., 75*: 6123 (1953).
93. C. A. Sperati, W. A. Franta, and H. W. Starkweather, *J. Am. Chem. Soc., 75*: 6127 (1953).

94. P. A. Small, *Adv. Polym. Sci.*, *18*: 1 (1975).
95. T. G. Scholte, in *Developments in Polymer Characterization*, Vol. 4 (J. V. Dawkins, ed.), Applied Science, London, p. 1 (1983).
96. B. H. Zimm and W. H. Stockmayer, *J. Chem. Phys.*, *17*: 1301 (1949).
97. P. J. Flory and T. G. Fox, *J. Am. Chem. Soc.*, *73*: 1904 (1951).
98. J. G. Kirkwood and J. Riseman, *J. Chem. Phys.*, *16*: 565 (1948).
99. H. Benoit, Z. Grubisic, P. Rempp, D. Decker, and J.-G. Zilliox, *J. Chim. Phys.*, *63*: 1507 (1966).
100. Z. Grubisic, P. Rempp, and H. Benoit, *J. Polym. Sci.*, *B5*: 753 (1967).
101. L. H. Tung, *J. Polym. Sci. A2*, 7: 47 (1969).
102. E. E. Drott and R. A. Mendelson, *J. Polym. Sci. A2*, 8: 1361, 1373 (1970).
103. A. Ram and J. Miltz, *J. Appl. Polym. Sci.*, *15*: 2639 (1971).
104. H. J. Cantow, E. Seitert, and R. Kuhn, *Chem. Ing. Techn.*, *38*: 1032 (1966).
105. A. C. Ouano and W. Kaye, *J. Polym. Sci.*, *12*: 1151 (1974).
106. A. C. Ouano, *J. Chromatogr.*, *118*: 303 (1976).
107. B. Millaud and C. Strazielle, *Makromol. Chem.*, *180*: 441 (1979).
108. L. D. Moore, G. R. Greear, and J. O. Sharp, *J. Polym. Sci.*, *59*: 339 (1962).
109. H. Matsuda, I. Yamada, and S. Kuroiwa, *Polym. J. (Jpn.)*, *8*: 415 (1976).
110. H. Matsuda, I. Yamada, M. Okabe, and S. Kuroiwa, *Polym. J. (Jpn.)*, *9*: 527 (1977).
111. G. N. Foster, *Polym. Prepr. (Am. Chem. Soc. Div. Polym. Chem.)*, *20*: 463 (1979).
112. W.-D. Hoffmann, G. Eckhardt, E. Brauer, and F. Keller, *Acta Polym.*, *31*: 233 (1980).
113. J. C. Randall, *ACS Symp. Ser.*, *142*: 93 (1980).
114. W. H. Stockmayer and M. Fixman, *Ann. N.Y. Acad. Sci.*, *57*: 334 (1953).
115. B. H. Zimm and R. W. Kilb, *J. Polym. Sci.*, *37*: 19 (1959).
116. T. Altares, Jr., D. P. Wyman, and Y. R. Allen, *J. Polym. Sci. A*, *2*: 4533 (1964).
117. W. W. Graessley and H. M. Mittelhauser, *J. Polym. Sci. A2*, *5*: 431 (1967).
118. R. A. Mendelson and E. E. Drott, *J. Polym. Sci.*, *B6*: 795 (1968).
119. G. C. Berry, *J. Polym. Sci. A2*, 9: 687 (1971).
120. C. D. Thurmond and B. H. Zimm, *J. Polym. Sci.*, *8*: 477 (1952).
121. H. Völker and F.-J. Luig, *Angew. Makromol. Chem.*, *12*: 43 (1970).
122. Q. A. Trementozzi, *J. Polym. Sci.*, *22*: 187 (1956).
123. J. E. Guillet, *J. Polym. Sci. A*, *1*: 2869 (1963).
124. E. Schröder and E. Winkler, *Plaste Kautsch.*, *21*: 269 (1974).
125. R. Prechner, R. Panaris, and H. Benoit, *Makromol. Chem.*, *156*: 39 (1972).
126. J. A. Cote and M. Shida, *J. Polym. Sci. A2*, *9*: 421 (1971).
127. J. Miltz and A. Ram, *Polymer*, *12*: 685 (1971).
128. L. Westerman and J. C. Clark, *J. Polym. Sci. Polym. Phys. Ed.*, *11*: 559 (1973).
129. L. H. Tung and G. W. Knight, *J. Polym. Sci. A2*, *7*: 1623 (1969).
130. D. E. Axelson and W. C. Knapp, *J. Appl. Polym. Sci.*, *25*: 119 (1980).
131. L. Wild and R. Guliana, *J. Polym. Sci. A2*, *5*: 1087 (1967).
132. W. W. R. Krigbaum and Q. A. Trementozzi, *J. Polym. Sci.*, *28*: 295 (1958).
133. L. D. Moore, Jr., *J. Polym. Sci.*, *20*: 137 (1956).
134. L. T. Muus and F. W. Billmeyer, *J. Am. Chem. Soc.*, *79*: 5079 (1957).
135. W. R. A. D. Moore and W. Millns, *Br. Polym. J.*, *1*: 81 (1969).
136. K. Yamaguchi, M. Kawaguchi, M. Kishi, N. Aki, and K. Sawanishi, *Kobunshi Kagaku*, *26*: 522 (1969).
137. T. Hama, K. Yamaguchi, and T. Suzuki, *Makromol. Chem.*, *155*: 283 (1972).
138. H. L. Wagner and F. L. McCrackin, *J. Appl. Polym. Sci.*, *21*: 2833 (1977).
139. L. Wild, R. Ranganath, and A. Barlow, *J. Appl. Polym. Sci.*, *21*: 3331 (1977).
140. G. N. Foster, A. E. Hamielec, and T. B. MacRury, *Liquid Chromatography of Polymers and Related Materials*, Vol. 2 (J. Cazes and X. Delamare, eds.), Marcel Dekker, New York, p. 143 (1980).

14

Environmental and Thermal Degradation of Low-Density Polyethylene Films

Febo Severini and Raffaele Gallo
Istituto di Chimica Industriale
dell' Università di Messina
Messina, Italy

FUNCTIONAL GROUPS

The macromolecules that constitute low-density polyethylene (LDPE) are made chiefly of sequences of the monomer, but they contain groups different from aliphatic groups, of great importance for the polymer's stability with respect to the action of outdoor agents. The most important groups are the following:

1. Tertiary carbon atoms due to branching reactions during polymerization.
2. Ketone groups formed through slow oxidation reactions during storage or in the course of hot-working processes to obtain manufactured goods, granules, or films.
3. Vinyl and vinylidene groups formed during synthesis or product manufacturing processes.

These groups have been identified by infrared spectrometry [1].

Moreover, the polymer can include hydroperoxide groups, whose production is supported by the presence of tertiary carbon atoms, and it can be polluted by foreign components which aid the oxidation reaction such as polynuclear aromatic hydrocarbons.

MORPHOLOGY

LDPE is a partially crystalline polymer; its crystallinity does not exceed 50%. The IR spectrum of this material shows the presence of peaks due to methyl groups which are to be considered prevalently the ends of short side chains formed during the polymerization reaction. The concentration of methyl groups is in the range 7–30 —CH_3 groups per 1000 carbon atoms. Table 1 summarizes the number of side chains per 1000 carbon atoms in an LDPE film [2].

The oxidation reaction of LDPE occurs mainly in the amorphous regions of this polymer [3, 4] and a growth of density of the material has been observed as the quantity of the absorbed oxygen increases [5] (Table 2). According to some authors [4, 5], a growth of the polymer crystallinity corresponds to this density increase, creating a "chemicrystallization effect" generated by the formation of segments shorter than those preexistent and thus more easily crystallizable, because of the oxidant degradation. Others [6]

Table 1 Frequency of Branching in an LDPE Film

Branching	Number per 1000 C atoms
—CH_3	0.0
—CH_2—CH_3	1.0
—CH_2—CH_2—CH_2—CH_3	9.6
—CH_2—CH_2—CH_2—CH_2—CH_3	3.6
C $\geq$ 6	6.6

Source: Data from Ref. 2.

Table 2 Density Change with Oxygen Uptake at 100°C of a Branched Polyethylene

Oxygen uptake (ml/g)	Density at 23°C
0	0.938
20	0.965
40	0.983
160	1.035
240	1.070

Source: Data from L. Reich and S. Stivala, *Autoxidation of Hydrocarbons and Polyolefins*, Marcel Dekker, New York, p. 388 (1969).

attribute the density increase to the formation of oxygenated functional groups, which replace the $-CH_2$ groups in the amorphous region, and the seeming crystallinity increase with the diminution of the amorphous area of the starting material. As a matter of fact, in the course of the natural weathering of an LDPE film it was not possible to verify crystallinity changes by means of differential scanning calorimetry [7, 8].

The oxidation rate of polymers containing amorphous and crystalline areas is mainly due to the amorphous regions for which a higher oxygen permeability is expected. In fact, high-density polyethylene (HDPE) shows a lower oxidizability than LDPE at 100°C [9]. This difference disappears at 140°C, above the fusion point [10, 11]. This behavior allows us to infer that LDPE can be considered a heterogeneous system as regards the distribution of the oxygen-reactive sites and the morphology of the polymer [12].

THERMAL DEGRADATION

The reaction of thermal degradation of LDPE in the absence of oxygen is notable at temperatures over 300°C only [9]. In the absence of oxygen the degradation proceeds with formation of fragments having molecular weight up to 700 and negligible amounts of the monomer. The degradation mechanism starts according to the following scheme [13]:

$$
\text{(opening)} \quad R-\underset{\underset{H}{|}}{\overset{\overset{H}{|}}{C}}-\underset{\underset{X}{|}}{\overset{\overset{H}{|}}{C}}-\underset{\underset{H}{|}}{\overset{\overset{H}{|}}{C}}-\underset{\underset{X}{|}}{\overset{\overset{H}{|}}{C}}-R' \rightarrow R-\underset{\underset{H}{|}}{\overset{\overset{H}{|}}{C}}-\underset{\underset{X}{|}}{\overset{\overset{H}{|}}{C}}\cdot + R'-\underset{\underset{X}{|}}{\overset{\overset{H}{|}}{C}}-\underset{\underset{H}{|}}{\overset{\overset{H}{|}}{C}}\cdot \quad (1)
$$

which foresees a random course for the scission. In fact, in the case of the strictly linear polymethylene obtained from diazomethane, the volatilization rate shows a maximum when plotted versus conversion, in accordance with the theory developed for random degradation [14].

The branched polyethylene does not follow this rule and has no maximum in its volatilization curve. The differences are attributed to the presence of branches in the molecule that are preferential points for the breaking reaction. The greater the branching, the greater the decomposition rate and the greater the deviation from random theory.

These connections are confirmed [15] by the greater formation of volatile products from the pyrolysis of the polypropylene in comparison with polyethylene.

The thermal degradation process can be described [14] according to the following scheme:

Initiation:

$$\text{polymer} \rightarrow R-CH_2-CH_2-CH_2\cdot \quad \text{(bond cleavage)} \quad (2)$$

Propagation by depolymerization:

$$R-CH_2-CH_2-CH_2\cdot \rightarrow R-CH_2\cdot + CH_2{=}CH_2 \quad (3)$$

(negligible formation of the monomer in the case of LDPE)

Propagation by transfer with random chain scission:

$$R-CH_2\cdot + R'-CH_2-CH_2-CH_2-R'' \rightarrow R-CH_3$$
$$+ R'-\overset{\cdot}{C}H-CH_2-CH_2-R'' \quad (4)$$

$$R'—\overset{\cdot}{C}H—CH_2—CH_2—R'' \rightarrow R'—CH{=}CH_2 + R''—CH_2\cdot \tag{5}$$

In the presence of branches in LDPE at least three transfer reactions which give double bonds, as identified by means of IR spectrometry, take place [14].

THERMAL OXIDATION

The thermal oxidation process of LDPE has been widely studied [16–19] and it is considered to proceed according to the scheme proposed by Bolland for the autooxidation of the hydrocarbons [20]:

Initiation:

$$RH \xrightarrow{\Delta,\, O_2} R\cdot,\ RO\cdot,\ \text{other radicals} \tag{6}$$

Propagation:

$$R\cdot + O_2 \rightarrow ROO\cdot \ (\text{fast}) \tag{7}$$

$$ROO\cdot + RH \rightarrow ROOH + R\cdot \tag{8}$$

$$ROOH \rightarrow RO\cdot + \cdot OH \tag{9}$$

$$RO\cdot + RH \rightarrow ROH + R\cdot \tag{10}$$

$$HO\cdot + RH \rightarrow H_2O + R\cdot \tag{11}$$

$$2ROOH \rightarrow RO\cdot + ROO\cdot + H_2O \tag{12}$$

Termination:

$$R\cdot + R\cdot \rightarrow R—R \tag{13}$$

$$R\cdot + ROO\cdot \rightarrow ROOR \tag{14}$$

$$ROO\cdot + ROO\cdot \rightarrow ROOR + O_2 \tag{15}$$

$$R\cdot + \cdot OH \rightarrow ROH \tag{16}$$

$$R\cdot + \cdot OR \rightarrow ROR \tag{17}$$

The factors affecting the thermooxidation are mainly tertiary carbon atoms due to side chains and allylic hydrogens which react very easily with radical agents. A similar behavior is predicted for hydrogen atoms in α to carbonyl groups. Table 3 shows the reactivity of the carbon–hydrogen bond as regards the attack of R—O—O· and R—O· radicals [21, 22].

The initiation reaction with formation of a macroradical requires the drawing of a hydrogen atom, which can take place according to a reaction of the following type:

1. Decomposition of a thermally unstable substance which gives radicals, such as a peroxide or an azoderivative.
2. Decomposition of hydroperoxide groups developed for slow oxidation of the polymer during storage or working processes of the melted material in the presence of air.

A study of the oxidation reaction under air of an LDPE film by means of thermogravimetry at 200°C [23] indicated the existence of a time of 10–12 min during which the material does not absorb oxygen and the weight of the sample does not change. The

Table 3 Relative Reactivities of C—H Bonds to ROO· and RO· Attack

Reaction	Relative rate	T (°C)
ROO· + R′—CH₂—H → ROOH + R′—CH₂· (H, H)	1	100
ROO· + R′—C(R″)H—H → ROOH + R′—C(R″)H·	10	100
ROO· + R′—C(R″)(R‴)—H → ROOH + R′—C(R″)(R‴)·	80	100
RO· + R′—CH₂—H → ROH + R′—CH₂·	1	45
RO· + R′—C(R″)H—H → ROH + R′—C(R″)H·	7	45
RO· + R′—C(R″)(R‴)—H → ROH + R′—C(R″)(R‴)·	20	45

Source: Data from Refs. 21 and 22.

weight of the sample subsequently increases about 1%, owing to absorption of oxygen, in a period of 20–30 min and a continuous diminution of the weight then is observed, probably corresponding to the initiation of the thermal oxidation process of the polymer (Figure 1).

The hydroperoxide groups that form are very weak because of the low value of the energy of the O—O bond, which is only 175.5 kJ/mol [24]:

$$P—C—OOH \rightarrow P—C—O· + ·OH \tag{18}$$

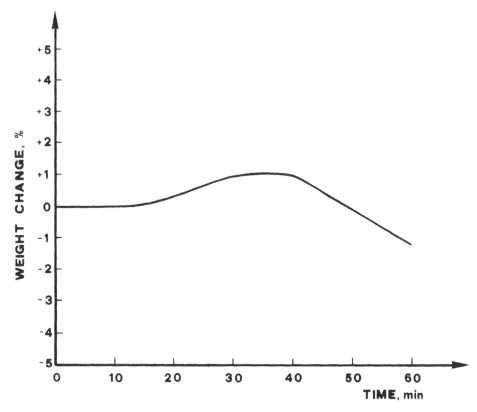

Figure 1   Weight change of an LDPE film at 200°C under air. (Data from Ref. 23.)

The two radicals can easily give transfer chain reactions activating the oxidation cycle, but the alkoxy radical can cause also a β-scission with polymer degradation. Obviously these reactions take place preferably in the amorphous region. The peroxide radicals shown in reaction 6 can react with a hydrocarbon group, producing an active center for the oxidation cycle, but they can also react with one another to yield oxygenated products, for example, dialkylperoxides, molecules containing carbonyl groups, and hydroxyl groups, according to the following schemes [25]:

$$-\underset{|}{\overset{|}{C}}-O_2\cdot \ + \ -\underset{|}{\overset{|}{C}}\cdot \quad \xrightarrow{\text{fast}} \quad -\underset{|}{\overset{|}{C}}-O-O-\underset{|}{\overset{|}{C}}- \tag{19}$$

$$-\underset{|}{\overset{|}{C}}-O_2\cdot \ + \ \cdot O_2-\underset{|}{\overset{|}{C}}- \quad \xrightarrow{\text{very slow}} \quad \left[-\underset{|}{\overset{|}{C}}-O\cdot \ + \ O_2 \ + \ \cdot O-\underset{|}{\overset{|}{C}}- \right] \tag{20}$$

$$H-\overset{|}{\underset{|}{C}}-O_2{}^{\bullet} \; + \; {}^{\bullet}O_2-\overset{|}{\underset{|}{C}}- \quad \xrightarrow{\text{slow}} \quad -C\overset{\displaystyle O}{\diagdown} \; + \; O_2 \; + \; HO-\overset{|}{\underset{|}{C}}- \tag{21}$$

Quantitative measurements of the kinetic constants have been carried out studying the behavior of model substances having low molecular weight. It is evident that, when we work with a semicrystalline material such as LDPE below its fusion temperature, the progress of the different reactions is determined by the morphology of the system and by the diffusion rate of the oxygen. Many studies have been carried out utilizing the material as a film and the influence of the thickness on the oxidation processes will be clear in the photooxidation.

The decomposition of hydroperoxide groups, initially produced through reaction 8, yields radicals such as RO· and ·OH (reaction 9) and ROO· (reaction 12), which are able to initiate transfer reactions leading to the formation of other R· macroradicals (reactions 10, 11, and 8).

Reactions 10, 11, and 12 indicate a branching of the kinetic chain and explain the autocatalytic character of the initial portion of the thermal oxidative curve at 110°C [9], as illustrated by Figure 2, which gives the oxygen uptake as a function of time in hours. The reaction proceeds with polymer degradation, and absorption of 1–2 cm^3 of O$_2$ per gram of polymer is sufficient to yield materials devoid of practical interest. A diminution of the oxygen absorption rate, due probably to the exhaustion of the accessible reactive centers, is observed later.

The crystallinity and the structural regularity distinguish the trend of the LDPE oxidation reaction from that of HDPE. In fact it can be observed that, in a curve of the absorbed

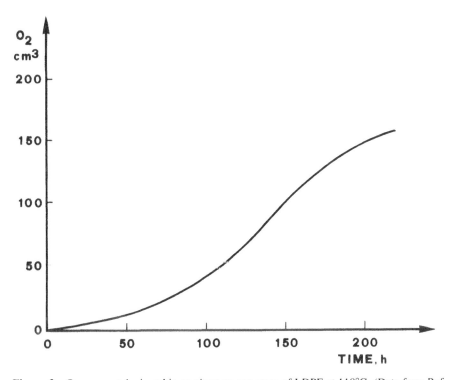

Figure 2 Oxygen uptake in cubic centimeters per gram of LDPE at 110°C. (Data from Ref. 9.)

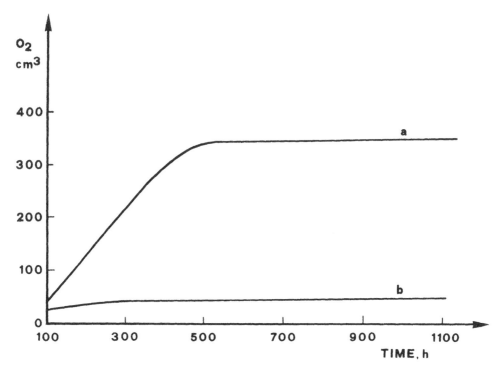

Figure 3 Oxygen uptake in cubic centimeters per gram of polymer at 110°C. Curve a, high-pressure polyethylene, density = 0.92; curve b, low-pressure polyethylene, density = 0.96. (Data from Ref. 9.)

oxygen at 110°C versus time, after 200 hr, a high-pressure polyethylene having density 0.92 absorbs an O_2 volume five times as large as that of a polyethylene having density 0.96 (Figure 3).

ENVIRONMENTAL DEGRADATION

We define environmental degradation of a polymer as the aging action performed on a polymer sample by all the stress factors present in the place under examination. This definition includes the main agents of polymer degradation [26, 27]: solar light, heat, moisture, oxygen, pollutants, and mechanical and biological factors.

In the case of LDPE films the agents of mechanical stress such as wind, rain, hail, suspended solid particles (dust and ash), and possible strains produced by the support system of the film can be of remarkable importance [28].

LDPE films are not generally blended with fillers and reinforcements [29] to avoid changing elasticity and optical and electrical properties of this material. Moreover, for some years, there have been attempts to produce polyethylene easily degradable in external applications in a way that would not leave pollutants after utilization.

BIODEGRADABILITY

Biodegradable polymers are those macromolecular materials which are degradable by means of microorganisms such as fungi and bacteria. The biodegradation of macromolecular materials such as paper and starch occurs generally because of enzymes secreted from microorganisms and brought into contact of the polymeric support by the water. We can say water is necessary for any biodegradation. Hydrophobic, waterproof, and semicrystalline materials such as polyethylene are biodegradable with difficulty, and the biological attack, if it is present, can be restricted to the superficial amorphous areas of the polymer. Many times photodegradable materials are improperly indicated as biodegradable materials for which the products of photodegradation are not proved to be truly biodegradable in acceptable times.

The most complete work on the biodegradation of polyethylene and linear alkanes is that of Potts et al. [30]. They have established the relations between molecular weight of linear hydrocarbons and the corresponding growth of microorganisms according to the ASTM method D-1924-63 (since 1980 designated ASTM G-21-70) [31]. The test consists of exposing, for 3 weeks, the hydrocarbon materials to a mixture of fungi *Aspergillus niger, A. flavus, Chaetonium globosum*, and *Penicillium funiculosum*. The growth rating is fixed as follows: 0 = no growth; 1 = traces (sample less than 10% covered); 2 = light growth (10–30% covered); 3 = middle growth (30–60% covered); 4 = strong growth (60–100% covered). Strips of filter paper act as a control in each experiment to ensure that an active mixture of fungi is used.

Data reported by Potts et al. indicate that linear alkanes having molecular weight up to 451 are easily utilized by the microorganisms, whereas products with higher molecular weight remain unaltered. Alkanes having low molecular weight, but branched, are resistant to biodegradation (Table 4). Potts et al. have also pointed out that the LDPE biodegradability, measured with the indicated ASTM method, is to be attributed to the presence of a small amount of fractions having molecular weight lower than 500. Johannessen [32] confirms that the microorganisms utilize the very degraded fractions only.

Tsuchii et al. [33] observed a partial degradation also for ethylenic oligomers with molecular weight in the range 600–800 by means of the Acinobacter.

The biodegradability of the alkanes having molecular weight lower than 500 only suggests that the enzyme attack can proceed through a process that starts from the ends of the paraffin chain. It is evident that polymers of the ethylene having high molecular weight contain very low concentrations of chain ends, which are not very accessible, owing also to the partially crystalline structure of the polymer and to the folded configuration of the macromolecules.

Another evaluation of the polymer biodegradation is given by the amount of carbon

Table 4 Biodegradability of Hydrocarbons

Compound	Molecular weight	Growth rate
Dodecane	170	4
Octadecane	255	4
Dotriacontane	451	4
Hexatriocontane	507	0
Tetratetracontane	620	0
Squalane	423	0

Source: Data from Ref. 30.

dioxide produced during the metabolic activity of the microorganisms. This method has been utilized, measuring the CO_2 development from polyethylene marked with ^{14}C, by Albertsson [34], who demonstrated, in the case of unaged HDPE, that the biodegradation is higher for the powder than for the film. This result indicates that the products of the photooxidation, if in powder, are more exposed than the film to attack by microorganisms. Moreover, the biodegradability of photooxidized powders can be helped by the presence of hydrophilic groups introduced by the oxidation processes.

PHOTOOXIDATION

Solar Activation

The main cause of polyethylene degradation is photooxidative attack due to the combined action of ultraviolet radiation of the solar light and oxygen on the polymer surface.

The composition of the solar spectrum is strongly influenced by gases, water, and particles present in the atmosphere. Oxygen and ozone absorb the UV radiation lower than 280 nm, while water and carbon dioxide attenuate the strength of the IR radiation. Scattering, due to the interaction of the solar light with the air and the suspended particles, contributes to absorb and attenuate a portion of the solar radiation. The extent of the phenomena described is determined by the air thickness crossed by the light and therefore it is a function of the seasonal and daily position of the sun. In the case of the vertical incidence of the sun, the intensity of the solar radiation is subdivided among the different spectrum ranges [35, 36] according to the amounts shown in Table 5. We deduce that about 6% of the total radiation lies in the field of the UV light and is effective for photooxidative degradation.

The geographical position of the exposure place of the material strongly influences the solar energy on the ground. The hours of daily insolation and the average annual energy per area unit change remarkably from one place to another on the earth. The changes observed are particularly important into the field of UV radiation. In fact, the seasonal change of energy due to UV light is nearly of no value at 0° latitude and increases as the distance from the equator increases [37].

The connection between the wavelength of the light and the irradiated energy is expressed by the equation

$$E = h\nu = h\frac{c}{\lambda} \tag{22}$$

Table 5 Intensity of Solar Radiation as a Function of Wavelength

Range of wavelength (nm)	Intensity (%)
280–400	6.1
400–800	51.8
800–1400	29.4
1400–3000	12.7

Source: Data from Ref. 36.

In this equation h is the Planck constant, ν is the frequency, c is the speed of light, and λ is the wavelength.

On the basis of Eq. 22 radiation in the range 270–340 nm has an energy content equal to or higher than the bond energies C—C or C—H, which are in the range 340–420 kJ/mol [38, 39]. But this radiation is not absorbed from the hydrocarbons and, in the photochemical processes, the absorbed radiation can become active only for subsequent chemical transformations.

In the case of the C—C or C—H bonds the absorption of energy to excite the electrons of the σ bonds to a higher level (σ^*) requires radiation having wavelength equal to or lower than 130 nm, not present in the natural environment.

So the absolutely pure polyethylene should be resistant to the photooxidative processes due to solar light. Indeed the photooxidative reactions of the polyethylene are due to the presence of functional groups (chromophores) able to absorb the atmospheric UV light. In the case of LDPE the chromophore groups consist of carbonyl functions and hydroperoxides which form during polymerization and film manufacturing processes in consequence of secondary or thermal-oxidative reactions [7, 40].

The ketone groups $\rangle C{=}O$ show an absorption peak which extends above 300 nm. The maximum of this band is at 270–280 nm [41]. The hydroperoxide groups absorb between 240 and 360 nm, but they have an absorption strength very much lower than that of the carbonyl groups [24, 41].

Finally, solar radiation with wavelength equal to 300 nm is the most efficient for polyethylene degradation [41].

The absorption of the ketone groups concerns an electron situated in a no-bonding orbital (n). The excitation due to UV light carries the electron to an antibonding orbital π^*. Whereas in the ground state the two electrons of the same orbital have opposite (paired) spin, after the $n \rightarrow \pi^*$ transaction the excited electron can have spin opposite or equal (unpaired) to that of the other electron. In the former—the most frequent—there is the singlet state, characterized by fast stages of decay to states having lower potential energy. In the latter the excited state is called the triplet. It has a relatively long lifetime (up to 10^{-3} sec) and so is more effective. It can be caused by the singlet state by means of a conversion ("intersystem crossing") which is slow but favored as regards the energy.

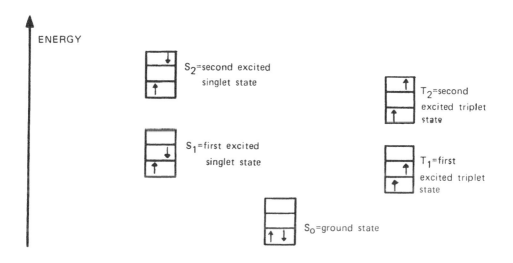

The energy absorbed by the carbonyl groups can be transferred to the hydroperoxide groups, which can decompose according to the reaction

$$ROOH \rightarrow RO\cdot + \cdot OH \qquad (9)$$

Yet the quenching of the excited state of the carbonyl groups can also be due to the molecular oxygen which thus proceeds to the singlet state [42]. Since commercial LDPE includes double bonds [7], in the presence of singlet oxygen allylic hydroperoxide can be formed [41]:

$$R{-}CH_2{-}CH{=}CH{-}R' \xrightarrow{{}^1O_2} R{-}CH{=}CH{-}\underset{\underset{H}{|}}{\overset{\overset{OOH}{|}}{C}}{-}R' \qquad (23)$$

A small amount of excited oxygen is directly produced by UV light. Under the effect of the light, charge-transfer complexes between polymer and atmospheric oxygen can be formed with subsequent production of hydroperoxides [43]:

$$RH + O_2 \xrightarrow{h\nu} RH^+ \cdots \cdot O_2^- \qquad (24)$$

$$RH^+ \cdots \cdot O_2^- \rightarrow R\cdot + \cdot OOH \qquad (25)$$

$$R\cdot + \cdot OOH \rightarrow ROOH \qquad (26)$$

The photooxidative cycle proceeds through a chain-radical mechanism similar to that Bolland [20] suggested for thermal oxidation. This mechanism can be subdivided into the steps of chain initiation, propagation, possible branching, and termination. Nevertheless, the initial stage of the photooxidation is different from the initiation of the thermal oxidation and includes all the phenomena of energy transfer described previously. Furthermore, the photooxidative degradation of LDPE is characterized by the Norrish I, Norrish II, and β-scission reactions. These reactions become particularly important to understand the global trend of the natural weathering of this polymer [7, 44].

Photooxidation Initiation

The opening reactions of photooxidation involve hydroperoxides, carbonyl groups, double bonds, catalyst residues, and oxygen. O—O bond energy of the hydroperoxides is 175.5 kJ/mol [24]; therefore these groups could be easily broken by the solar energy:

$$ROOH \xrightarrow{h\nu} RO\cdot + \cdot OH \qquad (9)$$

with formation of free radicals able to begin the degradation. Nevertheless, the decomposition of the hydroperoxides requires relatively high activation energies [45]. Therefore, the energy transfer carried out by the excited carbonyl groups, chiefly in polyolefin samples having high carbonyl/hydroperoxide ratios, is very important [46–48]. The hydroperoxides can also react for the possible presence of metals as catalytic residues. For example, some forms of TiO_2 catalyze the decomposition of the —ROOH groups [49–52]. A similar effect can be due to metals such as copper coming into contact with polyethylene during the exposure.

The ketone groups, besides active hydroperoxides, produce free radicals in the Norrish type I reaction:

$$R\text{—}\overset{\|}{\underset{O}{C}}\text{—}R' \overset{h\nu}{\rightarrow} R\text{—}\overset{\|}{\underset{O}{C}}\cdot + \cdot R' \tag{27}$$
$$\downarrow$$
$$R\cdot + CO$$

The beginning effect of the oxygen is attributed to the formation of charge-transfer complexes with subsequent generation of free radicals $R\cdot$ and $\cdot OOH$ and of saturated and allylic hydroperoxides (reactions 23–26).

Free radicals can be formed because of mechanical stress. In fact, the presence of alkyl macroradicals in polyolefins subjected to mechanical work, in the absence of oxygen and at low temperature, has been demonstrated [53]. In the presence of O_2 the alkyl macroradicals change in peroxy radicals.

Photooxidation Propagation

The propagation of the photooxidation is due to alkyl peroxy radicals $RO_2\cdot$, which can form through a fast reaction of the carbon macroradicals with oxygen:

$$R\cdot + O_2 \rightarrow ROO\cdot \tag{7}$$

This reaction shows that solubility and oxygen diffusion are important factors for the polyolefin's degradability. As the oxidation proceeds more easily in the amorphous materials, permeable to the oxygen, LDPE films, characterized by low crystallinity, are especially vulnerable to photooxidation. The importance of the oxygen diffusion into LDPE has been demonstrated by the change of the carbonyl index as a function of depth of the exposed sample [54]. Therefore, the photooxidation is, at least initially, a surface occurrence. Nevertheless, it has been observed that the simultaneity of thermal and photochemical processes produces the same oxidation degree on both sides of an LDPE film 145 μm thick after about a year of outdoor exposure [7]. The study of the oxidation reaction of different thick polyethylene films and of the importance of the oxygen diffusion as a factor determining the oxidation rate allows us to conclude that the diffusion effects are very important in the case of photooxidation of films ≥ 3 mm thick [54–56].

Expressions of the rate reaction changing thickness, diffusion constant, and oxygen solubility in the film have been obtained [54, 56–58]. The calculations show that for a film 3 mm thick, the photooxidation of the central portion of 1 mm is strongly reduced because of the low diffusion rate of the atmospheric oxygen. The calculated oxidation profiles agree with those found, as Figure 4 indicates [54]. It shows the carbonyl group concentration as a function of the film depth. In any case it is very important to remember that the photooxidation reaction of polyethylene films does not proceed with the same mechanism on all the areas, because of the structural heterogeneity of the system. Thus limited regions can exist in which there is a high concentration of oxidation products and spontaneous tearings of the exposed film are observed [8, 59, 60].

The alkyl peroxide radicals can also form with the following reaction caused by light, heat, or metallic ions [61]:

$$2ROOH \rightarrow ROO\cdot + RO\cdot + H_2O \tag{12}$$

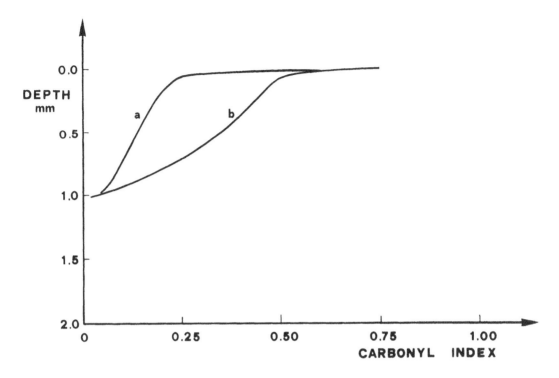

Figure 4 Carbonyl index as a function of depth. Curve a, LDPE exposed to fluorescent lamps in air; curve b, LDPE exposed in the tropics. (Data from Ref. 54.)

The ROO· radicals propagate the formation of hydroperoxides and alkyl macroradicals:

$$ROO· + R'H \rightarrow ROOH + ·R' \tag{8}$$

Chain Branching

The RO· and ·OH radicals cause the chain branching as a consequence of the following reactions and thus phenomena of reaction acceleration:

$$RO· + R'H \rightarrow ROH + ·R' \tag{10}$$

$$HO· + RH \rightarrow R· + H_2O \tag{11}$$

The presence of side chains in the polymer increases the LDPE degradability since the tertiary carbon atoms are preferential points which may be attacked by oxygen and help the formation of free radicals (see Table 3). Nevertheless, there are data showing that the attack points of the oxygen tend to lessen after prolonged exposure times of LDPE films [8]. The relative optical densities of the vinyl and carbonyl groups slacken their growth from about 10,000 hr of exposure.

Termination

The chain termination stage is due to reactions between two radicals, as in the following examples:

$$R\cdot + R\cdot \rightarrow \text{inactive products} \tag{28}$$

$$R\cdot + RO_2\cdot \rightarrow \text{inactive products} \tag{29}$$

$$RO_2\cdot + RO_2\cdot \rightarrow \text{inactive products} \tag{30}$$

After prolonged exposure times there is an increase in the percentage of crosslinking reactions [62] due to radical coupling, and this fact is indicated experimentally by the high content of material insoluble in orthodichlorobenzene [7, 8]. The presence in UV-irradiated polyethylene of absorption peaks in the range 1250–1170 cm^{-1} indicates the presence of ether bonds [63]:

$$
\begin{array}{c}
-CH_2-CH-CH_2- \\
| \\
O \\
| \\
-CH_2-CH-CH_2-
\end{array}
$$

Main Reactions

The main reactions characterizing the LDPE photooxidation process are Norrish I, Norrish II, β-scission, and macroradical breaking.

Norrish-Type Reactions

Norrish type I and II reactions are photolytic decomposition characteristic of the aliphatic ketones. The carbonyl groups are excited to singlet and triplet states by radiation in the range 270–330 nm. These excited carbonyl groups can give the Norrish-type reactions. Guillet and Norrish [64] have demonstrated that the polymeric ketones also give these reactions. The most reactive ketone groups are the most mobile, i.e., those near chain ends or in short side chains [65].

The type I reaction is a homolytic rupture process of the molecule giving two free radicals [66]:

$$
R-\underset{\underset{O}{\|}}{C}-CH_2-R' \xrightarrow{h\nu} R-\underset{\underset{O}{\|}}{C}\cdot + \cdot CH_2R' \tag{31}
$$

The R—C· radical can subsequently decompose to yield carbon monoxide and a macro-
$\quad\quad\;\|$
$\quad\quad\;O$
radical R·.

Some authors [8, 67, 68] believe the Norrish type I reaction to be of peculiar importance in the late stages of LDPE photooxidation. In fact this type of reaction can explain the decrease of carbonyl groups found in LDPE samples aged over 10,000 hours [8].

The Norrish type II reaction is an intramolecular process, nonradical, which produces a double bond and a methylketone:

$$
R-CH_2-CH_2-CH_2-\underset{\underset{O}{\|}}{C}-CH_2-R' \xrightarrow{h\nu} R-CH=CH_2 + CH_3-\underset{\underset{O}{\|}}{C}-CH_2-R' \tag{32}
$$

This reaction involves the rupture of the bond between the carbon atoms in α and β position with respect to the carbonyl group and is subsequent to the formation of a six-membered cyclic intermediate which involves the hydrogen on the γ carbon atom [69]:

$$R-CH \overset{\displaystyle H=\!=\!=0}{\underset{\displaystyle CH_2=\!=\!=CH_2}{}} C-CH_2-R'$$

Probably the type II reaction occurs prevalently in the first stage of the photooxidation [7, 67, 68], when a steady increase of the vinyl and carbonyl functions is observed. In this stage the stabilizing agents usually employed in LDPE films are still active enough to slacken the radical processes, giving a greater increase of vinyl and carbonyl groups and meaningful degradation phenomena [7].

β-*Scission*

The β-scission reaction becomes important when, in the absence of effective action of the antioxidants, free radicals to the tertiary carbon atoms can form. This fact emphasizes the importance of the branching in LDPE degradation, with short chains also. In fact, there is no experimental proof of meaningful amounts of aldehyde groups [70], obtainable from β-scission to secondary carbon atoms, in degraded LDPE. β-Scission of tertiary radicals has been considered by some authors [44] as the main reaction in the degradation of polypropylene as well as polyethylene. This reaction helps in explaining the greater increase [8] of carbonyl groups after the first 8000–9000 exposure hours of an initially stabilized LDPE:

$$
\underset{\underset{\textstyle R'}{\textstyle |}}{\overset{\overset{\textstyle O\cdot}{\textstyle |}}{R-C}}-CH_2-R'' \rightarrow \underset{\underset{\textstyle R'}{\textstyle \diagup}}{\overset{\overset{\textstyle R}{\textstyle \diagdown}}{C}}=O + \cdot CH_2-R'' \tag{33}
$$

The alkoxy radicals arising from decomposition of allylic hydroperoxides can give disproportion and crosslinking [71].

Macroradical Breaking

The breaking reactions of the macroradical $R\cdot$, which is remarkably unstable [72, 73], also involves the bond to the carbon atom adjacent to the radical site [74]. This reaction takes place owing to the energy gain obtained because of the formation of a new double bond:

$$R-CH_2-\dot{C}H-CH_2-R' \rightarrow R\cdot + CH_2=CH-CH_2-R' \tag{34}$$

As a matter of fact, measurements of the relative optical densities of the vinyl groups indicate that the ending double bonds reach very high concentrations in the latest stages of LDPE degradation, when there are low values of $\bar{M}_n$ and $\bar{M}_w$ [8, 68].

Other Reactions

The contemporaneous occurrence of the β-scission of the alkoxy radicals and the ruptures of the $\cdot R$ macroradicals leads the molecular weight $\bar{M}_w$ to reach the lowest value measurable under complete solubility conditions of the sample [8]. The observed reduction of the carbonyl groups in the later stages of exposition of LDPE films may be explained with the exhaustion of the points that may be attacked by oxygen and with the extension of the Norrish type I reaction [8].

Another possible cause of lessening of the carbonyl groups is their interaction with ·OH, RO·, and ROO· radicals [75]:

$$R\text{—}C\text{—}R' + \cdot OH \rightarrow R\text{—}C\text{—}OH + \cdot R' \tag{35}$$
$$\underset{O}{\overset{\|}{}} \qquad\qquad \underset{O}{\overset{\|}{}}$$

$$R'\text{—}C\text{—}R'' + RO\cdot \rightarrow R'\text{—}C\text{—}OR + \cdot R'' \tag{36}$$
$$\underset{O}{\overset{\|}{}} \qquad\qquad \underset{O}{\overset{\|}{}}$$

$$R'\text{—}C\text{—}R'' + ROO\cdot \rightarrow R'\text{—}C\text{—}OOR + \cdot R'' \tag{37}$$
$$\underset{O}{\overset{\|}{}} \qquad\qquad \underset{O}{\overset{\|}{}}$$

These reactions show as well that the aged samples can contain carboxylic acids, esters, and peroxyesters. Moreover, simultaneously to chain scissions, the formation of side chains is possible.

Quantum Yield

The stability of the polymers to photooxidation is often expressed from the quantum yield, i.e., from the ratio of the reacted moles to the energy amount as einstein. Generally the quantum yield of plastics materials for a generic chain scission is in the range of 10^{-2} to 10^{-5} [41]. Values of 0.025 have been found by Hartley and Guillet [69] for ethylene–1% CO copolymers in the solid state at 90°C.

Mechanical Properties

The degradation reactions decrease the average molecular weight of the polymer and change the physical and mechanical properties of the film up to values which can become a hundredth of the starting value. It is evident that the stabilized films lose their mechanical characteristics in the course of a longer time than that observed in the case of materials that are not stabilized [76]. A recent work [76] indicates that the time necessary for tensile strength and elongation at break of stabilized LDPE to be reduced to half of their starting values is four times greater than the LDPE free of stabilizers.

The trend of the change of mechanical properties compared with exposure time and reduction of the molecular weight of a photodegraded LDPE is shown in Figures 5 and 6, indicating that after 10,000 exposure hours there is a change of the mechanical properties that makes the material unutilizable [8].

The same authors [8] observed spontaneous tearing lines in the film after 10,000 exposure hours. The relative optical densities of the vinyl and carbonyl groups of samples taken near the tearing lines are higher than those measured on samples taken far from tearing lines. This result confirms that the LDPE films are heterogeneous materials as regards their oxidability, i.e., the distribution of the oxygen-reactive sites [12]. The useful life of a film changes, obviously, with the climatic conditions of the place of exposure.

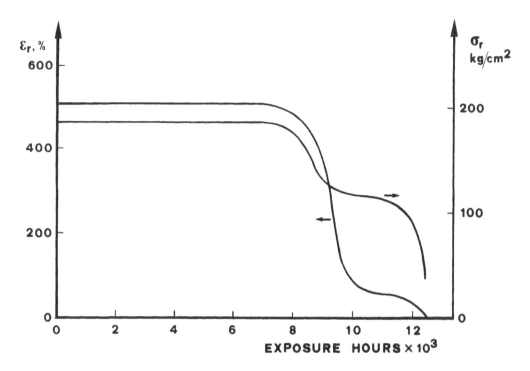

Figure 5   Time dependence of tensile strength (σ_r) and elongation at break (ϵ_r) of an LDPE film. (Data from Refs. 7 and 8.)

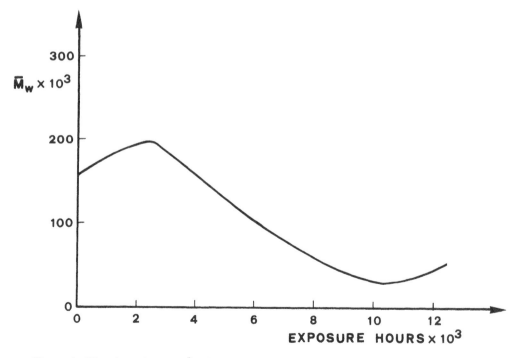

Figure 6   Time dependence of $\bar{M}_w$ of an LDPE film. (Data from Ref. 8.)

Controlled Photodegradability

Previously we stressed that the ketone groups present in the polymer chain aid its degradation on the basis of reactions of the Norrish type I and II. Guillet and Norrish [64, 77] prepared ethylene copolymers with unsaturated aliphatic ketones obtaining products degradable according to the following scheme:

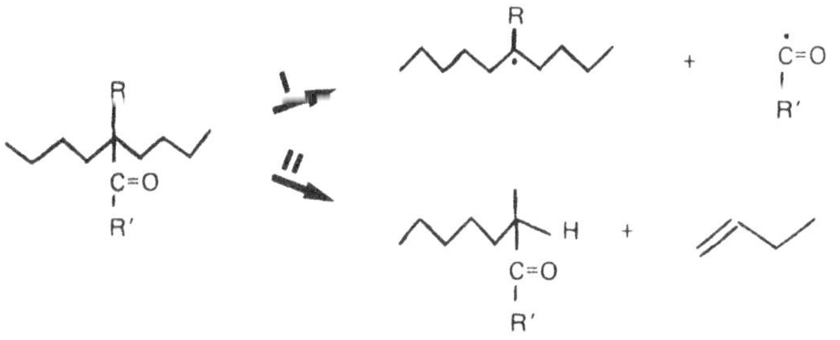

$$(38)$$

Since the product is generally kept in closed places, degradation does not occur during storage, because window glass is a very effective filter of UV radiation. Only during exposure to the atmosphere do the chromophore groups become active, in the absence of the direct solar radiation, since 50% of the UV radiation comes from scattering from the sky. The rate of the embrittlement process is due, under the same conditions, to the carbonyl concentration in the macromolecules.

Ethylene–carbon monoxide copolymers, characterized by an accelerated trend of photodegradation, are available on the market in the United States.

The presence of metallic ions aids the formation of radicals and thus the LDPE degradation processes according to the following scheme:

$$ROOH + M^{n+} \rightarrow ROO\cdot + M^{(n-1)+} + H^+ \tag{39}$$

$$ROOH + M^{(n-1)+} \rightarrow RO\cdot + M^{n+} + OH^- \tag{40}$$

Therefore, the addition of metal salts to LDPE can give rise to the formation of products having a controlled time of useful life. These systems are active in the absence of light too, and they give rise to unstable products. Scott [78, 79] first observed that metal complexes containing sulfur atoms act to destroy the hydroperoxides by means of nonradical reactions. In this way reactions of the type

$$ROOH \rightarrow RO\cdot + \cdot OH \tag{9}$$

which cause cycles of radical reactions giving carbonyl groups, activators of chain scissions (Norrish II), are inhibited. Metal complexes, such as metal–dialkyl–dithiocarbamate, act initially as catalysts for peroxide destruction and later as thermal stabilizers of the polymer during the hot-working or in the first exposure time of the material to solar light. The disruption of the peroxides takes place, however, with decomposition of the ligands and release of the metallic ion which subsequently develops its degrading action in the absence of light.

Scott additive allows the preparation LDPE films [80] characterized by a useful life predeterminable on the basis of the concentration of the metal complex added to the polymer.

REFERENCES

1. S. Krimm, *Advances in Polymer Sciences,* Vol. 2, Springer-Verlag, Berlin (1960/1961).
2. F. H. Bovey, *Chain Structure and Conformation of Macromolecules,* Academic Press, New York, p. 174 (1982).
3. W. L. Hawkins, W. Matreyek, and F. H. Winslow, *J. Polym. Sci., 41:* 1 (1959).
4. F. H. Winslow, C. I. Aloisio, W. L. Hawkins, W. Matreyek, and S. Matsuoka, *Chem. Ind. (London) 13:* 533 (1963).
5. F. H. Winslow and W. Matreyek, Paper presented at Chicago Meeting, ACS, Division of Polymer Chemistry, September 1954, Vol. 5, p. 552.
6. J. P. Luongo, *J. Polym. Sci., B1:* 141 (1963).
7. F. Severini, R. Gallo, S. Ipsale, and N. Del Fanti, *Polym. Deg. Stab., 14:* 341 (1986).
8. F. Severini, R. Gallo, S. Ipsale, and N. Del Fanti, *Polym. Deg. Stab., 17:* 57 (1987).
9. R. H. Hansen, in *Thermal Stability of Polymers* (R. T. Conley, ed.), Marcel Dekker, New York, chap. 6 (1970).
10. W. L. Hawkins, *SPE Trans., 4:* 187 (1964).
11. R. J. Martinovich, *Factors Affecting UV Degradation of Polyolefins,* Tech. Info., Marlex HDPE (1964).
12. N. M. Emanuel, G. E. Zaikov, and Z. K. Maizus, *Oxidation of Organic Compounds,* Pergamon Press, Oxford, Chap. 10 (1984).
13. R. Simha and L. A. Wall, *J. Polym. Sci., 6:* 39 (1951).
14. L. A. Wall, S. L. Madorsky, D. W. Brown, S. Straus, and R. Simha, *J. Am. Chem. Soc., 76:* 3430 (1954).
15. R. H. Hansen, Proceedings of the Symposium on Polypropylene Fibers, September 17–18, p. 130 (1964).
16. A. Hoff and S. Jacobson, *J. Appl. Polym. Sci., 26:* 3409 (1981).
17. A. Holmstrom and E. M. Sorvik, *J. Polym. Sci. Polym. Chem. Ed., 16:* 2555 (1978).
18. G. Scott, in *Developments in Polymer Stabilization* (G. Scott, ed.), Applied Science Publishers, London, p. 1 (1981).
19. Y. Kamiza and E. Niki, in *Aspects of Degradation and Stabilization of Polymers* (H. H. G. Jellinek, ed.), Elsevier, Amsterdam, p. 79 (1978).
20. J. L. Bolland, *Q. Rev. London, 3:* 1 (1949).
21. D. L. Allara, D. Edelson, and K. C. Irwin, *Int. J. Chem. Kinet., 4:* 345 (1972).
22. E. Niki and Y. Kamiya, *J. Org. Chem., 38:* 1403 (1973).
23. F. Severini and R. Gallo, unpublished results.
24. S. W. Benson, *J. Chem. Educ., 42:* 502 (1965).
25. E. Niki, C. Decker, and F. R. Mayo, *J. Polym. Sci. Polym. Chem. Ed., 11:* 2813 (1973).
26. A. Davis and D. Sims, in *Weathering of Polymers,* Applied Science Publishers, London, p. 5 (1983).
27. J. E. Potts, in *Encyclopedia of Chemical Technology,* 3rd ed., Supplement Volume, (M. Grayson, ed.), Wiley, New York, p. 626 (1984).
28. A. Davis and D. Sims, in *Weathering of Polymers,* Applied Science Publishers, London, pp. 54–57 (1983).
29. A. W. Bosshard and H. P. Schlumpf, in *Plastics Additive Handbook* (R. Gachter and H. Muller, eds.), Carl Hanser, Munich, p. 435 (1985).
30. J. E. Potts, R. A. Clendimming, W. B. Ackart, and W. D. Niegisch, in *Polymers and Ecological Problems* (J. Guillet, ed.), Plenum, New York, p. 61 (1973).
31. *Annual Book of ASTM Standards,* Part 27, ASTM-D-1924, p. 593 (1970).
32. M. S. Johannessen, *J. Appl. Polym. Sci. Appl. Polym. Symp., 35:* 415 (1979).
33. A. Tsuchii, T. Suzuki, and S. Fukuoka, *Biseibutsu Kogyo Giiutsu Keukyu Hokoku, 55:* 35 (1980). (In A. C. Albertsson, Advances in the Stabilization and Controlled Degradation of Polymers, Lucerne, Switzerland, 21–23 May 1986.)

34. A. C. Albertsson, "The Synergism between Biodegradation of Polyethylenes and Environmental Factors," in *Advances in the Stabilization and Controlled Degradation of Polymers*, Lucerne, Switzerland, 21–23 May 1986.
35. S. T. Henderson, *Daylight and Its Spectrum*, Adam Hilger, London (1970).
36. Publication by Commission Internationale de l'Eclairage (CIE), p. 20 (1972).
37. D. Kockott, 10th Colloquium of Danubian Countries for Natural and Artificial Aging of Polymers, Vienna, 1977, Farbe Lack 85, p. 102 (1979).
38. B. Ranby and J. F. Rabek, *Photodegradation, Photooxidation and Photostabilization of Polymers*, Wiley-Interscience, London (1975).
39. B. Dolezel, *Die Beständigheit von Kunstoffen und Gummi*, Carl Hanser, Munich (1978).
40. D. J. Carlsson and D. M. Wiles, *J. Macromol. Sci. Rev. Macromol. Chem.*, *C14*: 65 (1976).
41. F. Gugumus, in *Plastics Additives Handbook* (R. Gachter and H. Muller, eds.), Carl Hanser, Munich (1985).
42. A. M. Trozzolo and F. H. Winslow, *Macromolecules*, *1*: 98 (1968).
43. A. Garton, D. J. Carlsson, and D. M. Wiles, *Developments in Polymer Photochemistry*, Vol. 1, Applied Science Publishers, London, p. 93 (1980).
44. D. J. Carlsson and D. M. Wiles, *Macromolecules*, *2*: 587 (1969).
45. H. Muller, in *Plastics Additives Handbook* (R. Gachter and H. Muller, eds.), Carl Hanser, Munich, p. 75 (1985).
46. D. J. Carlsson and D. M. Wiles, *Macromolecules*, *7*: 259 (1974).
47. G. Geuskens and Q. Lu-Vink, *Eur. Polym. J.*, *18*: 307 (1982).
48. L. C. Stewart, D. J. Carlsson, D. M. Wiles, and J. C. Scaiano, *J. Am. Chem. Soc.*, *105*: 3605 (1983).
49. G. Frick, G. C. Newland, and R. H. S. Wang, *Am. Chem. Soc. Symp. Ser.*, *151*: 147 (1981).
50. H. G. Valz, G. Kaempf, H. G. Fitzky, and A. Klaeren, *Am. Chem. Soc. Symp. Ser.*, *151*: 163 (1981).
51. J. Lemaire, R. Arnaud, and J. L. Gardette, *Pure Appl. Chem.*, *55*: 1603 (1983).
52. R. Arnaud and J. Lemaire, *Dev. Polym. Photochem.*, *2*: 135 (1981).
53. J. Sohma, *Dev. Polym. Deg.*, *2*: 99 (1979).
54. G. C. Furneaux, K. J. Ledbury, and A. Davis, *Polym. Degrad. Stab.*, *3*: 431 (1981).
55. N. C. Billingham and P. D. Calvert, in *Development in Polymer Stabilization*, Vol. 3 (G. Scott, ed.), Applied Science Publishers, London, p. 139 (1980).
56. A. V. Cunliffe and A. Davis, *Polym. Degrad. Stab.*, *4*: 17 (1982).
57. J. E. Wilson, *J. Chem. Phys.*, *22*: 334 (1954).
58. J. E. Wilson, *Ind. Eng. Chem.*, *47*: 2201 (1955).
59. D. J. Carlsson, A. Garton, and D. M. Wiles, *Macromolecules*, *9*: 695 (1976).
60. M. N. Amin, G. Scott, and L. M. K. Tillekeratne, *Eur. Polym. J.*, *11*: 85 (1975).
61. A. J. Chalk and J. F. Smith, *Trans. Faraday Soc.*, *53*: 1214 (1957).
62. G. R. Cotten and W. Sacks, *J. Polym. Sci.*, *A1*: 1345 (1963).
63. H. C. Beachell and S. P. Nemphos, *J. Polym. Sci.*, *21*: 113 (1956).
64. J. E. Guillet and R. G. W. Norrish, *Proc. Roy. Soc.*, *A233*: 153 (1955).
65. D. J. Carlsson, A. Garton, and D. M. Wiles, in *Developments in Polymer Stabilization*, Vol. 1 (G. Scott, ed.), Applied Science Publishers, London, p. 219 (1979).
66. O. Cicchetti, *Adv. Polym. Sci.*, *7*: 70 (1970).
67. C. H. Chew, L. M. Gan, and G. Scott, *Eur. Polym. J.*, *13*: 361 (1977).
68. F. Severini, R. Gallo, S. Ipsale, and N. Del Fanti, *Mater. Plast. Elastom.*, *11*: 575 (1986).
69. G. H. Hartley and J. E. Guillet, *Macromolecules*, *1*: 165 (1968).
70. J. A. Adams, *J. Polym. Sci.*, *A1*, *8*: 1279 (1970).
71. G. Scott, in *Ultraviolet Light Induced Reactions in Polymers. ACS Symposium Series*, Vol. 25 (S. S. Labana, ed.), American Chemical Society, Washington, p. 340 (1976).
72. U. S. Pudov and A. L. Buchachenko, *Usp. Khim.*, *39*: 130 (1970).
73. V. Zakrevskii and V. E. Konshunov, *Vysokomol. Soedin.*, *A14*: 955 (1972).

74. A. Davis and D. Sims, in *Weathering of Polymers*, Applied Science Publishers, London, p. 111 (1983).
75. G. Geuskens and M. S. Kabamba, *Polym. Degrad. Stab.*, *4*: 69 (1982).
76. F. P. La Mantia, *Eur. Polym. J.*, *20*: 993 (1984).
77. J. Guillet, in *Polymers and Ecological Problems* (J. Guillet, ed.), Plenum, New York, p. 14 (1973).
78. G. Scott, Br. Pat. 1,356,107 (1974).
79. G. Scott, *J. Polym. Sci. Symp.*, *57*: 357 (1976).
80. D. Gilead and G. Scott, in *Developments in Polymer Stabilisation*, Vol. 5 (G. Scott, ed.), Applied Science Publishers, London, p. 71 (1982).

15

Properties of Isotactic Polypropylene

Vittoria Vittoria
*Istituto di Ricerche su Tecnologia
dei Polimeri e Reologia del CNR
Naples, Italy*

INTRODUCTION

Polypropylene is one of the most important of the hydrocarbon polymers and is the first synthetic stereoregular polymer to achieve considerable commercial and industrial importance.

Stereoregular polypropylene was first obtained by Natta and co-workers [1–4] in the 1950s at the Polytechnic of Milan. They reported the preparation of crystalline polypropylene using heterogeneous catalysts of the type discovered by Ziegler for the low-pressure polymerization of ethylene to high-density polyethylene. It was soon evident that the stereoregular polymer could be extensively used in the thermoplastic industry as well as in film and fiber production.

The molecular structure of polypropylene

$$\left(-CH_2-\underset{\underset{CH_3}{|}}{CH}- \right)_n$$

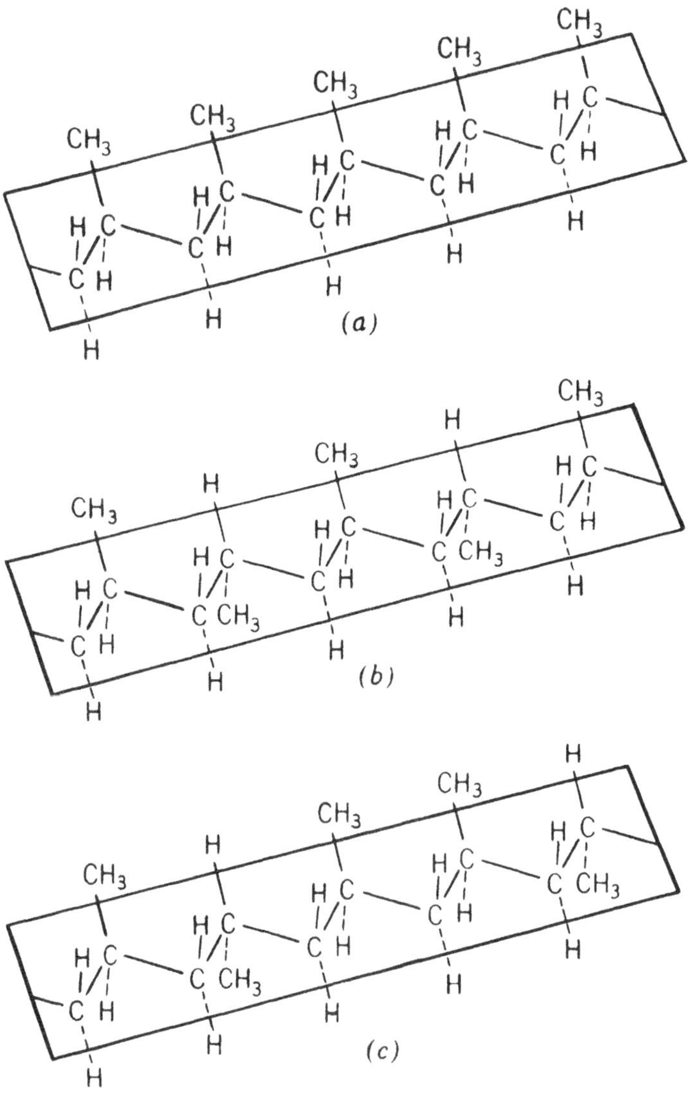

Figure 1 Representation of the spatial disposition of CH_3 in (a) isotactic, (b) syndiotactic, and (c) atactic polypropylene chain segments.

is formally derived from polyethylene by the substitution of one of the H atoms on alternate C atoms of the chain by a CH_3 group. Stereoregular configurations can be described as in Figure 1. Figure 1*a* is isotactic, in which all pendant CH_3 groups are attached on the same "side" of the chain, that is, all units have a spatially identical arrangement of atoms; Figure 1*b* is syndiotactic, in which the configuration of the C atom carrying the side group shows a regular alternation along the chain; Figure 1*c* is atactic, in which there is a random arrangement of pendant CH_3 groups. In addition to atactic, isotactic, and syndiotactic polypropylenes, it is also possible to obtain polymers in which blocks of isotactic sequences reverse their position along the chain. These structures are called stereoblock polymers. Figure 2 presents the models of isotactic, syndiotactic, and stereoblock polypropylene.

The stereoregularity of polymer molecules is of high importance for the properties of materials. It determines the possibility for adjacent molecules to fit together and therefore to crystallize. It also controls the strength of forces between molecules, on which the mechanical properties of the material depend.

Commercial polypropylenes generally have about 0.95 or higher isotactic indices. High isotactic index contributes to higher crystallization of the polymer and much improved mechanical properties of the products. It increases, for example, the yield stress, elastic modulus, hardness, and brittle point of the polymer [5].

In the case of polypropylene, among the possible stereoregular configurations, the isotactic form is the most important commercially.

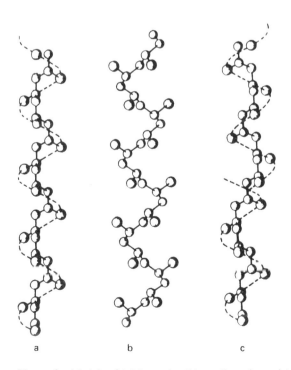

a   b   c

Figure 2   Models of (*a*) isotactic, (*b*) syndiotactic, and (*c*) stereoblock iPP.

STRUCTURE OF ISOTACTIC POLYPROPYLENE

Isotactic polypropylene (iPP) can crystallize in three different polymorphic forms, α, β, and γ [6]. The occurrence of these forms is dependent on many factors, among which the most important are the thermal treatment of the samples as well as the molecular weight and the degree of isotacticity. The infrared spectra of all three forms are closely similar and show the presence of ternary helices in all three.

α Modification

The principal isotactic form (the α form) was first observed by Natta [1, 2] in melt-crystallized material. It is normally observed also in solution-grown crystals [7, 8]. The structure determinations carried out on α phase isotactic polypropylene [1, 2, 6] showed that the crystal structure is monoclinic and the chains assume a helical conformation [Figure 2a]. From the x-ray data it follows that the identity in the main direction of the chain comprises three monomer units. The symmetry of the helix is characterized by a threefold screw axis. Consequently right- and left-handed helices are possible. They are arranged in a regular pattern, a left-handed helix always facing a right-handed one.

β Modification

The β form generally is formed in mixture with the α form by quenching the melt at temperatures between 100 and 120°C. In these conditions the relative amounts of α and β forms are dependent on the efficiency of the quenching conditions and on the tacticity and molecular weight of the specimen [6]. This form has been observed in melt-crystallized spherulites formed in particular conditions [9] and in thin films [10] or when crystallization has occurred in the presence of shearing forces [11, 12].

A number of additives, including the aluminum salt of 6-quinizarin sulfonic acid [13], a quinacridone dye [14, 15], and many others [16, 17], are effective in producing predominantly the β modification. Isophthalic and terephthalic acids also nucleate the β phase [18], but not as efficiently as the previously mentioned additives. Single crystals grown in low molecular weight diluent in the presence of 0.1% of a quinacridone dyestuff also show the β crystalline form [19]. The analysis of the crystal structure of this form [6, 19] suggests that the unit cell is hexagonal and the most likely packing of the helical molecules, within the unit cell, involves the incorporation of all left-handed or all right-handed helices within a given crystallite.

The different polymorphic forms α and β are generally organized into spherulites. Padden and Keith [9] characterized four types of spherulites. The α form is found in type I and II spherulites; the β form is found in types III and IV. Types I and II exhibit a well-defined Maltese cross and can be differentiated on the basis of birefringence, which for the first is positive and for the second is slightly negative. A more highly negative birefringence class of spherulites is that of type III. The birefringence of type IV is much more intense in its negative value and shows a distinctly banded structure [12, 20, 21].

γ Modification

A third crystalline modification of iPP is the γ form. It was first described [22] as observed only in polypropylene with low molecular weight having a degree of isotacticity not too low. It was later obtained from solvent fractionation [6]. A fraction of isotactic polypropylene precipitated from petrol ether or xylene in the temperature range 35–70°C, melted and slowly cooled from 190°C, under vacuum, at 6°/hr to room temperature, shows γ form crystallinity. It is therefore not necessarily associated with low molecular weight polymer crystallinity.

This crystalline modification has been obtained, free of α form, by melt crystallization under conditions of elevated pressure [18]. In this case a difference in behavior was found between specimens prepared at a large degree of supercooling ($\Delta T > 40°C$) and at a low degree of supercooling ($\Delta T < 40°C$). In the first case the samples, examined by low-angle x-ray diffraction, show long spacings, which are characteristic of chain-folded lamellar crystals. For the second one, no long spacings are obtained. These data suggest that for samples prepared at elevated pressure with low supercooling the amount of chain folding is minimal. It is suggested that in this case the γ phase is composed by extended chain crystallization.

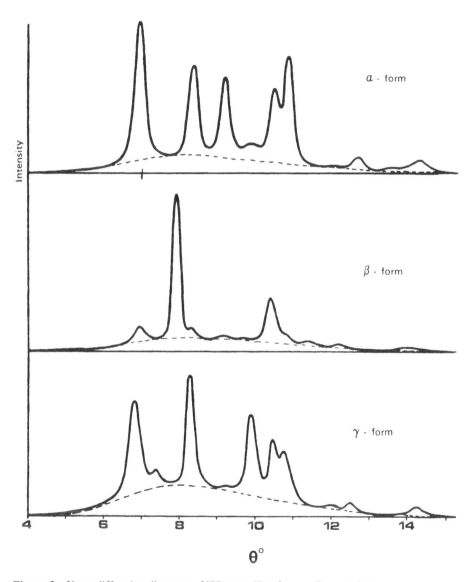

Figure 3 X-ray diffraction diagrams of iPP crystalline forms. (From Ref. 6.)

The most likely unit cell for the γ form is triclinic [6]. The method of crystallization and the molecular weight of the sample play an important role in determining the phases present in the sample.

Figure 3 presents the x-ray diffraction diagrams of the three crystalline forms of isotactic polypropylene.

Thermal and Mechanical Stability of β and γ Modifications

Figure 4 is a typical x-ray diffraction diagram showing both α and β form crystallinity. Turner Jones et al. [6] obtain the relative proportions of α and β forms by an empirical ratio K defined as

$$K = \frac{H\beta}{H\beta + (H\alpha_1 + H\alpha_2 + H\alpha_3)} \tag{1}$$

where $H\alpha_1$, $H\alpha_2$, and $H\alpha_3$ are the heights of the three strong equatorial α-form peaks (110), (040), and (130), and $H\beta$ the height of the strong single ($hk0$) peak at $d = 5.495$ Å, as shown in Figure 4. The K parameter is an empirical one. It tends to zero when no β form is present, and $K = 1$ when α form is absent, but it is not an absolute measure of the proportion of β crystallinity.

It has been found that the β form is not thermally stable and by heating it is transformed in the α monoclinic form. Turner Jones et al. [6] report the rates of conversion to the α form of a predominantly β-form specimen (initial value of $K = 0.85$) at several temperatures from 130 to 150°C (Figure 5). Conversion is very rapid at 150°C. This result is consistent with the observations made by Padden and Keith [9] on the melting behavior of type III or IV (β-form) spherulites using the optical birefringence method. At lower temperatures conversion proceeds more slowly and reaches a limiting value at each temperature. The authors attribute this to the melting of smaller or less perfect crystallites of melting point lower than the temperature studied.

Morrow [18] analyzed differential scanning calorimetry (DSC) data of the melting behavior of β-phase iPP prepared in the presence of 0.2% sodium phthalate. When the heating rate was as high as 80°C/min the material melted in the vicinity of 150°C. When the heating rate was reduced, the thermograms took a more complex aspect, explicable with the conversion of β phase to α phase.

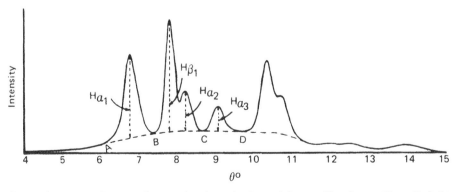

Figure 4 X-ray diffraction diagram showing mixed α and β crystalline forms. (From Ref. 6.)

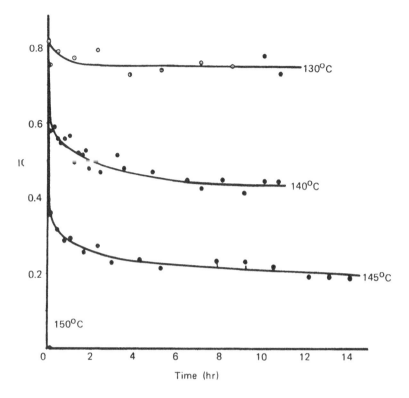

Figure 5 Conversion of β to α form at several temperatures of a sample with an initial value of $K = 0.85$. (From Ref. 6.)

Also, the γ-phase polypropylene crystallized at elevated hydrostatic pressure is metastable in character [18]. The DSC thermogram of a sample crystallized under pressure at a large degree of supercooling shows that the material transforms entirely to the α phase prior to melting. On the contrary, when the γ phase is prepared with a low degree of supercooling it is stable, showing no transformation to the α phase up to 173°C, temperature at which the x-ray scan obtained showed only the presence of γ phase.

Mechanical deformation of the β-form specimens [6, 18] occurs through neck propagation, but the drawn fibers show the smectic form oriented (discussed later). On drawing at higher temperatures the oriented α form is obtained. Therefore, it is not possible to obtain the β form oriented, free of the α form.

In the case of γ-form crystallinity all attempts to obtain oriented fibers in this crystalline form failed. Both cold drawing and drawing at higher temperatures produced only oriented α-form crystallinity.

Smectic Modification

When isotactic polypropylene is rapidly quenched from the melt it forms a phase of intermediate order (between amorphous and crystalline), which has been described as smectic [23], paracrystal [24], glass [25], or mesomorphic phase [26]. Recently Wunderlich and colleagues [27, 28] proposed the term "condis crystal" (conformationally disordered crystal) as possibly more appropriate for this semidisordered state. It is called here "smectic phase," which is the original and most frequently used name. Although the structure and the properties of the smectic form have been studied by many authors, the precise nature of this phase is still in debate.

The x-ray diffraction pattern of this form (seen in Figure 6) shows two broad diffuse peaks, suggesting that the quenched structure is in a disordered state. However, the presence of a second peak indicates a greater order than that existing in amorphous polypropylene samples. On the basis of IR spectra it was recognized that the individual chains maintain the threefold helical conformation [23, 24] and the chains are parallel. Furthermore, the packing of the chains perpendicularly to their axes is more disordered than along the axes, although the relative displacements and orientations of neighboring chains are not completely random.

Wyckoff [29], on the basis of the x-ray characterization of a wide series of samples obtained by quenching, cold drawing, and annealing of commercial isotactic polypropylene, found in the smectic form some correlation among the positions of adjacent helices, their Z coordinates, and their rotational coordinates, suggesting therefore the existence of some three-dimensional order in addition to the parallelism of axes.

The observation that the positions of the two most intense diffraction maxima of the partially ordered phase ($1/d = 0.17$ Å^{-1}, $1/d = 0.24$ Å^{-1}) are nearly coincident with the two most intense peaks of the hexagonal (β) crystalline form ($1/d = 0.18$ Å^{-1}, $1/d = 0.24$ Å^{-1}) led Gailey and Ralston [30] to propose that the quenched material is composed of very small (50–100 Å) hexagonal crystals. Takahara and Kawai [31], on the basis of orientation factors from x-ray data of melt-spun quenched polypropylene, also suggest a hexagonal crystalline form. From a morphological study of quenched iPP Gezovich and Geil also preferred the hexagonal crystal assumption [32]. These authors suggested the "ball-like" structures or nodules they observed are small, imperfect hexagonal crystallites that are the basic morphological units for quenched iPP.

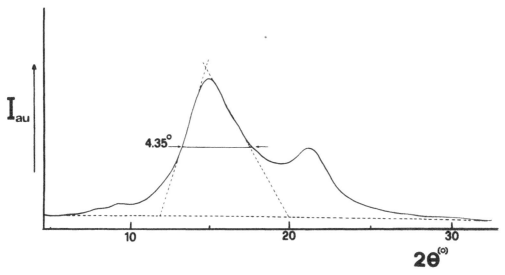

Figure 6 X-ray diffraction pattern of smectic polypropylene.

Bodor et al. [33] proposed that the quenched form is composed of microcrystals of monoclinic polypropylene and concluded that crystal size broadening is responsible for the typical x-ray diffraction pattern. The order existing in this form, according to Miller [24], might be that described by Hosemann as "paracrystalline" [34], defined as a deformation of the crystal structure obtained by replacing the constant cell edges by statistically determined vectors varying in both lengths and directions. On the basis of x-ray and IR data, Fujiyama et al. [35] agreed with the paracrystalline character.

McAllister et al. [36], from x-ray diffraction results, suggested the quenched state is 60% amorphous while the remainder of the sample has the helices arranged in a square array and a cubic tetragonal symmetry. It was later demonstrated that this model had to be discarded, since nearly all the diffraction maxima were interpreted as equatorial reflections, in contrast with the x-ray patterns of the oriented quenched samples [37]. To further clarify the structure of the partially ordered phase of iPP, Corradini et al. [26] compared the Fourier transforms of various kinds of aggregates of helices with the measured diffraction intensities. These authors draw some firm conclusions about the mesomorphic quenched form of iPP. It is built up of bundles of chains, having conformations very nearly threefold helices and parallel among themselves, as already suggested by other authors. The helices are probably terminated at helix reversal and other conformational defects, the mean dimension of each segment of helix of one sense being of the order of 40 Å. Lateral order does not seem as well developed as conformational order, although the relative heights of the neighboring chains within each bundle are mainly correlated. The local correlations between chains are probably nearer to those characterizing the crystal structure of the monoclinic form than to those characterizing the structure of the hexagonal form.

The conclusion of this study is that in mesomorphic iPP, short-range order is present within each chain and (to a lower degree) among chains. The chains are organized in bundles, for which any correlation about the relative position of the atoms is lost at distances of the order of 30–40 Å.

DETERMINATION OF PHASE COMPOSITION

It is well known that, for semicrystalline polymers, degree of crystallinity and morphology play a prominent role in influencing the macroscopic properties exhibited by the polymer. For example, drawing behavior, tensile strength, electrical and optical properties, and permeability of vapors are all directly related to the film's crystalline microstructure. The detailed study of the microstructure and how it can be altered during and after film fabrication by controlling, for example, the rate of cooling from the melt or the crystallization temperature can help improve some important properties of the film. In the case of isotactic polypropylene, in addition to degree of crystallinity and morphology, an important factor to be taken into account is the phase composition, that is, the possible presence of β or γ form crystallinity or a fraction of smectic form, in addition to the α form.

The predominant form of crystallinity in the melt crystallized at high temperatures is the α monoclinic form; on the other hand, in particular conditions, as reported earlier, β and γ crystallinity can be formed, mixed with the α form. Rapid quench cooling produces prevalently the smectic form. The lower the temperature of solidification, the larger is the fraction of smectic component formed. By rapid quenching at a sufficiently low temperature it is possible to obtain only the smectic phase without regular crystals. Since this form reverts to the monoclinic form when heated at temperatures above 70°C [38], by quench-

Table 1 Structural Organization of iPP Samples Crystallized in Different Conditions

Crystallization conditions	Structural organization
High temperature ($T > 120°C$)	α form
100°C $< T <$ 120°C and in the presence of particular additives	Mixed α and β form
Elevated pressure or sterically particular samples	γ form
Intermediate quenching temperature (0°C $< T <$ 70°C)	Mixed α and smectic form
Low quenching temperature ($T <$ 0°C)	Smectic form

ing at temperatures higher than 70°C it is possible to obtain only the crystalline phase without the smectic component. Therefore, a sample of iPP quenched at intermediate temperatures may contain different phases: the amorphous noncrystallizable phase, the crystalline phase, which is prevalently monoclinic but may contain β or γ crystallinity, and the intermediate smectic phase. The relative amounts of these phases are dependent either on the molecular characteristics of the polymer (molecular weight distribution and stereoregularity) or on quench conditions. Thickness of the sample, heat capacity, thickness and thermal conductivity of the substrate, and quenching medium and its temperature determine the relative amounts of each phase [39]. In Table 1 the structural organization corresponding to different crystallization conditions is reported.

To characterize a sample of quenched iPP it is necessary to determine the relative amounts of crystalline (α or β), smectic, and amorphous phases that are present. In fact, phase transformation ($\beta \rightarrow \alpha$ or smectic $\rightarrow \alpha$), due to change of temperature or to environmentals agents, can cause dimensional instability of the sample due to difference in density of the different phases. This is particularly relevant in the case of smectic $\rightarrow$ monoclinic transformation, since the densities of these two phases are quite different (0.916 g/cm^3 and 0.936 g/cm^3 respectively). Therefore, this problem is especially important in polymer technology, since, if a specimen of solid polypropylene is produced under rapid cooling conditions (e.g., in extrusion or molding processes), there will always be a fraction of material in an intermediate state of order, that is, in the smectic form.

Quantification of α Crystallinity Content

If a specimen of isotactic polypropylene has been obtained in conditions in which only the α modification is present, the crystallinity can be measured with the methods generally used for semicrystalline polymers. These methods include density, wide-angle x-ray scattering, and calorimetric measurements.

The wide-angle x-ray scattering determination of crystallinity was first published by Natta et al. [40] for unoriented specimens of iPP and later by Farrow for fibers of iPP [41]. The method consists of measuring the integrated area of the crystalline reflections (Ac) and the integrated area of the noncrystalline background (Aa) and comparing the two. The crystallinity is given by

$$\alpha_c = \frac{Ac}{Ac + kAa} \tag{2}$$

where k is a correction factor very near to 1 in the case of iPP.

The density method of obtaining the crystallinity in a polymer is based on the assumption that the material can be described as a two-phase system, amorphous and crystalline [41]. If the densities of a wholly crystalline and a wholly amorphous sample are known, then, by measuring the density of the sample of which crystallinity has to be determined and applying a simple proportion, the crystallinity of the specimen can be deduced. The equation that relates the density of the sample to the mass crystallinity is

$$\alpha_c = \frac{d_c(d - d_a)}{d(d_c - d_a)} \tag{3}$$

where d, d_c, d_a are the actual density and the densities of the crystalline and amorphous phases respectively.

In the case of polypropylene the density of the crystals, in the α form, can be calculated from the crystal structure [1] and has the value of 0.936 g/cm³. The range for the values of the amorphous density obtained by indirect methods is $0.85-0.87$ g/cm³ [41].

The calorimetric determination of crystallinity consists of obtaining, generally from measurements with a differential scanning calorimeter (DSC), the apparent enthalpy of fusion ΔH_f. This value compared with the enthalpy of fusion of crystals of iPP, ΔH_f°, gives the percentage of crystallinity:

$$\alpha_c = \frac{\Delta H_f}{\Delta H_f^\circ} \tag{4}$$

The reported value of ΔH_f° for iPP is 6943 J/mol [28].

Quantification of β Crystallinity Content

In the case of β crystallinity, the empirical ratio K (defined in Eq. 1), obtained by analyzing the x-ray diffraction spectra, has been generally used to quantify the amount of β form present [6]. Although this ratio varies from zero for no β form to unity for 100% β form, it is not an absolute measure of the proportion of β crystallinity. The overall percentage of crystallinity of the samples is estimated, as in the case of α form alone, by measuring the area under the crystalline diffraction peaks and dividing this by the total area under the crystalline and amorphous background curve. A new procedure for quantifying the volume fraction of β spherulites in a sample was developed recently [42]. It is based on the technique of selective solvent extraction. Samples with a high level of β form were prepared in the presence of a quinacridone dye, known as "permanent red" E3B. Because of the apparent high selectivity of hot toluene (95°C) in preferentially extracting the β phase, quantitative measurements of this extraction were performed. X-ray evidence for the selectivity of solvent extraction was obtained by recording the diffraction pattern of a film before and after toluene treatment.

One of the more interesting observations of this study is the variation of density with the β content of the sample, evaluated through the K parameter. Figure 7 is a plot of density versus K for compression-molded samples of two different series of samples. It is evident that there is an excellent linear correlation between these two properties. If this linear regression line is extrapolated, it predicts densities of 0.9023 and 0.9105 g/cm³,

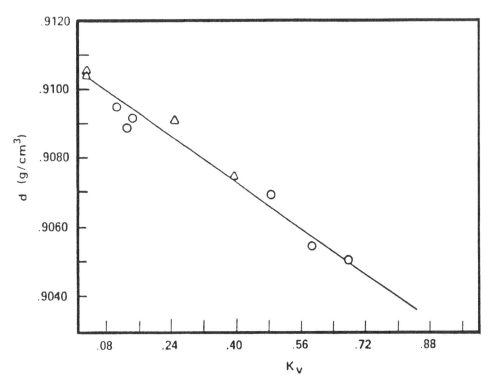

Figure 7 Density d of two series of compression-molded samples as a function of volume fraction of β crystals. (From Ref. 42.)

respectively, for K values of 1.0 (100% β content) and 0.0 (100% α content). Assuming a three-phase model for polypropylene, based on two crystalline phases (α and β) and one amorphous phase, from the measured densities the authors calculated the crystalline densities of α and β polymorphic forms. These crystal densities were in good agreement with those calculated from unit-cell dimensions.

Quantification of Smectic Content

The determination of the smectic content in a sample of iPP is even more important, because in polymer technology rapid cooling at low temperatures is a normal practice. In many cases the skin of a bulky specimen will be smectic, whereas the core will have a crystalline form. This indicates the presence of a three-phase system, crystalline–smectic–amorphous. It is therefore of interest to measure the composition quantitatively since it may have a considerable influence on the final properties of the product.

Methods have been developed for obtaining the fraction of each phase present by the use of x-ray diffraction techniques [32, 36, 43]. The method developed by Farrow [43] involves two estimates of crystallinity by x-ray methods. The first consists of the usual determination of crystallinity [40, 41] from x-ray spectra, that is, a comparison between the integrated intensity of the crystalline reflections and those of the noncrystalline background (Figure 8a, b). The second, which involves an additional measurement of crystallinity, is best illustrated with reference to Figure 8. In Figure 8c a trace of an unoriented

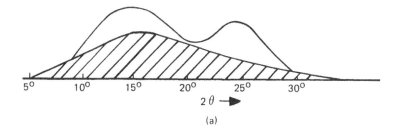

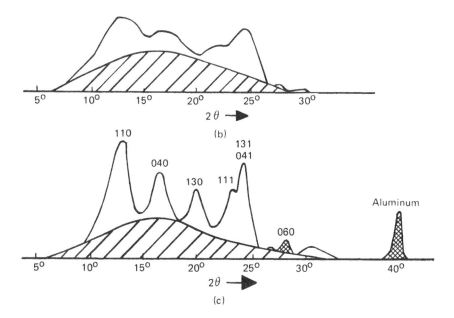

Figure 8 Microdensitometer traces of x-ray diffraction patterns. (*a*) Rapidly quenched polypropylene spun fiber; (*b*) rapidly quenched polypropylene spun fiber after heat treatment for ½ hr at 70°C; (*c*) rapidly quenched polypropylene spun fiber, after heat treatment for ½ hr at 140°C. (From Ref. 43.)

heat-crystallized sample of spun polypropylene is reported. On it are listed the indices of the principal peaks together with that of a minor peak, the 060. In Figure 8*a* the trace of a wholly smectic sample of polypropylene is reported. The absence of the 060 reflection is evident; the other peaks have merged into the two broad peaks characteristic of the smectic form. Therefore, the appearance of this peak is associated only with the presence of crystallinity due to the monoclinic phase and it can give a measure of the "true" crystallinity. It is then possible to obtain a measure of the smectic content from the difference between the value of crystallinity with the first method, which gives the total fraction of ordered phase (monoclinic + smectic), and the second, which gives a measure of monoclinic crystallinity only.

To use the 060 peak as a measure of crystallinity it is necessary to measure the integrated intensity of this peak, free of effect due to differences in exposure time to the x-ray beam, specimen size, absorption, etc. This was achieved by building into the focusing x-ray camera a standard reference sample in the form of a piece of aluminum foil 0.0005 in. in thickness. An x-ray diffraction pattern is obtained from this sample, most of which conveniently falls outside the pattern from a polypropylene sample, except for a line of reasonable intensity, which occurs at approximately $2\theta = 40°$ (Figure 8*c*). This line is used as a standard reference line, and the ratio between the integrated area of this and the 060 peak is obtained.

In this study, in addition to determining the fraction of smectic phase, the authors obtained the fraction of smectic component at different quenching temperatures. It was shown that the smectic content begins to change at about 40°C and decreases quite rapidly from 50°C onward, with a corresponding increase in the normal type of crystallinity. The smectic content reaches a minimum at about 85°C.

Gezovich and Geil [32], on the basis of Farrow's method, compared the measured wide-angle diffraction spectra for a given sample submitted to varying thermal treatment and crystallization conditions with spectra obtained by superposition of monoclinic and smectic x-ray diffraction patterns in different proportions. This method proved sufficiently accurate, but it is valid only for comparing different thermal treatments of a given sample. These authors obtain, with this method, the fraction of smectic form defined as

$$\frac{S}{S + C} \times 100 \qquad (5)$$

where S = fraction of smectic material and C = fraction of monoclinic material, for a series of samples quenched at different temperatures. In Figure 9 the percentage of smectic form is seen as a function of quenching temperature for two different substrates. The influence of the thermal conductivity of the substrate on the smectic content of the

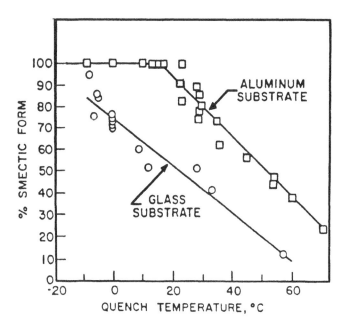

Figure 9 Effect of quenching temperature on percentage of smectic form for two different substrates. (From Ref. 32.)

sample is evident. At sufficiently low quenching temperature, using the aluminum substrate, the obtained samples contain 100% of smectic form; they are therefore biphase smectic–amorphous systems.

At the same low temperatures, with the glass substrate, lower fractions of smectic component are obtained, and the percentage of smectic form decreases as the quenching temperature increases, reaching a minimum at about 60°C.

Determination of the Structure of Smectic Form via Transport Properties

A different method for obtaining the fractions of smectic, monoclinic, and amorphous phase in samples quenched at different temperatures is based on the analysis of transport properties [44, 45]. Analysis of transport properties, sorption, and diffusion of gases and vapors has in many cases proved very successful in elucidating the amount and the role of the amorphous component, which is rather difficult to study with other techniques [46–48]. Since the crystals are assumed to be completely impermeable, they do not contribute directly to the sorption and impede the diffusion by compelling the penetrant molecules to detour through the amorphous component. For example, in the case of polyethylene, which is described as a two-phase amorphous crystalline system, it was found that the solubility of gases in a partially crystalline sample, K, is directly proportional to the amorphous volume fraction α_a, with the proportionality constant K^* being the solubility in completely amorphous polymer [49, 50].

The first problem dealt with was the determination of the transport parameters of a completely smectic–amorphous system, obtained by quenching the melt very rapidly at $-70°C$ (sample A) [44]. They were compared with the transport parameters of a polymer crystallized by rapid cooling to 110°C followed by slow cooling to room temperature. This sample resulted in a completely monoclinic–amorphous system, free of smectic phase (sample B). The penetrant used was CH_2Cl_2 at varying activity and the method used was the microgravimetric method, in which the increase of weight of a sample in contact with penetrant at a given activity is determined as a function of time. From this kind of measurement it is possible to obtain the equilibrium concentration of penetrant and, from the initial stage of sorption, the diffusion coefficient $\bar{D}$. This is a mean value in the range of activity explored and it is generally a function of concentration. To obtain the thermodynamic diffusion coefficient D_0, one has to know the dependence of $\bar{D}$ on concentration so as to extrapolate at zero penetrant concentration and obtain the thermodynamic D_0 [51]. The diffusion coefficient $\bar{D}$ (cm²/sec) as a function of the equilibrium bulk concentration C of penetrant (in grams of penetrant/100 g dry polymer) is reported in Figure 10 for samples A and B. All the experimental values for the two samples fall on the same straight line, with an extrapolated D_0 value at zero concentration of penetrant equal to 2.4×10^{-9} cm²/sec.

The concentration dependence of $\bar{D}$ is well represented by an exponential function

$$\bar{D} = D_0 \exp(\gamma_c) \tag{6}$$

where γ is a parameter characterizing the concentration dependence of the diffusion coefficient. The value of γ is 28.5 for both samples. Since the zero-concentration diffusion coefficient and γ are related to the fractional free volume (FFV) of the sorbing phase, this result is very important. For sample B, in fact, which is a two-component crystalline amorphous system, only the amorphous component can permit the passage of penetrant

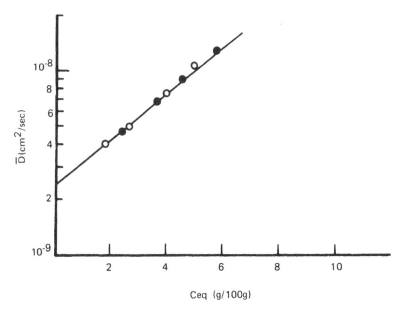

Figure 10 Logarithm of diffusion coefficient $\bar{D}$ of CH_2Cl_2 as a function of concentration of vapor in sample A (○) and sample B (●). (From Ref. 44.)

molecules, and the crystals do not contribute to the diffusion. For sample A, which is a smectic–amorphous system, the question of whether the smectic component contributes to the penetrant diffusion can be answered by the results of Figure 10. The coincidence of D_0 and γ for the two samples leads, in fact, to the conclusion that the smectic component does not contribute to the diffusion and the amorphous component in both samples behaves in a similar way.

For the crystalline sample B it is possible to derive the mass-fraction crystallinity from the density through the well-known Eq. 3, where $d_c = 0.936$ and $d_a = 0.856$ g/cm³ are the density of an ideal crystal and ideal supercooled melt respectively. This procedure seems to be correct for sample B, which is a two-component crystalline–amorphous system, without smectic component, but it yields a large error for sample A, which contains no crystalline fraction. In this case the smectic phase has a density surely lower than that of the ideal crystal; therefore, it is not possible to derive a reliable value of the phase composition. On the other hand, since the amorphous phase of the two samples behaves in a similar way with respect to the diffusion, it is possible to assume that the equilibrium sorption relative to the amorphous component is the same for the two samples. Thus the equilibrium concentration of a penetrant at a given activity is dependent on the amorphous fraction through the equation

$$C_{eq} = C_a(1 - \alpha_c) = C_a\alpha_a \tag{7}$$

where C_a is the equilibrium concentration of the amorphous phase and α_a the amorphous fraction. Assuming a constant C_a

$$C_a(A) = C_a(B) = \frac{C_{eq}(A)}{\alpha_a(A)} = \frac{C_{eq}(B)}{\alpha_a(B)} \tag{8}$$

Therefore,

$$R = \frac{C_{eq}(A)}{C_{eq}(B)} = \frac{\alpha_a(A)}{\alpha_a(B)} \tag{9}$$

Through this ratio R it is possible to derive the amorphous fraction in the sample A. In this case $R = 1.31$ and since $\alpha_a(B) = 0.4$, the smectic sample quenched at $-70°C$ was found to have an amorphous fraction $\alpha_a = 0.53$ and therefore a fraction of smectic component $\alpha_{sm} = 0.47$. By inserting 0.47 into Eq. 5, assuming for sample A a two-phase smectic–amorphous model, it is possible to calculate the density of the smectic phase:

$$\alpha_{sm} = \frac{d_{sm}}{d} \frac{(d - d_a)}{(d_{sm} - d_a)} = 0.47 \tag{10}$$

where $d = 0.883$ g/cm³ is the density of sample A, $d_a = 0.856$ g/cm³, and d_{sm} the value to be calculated. The calculations give 0.916 g/cm³, which is in very good agreement with the value reported in the literature [52].

In the study of phase composition in samples quenched at higher temperatures [45], two different series of samples have been obtained: the SM series obtained by quenching the melt with thinner substrate and the CR series with thicker substrate. The quenching temperature and the density of the obtained samples are listed in Table 2. For comparison, sample B of the previous investigation is reported as CLD. All these samples except CR100 contain both the crystalline and the smectic components in addition to the amorphous. They are therefore three-phase systems, and it is not possible to obtain from the density the exact distribution of three components.

If the mass fractions of the crystalline, smectic, and amorphous components are α_c, α_{sm}, and α_a, the density of this system has to be expressed as

$$\frac{1}{d} = \frac{\alpha_a}{d_a} + \frac{\alpha_{sm}}{d_{sm}} + \frac{\alpha_c}{d_c} \tag{11}$$

Table 2 Properties of Films Studied[a]

Sample	Quenching temperature (°C)	Density (g/cm³)	α_a	α_{sm}	α_c	A (°2θ)[b]	$1/A$	$\alpha_c/(\alpha_c + \alpha_{sm})$
SMR	−70	0.8830	0.53	0.47	0.0	4.4	0.23	0.00
SM0	0	0.8855	0.49	0.51	0.0	4.0	0.25	0.00
CR0	0	0.8915	0.42	0.48	0.10	3.8	0.26	0.17
CR10	10	0.8930	0.42	0.36	0.22	1.35	0.74	0.37
CR35	35	0.8953	0.42	0.26	0.32	0.77	1.30	0.56
CR50	50	0.8985	0.42	0.10	0.48	0.60	1.70	0.82
CR100	100	0.9030	0.39	0.00	0.61	0.45	2.20	1.00
CLD		0.9023	0.40	0.00	0.60	0.45	2.20	1.00

[a]α_a, α_{sm}, and α_c are the amorphous smectic and crystalline fractions calculated from Eqs. 11 and 12.
[b]A (°2θ) is the half-height broadening of the diffraction peak at about 2θ = 14° (x-ray CuK α radiation).
Source: Ref. 45.

The three fractions α_c, α_{sm}, and α_a are linked by the expression

$$\alpha_c + \alpha_{sm} + \alpha_a = 1 \tag{12}$$

Even if one knows the ideal density of all three components, i.e., d_{sm} in addition to d_c and d_a, one needs additional and independent information to derive the three fractions. For these samples, the amorphous fractions α_a have been obtained from sorption measurements, and the ratio R of Eq. 9 was derived for all the samples from measurements of C_{eq} as a function of vapor activity.

Figure 11 shows R calculated as a function of vapor activity for the different samples of Table 2. Only the sample CR100 shows a horizontal straight line; a deviation more and more marked going from CR50 to SMR clearly indicates that, above a given activity, the smectic phase also starts sorbing the penetrant. From the horizontal first part of the curve, in the range where the smectic phase does not sorb, it is possible to derive the amorphicity of the samples α_a through Eq. 9. From the densities of the samples this value can be used to determine the composition in terms of α_c and α_{sm} using Eqs. 11 and 12. In Table 2 the values of α_c, α_{sm}, and α_a obtained for all the samples quenched at different temperatures are listed.

A number of conclusions can be drawn from examination of Figure 11 and Table 2. CR100, quenched at 100°C, has no smectic component, because, inserting the obtained α_a and the measured density 0.9023 in Eq. 10, $\alpha_{sm} = 0$ is obtained. As a matter of fact there is no deviation from the horizontal straight line. The CR0, CR10, CR35, and CR50 samples have approximately the same amorphous fraction with a value of about 42%. It seems that as long as the crystalline and smectic phases are both present in the three-phase systems, the amorphous fraction has a fixed value, whereas the ratio of crystalline to smectic increases on increasing the quenching temperature. Since the density of the

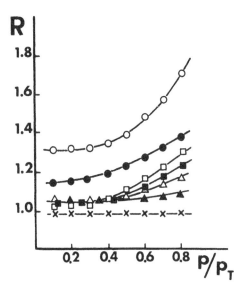

Figure 11 The ratio R between the equilibrium sorption of samples SMR ($\circ$), SMO ($\bullet$), CR0 ($\square$), CR10 ($\blacksquare$), CR 35 ($\triangle$), CR50 ($\blacktriangle$), and CR100 ($\times$) and the equilibrium sorption of sample CLD, as a function of vapor activity $a = p/p_T$. (From Ref. 45.)

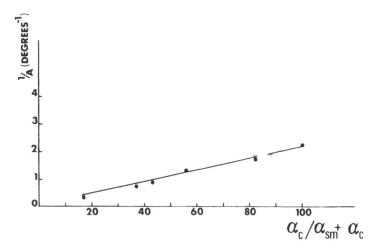

Figure 12 The reciprocal of the half-height broadening A of the (110) diffraction peak at about $2\theta = 14°$ (x-ray CuKα radiation), as a function of the percentage of monoclinic form. (From Ref. 45.)

crystalline phase (0.936 g/cm^3) is higher than that of the smectic phase (0.916 g/cm^3), an increase in the density of the sample in spite of the constant value of the (crystalline + smectic) sum is observed. For the CR50 sample, which has $\alpha_{sm} = 0.10$, there is only a small deviation of R at high activities, whereas the deviation increases with the increase of α_{sm} for the CR35, CR10, and CR0 samples.

The influence of the substrate thickness also is seen. In fact, the samples SM0 and CR0 are both quenched at 0°C with a thinner (SM0) or thicker (CR0) substrate. In the first case only the smectic and amorphous phases are present, whereas with a thicker substrate the system obtained is a three-phase system in which the crystalline phase is also present. The samples obtained with the thinner substrate SMR and SM0 have a different amorphous fraction, which is the more amorphous the lower the quenching temperature.

An interesting result of this work of Vittoria and Perullo [45] is the possibility to correlate the reciprocal of the half-height broadening A of the (110) diffraction peak at about $2\theta = 14°$ (x-ray, CuKα radiation) in the wide-angle spectrum, with the percentage of monoclinic form defined as

$$\% \text{ monoclinic} = \frac{\alpha_c}{(\alpha_c + \alpha_{sm})} \times 100 \tag{13}$$

In Figure 12 the $1/A$ parameter is reported as a function of the percentage of the monoclinic phase for three-phase CR systems. All the points fit the same straight line, indicating that there is very good agreement between the crystalline fractions derived from sorption experiments and the chosen order parameter. This ensures the reliability of the values obtained with sorption measurements.

CRYSTALLIZATION OF THE SMECTIC FORM

Thus far it has been assumed that the rapid quenching of iPP at low temperature produces the smectic phase. In reality it has been demonstrated that drastic quenching conditions produce the glassy polypropylene and that, during the warming to room temperature, the glassy polymer "crystallizes" in the smectic form.

Geil and colleagues recently showed that their ultraquenching technique is capable of quenching iPP into the glassy state [53, 54]. The wide-angle x-ray diffraction pattern of an 80-μm-thick ultraquenched iPP film, maintained at low temperature, shows a diffuse pattern similar to that obtained from atactic samples [24]. By increasing the temperature this diffuse halo becomes more intense and narrower by $-30°C$; when the temperature reaches $-20°C$, a new pattern, characteristic of the smectic structure, appears, consisting of two broad rings (as shown in Figure 6); i.e., iPP is transformed from the glass into the smectic structure.

This result agrees with the results of Hendra et al. [55]. These authors also find that the glass as originally produced at very low temperature is random, and local ordering occurs when it is warmed to room temperature. The thermal transitions of iPP quenched from the melt are clearly observable in the DSC thermogram of an ultraquenched film 0.1 mm thick shown in Figure 13, obtained by Hsu et al. [53]. The authors recognize in the thermogram an exothermic peak at approximately $-8°C$, due to crystallization of the glass into the

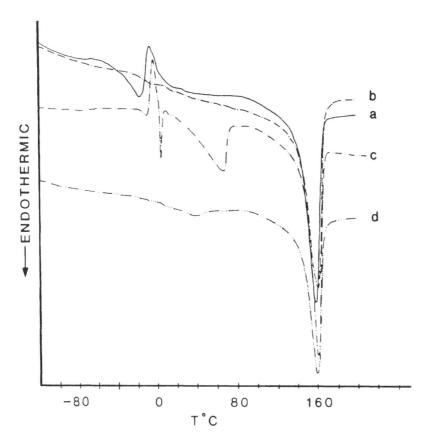

Figure 13 DSC thermograms of iPP. Curve a, as-ultraquenched sample 0.1 mm thick; curve b, second run; curve c, as-ultraquenched sample 0.3 mm thick; curve d, sample ultraquenched, warmed to room temperature before cooling back to liquid nitrogen temperature. (From Ref. 53.)

smectic phase. A small but broad exothermic peak occurring from -80 to $-30°C$ and centering at $-60°C$ is thought to be caused by some form of ordering taking place in glassy iPP in agreement with the x-ray diffraction results. Another small but broad exothermic peak centered at $\sim80°C$ is due to the transformation from smectic to monoclinic crystal structure. The largest peak, at $\sim160°C$, is due to the melting of monoclinic iPP crystals. A change in the baseline in the -30 to $-15°C$ range, just in front of the exothermic peak, is believed due to the glass transition in the glassy sample. In the second run (curve b), only a change of baseline in the glass transition region and a peak due to the melting of monoclinic iPP are observed. This is due to the crystallization of the sample when it was cooled after the first run. Curve c refers to an ultraquenched sample 0.3 mm thick and curve d is the thermogram of a sample that has been ultraquenched and warmed to room temperature; it shows only the broad exothermic peak centered at 80°C, caused by the smectic to monoclinic transformation, in addition to the normal melting at 160°C.

Thermal Crystallization

The range of temperature 60–80°C is generally indicated as the interval in which the smectic to monoclinic transition occurs [24, 38, 55–59]. The kinetics of crystallization from the smectic to the monoclinic phase have been studied by Zannetti et al. [38]. These authors also studied the dependence of the polymer order degree from the annealing conditions, i.e., how the different thermal treatments influence the fraction of the developed monoclinic form. The samples of smectic iPP were annealed at different temperatures between 60 and 160°C, for times from a minimum of 1 min to a maximum of 150 hr, and the x-ray diffraction spectra were obtained for each sample. The diffractometer scans registered after the different thermal treatments are presented in Figure 14. From this figure it is evident that the smectic to monoclinic transition takes place through a series of intermediate-order states that depend only on the annealing temperatures, when the saturation level is attained. The authors chose, as an index for the developed order degree during the transition, the reciprocal of the half-height broadening A of the (110) diffraction peak at about $2\theta = 14°$ (x-ray CuKα radiation) evaluated as is schematically represented in Figure 14. They report the variation of the parameter $1/A$ with the annealing time at different temperatures (Figure 15). One can observe that, at every temperature, an interesting time–saturation relation is detectable and that the maximum order level attainable at any annealing temperature is reached more quickly the higher the temperature is. An appreciable kinetic effect occurs only at the lower temperatures, whereas above 100°C the upper order limit is immediately attained. From the results the authors deduce that, since the structure of the ternary helix of the smectic form consists of short segments having opposite senses of spiralization, the smectic to crystalline transition occurs via intramolecular rearrangements of these short segments, right- or left-handed, having different length and winding themselves according to the prevalent sense of spiralization.

Infrared and Raman spectroscopy have also been used to study the smectic to monoclinic transition [25, 55]. Infrared spectroscopy is an excellent method for measurement of the helical content of the polymer. The principal bands in the spectrum of helical isotactic polypropylene are near 1220, 1163, 998, 900, 841, and 809 cm^{-1}. Of these, the band at 998 cm^{-1} has been widely used for estimating the helical content. The minimum number of monomer units necessary for observation of this band is about 10–11 [60]. Therefore, the increase in helix content is monitored as due to consequence of the progress of the crystallization process, according to the method proposed by Zerbi et al. [61].

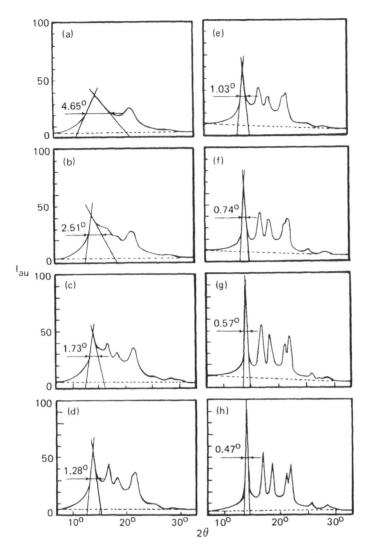

Figure 14 Diffractometer scans (CuKα radiation, proportional counter) of (a) the smectic form and of the same annealed up to saturation at (b) 70°C; (c) 80°C; (d) 90°C; (e) 100°C; (f) 120°C; (g) 140°C; and (h) 160°C. (From Ref. 38.)

Small-angle x-ray scattering (SAXS) [58] and transmission electron microscopy [62] have also been used to study this transition.

Solvent-Induced Crystallization

The presence of a solvent can cause a depression of both T_m and T_g of a polymer; the degree of this depression depends on the nature of the solvent and its interaction with the polymer [48]. In particular, the depression of T_g is due to an enhanced mobility of the amorphous chains plasticized by the solvent molecules.

In a crystallizable polymer obtained in the glassy state the presence of a solvent may

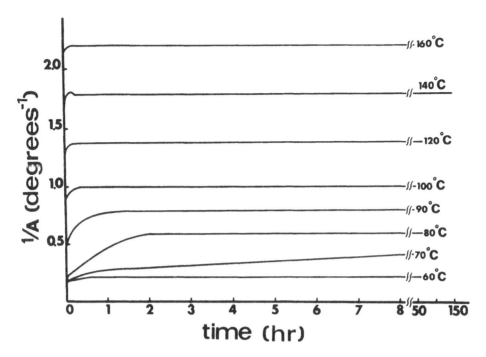

Figure 15 The reciprocal of the half-height broadening A of the (110) diffraction peak at about $2\theta = 14°$ (CuKα radiation) as a function of annealing time at different temperatures. (From Ref. 38.)

depress T_g to a temperature such that mobility will allow crystallization. This process, called solvent-induced crystallization (SINC), differs from thermal crystallization in that it takes place at lower temperatures and in the presence of solvent molecules. Vittoria and Riva [63] found this process to occur in smectic iPP exposed to different liquids. Solvent-induced crystallization is carried out at room temperature and provides an alternative mode of structure formation without increasing the temperature. Generally the ability of a solvent to interact and induce crystallinity in the amorphous polymers is correlated with its solubility parameter $\delta(cal/cm^3)^{1/2}$, defined as the square root of the cohesive energy density. It has often been found that the degree of interaction leading to SINC is maximum for solvents having a solubility parameter close to the value of the polymer [48].

Although for crystalline polymers the cohesive energy density is not well defined [64], the solubility parameter concept is useful in predicting swelling behavior and can help clarify the interactions that lead to crystallization. In the case of iPP, two different values for the solubility parameter are reported in the literature [65–67], 8.1 and 9.2–9.4. To study the SINC process, a smectic film of iPP obtained at $-70°C$ and equilibrated at room temperature, with a density of 0.883 g/cm^3 was used [63]. Different strips of the original film were immersed in various liquids at 25°C for different times. The liquids used in this study are reported in Table 3. They are all in the class of poorly hydrogen-bonded solvents, with zero or low polarizability and having a molar volume in a narrow range, except octane, with the greatest molar volume, and dichloromethane, with the smallest. It is therefore reasonable to compare the weight uptake and the ability to induce crystallization according to the solubility parameter of the solvents.

Table 3 Solubility Parameter (δ), Molar Volume (V), and Fractional Polarities (P) of Some Solvents

Solvent	$\delta(cal/cm^3)^{1/2}$	$V(cm^3/mol)$	P^a
n-Hexane	7.3	132	0
Octane	7.6	164	0
Cyclohexane	8.2	109	0
Carbon tetrachloride	8.6	97	0
Toluene	8.9	107	0.001
Benzene	9.2	89	0
Chloroform	9.3	81	0.002
Chlorobenzene	9.5	107	0.058
Dichloromethane	9.7	65	0.120

^{a}The fractional polarity P is the fraction of total interactions which are due to dipole–dipole attractions [64].
Source: Ref. 63.

The authors found that all the investigated liquids induce crystallization of the smectic form into the monoclinic form. They assume that in the liquids, with thermodynamic activity $a = 1$, the smectic phase becomes permeable to the penetrant molecules and the mobilizing presence of these molecules allows the process of crystallization. A fraction of solvent was rejected after the achievement of equilibrium sorption. In Figure 16 the volume of liquid sorbed and the weight of liquid sorbed by the polymer are reported as a function of the solubility parameters $\delta(cal/cm^3)^{1/2}$ of the solvents. The three curves refer to the first equilibrium concentration and to the value of liquid sorbed after 24 hr and 10 days of stay in the solvents. The greatest desorption occurs in the first 24 hr, but a further decrease on the subsequent days is observable. Since the desorption is indicative of crystallization, it means that this phenomenon continues for a long time. An interesting result is the presence of two maxima in the curves of liquid sorbed; generally the maximum of interaction occurs when the solubility parameters of polymer and solvent are close; in this case the observed maxima correspond to the two different solubility parameters reported in the literature [65–67] for polypropylene.

The type and the extent of crystallization are studied through measurements of wide-angle x-ray scattering (WAXS) and density. The WAXS spectra of iPP samples crystallized in the liquids show the presence of the monoclinic form (Figure 17). As an index of the order level attained by the iPP samples crystallized in the different solvents, the authors [63] chose the reciprocal of the width at half-height A for the (110) polypropylene strongest reflection ($2\theta = 14°$) as reported by Zannetti et al. [38]. The $1/A$ values for the samples immersed 10 days in the liquids are reported in Figure 18 as a function of the solubility parameter δ of the solvents. The starting smectic sample has a value of the $1/A$ parameter 0.22 (Figure 6); the solvent-crystallized samples show a much higher value, in the range 0.6–0.78; also on this curve there appear two clearly resolved maxima corresponding to the two zones of maximum interaction between polymer and solvent.

Figure 19 gives the values of the density for samples immersed 24 hr and 10 days in the liquids as a function of the solubility parameter. Since the density of the starting sample is 0.883 g/cm^3, a substantial increase in density, indicative of the crystallization phenomenon, is evident in all the samples. Moreover, the difference observed in many cases

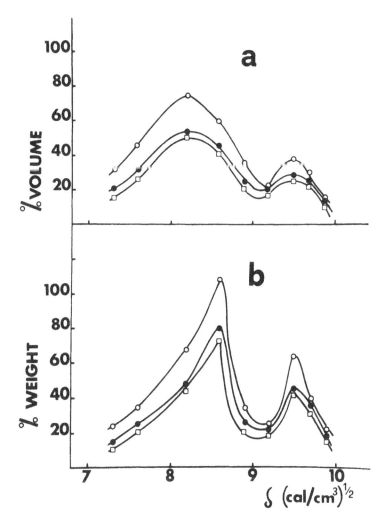

Figure 16 Volume of sorbed liquid as a percentage of the original volume of polymer (*a*) and weight of the sorbed liquid as a percentage of the original weight of polymer (*b*), corresponding to the maximum concentration sorbed (○) and to the value after 24 hr (●) and 10 days (□) as a function of the solubility parameter δ of the liquids. (From Ref. 63.)

between the 24-hr and 10-day values indicates that the process of crystallization went on for many days. Although the level of sorption and tendency to crystallization are not directly related, Figure 19 shows that with greater sorption and swelling, the density and hence the level of crystallinity increase. In fact, sorption is a measure of the interaction between the polymer and the liquid and the development of crystallinity may depend on such interaction. The degree of enhanced mobility and the depression of T_g are dependent on the uptake of solvent.

The $1/A$ values obtained for the solvent-crystallized samples if compared with thermally crystallized samples [38] would correspond to samples with lower density. This result suggests that the concentration of crystallites induced by SINC is high and that they are relatively small in size, as described already for other systems [48].

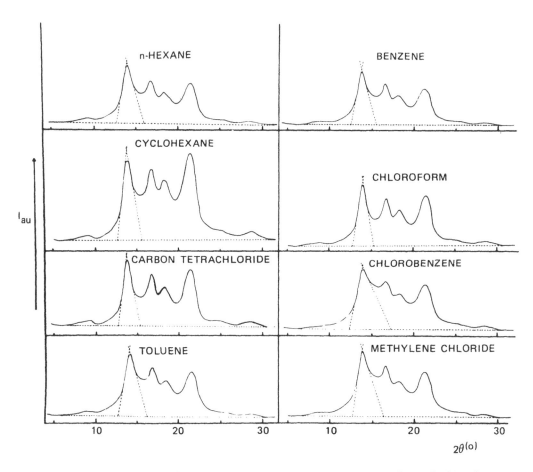

Figure 17 WAXS spectra of iPP smectic samples immersed 24 hr in the different liquids. (From Ref. 63.)

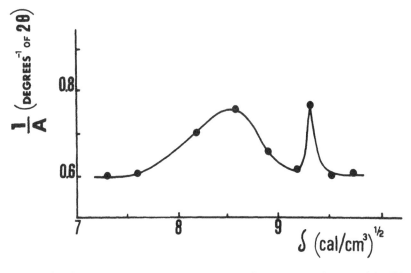

Figure 18 Reciprocal of the half-height width (A) for the (110) reflection of the iPP as a function of the solubility parameter of the liquids. (From Ref. 63.)

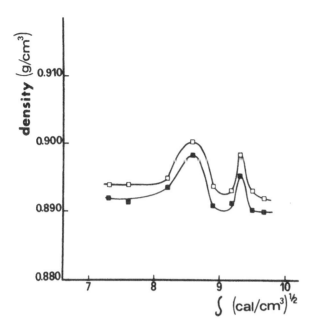

Figure 19 Density d for the samples immersed 24 hr (■) and 10 days (□) in the different liquids as a function of the solubility parameter of the liquids. (From Ref. 63.)

AGING PHENOMENA

The study of changes that can occur in a material with aging time is very important, since they can affect the application, performance, and lifetime of the materials.

Isotactic polypropylene quenched rapidly from the melt is in a nonequilibrium state and undergoes a strong enhancement of mechanical properties and an increase of density during its aging at room temperature [68–72]. Schael [68] found a linear increase of density with the logarithm of film age and a correspondent variation of impact strength and coefficient of friction. Gezovich and Geil [69] showed that some physical properties, such as yield stress and stress–strain curves of the smectic form of iPP, are dependent on aging time at room temperature. These properties were found to increase linearly with the logarithm of film age. The absence of changes in the x-ray diffraction pattern, infrared spectrum, electron diffraction pattern, or morphology led these authors to attribute the changes to a decrease in the concentration of randomly distributed voids, resulting from chain entanglements, crystal imperfections, and other irregularities. The decrease of holes with time occurs in the amorphous regions and can be explained by a mechanism in which molecules slowly reorient to relieve internal stresses imposed by rapid crystallization conditions.

Casper and Henley [70] reported a decrease in the permeation constant with time for C_2H_6 diffusing through iPP. The permeation constant reached a steady-state value after 50 hr, varying, from 5 to 50 hr, linearly with the logarithm of time.

Kapur and Rogers [39], starting from the assumption that structural changes which depend on aging occur only within amorphous and intercrystalline regions of the polymer, determined the transport behavior of small noninteracting penetrants. The transport properties can, in fact, be used as molecular probes sensitive to structural changes in the amorphous component on a scale comparable to the size of the penetrant molecule [46]. In considering the mechanical properties, these authors found an increase of yield stress by 16% of its initial value on aging for 70 hr. The initial modulus (reported in Figure 20), determined as the slope of the tangent to the stress–strain curve in the limit of zero strain, increased linearly with logarithmic film age to 185% of its initial value on aging for 100 hr. The drawing stress (reported also in Figure 20) increased with logarithmic time to 150% of its initial value on aging for 100 hr.

The resilience and mechanical damping showed more complex behavior. In Figure 21

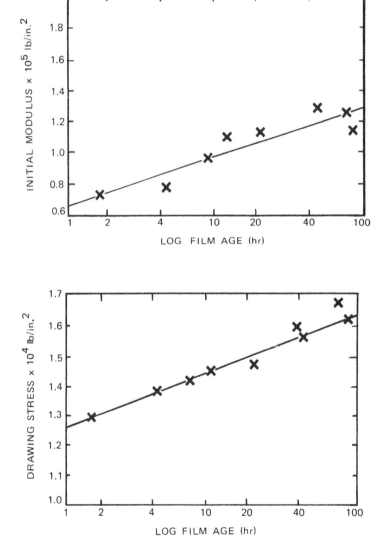

Figure 20 Initial modulus and drawing stress of iPP films as a function of logarithm of film age. (From Ref. 39.)

the diffusion coefficients for argon, neon, and helium are plotted as a function of aging time. The dependence of diffusion on the penetrant size immediately suggests that the structural changes accompanying the aging process involve domains of a size affecting the transport of helium and neon but small enough so that they do not affect the transport of argon. In light of the combined mechanical and transport measurements, the authors suggest that the aging process may involve structural changes associated with two different phenomena [39]. The first is a molecular rearrangement generally similar to secondary crystallization, accompanied by a decrease in segmental chain mobility of intercrystalline regions within domains comparable in size to that of the smaller penetrant species. The second is a relaxation of a residual stress distribution, related to nonuniform plastic flow during quenching. An important result of this work is the possibility of using transport of small penetrants as a molecular probe sensitive to structural phenomena in the amorphous and intercrystalline regions.

A study of the dependence of transport parameters on aging time has been recently reported [73]. In this case a vapor, dichloromethane, was used as penetrant. The interpretation of transport properties suggests two different mechanisms of aging. The first, very rapid, gives rise to a decrease of amorphous fraction corresponding to the increase of density. The second effect, which superimposes to the first, consists of a decrease of the mobility of the amorphous chains.

The reduction of mobility of the amorphous phase corresponding to a decrease of free volume agrees well with other results [74–76] and with the theory and the results of Struik [77, 78]. In the case of semicrystalline polymers above T_g, the aging phenomena can be explained from the model of Struik outlined in Figure 22. From this model it is evident

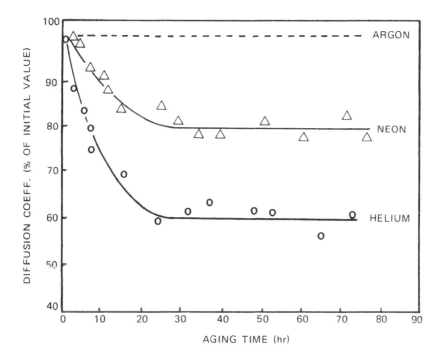

Figure 21 Diffusion coefficient of different gases as a function of film age. (From Ref. 39.)

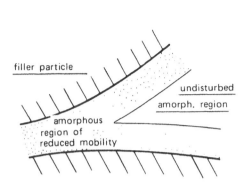

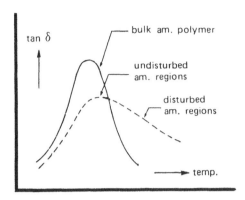

Figure 22 The extended glass transition in semicrystalline polymers and filled rubbers. (From Ref. 78.)

that the crystals disturb the amorphous phase and reduce its segmental mobility. The main consequence of this immobilization is that the glass transition will be extended toward the high-temperature side; therefore, even above the T_g of the bulk amorphous material some parts of the amorphous phase are rubbery, other parts are glassy, and still other parts are just passing their glass transition. Therefore, even above T_g, part of the amorphous phase of the semicrystalline polymer is still glassy. For such materials aging will persist at temperatures above T_g. This model can be successfully used for analyzing the problem of the aging of polypropylene [77, 78].

PLASTIC DEFORMATION OF ISOTACTIC POLYPROPYLENE

Polypropylene is one of the most important semicrystalline polymers used to obtain synthetic fibers; understanding of the plastic deformation mechanism of this material is thus of considerable interest. The mechanical characteristics of films and as-spun fibers of iPP are far from those required by the textile industry. They usually show neither a high anisotropy of physical properties nor a high elastic modulus and strength in the axial direction. A subsequent process of ''cold drawing'' imparts to the films and as-spun fibers their useful mechanical properties in the draw direction.

Cold drawing consists of irreversible elongation of the film in the solid state to many times its original length. The process of drawing can be undertaken at room temperature or at high temperature, in the interval between the glass transition, T_g, and the melting temperature T_m.

Determination of the stress–strain behavior of iPP is particularly important for studying mechanical properties such as elastic modulus, yield stress σ_y, and stress to break σ_b. These parameters are very useful in design considerations when the material is used in practical situations.

When a stress σ is applied to a film of iPP it will be deformed by a certain amount of strain ϵ. Under given conditions of rate of drawing and temperature, a stress–strain curve will be obtained similar to those shown in Figure 23. In this case the nominal stress σ is reported as a function of the strain ϵ. Curve a exhibits first a linear region, in which the stress is proportional to the strain and Hooke's law is obeyed. The tensile or Young's

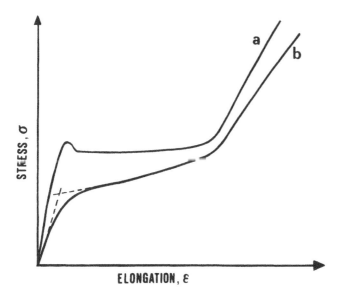

Figure 23 Stress–strain curves for inhomogeneous (curve a) and homogeneous (curve b) drawing. The nominal stress σ is reported as a function of the strain ϵ.

modulus can be obtained from the slope. When increasing the strain, the curve decreases in slope until it reaches a maximum, conventionally known as the yield point. During elastic deformation the cross-sectional area of the sample decreases uniformly as length increases, but at the yield point it starts to decrease more rapidly at one particular point along the specimen. At this point a "neck" starts to form, and it propagates all along the specimen. During this propagation the nominal stress σ remains very nearly constant or increases slightly. When the whole specimen is necked, strain hardening occurs and the stress rises until fracture intervenes. The draw ratio is the length of a fully necked or extended specimen divided by the original length. Since cold drawing takes place at approximately constant volume, it is also equal to the ratio of original sample cross-sectional area to that of the drawn sample.

Curve b represents the homogeneous extension of iPP. In this case drawing occurs without any visible neck formation. Homogeneous drawing can take place at higher temperature and lower draw rate; it is generally observed when smectic iPP is drawn in these conditions.

Drawing Behavior

The drawing of a semicrystalline polymer solid usually produces a fibrous material with high anisotropy of physical properties, e.g., elastic modulus [79], strength (stress to yield, stress to break) [80], and diffusivity [81]. This anisotropy is closely connected with the orientation of chains and amorphous regions and with the direction in which the two regions alternate [82–85]. Concomitant with these changes are corresponding changes in other properties such as refractive index (the polymer becomes birefringent) and coefficient of expansion [86].

Many physicochemical approaches to the measurement of anisotropy, including WAXS, sonic techniques, and spectroscopic methods, have shown that even at small draw ratios the chains in the crystal lattice are almost completely oriented in the draw direction and the orientation only slightly increases with further drawing [86]. The chain orientation of the amorphous regions is much less complete but increases steadily over the range of drawing.

In the drawing process there is a complex rearrangement of the structure and morphology of the sample. The more or less spherulitic starting material is composed of stacked folded chain lamellae; the fully drawn material is composed of highly oriented microfibrils. Many models have been proposed to characterize all stages of the deformation and the corresponding transition from a spherulitic to a microfibrillar morphology. The Peterlin model can better explain all the experimental evidences [87–90].

Figure 24 represents the very different basic elements of the starting lamellar and final fibrous material. The lamellae in the spherulitic sample (Figure 24*a*) are very thin, ~200 Å thick, and very wide, ~10,000 Å. The stacking of lamellae yields a regular grid with a long period $L = 200$ Å, as derived from SAXS (small-angle x-ray scattering) and electron microscopy (EM). The long period measures the sum of the thickness D of the crystal core and amorphous surface layers of each lamella. The amorphous surface layers contain chain folds, chain ends, and tie molecules (i.e., chain sections anchored at both ends in the crystal lattice of adjacent lamellae). The fraction of tie molecules is relatively small, about 1% of the total number of chain stems in the crystal lattice [88]. The microfibrils (Figure 24*b*), on the other hand, are very long, several micrometers, but have very small lateral thickness, ~100 Å. The microfibrils also show an axial alternation of crystalline and amorphous layers with a period L detectable by EM and SAXS.

Practically all molecular connections in the lamellar structure are within the same lamella, i.e., in the direction perpendicular to the chain, and in the fibrous structure are within the same microfibril, i.e., in the direction parallel to the chain. Hence it is possible to separate the lamellae in the former case and the microfibrils in the latter case without any large-scale cutting of molecules connecting these elements with the adjacent matrix.

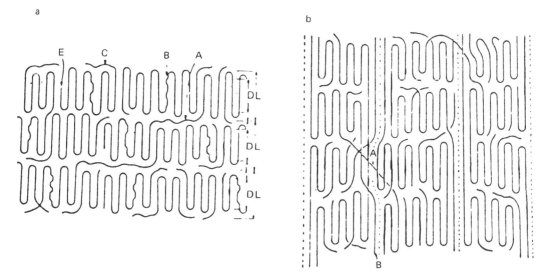

Figure 24 Schematic model of stacked parallel lamellae (*a*) and aligned microfibrils (*b*) as basic elements of lamellar and fibrous structure. (From Ref. 88.)

Uniaxial plastic deformation transforms the spherulitic solid into fibrous material of high mechanical anisotropy. The first stage of plastic deformation involves shear deformation and rotation of stacks of parallel lamellae [91]. Both modes are mainly the consequence of shear deformation of the amorphous layers between lamellae and stacks of lamellae. Very soon, however, the forces are sufficient for chain tilt and slip within the lamellae [92]. To a large extent these deformations are still recoverable at the temperature of the experiment and hence do not represent truly plastic deformation. But gradually the lamellae are so far deformed by chain slip that a crack develops. The partially unfolded chains bridging the crack lead to successive tearing off of crystal blocks from the lamellae and their incorporation into microfibrils connecting the opposite borders of the crack.

The microfibril consists of alternating crystal blocks and amorphous layers bridged by a great many taut tie molecules [87]. The unfolding of chains during the drawing of microfibrils yields a large number of intrafibrillar taut-tie molecules (TTM) connecting in the axial direction the crystalline blocks they are anchored in (Figure 24*b*). Their effective fraction, i.e., the fraction per amorphous layer, increases with the number of amorphous layers they cross. This number increases with the draw ratio of microfibril formation and the majority of them will be located on the outer boundary of microfibrils [93].

The most important effect of such axial connection of crystalline blocks across the amorphous layers by TTM is the enhancement of axial elastic modulus and strength of the microfibrils. The highly anisotropic fibrous material can deform plastically only by sliding motion of fibrils and microfibrils. The shear displacement of microfibrils enormously enhances the fraction of TTM connecting adjacent microfibrils, and such an increase of axial connectors, in turn, enhances the axial force transmission per unit area of fiber cross section and hence increases the axial elastic modulus. In the stress–strain curve (curve a in Figure 23) all the stages of the morphological transformation from spherulitic film to microfibrillar fiber are clearly distinguishable. The first linear part of the curve corresponds to the elastic deformation of the structure; very soon a plastic component of the deformation appears and after the yield, the neck, in which the morphological transformation is concentrated, starts to propagate all along the sample. When the transformation is completed, the deformation of the fibrous structure begins. This causes a relevant strain hardening, due to the increasing fraction of TTM, until fracture intervenes.

Influence of the Initial Structure on the Drawing Behavior

De Candia et al. [94] analyzed the monoaxial drawing at room temperature of films of iPP of different initial structure obtained by varying the quenching conditions. The samples prepared with thin (A samples) and thick (Q samples) substrate are identified by a number corresponding to the quenching temperature. The density of the samples is reported in Table 4. The structures of the films were analyzed in terms of phase composition and gross morphology. They show a different fraction of amorphous, smectic, and crystalline phases, the more crystalline the higher the temperature of quenching. The morphology of the analyzed samples is spherulitic, as is usually the case with iPP samples. A quantitative analysis of the dimensions of the spherulites was carried out using small-angle light scattering and detecting photographically the diffraction patterns. The average radius *R* of the spherulites was derived according to the theory of Stein and Rhodes [95].

The results show that the initial structure strongly affects the drawing behavior. The drawing occurs by neck propagation in all the samples, but the deformation in the neck depends on the gross morphology and crystallinity. Figure 25 shows the variation of the

Table 4 Mechanical Parameters of the Quenched Samples

Sample	d (g/cm^3)	σy (MPa)	σb (MPa)	$\Delta\lambda_{neck}$
A(-70)	0.8830	19.4	190.0	1.2–4.0
A0	0.8858	22.8	169.6	1.2–4.2
Q0	0.8915	28.7	204.0	1.2–5.5
Q35	0.8953	30.6	213.7	1.2–6.0
Q50	0.8985	34.7	206.7	1.2–6.5
Q100	0.9033	41.3	283.0	1.2–7.8

Source: Adapted from Ref. 94.

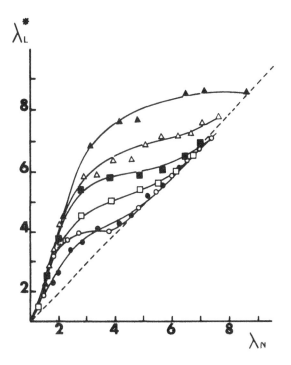

Figure 25 The local draw ratio, detected in the volume element where the virgin neck, indicated as λ_L^*, as a function of the local deformation λ_N for the quenched samples. Sample A, (-70), ○; A0, ●; Q0, □; Q35, ■; Q50, △; and Q100, ▲. (From Ref. 94.)

local deformation of the first volume element where the virgin neck appears, λ_L^*, as a function of the nominal deformation, λ_N, for all the different samples. The appearance of the virgin neck is followed by the neck propagation; in the first volume element the deformation rate drops again to a lower value, while the extension of the neck induces an acceleration of the deformation rate in the adjacent volume element, with a consequent increase in local deformation. This phenomenon is iterated to the next volume element, and so on, until the necking extends to all the sample. At this point the rate of deformation becomes homogeneous and local deformation in each volume element is equal to the

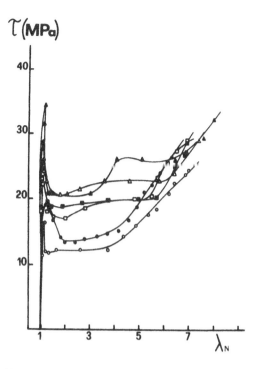

Figure 26 Load–elongation curves detected on the quenched samples. Sample A (−70), ○; A0, ●; Q0, □; Q35, ■; Q50, △; and Q100, ▲. (From Ref. 94.)

nominal deformation. The dashed line shows the condition $\lambda_L = \lambda_N$. This condition is achieved at $\lambda_N = 4$ in the samples quenched at −70 and 0°C, which are completely smectic–amorphous systems, while this value increases with increase in the quenching temperatures. As a matter of fact, the inhomogeneity in the drawing behavior increases on increasing the quenching temperature. The most crystalline sample, quenched at 100°C, at breaking point is still far from the conditions of homogeneous deformation.

In Figure 26 the stress–strain curves of the different quenched samples are reported; the tensile stress on the initial undeformed cross-sectional area τ is reported as a function of nominal deformation λ_N, which is linearly dependent on the drawing time. All the curves are conventional with an upper and lower yield point, both restricted in very sharp deformation range, which, at a macroscopic level, characterizes the appearance of a neck; the local deformation drastically increases in the neck, producing a significant reduction in the local cross section. The reduced cross section induces a stress concentration that stabilizes the neck, and the axial drawing takes place through the neck propagation over all the sample. This stage of the drawing process occurs at a constant load and therefore the plateau region identifies the deformation range characterized by the neck propagation. The most significant mechanical parameters for the different samples are listed in Table 4.

The general trend is an increase in σ_y and σ_b upon increasing the quenching temperature, which increases the fraction of crystalline material and spherulite radius, and upon passing from the A series to the Q series. This hardening effect is accompanied by an increase in $\Delta\lambda_{neck}$ as also shown in Figure 25. The mechanical work necessary to draw the sample increases on increasing the crystallinity while $\Delta\lambda_{neck}$ seems to be related to the spherulite dimensions.

STRUCTURAL CHANGES IN COLD DRAWING

SAXS and WAXS Studies

Results of SAXS and WAXS investigations on the plastic deformation of iPP films [96–101] support the three-step model of plastic deformation of polymers developed by Peterlin. In particular, it has been conclusively shown that there is a discontinuous transformation in the neck from the deformed spherulitic to the fiber structure.

Two samples of iPP, a quenched and an annealed sample, were drawn respectively at room temperature and at 100°C. WAXS and SAXS patterns were investigated in a range of values of draw ratio very closely spaced through the neck region and after the neck propagation. The contribution to the WAXS pattern of microfibrils could be clearly distinguished from that of deformed spherulites because of the better orientation parallel to the draw direction of the former compared to the latter.

WAXS photographs obtained for the sample annealed at 140°C and subsequently drawn at 100°C [99] show the presence of a spherulite deformation step up to $\lambda = 2$, followed by the formation of the fiber structure at higher draw ratios. For $\lambda > 2$, in fact, plastic deformation is accompanied by both microfibril formation and some spherulite deformation, as reflected by changes in both orientation and crystallite size. The onset of a discontinuous transformation of spherulitic to fiber structure was apparent from distinct changes in the WAXS pattern.

The SAXS study [97, 98] showed that the long period L changes discontinuously at the same draw ratio as indicated by WAXS and that the long period L_T of the fiber structure is a unique function of the temperature of drawing T_d and independent of the draw ratio λ

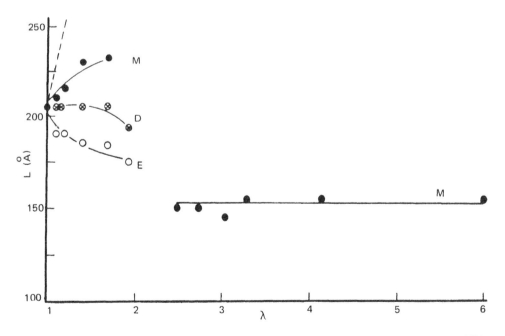

Figure 27 Meridional, diagonal, and equatorial long period of an iPP sample annealed at 140°C ($L_0 = 205$ Å) and drawn at 100°C as a function of draw ratio λ. (From Ref. 101.)

and of the initial long period L_0 of the microspherulitic film. This abrupt change from L_0 to L_T occurs in micronecks where each lamella is fractured in folded-chain blocks immediately incorporated into microfibrils, the basic element of the fiber structure.

The values of the meridional, diagonal, and equatorial long period as a function of the draw ratio are reported in Figure 27 for the sample annealed at 140°C, with $L_0 = 205$ Å, and subsequently drawn at 100°C [101]. The three different long spacings change in a different way up to $\lambda = 2$. At this draw ratio a new meridional maximum with a new long period L, which may differ appreciably from L_0, appears. This maximum gets stronger the closer it is to the section with well-developed fiber structure. It persists up to the maximum draw ratio and its value is dependent only on the drawing temperature.

In Figure 28 the long period L is plotted as a function of temperature for samples drawn, quenched, and annealed and drawn and annealed at different temperatures [101]. It is evident that within experimental errors all data are located on a single curve, showing

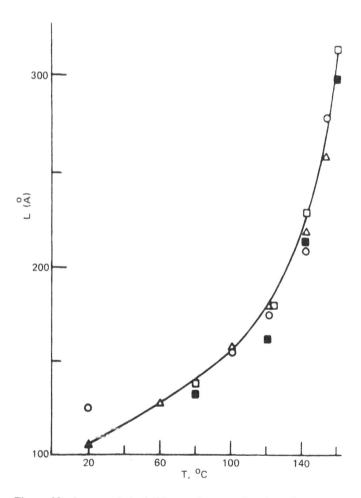

Figure 28 Long period of iPP samples as a function of temperature of treatment. △, drawn sample; ○, quenched from melt in ice water and annealed for 20 min; □, drawn at 20°C and annealed for 1000 min with free ends; ■, the same with fixed ends. (From Ref. 101.)

that the long period is a unique function of the temperature. If the drawing is performed at room temperature, the draw ratio at which the small-angle data show the beginning of fiber structure is anticipated as 1.4 with respect to the value of 2 for the drawing at 100°C. The WAXS photographs additionally show that microfibrils crystallize in the "smectic phase" as compared to the monoclinic phase for the initial sample and deformed spherulites. At this temperature plastic deformation proceeds through the spherulite deformation mechanism up to $\lambda = 1.4$, accompanied by an increase in chain orientation with increasing λ. For $\lambda > 1.4$ plastic deformation appears to occur exclusively through microfibril formation. Since the initial microspherulitic sample is largely in the α phase, the presence, in the fiber, of a large proportion of sample in the "smectic" phase indicates that chains initially packed in a relatively ordered manner are now packed in a somewhat more random fashion. This, together with the change to a smaller long period, indicates a considerable rearrangement in chain packing.

For the two samples drawn at room temperature and at 100°C, although the continuous plastic deformation of the spherulitic structure before necking is quite different in its WAXS manifestations, both WAXS and SAXS show a discontinuous transformation to the same type of new structure, the microfibrillar, after necking. This consists of molecules packed with their helical axes very nearly parallel to the draw direction in lamellae perpendicular to the drawing direction. In fact the main feature derived from the diffraction patterns is the gradual shift of orientation of the chain helices of the crystals toward the draw direction.

Many other authors have shown, through WAXS measurements, that cold drawing at room temperature produce an oriented but disordered structure, recognized as the "smectic" phase of iPP [29, 102, 103]. Annealing or drawing at higher temperatures produces the monoclinic α phase.

Orientation and Crystallinity

Any theoretical interpretation of the deformation mechanism requires a detailed knowledge of the crystallinity and morphology of deformed polymer. Therefore, many methods have been applied and developed to characterize the structure, morphology, and orientation of deformed samples [85, 86]. Isotactic polypropylene can be dealt with as a two-phase system [104] also in cases in which the smectic phase is present [44].

The two-phase model is excellent for representing the change in properties of a crystalline polymer during deformation. Each phase of the polymer is assumed to have intrinsic properties. In an unoriented polymer the phases are mixed at random and hence any difference in the observed properties will be a result of the relative amount of the two phases present:

$$P(\text{unoriented}) = \alpha_c P_c + (1 - \alpha_c)P_{\text{am}} \tag{14}$$

where P is the observed property of the polymer, P_c and P_{am} the intrinsic property of the crystalline and amorphous region respectively, and α_c and $(1 - \alpha_c)$ the fraction of crystalline and amorphous material. The orientation of the phases in a given direction, due to a deformation mechanism, will lead to a manifestation of the anisotropic character of the phases. Any observed anisotropic property of the oriented polymer $P(\text{oriented})$ will be a function of the properties of each phase and may be expressed, according to Samuels [85], as

$$\Delta P(\text{oriented}) = \alpha_c P_c f_c + (1 - \alpha_c)P_{\text{am}} f_{\text{am}} \tag{15}$$

where f_c and f_{am} are the orientation functions characterizing the average orientation of the crystalline and amorphous phases respectively. The measure of crystallinity and orientation is therefore the first important characterization of a drawn iPP sample.

The two generally used methods for obtaining the crystallinity of a drawn sample—density and x-ray techniques—give often conflicting results [105]. They are differently affected by many approximations and inaccuracies. The x-ray method fails to take into consideration the fraction of material in the smectic phase that forms on cold drawing. On the other hand, the calculation of crystallinity α_c from the density d (Eq. 3) is based on the extreme two-component model. It assumes that the density of the crystalline and amorphous components, even in drawn samples, is equal to that for the ideal crystal and ideal supercooled melt, respectively, and both components are clearly separated. Hence the derived mass-fraction crystallinity is an idealized parameter which characterizes the sample fairly well but does not correspond exactly to the mass fraction of the crystals. The drawing drastically distorts the crystal lattice and can produce a smectic fraction; hence it lowers the density of this component far below the value of the undrawn material. It also partially aligns the chains in the amorphous component. This favors a denser packing and hence a higher density than that of the relaxed supercooled melt.

Another source of error is the possible formation of voids during the drawing process, particularly for the highest draw ratios. These microvoids produced during the drawing of crystalline polypropylene fibers at low temperatures are generally of the order of magnitude of the wavelength of light. In particular drawing conditions the polypropylene fibers show a low apparent density and become opaque [106].

The values of crystallinity obtained from density measurements thus represent a lower limit. The density of the polymer sample may be determined by many experimental methods including flotation in mixtures of liquids and density gradient column. In many cases [103, 116] an increase of density, and hence of crystallinity, with drawing has been observed. Evidently when an iPP film with initially smectic structure is drawn at temperatures higher than 80°C, at which the transition from smectic to monoclinic phase occurs, the increase in density is due prevalently to this transition [99].

The measure of orientation is also a very important one. The orientation of the crystalline phase is obtained principally by studies of wide angle x-ray diffraction. Other methods, such as sonic modulus and birefringence, give a measure of the total orientation of the sample, related to f_c and f_{am} by Eq. 15. For example, the measured birefringence Δ_T may be defined as

$$\Delta_T = \alpha_c \Delta_c^\circ f_c + (1 - \alpha_c)\Delta_{am}^\circ f_{am} \tag{16}$$

where Δ_c° and Δ_{am}° are the intrinsic birefringences of the perfectly oriented crystal and perfectly oriented amorphous region respectively. Their values, derived by Samuels [85], are $\Delta_c^\circ = 29.1 \times 10^{-3}$ and $\Delta_{am}^\circ = 60.0 \times 10^{-3}$. Samuels determined the contribution of the separate phase to the total measured birefringence [109, 110]. These are plotted for a series of iPP samples in Figure 29 as a function of draw ratio. The crystalline contribution is $\Delta_c = \Delta_c^\circ f_c$ and the amorphous contribution is $\Delta_{am} = \Delta_{am}^\circ f_{am}$. The figure shows that the crystalline orientation contribution to birefringence is fairly high even at low elongations, whereas the orientation of the amorphous phase is steadily increasing with draw ratio. The birefringence measurement is thus a rapid and powerful tool for the study of morphological characteristics of deformed polycrystalline polymers. Since it is a measure of the total molecular orientation of the two-phase system, its examination in conjunction with other physical measurements (sonic modulus, x-ray diffraction, density, etc.) yields considerable insight into the molecular characteristics of bulk polymer.

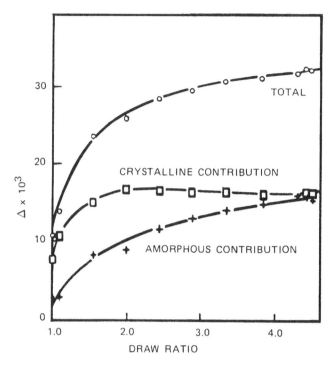

Figure 29 Crystalline and amorphous contributions to the total birefringence of the series of draw fibers, as a function of drawn ratio. (From Ref. 108.)

Generally the properties of a specimen are related to the draw ratio the polymer underwent during or after fabrication. Samuels showed that the average orientation, the crystalline orientation, and the amorphous orientation are all related to the fabrication draw ratio [109–111] and different processes and fabrication conditions can lead to equivalent states of average orientation. Hence the state of orientation and not the draw ratio should be used to characterize the mechanical properties of films and fibers. Samuels investigated many different mechanical properties, such as stress and strain fracture, true stress, tenacity, and yield stress, and interpreted them in terms of crystalline, amorphous, and average orientation. For example, yield stress increases with increasing molecular orientation and correlates directly with average molecular orientation.

Nadella and colleagues [103, 112] found the mechanical properties of the drawn fibers to depend on both the drawing and the melt spinning conditions. They investigated the behavior of a wide range of starting morphologies and different average molecular weights. The different starting morphologies have great effect on the drawing behavior of the filaments, but it was found that the mechanical properties of the spun and drawn fibers could all be correlated with the birefringence developed in them by the processing conditions, independent of the processing path. In Figure 30 the tensile strength of fibers obtained in different conditions is seen as a function of birefringence, showing the satisfactory correlation in spite of the different treatments. The same correlation was found for elastic modulus and elongation to break.

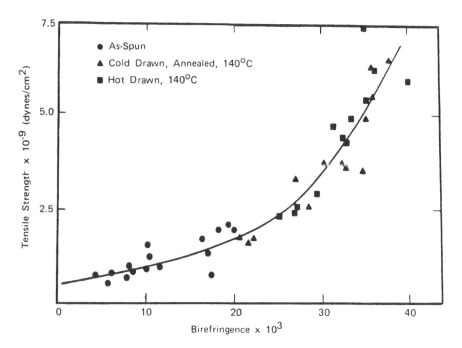

Figure 30 Tensile strength as a function of birefringence for melt-spun, hot-drawn, and cold-drawn and annealed fibers. (From Ref. 112.)

Transport Properties

The many techniques used to investigate the drawing of iPP show that, with cold drawing, the degree of orientation of the amorphous chains increases steadily up to the highest obtainable draw ratio. According to Peterlin, this implies a continuous increase of the fraction of taut molecules (TTMs) axially connecting crystalline blocks of the same or different microfibrils, which are responsible for the sharp rise in the axial elastic modulus [113–117].

Valuable information about the role of the amorphous component in the deformation process may be obtained from the study of sorption and diffusion of vapors [46]. Vittoria et al. [116] investigated quenched films of iPP drawn at 110°C up to draw ratio $\lambda = 18$. Uniformly deformed sections of the drawn sample were placed in an atmosphere of methylene chloride at varying activity $a = p/p_T$ of vapor and the transport parameters, sorption and diffusion, were gravimetrically measured. In Figure 31 the equilibrium-specific concentration c_{sp} at activity $a = 0.7$, the zero-concentration diffusion coefficient D_0, and the exponential concentration coefficient γ_D are reported as functions of the draw ratio. A continuous drop of sorption and diffusion with increasing λ is readily observable; the drop is accelerated between $\lambda = 8$ and 10. The same applies to the increase of γ_D with λ. This substantial decrease of transport parameters can be explained in terms of the structural changes caused by the drawing process.

According to Peterlin, the original unoriented lamellar material is gradually transformed into the highly oriented microfibrillar structure. The microfibrils consist of folded-chain crystalline blocks axially connected by a great many TTMs passing and compress-

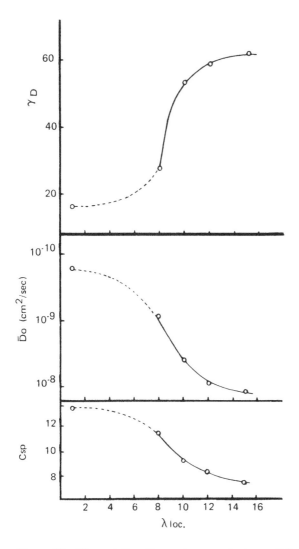

Figure 31 (*Bottom*) Specific sorption c_{sp} at $p/p_T = 0.7$ of methylene chloride at 25°C; (*middle*) logarithm of the zero-concentration diffusion coefficient D_0; and (*top*) concentration coefficient γ_D as a function of the draw ratio. (From Ref. 116.)

ing the amorphous layers separating the crystal cores of the subsequent blocks. The increase of the specific density of the amorphous layers causes a larger decrease of their FFV, which determines the decrease of transport properties in correspondence with the transformation from spherulitic to fibrillar morphology. After this transformation, although the drawing of the microfibrillar structure increases the fraction of TTMs by extension of the intermicrofibrillar tie molecules and thus increases the axial elastic modulus, it does not influence substantially the density of the amorphous layers of the microfibrils beyond the compression at the initial formation of the microfibrils out of the broken and destroyed lamellae. The achieved constancy or smaller rate of increase of γ_D clearly shows the smaller decrease or constancy of the FFV with increasing λ. The

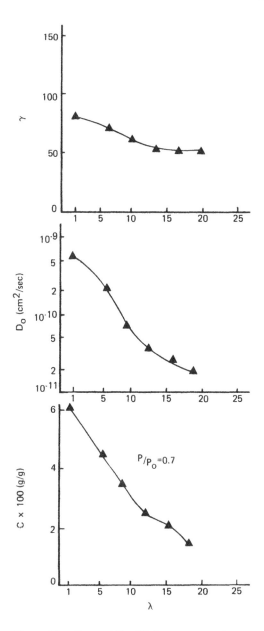

Figure 32 (*Bottom*) Equilibrium concentration c_{eq} at $p/p_T = 0.7$ of toluene at 30°C; (*middle*) the zero-concentration diffusion coefficient D_0; and (*top*) concentration coefficient γ_D as a function of draw ratio. (From Ref. 117.)

amorphous layers are close to a steady state, which does not change with increasing draw ratio.

The concept of the FFV explains very well the role of morphology in the observed specific transport properties. The sorption is proportional to the FFV, the zero-concentration diffusion coefficient D_0 is exponentially dependent on the inverse FFV, and the exponential concentration coefficient of diffusivity γ_D is inversely proportional to the square of the FFV of the polymer. A reduction of FFV yields a decrease of sorption ($S = c/p$), a much larger decrease of D_0, and a substantial increase of γ_D.

Highly drawn iPP was studied by Choy et al. [117], who investigated the sorption and diffusion of toluene vapor at 30°C in iPP, drawn at 130°C to draw ratios up to $\lambda = 18$. With increasing λ they found a decrease of sorption and zero concentration diffusion coefficient, as shown in Figure 32, in agreement with the results of methylene chloride. But in contrast with previous data on γ_D, which increases with λ, they found a decrease in the experimental concentration coefficient of the diffusivity. The authors explain this last result, which is at variance with data obtained with CH_2Cl_2 as penetrant in iPP drawn up to $\lambda = 15$, by the assumption that the several constrained chain elements in the drawn sample have much less freedom to mix with the molecules of the penetrant toluene. They derived an expression for γ_D similar to that of Fujita [118]:

$$\gamma_D = \frac{B\beta}{F^2} \tag{17}$$

where F is the fractional free volume, B is a constant, and β denotes the effectiveness of the penetrant molecule for increasing the free volume of the polymer–penetrant system. In their case β must decrease faster than F^2 in order to obtain a decrease of γ_D with λ. In the case of CH_2Cl_2 the factor β must remain constant or decrease more slowly than F^2 in order to yield an increase of γ_D with λ.

ULTRA–HIGH-MODULUS PRODUCTS

The possibility of obtaining ultra–high-modulus oriented iPP has been investigated by Ward and colleagues [119–121]. Their approach is based on deformation in the solid phase through the establishment of conditions for very high degrees of plastic deformation, involving both the crystalline and amorphous material.

This group undertook a systematic investigation to determine the suitable combination of processing parameters, i.e., draw temperature and strain rate, to overcome the limitation of the natural draw ratio. They found that draw temperature had by far the most significant effect over the strain rate. A threshold value of 100°C was found, above which draw ratio ratios far greater than 7–9 could be obtained. By drawing at high temperatures the process of the plastic deformation could be improved and draw ratios of 18 and higher obtained, as shown in Figure 33.

The modulus–draw ratio relationship was examined in samples prepared at different temperatures in order to define an upper limit for the draw temperature in terms of the mechanical properties of the oriented product. The results, seen in Figure 34, indicate an optimum draw ratio temperature in the vicinity of 110°C and showed that although higher draw ratios could be attained at temperatures well above 110°C, these are not characterized by an improvement in stiffness. Room temperature Young's moduli of 18 GPa were measured, representing a high fraction of the "theoretical" modulus proposed for iPP [86]. Low-temperature moduli as high as 25–27 GPa were also recorded; these are a still higher fraction of the theoretical modulus. Studying polymers of different molecular

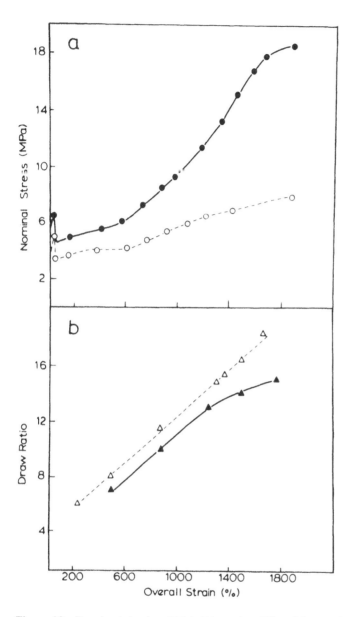

Figure 33 Drawing behavior of high ($M_w = 4 \times 10^5$, solid curves) and low ($M_w = 1.8 \times 10^5$, dashed curves) molecular weight PP. Draw temperature 110°C. The nominal stress (*a*) and the draw ratio in the necked region (*b*) are plotted as a function of overall strain. (From Ref. 119.)

weight and thermal history, Ward and colleagues found that the mechanical properties are to a good approximation insensitive to differences in these parameters, the stiffness of the final product being a unique fraction of its draw ratio. Only the analysis of the melting behavior of the samples indicates the presence of pronounced differences between samples of comparable draw ratio, depending on their molecular weight and thermal history.

The results of Ward's group are in agreement with the model of Peterlin in which the deformation of microfibrils is the principal mechanism involved in the achievement of high draw ratios.

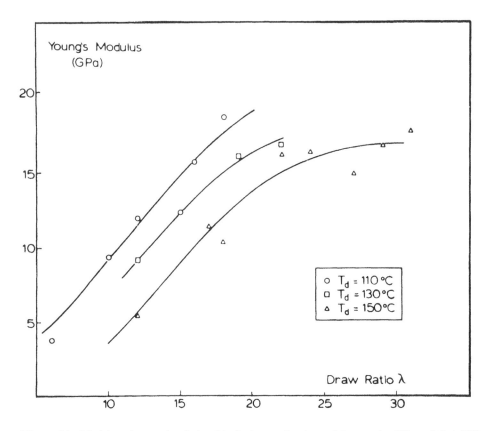

Figure 34 Modulus–draw ratio relationship for low molecular weight samples (M_w = 1.8 × 10^5) drawn at different temperatures. (From Ref. 119.)

REFERENCES

1. G. Natta, P. Corradini, and M. Cesari, *Rend. Accad. Naz. Lincei, 21*: 365 (1956).
2. G. Natta and P. Corradini, *Makromol. Chem., 16*: 77 (1955).
3. G. Natta and P. Corradini, *Nuovo Cim., 15*: 40 (1960).
4. G. Natta and P. Corradini, *J. Polym. Sci., 20*: 251 (1960).
5. M. Ahmed, *Polypropylene Fiber Science and Technology*, Elsevier, Amsterdam (1982).
6. A. Turner Jones, J. M. Aizlewood, and D. R. Beckett, *Makromol. Chem., 75*: 134 (1964).
7. D. R. Morrow, J. A. Sauer, and A. E. Woodward, *J. Polym. Sci., B3*: 463 (1985).
8. J. A. Sauer, D. R. Morrow, and G. C. Richardson, *J. Appl. Phys., 36*: 3017 (1965).
9. F. J. Padden, Jr., and H. D. Keith, *J. Appl. Phys., 30*: 1479 (1959).
10. P. H. Geil, *J. Appl. Phys., 33*: 642 (1962).
11. H. Dragaum, H. Hubeny, and H. Muschik, *J. Polym. Sci. Phys. Ed., 15*: 1779 (1977).
12. R. J. Samuels and R. Y. Yee, *J. Polym. Sci., A2, 10*: 385 (1972).
13. F. L. Binsbergen and B. G. M. de Lange, *Polymer, 9*: 23 (1968).
14. H. J. Leugering and G. Kirsch, *Angew. Makromol. Chem., 33*: 17 (1973).
15. H. J. Leugering, *Makromol. Chem., 109*: 204 (1967).
16. J. Garbarczyk and D. Paukszta, *Colloid Polym. Sci., 263*: 985 (1985).
17. J. Garbarczyk and D. Paukszta, *Polymer, 22*: 562 (1981).
18. D. R. Morrow, *J. Macromol. Sci. Phys., B3*(1): 53 (1969).

19. A. Turner Jones and A. J. Cobbold, *Polym. Lett., 6*: 539 (1968).
20. A. J. Lovinger, J. O. Chua, and C. Gryte, *J. Appl. Polym. Sci. Phys., 15*: 641 (1977).
21. H. D. Keith, F. J. Padden, Jr., N. M. Walter, and H. W. Wyckoff, *J. Appl. Phys., 30*: 1485 (1959).
22. E. J. Addink and J. Baintema, *Polymer, 2*: 185 (1961).
23. G. Natta, M. Peraldo, and P. Corradini, *Rend. Accad. Naz. Lincei, 26*: 14 (1959).
24. R. L. Miller, *Polymer, 1*: 135 (1960).
25. M. Glotin, R. R. Rahalkar, P. J. Hendra, M. E. A. Cudby, and H. A. Willis, *Polymer, 22*: 731 (1981).
26. P. Corradini, V. Petraccone, C. De Rosa, and G. Guerra, *Macromolecules 19: 2699 (1986)*
27. B. Wunderlich and J. Grebowicz, *J. Adv. Polym. Sci., 60/61*: 1 (1984).
28. J. Grebowicz, J. F. Lau, and B. Wunderlich, *J. Polym. Sci. Polym. Symp., 71*: 19 (1984).
29. H. W. Wyckoff, *J. Polym. Sci., 62*: 83 (1962).
30. J. A. Gailey and P. H. Ralston, *SPE Trans., 4*: 29 (1984).
31. H. Takahara and H. Kawai, *Rep. Prog. Polym. Phys. Jpn., 9* (1986).
32. D. M. Gezovich and P. H. Geil, *Polym. Eng. Sci., 8*: 202 (1968).
33. G. Bodor, M. Grell, and A. Kallo, *Faserforsch. Textiltech., 15*: 527 (1964).
34. R. Hosemann, *Acta Crystallog., 4*: 520 (1951).
35. M. Fujiyama, H. Awaya, and K. Azuma, *J. Polym. Sci. Polym. Lett. Ed., 18*: 105 (1980).
36. P. B. McAllister, T. J. Carter, and R. M. Hinde, *J. Polym. Sci. Phys. Ed., 16*: 49 (1978).
37. G. Guerra, V. Petraccone, C. De Rosa, and P. Corradini, *Makromol. Chem. Rapid Comm., 6*: 573 (1985).
38. R. Zannetti, G. Celotti, A. Fichera, and R. Francesconi, *Makromol. Chem., 128*: 137 (1969).
39. S. Kapur and C. E. Rogers, *J. Polym. Sci. Phys., 10*: 2107 (1972).
40. G. Natta, P. Corradini, and M. Cesari, *Rend. Accad. Naz. Lincei, 22*: 11 (1957).
41. G. Farrow, *Polymer, 2*: 409 (1961).
42. P. Jacoby, B. H. Bersted, W. J. Kissel, and C. E. Smith, *J. Polym. Sci. Phys. Ed., 24*: 461 (1986).
43. G. Farrow, *J. Appl. Polym. Sci., 9*: 1227 (1965).
44. V. Vittoria, *J. Polym. Sci. Phys., 24*: 451 (1986).
45. V. Vittoria and A. Perullo, *J. Macromol. Sci. Phys., B25*(3): 267 (1986).
46. A. Peterlin, *J. Macromol. Sci. Phys., B11*: 57 (1975).
47. A. S. Michaels, W. R. Vieth, and A. Barrie, *J. Polym. Sci., 41*: 53 (1959).
48. L. Rebenfeld, P. J. Makarewicz, H. D. Weighmann, and G. L. Wilkes, *J. Macromol. Sci. Rev., C15*(2): 279 (1976).
49. A. S. Michaels and R. B. Parker, *J. Polym. Sci., 41*: 53 (1959).
50. A. S. Michaels and H. J. Bixler, *J. Polym. Sci., 50*: 393 (1961).
51. J. Crank and G. S. Park, *Diffusion in Polymers*, Academic Press, London (1968).
52. J. Brandrup and E. H. Immergut, eds., *Polymer Handbook*, Wiley, New York (1975).
53. C. C. Hsu, P. H. Geil, H. Miyaji, and K. Asai, *J. Polym. Sci. Phys., 24*: 2379 (1986).
54. C. C. Hsu and P. H. Geil, *J. Appl. Phys., 56*: 2404 (1984).
55. P. J. Hendra, J. Vile, H. A. Willis, V. Zichy, and M. E. A. Cudby, *Polymer, 25*: 785 (1984).
56. G. Natta, M. Peraldo, and P. Corradini, *Rend. Accad. Naz. Lincei, 26*: 14 (1959).
57. R. Zannetti, A. Ferrero Martelli, and A. Fichera, *Gazz. Chim. Ital., 97*: 452 (1967).
58. R. Zannetti, G. Celotti, A. Fichera, and R. Martelli, *Eur. Polym. J., 4*: 399 (1968).
59. A. Fichera and R. Zannetti, *Makromol. Chem., 176*: 1885 (1975).
60. Y. Kissin and V. Rishina, *Eur. Polym. J., 12*: 757 (1976).
61. G. Zerbi, M. Gussoni, and F. Campielli, *Spectrochim. Acta, 23*: 301 (1967).
62. D. T. Grubb and D. Y. Yoon, *Polym. Comm., 27*: 84 (1986).
63. V. Vittoria and F. Riva, *Macromolecules, 19*: 1975 (1986).

64. J. L. Gardon, in *Encyclopedia of Polymer Science and Technology*, Wiley, New York, Vol. 12, 627 (1967).
65. W. Vieth and W. F. Wuerth, *J. Appl. Polym. Sci., 13*: 685 (1960).
66. R. A. Hayes, *J. Appl. Polym. Sci., 5*: 318 (1961).
67. J. F. Vocks, *J. Polym. Sci. A, 2*: 5319 (1964).
68. G. W. Schael, *J. Appl. Polym. Sci., 10*: 901 (1986).
69. D. M. Gezovich and P. H. Geil, *Polym. Eng. Sci., 8*: 210 (1968).
70. G. V. Casper and E. J. Henley, *Polym. Lett., 4*: 417 (1966).
71. J. J. Baum and J. M. Schultz, *J. Appl. Polym. Sci., 26*: 1579 (1981).
72. C. P. Buckley and M. Habibullah, *J. Appl. Polym. Sci., 26*: 2613 (1981).
73. V. Vittoria, *Polym. Comm., 28*: 199 (1987).
74. C. M. R. Dunn and J. Turner, *Polymer, 15*: 451 (1974).
75. J. M. Hutchinson and C. B. Bucknall, *Polym. Eng. Sci., 20*: 173 (1980).
76. C. K. Choi and N. G. McCrum, *Polymer, 21*: 706 (1980).
77. L. C. E. Struik, *Physical Aging in Amorphous Polymers and Other Materials*, Elsevier, Amsterdam (1978).
78. L. C. E. Struik, *Polymer, 28*: 1521 (1987).
79. G. Raumann and D. W. Saunders, *Proc. Roy. Soc. London, 77*: 1028 (1961).
80. Y. Tsunekawa, M. Oyane, and K. Kojima, *J. Polym. Sci., 50*: 35 (1961).
81. Y. Takagi and H. Hattori, *J. Appl. Polym. Sci., 9*: 2167 (1965).
82. S. L. Aggarwal and S. J. Sweeting, *Chem. Revs., 57*: 665 (1957).
83. W. Glenz and A. Peterlin, *J. Macromol., B4*: 473 (1970).
84. G. Parod, *Adv. Polym. Sci., 2*: 363 (1961).
85. R. J. Samuels, *Structured Polymer Properties*, Wiley, New York (1974).
86. I. M. Ward, *Structure and Properties of Oriented Polymers*, Applied Science Publishers, London (1975).
87. A. Peterlin, *J. Polym. Sci., C9*: 61 (1965).
88. A. Peterlin, *Colloid Polym. Sci., 253*: 809 (1975).
89. A. Peterlin, *J. Polym. Sci. A2, 7*: 1151 (1969).
90. A. Peterlin, in *Ultra High Modulus Polymers* (A. Ciferri and I. M. Ward, eds.), Applied Science Publishers, London (1979).
91. R. S. Stein, *Polym. Eng. Sci., 9*: 172 (1969).
92. H. Kiho, A. Peterlin, and P. H. Geil, *J. Polym. Sci., B3*: 257 (1965).
93. A. Peterlin, *J. Polym. Sci., C15*: 427 (1966).
94. F. de Candia, G. Romano, R. Russo, and V. Vittoria, *Colloid Polym. Sci., 265*: 696 (1987).
95. R. S. Stein and M. B. Rhodes, *J. Appl. Phys., 31*: 1873 (1960).
96. F. J. Baltà-Calleja and A. Peterlin, *J. Polym. Sci. A2, 7*: 1275 (1969).
97. F. J. Baltà-Calleja and A. Peterlin, *J. Mat. Sci., 4*: 722 (1969).
98. A. Peterlin and F. J. Baltà-Calleja, *J. Appl. Phys., 40*: 4238 (1969).
99. N. Morosoff and A. Peterlin, *J. Polym. Sci. A2, 10*: 1237 (1972).
100. F. J. Baltà-Calleja and A. Peterlin, *Makromol. Chem., 141*: 91 (1971).
101. F. J. Baltà-Calleja and A. Peterlin, *Plastic Deformation of Polymers*, Marcel Dekker, New York (1971).
102. M. F. Bottin, M. Bonduelle, and J. Guillet, *J. Polym. Sci. Phys., 21*: 401 (1983).
103. H. P. Nadella, J. E. Spruiell, and J. White, *J. Appl. Polym. Sci., 22*: 3121 (1978).
104. W. Ruland, *Polymer, 5*: 89 (1964).
105. P. G. Stern, *Polymer, 9*: 471 (1968).
106. S. Lee and D. R. Uhlmann, *J. Appl. Polym. Sci., 17*: 3747 (1973).
107. R. J. Samuels, *J. Polym. Sci., A3*: 1741 (1965).
108. R. J. Samuels, *J. Polym. Sci., C20*: 253 (1967).
109. R. J. Samuels, *J. Polym. Sci. A2, 6*: 2021 (1968).
110. R. J. Samuels, *J. Polym. Sci. A2, 7*: 1197 (1969).

111. R. J. Samuels, *J. Macromol. Sci. Phys.*, *B4*: 701 (1970).
112. H. P. Nadella, H. M. Henson, and J. E. Spruiell, *J. Appl. Polym. Sci.*, *21*: 3003 (1977).
113. F. de Candia, A. Perullo, V. Vittoria, A. Peterlin, in *Interrelations between Structure and Properties of Polymeric Materials* (J. C. Seferis and P. S. Theocaris, eds.), Elsevier, Amsterdam, p. 713 (1984).
114. M. Kamezawa, K. Yamada, and M. Takayanagi, *J. Appl. Polym. Sci.*, *24*: 1227 (1979).
115. W. P. Leung and C. L. Choy, *J. Polym. Sci. Phys. Ed.*, *21*: 725 (1983).
116. V. Vittoria, F. de Candia, V. Capodanno, and A. Peterlin, *J. Polym. Sci. Phys. Ed.*, *24*: 1009 (1986).
117. C. L. Choy, W. P. Leung, and T. L. Ma, *J. Polym. Sci. Phys. Ed.* *22*: 707 (1984).
118. A. Fujita, *Adv. Polym. Sci.*, *3*: 1 (1961).
119. G. Capaccio, A. G. Gibson, and I. M. Ward, in *Ultra High Modulus Polymers* (A. Ciferri and I. M. Ward, eds.), Applied Science Publishers, London, p. 1 (1979).
120. D. L. M. Cansfield, G. Capaccio, and I. M. Ward, *Polym. Eng. Sci.*, *16*: 721 (1976).
121. A. J. Willis, G. Capaccio, and I. M. Ward, *J. Polym. Sci. Phys. Ed.*, *18*: 493 (1980).

16

Physical and Chemical Crosslinking of Polyurethane

Jay Bhattacharyya
Gates Rubber Company
Denver, Colorado

INTRODUCTION

Polyurethanes are high-polymeric materials having —NHCOO— groups in the backbone of the polymer. This class of materials is not just the repetition of —NHCOO— groups as in polyethylene or polyvinyl chloride. The groups are present in the polymer as a result of reaction of an isocyanate (—NCO—) group at the ends of the monomers (OCN—R—NCO) with hydroxy (—OH—) groups present at the ends of other monomers (HO—R'—OH). The polyaddition reaction with diisocyanate was discovered by O. Bayer in the 1930s. Applications of the polyurethanes became apparent in the 1940s. This class of materials has become very important among the polymers because of the materials' superior tear strength and abrasion resistance among other properties and their use is steadily increasing. The total sale of polyurethane materials in the United States was 2333 million pounds in 1986, an increase of 0.65% from 1985 [1]. The global demand outlook was 9435 million pounds, which will grow to 13,051 million pounds in 1990, an estimated increase of about 7% for the decade [2].

Most polyurethanes are used as flexible foam in bedding, furniture, rug underlay, and transportation markets (1330 million pounds in 1985). The rigid foams are used in insulation and packaging markets and the volume is about 50% of that of flexible foams. The elastomers are used as specialty engineering materials; this is one of the most versatile materials among the elastomers, and the volume was 247 million pounds in 1985. The elastomers produced with urethane chemicals by reaction injection molding (RIM) and reinforced reaction injection molding (RRIM) have made major inroads in the automotive markets. Due to the high demand for smaller and lighter cars, automotive designers are looking for more plastic materials to be used in the cars replacing iron, at about half the weight, keeping satisfactory performance. The percentage of iron use has gone down from 17.9 for a 1978 car to 10.4 for a 1987 car, whereas the use of plastic has gone up from 5.4 to 9.1% during that time. The average use of plastics in the 1987 model year car is in the range of 255 pounds [3]. Metzger [4] gives further details on this subject.

CHEMISTRY

Over the past several years, research has shown that polyurethanes consist of segmented block copolymers of (AB)X type [5–12]. The hard segment consists of a reaction product of diisocyanate with a short-chain diol. The soft segment is a reaction product of a high molecular weight glycol with diisocyanate. Neither the hard segment nor the soft segment is suitable for engineering application as such, because of the lack of good physical properties. When diisocyanate, short-chain diol, and polymeric diol are reacted in the stoichiometric ratio, polyurethane is formed with hard and soft segments in the polymer backbone. The segments are incompatible, and as a result the hard segments cluster together by hydrogen bonding and thus serve as crosslinks between the soft segments. This is referred to as physical crosslinking. However, in the polymer matrix there are several crossovers and entanglements of the soft segments. The interconnection of the soft segments and the degree of phase separation are responsible for the physical properties of the polyurethanes [13–30]. The physical crosslinking is susceptible to breakage by thermal energy and solvation with good solvents. Thermal energy over a certain range will cause degradation of the polymer. Thus the application of the required amount of heat or a good solvent will make this class of polymer very useful for thermoforming processes and adhesive applications.

Figure 1 is a schematic representation of a typical polyurethane chain and the action of processing heat to the polymer. Wilkes and others reported that many of the segmented urethanes change the domain structure when the polymer is heated above 40°C [21, 23, 31]. When cooled to room temperature, the original domain texture is regained at a rate depending on the temperature and chemistry of the polymer involved and whether the system is linear or chemically crosslinked [32].

The chemical components for hard segments are predominantly linear difunctional molecules for elastomeric polyurethanes. The diisocyanate is typically diphenylmethane diisocyanate (MDI). The other symmetrical diisocyanates with or without aromatic molecules are being used for special applications. Aromatic diisocyanates are not light-stable. Aliphatic diisocyanates are used when light stability is desired, typically in coatings. Typical aliphatic isocyanates are 1,6-hexamethylene diisocyanate (HDI), hydrogenated MDI (RMDI), and isophorone diisocyanate (IPDI). The other component of the hard segment, also called the chain extender, is typically 1,4-butanediol or aromatic diamine. Other dihydric short-chain alcohols, with or without aromatic rings, are also being used but to a lesser extent.

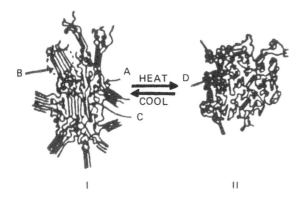

Figure 1 Schematic model depicting the morphology at both (I) long time (control) and (II) following heat treatment: (A) partially extended soft segment, (B) hard-segment domain, (C) hard segment, (D) "relaxed" soft segment, (E) lower order hard-segment domain. (From Ref. 23.)

The soft segment is a reaction product of a long-chain diol like polytetramethylene ether glycol, polycaprolactone diol, or polyadipate diol having molecular weight generally in the range of 650–2000, with structurally symmetrical diisocyanate like MDI. In polyurethane flexible foams toluene diisocyanate (TDI) with a mixture of 2,4- and 2,6-isomers is generally used for economic reasons. TDI, however, is more toxic than MDI. So there are systems available to use MDI instead of TDI in foam formulations, to overcome the toxicity problems. Polymeric diphenylmethane diisocyanate (PMPPI) is used in rigid polyurethane foams. Polyoxypropylene glycols (PPG) are the predominant choice as polyol in foam formulations for economic reasons. Polyoxyethylene–oxy-propylene glycols, commonly referred to as "tipped" PPG, and also PPG are being explored for applications as thermoplastic polyurethane [33]. Polyethers containing dispersed organic fillers like vinyl copolymer or polyurea are useful.

The chemical structures of the major components of the polyurethane systems are presented in Figure 2. The chain extender/resin (polyol) ratio, molecular weight of the polyol and the ratio of hard segment to soft segment, are the determining factors for the soft segment length. The percentage of the hard segment in the total polymer and also the length of the soft segment determine the physical and thermal properties of the thermoplastic polyurethane elastomers. The higher the ratio of the hard to soft segment, the higher is the polymer's mechanical and thermal properties in a given polyurethane system. The ratio of NCO/OH $\times$ 100, called the isocyanate index, is close to 100 in commercial thermoplastic polyurethanes. Generally the ratio is kept a little higher than 100 to compensate for the reactive groups in the impurities which will consume isocyanate molecules. Theoretically a polymer should have —OH and —NCO end groups at the same chain if a unit ratio NCO and OH is allowed to react. If the isocyanate index is less than 100, the end groups are —OH, and vice versa. If the NCO/OH ratio is much higher than 1, then a three-dimensional crosslinked polymer is formed with allophanate and biuret structures. The isocyanate molecules are also able to react with each other to form isocyanurates. The common chemical reactions are shown in Figure 3. The relative reaction rates of common urethane chemicals and impurities with isocyanates are amine > hydroxy > urea > urethane. Thus amine molecule present in the system will be consumed by isocyanates

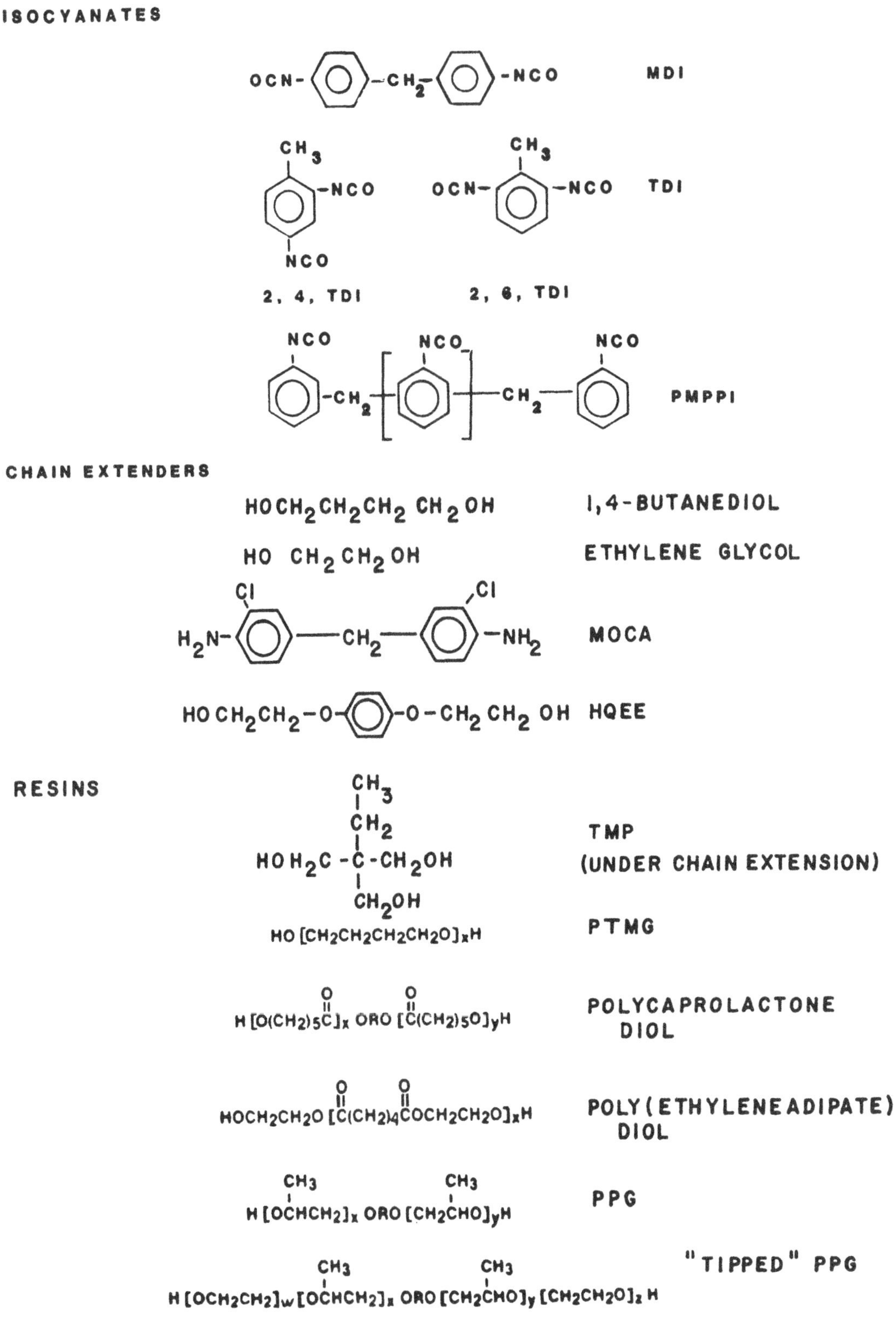

ISOCYANATES

OCN-⬡-CH₂-⬡-NCO   MDI

2, 4, TDI   2, 6, TDI   TDI

PMPPI

CHAIN EXTENDERS

HOCH₂CH₂CH₂CH₂OH   1,4-BUTANEDIOL

HO CH₂CH₂OH   ETHYLENE GLYCOL

H₂N-⬡-CH₂-⬡-NH₂   MOCA

HOCH₂CH₂-O-⬡-O-CH₂CH₂OH   HQEE

RESINS

HOH₂C-C(CH₃CH₂)(CH₂OH)-CH₂OH   TMP (UNDER CHAIN EXTENSION)

HO[CH₂CH₂CH₂CH₂O]ₓH   PTMG

H[O(CH₂)₅C(O)]ₓORO[C(O)(CH₂)₅O]ᵧH   POLYCAPROLACTONE DIOL

HOCH₂CH₂O[C(O)(CH₂)₄C(O)OCH₂CH₂O]ₓH   POLY(ETHYLENEADIPATE) DIOL

H[OCHCH₃CH₂]ₓORO[CH₂CHCH₃O]ᵧH   PPG

H[OCH₂CH₂]_w[OCHCH₃CH₂]ₓORO[CH₂CHCH₃O]ᵧ[CH₂CH₂O]_zH   "TIPPED" PPG

Figure 2   Chemicals commonly used in polyurethanes.

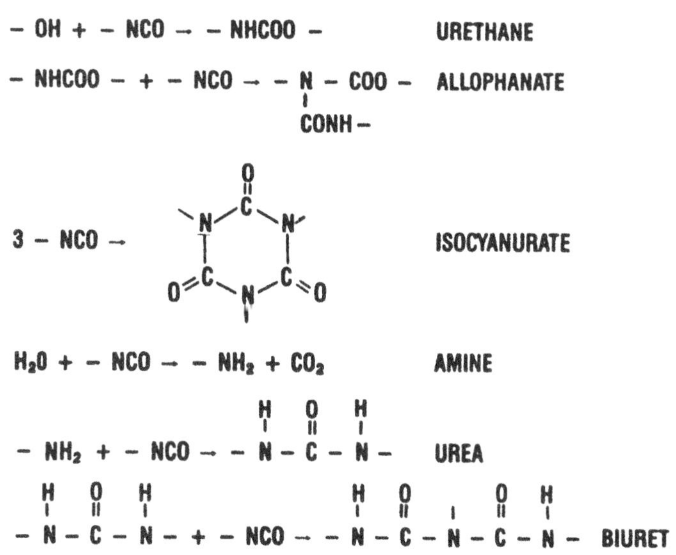

```
- OH  +  - NCO  →  - NHCOO -              URETHANE

- NHCOO -  +  - NCO  →  - N - COO -       ALLOPHANATE
                            |
                          CONH -
```

```
3 - NCO  →                                ISOCYANURATE
```

```
H₂O  +  - NCO  →  - NH₂  +  CO₂           AMINE
```

$$- NH_2 \; + \; - NCO \; \rightarrow \; - \overset{\overset{\textstyle H}{\textstyle |}}{N} - \overset{\overset{\textstyle O}{\textstyle \|}}{C} - \overset{\overset{\textstyle H}{\textstyle |}}{N} - \qquad \text{UREA}$$

$$- \overset{\overset{\textstyle H}{\textstyle |}}{N} - \overset{\overset{\textstyle O}{\textstyle \|}}{C} - \overset{\overset{\textstyle H}{\textstyle |}}{N} - \; + \; - NCO \; \rightarrow \; - \overset{\overset{\textstyle H}{\textstyle |}}{N} - \overset{\overset{\textstyle O}{\textstyle \|}}{C} - \overset{\overset{\textstyle H}{\textstyle |}}{N} - \overset{\overset{\textstyle O}{\textstyle \|}}{C} - \overset{\overset{\textstyle H}{\textstyle |}}{N} - \qquad \text{BIURET}$$

Figure 3 Isocyanate reactions.

first followed by hydroxy molecules, and so on. Unless the NCO/OH ratio is substantially higher than unity and the chemicals are pure, the chance of getting a crosslinked polymer is low. A wide variety of additives are used in polyurethanes to introduce certain properties or enhance the chemical reaction. These chemicals are catalysts, surfactants, fillers, blowing agents, antioxidants, light stabilizers, flame retardants, etc.

The thermoplastic polyurethanes show the characteristic properties of crosslinking due to the presence of physical crosslinking in the structure. These materials, however, can also be chemically crosslinked by generating free radicals in the ethylenic backbones by reacting with peroxides [34–37], which combine together to form a crosslinked structure. The classical method of sulfur vulcanization can also be applied to specially prepared polyurethanes having a pendent double bond with at least one methylene hydrogen adjacent to the double bond [38]. Crosslinking can also be introduced in the chain in various ways [39–42]. Unsaturated chain extenders are often used to incorporate pendent double bonds in the structure [43–45]. Taking the analogy of sulfur accelerator vulcanization of unsaturated rubber, Figure 4 represents the same for polyurethane millable gum. Another very useful way to crosslink specially formulated polyurethanes used in the coating application is by exposure to high-energy radiation such as UV or electron beam radiation. In this way a polyurethane can be formulated with no diluents but still liquid at room temperature and having good shelf life, thus capable of being easily applied as a coating material. When this type of material is exposed to high-energy radiation, free radicals are generated in the system, thus initiating the crosslinking process. Once the system is fully reacted the surface is hard and its properties can be varied by proper formulation. This has been found to be a very effective way to eliminate from the working place solvent, which is undesirable in most cases. This results in good cost savings since heating is not required to drive the solvent out and cure the system [46, 47]. The chemical reaction mechanism and other details can be found in the work of Guarino et al. [48] and Gupta et al. [49]. The crosslinked materials thus formed are insoluble and infusible, which is desired for many applications.

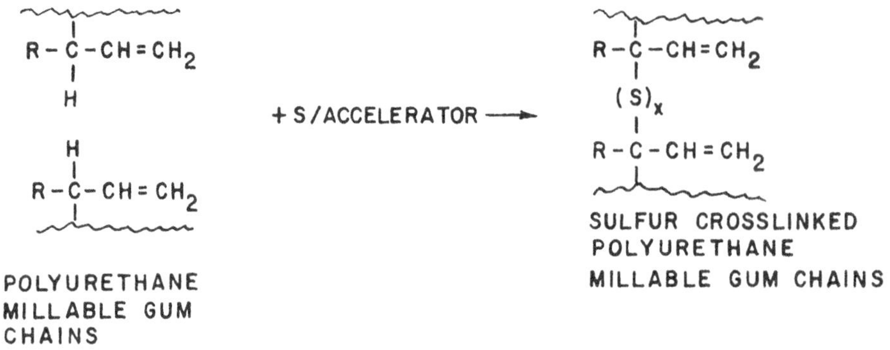

R – C̶ – CH = CH₂
     |
     H

H
|
R – C̶ – CH = CH₂

**POLYURETHANE
MILLABLE GUM
CHAINS**

+ S / ACCELERATOR ⟶

R – C̶ – CH = CH₂
     |
    (S)ₓ
     |
R – C̶ – CH = CH₂

**SULFUR CROSSLINKED
POLYURETHANE
MILLABLE GUM CHAINS**

Figure 4 Covalent chemical crosslinking of polyurethane millable gum chains by sulfur-accelerator systems. (Adapted from Ref. 30.)

A crosslinked material can also be made by incorporating a trifunctional material in the reaction mixture. This is common practice for the manufacture of polyurethane foams. When an isocyanate with an apparent functionality of more than 2.0 is allowed to react with other polyurethane chemicals, a crosslinked product is formed. A linear thermoplastic polyurethane with hydroxy end groups can be melted in processing equipment, under suitable conditions; it is also capable of reacting with excess isocyanate to form a crosslinked product. The amount of crosslinking can easily be detected by recording the torque required to run the equipment [50–52].

In recent years another important class of polymers is being studied including polyurethane as one of the components. These are called interpenetrating network (IPN) polymers [53–71]. These IPNs are made by two initial polymers in solution, dispersion, or bulk which are crosslinked in situ. In this way permanent entanglements will result, depending on the cohesive energies of the polymers [72]. True IPNs are homogeneous mixtures of the component polymers having no covalent bonds and grafts in the polymer chains. The polymers must be selected in such a way that there is no chemical reaction during crosslinking of the polymers. Ideal results are obtained when one of the polymers is rubbery and the other is glassy. The polymer chains are interlocked with each other; as a result there is virtually no phase separation. Thermodynamically incompatible polymers, when blended together, result in multiphase morphology. Semi-interpenetrating polymer networks (semi-IPNs) include one component that is a linear polymer. Lee and Kim [73–75] prepared IPNs and semi-IPNs under high pressure and at different temperatures and analyzed the morphology, dynamic mechanical properties, and density behavior to determine the effect of synthesis pressure and the effect of interlocking on the degree of intermixing of the component polymers. The IPNs made under high pressure were also studied with relation to the Gibb's free energy of mixing [76, 77], molecular weight, and polymer diffusion rate and crosslink density [78]. The existence of the interlocking of the polymer chains in the IPNs is a major factor for controlling properties like density [79, 80], synergistic effect in mechanical properties [81–83], finer dispersed domain structure [84], and increased thermal stability [85, 86].

Polyurethane chemistry offers significant utilization of wood and its by-products. Lignin is a naturally occurring resin present in the wood to hold the cellulosic materials. Isocyanate chemistry can be applied to bond wood products for variety of applications where hydroxy groups present in wood react with isocyanate groups. Several studies report the application of wood or wood components in polyurethane systems [87–98]. Glasser et al. [99] explored the possibilities of the use of lignin by-products from kraft and sulfite pulping and found that this is useful for polyol for polyurethane products despite the limitations of the lignin liquification process. Glasser and co-workers [100–102] synthesized and characterized the hydroxypropyl lignin derivatives. Several lignin-based solution polyurethanes were made and structure–property relationships were studied depending on the lignin type and composition [103] and crosslinking density. Saraf et al. [104] synthesized several hydroxypropyl lignin-based thermoset polyurethanes with hexamethylene diisocyanate and toluene diisocyanate and the polymers were characterized.

Polyurethanes are compatible with a number of other polymers and a number of polymer blends are available commercially. ABS can be used as a modifier of polyurethanes to decrease cost and improve processing characteristics; on the other hand polyurethanes can improve the impact strength and abrasion resistance of ABS [105, 106]. The thermal stability of a segmented elastomer depends in part of the hard-segment domains. The incorporation of a crystallizable, high-melting hard segment was achieved by the polymerization of MDI to an acid-terminated prepolymer and aliphatic dicarboxylic acid. The hard-segment content is controlled by the additional acid. The polymer produced by this system is capable of withstanding higher temperature than normal polyurethanes [107].

PROCESSING

The forming process for elastomeric polyurethanes is done by conventional methods like compression or injection molding and extrusion; in many cases, however, the finished shape is achieved while the components are still reacting. The manufacturing of bulk thermoplastic polyurethane can be done by a one-step or two-step process. In the one-step process the individual components, kept at proper reacting temperatures, are metered at a mixing head, where the components are mixed very rapidly and thoroughly and then dropped to a conveying system, which is also kept at a proper temperature, to complete the reaction. The polymer is then transferred to the proper shape and size according to the customer's desire. The proper selection of the urethane component metering system is a very important criterion for product uniformity as the shift of the isocyanate index will change the polymer properties to a great extent. Maroney and Griffin [108] discussed some criteria for proper selection of the metering pumps. In the two-step process, the polyol and the diisocyanate are made into a prepolymer where the end groups are isocyanates. In this process the ingredients are heated together in an inert atmosphere. The reaction is controlled by monitoring the —NCO number. The prepolymers are also sold commercially. The prepolymers are degassed before casting to eliminate any side reaction and imperfection in the finished products. This is done by heating the prepolymer and then applying vacuum. The curing agent is then added in a batch process or continuous process and the mix is injected in the heated mold of the desired shape. The viscosity of the mix and the reaction rate are controlled by the temperature of the prepolymer and the curing agent. The chemical reaction is not complete at this stage. The material has to be kept in the hot mold for some time before it can be removed. It is often necessary to postcure the items to have the full strength, which can be done by keeping the items at an elevated-temperature oven for few hours or by keeping them at ambient temperature for few weeks.

Although the prepolymer route is the preferred method for molding of articles, often it is more expensive than the one-shot process. Axelrood and co-workers [109, 110] have shown that by using proper catalyst comparable properties can be obtained with a polypropylene glycol–MOCA–TDI system. Table 1 shows the properties of the system. A polycaprolactone polyester casting system has been described by Magnus [111]. This system has better hydrolytic stability and low-temperature flexibility than the adipate polyester system and better strength and low-temperature flexibility than the polyether system. Polyurethane systems based on polybutadiene polyols have been described by Verdol and others [112–114]. Polyesters based on adipic acid, one of the earliest polyurethane systems developed, are described by Bayer and others [115–119]. Polybutylene and polypropylene glycols are two important raw materials for polyether polyurethane systems and have been described by Athey and co-workers [120–124]. The monomeric diisocyanate MDI requires special precaution since the monomer slowly dimerizes and has to be filtered to keep the correct ratio of NCO/OH. Storage temperature and time are the two most important factors for controlling dimer formation [125].

Polyurethane foams can broadly be classified as flexible and rigid. Polyether polyol and TDI are predominant materials in flexible foams, whereas in rigid foams PMPPI is also being used for crosslinking. The flexible foams are predominantly open-celled, whereas the rigid foams are close-celled. Polyethers produce materials with better hydrolytic stability and resiliency, whereas polyesters have better strength. The gas for blowing can be produced by the reaction of water with isocyanate [126] where carbon dioxide is generated. Methylene chloride can also be used to produce flexible foams [127]. The urethane reaction is exothermic, which allows the solvent to evaporate and get entrapped in the process of urethane polymer formation. Flexible slabstocks are produced by the one-shot continuous process where the ingredients are metered and mixed before pouring in the conveying system, yielding foam slabstocks. The slabstocks are then converted into different-sized products and shapes by variety of fabricating techniques. Saaty et al. [128] described foams made by various polyester polyols. Murai and others [129, 130] made polyether polyols for various other applications. Typical commercial slabstocks have a density range of 1–3 lb/ft^3. Properties can be tailored by the selection of suitable raw

Table 1 One-Shot Elastomers from Poly(propylene glycol)-MOCA-TDI

System	Prepolymer	One-shot	One-shot
Polyol	PPG 2010	PPG 2010	PPG 1010
NH$_2$/OH ratio	1 : 1	1 : 1	1 : 1
Stannous octoate	None	0.025	0.025
300% modulus, psi	900	920	2350
Tensile strength, psi	2800	2440	3900
Elongation, %	500	1020	
Hardness			
Shore A	75	82	
Shore D			49
Graves tear, pli	210	305	450
Bashore rebound	37	46	36

Source: Ref. 109.

materials and formulations. The molded foam is a one-shot discontinuous process where the mixed chemicals are introduced in the mold having generally a complex shape not easily attainable from slabstock. In the mold the mixture foams and fills the mold to attain the desired shape and properties. In the same mold different hardness-producing mixtures can also be introduced to give dual hardness by parallel introduction of the mixtures. Flexible foams are used in the furniture, bedding, and automotive industry for cushioning. Rigid foams are used for insulation for commercial and residential buildings. The fluorocarbon used as a blowing agent yields a foam product with very low coefficient of thermal conductivity. Since this is a crosslinked product, the usable temperature range is substantially higher than that for flexible foam and is chemical- and solvent-resistant. Rigid foams may be made in a variety of ways depending on the application. The components can be mixed and sprayed on, poured, sandwiched between metals, or made into a slab, which can be cut into standard sizes and shapes. Catalysts play a very important role in urethane foam manufacture, and the catalyst should be chosen very carefully before foam formulations are made [131].

Urethanes are often chosen as coating material for their superior abrasion resistance, tear strength, chemical and solvent resistance, and flexibility. TDI is generally used in this area. Color-stable coatings are made with aliphatic isocyanates. Coatings are available as a one-component moisture cure system or a two-component system where the polyol and the polyisocyanate are mixed before application. Both systems have limited pot life. Another system is available where the blocked isocyanate is premixed with polyol. When heated, the isocyanate is released to react with the polyol to make the polymer. Completely reacted urethane can also be dissolved in suitable solvents and can be used as coatings. Polyurethanes are very good adhesives when suitably formulated. The NCO groups are highly reactive and will react with any available hydrogen of the substrate. The types of the adhesives are similar to those of coatings. The recent trend is toward the use of 100% solids to meet the restrictions of different regulatory agencies. Crosslinking with UV light is a major step toward the utilization of 100% solids. In this process the urethane chemicals are mixed with some UV sensitizers and acrylates as comonomers. The system, still liquid at room temperature, when applied to the substrate and exposed to UV light becomes a hard solid crosslinked product. Further details can be obtained from Refs. 132–136.

STRUCTURE, PROPERTY, AND CHARACTERIZATION

Solvent Resistance

The polyurethane is a blocked copolymer consisting of hard and soft segments. The hard segments cluster together to act as filler particles to impart strength to the polymer. Solvents may act differently on hard and soft segments of the polymer [137, 138]. Solvents encapsulate the polymer molecules and break the intermolecular and intramolecular forces, thus increasing the distance between the molecules of the polymer; as a result the polymer goes into the solution phase, where the chains are separated and float free. This is true for linear and branched polymers. In the chemically crosslinked polymer, the chains cannot be totally separated from one another due to the presence of the chemical crosslinks. The degree of solvation will depend on the solvent and the crosslink density of a given system. Solvent swelling measurement of a crosslinked polymer can thus be used to calculate the crosslink density. Swelling can also result from the selective solvation of either hard or soft segments of the polyurethane, where one segment will hold the polymer together whereas the other segment will be separated by the solvent. Mixtures of solvents which are compatible with one another can be used to make a polymer solution where both the segments will go into solution by selective solvent properties. Polyurethane samples were made with BDO, MDI, and polytetramethylene adipate glycol (PTMA) in the ratio 0.3 : 1.0 : 1.3 by Schollenberger and Dinbergs [30] and tested for solvation resistance using ASTM standard D-741. The ratio of components was chosen to avoid any chemical crosslinking due to the presence of excessive isocyanate. The chemical crosslinking was achieved by adding 2 phr of organic peroxide and curing at a press at 210–220°C. The results are presented in Table 2. It can be seen from the table that with all the solvents the chemically crosslinked samples are less affected than the linear polymer. The strong solvents (solvents 12–17) have swelling action on the chemically crosslinked polymer but completely dissolve the linear polymer. Saraf et al. [104] prepared several polyurethanes based on a mixture of lignin-based polyol mixed with polyethylene glycols of different molecular weights, and hexamethylene diisocyanate (HDI) and TDI with varying NCO/OH ratios. Swelling studies in dimethylformamide were carried out. The results of the swelling as a function of NCO/OH ratio are seen in Figure 5. It can be seen that the swelling is lower with MW 4000 PEG and higher NCO/OH ratios, suggesting the crosslinking by allophanate and biuret linkages. Ershaghi et al. studied the structure of polyurethanes by swelling the polymer in dioxane solvent [139]. Frisch, Reegen, and Rumao determined the molecular weight of different polyurethanes by swelling in chlori-

Table 2 Solvation Resistance of Subject Thermoplastic Urethane Elastomer in VC and in CC-VC States (7 Days Immersion, 25°C)

Immersion liquid	VC polymer volume	Weight (% increase)	CC-VC polymer volume	Weight (% increase)
1. Water	1.31	1.00	1.27	1.16
2. Methanol	24.85	16.60	22.41	15.02
3. Ethylene glycol	1.96	1.82	0.78	0.82
4. ASTM Oil No. 1	0.87	0.77	0.29	0.25
5. ASTM Oil No. 3	7.49	5.16	1.21	0.96
6. JP4 Jet Fuel	5.08	3.49	2.85	1.84
7. Gasoline (Sunoco 200)	19.31	13.40	16.08	11.14
8. Benzene	169	126	101	74
9. Xylene	64	47	51	37
10. Chlorobenzene	273	255	135	125
11. Perchloroethylene	30.12	41.92	26.67	37.17
12. Methylene chloride	dissolves		259	295
13. Methyl ethyl ketone	dissolves		171	116
14. Cyclohexanone	dissolves		261	205
15. Tetrahydrofuran	dissolves		247	183
16. Ethyl acetate	dissolves		125	94
17. Nitromethane	dissolves		121	115

Key: VC = virtual crosslink; CC = covalent crosslink.
Source: Adapted from Ref. 30.

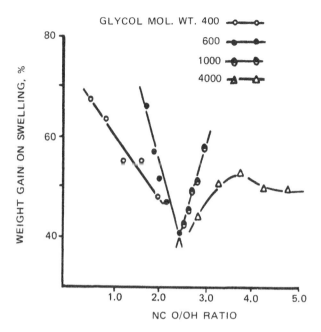

Figure 5 Swelling behavior of HDI-based polyurethanes in relation to NCO/OH ratio; films. Glycol MW: (○) 400; (●) 600; (◑) 1000; (△) 4000 in DMF. (Adapted from Ref. 106.)

nated solvents [140]. Sung and Mark studied the network structure of polyurethane by swelling in benzene [141]. Seefried et al. established the relationship of intrinsic viscosity and molecular weight by dissolving a polyurethane in dimethylformamide, dimethylacetamide, tetramethylurea, *N*-methylpyrrolidone, and meta-cresol [142].

Stress–Strain Properties

In a given polyurethane system, a good correlation exists between the stress–strain properties and the content of the urethane hard segment. As the hard-segment content increases, the Young's and secant moduli will also increase, whereas the elongation at break will decrease. The hard segments act as a filler in the rubbery soft polyurethane system, and increasing the amount will naturally increase the strength of the material up to a certain point. This is true for both polyether and polyester systems. The modulus values are generally higher in the polyester systems than in the polyether system at the same composition and tested at the same temperature. Figures 6 and 7 show the stress and strain properties of polyether and polyester systems at different compositions. It can be seen that the modulus values are higher in the polyester systems than the polyether systems at similar compositions. Lilaonitkul and Cooper [143] explained this phenomenon through the viscoelastic state of the polymers. The polyether polymer is still in the rubbery plateau, whereas the polyester polymer is in the leathery region. When the polyester crystallizes a high modulus is obtained. In a polyurethane chain there is extensive hydrogen bonding, which also influences the stress–strain properties. The NH group of the urethane linkage will bond with oxygen of the hard segment and the ester carbonyl in the polyesters or ether oxygen in case of polyether polymers.

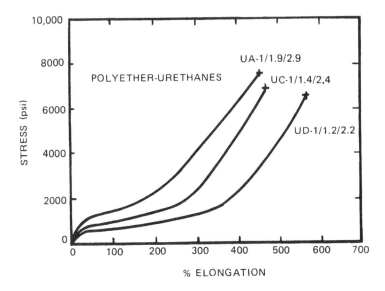

Figure 6 Stress–strain curves for polyether urethanes of varying urethane content. (Adapted from Ref. 143.)

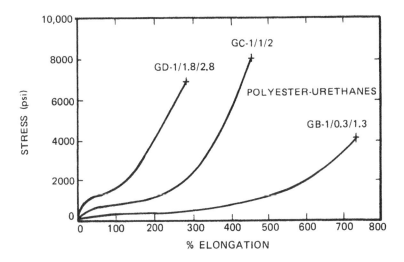

Figure 7 Stress–strain curves for polyester-urethanes of varying urethane content. (Adapted from Ref. 143.)

The nature of hydrogen bonding has been the subject of several investigations [9, 91, 144–152]. Schollenberger and Dinberg [30] tested the stress–strain properties of MDI, PTMA, and BDO systems, linear and crosslinked with peroxide, at room temperature and at 100°C. Their results are presented here in Figure 8. It can be seen that the chemically crosslinked polymer has much higher modulus at the same elongation at the same temperature than the linear polymer. The elongation of the crosslinked polymer is lower

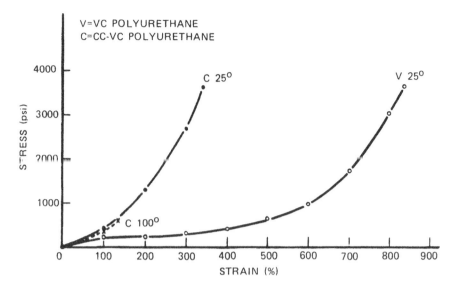

Figure 8 Stress–strain values for VC and CC–VC polyurethanes at 25 and 100°C (Instron, 20 in./min, 25 mil ⅛-in. dumbbell). (Adapted from Ref. 30.)

because the chemical bonds developed between polymer chains make it more resistant to deformation. At higher temperature the integrity of the crosslinked polymer is maintained and the polymer could be tested, whereas the linear polymer could not be tested. In certain polyurethane compositions the stress–strain properties can change with postcuring. In this process the polymer is either kept at room temperature or at elevated temperature before use to get the full strength. This process enables the chemicals to react completely. In certain compositions there may be crosslinking reactions if there are excess isocyanates.

Rearrangement of the crystalline structure has also been given as an explanation of this enhancement of physical properties [153]. Figure 9 illustrates this for a polyester system.

Chiu [154] studied the stress–strain properties at different times and temperatures of aging of polyurethane ureas made from PTMEG, substituted diamines, and saturated MDI. The data are presented in Figure 10. It can be observed that at 75°C the tensile properties are higher with more aging, indicating the postcuring effect. At 100°C the tensile properties are lower at higher aging, indicating the chain scission reaction.

Paik Sung [155] studied the mechanical properties of polyether polyurethane ureas and concluded that the excellent properties of the polyurethane ureas are due to better phase segregation and domain structure, which are derived from three-dimensional hydrogen bonding. The polyurethane ureas have lower T_g of the soft segments and higher T_g of the hard-segment domains, even at lower urea content. The T_g of the soft and hard segments of the PTMO-2000 polymers are found to be independent of urea or urethane content, whereas for PTMO-1000 this is dependent on the urethane content but independent of the urea content, as can be seen from Figures 11 and 12.

Boenig [156] pointed out the importance of the perfection of the microphase separation which is associated with the symmetrical structure required for the rigid segment of the polyurethane ureas with amines as chain extenders. Muller and co-workers [116] showed that the bulky and rigid diisocyanates gave products of high modulus and elasticity (Table

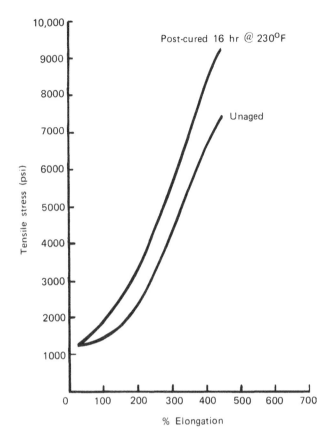

Figure 9 Effect of postcuring on stress–strain properties of 0.080-in. thickness samples stressed at 2 in./min, Roylar RD52. (Reprinted with permission from Van Nostrand Reinhold Company.)

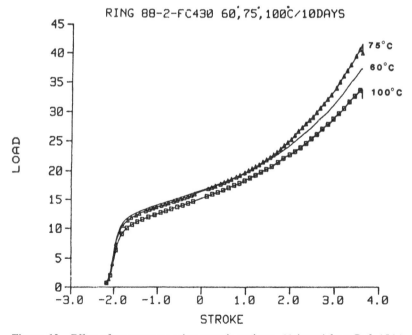

Figure 10 Effect of temperature aging on polyurethane. (Adapted from Ref. 154.)

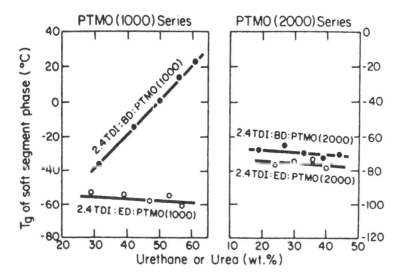

Figure 11 Comparison of T_g of the soft-segment phase in polyether polyurethanes and polyether polyurethane ureas. (Adapted from Ref. 155.)

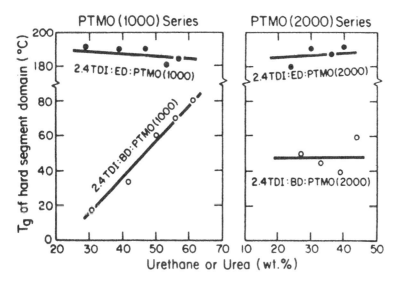

Figure 12 Comparison of T_g of the hard-segment domain in polyether polyurethanes and poly-ether polyurethane ureas. (Adapted from Ref. 155.)

Table 3 Effect of Diisocyanates on Polyurethane Elastomer[a] Properties

Diisocyanate	Tensile strength (kg/cm³)	Elongation at break (%)	300 modulus (kg/cm³)	Tear strength (kg/cm³)	Hardness shore B
1,5-Naphthalene	300	500	210	36	86
1,4-Phenylene	450	600	161	53.5	72
2,4- and 2,6-Toluene	320	600	25	27	40
4,4'-Diphenyl methane	555	600	112	48	61
3-3'-Dimethyl 4-4'-diphenyl methane	370	550	42	7	47
4-4'-Diphenyl propane	245	700	21	16	56
3-3'-Dimethyl-4-4' biphenyl	280	400	161	32	72

[a]Composition molar ratio poly(ethylene adipate) MW 200/1,4-butanediol/diisocyanate. 1 : 2 : 3.2.
Source: Adapted from Ref. 116.

3). Joseph, Wilkes, and Park [157] crosslinked model polyurethanes with electron beams and measured the mechanical properties. They found that up to a 10-Mrad dose there is an increase in the modulus but above this the modulus decreases, probably due to chain scission. This is presented in Figure 13. Agrawal and Rubner [158] introduced diacetylene group in model polyurethane and irradiated the molded specimen. The stress increases with strain as the radiation dose is increased.

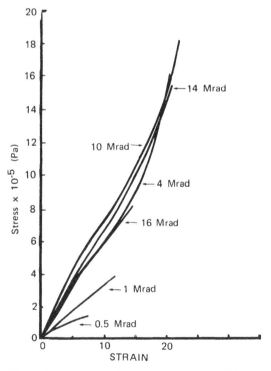

Figure 13 Stress–strain curves above the melting temperature of the 0.5-, 1-, 4-, 10-, 14-, and 16-Mrad materials which were crosslinked at room temperature. (Adapted from Ref. 157.)

Table 4 Urethane Thermoplastic Elastomers Prepared by Reacting NIAX Polyol D560/MDI/Curing Agent at a 1 : 2 : 1 Molar Ratio with the Curing Agents Differing in Methylene Group Concentration

% Lactone in elastomer:	88.6	77.6	77.2	76.7	76.3	75.9	75.5	75.1	74.9	74.3	73.5	
Curing agent:	None	Ethylene glycol	1,3-Propanediol	1,4-Butanediol	1,5-Pentanediol	1,6-Hexanediol	1,7-Heptanediol	1,8-Octanediol	1,9-Nonanediol	1,10-Decanediol	1,12-Dodecanediol	Test procedure
Physical properties												
Hardness, Shore A	63[a]	63[b]	65	64	65	67	58	65	67	65	72	ASTM D2240
100% modulus, psi	770	245	275	290	280	255	200	365	290	340	425	ASTM D412
300% modulus, psi	836	435	665	610	460	420	350	585	550	650	850	ASTM D412
Tensile strength, psi	4200	4400	4500	4400	3400	4200	3600	4100	4300	4200	4900	ASTM D412
Elongation, %	835	685	590	705	650	705	760	700	665	610	645	ASTM D412
Graves tear, psi	445	340	300	320	275	235	225	285	255	270	300	ASTM D412
B compression set, %	93	63	68	73	70	85	94	68	71	72	68	ASTM 624C
Zwick rebound	58	66	70	69	65	69	63	65	62	70	66	ASTM D395B
$T_g(G')$, °C	−40	−38	−38	−40	−40	−42	−42	−42	−40	−43	−43	DiN 53-512
η_{sp}/C 30°C, DMF, 0.2 gm/dl	0.89	0.85	1.19	1.05	0.98	1.26	0.89	0.88	1.08	1.01	0.99	

[a]Crystallized in less than one week. Properties are for the partially crystalline polymer.

[b]Crystallized on long standing.

Source: Adapted from Ref. 59.

Critchfield et al. [159] studied the effect of the short-chain diol on the properties of polycaprolactone urethanes. They selected C-2 to C-12 straight-chain diols and used them with MDI–polycaprolactone diol in the same molar ratios. The results are presented in Table 4. They concluded that the molecular weight of the glycols can affect the properties of the polymer. Increasing the molecular weight of the glycol does improve the low-temperature flexibility of the polymer. Use of 1,6-hexanediol and 1,7-heptanediol made polymers with low compression set, which is thought to be associated with decrease in molecular association.

Dieter et al. [160] studied the structure–property relationship with polyurethanes made with different polyesters, polycaprolactone, and polyoxymethylene glycol. They concluded that the azelates are better in tensile strength, torsional flexibility, and elongation than the adipates at lower temperatures. The azelates also have better hydrolytic stability than the adipates. The hydrolytic stability of the polyoxymethylene glycols are superior to any other polyglycols in this study, as is also well documented in the literature [161–163]. The hydrolytic stability as measured by the retention of mechanical properties after humid aging are polyoxymethylene glycol > polycaprolactone > polyazelates > polyadipates. The addition of polycarbodiimides has been found to be very effective in improving the hydrolytic stability of the azelates but the physical properties are somewhat lower than unmodified polymer.

The kinetics of the hydrolytic aging of polyester urethanes were studied by Brown and co-workers [164], who found that the hydrolytic degradation is due to the acid-catalyzed hydrolysis of the ester group.

Molecular Weight–Property Relationship

It is well recognized that the mechanical properties of the urethane polymer increase with the increase of molecular weight to a certain level. The molecular weight can be determined by several methods. The number average molecular weight $(\bar{M}_n)$ can be measured by the osmometry method where the osmotic pressure depends on the molar concentration of the solute. The weight average molecular weight $(\bar{M}_w)$ can be determined by gel permeation chromatography.

There are other methods for determining molecular weight. One common method is by measuring the viscosity of the polymer solution in a very dilute solution. The molecular weight can be expressed by the Mark–Houwink equation:

$$[\eta] = KM^a$$

where the $[\eta]$ is the intrinsic viscosity, M is the molecular weight, and K and a are constants depending on the solvent, polymer, and temperature. Seefried and co-workers [142] determined the values of K and a for a number of solvents for polyurethanes. The value of a is generally between 0.6 and 0.8.

The molecular weight distribution (MWD) is the ratio of M_w/M_n, which is always greater than unity. The higher values mean broader distribution of the molecular weight. Schollenberger and Dinbergs [165, 166] studied different properties of MDI/BDO/PTAd polyurethane and found that with increase of polymer molecular weight, tensile strength, abrasion resistance, specific gravity, low-temperature modulus, and Clash-Berg T_f, T_g, processing temperature, and dynamic extrusion rheometer will increase. Melt index, hysteresis, extension set, stress relaxation, flex life, and deMattia index decrease with the increase of the molecular weight. They also noted the inflection molecular weight, after which the properties tend to level off.

Wolgemuth [167] prepared very high molecular weight polyurethanes from adipo-dinitrile carbonate and measured the tensile strength and molecular weight. The results are presented in Figure 14. It can be seen from the figure that for polyester polymer, tensile strength increases at a faster rate than that of the polyether polymer.

Thermal Analysis

Differential scanning calorimetry (DSC) is an important tool to character the thermal properties of polyurethanes. In DSC experiments the endotherms or exotherms are measured as a function of temperature. In thermogravimetric analysis (TGA) the rate of weight loss measured as a function of temperature and can be applied to characterize the degradation of the polymer.

Lilaonitkul and Cooper [143] studied the thermal properties of commercial polyurethanes of different but essentially linear composition. Their results are presented in Figures 15 and 16. In Figure 15 the trace of sample QA-1/0.9/1.9, the first endotherm indicates the glass transition (T_g) of the soft segment followed by the crystallization of the soft segment showing the exotherm. The next endotherm is the melting of the just-formed crystals. The T_g's of different polymers as measured by these authors are presented in Table 5. It can be seen that the T_g's of the soft segment of polyester urethanes are higher and more dependent on the hard-segment content than the polyether polymers, suggesting that in the polyether polymer there is better microphase separation. The same observation has been made by Clough and colleagues [5, 6] and also by Seymour et al. [145] by infrared spectroscopy.

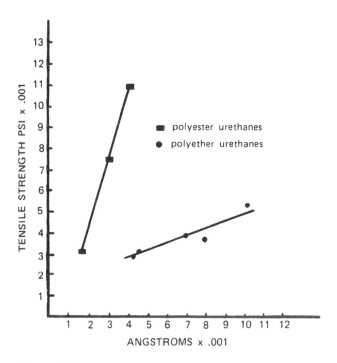

Figure 14 Tensile strength vs. angstrom size for thermoplastic urethanes. (Adapted from Ref. 167.)

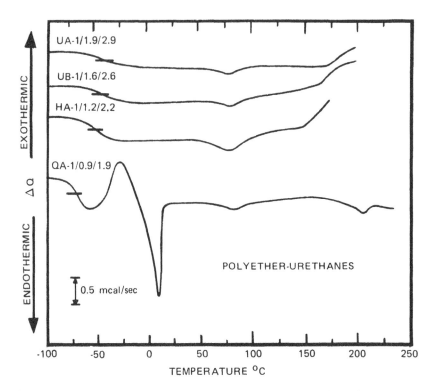

Figure 15 Thermal analysis curves for polyether–urethanes of varying urethane content and segment length. (Adapted from Ref. 143.)

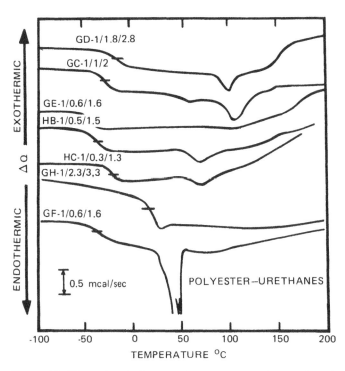

Figure 16 Thermal analysis curves for polyester–urethanes of varying urethane content and segment length. (Adapted from Ref. 143.)

Table 5 Thermal Properties of Polyurethanes

Sample	T_g (°C)	Endothermic transition (°C)[a]
UA-1 : 1.9 : 2.9	−41	79 (W), 170 (B)
UB-1 : 1.6 : 2.6	−44	80 (W), 165 (B)
UC-1 : 1.4 : 2.4	−40	77 (W), 150 (B)
UD-1 : 1.2 : 2.2	−44	76 (W), 162 (B)
QA-1 : 0.9 : 1.9	−70	(−28),[b] 7 (S), 81 (W), 203 (W)
QB-1 : 1.4 : 2.4	−20	70 (W), 136 (B)
QC-1 : 1.5 : 2.5	−19	70 (W), 139 (B)
QD-1 : 0.3 : 1.3	−44	37 (Sh), 49 (VS)
HA-1 : 1.2 : 2.2	−50	77
HB-1 : 0.5 : 1.5	−37	70
HC-1 : 0.3 : 1.3	−21	70
HD-1 : 0.4 : 1.4	−46	35 (Sh), 41 (Sh), 50 (VS)
HE-1 : 1.1 : 2.1	−23	30 (Sh), 45 (VS)
HF-1 : 2 : 3	+9	
HG-1 : 1.2 : 2.2	+4	76 (W)
GA-1 : 1.3 : 2.3	−53	116, 150 (B)
GB-1 : 0.3 : 1.3	−32	58 (B), 80
GC-1 : 1 : 2	−28	58 (B), 107
GD-1 : 1.8 : 2.8	−22	105, 153 (BW)
GE-1 : 0.6 : 1.6	−42	103 (BW), 165 (BW)
GF-1 : 0.6 : 1.6	−38	46 (VS), 72 (W)
GG-1 : 1 : 2	−24	109, 218 (VBW)
GH-1 : 2.3 : 3.3	+19	
GI-1 : 1.8 : 2.8	−23	71, 156 (BW)

[a](B), broad; (S), strong; (Sh), shoulder; (V), very; (W), weak.
[b]Exothermic crystallization peak.
Source: Adapted from Ref. 143.

Hoffman [168] made polyether polymers based on PTMEG, MDI, and BDO. A few samples were made with a mixture of tetrafunctional alcohol and BDO to get chemical crosslinking. Polyurethane ureas were made with mixtures of diamines with PTMEG and BDO. The DSC traces are given in Figures 17 and 18. It can be seen from the figures that the T_g of the low molecular weight soft segments of the samples of 73 series are higher than those of the 87 and 88 series high molecular weight soft-segment samples. This is thought to be due to the more frequent placement of the hard segments in the 73 series samples causing less mobility in the 73 series samples [169, 170]. The addition of the tetrafunctional alcohol to the 73 series also raises the T_g's, probably due to the change in the morphology of the hard and soft segments. The transitions of the hard segments are difficult to identify due to low concentration and the small amount of crystallinity that exhibits broad and weak melting isotherms [12, 171]. The endotherms are smaller in the crosslinked hard segments than in the linear hard segments. The 87 and 88 series poly-ureas are not totally polymerized at room temperature; as a result, a curing endotherm is observed at about 100°C with polyurethane urea polymers.

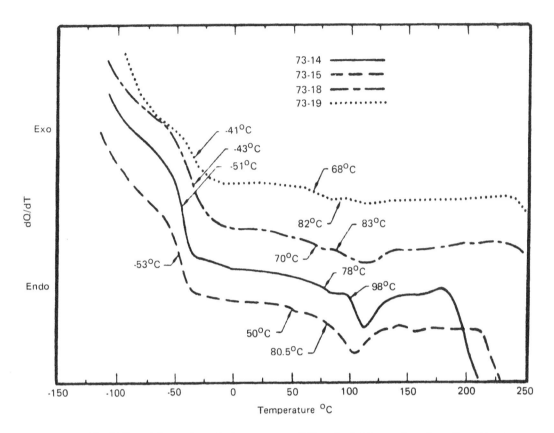

Figure 17 Differential scanning calorimeter traces of 73-series halthanes show the soft- and hard-segment glass transitions and a melting endotherm. The temperatures of these transitions are consistent with the dynamic mechanical measurements. (Adapted from Ref. 168.)

Kong and Wilkes [172] have studied the effect of the T_g's on the stretching of the heated samples. They showed that there is lowering of T_g's for the heated samples, whereas the opposite is true for well-aged samples. Figures 19 and 20 present the TGA traces of 73, 87, and 88 samples as prepared by Hoffman. It can be seen that the degradation of polyurethanes is bimodal. The degradation starts at about 250°C. The first mode is thought to be of hard segments where polyurethane ureas are more stable than polyurethanes, as can be seen from the figures.

Dynamic Mechanical Properties

The measurement of the dynamic mechanical properties provides a very important tool to study the first- and second-order transitions, secondary molecular motion, phase separation, and modulus–temperature relationship. Figure 21 presents the thermomechanical properties of a polyurethane made out of PTMEG, BDO, and MDI. The trace of the loss modulus shows two peaks before T_g, which are thought to be associated with the secondary molecular relaxations association with the soft segments. These are associated with the molecular motion of the urethane [14] and the polyether of the soft segments [173]. In

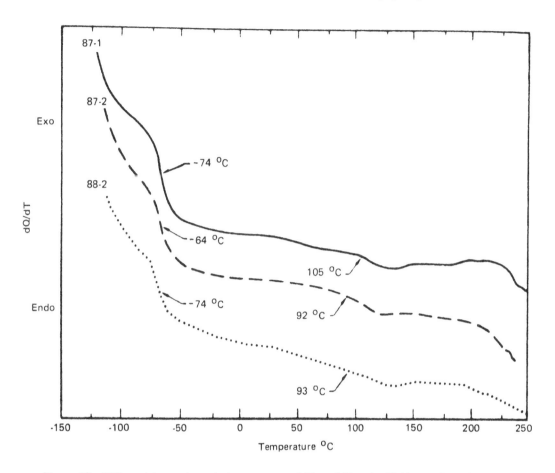

Figure 18 Differential scanning calorimeter traces of 87- and 88-series Halthanes show the soft-segment glass transition and a curing endotherm at about −70 and 100°C, respectively. (Adapted from Ref. 168.)

polyester urethanes an additional peak can be seen which is related to the water absorption. Three temperature ranges characterize a typical polymer: a glassy state below T_g with a very high modulus; a transition state, where the modulus decreases; and a rubbery state above T_g. Hoffman [168] measured the dynamic mechanical properties of linear and crosslinked polyurethanes and polyurethane ureas. The results of the transition temperatures of the studied systems are presented in Tables 6 and 7. From the tables it can be seen that in the polyurethane ureas, the T_g's of the soft segments are much lower than the T_g's for polyurethanes, and the T_g's of the hard segments are much higher. The difference in the dynamic mechanical properties of the linear and crosslinked polymers is presented in Figure 22. It can be noted from this figure that the storage modulus of the linear polymer decreases with increase in temperature, whereas the storage modulus for crosslinked polymer decreases after degradation. Figure 23 presents the dynamic mechanical properties of polyurethane ureas where the hard segments did not cure at room temperature. As the temperature is raised further curing occurs, causing the increase of the storage and lost moduli.

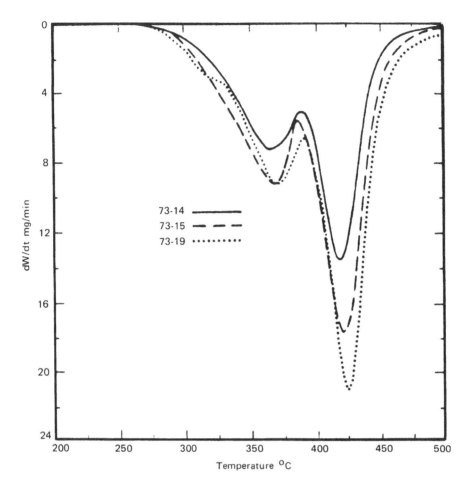

Figure 19 Thermogravimetric analysis curves of 73-series Halthanes show bimodal pyrolysis behavior starting at about 250°C. The rate of change in weight of the adhesive with time, dW/dt, is plotted against temperature for a programmed heating rate of 12°C/min. (Adapted from Ref. 168.)

Lilaonitkul and Cooper [143] measured the dynamic properties of a series of commercially available polyester and polyether urethanes. Some of their findings are presented in Figures 24 and 25. It can be seen that the aromatic content increases the T_g and the plateau modulus. Killis et al. [174] studied the dynamic mechanical properties of polyurethanes containing polypropylene oxide, triisocyanate, and sodium tetraphenylborate. They concluded that the variation of the storage modulus with salt concentration or network chemical structure is directly correlated with the variation in the glass transition temperature. Critchfield et al. [159] studied the effect of the short-chain glycols on dynamic mechanical properties. Lee and Kim [73] studied the dynamic mechanical properties of IPNs made under different pressure and concluded that the degree of intermixing increased with increasing synthetic pressure. Dynamic mechanical analysis of the lignin-based polyurethanes was carried out by Saraf et al. [104].

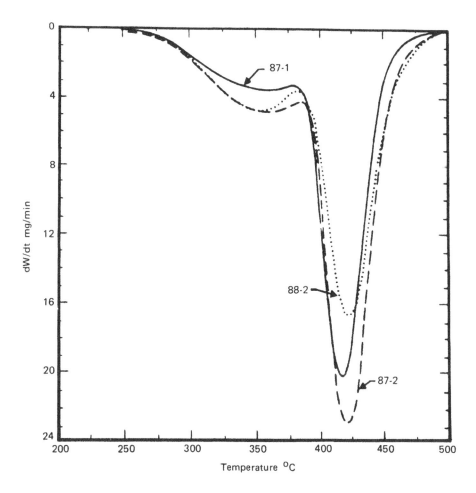

Figure 20 Thermogravimetric analysis curves of 87- and 88-series Halthanes show degradation behavior similar to 73-series Halthanes (Figure 19). (Adapted from Ref. 168.)

Torque Rheology

Torque rheology has been used in the rubbery industry for a long time to measure the increase in viscosity during vulcanization of rubber compounds. The applications of torque rheology in thermoplastics are limited. This method has been applied to measure the crosslinking of polyethylene [175], curing of thermoset polymer [176], and gelling of plastisols [177]. Viscosity of a polymer increases when molecular weight increases and also when a three-dimensional network structure is formed. In torque rheometer, the principle is that the mixer sensor is able to determine the energy required to perform the task, which in turn is dependent on the viscosity of the material being processed. Hartley and Williams [178] measured the dynamic viscosity with conversion of different polyurethanes.

Schollenberger and co-workers [179] used torque rheology to measure the polymer

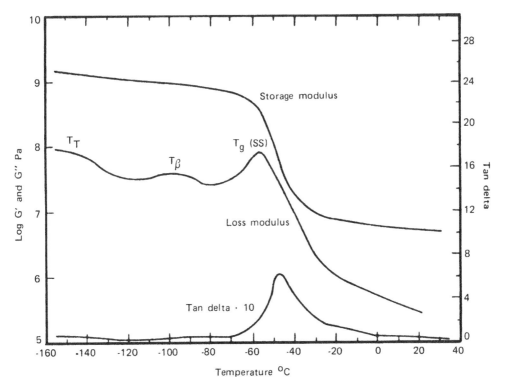

Figure 21 The low-temperature dynamic mechanical spectrum of Halthane 73–14 is typical of the 73-series polyurethane adhesives. Two secondary relaxations, T_β and T_γ, are shown as peaks in the loss modulus at -100 and $-150°C$. The soft-segment glass transition, $T_g(SS)$, occurs at about $-50°C$. The frequency of oscillation was held constant during the measurement of 0.1 Hz. (Adapted from Ref. 168.)

Table 6 Transition Temperatures of 73 Series Halthanes[a]

Transition[b]	73–14	73–15	73–18	73–19
T_γ	-155	-160	-160	-160
T_β	-100	-100	-99	-100
E_{ACT} (kJ/mol)			57.3	62.3
T_g (SS)	-56.7	-53.8	-49.2	-49.4
E_{ACT} (kJ/mol)	339	323	298	195
T_g (SS)[c]	-51	-53	-43	-41
T_m (HS)[d]	92	81	85	85
T_g (HS)[c]	78	50	70	68
T_m (HS)[c]	98	80	83	82

[a]Dynamic mechanical transition temperatures were measured at the maximum in the loss modulus at 0.1 Hz.

[b]Temperatures in °C.

[c]Differential scanning calorimeter measurements from Figure 17.

[d]This transition shifts from sample to sample characteristic of urethane hard segment multiple melting behavior.

Source: Adapted from Ref. 168.

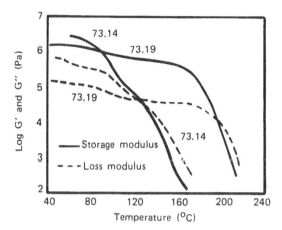

Figure 22 The high-temperature shear storage and loss moduli of Halthane 73–14 and 73–19 adhesives are controlled by the presence or absence of the crosslinking agent quadrol in the hard segments. In the linear urethane (73–14), viscous flow follows the melting of the hard segments, whereas in the crosslinked urethane (73–19) the modulus drops only when the polymer begins to degrade. (Adapted from Ref. 168.)

Table 7 Transition Temperatures of 87 and 88 Series Halthanes[a]

Transition[b]	87–1	87–2	88–2
T_γ	−150	−150	−150
T_g (SS)	−79	−77	−76
E_{ACT} (kJ/mol)	189	162	256
T_c (postcure)	105	92	93
T_g (HS)	185	196	188

[a]Dynamic mechanical transitions were measured at the maximum in the loss modulus at 0.1 Hz.

[b]Temperatures in °C.

[c]Differential scanning calorimeter measurements from Figure 18.

Source: Adapted from Ref. 168.

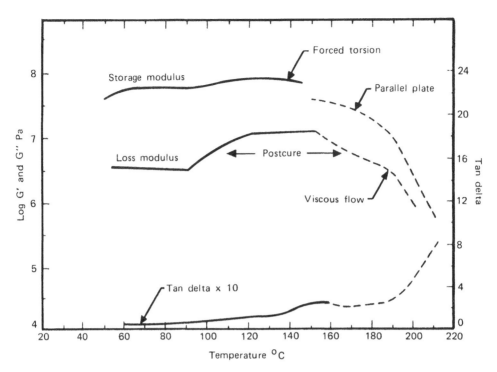

Figure 23 The high-temperature dynamic mechanical spectrum of Halthane 88–2 shows that some further curing is occurring above 100°C because both storage and loss modulus increase over a broad range of temperatures. The onset of viscous flow above 200°C indicates that the hard-segment glass transition temperature has been exceeded. Solid-line data were obtained using the RMS forced-torsion fixture and dashed-line data using the parallel-plate fixture. (Adapted from Ref. 168.)

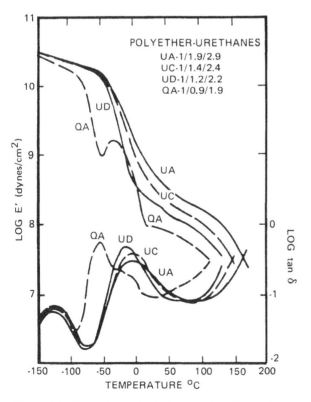

Figure 24 Dynamic mechanical properties of polyether–urethanes of varying urethane content and segment length. (Adapted from Ref. 143.)

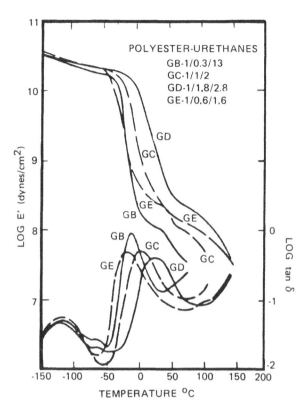

Figure 25 Dynamic mechanical properties of polyester–urethanes of varying urethane content. (Adapted from Ref. 143.)

viscosity to follow the course of the polymerization of polyurethane elastomers. High molecular weight thermoplastic urethane can have a hydroxy or isocyanate group at one or both ends, depending on the isocyanate index used for polymerization. A polymer with hydroxy groups at ends is formed when NCO/OH is less than 1. This hydroxy-terminated polymer is capable of reacting with isocyanate under suitable conditions. When bifunctional isocyanate is allowed to react with a polyurethane with hydroxy groups at both ends, the molecular weight increases, further increasing the viscosity. An excess of bifunctional isocyanate is capable of making a crosslinked polymer with allophanate linkages. When an isocyanate with apparent functionality of more than 2.0 is reacted with an OH-terminated polyurethane, a three-dimensional structure is formed with higher molecular weight and viscosity.

Bhattacharyya [50–52] used torque rheometry to study the effects of isocyanates of different structures and functionalities on a hydroxy-terminated polyurethane. In this study, 60 g of polymer was mixed with the desired amount of isocyanate in the mixing bowl of the machine. The materials were mixed at 80 rpm for 90 min under a 5-kg load. The torque and the total torque were measured, plotted, and printed against time by the machine. In these studies MDI, PPDI (*p*-phenylene diisocyanate), CHDI (1,4-cyclohexane diisocyanate), and PMPPI (polymethylene polyphenyl isocyanate) of functionalities 2.3 and 2.6 were used. The total torques (m-kg-min) against time of different materials are presented in Table 8. Figure 26 plots the individual torques against time of

Table 8 Total Torque (m-kg-min) against Time of Different Isocyanates

Polymer	Isocyanate	% w/w	30	60	90
TPU			24.3	44.3	64.2
TPU	MDI	1.39	35.8	67.9	99.4
		2.00	37.9	73.5	108.2
		3.48	46.6	95.9	145.7
		4.34	37.8	83.3	133.9
		6.51	21.8	44.5	85.8
		8.68	13.3	21.0	32.3
TPU	PMPPI	1.45	37.1	67.7	96.3
		1.82	51.3	99.3	144.2
		2.73	62.6	119.0	172.1
		3.64	80.2	152.6	223.2
		6.80	55.0		
		9.10	31.1	79.9	95.7
TPU	PPDI	1.11	31.7	67.7	96.3
		1.28	33.5	60.6	86.5
		2.22	45.2	92.8	141.3
		2.78	42.1	82.4	120.2
		4.17	13.9	27.0	49.8
		5.56	8.4	14.0	23.5
TPU	CHDI	1.15	24.7	50.3	78.7
		1.33	29.0	62.1	97.2
		2.29	30.2	64.5	104.2
		2.88	27.5	61.2	104.6
		4.33	21.8	42.8	75.4
		5.77	13.6	20.5	28.8

Source: Adapted from Ref. 51.

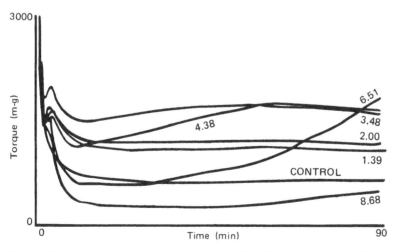

Figure 26 Torque vs. time as a function of % MDI. (Adapted from Ref. 51.)

polymers mixed with different amounts of MDI. The amount of MDI was varied from 1.39 to 8.68% w/w of polymer. The initial peaks are believed to be due to formation of high molecular weight products at the beginning, which take longer to melt and react. It can also be noted that the heights of the peaks are inversely proportional to the amounts of MDI. This is believed to be due to simultaneous reaction of MDI to form high molecular weight products and at the same time the plasticization effect of the excess isocyanates. MDI is liquid at reaction temperatures, and it is expected that the MDI will plasticize the reaction mixture at that temperature. Torque increases steadily when the amount of MDI increased from 1.39 to 3.48%. When 4.38% MDI is added, the initial torque drops below that of 1.39%, then rises even higher than that of 3.48%. The isocyanate index used for the polymer is not known. It is likely that 4.38% MDI is much higher than what is required for a linear polymer. The excess amount has the plasticization effect first before being reacted. At 6.51% a similar effect is observed, but this time the lowest torque value is even lower than that of the control value. The 90-min torque value is the highest in this series. At 8.68% MDI the plasticization effect is so dominant that the torque value never went above the control value. The reactions with 6.51 and 8.68% MDI are not complete even at 90 min under the reaction conditions, as evident from the upward movement of the torque traces.

Figure 27 shows the torque traces with different amounts of PPDI. The amounts used are equimolar with corresponding amounts of MDI. The trends of the traces are similar to those of MDI but the corresponding torques are lower compared to those of MDI, as is also evident from the total torque values listed in the table. PPDI has one benzene ring in the molecule, whereas MDI has two. Thus it is expected that in a polymer with MDI in the backbone, the molecule should be tougher than with an equimolar amount of PPDI. It is interesting to note that polymer viscosity with 2.22% PPDI remained highest in this series, which is contrary to the trend with MDI. CHDI has a cyclohexane ring in the structure. The absence of double bonds in the structure will make a polymer softer with the same molecular equivalents of PPDI or MDI. The same can be observed from the torque traces from Figure 28. These experiments were run with 1.15–5.77% CHDI. The

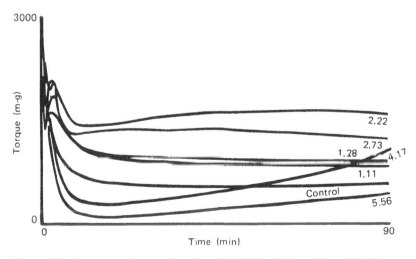

Figure 27 Torque vs. time as a function of % PPDI. (Adapted from Ref. 51.)

general nature of the traces is similar to that of MDI and PPDI but the torque and total torque values are lower.

PMPPI used for this study is commercially available and is believed to be a mixture of high molecular weight isocyanates of different functionalities. The apparent functionality of this PMPPI is 2.3. If there is any monofunctional isocyanate present in the mixture, it will terminate the polymer chain. The lower molecular weight bifunctional isocyanate will have greater mobility than the high molecular weight multifunctional isocyanates. Thus the reactions possible with the hydroxy-terminated polyurethane with PMPPI are to increase the molecular weight by reacting with bifunctional isocyanates and the formation of a three-dimensional network structure by reacting with tri- or higher functional isocyanates. The crosslinked polymer viscosity is higher than the linear polymer. The torque traces with PMPPI (f = 2.3) are presented in Figure 29. In this series of experiments 1.45–9.45% w/w PMPPI were used. It can be observed from the figure that the increase in viscosity is directly proportional to the amount of PMPPI up to 3.64%. The viscosities with 2.73 and 3.64% PMPPI are so high that the initial torque traces went out of scale. It is interesting to note that the viscosity values have more spread at any given time than any other isocyanate in this study. With 6.8% PMPPI the polymer degraded at about 40 min, pushing the ram out of the bowl. With an additional 5 kg the experiment with 9.10% PMPPI was continued but the polymer degraded at about 50 min, as can be seen from the figure. There are second peaks present with 2.73 and 3.64% PMPPI, the reason for which is not clear. PMPPI with an apparent functionality of 2.6 was also included in the study. The torque traces are presented in Figure 30. It can be observed from the last two figures that with PMPPI with functionality of 2.6, the polymer viscosities are higher compared to that of PMPPI with functionality of 2.3. The appearance of the traces otherwise is similar. Figure 31 shows the torque traces with different isocyanates at equimolar amounts. The viscosities with MDI and PPDI are very similar to each other, PPDI being slightly higher, although the total torque with MDI is slightly higher. The polymer viscosity with CHDI is lowest in this series, although it is rising even at 90 min, suggesting the incompletion of the reaction. The viscosity with PMPPI is highest with a gradual decrease starting at about

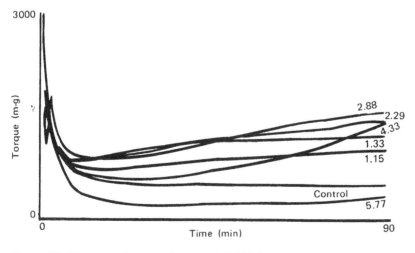

Figure 28 Torque vs. time as a function of % CHDI. (Adapted from Ref. 51.)

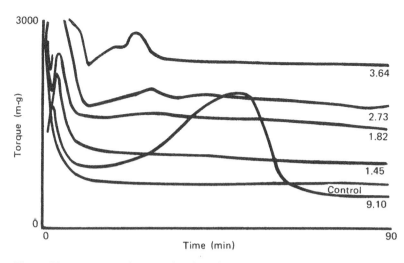

Figure 29 Torque vs. time as a function of % PMPPI ($f = 2.3$). (Adapted from Ref. 51.)

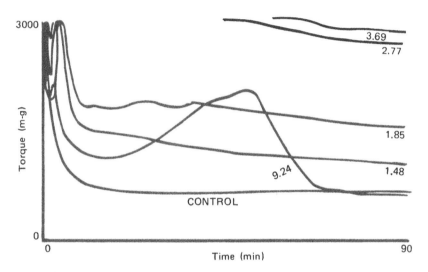

Figure 30 Torque vs. time as a function of % PMPPI ($f = 2.6$). (Adapted from Ref. 52.)

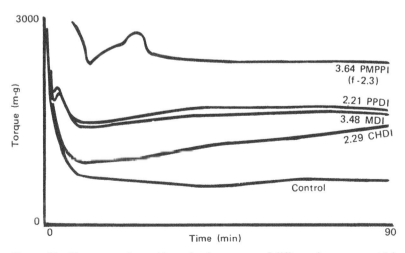

Figure 31 Torque vs. time with equimolar amount of different isocyanates. (Adapted from Ref. 51.)

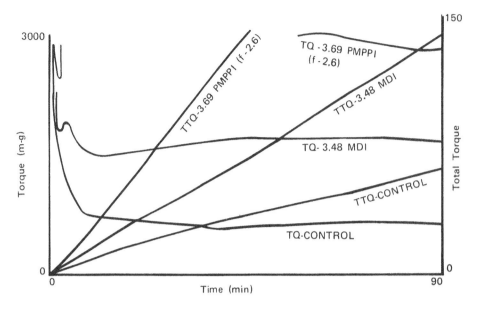

Figure 32 Torque and total torque vs. time with equimolar MDI and PMPPI ($f = 2.6$).

30 min, suggesting polymer degradation. The comparison of torques and total torques of MDI and PMPPI with functionality 2.6 at equimolar amounts is presented in Figure 32. It is obvious from the figure that the viscosity with PMPPI is much higher, almost twice that with MDI, and both of them are considerably higher than that of the control polymer.

Optical Methods

Lilaonitkul and Cooper [143] used proton and carbon-13 nuclear magnetic resonance (^{13}C NMR) to perform the chemical analysis of 24 polyurethane samples of different compositions and structures. They also used infrared (IR) spectrophotometry for analysis of polyurethane samples. This technique can be applied to distinguish the polyether from polyester. The identification of different structures within a given family can be difficult, although some assessment can be made. The results of their findings are presented in Table 9. Figure 33 shows typical IR traces of polyether and polyester urethane samples. Paik Sung [155] used IR studies to have a detailed understanding of the composition, transition behavior, and properties of polyurethane and polyurethane ureas. He suggested the presence of three-dimensional hydrogen bonds between the urea–carbonyl group and two NH groups in the hard-segment domains. Wolgemuth [167] used IR observations to characterize the polyurethanes made from adipodinitrile carbonate. Thermal degradation of polyurethane was studied by IR spectrophotometry by Gaboriaud and Vantelon [180], who have also used UV spectra for their analysis. Foti et al. [181] studied the thermal degradation of totally aromatic polyurethane by mass spectrometry. Small angle x-ray scattering (SAXS) investigations of linear polyurethanes were made by Ophir and Wilkes [182]. Paik Sung and co-workers [183] investigated TDI-based polyurethane by different x-ray techniques. Joseph, Wilkes, and Park [157] studied the electron beam–cured polyurethanes by SAXS and other optical methods.

Table 9 IR Spectra of Polyurethane Elastomers

Frequency (cm⁻¹)	Relative intensity[a]	Urethane	Assignment[b]	
			PTMO	PTMA
3420	W, Sh	ν(N—H) free		
3330–3340	S	ν(N—H)-bonded		
2960	S			ν_a(CH$_2$)
2940	S		ν_a(CH$_2$)	
2890	M, Sh			ν_s(CH$_2$)
2870	M, Sh			ν_s(CH$_2$)
2860	S		ν_a(CH$_2$)	
2795	M		ν_s(CH$_2$)	
1730	VS	ν(C=O) free		
1703	VS	ν(C=O) bonded		
1595	S	ν(C=C) ring		
1530	VS	δ(N—H) + ν(C—N)		
1410	S	ν(C—C) ring		
1360	M		W(CH$_2$)	W(CH$_2$)
1305	S	δ(N—H) + ν(C—N), β(C—H) ring		
1250	S, Sh	ν(C—O—C)	W(CH$_2$)	W(CH$_2$)
1220	VS	δ(N—H) + ν(C—N)		
1170	S			ν(C—O—C)
1105	VS		ν(C—O—C)	
1065–1075	S	ν(C—O—C)		
1030	W	β(C—H) ring		
985	W		ν_a(C—O—C)	
915	W	γ(C—H) ring		
815	W	γ(C—H) ring		
768		γ(—C—O—)— with C=O above		

[a]Refer to samples UB-1:1.6:2.6 and GD-1:1.8:2.8 both of which contain soft segment of MW 1000. W, weak; M, medium; S, strong; VS, very strong; Sh, shoulder.

[b]ν, stretching; ν_s, symmetric stretching; ν_a, asymmetric stretching; δ, bending; β, in-plane bending; γ, out-of-plane bending; W, wagging.

Source: Adapted from Ref. 143.

General

Miyabayashi and Kennedy [184] prepared polyurethanes with dihydroxy polyisobutylenes with triphenylmethyl isocyanate and characterized by extraction, the Flory–Rehner swelling, and Mooney–Rivlin equilibrium methods. IR spectroscopy and ¹³C NMR were used by Bauer and co-workers [185] to follow the reactions of the coatings made with acrylic copolymers crosslinked with aliphatic isocyanates. The extent of the isocyanate reaction was followed as a function of time and temperature by IR. Tensile testing, sonic velocity measurements, and dynamic mechanical analysis were used to characterize the IPNs made

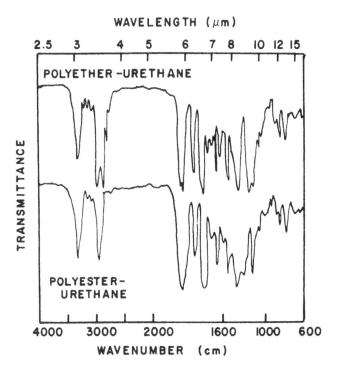

WAVELENGTH (μm)

2.5   3   4   5   6   7   8   10 12 15

POLYETHER –URETHANE

TRANSMITTANCE

POLYESTER–
URETHANE

4000   3000   2000   1600   1000   600

WAVENUMBER (cm)

Figure 33   Survey infrared spectra for typical polyether–urethane and polyester–urethane samples studied. (Adapted from Ref. 143.)

with polymethyl acrylate and polyurethane by Houston and McCluskey [186]. Hsu and Lee [187] followed the kinetics of formation of simultaneous interpenetrating networks composed of unsaturated polyester and polyurethane. They showed that increasing the polyester content enhanced the polymerization of polyurethane due to the solvent effect of polyester. When polyurethane content is increased, the polymerization of polyester is retarded by the cage effect. When an excess of MDI was reacted with polyoxypropyltriol to make a polyurethane the very high values of equilibrium modulus found were thought to be due to formation of allophanate crosslinking [188]. A series of fluorinated polyurethanes were made by Keller [189] by reacting with hexamethylene diisocyanate and 1,3-bis(2-hydroxyhexafluoro-2-propyl) benzene. Born and Hesoe [190] used crystallographic analysis to distinguish the polyurethanes made with diamines and glycols. Their most significant observation is that the neighboring molecules in one and the same plane are linked to one another by means of bifurcated hydrogen bonds. Polyurethanes with different polyols made with isocyanatoethyl methacrylate were studied for the effects of polyol type and molecular weight and crosslink content on the mechanical properties by DSC, dynamic mechanical spectroscopy, and tensile testing [191]. Usmani [192] used dynamic mechanical analysis and IR spectroscopy to study the curing kinetics of polyurethane coatings. It was found that the reduced concentration of isocyanate groups resulted by heat and moisture produced a hardened urethane surface.

REFERENCES

1. *Modern Plastics*, January: 55 (1987).
2. P. J. Manno, *ACS Symp. Ser.*, *172*: 9 (1981).
3. D. G. Leis, *ACS Symp. Ser.*, *172*: 33 (1981).
4. S. H. Metzger, "Metal Replacement Opportunities for Urethane Systems," *ACS Symp. Ser.*, *172*: 69.
5. S. B. Clough and N. S. Schneider, *J. Macromol. Sci.*, *B2*: 553 (1968).
6. S. B. Clough, N. S. Schneider, and A. O. King, *J. Macromol. Sci. Phys.*, *B2*: 553 (1968).
7. G. M. Estes, S. L. Cooper, and A. V. Tobolsky, *J. Macromol. Sci. Rev. Macromol. Chem.*, *C4(2)*: 313 (1970).
8. S. L. Samuels and G. L. Wilkes, *J. Polym. Sci.*, *43*: 149 (1973).
9. R. W. Seymour, A. E. Allegrezza, Jr., and S. L. Cooper, *Macromolecules*, *6*(6): 896 (1973).
10. Y. P. Chang and G. L. Wilkes, *J. Polym. Sci.*, *13*: 455 (1975).
11. J. Blackwell and K. H. Gardner, *Polymer*, *20*: 13 (1979).
12. R. Bonart, L. Morbitzer, and G. Hentze, *J. Macromol. Sci. Phys.*, *3*(2): 337 (1969).
13. L. L. Harrell, Jr., *Macromolecules*, *2*(6): 607 (1969).
14. D. S. Huh and S. L. Cooper, *Polym. Eng. Sci.*, *11*(5): 369 (1971).
15. H. N. Ng, A. E. Allegrezza, R. W. Seymour, and S. L. Cooper, *Polymer*, *14*: 255 (1973).
16. T. L. Smith, *J. Polym. Sci. Phys.*, *12*: 1825 (1974).
17. R. J. Zdrahala, R. M. Gerkin, S. L. Hager, and F. E. Critchfield, *J. Appl. Polym. Sci.*, *24*: 2041 (1979).
18. C. G. Seefried, Jr., J. V. Koleske, and F. E. Critchfield, *J. Appl. Polym. Sci.*, *19*: 2493, 2503, 3185 (1975).
19. S. L. Samuels and G. L. Wilkes, *J. Polym. Sci.*, *11*: 807 (1973).
20. R. W. Seymour and S. L. Cooper, *Macromolecules*, *6*: 48 (1973).
21. G. W. Wilkes and R. Wildnauer, *J. Appl. Phys.*, *46*: 4148 (1975).
22. G. W. Wilkes and J. A. Emerson, *J. Appl. Phys.*, *47*: 4161 (1976).
23. G. W. Wilkes, S. Bagrodia, W. Humphries, and R. Wildnauer, *J. Polym. Sci. Polym. Lett. Ed.*, *13*: 321 (1975).
24. G. W. Wilkes, B. Morra, J. A. Emerson, and R. Wildnauer, *Proceedings VIIth International Congress on Rheology*, pp. 224–226 (August 1976).
25. J. C. West and S. L. Cooper, in *Science and Technology of Rubber* (F. R. Eirich, ed.), Academic, New York, pp. 531–567 (1978).
26. S. L. Cooper and G. M. Estes, eds., *Multiphase Polymers* (Advances in Chemistry Series, Vol. 176), American Chemical Society, Washington (1979).
27. C. S. Schollenberger and K. Dinbergs, *J. Elastoplast.*, *5*: 222 (1973).
28. C. S. Schollenberger and K. Dinbergs, *J. Elastom. Plastics*, *7*: 65 (1975).
29. C. S. Schollenberger and L. E. Hewlett, *Polym. Prepr.*, *19*(1): 17 (1978).
30. C. S. Schollenberger and K. Dinbergs, *Adv. Urethane Sci. Technol.*, *6*: 60 (1978).
31. R. A. Assink and G. L. Wilkes, *Polym. Eng. Sci.*, *17*: 606 (1977).
32. Z. H. Ophin and G. L. Wilkes, in *Multiphase Polymers* (S. L. Cooper and G. M. Estes, eds.), American Chemical Society, Washington, pp. 53–67 (1979).
33. F. X. O'Shea, *ACS Symp. Ser.*, *172*: 243 (1981).
34. Farbenfabriken Bayer, Leverkusen, Germany, Technical Bulletin, "Urepan E, an Ester Based Urethane Rubber for Crosslinking with Peroxides" (July 1, 1961).
35. E. E. Gruber and O. C. Keplinger, *Ind. Eng. Chem.*, *51*(2): 151 (1959).
36. The General Tire and Rubber Co., Akron, Ohio, Technical Bulletins: "Genthane S" (GT-S3); "Genthane SR" (GTOSR1).
37. Naugatuck Chemical Co., Division of United States Rubber Co. (Uniroyal), Naugatuck, Conn., Technical Bulletins: "Vibrathane 5003"; "Vibrathane 5004."

38. D. B. Patterson, U. S. Pat. 2,808,391 (1957); E. I. duPont de Nemours and Co.
39. E. F. Cluff and E. K. Gladding, *J. Appl. Polym. Sci., 3*: 290 (1960).
40. Thiokol Chemical Corporation, Trenton, N.J., Technical Bulletin, "Thiokol Facts," 5(2), (1963).
41. M. M. Swaab, *Rubber Age, 92*(4): 567 (1963).
42. E. I. duPont de Nemours & Co. Elastomers Chemicals Dept., Wilmington, Delaware, Technical Bulletins: "Adiprene C, A Urethane Rubber," Development Products Report No. 4 (July 15, 1957); "Adiprene CM, A Sulfur Curable Urethane Rubber," by S. M. Hirsty, Report A-66693 (1969).
43. W. Kallert, *Kautsch Gummi Kunstst., 19*: 363 (1966).
44. H. S. Kincaid, G. P. Sage, and F. E. Critchfield, "Polycaprolactone Millable Urethane Elastomers," American Chemical Society (Rubber Division) Meeting, Montreal (May 1967).
45. P. Wright and A. P. C. Cumming, *Solid Polyurethane Elastomers,* Gordon and Breach, New York (1969).
46. W. K. Walsh and B. S. Gupta, *J. Coated Fabrics, 7*: 253 (1978).
47. F. J. Gasparrini, *J. Coated Fabrics, 9*: 6 (1979).
48. J. P. Guarino and E. P. Tripp, *Polym. Prepr. Am. Chem. Soc. Div. Organ. Coat. Plast. Chem., 39*: 9 (1978).
49. B. S. Gupta, W. S. McPeters, and W. M. Walsh, *J. Coated Fabrics, 9*: 12 (1979).
50. J. Bhattacharyya, "Application of Computer for Studying Crosslinking of Thermoplastics Polyurethane," Pittsburgh Conference and Exposition on Analytical Chemistry and Applied Spectroscopy, No. 425 (1985).
51. J. Bhattacharyya, *J. Elastomers Plastics, 18*: 85 (1986).
52. J. Bhattacharyya, *Am. Lab.,* February: 148 (1987).
53. Y. Lipatov and L. Sergeeva, *Vzaimopronikayushehie Setki* (Interpenetrating Polymer Networks), Naukova Dumka, Kiev (1979).
54. J. R. Miller, *J. Chem. Soc., 1960*: 1311.
55. K. Shibayama and Y. Suzuki, *Kobunshi Kagaku, 23*: 249; *24* (1966).
56. K. Shibayama, *Zairyo, 12*: 362 (1962).
57. K. Shibayama, *Kobunshi Kagaku, 19*: 219 (1962).
58. K. Shibayama, *Kobunshi Kagaku, 20*: 221 (1963).
59. H. L. Frisch, D. Klempner, and K. C. Frisch, *J. Polym. Sci. Polym. Lett., 7*: 775 (1969).
60. H. L. Frisch, D. Klempner, and K. C. Frisch, *J. Polym. Sci. A2, 8*: 921 (1970).
61. M. Matsuo, T. K. Kwei, D. Klempner, and H. L. Frisch, *Polym. Eng. Sci., 10*: 327 (1970).
62. D. Klempner and H. S. Frisch, *J. Polym. Sci., B8*: 525 (1970).
63. L. H. Sperling and D. W. Friedman, *J. Polym. Sci. A2, 7*: 425 (1969).
64. L. H. Sperling, D. W. Taylor, M. L. Kirkpatrick, H. F. George, and D. R. Bardman, *J. Appl. Polym. Sci., 14*: 73 (1970).
65. L. H. Sperling, H. F. George, V. Huelck, and D. A. Thomas, *J. Appl. Polym. Sci., 14*: 2815 (1970).
66. H. L. Frisch and D. Klempner, *Adv. Macromol. Chem., 2*: 149 (1970).
67. K. C. Frisch, D. Klempner, and H. L. Frisch, *Polym. Eng. Sci., 22*: 1143 (1982).
68. D. Klempner, *Angew. Chem., 17*: 97 (1978).
69. A. J. Curtius, M. J. Covitch, D. A. Thomas, and L. H. Sperling, *Polym. Eng. Sci., 12*: 101 (1972).
70. V. Huelck, D. A. Thomas, and L. H. Sperling, *Macromolecules, 5*: 340 (1972).
71. K. C. Frisch, D. Klempner, S. Migdal, and H. L. Frisch, *Polym. Eng. Sci., 1*: 76 (1974).
72. K. C. Frisch, D. Klempner, and S. Migdal, *J. Polym. Sci. Polym. Chem., 12*: 885 (1974).
73. D. S. Lee and S. C. Kim, *Macromolecules, 17*: 2222 (1984).
74. D. S. Lee and S. C. Kim, *Macromolecules, 17*: 268 (1984).
75. D. S. Lee and S. C. Kim, *Macromolecules, 17*: 2193 (1984).

76. L. H. Sperling, *Interpenetrating Polymer Networks and Related Materials*, Plenum, New York (1981).
77. O. Olabisi, L. M. Robeson, and M. T. Shaw, *Polymer–Polymer Miscibility*, Academic Press, New York (1979).
78. A. A. Donatelli, L. H. Sperling, and D. A. Thomas, *J. Appl. Polym. Sci.*, *21*: 1189 (1977).
79. S. C. Kim, D. Klempner, K. C. Frisch, H. L. Frisch, and H. Ghiradella, *Polym. Eng. Sci.*, *15*: 339 (1975).
80. S. C. Kim, D. Klempner, K. C. Frisch, and H. L. Frisch, *Macromolecules, 9*: 263 (1976).
81. S. C. Kim, D. Klempner, K. C. Frisch, and H. L. Frisch, *Macromolecules, 10*: 1187 (1977).
82. N. Devia, L. H. Sperling, J. A. Manson, and A. Conde, *Macromolecules, 12*: 360 (1979).
83. N. Devia, L. H. Sperling, J. A. Manson, and A. Conde, *Polym. Eng. Sci.*, *19*(12): 869, 878 (1979).
84. S. C. Kim, D. Klempner, K. C. Frisch, W. Radigan, and H. L. Frisch, *Macromolecules, 9*: 258 (1976).
85. S. C. Kim, D. Klempner, K. C. Frisch, and H. L. Frisch, *J. Appl. Polym. Sci.*, *21*: 1289 (1977).
86. V. K. Belyakov, A. A. Berlin, I. I. Bukin, V. A. Orlov, and O. G. Tarakanov, *Polym. Sci. USSR, 10*: 700 (1978). (Engl. trans.)
87. R. Senzyn and H. Ishikawa, *J. Agric. Chem. Soc. Jpn.*, 1948, 22, 72.
88. H. Ishikawa, T. Oki, and F. Fugita, *Nippon Mokuzai Gakkaishi*, 7: 85 (1961).
89. K. Kratzl, K. Buchtela, J. Gratzl, J. Zauner, and O. Ettingshausen, *Tappi, 45*(2): 113 (1962).
90. G. G. Allan, in *Lignins—Occurrence, Formation, Structure and Reactions* (K. V. Sarkanen and C. H. Ludwig, eds.), Wiley-Interscience, New York, pp. 511–573 (1971).
91. J. H. Saunders and K. C. Frisch, *Polyurethanes—Chemistry and Technology*, Part 1, Interscience, New York (1962).
92. H. H. Moorer, W. K. Dougherty, and F. F. Ball, U.S. Pat. 3,519,581 (1970).
93. G. G. Allan, U.S. Pat. 3,476,795 (1969).
94. D. T. Christian, M. Look, A. Nobell, and T. S. Armstrong, U.S. Pat. 3,546,199 (1970).
95. W. G. Glasser and O. H.-H. Hsu, U.S. Pat. 4,017,474 (1977).
96. O. H.-H. Hsu and W. G. Glasser, *Appl. Polym. Symp.*, *28*: 297 (1975).
97. O. H.-H. Hsu and W. G. Glasser, *Wood Sci.*, *9*(2): 97 (1976).
98. W. Lange and W. Schweers, *Wood Sci. Technol.*, *14*(1): 1 (1980).
99. W. G. Glasser, O. H.-H. Hsu, D. L. Reed, R. C. Forte, and L. C.-F. Wu, *ACS Symp. Ser.*, *172*: 311 (1981).
100. L. C.-F. Wu and W. G. Glasser, *J. Appl. Polym. Sci.*, *29*: 1111 (1984).
101. W. G. Glasser, L. C.-F. Wu, and J. F. Selin, in *Wood and Agricultural Residues: Research on Use for Feed, Fuels, and Chemicals* (E. J. Soltes, ed.), Academic, New York, pp. 149–166 (1983).
102. W. G. Glasser, C. A. Barnett, T. G. Rials, and V. P. Saraf, *J. Appl. Polym. Sci.*, *29*: 181 (1984).
103. V. P. Saraf and W. G. Glasser, *J. Appl. Polym. Sci.*, *29*: 1831 (1984).
104. V. P. Saraf, W. G. Glasser, G. L. Wilkes, and J. E. McGrath, *J. Appl. Polym. Sci.*, *30*: 2207 (1985).
105. C. N. Georgacopoulos and A. A. Sardanopoli, *Modern Plastics*, May: 76 (1982).
106. J. V. Koleske, in *Polymer Blends*, Vol. 2, Academic, New York (1978).
107. R. G. Nelb II, A. T. Chen, W. J. Farrissey, Jr., and K. B. Onder, "Poly(Esteramide): Thermoplastic Elastomers for High Temperature Applications," SPE 39th Annual Technical Conference (1981).
108. G. E. Maroney and T. R. Griffin, *Elastomerics*, June: 26 (1982).

109. S. L. Axelrood, C. W. Hamilton, and K. C. Frisch, *Ind. Eng. Chem., 53*(11): 889 (1961).

110. S. L. Axelrood, L. C. Smith, and K. C. Frisch, *Rubber Age, 96*(2): 233 (1964).

111. G. Magnus, *Rubber Age, 97*(4): 86 (1965).

112. J. A. Verdol, P. W. Ryan, D. J. Carrow, and K. L. Kuncl, *Rubber Age, 98*(7): 57 (1966).

113. J. A. Verdol, P. W. Ryan, D. J. Carrow, and K. L. Kuncl, *Rubber Age, 98*(8): 62 (1966).

114. R. A. Moore, K. L. Kuncl, and B. G. Gower, *Rubber World, 159*(5): 55 (1969).

115. O. Bayer, E. Muller, S. Peterson, H. F. Piepenbrink, and E. Windemuth, *Rubber Chem. Technol., 23*: 812 (1950).

116. E. Muller, O. Bayer, S. Peterson, H. F. Piepenbrink, W. Schmidt, and E. Weinbrenner, *Rubber Chem. Technol., 26*: 493 (1953).

117. K. A. Piggot, B. F. Frye, K. R. Allen, S. Steingiser, W. C. Darr, J. H. Saunders, and E. E. Hardy, *Chem. Eng. Data, 5*: 391 (1960).

118. J. H. Saunders, *Rubber Chem. Technol., 33*: 1259 (1960).

119. J. H. Saunders, *J.I.R.I., 2*(1): 21 (1968).

120. R. J. Athey, *Rubber Age, 85*: 77 (1959).

121. R. J. Athey, *Ind. Eng. Chem., 52*: 611 (1960).

122. R. J. Athey, J. G. DePinto, and J. M. Keegan, "Adiprene L-100," Bulletin No. 7, Elastomer Chemicals Department, E. I. DuPont de Nemours & Co.

123. R. J. Athey, "Adiprene L-167," Bulletin No. 12, Elastomer Chemicals Department, E. I. DuPont de Nemours & Co.

124. R. J. Athey and J. G. DePinto, "Adiprene L-315," Bulletin No. 1, Elastomer Chemicals Department, E. I. DuPont de Nemours & Co.

125. The Isonate Resource Library, Upjohn Polymer Chemicals, Laporte, Texas.

126. R. P. Kane, *Rubber World, 147*: 35 (1963).

127. D. Bianca and R. E. Knox, *Rubber Age, 98*(5): 76 (1966).

128. N. N. Saaty, L. W. Abercrombie, H. E. Reymore, Jr., and A. A. R. Sayigh, *J. Elastoplast., 1*: 170 (1969).

129. K. Murai and K. Fukuda, *J. Elastoplast., 1*: 150 (1969).

130. W. K. Fischer, *J. Elastoplast., 1*: 241 (1969).

131. T. E. Rusch and D. S. Raden, *Plastic Compound*, July/August (1980).

132. J. C. Chuang, I. S. Lin, and M. M. Hashem, "1,4-Butanediol Based Oligomers and *N*-Vinyl-2-Pyrrolidone in UV Curing," Proceedings of the SPI 29th Annual Technical Marketing Conference, Reno (1985).

133. W. K. Walsh, A. Makati, and E. Bittencourt, *Printing Symposium: Meeting the Challenge of '80*, p. 54 (1978).

134. R. Dowbenko, C. Friedlander, G. Gruber, P. Prucnal, and M. Wismer, *Prog. Org. Coatings, 11*: 71 (1983).

135. J. Chuang, *Am. Paint Coatings J.*, Jan. 16: 39 (1984).

136. G. E. Green, B. P. Stark, and S. A. Zahir, *Macromol. Sci. Rev. Macromol. Chem., C21*: 187 (1982).

137. S. L. Cooper and A. V. Tobolsky, *J. Appl. Polym. Sci., 10*: 1837 (1966).

138. S. L. Cooper, *J. Polym. Sci., A1, 7*: 1765 (1969).

139. B. Ershaghi, A. J. Chompff, and R. Salovey, *ACS Symp. Ser., 172*: 373 (1981).

140. K. C. Frisch, S. L. Reegen, and L. P. Rumao, *Effect of Isocyanate Structure on Properties* (Advances in Urethane Science and Technology, Vol. 1), Technomic Publishers, Lancaster, Pa., p. 49 (1971).

141. P. H. Sung and J. E. Mark, *J. Polym. Sci. Polym. Phys., 19*: 507 (1981).

142. C. G. Seefried, Jr., J. V. Koleske, F. E. Critchfield, and C. R. Pfaffenberger, *J. Polym. Sci. Polym. Phys., 18*: 817 (1980).

143. A. Lilaonitkul and S. L. Cooper, *Properties of Thermoplastic Polyurethane Elastomers* (Advances in Urethane Science and Technology, Vol. 7), Technomic Publishers, Lancaster, Pa., p. 163 (1979).

144. W. J. Macknight, and M. Yang, *J. Polym. Sci. Symp. Ser., 42*: 817 (1973).

145. R. W. Seymour, G. M. Estes, and S. L. Cooper, *Macromolecules, 3*: 579 (1970).
146. G. M. Estes, R. W. Seymour, and S. L. Cooper, *Macromolecules, 4*: 452 (1971).
147. R. W. Seymour and S. L. Cooper, *J. Polym. Sci. Symp. Ser., 46*: 69 (1974).
148. L. B. Weisfeld, J. R. Little, and W. E. Wolstenholme, *J. Polym. Sci., 56*: 455 (1962).
149. J. F. Terenzi, Jr., University Microfilms, L.C. Card No. Mic. 59-5231 (1957); D. S. Trifan and J. F. Terenzi, *J. Polym. Sci., 28*: 443 (1958).
150. Yu. M. Boyarchuk, L. Ya. Rappoport, V. N. Nikitin, and N. P. Apukhtina, *Polym. Sci. U.S.S.R., 7*: 859 (1965); *Vysokomol. Soedin., 7*: 778 (1965).
151. T. Tanaka, T. Yokoyama, and K. Kaku, *Mem. Fac. Eng., Kyushu Univ., 23*: 113 (1963).
152. T. Tanaka, T. Yokoyama, and Y. Yamaguchi, *J. Polym. Sci. A1, 6*: 2153 (1968).
153. K. W. Rausch, Jr., and W. J. Farrissey, Jr., *J. Elastoplast., 2*: 114 (1970).
154. I. L. Chiu, Lawrence Livermore National Lab, California, Report No. UCID-19959.
155. C. S. Paik Sung, *Polymer Alloys*, Vol. 2, Plenum, New York, p. 119 (1980).
156. H. V. Boenig, *Elastomerics*, February: 39 (1981).
157. E. Joseph, G. Wilkes, and K. Park, *J. Appl. Polym. Sci., 26*: 3355 (1981).
158. R. Agrawal and M. F. Rubner, "Mechanical Properties of Segmented Polyurethanes Containing Reactive Diacetylene Groups in the Hard Segments," Proceedings of the ACS Div. of Polymeric Materials, Science and Engineering, Vol. 56 (1987).
159. F. E. Critchfield, J. V. Koleske, G. Magnus, and J. L. Dodd, in *Advances in Urethane Science and Technology*, Vol. 2, Technomic Publishers, Lancaster, Pa., p. 141 (1973).
160. J. A. Dieter, K. C. Frisch, G. K. Shanafelt, and M. T. Devanney, in *Advances in Urethane Science and Technology*, Vol. 3, Technomic Publishers, Lancaster, Pa., p. 197.
161. R. J. Athey, *Rubber Age, 96*(5): 705 (1985).
162. G. Magnus, R. A. Dunleavy, and F. E. Critchfield, *Rubber Chem. Technol., 39*(4): 1328 (1966).
163. C. S. Schollenberger and F. D. Stewart, in *Advances in Urethane Science and Technology*, Vol. 1 (K. C. Frisch and S. L. Reegen, eds.), Technomic Publishers, Lancaster, Pa., Chap. 4 (1971).
164. D. W. Brown, R. E. Lewry, and L. E. Smith, *Macromolecules 13*: 248 (1980).
165. C. S. Schollenberger and K. Dinbergs, in *Advances in Urethane Science and Technology*, Vol. 3, Technomic Publishers, Lancaster, Pa., p. 36.
166. C. S. Schollenberger and K. Dinbergs, in *Advances in Urethane Science and Technology*, Vol. 7, Technomic Publishers, Lancaster, Pa., p. 1 (1979).
167. L. G. Wolgemuth, in *Advances in Urethane Science and Technology*, Vol. 2, Technomic Publishers, Lancaster, Pa., p. 153 (1973).
168. D. M. Hoffman, *ACS Symp. Ser., 172*: 343 (1980).
169. J. L. Illinger, N. S. Schneider, and F. E. Karasz, *Polym. Eng. Sci., 12*: 25 (1972).
170. Y. Minoura, S. Yamashita, H. Okamoto, T. Matsuo, M. Izawa, and S. Kohmoto, *J. Appl. Polym. Sci., 22*: 1817 (1978).
171. R. Benart, *J. Macromol. Sci. Phys., 132*: 115 (1968).
172. E. S. W. Kong and G. L. Wilkes, *J. Polym. Sci. Polym. Lett., 18*: 369 (1980).
173. R. E. Wetton and G. Allen, *Polymers, 7*: 331 (1966).
174. A. Killis, J. F. Lenest, A. Gandini, and H. Cheradame, *J. Polym. Sci. Polym. Phys., 19*: 1073 (1981).
175. J. E. Hager, *Plastics Technol., 15*(7): 55 (1969).
176. M. Roller, "Characterization of the Time-Temperature-Viscosity Behavior of Curing B-Staged Epoxy Resin," Society of Plastics Engineers, Technical Papers, Vol. 21, pp. 212–216 (1975).
177. R. J. Zietlin, "Reaction at the Terminal Vinyl Groups in Phillips Tire Polyethylene," Proceedings of the ACS Div. of Polymer Chemistry, Miami Beach Meeting, Vol. 8, No. 1, pp. 823–829 (1967).
178. M. D. Hartley and H. L. Williams, *Polym. Eng. Sci., 21*(3): 135 (1981).
179. C. S. Schollenberger, K. Dinbergs, and F. D. Stewart, *ACS Symp. Ser., 172*: 433 (1980).

180. F. Gaboriaud and J. P. Vantelon, *J. Polym. Sci. Polym. Chem.*, *19*: 139 (1981).

181. S. Foti, P. Maravigna, and G. Montaudo, *J. Polym. Sci. Polym. Chem.*, *19*: 1679 (1981).

182. Z. Ophir and G. L. Wilkes, *J. Polym. Sci. Polym. Phys.*, *18*: 1469 (1980).

183. C. S. Paik Sung, C. B. Hu, and C. S. Wu, *Macromolecules*, *13*: 111 (1980).

184. T. Mivabavashi and J. P. Kennedy, *J. Appl. Polym. Sci.*, *31*(8): 2523 (1986).

185. D. R. Bauer, R. A. Dickie, and J. L. Koenig, *IEC Prod. Res. Dev.*, *25*(2): 289 (1986).

186. D. J. Houston and J. A. McCluskey, *J. Appl. Polym. Sci.*, *31*(2): 645 (1986).

187. T. Hsu and L. J. Lee, *Polym. Eng. Sci.*, *25*(15): 951 (1985).

188. M. Liavsky, K. Bouchal, and K. Dusek, *Polym. Bull.*, *14*(3): 295 (1985).

189. T. M. Keller, *J. Polym. Sci. Polym. Chem.*, *23*(9): 2557 (1985).

190. L. Born and H. Hesoe, *Coll. Polym. Sci.*, *263*(4): 335 (1985).

191. T. A. Speckhard, K. K. S. Hwang, S. Y. Tsay, S. B. Lin, and S. L. Cooper, *Polym. Prepr.*, *25*(1): 125 (1984).

192. A. M. Usmani, *J. Coatings Technol.*, *56*(716): 99 (1984).

17

Polyisobutylene-Based Polyurethanes

Joseph P. Kennedy
Institute of Polymer Science
The University of Akron
Akron, Ohio

INTRODUCTION

Polyurethanes (PUs) are a large and versatile family of elastic and rigid engineering materials containing the urethane unit: —NH—CO—O—. PUs are customarily prepared by reacting di- or multiisocyanates with —OH-capped polyesters in the presence of a curing agent such as an amine or a polyol. These polymers exhibit an outstanding combination of properties and are being used in a large variety of rigid–flexible, elastomeric, thermoplastic–elastomeric, coating, foam, etc., applications.

Polyisobutylene-based (PIB-based) PUs belong to the class of elastomeric PUs. The earliest PU elastomers were prepared during the decade 1946–1956 by Pinten [1] and Bayer et al. in Germany [2, 3], Harper et al. in England [4], and Dinsmore [5], Hill et al. [6], and Rugg and Scott in the United States [7]. The raw materials for conventional elastomeric PUs are polyester or polyether glycols in combination with diisocyanates such as 4,4′-diphenylmethane diisocyanate (MDI) or the less expensive tolylene diisocyanate (TDI). The polyesters are usually prepared from adipic acid with ethylene, propylene, butylene, or diethylene glycols. The polyether glycols may be obtained by polymerizing ethylene oxide or propylene oxide with basic catalysts in the presence of a glycol or water. High-quality polyethers can be obtained from tetrahydrofuran by the use of cationic initiators. The molecular weights of the polyester or polyether glycols are preferentially in the 1000–2000 $\bar{M}_n$ liquid range to ensure low initial charge viscosities and rapid mixing of the components.

In addition to these conventional polyester- and polyether-based PUs, specialty elastomeric PUs can also be prepared by the use of liquid hydroxyl-terminated polybutadienes;

however, the number average terminal functionality $\bar{F}_n$ of these products is usually not exactly 2.0, which leads to compounding difficulties.

Justification for interest in polyisobutylene PIB-based PUs stems from certain material deficiencies of the aforementioned PUs. Major shortcomings of conventional polyester- and polyether-based PUs are relatively low acid, base, hydrolytic, steam, and environmental stability and a maximum service temperature of ~105°C. Polybutadiene-based PUs, in addition, possess low oxidative and aging resistance. It is anticipated that PUs prepared from hydroxyl-terminated PIBs will alleviate these shortcomings.

THE SYNTHESIS OF HYDROXYL-TERMINATED PIBs

The synthesis of hydroxyl-terminated (hydroxyl-telechelic) PIBs can occur by indirect or direct routes. The indirect route involves the ozonization (or oxidation) of internal double bonds of isobutylene-conjugated diene copolymers (e.g., butyl rubbers) followed by reduction of the oxidation products to —OH end groups. Thus Marvel and colleagues ozonized copolymers of isobutylene with isoprene [8] and 2,5-dimethyl-2,4-hexadiene [9] and reduced the reaction products by $LiAlH_4$; however, the final polymers were rather ill-characterized mixtures. Zapp et al. [10] ozonized isobutylene-1,3-pentadiene copolymers and reduced the carboxyl-terminated intermediates with $LiAlH_4$ in ethyl ether (conversion >96%) or converted them by reaction with propylene oxide to form —COO—$CH_2CH(OH)CH_3$ end groups with $\bar{F}_n \sim 2.0$.

Similarly, Speckhard et al. prepared hydroxyl-terminated PIB by ozonization of isobutylene–isoprene copolymers followed by catalytic hydrogenation and obtained a product with $\bar{M}_n = 1439$, $\bar{M}_w/\bar{M}_n = 1.9$, and $\bar{F}_n = 1.9$ [11].

The state of the art direct route to hydroxyl-telechelic PIBs is by the inifer technique [12–14]. In this process isobutylene is polymerized in the presence of bi- or trifunctional initiating-transfer agents (so-called inifers) in conjunction with Friedel–Crafts acids (preferentially BCl_3). The structures seen in Scheme 1 can be obtained [15–18]. The molecular

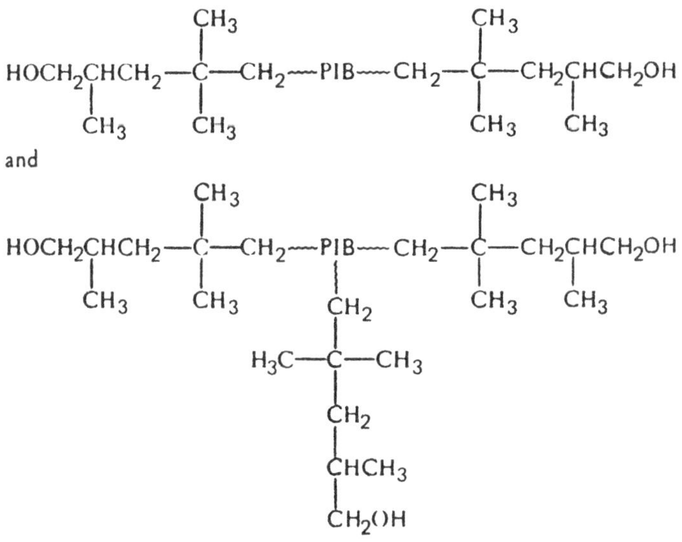

Scheme 1

weights are readily controlled from ~1000 to 5000 or to much higher values [15, 16], and $\bar{F}_n = 2.0 \pm 0.1$ and 3.0 ± 0.1. The molecular weight distributions ($\bar{M}_w/\bar{M}_n$) of the linear glycol is ~1.5 and that for the three-arm star triol is 1.33. The products are odorless, colorless, and tasteless liquids. Due to the absence of tertiary hydrogens PIB has excellent chemical and environmental stability and exhibits outstanding gas and liquid barrier properties. The linear and the three-arm star PIBs have recently become available commercially (Akron Cationic Polymer Development Co., Akron, Ohio).

FLEXIBLE PU NETWORKS PREPARED FROM HYDROXYL-TERMINATED PIBs

The first PU networks from hydroxyl-terminated PIBs were prepared by Zapp et al. [10]. These workers extended linear PIBs carrying —COO—CH$_2$CH(OH)CH$_3$ termini ($\bar{M}_n$ = 1800) with TDI and 1,4-butanediol and obtained excellent fully cured (negligible sol) networks exhibiting up to ~5000 psi tensile strength with 80% elongation. Unfortunately only some tensile/modulus/elongation data are given, and this work has not been continued beyond the feasibility stage.

A more detailed study of PIB-based PUs was carried out by Speckhard et al. [11]. These authors compared the properties of polyester-, polyether-, and polybutadiene-based PUs with those of PIB-based products. They found good phase separation, but network synthesis was hampered by difficulties due to incompatibility between the PIB–diol and diol chain extender. According to environmental tests PIB-based PUs exhibited improved hydrolytic stability and moisture permeability compared to polyester- and polyether-based PUs and improved oxidative stability compared to polybutadiene-based products. Unfortunately the $\bar{F}_n$'s of the starting hydroxyl-terminated PIBs were less than 2.0, which impaired ultimate mechanical properties of the PUs [11].

Improved PUs prepared from PIB–glycols and MDI with $\bar{F}_n = 2.0$ by the inifer technique have also been studied [19]; the mechanical properties of these materials, however, were still lower than those of conventional PUs. Lesser mechanical properties may be due to the absence of strain-induced crystallization of the soft PIB segment and compositional heterogeneity on account of reactant incompatibility [19]. Somewhat improved ultimate properties could be obtained from the same PIB–glycols with 4,4'-dicyclohexylmethane diisocyanate [20]. Increasing hard-segment concentration gave improved dynamic and tensile modulus while elongations at break were unaffected.

Extensive fundamental investigations have been made of the properties of unextended pure gum PIB-based PUs, i.e., PUs prepared by reacting PIB–diols and triols with tri- and diisocyanates, respectively (absence of low molecular weight glycol extenders) [11, 12, 21]. These transformations, seen in Scheme 2, yield "perfect" or "model" networks in

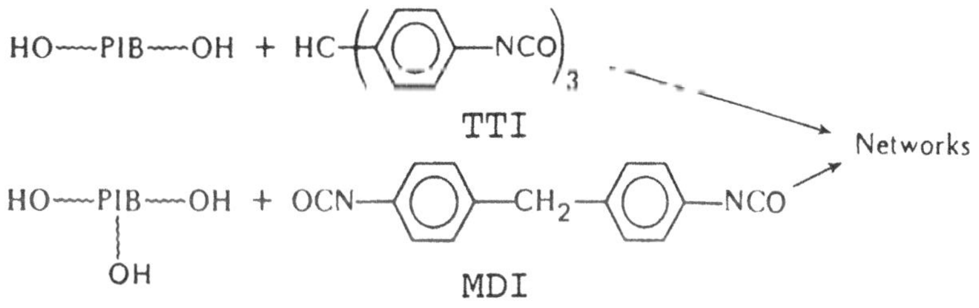

Scheme 2

Table 1 Hot-Water and Hot-Air Resistance of PIB-Based Polyurethanes

Before degradation testing			After exposure to hot water[b]			After air-oven aging[c]		
$\bar{M}_c{}^a$	$\sigma_{t,b}$ (Pa)	E_b (%)	$\sigma_{t,b}$ (Pa)	E_b (%)	Change in $\sigma_{t,b}$ (%)	$\sigma_{t,b}$ (Pa)	E_b (%)	Change in $\sigma_{t,b}$ (%)
2100	1.85×10^7	227	1.76×10^7	222	-4.86	1.81×10^7	224	-2.16
3700	1.81×10^7	330	1.73×10^7	298	-4.42	1.76×10^7	290	-2.76
7700	9.40×10^6	510	9.34×10^6	492	-0.64	9.11×10^6	494	-3.09

[a]Based on $\bar{M}_n$ of prepolymer.
[b]ASTM-D3137.
[c]ASTM-D537.

which the molecular weight between crosslinks can be calculated from that of the PIB–diol or triol. The stress–strain and other properties of these pure gum networks have been investigated [24, 25]. For example, a model network prepared from PIB–diol of $\bar{M}_n =$ 1400 exhibited ~5.5 MPa with ~300% elongation [24]. According to ASTM D-3137 tests the tensile strength and elongation of networks made with $\bar{M}_n = 3400$ and 11,500 PIB–diols remained essentially unaffected upon treatment with 85°C steam for 4 days [26]. Water absorption of the networks by ASTM-D570 was negligible. The high hydrolytic stability of these PUs is due to the low water absorption and moisture permeability of the well-segregated PIB domains, which protect the urethane linkages. Resistance to hot-air aging (circulating air oven at 128°C for 48 hr, ASTM-D513) was also outstanding, which is not too surprising in view of the saturated nature of these largely hydrocarbon networks. These rubbers also exhibited excellent gas barrier properties [26].

Unextended PU networks prepared from PIB–triols and MDI also showed outstanding hydrolytic/aging/environmental characteristics combined with adequate mechanical properties [25]. For example, a PIB–triol of $\bar{M}_n = 7700$ showed ~10 MPa stress at ~500% elongation at break. The hydrolytic and hot-air stability of representative PIB–triol-based networks are shown in Table 1. The loss in properties in these demanding tests was negligible. Moisture absorption was also negligible [25].

While the excellent hydrolytic stability, aging resistance, and moisture absorption can be explained in view of the saturated hydrophobic nature of the networks, the high thermal resistance is more surprising. Deblocking (reversion) of the urethane group to alcohol and isocyanate is thought to start at ~120°C [27], and in conventional PUs the isocyanate group may react with moisture present on account of the relatively polar ester or ether groups, resulting in permanent chain breaking. In contrast, in hydrophobic PIB-based PUs moisture is absent and the deblocked isocyanate can react only with the deblocked alcohol; i.e., deblocking is reversible and permanent damage does not occur.

The engineering applicational potential of these new materials is very large both for biomedical (PIB-based PUs are nonthrombogenic [28]) and conventional use.

REFERENCES

1. H. Pinten, Germ. Pat. Appl. (March 1942); see O. Bayer, *Angew. Chem.*, A59: 275 (1947).
2. O. Bayer, E. Müller, S. Petersen, H. F. Piepenbrink, and W. Windemuth, *Angew. Chem.*, 62: 57 (1950).

3. E. Müller, O. Bayer, S. Petersen, H. F. Piepenbrink, F. Schmidt, and E. Weinbrenner, *Angew. Chem.*, *64*: 523 (1952).

4. D. A. Harper, W. F. Smith, and H. G. White, *Rubber Chem. Tech.*, *23*: 608 (1950).

5. R. P. Dinsmore, Washington Rubber Group, Washington (March 18, 1953).

6. F. B. Hill, C. A. Young, J. A. Nelson, and R. G. Arnold, *Ind. Eng. Chem.*, *48*: 927 (1956).

7. J. S. Rugg and G. W. Scott, *Ind. Eng. Chem.*, *48*: 930 (1956).

8. E. B. Jones and C. S. Marvel, *J. Polym. Sci.*, *A2*: 5313 (1964).

9. J. K. Hecht, C. S. Marvel, and T. W. Campbell, *J. Polym. Sci. A1*, *5*: 1486 (1967).

10. R. L. Zapp, G. E. Serniuk, and L. S. Minkler, *Rubber Chem. Tech.*, *43*: 1154 (1970).

11. T. A. Speckhard, G. VerStrate, P. E. Gibson, and S. L. Cooper, *Polym. Eng. Sci.*, *23*: 337 (1983).

12. J. P. Kennedy, R. A. Smith, and L. R. Ross, U.S. Pat. 4,276,394 (1981).

13. J. P. Kennedy, U.S. Pat. 4,316,973 (1982).

14. J. P. Kennedy, U.S. Pat. 4,342,849 (1982).

15. V. S. C. Chang, J. P. Kennedy, and B. Ivan, *Polym. Bull.*, *3*: 339 (1980).

16. J. P. Kennedy and R. A. Smith, *J. Polym. Sci. Polym. Chem. Ed.*, *18*: 1523 (1980).

17. J. P. Kennedy, L. R. Ross, J. E. Lackey, and O. Nuyken, *Polym. Bull.*, *4*: 67 (1981).

18. A. Fehervari, J. P. Kennedy, and F. Tüdös, *J. Makromol. Sci. Chem.*, *A15*: 215 (1981).

19. T. A. Speckhard, P. E. Gibson, S. L. Cooper, V. S. C. Chang, and J. P. Kennedy, *Polymer*, *26*: 55 (1985).

20. T. A. Speckhard, K. K. S. Hwang, S. L. Cooper, V. S. C. Chang, and J. P. Kennedy, *Polymer*, *26*: 70 (1985).

21. V. S. C. Chang and J. P. Kennedy, *Polym. Bull.*, *8*: 69 (1982).

22. T. Miyabayashi and J. P. Kennedy, *J. Appl. Polym. Sci.*, *31*: 2523 (1986).

23. J. P. Kennedy and J. E. Lackey, *J. Appl. Polym. Sci.*, *33*: 2449 (1987).

24. T. Miyabayashi and J. P. Kennedy, *J. Appl. Polym. Sci.*, *31*: 2523 (1986).

25. J. P. Kennedy and J. Lackey, *J. Appl. Polym. Sci.*, *33*: 2449 (1987).

26. V. S. C. Chang and J. P. Kennedy, *Polym. Bull.*, *8*: 69 (1982).

27. P. Wright and P. C. Cummings, *Solid Polyurethane Elastomers*, Maclaren and Sons, London (1969).

28. P. Giusti, M. Palla, F. Artigiani, and G. Soldani, IUPAC Prepr., p. 371 (July 1982).

18

Photodegradation of Polystyrene

Ayako Torikai
Nagoya University
Nagoya, Japan

Recent progress in the photodegradation of polystyrene is reviewed in this chapter. A wide range of current studies concerning the photodegradation of polystyrene have been published and it is necessary to compile the information from various papers. This chapter covers fields such as analytical techniques in photodegradation applied for polystyrene, photoinitiation and elementary processes of photodegradation of polystyrene, the photodegradation mechanism in solid films and in solutions, photosensitized degradation, factors affecting photodegradation, and photodegradation of copolymers and blends of polystyrene with other polymers.

INTRODUCTION

Polystyrene is one of the important commercial polymers widely used in various industrial fields and often subjected to the irradiation of sunlight on outdoor exposure. The ultraviolet absorption spectrum of polystyrene has an absorption maximum at 250 nm due to the $S_0 \rightarrow S_1$ transition of benzene ring [1, 2], and this absorption band extends to longer wavelengths. The end groups of polystyrene [3, 4] or ketonic impurities [5] also absorb the light of wavelength longer than 300 nm. As the terrestrial sunlight consists of the light of wavelength longer than 290 nm, photodegradation of polystyrene (e.g., embrittlement and color change) can take place upon irradiation with a portion of UV light which is contained within sunlight.

Studies on the photodegradation of polystyrene are important to develop methods to prevent photodegradation. These studies offer the key to obtaining light-stable polystyrene. Photodegradation and photooxidative degradation of polystyrene have long been studied and many reviews have been published [6–14]. The last review by Weir summarizes the recent progress and unsolved problems in photoreactions of polystyrene. Although much information about photodegradation of polystyrene has been accumulated in the last decade, the reaction mechanism of photodegradation of polystyrene is not always clear. The reasons for the controversy on this subject are considered in Ref. 14. The lack of standardization of preparation and reaction conditions makes the problem difficult. The relatively low photoconversion of polystyrene also contributes to this problem. Although the terrestrial sunlight consists of light of wavelength longer than 290 nm, most researchers have studied the photodegradation using low-pressure (254 nm) or medium pressure (>250 nm) mercury lamps. Detectable changes can be obtained under these conditions. The changes induced by irradiation with sunlight can be deduced from experimental results.

The purpose of this chapter is to review recent studies on the photodegradation of polystyrene.

ANALYTICAL TECHNIQUES IN PHOTODEGRADATION

Photodegradation of polystyrene results in chain scission, crosslinking, and marked discoloration (yellowing). These reactions are accompanied by change of chemical structure, molecular weight, solubility in solvents, and various characteristics and mechanical properties of the polymer. Analytical methods for studying these changes have been developed and much information about the photodegradation processes of polystyrene has been obtained. Ultraviolet (UV) and infrared (IR) absorption spectroscopy are used to detect the formation or disappearance of functional groups in the polymer, i.e., chemical changes of the polymer. Recently Fourier transform infrared (FTIR) spectroscopy, due to its excellent sensitivity, has been applied for analysis of a small amount of the products formed by photoirradiation. Since polystyrene generally has relatively low reactivity to photoirradiation, FTIR spectroscopic analysis serves as a powerful tool for the detection of very small chemical changes in the polymer. In Figure 1 the detection of hydroperoxides by FTIR spectroscopic measurement is shown [15]. These spectral changes cannot be detected by using conventional dispersion-type infrared spectrophotometry under the same experimental conditions.

Nuclear magnetic resonance (NMR) also is useful in analyzing structural changes. Changes of molecular weight are followed by gel permiation chromatography (GPC), viscosity, osmotic pressure, and light scattering. Among these techniques, GPC and viscosity measurements are very simple and frequently applied in molecular weight determination. Changes in the characteristics of polystyrene are followed by gel and density measurements. The macroscopic deterioration of polystyrene can be followed directly by measurements of mechanical properties. Recently studies on the effect of change in

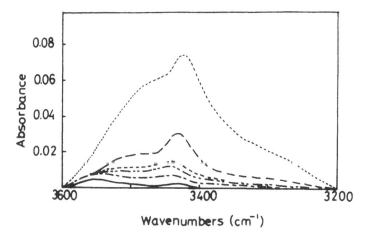

Figure 1 FTIR spectra of photoirradiated polystyrene–polymethyl methacrylate blend (1 : 1). Irradiation time: ——, 0 hr; – – –, 1 hr; – – – –, 2 hr; – – –, 3 hr; – –, 4 hr; · · · ·, 5 hr.

chemical structure (microscopic change) and on the mechanical properties of the polymer (macroscopic change) have started and information in this field is now accumulating. Reaction intermediates in the photodegradation processes play an important role in producing chemical and physical changes in photoirradiated polystyrene.

Reaction intermediates—radicals—have been detected and identified by using electron spin resonance (ESR) spectroscopy. Radicals produced immediately after photon absorption can be detected by measurement at low temperature ($-196°C$, liq. N_2), where the radical decay is almost suppressed. Generally some polymer radicals can be detected at room temperature owing to the rigidity of the polymer. The methods mentioned have been applied to elucidate the photodegradation mechanism of polystyrene.

Additional information on analytical techniques can be obtained in some detail from Refs. 16–20.

PHOTOINITIATION AND ELEMENTARY PROCESSES

The initiation of photochemical reactions is the formation of an electronically excited state. In the case of polystyrene, absorption of light of wavelength less than 300 nm results in the formation of the excited singlets (S_1^*) state of the phenyl groups [3]:

This electronically excited state can be detected by luminescence (fluorescence and phosphorescence) [21–24] and by light absorption [25, 26] methods. Fluorescence spectra of polystyrene obtained in CH_2Cl_2 solution in the absence of oxygen are shown in Figure 2 [23]. The emission from monomer singlet state was detected at the end of the pulse. After 45 ns the spectrum mainly consists of the emission from excimer (complexes of excited and nonexcited states). Excimer formation has also been detected by absorption spectro-

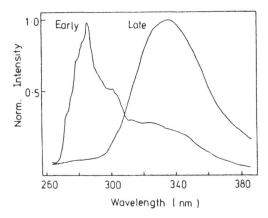

Figure 2 Time-resolved emission spectra for polystyrene obtained in CH₂Cl₂ solution in the absence of oxygen. Early, recorded at the end of the flash; late, recorded 45 ns after the flash. (From Ref. 23.)

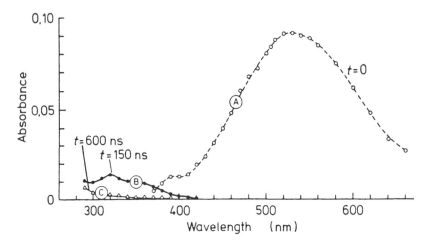

Figure 3 Photolysis of polystyrene in 1,4-dioxane solution (1×10^{-3} basemol/L) saturated with argon. Optical absorption spectra recorded at the end of the flash (A) and 150 ns (B) and 600 ns (C) after the 15-ns flash. (From Ref. 25.)

scopy as shown in Figure 3 [25]. These spectra are due to the direct excitation of a benzene ring of polystyrene. Polystyrene also degrades by photoirradiation with light of wavelength longer than 300 nm owing to the end groups of polystyrene and ketonic and occasionally incorporated impurities. Recent studies on the photoinitiation processes with light of wavelength longer than 300 nm are summarized in Ref. 14.

The electronically excited state of the polystyrene molecule then dissociates into radicals. The radicals produced by the photoirradiation of polystyrene at −196°C and detected by ESR spectrometry at the same temperature are assigned to the polystyryl radical (**I**) and phenyl radical (**II**) [27]. Hydrogen radicals formed as the counter part of radical (**I**) are mobile and can diffuse into the polymer matrix and recombine to form hydrogen molecules [28]. Reactions of the hydrogen radicals escaped from the recombination will be discussed in the next section.

$$\sim CH_2 - \overset{\bullet}{C} \sim \qquad (I)$$

$$(II)$$

The phenyl radicals cannot diffuse out and may abstract hydrogen atom from adjacent polymer molecules to form polystyryl radical. These elementary processes are summarized as follows:

$$\sim CH_2 - CH - CH_2\sim \quad \overset{h\upsilon}{\longrightarrow} \quad \sim CH_2 - CH - CH_2\sim * \qquad (1)$$

$$\sim CH_2 - CH - CH_2\sim * \quad \longrightarrow \quad \sim CH_2 - \overset{\bullet}{C} - CH_2\sim + H\cdot \sim \qquad (2)$$

$$\sim CH_2 - \overset{\bullet}{C} - CH_2 \sim + \qquad (3)$$

$$\sim CH_2 - CH - CH_2\sim \; + \qquad \sim CH_2 - \overset{\bullet}{C} - CH_2\sim + \qquad \longrightarrow \qquad (4)$$

The polystyryl radicals thus produced are the important precursors of photoinduced chain scission, crosslinking, and the formation of unsaturation (yellowing).

The elementary processes of photodegradation of polystyrene in chlorine-containing solvents have been investigated [26, 29]. Charge transfer complexes ($PSt^{+\delta} \cdot Cl^{-\delta}$) having λ_{max} at 320 nm and at 500 nm have been observed as transient absorption spectra in $CHCl_3$ and CCl_4 solution. This charge transfer complex degrades into polymer radical, according to

$$(P^{+\delta} \cdot Cl^{-\delta}) \rightarrow P\cdot + HCl \qquad (5)$$

where P indicates the polystyryl radical (**I**). Charge transfer complexes are precursors of

(I). Polystyryl radical produced by reaction 5 contributes to the degradation of the polymer. In fact, polystyrene decomposes effectively in these solutions.

Recently studies of photophysical processes (energy migration) in polystyrene have been conducted by Blonski and Sienicki [30] using the Monte Carlo method. They estimated the quantum yields of monomer and excimer fluorescence and assumed resonance and exchange mechanisms of energy migration in the simulation of photophysics of polystyrene. They concluded that an exchange mechanism is more appropriate in a description of the photolysis of polystyrene on the basis of the obtained values of quantum yields of monomer and excimer fluorescence and respective decay time.

PHOTODEGRADATION MECHANISM

Polystyrene Films

The main precursor of photodegradation of polystyrene is polystyryl radical, as discussed in the previous section. The decay rate of polystyryl radical has been estimated by ESR spectrometry [12]. The relative ESR peak height of polystyryl radical, which is the measure of radical concentration, stored for 10 days, is about 75% of the initial value. Thus the decay rate of the radical is rather slow in this case. This means the photoreaction of solid polystyrene in vacuum may occur slowly. The main reactions in photoirradiated polystyrene in vacuum are crosslinking, chain scission, and discoloration (yellowing) [31–33]. The only gaseous product produced at the same condition is hydrogen. Reaction 2 has been thought to be the main source of hydrogen production [34, 35]. It has been suggested that reaction 2 is a minor source, and that most of the H_2 results from the abstract reaction between H atoms and the polymer [31]. Gel formation in photoirradiated polystyrene is predominant in vacuum, and viscosity average molecular weight ($\overline{M}_v$) of polystyrene decreases slightly with photoirradiation [12]. These experimental results show the main reaction taking place on photoirradiation in vacuum is crosslinking, and a small amount of chain scission also participates in this system, as seen in Figure 4.

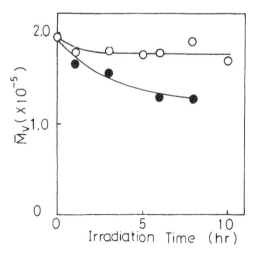

Figure 4 Changes in the viscosity average molecular weight in photoirradiated polystyrene films in vacuum ($\bigcirc$) and in air ($\bullet$) with the irradiation time. (From Ref. 12.)

The following reaction mechanism can reasonably explain the photoinduced main reactions of polystyrene in vacuum [12, 14, 31, 32]:

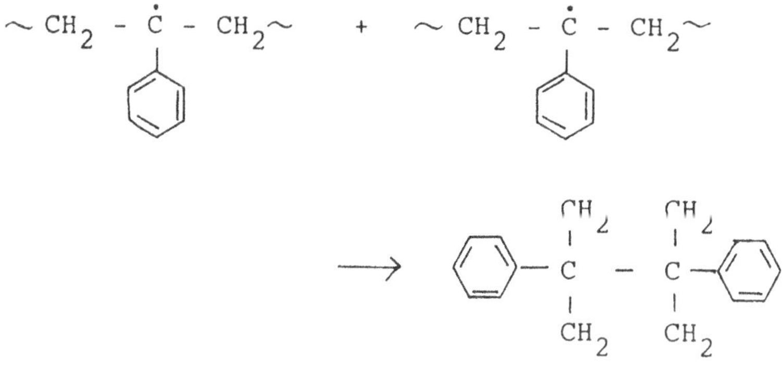

$$\sim CH_2 - \overset{\bullet}{C} - CH_2\sim \quad + \quad \sim CH_2 - \overset{\bullet}{C} - CH_2\sim \tag{6}$$

$$\longrightarrow \quad \text{(Crosslinking)}$$

$$\sim CH_2 - \overset{\bullet}{C} - CH_2 - CH\sim \quad \sim CH = CH \quad + \quad \overset{\bullet}{C}H_2 - CH\sim \tag{7}$$

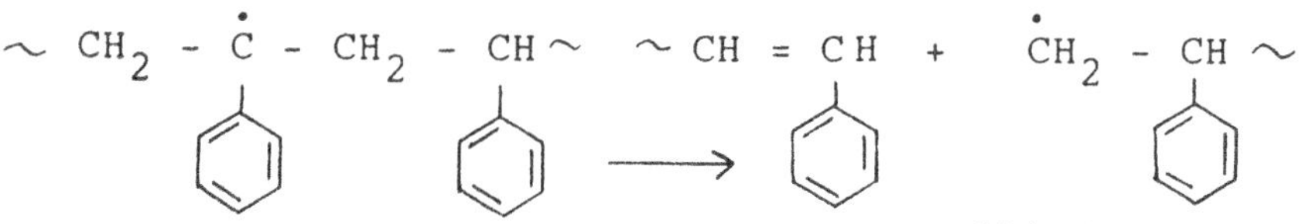

$$\longrightarrow \quad \text{(Main-chain scission)}$$

$$H \ (\text{reaction}(2)) \quad + \quad \cdot H \longrightarrow H_2 \tag{8}$$

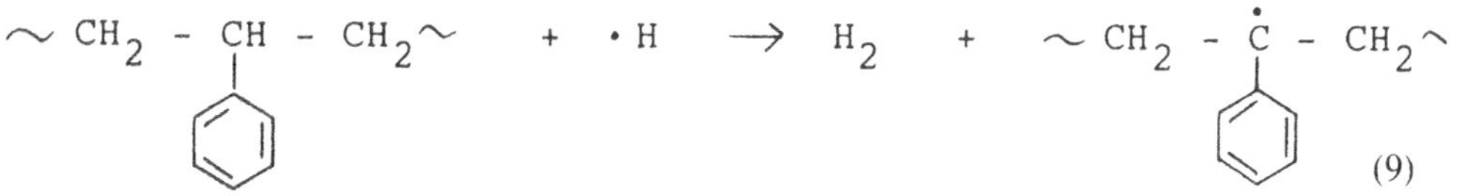

$$\sim CH_2 - CH - CH_2\sim \quad + \quad \cdot H \longrightarrow H_2 \quad + \quad \sim CH_2 - \overset{\bullet}{C} - CH_2\sim \tag{9}$$

$$\sim CH_2 - \overset{\bullet}{C} - CH_2\sim \quad + \quad \cdot H \longrightarrow H_2 \quad + \quad \sim CH_2 - C = CH \sim \tag{10}$$

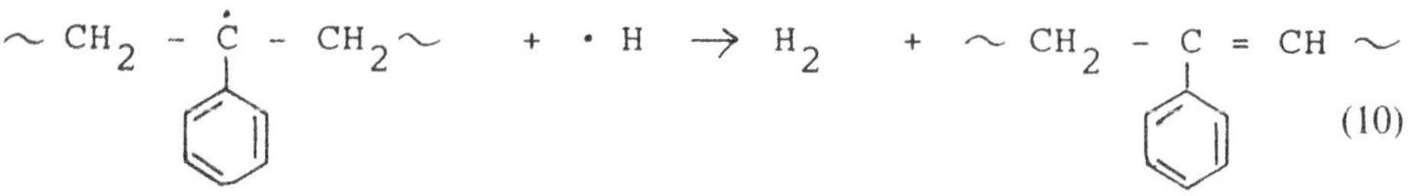

Reactions 8, 9, and 10 result in the formation of hydrogen.

Photodegradation of polystyrene in the presence of air (photooxidation) results in chain scission, some crosslinking, and discoloration. On UV irradiating polystyrene in air, the ESR spectrum shown in Figure 5 was obtained. The spectral shape and the line width are typical for the peroxy radical [12], so this spectrum can be assigned to the peroxy radical of polystyrene (**III**).

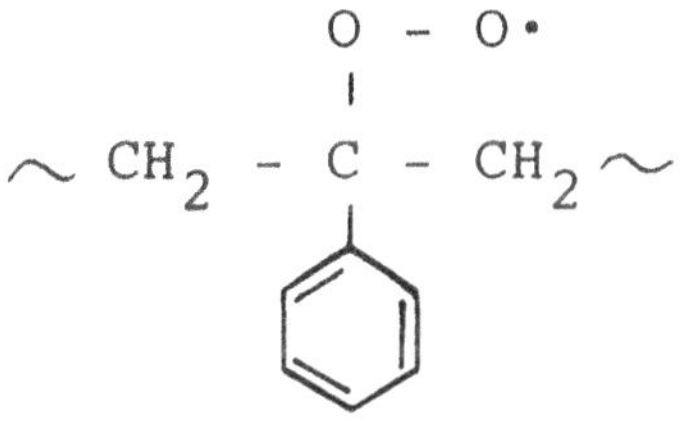

$$\sim CH_2 - \overset{\overset{\textstyle O - O\cdot}{\textstyle |}}{C} - CH_2\sim \quad (\text{III})$$

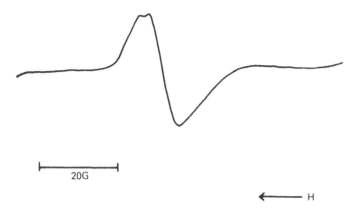

Figure 5 ESR spectrum of photoirradiated polystyrene film in air. Irradiation time, 7 hr. (From Ref. 12.)

It is found by the analysis of radical formation that oxygen attack on the polystyryl radical is faster during the early stages of irradiation. This may be explained by assuming that oxygen penetrates into polystyrene and the rate of penetration becomes slower in the interior of the polystyrene. The $\overline{M}_v$ decreases with the irradiation time, as shown by closed circles in Figure 4. This decrease is complementary with the increase of peroxy radical formation.

Infrared absorption spectra of polystyrene photoirradiated in air indicate an increase in the absorption band at $\lambda_{max} = 1740$ cm^{-1}. This band is attributable to a C=O stretching vibration. This absorption band increases in its intensity with an increase in the irradiation time. The increase in the intensity at 1740 cm^{-1} parallels that of the peroxy radical. These results suggest the degradation reaction may proceed via peroxide intermediates and results in the main chain scission of polystyrene, leading to the formation of carbonyl compounds. The following mechanism can explain the photooxidation reaction in polystyrene [4, 9, 12, 36]:

$$(11)$$

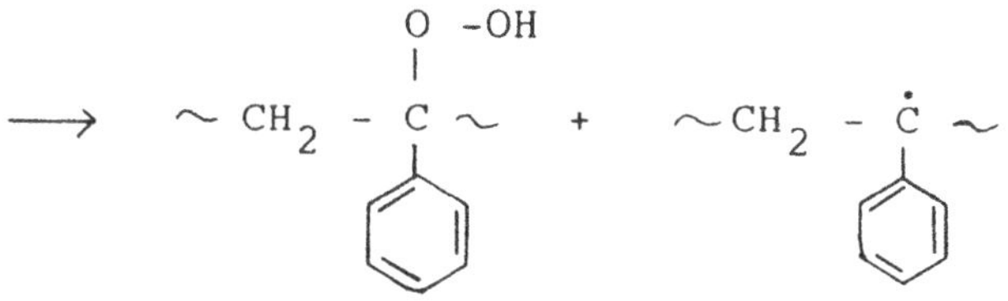

$$\longrightarrow \quad \sim CH_2 - \overset{\overset{\displaystyle O\ -OH}{|}}{\underset{\underset{\bigcirc}{}}{C}} \sim \quad + \quad \sim CH_2 - \overset{\overset{\displaystyle \cdot}{}}{\underset{\underset{\bigcirc}{}}{C}} \sim \qquad (12)$$

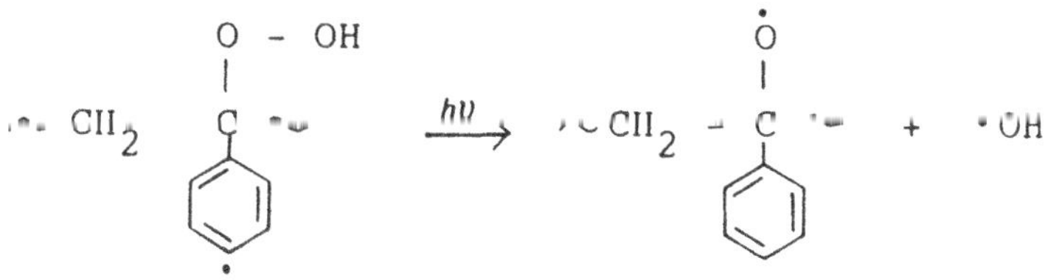

$$\sim CH_2 \quad \overset{\overset{\displaystyle O\ -\ OH}{|}}{\underset{\underset{\bigcirc}{}}{C}} \sim \quad \overset{h\upsilon}{\longrightarrow} \quad \sim CH_2 - \overset{\overset{\displaystyle \overset{\cdot}{O}}{|}}{\underset{\underset{\bigcirc}{}}{C}} \sim \quad + \quad \cdot OH \qquad (13)$$

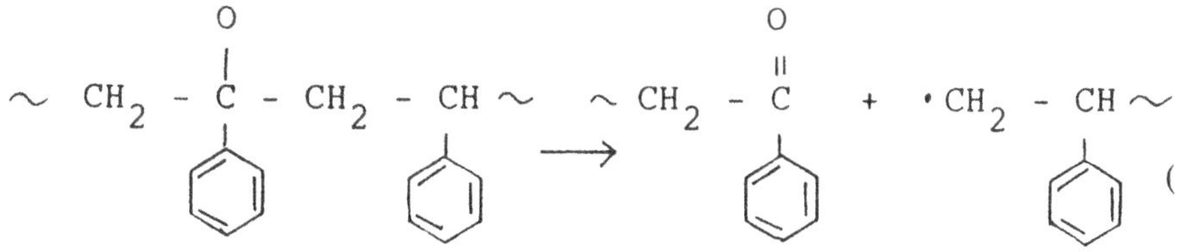

$$\sim CH_2 - \overset{\overset{\displaystyle \overset{\cdot}{O}}{|}}{\underset{\underset{\bigcirc}{}}{C}} - CH_2 - \underset{\underset{\bigcirc}{}}{CH} \sim \quad \longrightarrow \quad \sim CH_2 - \overset{\overset{\displaystyle O}{\|}}{\underset{\underset{\bigcirc}{}}{C}} \quad + \quad \cdot CH_2 - \underset{\underset{\bigcirc}{}}{CH} \sim \qquad (14)$$

(Chain scission)

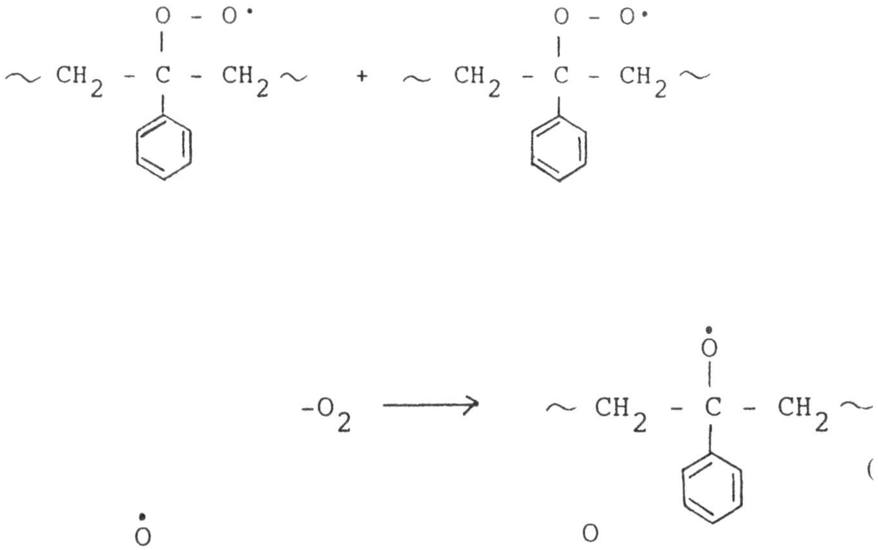

$$\sim CH_2 - \overset{\overset{\displaystyle O\ -\ O\cdot}{|}}{\underset{\underset{\bigcirc}{}}{C}} - CH_2 \sim \quad + \quad \sim CH_2 - \overset{\overset{\displaystyle O\ -\ O\cdot}{|}}{\underset{\underset{\bigcirc}{}}{C}} - CH_2 \sim$$

$$-O_2 \quad \longrightarrow \quad \sim CH_2 - \overset{\overset{\displaystyle \overset{\cdot}{O}}{|}}{\underset{\underset{\bigcirc}{}}{C}} - CH_2 \sim \qquad (15)$$

$$\sim CH_2 - \overset{\overset{\displaystyle \overset{\cdot}{O}}{|}}{\underset{\underset{\bigcirc}{}}{C}} - CH_2 \sim \quad \longrightarrow \quad CH_2 - \overset{\overset{\displaystyle O}{\|}}{\underset{\underset{\bigcirc}{}}{C}} \quad + \quad \overset{\cdot}{C}H \qquad (16)$$

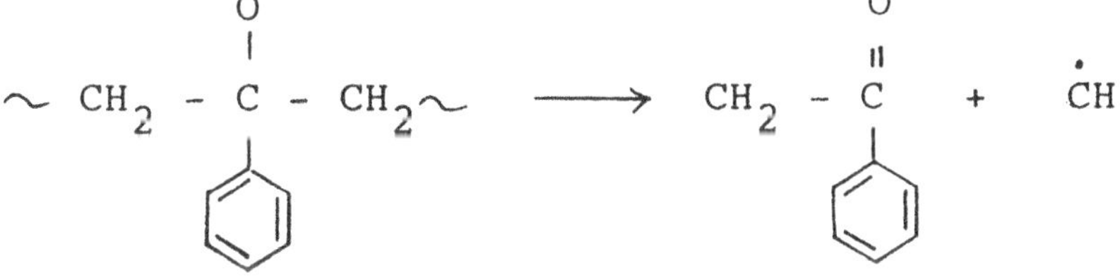

(Chain scission)

The yellowing of photoirradiated polystyrene in vacuum has been attributed to the conjugated double bond formation in the polymer backbone [34]. It has also been suggested that the yellowing is due to the fulvene formed by the photoinduced isomerization of benzene rings of polystyrene [7].

Analogous to the photodegradation of various types of polymers such as polyvinyl chloride [37], polyethylene [38], and polypropylene [39], the evolution of conjugated double bond sequences in the polymer backbone has been suggested as follows [40]:

$$\sim CH_2 - \overset{\bullet}{C} - CH_2 - CH - CH_2 \sim$$

$$\downarrow$$

$$\sim CH_2 - \overset{\bullet}{C} - CH = CH - CH_2 \sim \qquad (17)$$

$$\downarrow$$

$$\sim CH_2 - \overset{\bullet}{C} - (CH = CH)_n - CH_2 \sim \qquad (18)$$

The plausible mechanism for yellowing of the polymer is the formation of conjugated double bond in the polymer backbone. Main chain scission, structural change, and quantum yield of photodegradation of polystyrene have been reported recently [13, 41].

Radical species of photoirradiated polystyrene have also been studied by ESR at 90 K [42].

Irradiation of polystyrene with the light of wavelength longer than 300 nm causes the excitation of impurity chromophores, which are incorporated in the chain during the polymerization and/or subsequent processing. The roles in photoinitiation of impurity chromophores such as side chain hydroxyl groups and in-chain peroxide linkage [43], acetophenone end group, and oxygen–polymer complex are summarized in Ref. 33. Geuskens and Lu-Vinh reported on the role of aromatic ketones in the photodegradation of low molecular weight model compounds of polystyrene irradiated with 365-nm light [44].

Polystyrene in Solutions

The rate of photodegradation of polystyrene is much affected by the nature of the solvent. Price and Fox [45] reported that a low concentration of carbon tetrachloride in cyclohexane solution exhibits marked sensitization, whereas a low concentration of cyclohexane in carbon tetrachloride solution inhibits rapid scission. For example, the quantum yield for chain scission of polystyrene (ϕ_s^p) in carbon tetrachloride is 560 times greater than that in cyclohexane. Relatively less attention has been paid, however, to the behavior of polystyrene in solution. The studies in this field have been developed rather recently [7, 46–48].

Ishii et al. [47] studied the variation of excimer fluorescence intensity with number average molecular weight for anionic polystyrene in 1,2-dichloroethane and observed an increase in the ratio of excimer emission with increased molecular weight. Excimer formation in the same solvent has also been detected by laser flash photolysis, as already mentioned in the previous section [25].

Bortolus et al. [29] studied the photooxidative degradation of polystyrene in halomethanes and found the rate of degradation to increase in the order of increasing electron affinity of the solvent (in methylene chloride, chloroform, carbon tetrachloride). They attributed this behavior to the charge-transfer complex formation between the halomethane solvent and ground and/or electronically excited polystyrene molecules, which is an active intermediate in the photolysis of polystyrene in halomethane. These reactions are summarized as follows:

$$\sim CH_2—CH\sim (PS) + RCl \rightleftarrows (PS\ RCl) \tag{19}$$

$$PS \xrightarrow{h\nu} PS^* \tag{20}$$

$$PS^* + RCl \rightleftarrows (PS\ RCl)^* \rightleftarrows PS^+ + RCl^- \rightarrow R\cdot + Cl^- + PS^+ \tag{21}$$

$$PS^+ \rightarrow PS\bullet\left(\sim CH_2—\dot{C}\sim\right) + \cdot H \tag{22}$$

$$(PS\cdot RCl) \xrightarrow{h\nu} (PS\cdot RCl)^* \rightleftarrows PS^+ + RCl^- \rightarrow R\cdot + Cl^- + PS^+ \tag{23}$$

$$PS + \cdot R \rightarrow PS\left(\sim CH_2—\dot{C}\sim\right) + RH \tag{24}$$

in the presence of air,

$$R\cdot + O_2 \rightarrow R—O—O\cdot \tag{25}$$

$$R—O—O\cdot + PS \rightarrow R—O—O—H + PS\cdot\left(\sim CH_2—\dot{C}\sim\right) \tag{26}$$

$$PS + O_2 \rightarrow PS—O—O\cdot \tag{27}$$

Ikada et al. [48] recently studied the main chain scission in photodegradation of low concentration of polystyrene in benzene, carbon tetrachloride, and chloroform, avoiding the crosslinking reaction. The number of chain scissions determined by viscosity measurement and GPC analysis is much faster in chloroform and carbon tetrachloride than in benzene. The hydrogen abstraction reaction by the radical species formed by the photoirradiation of polystyrene in solvents produces polystyryl radical, which is the precursor of photodegradation of polystyrene:

$$CCl_4 \xrightarrow{h\nu} \cdot CCl_3 + \cdot Cl \tag{28}$$

$$CHCl_3 \xrightarrow{h\nu} \cdot CCl_3 + \cdot H \tag{29}$$

$$\sim CH_2—CH\sim + \cdot H(\cdot Cl, \cdot CCl_3) \rightarrow \sim CH_2—\dot{C}\sim + H_2 \tag{30}$$

In any case, reaction intermediates produced from the solvents participate with the formation of polystyryl radical. Other studies on the photodegradation of polystyrene in solutions have been carried out primarily in halomethanes [49–51].

The photosensitized degradation of polystyrene in benzene solution containing benzophenone has also been studied [52]. The ketyl radical has been detected as a transient intermediate by absorption spectroscopy. This fact implies the formation of ketyl radicals in the primary process of polymer degradation, involving hydrogen abstraction from the polymer by excited benzophenone molecules. Polystyryl radical has also been detected by the spin trapping method in ESR spectrometry [53].

PHOTOSENSITIZED DEGRADATION

Photosensitized degradation of polystyrene in chlorine-containing solvents such as chloroform and carbon tetrachloride has been reported by Price and Fox [45] as described in the preceding section. The reaction mechanism has recently been studied by many workers [7, 46–54]. In this section photosensitized degradation in the solid phase is described. In view of the efficient utilization of absorbed photon, the accelerated photoreaction is an interesting subject for photochemists, and attention has been paid to the reaction in the solid phase, especially concerning the industrial applications, and this study also is devoted to elucidation of specific behavior in photoirradiated bulk polymers. Studies of sensitized photodegradation have also been applied to develop light-sensitive photo-degradable plastics combined with the problem of environmental pollution by plastic litter.

Ketonic compound such as quinones and benzophenone have long been recognized as effective sensitizers in photodegradation [2]. Halocarbon-sensitized degradation of polystyrene has been studied mostly in the liquid phase. Studies of photodegradation of polystyrene in the solid phase are relatively scant. Horie et al. [55] studied the interaction of benzophenone triplet with polymer matrix in connection with polymer degradation by analyzing the decay curve of phosphorescence of benzophenone in polystyrene. The decay curve $\Phi(t)$ did not obey exponential relations in the range of $T_\beta < T < T_g$. The authors attributed this phenomenon to the quenching of phosphorescence by the phenyl group of polystyrene. T_β is defined as the temperature of the initiation of local motion of polystyrene molecule concerning the phenyl groups and estimated to be $-75°C$ from the point of inflection in the decay curve. Polystyrene films containing benzophenone as an additive have been irradiated with UV light and sensitized degradation was studied by ESR and viscosity measurements [12].

Benzophenone is a well-known sensitizer in photochemical reactions forming a ketyl radical by the following reactions [52]:

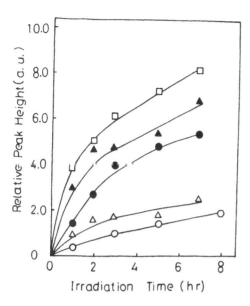

Figure 6 Relative ESR peak heights of photoirradiated polystyrene films containing benzophenone in vacuum against the irradiation time. ○, pure polystyrene; polystyrene containing 0.14 mol% (△), 0.48 mol% (●), 0.69 mol% (▲), and 0.7 mol% (□) of benzophenone. (From Ref. 12.)

$$Ph_2CO \rightarrow Ph_2CO^* \tag{31}$$
$$PhCO^* + RH \rightarrow Ph_2\dot{C}OH + \cdot R \tag{32}$$

The ESR spectrum of photoirradiated polystyrene film containing benzophenone in vacuum consists of the polystyryl radical and radical arising from the additive [18]. When the irradiation was carried out in the presence of air, only the peroxy radical was produced. The concentration of polystyryl radical increases with an increase in the additive concentration at any time of the irradiation, as shown in Figure 6. As polystyryl radical is considered to be a precursor of the photodegradation of polystyrene, benzophenone is seen to act as a sensitizer in the polymer radical formation. The sensitized degradation of polystyrene has been confirmed by viscosity and IR spectroscopic measurements.

The following reaction mechanism has been proposed for the photodegradation of polystyrene in the presence of benzophenone [12]:

PhCO $\xrightarrow{h\upsilon}$ PhCO*

PhCO* + $\sim$ CH$_2$ – CH – CH$_2$ $\sim$

$$
\text{Ph}\overset{\bullet}{\text{C}}\text{OH} \quad + \quad \sim \text{CH}_2 -\overset{\bullet}{\underset{\text{Ph}}{\text{C}}} - \text{CH}_2\sim \qquad (33)
$$

$$
+ \text{O}_2 \swarrow \qquad \searrow \text{recombination}
$$
$$
\text{chain scission} \qquad\qquad \text{crosslinking} \qquad (34)
$$

p-p'-Dibenzoylphenyl ether-sensitized photoinitiation of polystyrene has been applied to thermal degradation of polystyrene by Mita et al. [56]. They also studied the reaction of aromatic ketone triplet with polystyrene and concluded that the rate of deactivation of the triplet(k_d^{PS}) is faster than that of the hydrogen abstraction reaction from polystyrene(k_α^{PS}). Thus the quantum yield of the radical formation by S is represented by $\phi = k\alpha/kd$. ϕ estimated from the disappearance of S is on the order of 10^{-2} [57].

Torikai et al. [40] recently studied the carbon tetrachloride-sensitized degradation of polystyrene in the solid phase. Carbon tetrachloride-doped polystyrene films were irradiated with UV light and the effect of the additive on photodegradation was analyzed by UV and FTIR spectroscopic techniques and by gel and viscosity measurements. On photo-irradiation of CCl$_4$-doped polystyrene film, the amount of oxygenated products and unsaturation increased with the concentration of carbon tetrachloride up to 5 mol%. The same behavior was found in the change of gel fraction and number of chain scission. These results support the photosensitized degradation of polystyrene by carbon tetrachloride. Plausible mechanisms for the processes are suggested as follows.

$$
\sim \text{CH}_2 - \underset{\text{Ph}}{\text{CH}} - \text{CH}_2\sim \quad \overset{h\nu}{\longrightarrow} \quad \sim \text{CH}_2 - \underset{\text{Ph}}{\text{CH}} - \overset{*}{\text{CH}_2}\sim \qquad (35)
$$

$$
\text{CCl}_4 \overset{h\nu}{\longrightarrow} \text{CCl}_4^* \rightarrow \cdot\text{CCl}_3 \quad + \quad \cdot\text{Cl} \qquad (36)
$$

$$
\sim \text{CH}_2 - \underset{\text{Ph}}{\text{CH}} - \overset{*}{\text{CH}_2}\sim \quad \longrightarrow \quad \sim \text{CH}_2 - \underset{\text{Ph}}{\overset{\bullet}{\text{C}}} - \text{CH}_2\sim \quad + \cdot\text{H} \qquad (37)
$$

$$
\sim \text{CH}_2 - \underset{\text{Ph}}{\text{CH}} - \overset{*}{\text{CH}_2}\sim \quad + \quad \text{CCl}_4
$$
$$
\longrightarrow \quad \sim \text{CH}_2 - \underset{\text{Ph}}{\text{CH}}\sim \quad + \quad \text{CCl}_4^* \qquad (38)
$$

$$CCl_4^* \longrightarrow \cdot CCl_3 + \cdot Cl \tag{39}$$

$$\sim CH_2 - \underset{\underset{\bigcirc}{|}}{CH} - CH_2 \sim + \cdot Cl$$

$$\longrightarrow \sim CH_2 - \overset{\cdot}{\underset{\underset{\bigcirc}{|}}{C}} - CH_2 \sim + HCl \tag{40}$$

chain scission

crosslinking $\tag{41}$

Other factors affecting these processes may be interpreted as follows: The softening of the polymer matrix (polystyrene) by the addition of CCl_4 as a plasticizer allows the radical reaction to occur more easily, leading to crosslinking, chain scission, and unsaturation in the polystyrene molecules. Excitation of impurity chromophores such as hydroperoxides, carbonyl group, and oxygen–polymer charge-transfer complexes may initiate the impurity-sensitized photodegradation of polystyrene [7, 8, 10, 31, 58]. Three major mechanisms have been proposed to account for the photooxidation of polystyrene. These reactions are summarized in detail in Ref. 33. The adventitious impurities incorporated in the chains and end groups during polymerization and/or subsequent processing also initiate photodegradation [27]. The photoinitiation mechanisms by these impurities have also been discussed by Schnabel [2].

FACTORS AFFECTING PHOTODEGRADATION

Efficiency of photon absorption in polystyrene films has been discussed by some authors. Weir [14] showed that the rapid attenuation in the photon absorption on irradiation of polystyrene film with 254-nm light has the effect of confining the effective zone to the surface layers. The efficiency of photon absorption is represented by the equation $I_L = I_0 \exp(-\beta L)$ in which I_L, I_0, and β are the intensity at depth L, the incident 254-nm intensity, and an absorption coefficient, respectively. Konuma et al. [59] studied the effect of photodegradation of multilayer polystyrene films and observed the same type of attenuation of photon absorption in polystyrene film. They also reported that the photodegradation of polystyrene takes place on the surface of films by analyzing the changes in chemiluminescence spectra of polystyrene films during photoirradiation.

Oxygen penetration into the polymer film also affects the photoreaction of polystyrene [12]. On the surface of polystyrene film, the rate of oxygen attack to polystyryl radical formed by photoirradiation much faster than that in the interior of the polystyrene film. As a result, chain scission takes place mostly on the surface of the film while crosslinking occurs in the interior of the film in photooxidation processes. Polystyryl radical has been considered to be a precursor of photoinduced main chain scission and crosslinking.

It is well known that the molecular mobility in the amorphous part of any polymer begins to increase at the glass transition temperature (T_g) due to the segmental motion. Torikai et al. [60] carried out photodegradation of polystyrene at 30°C ($<T_g$), 100°C ($\simeq T_g$), and 120°C ($>T_g$) and analyzed the results by ESR, FTIR, and gel fraction measurements. The concentration of polystyryl radical decreased and oxygenated product formation was accelerated by irradiation at higher temperatures (Figure 7), reflecting the segmental motion of the polymer chain. Gel formation is favored by photoirradiation near

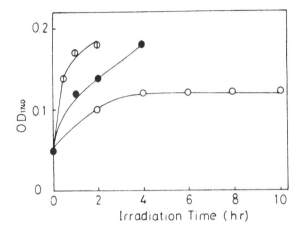

Figure 7 Effect of temperature on the formation of carbonyl groups in polystyrene films photo-irradiated in air. Irradiation temperatures, ⊕, 30°C; ●, 100°C; ○, 120°C. (From Ref. 12.)

T_g because of the initiation of the segmental motion of the polymer near T_g, whereas this reaction is suppressed above T_g because the reaction of polystyryl radical with oxygen takes place favorably at higher irradiation temperatures. The last aspect was also confirmed by GPC measurement in the photodegradation of polystyrene at 180 or 250°C [61]. Mita et al. [56] studied photosensitized thermal degradation of polystyrene and reported that by using p-p'-dibenzoylphenyl ether as a sensitizer, thermal-induced main chain scission of polystyrene takes place at relatively lower temperature than that in the absence of the sensitizer.

COPOLYMERS AND POLYMER BLENDS OF POLYSTYRENE

The usefulness of composite materials in polymeric engineering materials is now increasing because various characteristics of polymeric materials can be improved by blending the polymeric materials with other materials and/or low molecular weight compounds. Various characteristics of homopolymers can also be improved by copolymerization. Although the photodegradation of polystyrene homopolymer has been studied, as mentioned in the preceding sections, much less information is available for copolymers and blends than for homopolymers.

Copolymers

The investigations of energy transfer in the elementary processes of photodegradation of polystyrene copolymers is important both from the scientific and practical point of view. David et al. [62] indicate that excimer emission (I_d) in styrene copolymers with methyl methacrylate, poly (St-co-MMA), and methacrylate, poly(St-co-MA), relative to monomer emission (I_m), depends linearly on the concentration of the styrene diad fraction (fss). The authors suggested that excimer formation in these copolymers is associated with energy migration. The later work for poly(St-co-MMA) by Reid and Sautar [63] confirmed the conclusions of David et al. and showed that the concentration of excimer-forming sites is proportional to the fraction of linkages between aromatic species. Sienicki and

Bojarski [64] measured the quantum yields of monomer and excimer fluorescence of statistical styrene copolymers with methyl methacrylate, ethyl acrylate, and acrylonitrile in 1,2-dichloroethane and suggested that the numbers of excimer-forming sites in these copolymers were proportional to the fraction of linkages between styrenes.

Recently Semerak and Frank [65] studied the electronic energy migration in the copolymer of polystyrene with poly-2-vinylnaphthalene and suggested that (1) migration is not limited to nearest neighbor rings, (2) the morphology of the dilute blends does not consist of isolated chains of the aromatic vinyl polymers, and (3) short-range electronic interactions other than dipole–dipole are involved in energy migration.

It has been reported by Fox et al. [66] that the MMA unit acts as an energy sink in poly(St-co-MMA) in photodegradation and that the alternating copolymer is most subject to photodegradation. Quantum yields of chain scission of the copolymer changing the ratio of styrene to methyl methacrylate have been estimated by the same authors.

Torikai et al. [67] recently investigated the photodegradation of a series of poly(St-co-MMA) which are prepared by γ-ray–induced polymerization of styrene and methyl methacrylate, avoiding the inclusion of impurities. The ESR spectrum of the photoirradiated copolymer consists of the mixture of radicals of polystyrene and polymethyl methacrylate and the concentration of polymethyl methacrylate radical decreases with the increase of styrene fraction in the copolymer. The same trend was also detected by analyzing oxygenated products in the 3450-cm^{-1} region of extended FTIR spectra. Decrease in molecular weight of the copolymer was found by the incorporation of methyl methacrylate in the styrene unit (Figure 8). The energy migration from excited styrene unit to methyl methacrylate unit has been proposed by estimating the probability of chain sessions of methyl methacrylate component changing the mol fraction of styrene in the copolymers. The ESR studies of photoirradiated poly(St-co-MMA) supports the conclusion that the energy migration takes place in the copolymer [68].

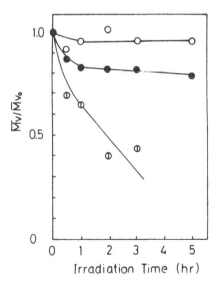

Figure 8 Changes in the viscosity average molecular weight of photoirradiated polystyrene (○), poly(St-co-MMA) (●), and PMMA (◐) films at 30°C in vacuum with irradiation time (From Ref. 67.)

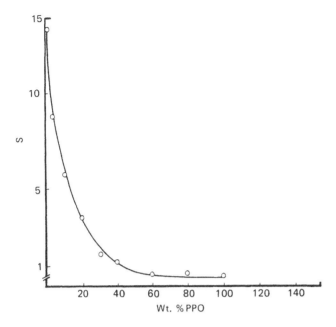

Figure 9 Number of chain scissions per macromolecule, s, for various polystyrene–polyphenylene oxide blends (9.6×10^{-4} ml^{-1} polystyrene monomer segments) in tetrahydrofuran at 298 K. (From Ref. 70.)

Polymer Blends

Gelles and Frank [69] treated the energy migration in miscible polystyrene–poly(vinylmethyl ether) blend as a three-dimensional random walk on a spatially periodic lattice.

Kryszewski et al. [70] examined the photodegradation of polystyrene–poly(2,6-dimethyl phenylene oxide) (PS–PPO) miscible blends. The main chain scission of polystyrene is well suppressed by adding PPO as shown in Figure 9. PS–PPO is an almost completely miscible blend which has recently been applied in various industrial fields such as engineering plastics [71].

CONCLUSIONS

Recent progress in the photodegradation of polystyrene has been reviewed in this chapter. Although polystyrene is widely used in various industrial fields, the photodegradation studies have not been well established compared with those of polyolefines. The studies on photodegradation of polystyrene in the solid phase are very important because this polymer has been and will be used in the solid phase in industrial applications.

Some photoreactions of polystyrene have been reviewed by Weir [14], who discussed the progress and unsolved problems in the photodegradation of polystyrene.

Studies on photodegradation of polystyrene are now accumulating by many researchers and deeper understanding of this problem is expected.

REFERENCES

1. G. Loux and G. Weill, *J. Chem. Phys.*, *61*: 484 (1964).
2. W. Schnabel, in *Polymer Degradation*, Macmillan, New York, p. 98 (1981).
3. W. Klopffer, *Eur. Polym. J.*, *11*: 203 (1975).
4. G. A. George and D. K. C. Hodgeman, *Eur. Polym. J.*, *13*: 63 (1977).
5. G. Geuskens and C. David, *Pure Appl. Chem.*, *51*: 233 (1979).
6. N. Grassie and N. A. Weir, *J. Appl. Polym. Sci.*, *9*: 999 (1965).
7. J. F. Rabek and B. Ranby, *J. Polym. Sci. Polym. Chem. Ed.*, *12*: 273 (1974).
8. P. J. Burchill and G. A. George, *J. Polym. Sci. Polym. Lett. Ed.*, *12*: 497 (1974).
9. G. A. George and D. K. C. Hodgeman, *J. Polym. Sci.*, *C55*: 195 (1976).
10. G. A. George and D. K. C. Hodgeman, *Eur. Polym. J.*, *13*: 63 (1977).
11. N. A. Weir, *Dev. Polym. Degrad.*, *4*: 143 (1982).
12. A. Torikai, T. Takeuchi, and K. Fueki, *Polym. Photochem.*, *3*: 307 (1983).
13. V. Kovacevic, *Polym. Photochem.*, *4*: 459 (1984).
14. N. A. Weir, in *New Trends in the Photochemistry of Polymers* (N. S. Allen and J. F. Rabek, eds.), Elsevier Applied Science Publishers, London, p. 161 (1985).
15. A. Torikai, Y. Sekigawa, and K. Fueki, unpublished results.
16. G. M. Kline, ed., *Analytical Chemistry of Polymers*, Interscience, New York (1962).
17. B. Ranby and J. F. Rabek, *ESR Spectroscopy in Polymer Research*, Springer, Berlin (1977).
18. C. David, W. Piret, M. Sekiguchi, and G. Geuskens, *Makromol. Chem.*, *179*: 181 (1978).
19. H. W. Siesler, *J. Molec. Struct.*, *59*: 15 (1980).
20. M. A. Winnik, ed., *Photophysical and Photochemical Tools in Polymer Science, Conformation, Dynamic, Morphorogy* (NATO ASI Series, Vol. C182), Reidel, Dordrecht (1986).
21. M. T. Vala, Jr., J. Haebig, and S. Rice, *J. Chem. Phys.*, *43*: 886 (1965).
22. C. David, N. Putman de Lavarielle, and G. Geuskens, *Eur. Polym. J.*, *19*: 617 (1974).
23. K. P. Ghiggino, R. D. Wright, and D. Phillips, *J. Polym. Sci. Polym. Phys. Ed.*, *16*: 1499 (1978).
24. J. R. MacCallum, *Eur. Polym. J.*, *17*: 797 (1981).
25. S. Tagawa and W. Schnabel, *Makromol. Chem. Rap. Commun.*, *1*: 345 (1980).
26. S. Tagawa, W. Schnabel, M. Washio, and Y. Tabata, *Rad. Phys. Chem.*, *18*: 1087 (1981).
27. B. Ranby and J. F. Rabek, *Photodegradation, Photooxidation and Photostabilization of Polymers*, Wiley-Interscience, New York (1975).
28. J. Wilske and H. Heusinger, *J. Polym. Sci. A1*, *7*: 995 (1965).
29. P. Bortolus, F. Minto, S. Loa, M. Gleria, and G. Beggiato, *Polym. Photochem.*, *4*: 45 (1984).
30. S. Blonski and K. Sienicki, *Polym. Commun.*, *27*: 130 (1986).
31. G. Geuskens, D. Baeyens-Volant, G. Delaunois, Q. Lu-Vinh, W. Piret, and C. David, *Eur. Polym. J.*, *14*: 291 (1978).
32. C. David, D. Baeyens-Volant, G. Delanois, Q. Lu-Vinh, W. Piret, and G. Geuskens, *Eur. Polym. J.*, *14*: 501 (1978).
33. J. F. Mckeller and N. S. Allen, *Photochemistry of Man Made Polymers*, Applied Science Publishers, London, p. 103 (1979).
34. N. Grassie and N. A. Weir, *J. Appl. Polym. Sci.*, *9*: 975 (1965).
35. I. K. Chernova, V. P. Golikov, S. S. Leschenko, and V. L. Kaplov, *Khim. Visokikh Energ.*, *8*: 265 (1975).
36. P. J. Burchill and G. A. George, *J. Polym. Sci. Polym. Lett. Ed.*, *12*: 497 (1974).
37. A. Torikai, H. Tsuruta, and K. Fueki, *Polym. Photochem.*, *2*: 227 (1982).
38. A. Torikai, S. Asada, and K. Fueki, *Polym. Photochem.*, *7*: 1 (1986).
39. A. Torikai, K. Suzuki, and K. Fueki, *Polym. Photochem.*, *3*: 379 (1983).
40. A. Torikai, S. Kato, and K. Fueki, *Polym. Degrad. Stab.*, *17*: 21 (1987).
41. I. Kuzina, *Visokomol. Soedin*, *B26*: 788 (1984).

42. T. Hill, *J. Macromol. Sci. Chem.*, *A22*: 403 (1985).

43. D. Adams and D. Braun, *J. Polym. Sci. Polym. Lett. Ed.*, *18*: 629 (1980).

44. G. Geuskens and Q. Lu-Vinh, *Eur. Polym. J.*, *18*: 307 (1982).

45. T. R. Price and R. B. Fox, *J. Polym. Sci. Polym. Lett. Ed.*, *4*: 771 (1966).

46. J. B. Lawrence and N. A. Weir, *J. Polym. Sci. Polym. Chem. Ed.*, *11*: 105 (1973).

47. T. Ishii, T. Hanada, and S. Matsunaga, *Macromolecules*, *11*: 40 (1978).

48. H. Ikada, M. Kimura, and M. Ashida, *Kobunshi Ronbunshu*, *43*: 405 (1986).

49. J. Kowal and M. Nowakowska, *Polymer*, *23*: 281 (1982).

50. L. Wolinski, *Makromol. Chem.*, *183*: 3089 (1982).

51. L. Wolinski, *Makromol. Chem.*, *186*: 1001 (1985).

52. H. Yamaoka, T. Ikeda, and S. Okamura, *Macromolecules*, *10*: 717 (1977).

53. T. Ikeda, H. Yamaoka, T. Matsunaga, and S. Okamura, *J. Phys. Chem.*, *82*: 3329 (1978).

54. R. Ssatre and F. Gonzalez, *Polym. Photochem.*, *1*: 151 (1981).

55. K. Horie, K. Morishita, and I. Mita, *Polym. Prepr. Jpn.*, *31*(7): 1649 (1982).

56. I. Mita, K. Horie, A. Okamoto, and T. Hisano, *Polym. Prepr. Jpn.*, *33*(2): 287 (1984).

57. I. Mita, T. Takagi, K. Horie, and Y. Sindo, *Macromolecules*, *17*: 2256 (1984).

58. J. R. MacCallum and D. A. Ramsay, *Eur. Polym. J.*, *13*: 945 (1977).

59. F. Konoma, Song Wu, and Z. Osawa, *Polym. Prepr. Jpn.*, *35*(2): 376 (1986).

60. A. Torikai, A. Takeuchi, and K. Fueki, *Polym. Degrad. Stab.*, *14*: 367 (1986).

61. I. Mita, private communication.

62. C. David, M. Lempereur, and G. Geuskens, *Eur. Polym. J.*, *9*: 1315 (1973).

63. R. Reid and J. Sautar, *J. Polym. Sci. Polym. Phys. Ed.*, *16*: 231 (1978).

64. K. Sienicki and C. Bojarski, *Macromolecules*, *18*: 2714 (1985).

65. S. N. Semerak and C. W. Frank, *Can. J. Chem.*, *63*: 1328 (1985).

66. R. B. Fox, T. R. Price, R. R. Cozzens, and H. Echols, *Macromolecules*, *7*: 937 (1974).

67. A. Torikai, A. Hozumi, and K. Fueki, *Polym. Degrad. Stab.*, *16*: 13 (1986).

68. A. Torikai and K. Fueki, *Polym. Degrad. Stab.*, *6*: 81 (1984).

69. R. Gelles and C. W. Frank, *Macromolecules*, *15*: 747 (1984).

70. M. Kryszewski, B. Wandelt, D. J. S. Birch, R. E. Imhof, A. M. North, and R. A. Pethrick, *Polymer*, *23*: 294 (1982).

71. U.S. Pat. 338,345 (1968); General Electric.

19

Liquid Crystal Polymers and Their Applications

Tai-Shung Chung, Gordon W. Calundann,
and Anthony J. East
Celanese Research Company
Summit, New Jersey

INTRODUCTION

During the last two decades a new class of polymeric materials has been developed which received a great deal of attention from both industry and academia. These materials are called liquid crystal polymers (LCPs). In the liquid state, either as a solution (lyotropic) or a melt (thermotropic), they lie between the boundaries of solid crystals and isotropic liquids. This polymeric state is also referred to as a mesomorphic structure, or a mesophase, a combined term adopted from the Greek language (mesos = intermediate; morphe = form). This state does not meet all the criteria of a true solid or a true liquid, but it has characteristics similar to both a solid and a liquid. For instance, the anisotropic optical properties of LC polymeric fluids are like those of crystalline solids, but their molecules are free to move as in liquids. The main difference between these polymers and the conventional liquid crystals used in electrical display devices is the molecular weight. LCPs have a much higher molecular weight. This difference provides unique features. Especially important is the existence of a glass transition point which permits freezing of an LC phase for use over a wide range of temperatures.

The earliest recognition of liquid crystalline behavior, observed on a thermotropic cholesteric small molecule, is attributed to Reinitzer in 1888. In the 1940s and 1950s polymeric liquid crystallinity was recognized. The systems studied were mostly natural and synthetic lyotropic biopolymers such as tobacco mosaic virus, collagen, and poly(γ-benzyl-1-glutamate). Paul Flory proposed in 1956, in essence, that a solution of a rigid rodlike synthetic polymer above a critical solute concentration would behave as a lyotropic system [1]. In other words, the LC state is a unique condition in which long-range molecular orientational order persists in the absence of various types of short-range translational order. This class of ordering is a direct result of specific types of intermolecular interactions and manifests itself in unique macroscopic physical properties. This theoretical prediction was well demonstrated when, in 1965, Stephanie Kwolek of the DuPont Company discovered that certain rigid wholly aromatic polyamides gave anisotropic solutions in alkylamide and alkylurea solvents, an observation that led ultimately to the development of Kevlar aramid fiber [2]. This was a milestone as it raised the question in the fibers industry: Could a similar phenomenon occur in the melt? Were thermotropic polymers possible? They were possible, and indeed the number reported to date by academic and industrial researchers is nearly beyond count. In addition, as it later turned out, aromatic thermotropic polymers were found to offer a great many more useful properties than just their now well-known tensile capabilities.

A 1972 patent [3] to Carborundum showed data comparing the melting point–composition curves of two wholly aromatic copolyesters: (1) the copolymer of p-hydroxybenzoic acid (HBA), terephthalic acid (TA), and hydroquinone (HQ) and (2) the copolyester made from HBA, TA, and 4,4'-biphenol (BP), the Ekkcel I-2000 polymer (Figure 1). The minimum melting temperature of the former copolymer is about 420°C, while the biphenol-based injection-moldable copolyester has a minimum melting temperature of approximately 395°C. These polymers are injection moldable, albeit at temperatures in the vicinity of 400°C, a temperature not compatible with common melt-spinning equipment. The rate of thermal degradation of such a polyester at 400°C makes stable fiber production particularly difficult. Moreover, most conventional injection-molding equipment requires modification to operate at the high temperatures needed to ensure reasonable processing of this polymer. In 1976, in another patent to Carborundum [4], an HBA/TA/BP copolymer system, modified and "softened" with isophthalic acid (IA), was reported to be melt spinnable and to produce fiber with as-spun properties of 5 g/d tenacity and 390 g/d modulus. These researchers did not specifically report thermotropic behavior, but this does manifest itself for at least some compositions of this terpolyester. (Part of this patent property was subsequently sold to Dart-Kraft and has become the basic technology supporting Dartco Manufacturing's engineering resin, Xydar.) Furthermore, it is now known that unpublished works at ICI Fibers, Harrogate, England, noted thermotropic behavior in aromatic copolymers as early as 1963.

Researchers at Eastman Kodak reported on and produced the first well-characterized thermotropic aromatic–aliphatic copolyesters by the reaction of p-acetoxybenzoic acid (ABA) and poly(ethylene terephthalate) (PET) [5, 6], and later Calundann and his co-workers at Celanese invented variously tractable wholly aromatic thermotropic polyesters

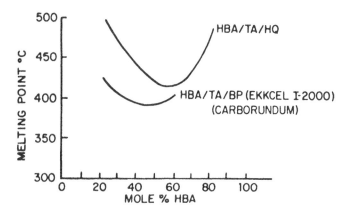

Figure 1 Effect of composition on polymer melting point, Ekkcel I-2000. (Data from Ref. 10.)

using 2,6-functionally disubstituted naphthalene monomers [7–11]. Numerous studies on these two companies' products have been reported in the literature.

The existing lyotropic Kevlar and thermotropic polyester fibers have very attractive mechanical properties. Their tensile moduli are in the range of 9–17 Mpsi, depending on the chemical compositions and process conditions, compared with values of 0.9–1.8 Mpsi for conventional nylon and PET. In an injection-molded part, the thermotropic liquid crystal copolyesters have very high tensile strength and stiffness in the flow or machine direction of the polymer melt. For example, the tensile modulus of an injection-molded thermotropic polyester formed from hydroxybenzoic acid and hydroxynaphthoic acid is at least three times that of conventional polyester (2.5 versus 0.5 Mpsi), and the flexural strength of the former is twice that of the latter. The coefficient of thermal expansion of these polymers is significantly lower than that of conventional isotropic polymers, particularly in the machine direction. Therefore, the shrinkage of molded parts after molding is negligible. In addition, due to its low viscosity and high yield melt stress, the molding cycle is much faster for conventional polymers. Such impressive properties are due to the nearly rigid and linear mesogenic units in the chain backbones of these two materials. These polymers show spontaneous anisotropy and tend to remain oriented in domains, even in quiescent conditions, for long periods.

LC polymer phases also possess attractive optical properties. For example, smectic side chain LC polymers are able to form optically clear monodomain samples, free of defects. By selectively heating a region of the material with a laser beam, the material may be locally disordered and becomes a turbid polydomain. The material is therefore useful for optical storage of information, which is another example of the unique applications for which LC polymers are potentially suited.

CLASSIFICATION OF LIQUID CRYSTAL POLYMERS

In general, there are two types of chemical structure (Figure 2) for LC polymers [12]: main-chain LCPs (MCLCPs) and side-chain LCPs (SCLCPs). In the MCLCPs the mesogenic groups form part of the backbone of the molecular chains, whereas in the SCLCPs the mesogenic units are linked as pendant side chains to a polymer backbone. Both can form thermotropic or lyotropic LC states. Examples of both structures are illustrated in

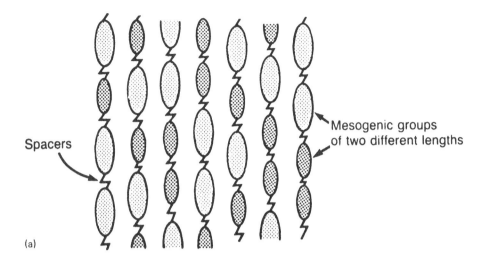

Spacers

Mesogenic groups
of two different lengths

(a)

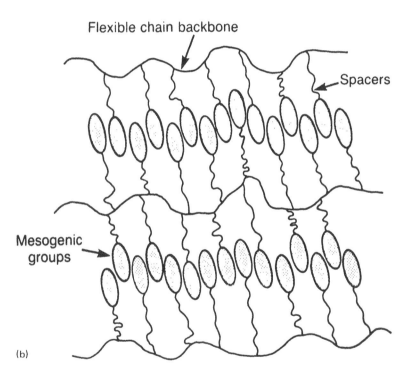

Flexible chain backbone

Spacers

Mesogenic
groups

(b)

Figure 2 (a) The nematic structure for MCLCPs and (b) the smectic C structure for SCLCPs. (From Ref. 12.)

Figure 3 [12]. Flexible spacers (e.g., alkyl chains) between mesogenic groups in MCLCPs or between the backbone and mesogenic units in SCLCPs are sometimes necessary to form a particular LC phase. Mesogenic units are typically rigid structures with delocalized electrons. The chemical constituents of LCP mesogens are usually two or more linear-substituted cyclic units which may be linked by a short, rigid central bridging group. The

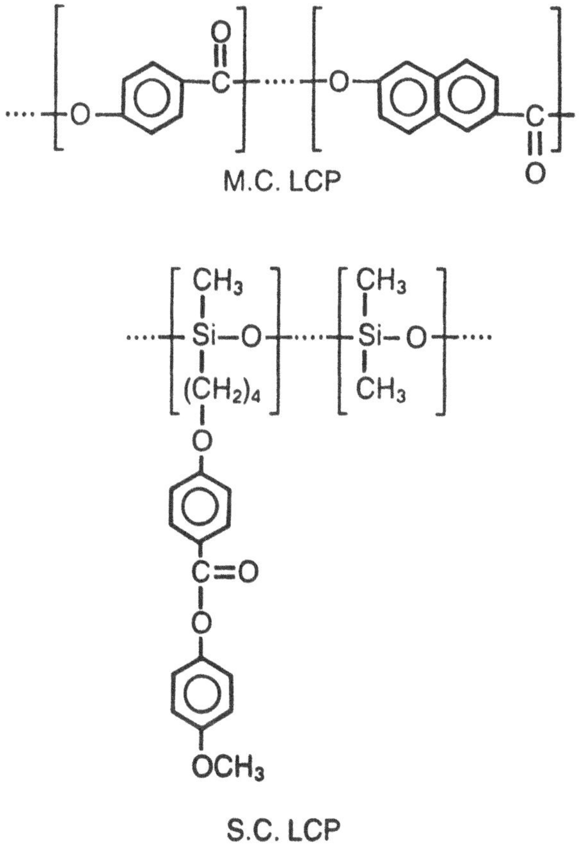

M.C. LCP

S.C. LCP

Figure 3 Typical main-chain and side-chain LCPs. (From Ref. 12.)

spacer is typically an alkyl chain or an aliphatic ether segment, although any flexible unit is satisfactory. The multiple-bond character in both the central linkage group and the spacer is of great importance in maintaining polymer rigidity and linearity. Although there is no formal double bond in the ester or amide groups, the electron-rich aryl or phenyl group slightly shifts its bonding electrons to the carbonyl group, imparting to the ester or amide linkages some double-bond character. It therefore exhibits a more rigid structure than might be expected. Furthermore, the amide group is coplanar with adjacent rings, which contributes to this effect.

POLYMER SYNTHESIS

Lyotropic LCPs

For a typical LC polyamide, such as Kevlar, polymerization proceeds by the reaction of *p*-phenylene diamine and terephthloyl chloride at low temperature in a suitable solvent (Figure 4) [10]. The degree of polymerization is increased by the presence of tertiary amine salts. The polyamide is separated from the solvent and redissolved in a sulfuric acid to form a lyotropic "dope" for further dry-jet wet spinning. The polycondensation reac-

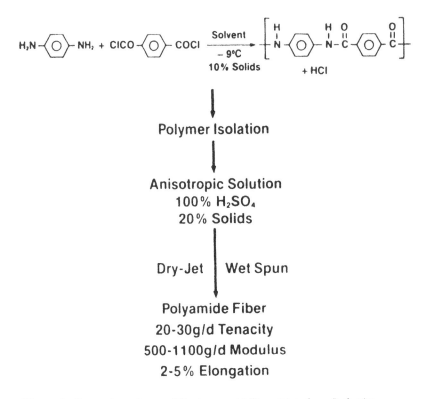

Figure 4 Processing scheme of Kevlar aramid fiber. (Data from Ref. 10.)

tion is affected by monomer and solvent purity, reaction temperature, concentration, solubility, and mixing.

Thermotropic Main-Chain LCPs

Most industrial main-chain aromatic LC polyesters are formed via a polycondensation (mostly transesterification) reaction. The reaction equations for two well-known thermotropic copolyesters are illustrated in Figure 5 [11–13]. There are two main routes. One uses a transesterification reaction between the phenyl esters of the diacids with aryl diols evolving phenol. The other uses an acidolysis reaction between the free diacids and the acetylated aryl diols at relatively high temperatures. In general, anhydrous sodium or potassium acetate is added in the reactor as a catalyst. Other catalysts are tin salts, tertiary amines, phosphines, and in some cases the polymerization may be carried out without any catalyst. Catalysts affect the color and the thermal stability of the final products, but their exact role is still unknown.

To reduce oxidation, the polymerization is conducted in an inert atmosphere. Since in the later stage the condensation reaction is a diffusion-controlled process, a noncorrosive metal stirrer is needed to improve the mixing and to accelerate the evolution of volatile by-products such as acetic acid. Reaction temperatures are generally set 50–80°C above the highest melting point of the monomers. When a low melt viscosity prepolymer is ob-

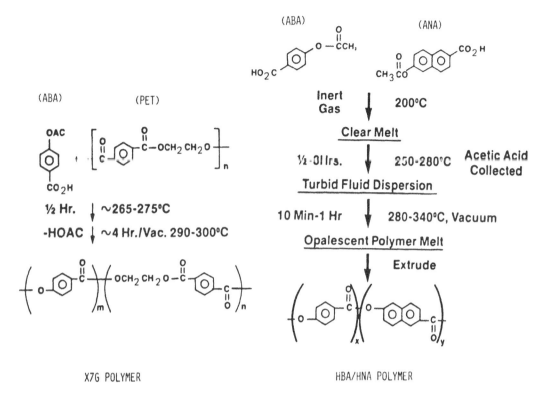

Figure 5 Polymerization of two main-chain LCPs. (Data from Refs. 10–13.)

tained, the melt rapidly becomes turbid since the fluctuations of the growing mesogenic domain orientations are in the range of the wave length of visible light. A vacuum is then applied to the reactor to further remove volatiles from the melt and to increase the molecular weight of the polymer very much, as in standard PET manufacture.

To obtain a still higher molecular weight, solid-state polymerization of ground polymer has been demonstrated under a reduced pressure or in a nitrogen environment. A multistep change in the temperature profile is the key to facilitate the molecular weight enhancement, but the final temperature is about 10–30°C below the original melting point. LC polyester-amides can be prepared by this method. In general, the mesogenic ester-amides show increased thermal stability compared to esters in similar compounds [9].

Side-Chain LCPs

Two approaches are frequently employed to synthesize SCLCPs. Figure 6 illustrates typical examples of the reaction scheme [13–15]. The first example describes the monomer preparation before polymerization. By stepwise synthesis, a reactive monomer is prepared. This monomer has an anisodiametric group or rigid, rodlike unit attached to a polymerizable group through a flexible spacer. The reactive (vinyl) group is capable of addition polymerization. Acrylates and methacrylates are commonly used as reactive vinyl groups. The second example is a grafting of vinyl-substituted mesogenic monomers

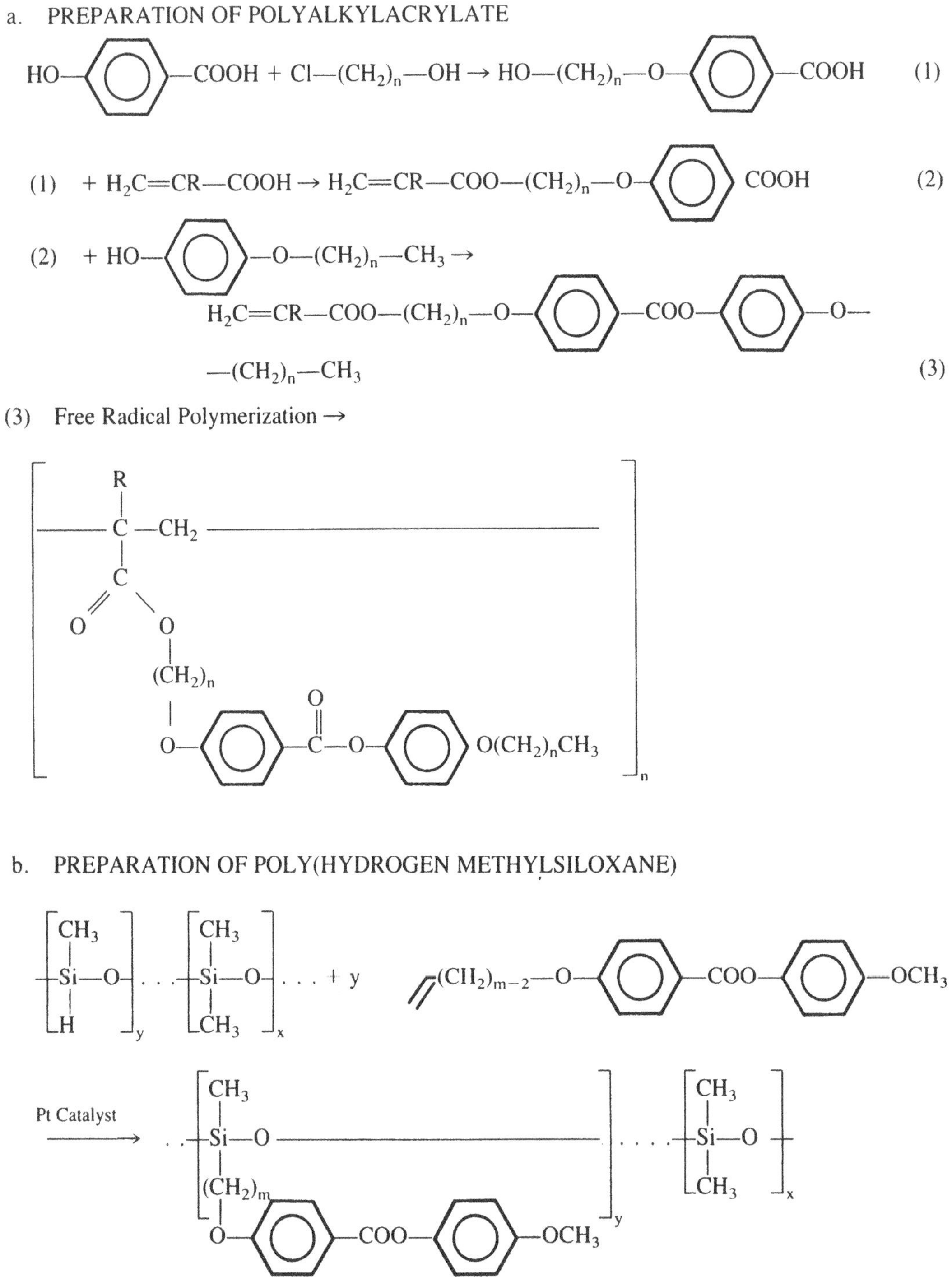
a. PREPARATION OF POLYALKYLACRYLATE
HO—
—COOH + Cl—(CH2)n—OH → HO—(CH2)n—O—
—COOH
(1)
(1) + H2C=CR—COOH → H2C=CR—COO—(CH2)n—O—
COOH
(2)
(2) + HO—
—O—(CH2)n—CH3 →
H2C=CR—COO—(CH2)n—O—
—COO—
—O—
—(CH2)n—CH3
(3)
(3) Free Radical Polymerization →
R
C
CH2
C
O
O
(CH2)n
O
O
C
O
O(CH2)nCH3
n
b. PREPARATION OF POLY(HYDROGEN METHYLSILOXANE)
CH3
Si
O
H
y
CH3
Si
O
CH3
x
+ y
(CH2)m−2—O—
—COO—
—OCH3
Pt Catalyst
CH3
Si
O
(CH2)m
O—
—COO—
—OCH3
y
CH3
Si
O
CH3
x

Figure 6 Polymerization of (a) polyalkylacrylate and (b) poly(hydrogen methylsiloxane). (Data from Refs. 12–14.)

onto reactive poly(methylhydrosiloxane), whose Si–H groups add onto the vinyl unit. In practice, the first approach uses either radical or ionic initiations of vinyl polymerization in solution. The polymer is then purified through reprecipitation. Reaction temperatures are relatively low compared to those for MCLCPs, and polymerization below 0°C has been reported [16]. As with MCLCPs, monomer preparation and purification before polymerization are very critical. Grafted polysiloxanes have also been polymerized in solvents and were reprecipitated and dried [15, 17]. The catalysts utilized in the grafting reaction are platinum and hexachloroplatinic acid.

It is a fact that monomers for both MC and SCLCPs do *not* have to exhibit LC states themselves, but should possess some of the potential characteristics of anisotropy and rigidity. On the other hand, LC monomers are not guaranteed to form LCPs. At this moment, the literature has indicated that most MCLCPs have been prepared from non-mesomorphic monomers, whereas most SCLCPs have been synthesized from meso-morphic monomers. In addition, experimental data have demonstrated that nematic mono-mers may produce smectic-phase polymers. Polymerization of SCLCPs under electrical and magnetic fields has been conducted. The fields changed the polymerization rate and the properties of final products [18, 19]. Perplies et al. [19] observed that the polymeriza-tion rate increased with an increase in the magnetic field due to a decrease in the chain termination step. Well-oriented products of SCLCPs have been prepared in this way. The crystallization kinetics of LC structural transformations obey the Avrami equation [20].

BASIC STRUCTURES

LC phases are divided into three broad categories: nematic, cholesteric, and smectic. Briefly, the nematic phase (Figure 2a) consists of parallel molecules with no translational order. Unoriented nematics show correlated orientation over a micrometer-length scale. Nematics may be oriented over large areas; however, the orientational order parameter usually does not exceed 0.65 due to molecular fluctuations. The cholesteric phase is a twisted nematic in which the order parameter (pointed along a molecular axis) exhibits a helical twist with a pitch on the order of 0.4 μm to several micrometers. The smectic phase (Figure 2b) is a layered structure consisting of sheets of molecules with orienta-tional order and translational order between the smectic layers. Many different types of smectics exist displaying varying degrees of order both within an individual smectic layer and between layers. Unoriented smectics show a domain texture in which the layers are aligned within each domain on a micrometer or submicrometer scale. Smectics obtained from oriented nematics are able to form large monodomains in which very high orienta-tional order approaching that of a single crystal is achieved (i.e., orientational order parameters exceeding 0.95) [12].

Similarly to conventional polymers, LCPs have phase transitions. Polymeric crystals may form either smectic or nematic structures upon heating, but they may become a completely nematic phase once the temperature is further increased. If the temperature is extremely high, an isotropic melt is formed. However, because LCPs decompose at high temperature, the isotropic transition may not take place before the decomposition. As a result, it is sometimes very difficult to determine the exact structure of potential LCPs. For example, solid poly(p-hydroxybenzoic acid) (HBA) has shown microfibril and slab-like structures depending on the polymerization conditions [21]. However, due to thermal degradation at high temperatures, its exact melting phase is still debatable.

STRUCTURAL INVESTIGATION OF LIQUID CRYSTAL POLYMERS

Numerous techniques (e.g., microscopy, x-ray diffraction, NMR spectroscopy) have been used to determine the structure of LCPs [13]. Frequently a hot-stage polarizing microscope is required for the examination of a thin LC polymeric film at various temperatures. The sample is held between a slide and a cover slip. The sample appears isotropic if the molecules align perpendicular to the surface (homeotropic), while it is optically birefringent if the molecular alignment is parallel to the surface. Once the temperature of the hot stage reaches phase transition, the cover slip can be slightly depressed to investigate the birefringency. Hartshorne [22] and Demus and Richter [23] summarized test methods and the means of identification. In short, nematic fluids show threaded schlieren and satinlike textures; cholesterics exhibit oily streaks and striated textures; and smectics have a focal–conic or fan-shaped appearance.

Both low molecular weight LCPs and high molecular weight LCPs have similar colorful textures under the polarizing microscope. These colors are produced by the interference of the transmitted polarized light with the regular molecular structure. On a micrometer-length scale, the structure of LCPs are therefore the same as those found for small-molecule LCPs; namely, all of the major phase types found for LCPs (nematic, cholesteric, and smectic) are also found for LCPs.

When employing x-ray diffraction techniques to characterize molecular structure of LCPs, researchers found that there are two types of diffusion halos for a nematic phase [24–28]. One is a weak inner ring; the other, a strong outer ring. The former depends on the length of a repeating unit along the chain axis for MCLCPs or the length of the side group in SCLCPs, while the latter is a measure of lateral packing and the distance between neighboring molecules. On a molecular scale, LCPs present some novel features. For example, MCLCPs often display a nematic structure in which long polymer molecules are packed in a parallel array (see Figure 2a). The individual chains are arranged on a two-dimensional lattice perpendicular to the molecular axis (e.g., a hexagonal lattice with unit cell dimensions such that the interchain distance is approximately 4.5 Å). However, there is little or no axial registration between neighboring polymer molecules. (This is as if one held a bundle of pencils but did not align the ends.) Thus the nematic polymer molecules are highly parallel but nearly two-dimensionally crystalline.

The molecular architecture of SCLCPs is not as clearly understood. The structural arrangement of side-chain mesogens clearly follows that found in a small molecule LC of the same phase. For example, a smectic C polymer contains sheets of mesogens in which the mesogen long axis is tilted with respect to the sheet and for which there is no regular crystalline arrangement of mesogens within the layer or between layers (see Figure 2b). Thus the LCP is envisaged as forming a comblike structure with neighboring molecules. The nature and degree of ordering of the chain axis to which the mesogenic units are attached is not presently understood.

The use of NMR spectroscopy to characterize thermotropic LCPs was initiated by researchers at Eastman Kodak and at Celanese. Blumstein and his group [29] and Samulski [30] further extended it using proton and deuterium NMR to measure the nematic order parameter S and the fraction of nematic (F_n) phase. The order parameter was calculated from the dipolar or quadrupolar splittings of the NMR spectrum when cooling from the isotropic stage. The fraction of the mesogenic core aligned was deduced from the NMR free-induction-decay curve. Figure 7 gives a typical example of their data [29] on cooling from the isotropic state. The high values of S in the nematic melt, and even at the

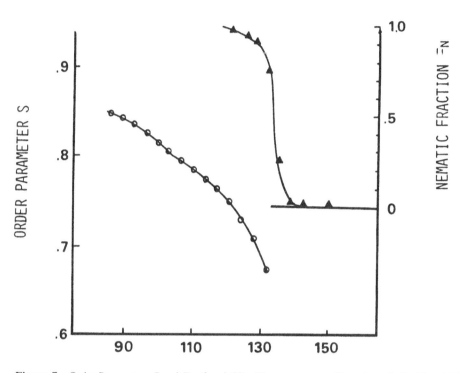

Figure 7 Order Parameters S and F_n of an LCP with temperature. (Data from Refs. 13 and 29.)

transition temperature, result from the inherent characteristics of LCPs and indicates that the C_{10}-flexible spacer is surprisingly in a largely extended configuration. F_n is more temperature sensitive than is S; its value drops remarkably at the transition temperature.

However, the x-ray and NMR approaches generally are time-consuming and expensive. There are some simple qualitative identification methods:

1. Most LCPs have a fibrous, almost woodlike bulk texture.
2. The fracture surface of an extruded fiber or strand shows the highly fibrillar morphology.
3. Both extruded and single-gate injection-molded articles are highly anisotropic.
4. The color of molten LCPs is almost opaque or pearlescent, whereas most plastics are either transparent or translucent.
5. The viscosity of LCPs is often much lower than that of conventional polymers.

If an unknown polymer shows similar characteristics, it may be worthwhile to study it further.

CRITERIA FOR FORMING A LIQUID CRYSTAL POLYMER

Up to the present, no firm statements regarding the criteria for forming an LCP have been established. However, after surveying the existing LCPs, one could conclude that LCPs have at least the following common characteristics [1–13, 30–38]:

1. Molecular shape has a large length (or diameter) to width (or thickness) ratio (i.e., aspect ratio is high).
2. Polarizability *along* the rigid chain axis is substantially larger than that in the transverse direction.
3. Good molecular parallelism of rigid units prevails in the structure.

To meet these requirements, an LCP probably will possess a rigid molecular structure similar to a combination of successive *p*-arylene groups. Each *p*-arylene group is an almost flat molecule (a large length-to-thickness ratio) and has a highly polarized electron cloud along the main chain. This results in the aryl ring being symmetrical and rigid along the main axis, and the overall molecules fit well into a crystal lattice. A *p*-arylene type of LCP

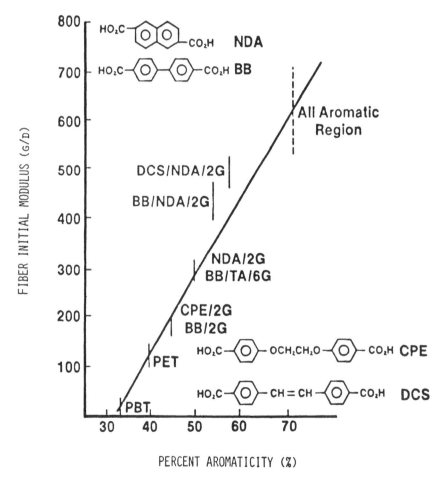

Figure 8 Effect of polyester aromaticity on fiber tensile modulus. (Data from Ref. 10.)

can easily retain a fair parallelism by means of chain connections with other *p*-arylene groups, rigid double-bond spacers, or amide and ester linkages. However, a polymer filled with *p*-arylene groups is not necessarily an LCP; the electronic arrangement and stereochemistry within the polymer are also important.

The parallelism criterion implies that spacer units should have an adequate length. For an SCLCP, the spacer reduces the influence of the molecular movements at the flexible main-chain backbone on the mesogenic unit, whereas for an MCLCP it restrains the chain flexibility and entanglement.

Chung [13] proposed that the foregoing criteria might mathematically be described using the following parameters. (1) the Mark–Houwink exponent, (2) the persistence length a', chain length L, Kuhn segment A, and radius of gyration $\langle R*R \rangle$; (3) the aspect ratio (asymmetry) of the domain; and (4) chain flexibility f. However, Calundann and Jaffe [10] discovered a useful empirical parameter for characterization of LCPs. They related the fiber tensile modulus of aliphatic–aromatic polyesters with an empirical index, called the degree of aromaticity of polymers. This index is defined as the ratio of the number of sp 2 hybridized carbons to total number of atoms in the repeat unit. Figure 8 illustrates their results and indicates that "wholly" aromatic polyester would be required to obtain fiber moduli greater than 600–700 g/d.

THE EFFECT OF POLYMER COMPOSITION ON LIQUID CRYSTAL POLYMER BEHAVIOR

Although a highly rigid rodlike MCLCP offers attractive properties, its high melting point causes plastics engineers difficulties. For example, the current Dartco's Xydar LCP (polyester from TA, HBA, and *p,p'*-biphenol) has a melting point of 793°F (423°C). The injection molders have to install high-temperature ceramic heaters to their extruder barrels and also heat molds to 465–535°F in order to process the resin [39]. This disadvantage may offset the unique properties which this LCP has to offer. On the other hand, reducing the influence of the chain backbone of an SCLCP on mesogenic units is essential to form the LC order and improve its thermal stability.

Figure 9 summarizes the molecular architecture used to promote polyester melt anisotropy at reasonable temperatures. The Celanese, DuPont, and Eastman Kodak laboratories have outlined most of the key approaches to lower polymer melting point via crystalline order disruption. This is accomplished while maintaining sufficient molecular symmetry to preserve the melt anisotropy inherent in linear aromatic polyesters; that is, polyesters derived typically from symmetric monomers such as *p*-hydroxybenzoic acid, terephthalic acid, hydroquinone, 4,4'-biphenol, and the like. Pioneering work at Eastman Kodak has already been reviewed; the approach used there involved introduction of aromatic comonomers into aliphatic–aromatic polyesters such as PET. The increased symmetric ring content imparts melt anisotropy to the copolyester but, as mentioned, the aliphatic–aromatic structure does not give the mechanical property levels achieved for the wholly aromatic polyesters.

Recent studies by academic scholars have shown that three accompanying effects might occur when aliphatic units are added into an MC or SCLCP [13, 33–35]: (1) a decrease in the polymer transition temperatures; (2) an odd–even zigzag profile for transition temperatures and isotropization entropy; and (3) a change in the micromolecular packing structure. Figure 10 shows the dependence of the melting point T_m and the isotropization temperature T_i of two MCLCPs on the number of methylene units [40]. An

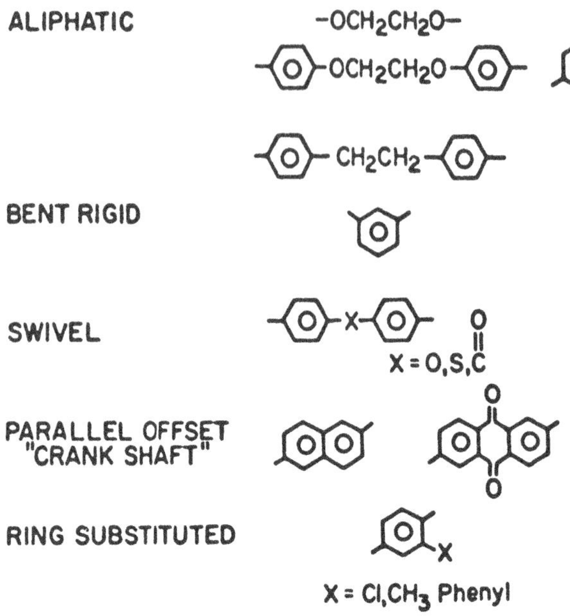

ALIPHATIC

$-OCH_2CH_2O-$

$-OCH_2CH_2O-$ (with phenyl rings)

(thiophene/S ring)

$-CH_2CH_2-$ (with phenyl rings)

BENT RIGID

(bent ring structure)

SWIVEL

$-X-$ (with phenyl rings)

$X = O, S, \overset{O}{\overset{\|}{C}}$

PARALLEL OFFSET "CRANK SHAFT"

(fused ring structures)

RING SUBSTITUTED

X

$X = Cl, CH_3$ Phenyl

Figure 9 Structures providing tractability in aromatic thermotropic polyesters. (Data from Ref. 10.)

increase in the length of spacers results in a decrease in the T_i in a zigzag fashion. The temperature difference (DT) between T_i and T_1 implies the stability of the mesophase. A variety of MC and SCLCPs has also been developed in academic circles using more rigid anisodiametric bridging units [34, 35, 50–52]. The resulting LCPs seem to have better thermal and chemical stability than those from the aliphatic bridging units. The effectiveness of these bridging units in extending the mesogenic behavior of chiral LC polyesters follows the order [50, 51]

$$-CH=C(CH_3)- > -N=N(O)- > -N=N- > -CH=CH-COO-$$

and the stability of the resultant mesogenic phase is in the order

$$-N=N(O)- > -CH=C(CH_3)- > -N=N- > -CH=CH-COO-$$

alternatively

$$-CH=CH- > -N=N(O)- > -CH=N- > \text{none}$$

These orders are in agreement with the trend observed for low molecular weight LCs [48] and can be explained from the stereochemical viewpoint.

Introduction of bent rigid units is an obvious route toward increasing polymer tractability. Morgan [41] at DuPont synthesized poly(1,3-phenyleneisophthalamide; trademark Nomex) using this approach. A comparison of the mechanical properties of Nomex with Kevlar (1,4-phenyleneterephthalamide) shows that the former has a much lower tensile strength (645 MPa) and modulus (17.0 GPa) than those of the latter (3620 MPa and 83 GPa). The elongation at break of Nomex is also increased to 22%, while that of Kevlar 29 is only 4.4% [42]. These results clearly demonstrate that the kink arrangement reduces

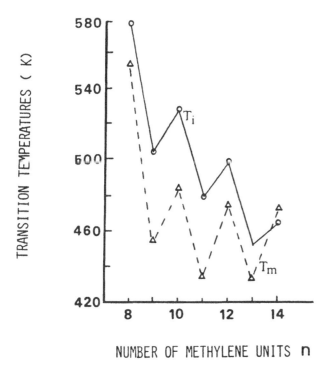

Figure 10 Depending of the melting point T_m and the isotropization temperature T_i on the number of methylene units. (From Refs. 13 and 40.)

the rigidity of the structure. Nomex is not a lyotropic polymer. Therefore, there are problems associated with the meta linkage in many thermotropic systems. Incorporation of low levels of isophthalic acid increases polymer tractability, but increasing isophthaloyl content beyond a certain point tends to offset this gain by reducing polymer melt anisotropy with consequent negative impact on polymer rheology and ultimately on fiber (film, molding, etc.) properties. An example of this effect will be seen later. Resorcinol, even at rather low levels, generally results in amorphous polyesters with isotropic melts.

DuPont workers have described [43] a wide variety of tractabilizing molecules and the thermotropic polyesters derived therefrom. Their research has focused for the most part on ring-substituted monomers such as chloro-, methyl-, or phenyl-substituted hydroquinone and "swivel" or linked-ring molecules, examples of which are 3,4'- or 4,4'- functionally (hydroxy- or carboxy-) disubstituted diphenyl ether, sulfide, or ketone monomers. Their data clearly indicated that these substituents strongly affected the mechanical properties of heat-treated fibers. This implies the growth of LC order and packing of molecular chain during annealing are associated with the type of substituent. Lenz and his co-workers [33,

44–46] have introduced halogen and alkyl groups into one aromatic ring of a mesogenic unit and found that they depress the glass temperature, melting point, and clearing point. Polymers lose LC order if they contain long alkyl ring substituents. Kapuscinska and colleagues found that halogen substitution significantly reduced both the thermal properties and flammability of a Kevlar-type fiber in the following order [47]:

$$Br > Cl > F > H$$

A similar order of effectiveness was found for halogens on the transition temperatures of low molecular weight crystals [48].

The effects of ring substitution on SCLCPs have been investigated by many workers [34, 35]. Ring substitution usually occurs at the para position and results in a nematic phase with short substituents, whereas long substituents form smectic structures. These results are also in agreement with the behavior of low molecular weight liquid crystals.

Use of the parallel offset or "crankshaft" geometry provided by 2,6-disubstituted naphthalene monomers has been the major thrust of the Celanese Corporation's development of easily processible, high-performance, wholly aromatic thermotropic polyesters. Specifically, Celanese has defined families of thermotropic polyesters derived from 2,6-naphthalene dicarboxylic acid (NDA), 2,6-dihydroxynaphthalene (DHN), and 6-hydroxy-2-naphthoic acid (HNA). Indeed, the Vectra LCP engineering resins of Celanese Corporation are based on these monomers. This approach was also adopted by Monsanto. Simoff and Porter [49] have summarized the properties of Monsanto's thermotropic poly(bisphenol-E isophthalate-co-naphthalate). The polymer (BECN) is made from bisphenol-E diacetate, IA, NDA. Without the NDA, this polymer becomes a non-LCP.

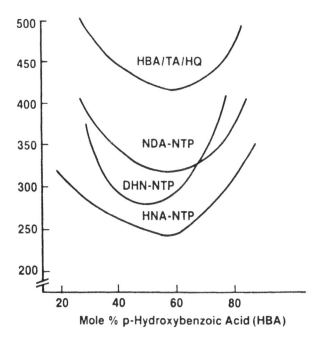

Figure 11 Composition vs. T_m relationships of naphthalene-derived thermotropic polyesters (NTP). (Data from Ref. 10.)

To illustrate the dramatic effect on polymer melting point depression of the 2,6-naphthalene nucleus, Figure 11 compares the melting point–composition curves of NDA-, DHN-, and HNA-based polymers with those of the copolyester, HBA/TA/HQ [7]. Replacement of the TA with 2,6-NDA in the polymerization charge gives a series of copolyesters with a melting point minimum at about 325° (near 60 mol% HBA). Replacing hydroquinone with 2,6-DHN produces a series with a polymer T_m minimum near 280°C (50 mol% HBA). Finally, the composition range of the two-component polyester of *p*-hydroxybenzoic acid and 6-hydroxy-2-naphthoic acid (HNA/HBA) falls largely within the industrially convenient melt temperature zone of 250–310°. A great many compositions of the three groups of copolyesters shown were investigated in detail; all formed anisotropic melts and all were melt spun to high strength, high modulus fibers. X-ray diffraction data indicated that the diffraction patterns of these LCPs can be predicted by employing a completely random sequence model [25, 26]. Due to the nonlinearity of the 2,6-naphthalene linkage, the x-ray diffraction patterns showed that the directions of successive ester-oxygen–ester-oxygen central linkage groups are no longer parallel to the chain axis but off the axis in the range of ±20 degrees.

Replacement of the *p*-oxybenzoyl fraction with *p*-phenylene terephthloyl units in the two-component polyester resulted in a more symmetric, higher melting copolyester series, shown as polymer II, HNA/TA/HQ, in Figure 12. Similarly, all of the compositions of II that were examined were thermotropic and ultimately gave high-strength, high-modulus fibers. As an example of a more complex system, isophthalic acid was interpolymerized with HBA, NDA, and HQ to form a four-component terpolyester, HBA/NDA/IA/HQ [8] with increased tractability over the parent copolymer, HBA/NDA/HQ.

Figure 13, the ternary T_m–composition diagram of this terpolyester, shows contour plots representing groups of terpolyester compositions with the same crystalline melting

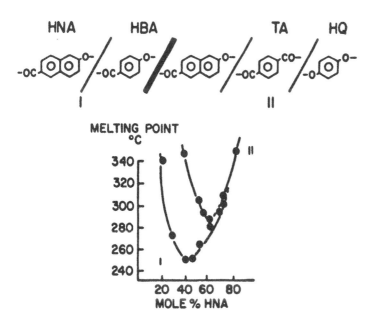

Figure 12 Composition vs. T_m relationship of HNA-based copolyesters. (Data from Ref. 10.)

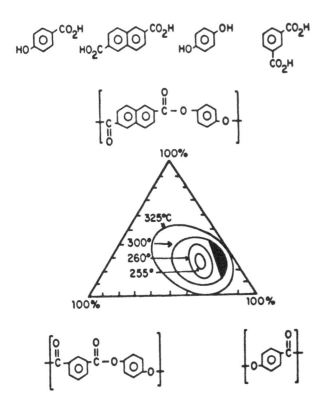

Figure 13 Composition vs. T_m relationship of NDA-based terpolyesters. (Data from Ref. 10.)

point; note the small inner contour with compositions melting near the minimum for this series at about 255°. Melt rheology, spinning behavior, and fiber properties were by no means similar across this diagram. The shaded area encloses those terpolyesters with an optimum processability/fiber property profile; that is, systems melting at 300° or less, with equal to or less than 15 mol% isophthaloyl units, and from 55 to 75 mol% *p*-oxy-benzoyl units. Above 300°, stable fiber spinning becomes increasingly difficult. Lower melting terpolyesters of increased isophthalate content tend to have lower melt anisotropy and thus less favorable rheology and generally become more difficult to process. Further, fibers from terpolyesters with a high IA content show poorer mechanical properties and considerably lower chemical stability.

Work with NDA, DHN, and HNA monomers has also led to the development of thermotropic aromatic poly(ester-amides). The question arose as to whether the introduction of some hydrogen-bonding capability into the chain might result in some useful property improvements. The greater interchain interaction could result in increased fiber fatigue resistance, higher filament shear modulus (and tensile modulus), and improved fiber surface adhesion. Early data are promising, but more research remains to be done before the potential of the ester–amide variants can be fully defined.

The introduction of a mixed linkage does not appear to offer any gain in system tractability; specifically, an increase in the range of compositions melting below 320°. Figure 14 compares the T_m–composition curves of the HNA/TA/HQ copolyester and an analogous poly(ester-amide) in which *p*-acetoxyacetanilide replaced hydroquinone diace-tate in the initial polymerization charge. In fact, amide linkage incorporation resulted in a

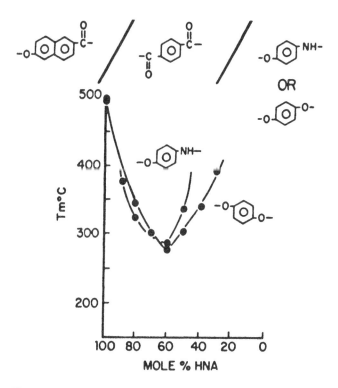

Figure 14 Composition vs. T_m relationship of HNA-based poly(ester amide). (Data from Ref. 10.)

reduced tractable zone, although minimum melting points for both systems were nearly the same at about 275–280°C (60 mol% HBA). The maximum amide linkage tolerable from a tractability standpoint was about 25 mol%. However, both fibers and moldings from such polyesteramides show excellent mechanical properties [9].

Thus factors critical in the molecular design of a commercially viable thermotropic polymer for fiber/resin applications include:

1. Processability—ideally the polymer should process in the range of 250–350°C.
2. Melt anisotropy—a careful balance of molecular symmetry is required.
3. End-use properties—considerable structure/property parameters are needed to achieve optimum results.
4. Minimum monomer cost—obviously important.
5. Ease of preparation—melt-polymerized thermotropic polymers are particularly favored.

All of the polymers described thus far may be made in a conventional melt acidolysis process as previously described.

RHEOLOGICAL BEHAVIOR OF LIQUID CRYSTAL POLYMERS

Liquid crystal polymers generally have a lower viscosity than that of isotropic polymers due to the partially oriented nature of mesophases. Onogi and Asada [53] described the rheological behavior of LCPs using three flow regions, as shown in Figure 15. The shear-

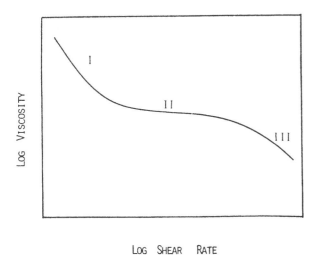

LOG SHEAR RATE

Figure 15 Three regions of flow behavior. (Data from Ref. 53.)

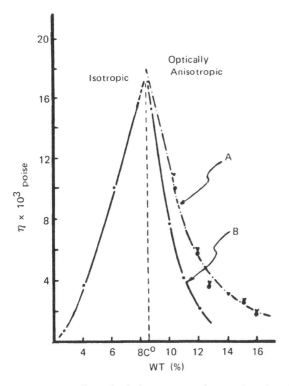

Figure 16 Effect of solution concentration on viscosity of solution. Curve A, IV 3.85 at 50°C in 100% sulfuric acid; curve B, IV 2.73 at 18°C in 100% sulfuric acid. (From Ref. 57.)

thinning viscosity at *low* shear rate in the first region is a unique feature of LCPs and has not been observed for conventional polymers. It occurs due to yield stress, deformation-history dependence, and structure change [53–63]. The second and third regions seem to be similar to the conventional polymers; however, LCPs do show some strange characteristics. According to the results published by previous authors [54–62] and partially summarized by Wissbrun et al. [62, 63], they can be described as follows:

1. The solution viscosity of lyotropic LCPs does not increase monotonically with an increase in polymer concentration. With an increase in the polymer concentration, the solution first transforms from normal isotropic liquid structure to two-phase systems containing an anisotropic phase; it then becomes a single anisotropic phase at higher concentrations. Figure 16 illustrates this phenomenon and shows that the viscosity η of PPTA solution in 100% sulfuric acid increases rapidly until a critical concentration. It then drops rapidly with an increase in concentration [59]. The rise in viscosity with concentration is expected; the reverse trend is unique and is a characteristic of LCPs, and it was forecast by Flory in 1956. IV is the inherent viscosity of polymer.

2. Thermotropic LCPs generally have a high melt elasticity but exhibit a *small* die-swell. Their relaxation time is much longer than that of conventional molten polymers.

3. Both lyotropic solutions and thermotropic melts can be oriented more effectively by elongational stress rather than shear stress.

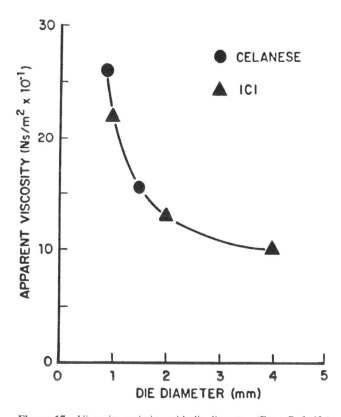

Figure 17 Viscosity variation with die diameter. (From Ref. 63.)

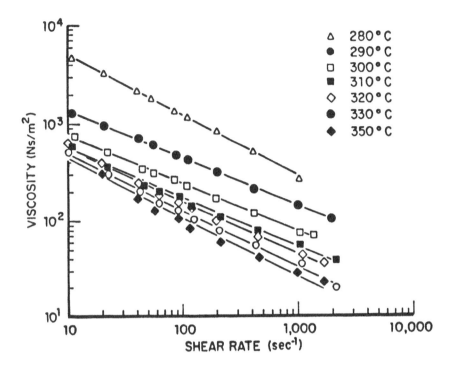

Figure 18 Viscosity variation with shear rate. (From Ref. 63.)

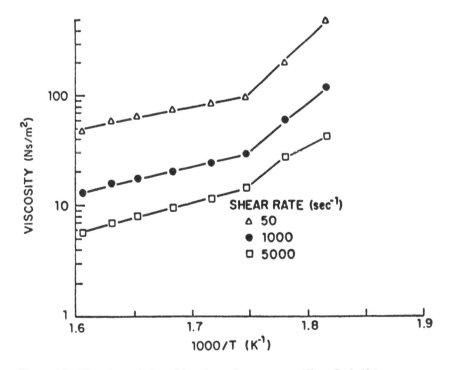

Figure 19 Viscosity variation with reciprocal temperature. (From Ref. 63.)

4. The melt viscosity is dependent on the gap dimensions of the test apparatus. This phenomenon is enhanced at low shear rates. Figure 17 gives an example of this effect. Data were obtained by various capillary viscometers at a shear rate of inverse seconds [63].

5. The flow activation energy of thermotropic LCPs, E_l, is temperature and shear dependent and is much greater at low temperatures than at high temperatures. For example, Figures 18 and 19 show the viscosity of an HBA/HNA copolyester as a function of shear rate and reciprocal temperature, respectively. E_l is independent of shear rate and is about 11 ± 1.5 kcal/mol for temperatures above 300°C, while it has values of 45, 41, and 29 kcal/mol at shear rates of 30, 1000, and 5000 sec^{-1} at temperatures below 300°C [63].

6. The first normal stress difference may be negative for some LCPs. Due to these unique characteristics, the thermal and mechanical history of LCPs also affect the rheological and mechanical properties of molded LCP articles. Samples have to be properly dried and certain test parameters such as the aspect ratio and die diameter of a capillary rheometer have to be defined before meaningful tests can be conducted.

PROCESSES AND APPLICATIONS

Fibers

Jaffe and Jones [64] reported the detailed spinning process for lyotropic fibers such as Kevlar and poly(1,4-benzamide). Because Kevlar (poly[1,4-phenyleneterephthalamide]; PPTA) decomposes before its melting point, solution spinning is the only way to fabricate it into fibers. As depicted in Figure 16, once the PPTA dope is over the "critical concentration," its viscosity drops due to the formation of a nematic lyotropic phase. In general, a dope having 18–23% PPTA has the advantage of low viscosity and of forming fibers with high mechanical properties. However, this dope is solid and has been warmed up to 80°C in order to be spun. Dry-jet wet spinning is employed for producing high-modulus Kevlar fibers.

Researchers at Celanese studied the spinning of PPTA fibers and reached the following conclusions [64]: (1) fiber tensile strength (tenacity) increases with an increase in drawdown ratio [or spin stretch factor (SSF): the ratio of fiber diameter at the spinneret to the final fiber diameter] and inherent viscosity (IV); and (2) fiber modulus increases with increasing drawdown ratio but seems to be independent of IV and the air gap (AG) between the spinneret and the coagulation bath. Figures 20–22 illustrate these results. The mechanical properties of aramid fibers can be further enhanced by means of heat treatment under tension. However, their final properties are strongly dependent on the structure of the as-spun fibers and the spinning techniques. Annealing of dry-jet wet-spun fibers can transform a low-modulus yarn to a higher modulus yarn than that obtained for conventional wet-spun or dry-spun fibers.

Thermotropic LCPs can be converted to fiber using conventional melt-spinning facilities, as illustrated in Figure 23. Owing to the ease of orientating LC domains in an elongational flow field, mechanical properties develop very quickly with increased drawdown ratio, as shown in Figure 24. The freeze point (position of final diameter or final velocity attainment in the threadline) is reached very quickly in the molten threadline. Within 10 cm of the spinneret, all orientational and structure-forming events occurring in the developing fiber are complete. The tensile strength (tenacity) of as-spun fibers is a function of polymer molecular weight as well as orientation, as shown in Figure 25. The

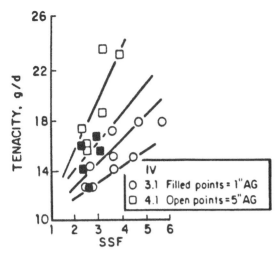

Figure 20 PPT fiber tenacity (tensile strength) development in dry-jet wet-spinning as a function of SSF. (Courtesy of the Hoechst-Celanese Corporation.)

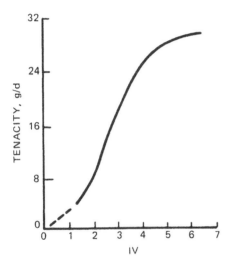

Figure 21 Fiber tenacity development as a function of PPT molecular weight (IV). (Courtesy of the Hoechst-Celanese Corporation.)

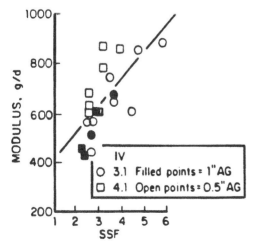

Figure 22 PPT fiber modulus development in dry-jet wet-spinning as a function of SSF. (Courtesy of the Hoechst-Celanese Corporation.)

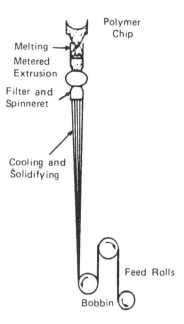

Figure 23 Development of NTP (naphthalene-derived thermotropic polyester) fiber properties with drawdown. (Data from Ref. 10.)

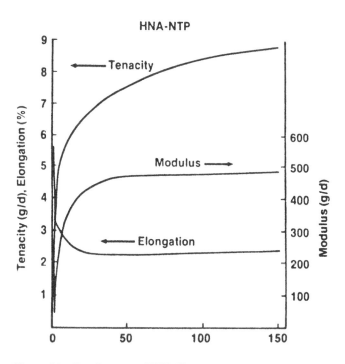

Figure 24 Development of NTP fiber properties with drawdown. (Data from Ref. 10.)

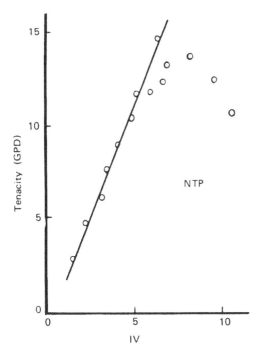

Figure 25 Relationship between NTP molecular weight (IV) and fiber tensile strength. (Data from Ref. 10.)

fall-off in tenacity at high molecular weights indicates the point at which the spinning process employed becomes nonoptimal rather than any inherent tenacity-lowering at high IVs.

As-spun fibers are not drawable in the conventional sense (e.g., PET or nylon), but to improve fiber properties beyond what is achieved in spinning, a heat treatment or annealing step in the absence of tension may be performed. In a typical heat treatment, the fibers are subjected to a high temperature (typically 10–20°C below the melting temperature) in an inert environment for a prolonged time (10 min to several hours). As a result of heat treatment, the tenacity shows a dramatic increase from about 10 g/d for as-spun fibers to values in excess of 20 g/d, occasionally reaching values as high as 40 g/d. In addition, chemical resistance, melting point, elongation, thermal retention of tensile properties, and end-use temperature also increase. Many thermotropic copolyester fibers do not show significant increases in fiber *modulus* (typically about 500–650 g/d) with heat treatment, due to the absence of orientational changes during this process. However, this is not true of all copolyesters—some may more than double their modulus on heat treatment. Figure 26 schematically summarizes the effects noted during the heat treatment.

Figure 27 compares the tenacity–temperature relationship of a typical thermotropic LC fiber to a conventional fiber such as poly(ethylene terephthalate). The LCP fiber retains about 65% of its ambient temperature properties at 150°C. This unique property occurs despite the fact that the fibers are *not* highly crystalline in the conventional sense and have a glass transition temperature in the range of 100–120°. The rigid rodlike nature prevents the LC chains from disorienting conformations above the glass temperature and results in excellent mechanical property retention at high temperatures.

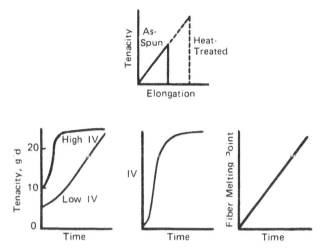

Figure 26 Summary of the property change observed after the high-temperature annealing of fibers spun from NTPs. (Data from Ref. 10.)

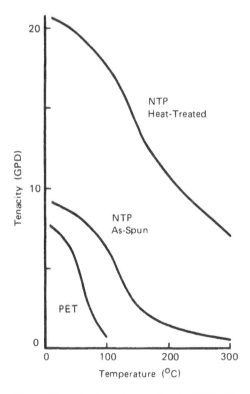

Figure 27 The dependence of typical NTP fiber tensile strength on temperature. (Data from Ref. 10.)

The observed changes in LCP fiber mechanical properties with temperature can be followed from their dynamic mechanical behavior. Figure 28 shows a plot of E' and tan δ as a function of temperature for an HBA/HNA copolymer fiber (11 Hz, Rheovibron) [65]. Three loss processes, called γ, β, and α transitions, are observed at -50, $+41$, and $110°C$, respectively, with changes in E' corresponding to these events. The γ transition is due to the reorientational motion of *p*-phenylene groups; the β transition is due to the reorientational motion of 2,6-naphthalene groups; and the α transition is cooperative in nature and may be minimized by annealing, i.e., it behaves like a glass transition. Since the β transition originates from the presence of 2,6-naphthylene units in the polymer chain, it is absent in polymers that do not contain naphthalenic units.

In comparing Figure 28 to a plot of conventional polymers [66], there are at least two differences: (1) LCP fibers have a greater E' value and the change of E' from the glassy state to the rubbery plateau state is much smaller than that of conventional polymers (10^{-1} vs. $10^{-2} \sim 10^{-5}$); and (2) the change of tan δ at the α transition is much smaller in LCP fibers than in conventional polymers. Usually the greatest change in tan δ for most isotropic polymers occurs at the α transition, but this is no longer true for LCPs. As a consequence, the energy changes for LCPs in both T_g and T_m during a DSC measurement are much smaller than that of conventional polymers.

Both Kevlar and thermotropic HBA/HNA-type polyesters have excellent chemical resistance, as illustrated in Table 1. Significant property dropoffs are noted only under strongly alkaline conditions. A comparative analysis of recently developed high-performance LCP fibers, such as Twaron (Enka), Technora (Teijin), Ekonol (Sumitomo), and Vectran (Celanese–Kuraray), has been summarized by Schaefgen [67]. In general, these commercial fibers have a tenacity of 20–30 g/d and a modulus of 500–1100 g/d. Schaefgen also summarized their applications, as shown in Table 2 [67]. The end-uses of LCP fibers can be divided into three categories: industrial fibers, fabrics–composites, and short-fiber composites.

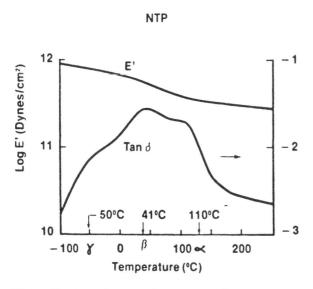

Figure 28 Dynamic mechanical behavior of a typical NTP fiber. (Data from Ref. 10.)

Table 1 Chemical Resistance of Typical Naphthalene Thermotropic Polyester (NTP) Fibers

Solvent	% modulus retention	% strength retention	% modulus retention	% strength retention
Original	100	100	100	100
Water	103	106	95	95
Gasoline	96	95	101	107
Motor oil	96	89	97	92
50% antifreeze	103	98	99	104
5% sodium hypochlorite	84	75	Dissolved	
10% sodium hydroxide	88	40	105	69
20% sulfuric acid	94	88	101	72
20% hydrochloric acid	108	117	104	65

Note: Percent modulus and strength retention of fibers aged in solvents for 1 month at 50°C.
Source: Data from Ref. 10.

LCP fibers are lightweight; 1 lb of industrial LCP fibers may replace 2 lb of glass or nylon fibers or 5 lb of steel wire while yielding the same tensile strength. Ribbon parachutes made from Kevlar have the same dimension and strength but weigh only half as much as their nylon counterparts [68, 69]. Since Kevlar and HBA/HNA fiber have excellent creep and fatigue resistance, they meet the requirements for industrial fibers.

Fabrics of LCP fibers have been used as ballistic garments, in helmets and military flak jackets. Woven fabrics were commonly preimpregnated with epoxy resins and then cured and fabricated in an autoclave, press, or vacuum bag. Due to the unique properties of LCP fibers, the helmet can stop a flying bullet (2000 ft/sec from a 0.22-caliber projectile) in a very short distance. Experimental results showed that although the exterior of a tested HBA/HNA helmet was damaged by a bullet, no transfixion was observed. Since a helmet has a cradle-type suspension system, it has interior space for the deformation and protects the human carrier from damage. Excellent cut/tear resistance and thermal insulation also make Kevlar and other LCP fibers desirable for gloves and protective clothing.

As described by DuPont researchers [68–70], significant weight saving is the primary driving force behind the adoption of Kevlar by the aircraft industry. Kevlar–epoxy composites usually are 25–50% lighter than those from glass or aluminum and have greater tensile properties. The fabrication processes, such as wet layup, prepregging, and filament-winding pultrusion, for both Kevlar and glass fibers, are quite similar; therefore, only minor modification is needed for a glass–fiber plant to be adapted for Kevlar. However, the matrices and curing cycles for these two composites may be different. Table 3 summarizes the mechanical properties of unidirectional laminates made with Kevlar 49, carbon fiber, and E-glass [69]. Kevlar 49 shows significant advantages in density and stiffness over glass. However, weak compression strength limits Kevlar from being used where compression strength is critical. To overcome this, Kevlar–carbon hybrid composites were invented which resulted in good properties [70]. It is also important to note that predrying Kevlar yarns and rovings (e.g., at 105°C for 16 hr) to below 1% moisture is essential for the production of high-quality panels; otherwise, moisture on the yarn surface reduces interlaminar shear properties.

Table 2 Applications of LCP Fibers

Sample	Property
Protective fabrics	
Ballistic vests	Strength, thermal resistance
Gloves	Cut, thermal resistance
Clothing	Strength, thermal resistance, comfort
Strong fabrics	
Tarpaulins	Strength, weatherability
Conveyer belts	Strength, stiffness, fatigue resistance
Coated fabrics	
Inflatable boats	Strength, fatigue resistance
Sails	Strength, fatigue resistance
Industrial fibers	
Ropes, cables (oil rig mooring)	Strength, low density, low creep
Ropes (marine uses)	Strength, low density, low creep
Filament wound pressure vessels	Strength, low density, low creep
Sails	Strength, weatherability, low fatigue
Sewing threads	Strength, weatherability, low fatigue
Rubber reinforcement	
Radial tires (trucks, automobiles)	Strength, stiffness, low fatigue, adhesion, low density, oxidation resistance
Belts (conveyor, power transmission)	Strength, stiffness, hydrolytic stability
Plastics reinforcement	
Space applications	Tensile properties, low density, thermal resistance
Aircraft, interior	Tensile properties, low density, thermal resistance
Aircraft, propeller	Tensile properties, low density, thermal resistance, vibration damping
Boats, canoes, kayaks	Tensile properties, low density, low fatigue, vibration damping
Military (helmets)	Tensile properties, low density
Sporting goods	Modulus, compression strength
Cement reinforcement	
Building materials, pipes	Tensile properties, hydrolytic stability, thermal properties
Friction uses, asbestos replacement	
Brake linings, clutch facings, gaskets, packing	Modulus, nontoxicity, nonaggressive wear, thermal stability

Source: Based on Ref. 67.

By analogy with the aerospace experience, boats made from Kevlar have demonstrated outstanding durability and maneuverability. Various fishing boats and motor yachts have been built and tested [68, 69]. Since Kevlar is lighter than glass, boats reinforced with Kevlar also have greater speed, quicker response, and easier portage than those reinforced with glass. Although the initial cost of a boat built from Kevlar is higher, calculations indicate that the extra cost would be recovered in two years [68].

Because Kevlar and LC polyester are strong, nonabrasive, dimensionally and thermally stable, they have been used to reinforce brake linings. Kevlar is not a friction

Table 3 Unidirection Composition Lamina Properties

Property	Kevlar® 49 Aramid	E-Glass	Graphite
Density, kg/*l*	1.38	2.08	1.52
Tensile strength 0°, MPa	1,380	1,100	1,240
Compressive strength 0°, MPa	276	586	1,100
Tensile strength 90°, MPa	27.6	34.5	41.4
Compressive strength 90°, MPa	138	138	138
In-plane shear strength, MPa	44.1	62.0	62.0
Interlaminar shear strength, MPa	48.69	83	97
Poisson's ratio	0.34	0.30	0.25
Tensile and compressive modulus 0°, MPa	75,800	39,300	131,000
Tensile and compressive modulus 90°, MPa	5,500	8,960	6,200
In-plane shear modulus, MPa	2,070	3,450	4,830

Source: Ref. 69.

controller but provides uniform wear and low creep and reduces irregular cracking. Experimental data indicate that 5% by weight of chopped Kevlar fibers, plus pulp, can replace 60% or more asbestos [68]. With a better formulation, brake performance can be significantly improved with the addition of Kevlar and LCP fibers. Kevlar provides the following advantages when used as clutch facings: (1) greater wear resistance; (2) higher peak strength; and (3) less weight [68]. These advantages arise from the unique properties of Kevlar fibers.

LCP Rods, Sheets, Extrusions, and Composites

Large-diameter melt-extruded LCP (30–70 mil) rods have been used to replace steel wire; Kevlar ropes are even used as strength members in optical cable applications. This is because (1) LCP rods are lightweight and flexible so that the handling is easier; (2) LCP rods have excellent tensile properties, which prevent optical fibers from breaking during the laydown process; (3) LCP rods have a very small negative coefficient of thermal expansion, which minimizes any external stress (expansion) on optical fibers due to the changes in the weather; (4) good chemical resistance and almost zero water regain of LCP rods prolongs the cable lifetime; and (5) the fabrication process of thermotropic LCP rods is easier and more straightforward than that of dry-wet spinning of Kevlar. Finally, both types of LCP are dielectric, which precludes lightning strikes in optical cables aboveground.

However, engineers should be aware of four key characteristics of LCPs during extruding [71]: (1) little drawdown is recommended because LCPs set up quickly on leaving the die; (2) the extrusion line should have a sizing ring and a short waterbath set close to the die; a 2-ft trough set 7–9 in. from the die has worked well in extruding 10- to 100-mil rod of Vectra A900; (3) LCP melts have a very high "memory"; and (4) materials are very stiff and highly oriented. Chung [72] noted that the distance from die to water bath has a strong effect on extrudate diameter uniformity and concentricity. For example, if the roundness of a rod is defined by the ratio of the major to the minor diameter and the air gap denotes the distance between the die exit and the sizing ring, Figure 29 shows the roundness of extruded rods as a function of air gap. Surprisingly, an air gap of 6 in. was required for HBA/HNA polymeric rods to reach 95% of roundness. Using the same die, polypropylene and nylon-12 yielded rods with a roundness exceeding 95% when the air gap was less than 2 in. After examining the extrudate we found it seemed to shrink unevenly after exiting from the die, and it continued rotating until it was solidified. The rotation speed was fastest at the die exit. Thus its cross section seemed to be elliptical at the die exit and gradually became round at a distance from the die. In addition, these two anomalous flow phenomena were enhanced when the screw rotational speed was increased. The rotational phenomenon is most likely due to the long relaxation time of LCP melts. The uneven shrinkage probably arises from two factors, one being the uneven stress relaxation of LCPs and the other the negative normal stress differences.

Figure 30 illustrates the Hermans orientation profiles for two different LCP rods and shows that orientation is highest near the skin and lowest at the center. However, the variation in the LCP orientation across a rod is much less than that of a conventional polymeric rod. Upon examination of fracture surfaces of LCP rods, fibrils occurred near the skin. Figure 31 gives the relationship between the initial moduli of extruded rods and the Hermans orientation. The rod diameters were in the range of 40–100 mil and were extruded with various drawdown ratios. Since the slope of this curve is increased significantly with an increase in the Hermans function, it indicates that the modulus of a highly oriented rod can be remarkably enhanced if even a slight improvement is made on the domain alignment. Polyester amides perform better than polyester LCPs in this respect, as they orient under lower extentional flow field conditions.

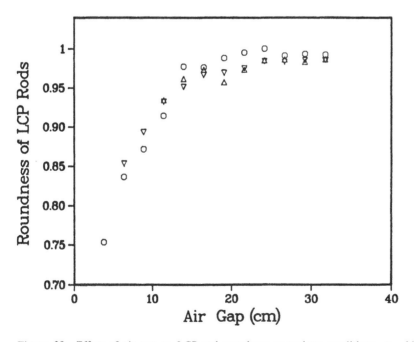

Figure 29 Effect of air gap on LCP rod roundness at various conditions. ○, 11.4 mm, 24.8 m/sec; △, 15.2 mm, 19.1 m/sec; ▽, 22.9 mm, 13.4 m/sec. (Data from Ref. 72.)

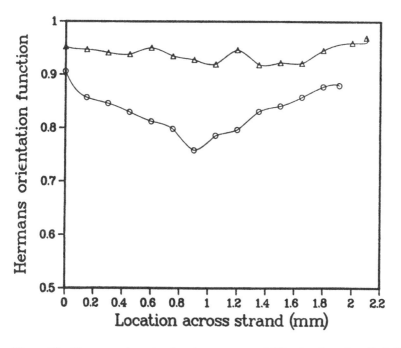

Figure 30 Hermans orientation function across two LCP rods. (Data from Ref. 72.)

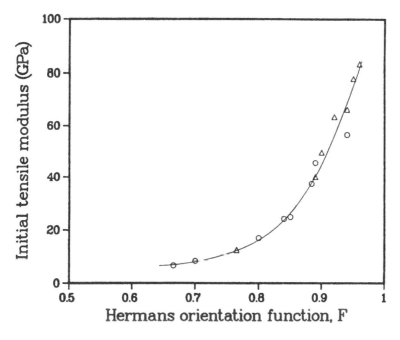

Figure 31 Initial modulus as a function of Hermans orientation function of two main-chain LCPs. (Data from Ref. 72.)

For similar reasons, wire coating of LCP over silicone-buffered optical fibers has resulted in jacketed fiber capable of withstanding tensile loads 10 times higher than that sustainable by conventional thermoplastic-jacketed optical fibers [73]. In addition, LCP is more resilient than a pultruded fiber because LCP is stiffer than conventional polymers. Bending tests on LCP-jacketed fiber indicate that it has smaller bend radii than that of conventional polymeric jackets. Figure 32 compares attenuation loss between a conventional nylon 12 and an LCP-jacketed optical fiber [74]. The LCP-jacketed fiber demonstrates no signal loss increase down to −60°C upon cooling, whereas the nylon-jacketed fiber shows a significant loss near −20°C. This difference in signal loss is primarily due to the nylon having a 50 times greater CTE than the LCP, resulting in excess compression strain (shrinkage) on fiber upon cooling.

Sheet products made from mineral-filled LCP variants have been used for thermoform-

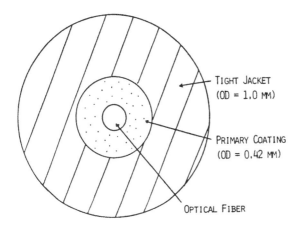

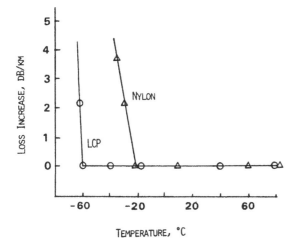

Figure 32 Transmission-loss change of jacketed fibers at 1.3 μm, heated to 80°C and then cooled to low temperatures. (From Ref. 74.)

ing and electroplating for printed circuit boards. During the sheet fabrication, calendering rolls should be placed up against the die. The stiffness of LCPs requires a wider spool— about 200 times the product diameter. Figure 33 illustrates the typical mechanical properties as a function of drawdown ratio (defined as the ratio of cross section area of the die to the cross section area of the extruded article) in both machine direction (M.D.) and transverse direction (T.D.) for unfilled LCP sheets. It is evident from the data that, in thin tapes, mechanical properties in the machine direction are equivalent to fiber properties. Figures 34 and 35 show that the angular dependence of the tensile strength and modulus fit very well with the Tsai–Hill and Lees' carbon fiber–epoxy composite theory [75]. In other words, the mechanical properties and performance of the extruded LCP sheets can be predicted using the existing composite theories developed for long- or short-fiber-reinforced composites, such as epoxy–glass fiber and epoxy–carbon fiber systems. From a molecular standpoint this unique result is mainly due to the highly oriented mesogenic domains having a tendency to remain parallel in the flow direction and act like fiber-reinforced elements in the fabricated articles.

Since LCPs offer a low viscosity for the impregnation of carbon fibers and excellent chemical resistance, LCP–carbon fiber composites have been developed and used as secondary composites for the aerospace industry. The impregnation of carbon fibers was carried out in a crosshead tape die. Composite panels were prepared by compression molding stacked layers of the prepreg. Tensile and flexural properties of a LCP/CF composites are given in Tables 4 and 5 [76]. The wet "conditions" in these tables refer to samples immersed in water at 160°F for two weeks prior to testing at the indicated temperature. Both tensile and flexural properties are comparable to those of commercial

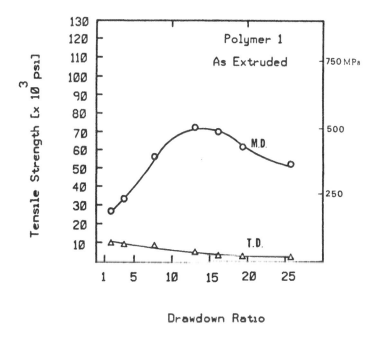

Figure 33 Tensile strength of as-extruded LCP sheet as a function of drawdown ratio. (Data from Ref. 75.)

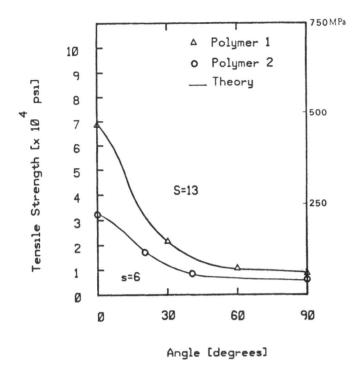

Figure 34 Angular dependence of tensile strength for two LCP-extruded sheets. (Data from Ref. 75.)

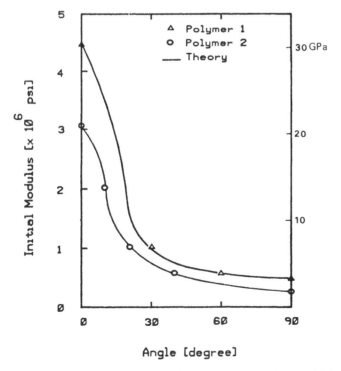

Figure 35 Angular dependence of Young's modulus for two LCP-extruded sheets. (Data from Ref. 75.)

Table 4 Tensile Properties of Unidirectional Composites

CF volume (%)	Conditions	Tensile strength (MPa)	Tensile modulus (Gpa)	Tensile strain (%)
56.5	Dry	1492	143	1.045
56.5	Wet	1297	121	1.071

Source: Data from Ref. 76.

Table 5 Flexural Properties of LCP–CF Composites

Fiber content (vol. %)	Conditions	Strength (MPa)	Modulus (GPa)
51	RT (dry)	1515	106.5
51	RT (wet)	1447	106.5
51	200°F (dry)	1054	106.2
51	200°F (wet)	904	105.3
51	250°F (dry)	744	98.8
51	250°F (wet)	854	97.8

Source: Data from Ref. 76.

Table 6 Mechanical Properties of Isotropic Laminated Panel (LCP-Celion 6000)

Property	CF (vol %)	Result
Open-hole tensile strength	52	313.5 PMa
45° tensile strength	53	134.5 MPa
Impact test		
Maximum load	52	2420 N
Energy at maximum load	52	2.76 J

Source: Data from Ref. 76.

epoxy–CF composites at the same fiber volume content. Flexural modulus retention at elevated temperatures is extremely good. However, the flexural strength retention at 200°F is only fair (67%), and it becomes poor (54%) at a test temperature of 250°F. This poor retention may be due to poor interface between fiber and matrix. Thus the adhesion between fibers and matrix fails long before the composites fails. The mechanical properties of an isotropic laminate are given in Table 6. Both open-hole and 45° tensile strength are comparable to those of epoxy–carbon fiber composites. Impact test results indicate that the isotropic panel has a maximum load of 2420 N, which is clearly superior to thermoset matrix systems (typically 1480 N).

LCP Plastics and Applications

When unfilled thermotropic copolyesters were used as injection-molding resins, several distinct phenomena were reported: (1) the tensile and flex properties of a molded LCP article are comparable to those of short-fiber glass-filled thermoplastics; (2) mechanical properties are anisotropic across the part—Figure 36 shows that an unfilled LCP part is more anisotropic than that of a 30% glass filled PBT part; (3) physical properties are strongly dependent on the thickness of a molded part: parts containing from 7 to 11 layers have been reported due to various flow fields [10, 77]; and (4) unfilled LCP parts have very poor weld-line strengths. To maintain mechanical strength while eliminating these drawbacks, Celanese and Dartco have discovered that the addition of fillers or fiber-reinforced elements into neat LCP resins may be the best approach. Thickness effect is less pronounced and weld-line strength is enhanced.

There are currently two major families of LCPs on the market: Vectra from Celanese and Xydar from Dartco. The two materials differ significantly in terms of processing and properties. Figure 37 compares melt viscosity for LCPs and conventional polymers [78]. The key advantages of Vectra are its fast cycling and ease of processing compared with any other LCPs that have been tried. Typical melt temperatures for both injection and extrusion grades are around 285°C, with injection-molding tool temperatures held at about 100°C. The exceptionally low melt viscosity of Vectra ensures easy filling of complex,

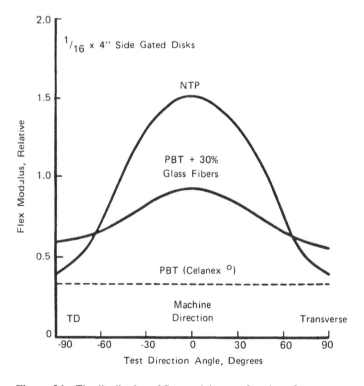

Figure 36 The distribution of flex modulus as a function of measurement direction in an injection-molded 0.1587- × 10.16-cm disk. (Data from Ref. 10.)

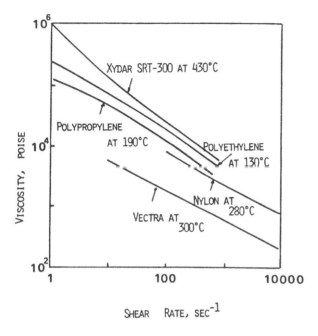

Figure 37   Viscosity of Vectra and Xydar polymers. (From Ref. 78.)

small-part geometries. As an indication, one part for a metering instrument has been molded successfully 0.56 mm thick and some 125 mm long. No modifications are re-quired to processing equipment and the material is noncorrosive to molds and is clean-running. As with filled or reinforced polymers, tools should be hardened to minimize the effect of abrasion. Overall, tool wear should be similar to that for molding with 30% glass-filled nylon. The inherent thermal stability allows the reuse of reground Vectra LCP for injection molding.

Xydar, by contrast, has a substantially higher melting point and requires that certain modifications be made to processing equipment. It also has a melt viscosity two orders of magnitude higher, making it much more difficult than Vectra to mold in thin sections. However, Xydar offers higher retention of mechanical properties at extreme application temperatures, such as those in excess of 300°C. Its nominal melting point is 421°C and the HDT (heat deflection temperature) is around 316–355°C. Dartco offers unfilled Xydar in two flow ranges, plus a 50% talc-filled version and a 50% glass-filled compound. All four grades are suitable for injection molding.

Neither of these LCPs reorients when they solidify from the melt, which ensures that molded parts can be produced with minimum warpage and distortion as illustrated by Table 7 [78]. Parts can be ejected from the hot mold and no mold release agent is needed. Gating is a key factor in mold design for LCPs, as the direction and degree of orientation directly influence physical properties. The multilayered structure of injection-molded parts, as shown in Figure 38, is due to the effect of various flow fields imposed on the LC mesophase. As a result, each layer has different structure and degree of orientation. X-ray diffraction data studied by Stamatoff [79], Duska [77], Joseph et al. [80], and Kenig [81] found that the number of color bands (layers) and their color depend on the mold thick-

Table 7   LCP Mold Shrinkage

Dartco Xydar (ASTM Bar: 12.7 × 1.27 × 0.3175 cm)		
Filler (%)	Length (mm/m)	Width (mm/m)
0	8	2
30	2	−2
40	0	−1
50	−3	−2

Celanese LCP		
Filler	Flow	Transverse
15% mineral	2	6
30% glass	0.4	1–1.5

Other Engineering Resins (Published Values)

Resin	Glass (%)		
	0	30	40
PEEK	11	2	
Polyetherimide	5–7	2	
Polysulfone	7	1–3	
Polyamide-imide	6–8	3–5	
Polyphenylene sulfide	6–8		2–4

Source: Ref. 78.

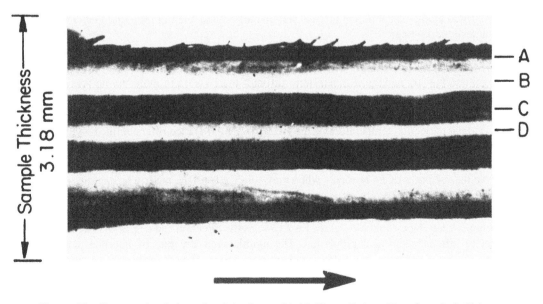

Figure 38   Cross-sectional view of an injection-molded LCP tensile bar. (Data from Ref. 10.)

ness, location, and direction of the testing sample with respect to the machine direction. Mold thickness controls both the fountain flow pattern and elongational stresses at the melt front and the degree of shear stress and viscous dissipation energy. Location decides if there is a radial (diverging) or converging flow. A diverging flow results in disorientation of the LCP domains from the machine (flow) direction (sometime perpendicular to each other), while a converging flow parallels domains with the flow direction. Experimentation reported by the previous authors indicates that LCP orientation changes over 90° from the surface to the core; it is parallel to the flow direction near the surface. Whenever there is color change, the orientation and morphology change. The slight disorientation of region B in Figure 38 may be due to the effect of diverging flow of the fountain flow at the melt front. The orientation is almost frozen because it is very close to the mold surface. Due to the effect of shear stresses, region C has a slightly higher orientation than that of region B.

Table 8 summarizes mechanical properties of some Vectra variants. Some highlights of the resin properties are the flexural and tensile strengths, which are typically in the range of 20–26 kpsi (140–180 MPa), and flex and tensile moduli, which are in the range of 1.5–2.8 Mpsi (10–19 GPa). For comparison, these values for lightly filled variants are about equal to a 30% fiber-reinforced engineering plastic. When highly fiber reinforced, they substantially exceed the stiffness of commonly available engineering polymers with comparable fiber levels. Both Vectra and Xydar are very resistant to chemical and solvent attack. Tables 9 and 10 show their properties [83]. In addition, LCPs are little affected by radiation. Vectra shows no change in properties when exposed to 500 Mrad of radiation (cobalt-60), when weathered in a weatherometer for 2000 hr; it absorbs little water, 0.04% at equilibrium, and is hydrolytically stable, resisting 230°F (110°C) water for 200 hr with no change in appearance or properties [82]. Similar properties have been reported for Xydar [83].

Table 8 Injection Molding Grades of Vectra

Property	ASTM test	Glass fiber A130	High flow A515	High temperature C130	Wear resistance A420	Chemical resistance A625
Tensile strength, kpsi (MPa)	D638	26(180)	26(180)	23(160)	20(140)	24(170)
Tensile modulus, Mpsi (GPa)	D638	2.2 (15)	1.7 (12)	2.2 (15)	3.0 (20)	1.5 (10)
Tensile elongation, %	D638	2.3	4.5	2.0	1.3	6.9
Flex modulus, Mpsi (GPa)	D790	2.0 (14)	1.5 (10)	2.0 (14)	2.8 (19)	1.5 (10)
Notched izod, ft-lb/in. (J/m)	D256	2(110)	6(320)	2.2(120)	1.8(100)	2.5(130)
HDT, 264 psi, °F (1.82 MPa, °C)	D648	445(230)	365(185)	465(240)	440(225)	365(185)
Specific gravity	D792	1.6	1.5	1.6	1.9	1.5

Source: Courtesy of the Hoechst-Celanese Corporation.

Table 9 Chemical Resistance of Vectra

Sample	Temperature (°F)[a]	Rating[b]
Acids		
Acetic, glacial	216 (118)	A
Formic, 80%	216 (104)	A
Sulfuric, 70%	375 (190)	A
Sulfuric, 93%	375 (190)	B
Chromic, 70%	190 (88)	A
Chromic, 70%	250 (121)	B
Hydrochloric, 37%	190 (88)	A
Bases		
Sodium hydroxide, 10%	190 (88)	A
Sodium hydroxide, 50%	190 (88)	B
Diphenylamine	150 (66)	A
Organic solvents		
Acetone	125 (52)	A
Methanol	125 (52)	A
Methylene chloride	150 (66)	A
Trichloroethane	150 (66)	A
Ethyl acetate	170 (77)	A
Others		
Gasoline	122 (50)	A
Methanol fuel	122 (50)	A
Skydrol	122 (50)	A
Fluorinert FC-70	419 (215)	A
Water	110 (230)	A

[a]°C given in parentheses.
[b]Rating: A = essentially no effect. Less than 5% change in mechanical properties and less than 2% change in weight and dimensions. B = some change.
Source: Courtesy of the Hoechst-Celanese Corporation.

The main application of Xydar is for dual-oven cookware; Celanese intends to formulate various grades for a variety of applications, as listed in Table 11 [82]. Most of these are for high technology and high value-added applications. Some of them will be discussed next.

Optical couplers are used to passively split, combine, and distribute optical signals to or from an optical fiber. The key requirements for an optical coupler are the following: (1) its CTE must closely match the CTE of optical fibers; (2) it must have excellent dimensional stability at elevated temperatures; and (3) it must have high flexural modulus. Otherwise, the stresses accumulated at the joint (where the optical fibers fuse together or just touch) and the nearby tapering area may separate the fibers, resulting in increasing attenuation loss or even a break in communication. The dynamic breaking strength in these weak regions is less than 0.5 N (0.1 lb). The Vectra variant developed for this application consists of a compound reinforced with less than 20% glass and colored with

Table 10 Chemical Resistance of Xydar SRT-300

Substance	Test condition	Retention of tensile strength (%)
60% relative humidity	5 weeks at ambient temp.	100
Water	200 hours at 110°C (230°F)	82
Water	30 days at 82°C (180°F)	95
Acetic acid	1 week at reflux	100
Ethyl acetate	1 week at reflux	97
20% sulfuric acid	1 month at 50°C (122°F)	100
37% sulfuric acid	1 week at 50°C	93
70% nitric acid	11 days at ambient temp.	98
Methyl ethyl ketone	1 week at reflux	96
Isopropyl alcohol	1 week at reflux	100
Trichloroethylene	1 week at 50°C	100
Detergent	1 month at 50°C	100

Source: Data from Ref. 83.

Table 11 Possible Applications of LCPs

Market areas	Applications
Electrical/electronic	Sockets, chip carriers, bobbins, high-temperature connectors, switches, vapor phase–soldered components, encapsulation of hybrid and micro-circuitry
Chemical	Tower packing, pumps, meters, down-hole oil well components, valves
Fiber optics	Connectors, couplers, secondary reinforcing members, loose and tight buffers
Industrial/mechanical	Pulleys, bushings, bearings, seals, wear blocks, mechanical components
Transportation	Fuel and electronic components, mechanical and under-hood components
Aircraft/aerospace	Interior components, fuel systems, radoms, brackets, fasteners, electrical components
Other	Appliances, housewares, business equipment, sports and recreation

Source: Courtesy of the Hoechst-Celanese Corporation.

less than 5% carbon black. Its CTE is about 1.4 ppm/°C and the CTEs of most optical glass fibers vary from 0.8 to 1.9 ppm/°C, dependent on the producers. Its percentage of dimensional change after aging over 125°C for 24 hr is about 0.015%, after which it showed almost no change at the maximum operating temperature of 85°C for extended periods [84]. The coupler also passed severe environmental tests, such as temperature cycling (−50 to 85°C), humidity (95% at 65°C), and vibration/shock (according to MIL-STD 810C specification).

Ceramic saddle packing had been used as tower packing in chemical distillation columns for years. It provides high-temperature corrosion resistance, but it is brittle. This leads to pressure surges and other disturbances after operating the column for one month, due to breakage of the saddles. The loss in packed-bed height decreases the mass-transfer efficiency within the column, and ceramic packing debris can plug lines, damage seals, and cause leaks. To overcome this problem, Celanese invented a Vectra–graphite flake tower packing to replace ceramic ones. As previously described, LCP provides a unique combination of chemical and mechanical properties; the new packing was tested in a Celanese plant distilling formic acid. The following advantages were reported: (1) loading the column is easier than with ceramic packing because LCP is lightweight; (2) no breakup of the polymer packing was noted in the first year of service in the tower; (3) turnarounds were reduced from 6–7/yr to 3–4/yr, producing significant annual savings in maintenance; and (4) even when the packing did start to break up, its debris did not damage the system because of its nonabrasive nature. As a result of having fewer turnarounds and improved formic acid recovery, it is estimated that LCP tower packing saves \$110,000/yr/column and the column performance can be more accurately predicted than with ceramic packing.

Electrical properties of LCP and polybutylene terephthalate (PBT) resins are comparable; however, LCP offers at least a few advantages over PBT in electronic applications: (1) low mold shrinkage; (2) fast cycling time; (3) capability of molding thin parts; (4) low

Table 12 CTE of Typical Electronic Components

Material	CTE (10^{-6} m/m°C)
FR4 epoxy glass board	−13
Invar-clad epoxy/glass board	−6–7
Ceramic substrates/components	−6–7
40% glass/PPS	−22.5
Vectra A-130	−11.3

Source: Courtesy of the Hoechst-Celanese Corporation.

Table 13 Dimensional Test Results—Connectors Exposed to Vapor Pressure Soldering

Material	Length (L) 5.626 cm		
	Mold (cm)	Due to vapor phase soldering (cm)	Total (cm)
PBT (30% glass)	0.0061	0.0053	0.0114
PEI (30% glass)	0.0003	0.0064	0.0067
PPS (40% glass)	0.0041	0.0038	0.0079
Vectra A-130	0.0002	0.0008	0.0110

Source: Courtesy of the Hoechst-Celanese Corporation.

moisture regain; (5) better chemical and mechanical properties. Tables 12 and 13 give a comparison in CTE and dimensional stability in vapor-phase soldering processes between LCP and other electrical-grade engineering resins and clearly show that LCP is superior to its competitors. However, it is important to point out that the electrical-grade LCP variants must contain fillers or additives in order to improve overall properties and broaden its applications. This phenomenon may be explained by Figure 39, in which the flexural properties of two unfilled LCPs were remarkably improved by the addition of either mineral or glass–carbon fibers. Mold gate design is of particular importance in this case. A submarine gate 50–100% of the wall in thickness is recommended with the gate at least as wide as it is high. It should be located in the ejector side of a mold and should be

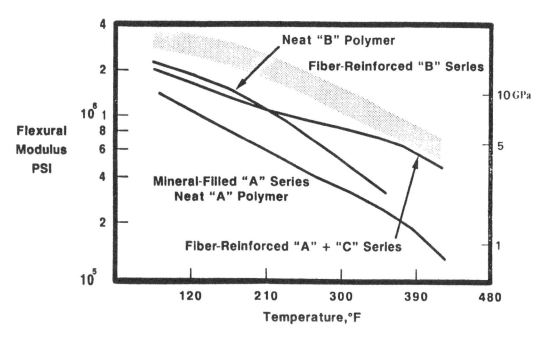

Figure 39 Flexural modulus of LCPs vs. temperature. (Courtesy of the Hoechst-Celanese Corporation.)

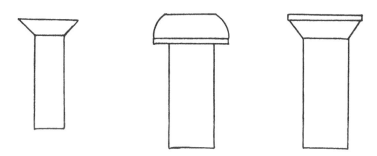

Figure 40 Available LCP rivet configurations. (Courtesy of the Hoechst-Celanese Corporation.)

Table 14 Shear Strength and Specific Gravity of Selected Vectra Compounds

Sample	Specific gravity	Shear strength ASTM B565-76 (MPa)	Specific shear strength[a] (MPa)
Vectra grade			
A950	1.4	104.8	75.2
A130	1.6	123.4	77.2
A230	1.5	121.3	80.7
A515	1.5	101.4	68.2
A950	1.4	128.9	92.4
A130	1.6	132.4	82.7
A150	1.8	144.8	80.7
Metals			
Aluminum	2.7	193–310	71.7–115
Titanium alloy	6.2	303–372	48.9–60

[a]Shear strength/specific gravity.

Source: Courtesy of the Hoechst-Celanese Corporation.

designed to be suitable for high-modulus materials. Vents should be located in all sections of a mold where air may become trapped by the molten plastic.

LC polymeric rivets offer an attractive alternative to metal for fasteners. For example, rivets have been shown to securely fasten carbon fiber–epoxy composite panels without the high cost and heavy weight of titanium or the corrosion problems associated with aluminum. These fasteners have also been demonstrated to be suitable for thin aluminum panels where it is desirable to spread loads over a larger surface area. Figure 40 illustrates three popular types of rivets, with Table 14 summarizing their properties. Although LC rivets are not as strong, on a size basis, as aluminum or titanium, their specific shear strength is comparable to aluminum and much superior to titanium.

OPTICAL APPLICATIONS

The use of LC polymers for electrooptical applications is in the early stage of development. Coles and Simons [85] and Platé and Shibaev [86] have summarized most of the previous work. The main application of LCPs is for data storage; however, compared with a monomeric system there are a few drawbacks with currently developed LCPs [85–89]: (1) The operating parameters are not practical, i.e., the operating temperature and threshold voltages are high and the response times are too slow; and (2) the viscosity of LCP is still too high. Figure 41 is a schematic drawing of the measuring arrangement designed by Coles and Simons [90]. The testing sample (center box) is located on a temperature-controlled hot stage. Fields are applied across the sample with various voltages. A transmission polarizing microscope is used to record the signal. Many phenomena were

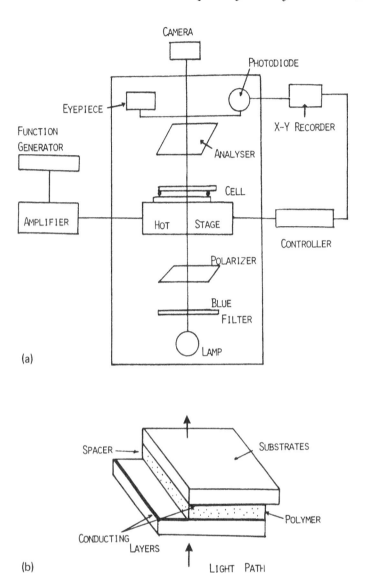

Figure 41 (a) Schematic diagram of experimental apparatus and (b) assembly of polymer liquid crystal cell. (From Ref. 90.)

observed; one of them is the Freedericksz transition, where the LC orientation changes through 90° and the sample changes from clear to opaque or vice versa. The time for this change depends on the sample preparation, viscosity, temperature, voltage, and field frequency. Figures 42 and 43 illustrate some of these effects and the response time is defined at 50% change in light transmission. At present, the response times for most existing LCPs are too long to compete with existing commercial dyes (10–0.1 sec vs. 100–1000 nsec). More work is needed in this area.

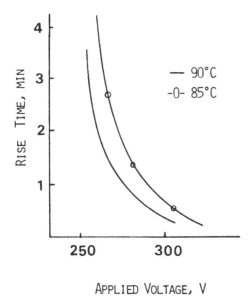

Figure 42 Rise time as a function of applied voltage at temperature just below T_m. (From Ref. 90.)

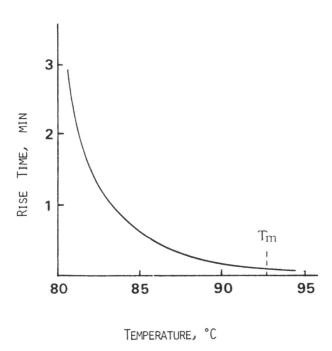

Figure 43 Optical rise time as a function of temperature. (From Ref. 90.)

ACKNOWLEDGMENTS

The authors wish to express their gratitude to Drs. M. Jaffe, R. S. Jones, H. N. Yoon, K. Wissbrun, J. A. Flint, K. T. O'Brien, G. Farrow, and Mr. M. Braeckel at Celanese for their useful comments and corrections. Thanks are also due to Dr. A. Buckley, Mr. P. Garrett, Mr. W. C. Brinegar, and the Celanese LCP Department for their support and permission to publish this work.

REFERENCES

1. P. J. Flory, *Proc. R. Soc. London, A234*: 73 (1956).
2. S. L. Kwolek, U.S. Pat. 3,600,350 (1971); (DuPont).
3. S. G. Cottis, J. Economy, and B. E. Nowak, U.S. Pat. 3,637,595 (1972); Carborundum.
4. S. G. Cottis, J. Economy, and L. C. Wohrer, U.S. Pat. 3,975,487 (1976); Carborundum.
5. F. E. McFarlane, lecture, Gordon Research Conference on Polymers (1974); H. F. Kuhfuss and W. J. Jackson, Jr., U.S. Pat. 3,778,410 (1973); Eastman Kodak.
6. W. J. Jackson, Jr., and H. F. Kuhfuss, *J. Polym. Sci. Polym. Chem. Ed., 14*: 2043 (1976).
7. G. W. Calundann, U.S. Pat. 4,067,852 (1978); 4,161,470 (1979); 4,185,996 (1980), 4,256,624 (1981); Celanese.
8. G. W. Calundann, H. L. Davis, F. J. Gorman, and R. M. Mininni, U.S. Pat. 4,083,829 (1978); Celanese.
9. A. J. East, L. F. Charbonneau, and G. W. Calundann, U.S. Pat. 4,330,457 (1982); Celanese.
10. G. W. Calundann and M. Jaffe, "Anisotropic Polymers, Their Synthesis and Properties," Proceedings of The Robert A. Welch Conferences on Chemical Research, XXVI, Synthetic Polymers, Houston, Texas, p. 247 (1982).
11. G. W. Calundann, in *High Performance Polymers: Their Origin and Development* (R. B. Seymour and G. S. Kirshenbaum, eds.), Elsevier, New York, p. 235 (1986).
12. G. W. Calundann, T. S. Chung, and J. B. Stamatoff, in *McGraw-Hill Yearbook of Science and Technology*, McGraw-Hill, New York (1987).
13. T. S. Chung, *Polym. Eng. Sci., 26*: 901 (1986).
14. R. Zentel and H. Ringsdorf, *Makromol. Chem. Rapid Commun., 5*: 393 (1984).
15. H. Finkelmann, H. J. Kock, and G. Rehage, *Makromol. Chem. Rapid Commun., 2*: 317 (1981).
16. A. Blumstein, Y. Osada, S. B. Clough, E. C. Hsu, and R. B. Blumstein, in *Mesomorphic Order in Polymers and Polymerization in Liquid Crystalline Media* (A. Blumstein, ed.), ACS Symposium Series 74, Washington, p. 56 (1978).
17. H. Finkelmann and G. Rehage, *Makromol. Chem. Rapid Commun., 1*: 31 (1980).
18. L. Strzelecki and L. Liébert, *Bull. Soc. Chim. Fr., 2*: 597, 603 (1973).
19. E. Perplies, H. Ringsdorf, and J. H. Wendorff, *J. Polym. Sci. Lett., 13*: 243 (1975).
20. Y. S. V. Lipatov, V. V. Tsukruk, and V. V. Shilov, *J. Macromol. Sci. Rev., C24*: 173 (1984).
21. R. Geiss, W. Volksen, J. Tsay, and J. Economy, *J. Polym. Sci. Polym. Lett. Ed., 22*: 433 (1984).
22. N. H. Hartshorne, *The Microscopy of Liquid Crystals*, Microscope Publications, London (1974).
23. D. Demus and L. Richter, *Textures of Liquid Crystals*, Verlag Chemie, Weinherim (1978).
24. J. H. Wendorf, in *Liquid Crystalline Order in Polymers* (A. Blumstein, ed.), Academic Press, New York, p. 1 (1978).
25. R. A. Chivers, J. Blackwell, G. A. Gutierrez, J. B. Stamatoff, and H. N. Yoon, *Polym. Prep., 24*: 288 (1983).
26. J. B. Stamatoff, *Mol. Cryst. Liq. Cryst., 110*: 75 (1984).

27. H. Finkelmann, H. J. Kock, W. Gleim, and G. Rehage, *Makromol. Chem. Rapid Commun.*, 5: 287 (1984).

28. C. Noel, C. Friedrich, F. Laupretre, J. Billard, L. Bosio, and C. Strazielle, *Polymer, 25*: 263 (1984).

29. F. Volino, J. N. Allonneau, A. M. Giroud-Godquin, R. B. Blumstein, E. M. Stickles, and A. Blumstein, *Mol. Cryst. Liq. Cryst., 102*: 21 (1984).

30. E. T. Samulski, *Polymer, 26*: 177 (1985).

31. J. L. White and J. F. Fellers, *J. Appl. Polym. Sci. Symp., 33*: 137 (1978).

32. J. I. Jin, S. Antoun, C. Ober, and R. W. Lenz, *Br. Polym. J., 1980*: 132 (December).

33. C. K. Ober, J. I. Jin, and R. W. Lenz, *Adv. Polym. Sci., 59*: 103 (1984).

34. H. Finkelmann and G. Rehage, *Adv. Polym. Sci., 60/61*: 99 (1984).

35. V. P. Shibaev and N. A. Platé, *Adv. Polym. Sci., 60/61*: 173 (1984).

36. P. J. Flory, *Adv. Polym. Sci., 59*: 1 (1984).

37. L. L. Chapoy, *Recent Advances in Liquid Crystalline Polymers*, Applied Science Publishers, London (1985).

38. H. Finkelmann, H. Ringsdorf, and J. H. Wendorff, *Makromol. Chem., 179*: 273 (1978).

39. See for example, *Plastics Tech., 1984*: 82 (December).

40. A. Roviello and A. Sirigu, *Makromol. Chem., 183*: 895 (1982).

41. P. W. Morgan, *Macromolecules, 10*: 1381 (1977).

42. P. G. Riewald, *J. Ind. Fabr., 1983*: 7.

43. J. J. Kleinschuster, U.S. Pat. 3,991,014 (1976); T. C. Pletcher, U.S. Pat. 3,991,013 (1976); J. J. Kleinschuster and T. C. Pletcher, U.S. Pat. 4,066,620 (1978); J. R. Schaefgen, U.S. Pat. 4,075,262 and 4,118,372 (1978); C. R. Payet, U.S. Pat. 4,159,365 (1979); R. S. Irwin, U.S. Pat. 4,232,143 and 4,232,144 (1980) and 4,245,082 (1981); DuPont.

44. Q. F. Zhou and R. W. Lenz, *J. Polym. Sci. Polym. Chem. Ed., 21*: 3313 (1983).

45. C. K. Ober, J. I. Jin, and R. W. Lenz, *Polym. J., 14*: 9 (1982).

46. G. Chen and R. W. Lenz, *J. Polym. Sci. Polym. Chem., 22*: 3189 (1984).

47. M. Kapuscinska and E. M. Pearce, *J. Polym. Sci. Polym. Chem., 22*: 3989 (1984).

48. G. W. Gray, in *The Molecular Physics of Liquid Crystal* (G. R. Luckhurst and G. W. Gray, eds.), Academic Press, New York, p. 1 (1979).

49. D. A. Simoff and R. S. Porter, *Mol. Cryst. Liq. Cryst., 110*: 1 (1984).

50. E. Chiellini and G. Galli, *Faraday Disc., 79*: 241 (1985).

51. A. S. Angeloni, M. Laus, C. Castellari, G. Galli, P. Ferruti, and E. Chiellini, *Makromol. Chem., 186*: 977 (1985).

52. A. Blumstein, S. Vilasagar, S. Ponrathnam, S. B. Clough, and R. B. Blumstein, *J. Polym. Sci., 20*: 887 (1982).

53. S. Onogi and T. Asada, in *Rheology I* (G. Astarita, G. Marrucci, and L. Nicolais, eds.), Plenum, New York, p. 127 (1980).

54. Y. Onogi, J. L. White, and J. F. Fellers, *J. Non-Newtonian Fluid Mech., 7*: 121 (1980).

55. R. S. Porter and J. F. Johnson, in *Rheology IV* (F. R. Eirich, ed.), Academic Press, New York, p. 317 (1967).

56. D. G. Baird, in *Liquid Crystalline Order in Polymers* (A. Blumstein, ed.), Academic Press, New York, p. 237 (1978).

57. N. Yamazaki, M. Matsumoto, and F. Higashi, *J. Polym. Sci. Chem. Ed., 13*: 1373 (1975).

58. T. I. Bair, P. W. Morgan, and F. L. Killian, *Macromolecules, 10*: 1396 (1977).

59. B. Jingsheng, Y. Anji, Z. Shengging, Z. Shufan, and H. Chang, *J. Appl. Polym. Sci., 26*: 1211 (1981).

60. A. D. Gotsis and D. G. Baird, *Rheol. Acta, 25*: 275 (1986).

61. Y. Ide and Z. Ophir, *Polym. Eng. Sci., 23*: 261 (1983).

62. K. F. Wissbrun, *J. Rheol., 25*: 619 (1981).

63. K. F. Wissbrun, G. Kiss, and F. N. Cogswell, *Chem. Eng. Commun.* (1987).

64. M. Jaffe and R. S. Jones, in *Handbook of Fiber Science and Technology*, Vol. 3 (M. Lewin and J. Preston, eds.), Marcel Dekker, New York, p. 349 (1985).
65. H. N. Yoon, forthcoming.
66. A. V. Tobolsky and H. Mark, *Polymer Science and Materials*, Wiley, New York (1974).
67. J. R. Schaefgen, "A Comparative Analysis of High-Performance Fibers," Proceedings of Fiber Producer Conference, Greenville, South Carolina, p. 5A-1 (1986).
68. DuPont Technical Information, Kevlar High Strength Fiber.
69. DuPont Technical Information, Kevlar Aramide, Bulletin K-5 (September 1981).
70. R. E. Wilfong and J. Zimmerman, *J. Appl. Polym. Sci. Symp.*, *31*: 1 (1977).
71. M. H. Naitove, *Plast. Technol.*, p. 23, 1986 (February).
72. T. S. Chung, *J. Polym. Sci. Lett.*, *24*: 299 (1986).
73. F. E. Martin, U.S. Pat. 4,553,815 (1985).
74. S. Yamakawa, Y. Shuto, and F. Yamamoto, *Electron. Lett.*, *20*: 199 (1984).
75. Y. Ide and T. S. Chung, *J. Macromol. Sci. Phys.*, *B23*: 497 (1984–1985).
76. T. S. Chung and P. E. McMahon, *J. Appl. Polym. Sci.*, *31*: 965 (1986).
77. J. J. Duska, *Plast. Eng.*, *12*: 39 (1986).
78. M. H. Naitove, *Plast. Technol.*, p. 85, 1985 (April).
79. J. B. Stamatoff, personal communication.
80. E. J. Joseph, G. L. Wilkes, and D. G. Baird, *Polym. Eng. Sci.*, *25*: 377 (1985).
81. S. Kenig, Celanese Research Report No. 2795 (1983).
82. J. C. Chen, D. V. Gorky, R. C. Haley, F. C. Jaarsma, and C. E. McChesney, "Vectra: A Unique New Liquid Crystal Polymer Resin for High Strength/Stiffness Injection Molding and Extrusion Applications," Proceedings of 5th SPE RETEC, Ohio University, Ohio, p. 61 (1985).
83. M. M. Fein, "Introducing Xydar," Proceeding of Food Plastics Conference, Atlantic City, New Jersey (1984/1985).
84. K. Rogers, F. T. McDuffee, A. Holt, and A. Mekkaoui, "Improvements in Fused Optical Coupler Technology," Proceedings of SPIE Fiber Optic Couplers, Connectors, and Splice Technology, San Diego, California, *574*: 29 (1985).
85. H. J. Coles and R. Simons, in *Polymeric Liquid Crystals* (A. Blumstein, ed.), Plenum, New York, p. 351 (1985).
86. N. A. Platé and V. P. Shibaev, *Makromol. Chem. Suppl.*, *6*: 26 (1984).
87. H. Finkelmann, U. Kiechle, and G. Rehage, *Mol. Cryst. Liq. Cryst.*, *94*: 343 (1983).
88. H. Ringsdorf and R. Sentel, *Makromol. Chem.*, *183*: 1245 (1982).
89. H. Ringsdorf and H. W. Schmidt, *Makromol. Chem.*, *185*: 1327 (1984).
90. H. J. Coles and R. Simons, in *Recent Advances in Liquid Crystalline Polymers* (L. L. Chapoy, ed.), Applied Science Publishers, London, p. 323 (1985).

20

Dimensional Stability of Thermoplastic–Polymeric Liquid Crystal Blends

A. Apicella, L. Nicodemo, and L. Nicolais
University of Naples
Naples, Italy

R. A. Weiss
Institute of Materials Science
University of Connecticut
Storrs, Connecticut

INTRODUCTION

Processing of thermoplastic polymers often leads to parts which are partially oriented and contain frozen-in stresses. Nevertheless, molecular orientation is often imparted to the processed part in order to improve mechanical properties. The resulting objects present a very high molecular orientation and show a great degree of springback once subjected to temperatures higher than their own glass transition. At sufficiently high temperatures, these oriented thermoplastic materials partially recoil, modifying the shape of the formed parts (see Figure 1). The total deformation is generally described as formed by instantaneous elastic, viscoelastic, and viscous (plastic) components [1–3]. This poor dimensional stability reduces the range of application for oriented thermoplastic materials and remains one of the major problems in the forging of plastics. In this regard, stiff fibers can be added to a polymer to improve dimensional stability and to minimize the shrinkage normally encountered in common polymer-processing operations upon cooling the material from the melt.

The effect of reinforcing fibers and fillers on the mechanical properties of polymers is

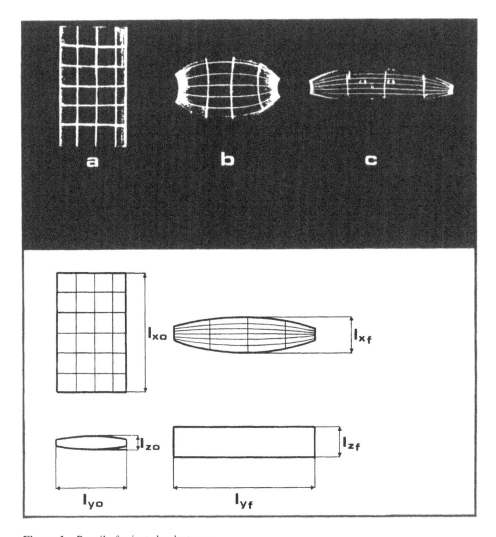

Figure 1 Recoil of oriented polystyrene.

well known. Structural fiber composite materials with very high specific properties have been successfully utilized in many engineering applications. Due to the higher strength of the fibers compared to that of ordinary polymeric materials, they normally impart strength to whatever matrix they are in. When the fibers are all aligned in one direction, maximum strength is achieved in the composite material along the direction of the fiber length.

It thus becomes possible to tailor-make specific properties in specific directions and thereby satisfy the design requirements of structures subjected to a variety of multiaxial stresses. Moreover, although these fillers are usually used to increase the stiffness of the composite, an improvement of the dimensional stability is also achieved. The extent of reinforcement and stabilization depends on factors such as the shape and the amount of filler, interfacial filler–matrix adhesion, and filler surface treatments. Nevertheless, the addition of rigid fibers to a polymer, which is a benefit in terms of mechanical and stability properties, has an adverse effect on its processability. For reasons similar to those accounting for the increase in the modulus and the decrease in the coefficient of expansion, the presence of the fibers in a polymer melt increases its apparent viscosity. The consequences of this are an increase in the power consumption of the plastics processing machinery, which is due to the too-high viscosity, and a breakage of the fibers, which is due to the high shear stresses generated during processing and drawing.

Thus there is an incentive to find alternative polymeric systems which exhibit improved properties without sacrificing the melt viscosity characteristics. A possible alternative is to use a system that behaves as homogeneous melt when processed at elevated temperatures but develops a rigid reinforcing phase once cooled down to the solid state. This goal can be achieved by adding to a polymer organic liquid crystal (PLC) material that has the potential of both reinforcing the polymer in one temperature interval and plasticizing the polymer in another temperature interval [4].

In general, a liquid crystalline material is one that may exhibit over a well-defined range of temperatures either isotropic or anisotropic liquid phases. As a consequence of their anisotropy, the molecules can be oriented in the shear fields developed during melt processing operations leading to preferred orientation of the crystalline molecules once the composite is cooled from the melt. These particulate fillers introduce a highly oriented second phase into the material, reducing the mobility and deformability of the matrix.

However, although much attention has been devoted in the scientific literature to mechanical properties of oriented and reinforced polymers, only a limited number of papers have been published on the fundamental studies of the recoil phenomenon active in such oriented thermoplastics [3, 5–7].

POLYMER PROCESSING AND DIMENSIONAL STABILITY

Orientation in Amorphous Polymers

From the morphological point of view, amorphous polymers are considered orderless structures composed of coiled and entangled molecules. Due to the high flow hindrance induced by the presence of such entanglements, it is only well above their glass transition, where interchain slippage does not require very high stresses, that an amorphous polymer can be considered a processable melt.

When deformation, or flow, is imposed to the melt, polymer chains acquire a preferential orientation, which increases their end-to-end mean distance and induces a mechanical and optical anisotropy either in the melt or in the solid glass once quenched below its glass transition. Several theories based on the analysis of forces such as the hydrodynamic drag, which orients the chains in the direction of the flow, and of Brownian type, which randomizes the chain, provide information about the behavior of the polymer chains in flow fields. An increase in the rate of deformation then results in a more or less pronounced chain elongation according to the material characteristic relaxation times. The driving force of such relaxation is the Brownian motion, whereas the flow drag, which is governed by the melt viscosity, is the retarding opposing force. Processing temperature therefore can have a dramatic influence on the rate of orientation loss.

During the application of a constant stress an elastic, a delayed elastic, and a viscous deformation are produced, but it is only the delayed elastic one that will produce frozen-in strains upon quenching. The elastic strain component, in fact, immediately relaxes, while the viscous one has no tendency to recover. This orientation (which is often present in injection-molded pieces) is stable below glass transition and is also accompanied by the development of birefringence and enhancement of strength in the direction of the applied stress. Any chain orientation present in the glassy state is practically permanent, unless the polymer is brought near and above glass transition. For oriented amorphous polymers annealed above the glass transition temperature the molecular randomization results in dimensional changes.

Orientation and Order in Liquid Crystal Polymers

A liquid crystal polymer (LCP) may exhibit both the disordered isotropic fluid state and the ordered crystalline solid state. The mesophase structure consists of highly anisotropic rodlike molecules.

The order of the LCP can be achieved under temperature or concentration gradients, electrical field, or mechanical shear stresses. The liquid crystalline materials show three principal mesophase transitions associated with the orientation and relative position of the macromolecules, alignment of molecular layers, and backbone morphology: smectic, nematic, and cholesteric phases.

In the smectic mesophase, the centers of the elongated molecules are equally arranged in ordered planes showing molecular orientation and bond orientation order. The nematic mesophase differs from the smectic in that it contains only molecular orientation. The cholesteric mesophase displays a twisted type of nematic phase with short-range positional order of the molecules varying helically along an axis perpendicular to the planes of orientation.

In some processing operations the orientation of such LCPs can be achieved under a stress field generated by a shear and elongational melt flows.

A temperature change can induce a mesophase transition in so-called thermotropic LCPs. At constant temperature, the rheological characteristics of thermotropic liquid crystals as a function of the shear rate are generally represented [8] by three regions: a low-shear-rate thinning, a plateau, and a high-shear-rate thinning. The liquid crystal anisotropic molecule behaves in a different manner under the melt flow conditions encountered under ordinary polymer processing. A rodlike structure, in fact, can be better oriented in an elongational flow than in a shear flow.

When these crystalline polymers are crystallized in the absence of external forces, there is no preferred direction in the specimen along which the polymer chains lie. If such unoriented crystalline polymer is subjected to an external stress, it undergoes a rearrangement of the crystalline material. During ordinary extrusion operations, and after a proper postdie treatment, which includes melt stretching and cooling, mechanically very strong crystalline fibers of reduced diameters and high anisotropy can be formed [9].

Their structure, as detected from x-ray techniques, is formed by aligned molecules where a long periodicity is present [9].

Rheology of Multicomponent Polymeric Systems

Although shear flows are the predominant forms encountered in most polymer processing, elongational (extensional or shear free) flows can be developed under particular polymer technologies such as uniaxial or biaxial stretching of molten films or the multiaxial stretching of a parison during blow molding.

These elongational flows are more complex to analyze than those involving shearing. In steady elongational flows a material function, the elongational viscosity, is defined to describe the normal stress difference. In this regard, stiff fibers can be added to a polymer to minimize the shrinkage normally encountered upon cooling the material from the melt in common polymer processing operations.

For example, the addition of small amounts of chopped glass filaments to a thermoplastic such as polystyrene can result in a significant reduction in film shrinkage during film extrusion or in the subsequent thermal treatments above glass transition, as indicated in Figure 2, where the equilibrium degree of recoil for highly oriented amorphous polymer is reported [6].

Figure 3 is a schematic of the changes in orientation of rigid fibers occurring during shear and elongational flows. In a shear flow, the stress field induces the rotation of the initially randomly distributed fibers rather than their orientation in the flow direction, such is the case of the alignment reached in an elongational flow. In both cases, although of different intensities, a great increase of the apparent viscosity, which is greatly influenced by the aspect ratio of the filler (see Figure 4), is attained. An increase of the length-to-diameter ratio is accompanied by a greater increase of the apparent viscosity as the filler content is increased.

With multicomponent polymeric systems containing a liquid second phase as indicated in Figure 3, the application of an elongational flow to the amorphous melt results in the

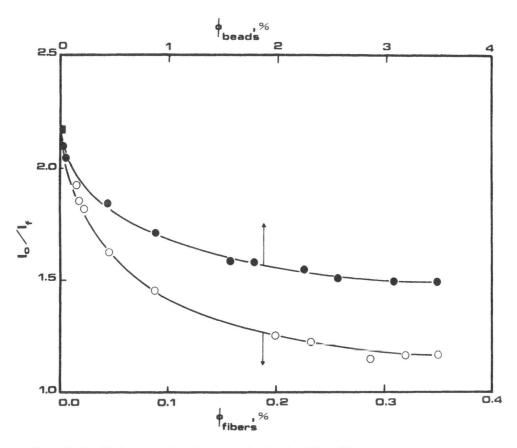

Figure 2 Equilibrium retraction of oriented glass bead and fiber-filled polystyrene.

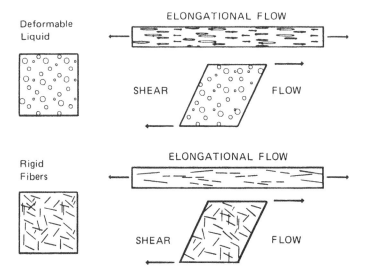

Figure 3   Elongational and shear flows for multicomponent polymer system.

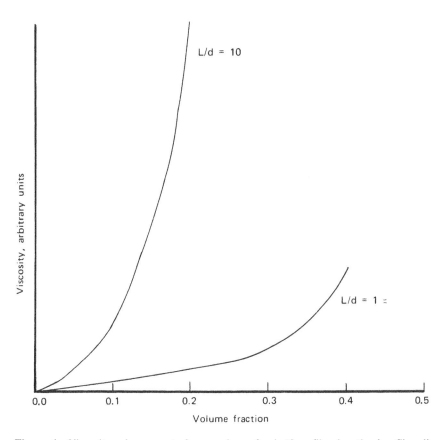

Figure 4   Viscosity enhancement of suspensions of rods (L = fiber length, d = fiber diameter).

development of shear stresses at the boundary between the phases, and to a stretching and deformation of the liquid inclusion with no enhancement of the viscosity of the system. In a shearing flow, also, the use of deformable liquid inclusions, such as the insoluble liquid crystals, which preserve an aspect ratio near unity, results in less pronounced viscosity enhancements, as seen in Figure 4.

Influence of Melt Orientation and Flow on the Properties of Composites

With amorphous polymers such as polystyrene, uniaxial orientation will increase the strength and the modulus of the polymer in the direction of orientation, as indicated in Figure 5, while reducing the same properties in the transverse direction [10]. To obtain the maximum performance from these materials, biaxial or multiaxial orientation is required. Once oriented thermoplastic materials are placed at sufficiently high temperatures, partial recoil will occur, modifying the shape of the parts as seen in Figure 1.

The effects of extrusion condition and subsequent drawing and cooling conditions on

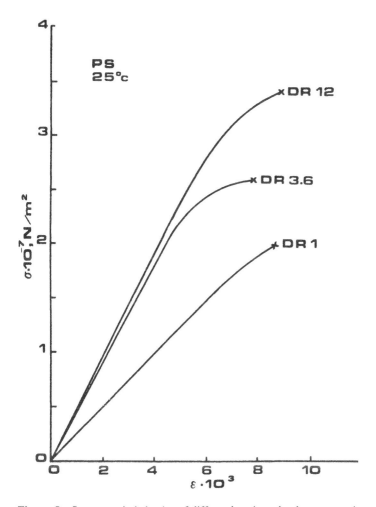

Figure 5 Stress–strain behavior of differently oriented polystyrene strips.

the fiber orientation, on the extent of damage done to the fibers during processing, and on the mechanical properties of the resulting composite sheet were presented earlier. The hot-drawing of extruded composite sheets induces an elongational flow which controls the orientation of both matrix and reinforcing fibers. A study was made on the effect of draw ratio on the properties of an extruded polystyrene sheet containing 0–1% of short glass fibers. An increase in draw ratio resulted in an increase in fiber orientation.

A model of rigid fiber rotating in an elongational flow field was used to describe the effect of the draw ratio on the final orientation distribution. The increase in the draw ratio resulted in either fiber orientation or an increase in the amount of fiber breakage. A shear lag analysis was used to estimate the extent of damage as a function of the draw ratio. When the total elongation of the molten matrix parallel to the fiber axis equals the elongation to break of the fiber there must be a rupture of either the fiber or the interface between the two phases. If the interface is strong enough to withstand the tensile and shear stresses generated at the interface, the fiber must break as a result of the imposed stress.

This would account for the fiber damage observed during the hot-drawing process. The higher the degree of orientation, the lower, independently of the initial size, is the length of the short fibers in the final product. The fiber breakage during drawing is principally a result of a shear stress developed at the fiber interfaces. A quantitative analysis of the latter effect is quite difficult unless the fiber is already aligned along the axis of elongation. In this case, the classical "shear lag analysis" of Kelly and Tyson (11) can be utilized. If the tensile load transmitted to the encapsulated fiber by means of a shear stress reaches the ultimate strength of the fiber, this will break at the weakest point. The length of the fiber for which the tensile stress reaches the fiber ultimate strength at its midpoint (Figure 6) is called the critical length L_c. Encapsulated fibers smaller than the critical length cannot be fractured in uniaxial tension. For a matrix that has a constant shear yield stress τ_y, the critical length is

$$L_c = \frac{\sigma_f d_f}{2\tau_y} \tag{1}$$

where σ_f and d_f are the ultimate strength and diameter of the fiber, respectively, and τ_y is the yield strength of the matrix.

From reported data [10, 11] it has been estimated that the elongational stresses developed during drawing of polystyrene are in the range of ~ 3 MN/m^2 for low draw ratios to ~ 7 MN/m^2 for higher orientations. From the von Mises criterion, the maximum shear stress developed should be of the order of 1.7–4.0 MN/m^2. For tensile strength of glass fibers of about 2100 MN/m^2 and fiber diameter of 0.01 mm, the critical fiber length varies from 6 to 2.6 mm. This was in excellent agreement with the maximum length found in the drawn sheets.

The Measure of Orientation and Dimensional Stability

As previously discussed, the randomization of the frozen-in strains induced by the increase of the temperature above the material glass transition is accompanied by dimensional changes. The recoil kinetics of polystyrene sheets obtained from extrusion and subsequent drawing have been extensively investigated. The length reversion ratio, defined as the ratio of the initial length of a drawn sample to the length of the same sample, has been measured after exposure at temperature higher than T_g for a defined length of time. Typical recovery experiments above T_g performed on highly oriented polystyrene sheets obtained from extrusion and subsequent hot-drawing [3–7] are reported in Figure 7.

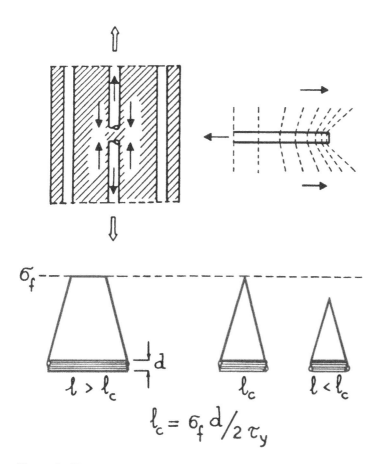

Figure 6 Shear-lag analysis.

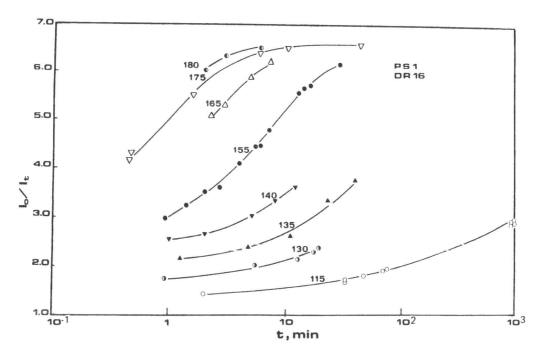

Figure 7 Recoil kinetics of oriented polystyrene strips as a function of temperature.

A shift procedure was successfully applied to the recoil data using the William–Landel–Ferry (WLF) equation on differently oriented samples. The shrinkage phenomenon proceeds at a constant volume since the material density does not change during the heat treatment. At constant volume $\epsilon_x\epsilon_y\epsilon_z = 1$ where $\epsilon_i = l_f/l_0$ and the subscripts 0 and f refer to the initial and final dimension of the sample according with the definitions given in Figure 1.

The inhomogeneous deformation in the x–y plane is due to the different degree of shrinkage between the central and lateral parts of the samples. It can be reasonably assumed that the region around the center line of the sample is in uniaxial extension during the extrusion, drawing, and recoil processes.

The recoil kinetics of stretched and oriented systems thus can be analyzed by measuring length along a symmetry line as a function of time. A typical kinetic curve of recoil is reported in Figure 7, where the ratio of the length at the time t, l_t, to the initial length, l_0, is plotted versus time for various test temperatures.

At the higher temperatures the asymptotic equilibrium length is reached after a relatively short time. Similar results were obtained for different drawing ratios. Assuming that the recoil is a viscoelastic process, it is expected that all the curves obtained at the different temperatures can be superimposed to form a master curve through a shifting procedure. The data are superimposed using horizontal shift factors a_t that produced the smoothest possible master curve and are reported as l_0/l_t versus t' in Figure 8 where t' is given by $t' = t/a_t$. The values of the shift factors are fitted by the curve obtained using the WLF equation:

$$\log a_t = \left(\frac{C_1(T - T_g)}{C_2 + (T - T_g)}\right) - \log a_{r0} \tag{2}$$

where T_g is the matrix glass transition, C_1 and C_2 are the WLF constants, and $\log a_{r0}$ is a constant which accounts for the chosen reference temperature.

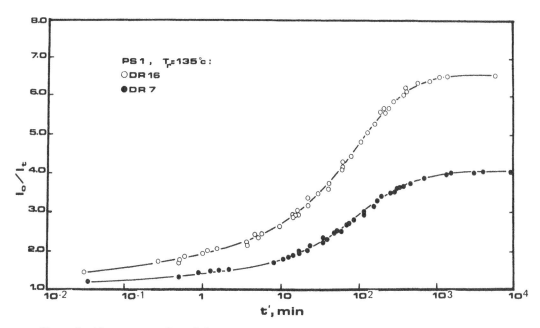

Figure 8 Master curves of recoil data.

The fact that the experimental shift factors follow very closely the WLF equation indicates that the recoil kinetic can be modeled as a viscoelastic process. Moreover, assuming that the length reversion ratio l_f/l_0 represents the maximum value of the shrinkage that can be obtained, a variable defined as the fractional distance from equilibrium ($\lambda = [(l_0 - l_t) - 1]/[(l_0 - l_f) - 1]$) should reduce all the data obtained on sheets drawn at different draw ratios to a single master curve (see Figure 9).

The experiments are well correlated by a single master curve and the value of the shift factors are fitted by the WLF equation. Any variation of the molecular weight or the presence of plasticizers that may alter the glass transition of the system does not produce a modification of the constant used in the WLF equation.

The kinetics of the retraction phenomenon is therefore a viscoelastic process that can be modeled once the glass transition temperature of the system and the equilibrium value of the shrinkage are determined. The presence of rigid fillers, such as glass beads and fibers, significantly enhances the dimensional stability of the drawn thermoplastic matrices. The second rigid phase does not influence the recoil kinetics of the polymeric matrix (see Figure 10, where the normalized curve of filled and unfilled stretched polystyrenes is reported), but has a strong influence on the equilibrium value of retraction, l_f, as already reported in Figure 2, where the length reversion ratio of glass beads and short

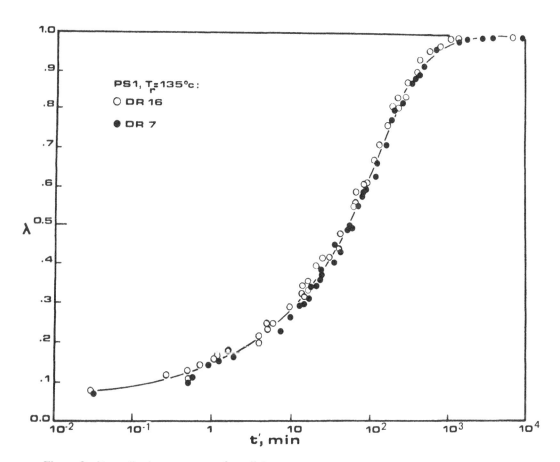

Figure 9 Normalized master curve of recoil data.

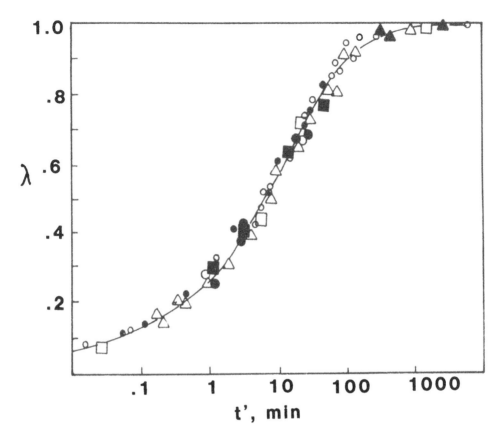

Figure 10 Normalized master curve, $T_r = 135°C$, for differently filled polystyrene. $\mathbb{O}$, $\phi = 0\%$; ▲, $\phi = 0.017\%$ f; $\bigcirc$, $\phi = 0.017\%$ b; ▼, $\phi = 0.35\%$ f; ●, $\phi = 3.5\%$ b.

fiber composites containing different amounts of inclusions are reported. In particular, very small fiber concentration (less than 0.35%) is sufficient to reduce l_0/l_t from 2.3 of the unfilled system (i.e., 130% of deformation) to a value of 1.15 (i.e., 15% of deformation).

The influence of spherical fillers on the dimensional stability is much less evident. In fact, different mechanisms are responsible for the improved dimensional stability of the short fiber and bead composites.

It is possible to assume that the fibers act as constraints for a certain amount of polymer surrounding them, leading to a partial relaxation of the internal stress and then to a randomization of the molecular orientation, which is not accompanied by strain recovery. Consequently, the fiber-filled systems partially creep and partially relax, and the contribution of the stress relaxation overcomes the creep behavior at very low filler concentration. The reduction of retraction in systems containing spherical fillers is due only to the presence of an inclusion which reduces the actual relaxing volume.

THE USE OF LIQUID CRYSTALS IN POLYMER PROCESSING

As previously discussed, the use of a small percentage of rigid fibers in oriented polymers will strongly contribute to the improvement of the dimensional stability of formed objects.

However, the processing operations induce severe damage on the fibers, reducing their length below a critical value, especially if high orientation is achieved. Moreover, the presence of the rigid fillers enhances dramatically the viscosity of the melt, suggesting the possibility of using incompatible thermotropic polymeric liquid crystals to improve the dimensional stability and to influence the rheological properties of the melt.

It is conceivable that one may exploit the isotropic liquid characteristics to plasticize the polymer melt at processing temperatures, and at the same time utilize the ability to induce order into the plasticizer by simply allowing the melt temperature to fall below the liquid crystal–isotropic liquid phase transition.

In very general terms, a liquid crystalline material is one that in addition to exhibiting an isotropic liquid phase also exhibits one or more anisotropic liquid phases. Although each of these phases represents thermodynamically stable liquids, the liquid crystal possesses a molecular order not normally associated with conventional liquids. Because molecules that exhibit liquid crystalline characteristics are elongated in shape, orientation of these molecules in a stress field, as encountered in polymer processing, may result in improvements in the composite mechanical properties similar to those attained by using reinforcing fibers. Compared with the other polymeric materials, these PLCs have very high unidirectional mechanical properties. They also can have outstanding chemical resistance and high dimensional stability, and they are easily processed, depending on the monomers used to prepare the polymers. These PLCs then act as fibers in hindering the recoil of polystyrene, as long as they exist as a separate solid phase above its glass transition temperature. They are therefore more useful for practical purposes the higher their melt temperature is.

The structures of typical liquid crystalline polymers are given in Figure 11. Commercial melt–processable polymeric liquid crystals include Vectra from Celanese and Xydar from Dartco Manufacturing Co. These are primarily aromatic polyesters based on *p*-hydroxybenzoic acid and hydroxynapthoic acid monomers or on terephthalic acid, *p,p'*-dihydroxybiphenyl, and *p*-hydroxybenzoic acid.

The elongational flow induces stretching of the molecules aligning the crystalline domains along the flow direction. Induced orientation effects of rigid molecules can be advantageous in the processability and the mechanical properties of these melts. The shape of the inclusion is also an important factor in influencing the material properties. The aspect ratio, the ratio of the length to the diameter of the filler, is crucial for the improvement of mechanical properties. As the fiber length increases, the portion of the fiber not carrying load decreases, improving strength, toughness, and modulus. Fibrous fillers with aspect ratio greater than 150 behave as continuous fibers in their ability to carry load transferred from the matrix. On the other hand, a plasticizing effect is expected when the LCP is in its isotropic form in the molten polymer. At elevated temperatures the additive can be completely dissolved in the polymer, leading to an increase in the flexibility and softness of the mixture, which favors the plastic flow by weakening the intermolecular bonding forces and the separation of the polymer chains. As the cooling of the material proceeds, however, the additive solubility should significantly decrease, finally becoming insoluble and crystallizing in the still molten matrix. It is therefore crucial to select LCPs with well-defined melting and solubility characteristics.

Compared with fiber-reinforced resins, thermotropic liquid crystal systems exhibit comparable mechanical properties. Typical behavior of thermotropic liquid crystal on cooling is characterized by a sudden viscosity drop, associated with the transformation from isotropic to nematic states, followed by an increase as the temperature further

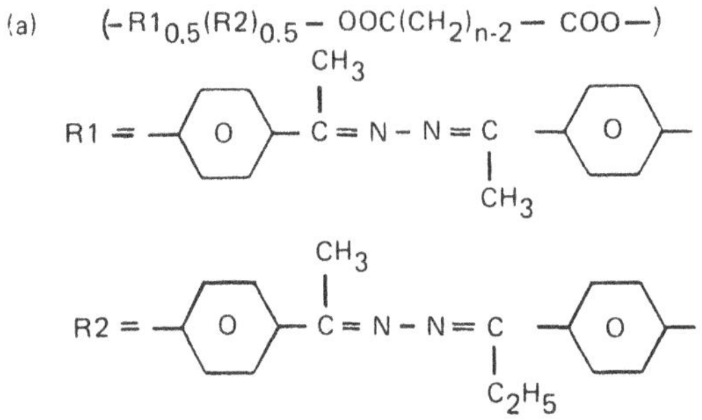

(a) $(-R1_{0.5}(R2)_{0.5} - OOC(CH_2)_{n-2} - COO-)$

$$R1 = -\!\!\left\langle\!\!\bigcirc\!\!\right\rangle\!\!- O -\!\!\left\langle\!\!\bigcirc\!\!\right\rangle\!\!- \overset{\overset{\displaystyle CH_3}{|}}{C} = N - N = \overset{\underset{\displaystyle CH_3}{|}}{C} -\!\!\left\langle\!\!\bigcirc\!\!\right\rangle\!\!- O -\!\!\left\langle\!\!\bigcirc\!\!\right\rangle\!\!-$$

$$R2 = -\!\!\left\langle\!\!\bigcirc\!\!\right\rangle\!\!- O -\!\!\left\langle\!\!\bigcirc\!\!\right\rangle\!\!- \overset{\overset{\displaystyle CH_3}{|}}{C} = N - N = \overset{\underset{\displaystyle C_2H_5}{|}}{C} -\!\!\left\langle\!\!\bigcirc\!\!\right\rangle\!\!- O -\!\!\left\langle\!\!\bigcirc\!\!\right\rangle\!\!-$$

with $n = 8$ for the PLC indicated as P_1 and with $n = 10$ for the PLC indicated as P_2 in the text.

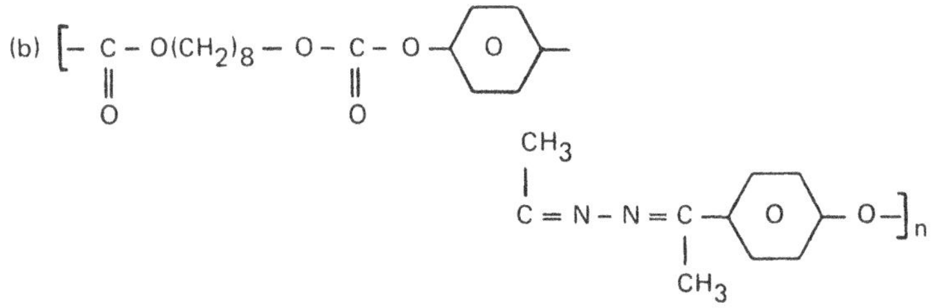

(b)

referred as P_3, and

(c) $C_4H_9 -\!\!\left\langle\!\!\bigcirc\!\!\right\rangle\!\!- O -\!\!\left\langle\!\!\bigcirc\!\!\right\rangle\!\!- N = C -\!\!\left\langle\!\!\bigcirc\!\!\right\rangle\!\!- C = N -\!\!\left\langle\!\!\bigcirc\!\!\right\rangle\!\!- C_4H_9$

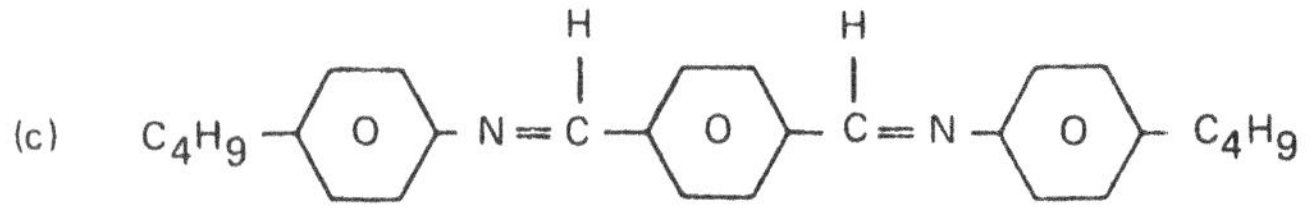

referred as TBBA (terephthal bis-4-n-butylaniline)

Figure 11 Polymeric liquid crystals.

decreases. The shear viscosity of the isotropic phase at high temperature is, in fact, higher than the viscosity of the nematic phase at lower temperatures. During flow, the mesophase domains easily slide by each other when the anisotropic phase is oriented along the flow direction, reducing viscosity and improving the melt processing characteristics. Moreover, the liquid crystal polymers possess relaxation times up to four times longer than normal polymers, presumably as a consequence of the more difficult cooperative motion and rotation of the molecular units. The morphology and the behavioral response of an LCP in shear and elongational flows are strongly dependent on the anisotropic nematic or isotropic state. Shearing of isotropic melts containing rod-shaped molecular units induces their rotation while the elongational flow induces local molecular orientation. Conversely, elongation of an anisotropic melt containing domains of local orientation induce their stretching and alignment. Evidence of the extent of orientation is particularly evident in the skin sections of injection-molded parts which undergo highly oriented flow field. In some cases the physical mixing of a polymer with a liquid crystal results in the plasticization of the matrix. Phase separation, plasticization, and mesophase transition temperature depressions have been analyzed by means of the classical Flory–Huggins equation of state for compatible systems. The plasticizing effect was identified from a decrease of the melt viscosity, a decrease of the glass transition, and clearing temperature of the system.

Thermotropic LCPs are often referred to as self-reinforcing thermoplastics. The aromatic backbone, in fact, stiffens the network structure, especially in the direction of the polymer chains orientation developed during processing. To improve the mechanical properties of an ordinary polymer, such self-reinforcing liquid crystals have been used in melt-processable polymer composites. During extrusion and subsequent drawing processes, the polymeric liquid crystals are molten and drawn in the direction of the extensional flow. In these stress fields oriented fibrils are formed and quenched during the cooling procedures. After a drawing process, liquid crystal domains are aligned along the flow direction and the in situ formation of fiberlike liquid crystal phase acts as a reinforcing filler for the blend. Morphological studies indicate the presence of layered structures, accounting for the high orientation of the liquid crystal phase. The mechanical properties and heat stability may be similar to those of fiber-reinforced composite materials.

The addition of a polymeric liquid crystal to a thermoplastic matrix can lead to different results, depending on the characteristics of the single components. If the PLC is compatible with the polymer, only a plasticization effect can be expected and only the recoil kinetics will be influenced. In contrast, if the PLC forms a second phase, it can be oriented during the drawing process and can act like short fibers, reducing the equilibrium value of retraction of drawn samples for temperatures between the glass transition of the matrix and the melting point of the crystalline phase. The P_1, P_2, and P_3 seen in Figure 11, for example, are immiscible in the polystyrene matrices investigated in Refs. 8 and 9, whereas the TBBA was found miscible with polystyrene at all the temperatures investigated and up to concentrations greater than 10%. The presence of a second phase and of orientation both in the matrix and the liquid crystalline phase can be evidenced by using polarized light microscopy and x-ray diffraction techniques. The x-ray diffraction patterns of unoriented crystalline polymer are characterized by well-defined rings that break into arcs and then to more defined and sharp patterns as orientation is increased. Figure 12 reports the diffraction patterns of a single PLC fiber and of amorphous stretched polystyrene containing 5% of P_3. It is evident that the PLCs form fibrils well oriented in the drawing direction, as also indicated by the optical observation in polarized light (see Figure 13a). In the temperature region of existence of the crystalline phase, the PLC acts as a short fiber filler, enhancing the dimensional stability of the oriented polymer. The same optical observation (see Figure 13b) indicates that after a heat treatment above the glass transition temperature of the glassy matrix but below the melting temperature of the crystalline phase (i.e., 130°C in the case reported here), the orientation and fibril structure of the PLC guest is still preserved. Figure 14 reports the length reversion ratio, l_0/l_f, for unfilled polystyrene and for blends containing PLC characterized by different melting behavior and temperature. The reported data for temperatures lower than 125°C cannot be considered equilibrium values because residence experimental times too long are needed in order to reach the equilibrium retraction (in this case the tests were stopped after one week). Therefore, the low temperature dependence of the length reversion ratio with the temperature for the unfilled polystyrene is due to the different degree of springback achieved when the experiment was interrupted. For higher temperatures, in fact, the equilibrium retraction is, as expected, temperature independent. The data of Figure 14 relative to the PLC-filled samples clearly indicate the positive influence of the fillers on the dimensional stability of the oriented matrix. In the range of temperatures in which the

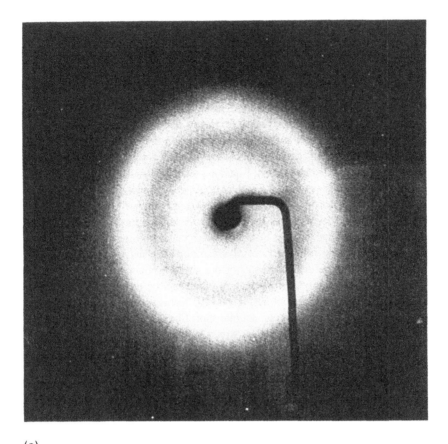

(a)

Figure 12 X-ray diffraction pattern of (a) P_3-filled polystyrene and (b) P_3-(fiber)-CuKα, flat film camera.

crystalline phase is stable, the blends behave similarly to the glass–short-fiber composites previously discussed. While the presence of the incompatible liquid crystalline filler has strongly affected the extent of the recovery, the recoil kinetics have not been altered by the presence of the PLC, as already observed for the glass–short-fiber composites.

The reinforcing effect of the PLC is then preserved up to temperatures close to the melting zone of crystals, where the filler loses its shape and effectiveness. Once the melting temperature is reached, the PLC cannot act as a fiber and previously hindered springback takes place. As a consequence, the equilibrium length reversion ratio of the blend becomes similar to that obtained by the unfilled polystyrene. The presence of a second insoluble phase or the plasticization effect of the LCP can be evidenced by using differential scanning calorimetry techniques. A DSC thermogram of a multicomponent polymeric system will present the thermal events characteristic of the single components, i.e., the glass transition of amorphous matrix and the melting of the liquid crystalline filler.

If the liquid crystal is compatible, and hence soluble in the matrix, the DSC thermogram will present only the reduced glass transition temperature of the system. The thermal

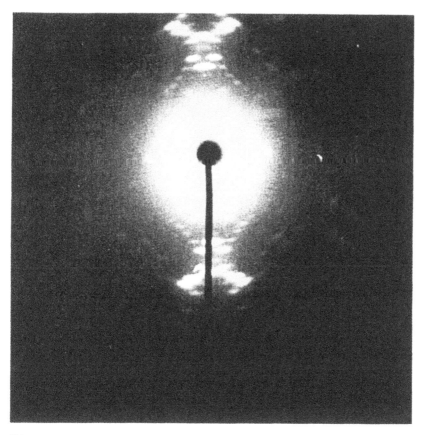

(b)

behavior of the blends of polystyrene and highly soluble TBBA is the same as that of the unfilled polystyrene. The high solubility of the TBBA in polystyrene makes this PLC act as a plasticizer and not as a reinforcing filler because the liquid crystal is dissolved in the polymer and cannot form a second phase. Figure 15 shows the DSC thermograms relative to the type 1 and 3 liquid crystalline polymers and of the drawn polystyrene samples blended with the type 2 PLC. The low-temperature peak around 85°C, which can be seen on the thermograms relative to the composites, is due to the high degree of orientation of the thermoplastic matrix. These subglass transition thermal events have been also observed in stretched polystyrene [4–7]. These DSC thermograms can be used to determine the processing and application temperature limits of the blends. As indicated in Figure 14 and Figure 15, the PLC–polystyrene composite will be stable up to the melting point of the crystalline phase, i.e., 160 and 180°C for type 1 and 3 liquid crystals.

The sharper the melting behavior of the crystalline phase, the more defined is the limit between the zones of dimensional stability and springback. Moreover, the use of liquid crystals in polymer processing can significantly reduce the viscosity of the melt, improving the processability of the blends. Cogswell et al. [12] and Nicolais et al. [13] have shown that the melt viscosity of such a blend is much lower than that of unfilled polymer, particularly at high shear rates, which are the conditions similar to those encountered during the molding and extrusion operations.

(a)

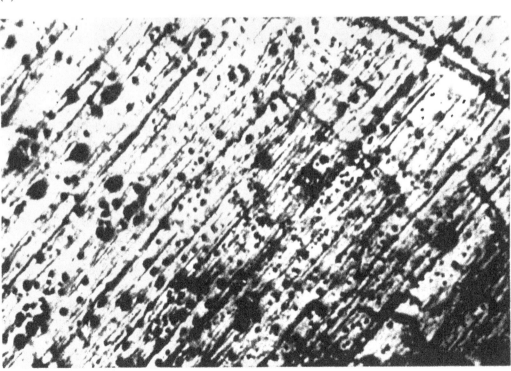

(b)

Figure 13 (a) P_1-filled polystyrene; (b) partially recoiled P_1-filled polystyrene. Both are observed with polarizing microscopy.

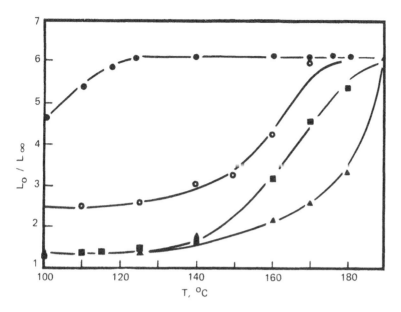

Figure 14 Equilibrium recoil vs. temperature. ●, unfilled polystyrene; ○, P_2-filled polystyrene; ■, P_1-filled polystyrene; ▲, P_3-filled polystyrene.

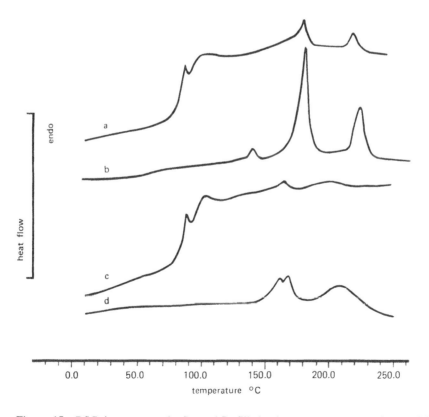

Figure 15 DSC thermograms for P_3- and P_1-filled polystyrene (curves a and c), and for pure P_3 and P_1 (curves b and d).

REFERENCES

1. A. Peterlin, ed., *Plastic Deformation of Polymers*, Marcel Dekker, New York (1971).
2. R. J. Samuels, *Structured Polymer Properties*, Wiley-Interscience, New York (1974).
3. A. Ram, Z. Tadmor, and M. Schwartz, *Int. J. Polym. Mat.*, *6*: 57 (1977).
4. Wansoo Huh, R. A. Weiss, and L. Nicolais, *Polym. Eng. Sci.*, *23*: 779 (1983).
5. A. Apicella, L. Nicodemo, and L. Nicolais, *Rheol. Acta*, *19*: 291 (1980).
6. L. Nicolais, A. Apicella, and L. Nicodemo, *Polym. Eng. Sci.*, *21*: 151 (1981).
7. L. Nicodemo, L. Nicolais, and A. Apicella, *J. Appl. Polym. Sci.*, *26*: 129 (1981).
8. S. Onogi and R. S. Asada, in *Rheology*, Vol. 1, Plenum, New York, p. 127 (1980).
9. A. Apicella, P. Iannelli, L. Nicodemo, L. Nicolais, A. Roviello, and A. Sirigu, *J. Appl. Polym. Sci.*, *26*(9): 600 (1986).
10. L. Nicolais, L. Nicodemo, P. Masi, and A. T. DiBenedetto, *Polym. Eng. Sci.*, *19*(14): 1046 (1979).
11. A. Kelly and W. R. Tyson, *J. Mech. Phys. Solids*, *13*: 329 (1965).
12. F. N. Cogswell, B. P. Griffin, and J. B. Rene, U.S. Pat. 4,433,083 (1984).
13. L. Nicolais et al., unpublished data (1987).

21

Molecular Aggregation of Solid Aromatic Polymers

Masakatsu Kochi
University of Tokyo
Tokyo, Japan

INTRODUCTION

During the late 1950s and early 1960s aromatic polymers were produced for the development of heat-resistant polymers. More recently, rodlike aromatic polymers have attracted considerable attention as ultra–high-strength, high-modulus fibers. Because of the conjugated moieties in main chains, aromatic polymers are also expected to be highly electrically conductive or photosensitive materials.

Aromatic and/or heteroaromatic nuclei are planar and rigid, and if suitable substituents are correctly positioned in the aromatic rings, intermolecular forces should operate strongly enough for elongated molecules to form liquid crystals [1]. Therefore, we can expect some molecular order in the molecular aggregation of rigid aromatic polymers. For example, polybisbenzimidazobenzophenanthroline-dione (BBB) [2], a heat-resistant heterocyclic polymer, shows extensive intermolecular association, giving a supermolecular structure with nearly planar repeat units stacked in a "graphitelike" array. In thermoplastic aromatic polymers with flexible main chains, the intermolecular ordering should be generally weak, as is the case of poly(bisphenol-A carbonate) [3].

This chapter is concerned with the molecular aggregation of typical heat-resistant polymers, polyimide and polyamideimide, and the relation between their molecular aggregation and mechanical properties.

MOLECULAR AGGREGATION OF SOLID AROMATIC POLYIMIDE AND POLYAMIDEIMIDE

Polypyromellitimide

Polypyromellitimide (PI) shows excellent chemical, mechanical, and electrical performance over a wide temperature range.

As seen in Figure 1, PI has rigid planar aromatic and imide rings in the main chain, and

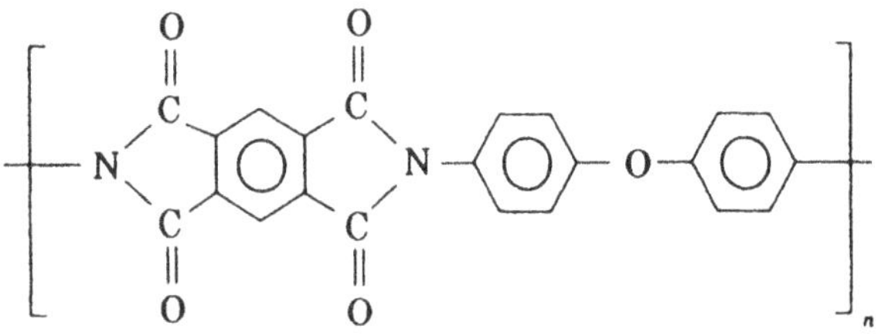

Figure 1 Chemical structure of PI.

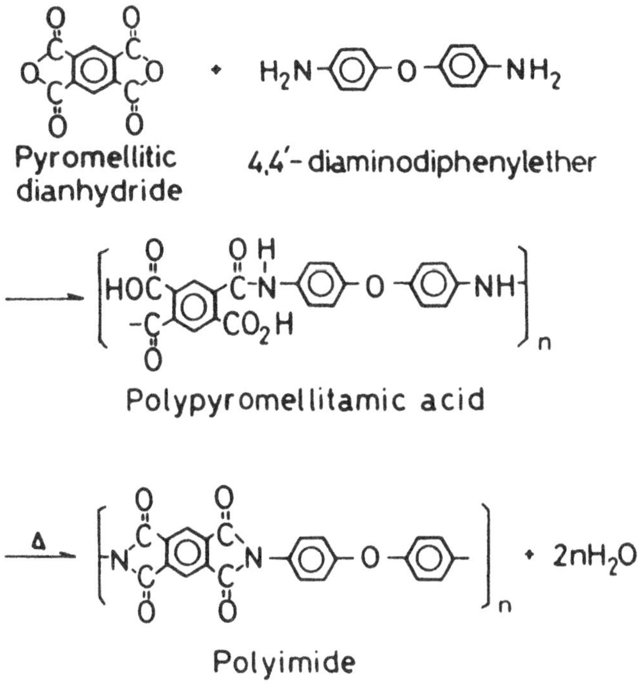

Pyromellitic dianhydride

4,4'-diaminodiphenylether

Polypyromellitamic acid

Polyimide

Figure 2 Preparation procedure of PI.

the length of the rigid moiety corresponding to the distance between ether linkages is ~18 Å. Conformational changes in this chain occur only by internal rotation around ether linkages.

Some authors indicate that the molecular aggregation of PI gives only an amorphous state [4, 5], whereas other authors refer to the crystallizability of PI [6–8]. The inconsistencies mean that the solid state of PI is sensitive to thermodynamic processes, including imidization temperature.

By wide-angle x-ray diffraction (WAXD) analysis of highly oriented PI fibers and films, Kazaryan et al. [7] showed that the space lattice of PI has orthorhombic symmetry with lattice parameters $a = 6.31$, $b = 3.97$, and $c = 32$ Å (fiber axis), and that the molecule has a second-order helix axis passing along the fiber axis.

Isoda et al. [9] analyzed the molecular aggregation of PI film by small-angle x-ray scattering (SAXS). PI films were prepared in three steps, as shown in Figure 2:

1. Polyamic acid (PAA), the precursor of PI, was polymerized in dimethylacetamide (DMAc) solution by dissolving 4,4'-diaminodiphenylether and pyromellitic dianhydride. The resulting prepolymer solution contained approximately 10 wt% solids. The inherent viscosity of the prepolymer was measured in DMAc solution and the molecular weight was estimated to be about 30,000.

2. The prepolymer solution was cast onto a glass plate and leveled off to a thickness of approximately 70 μm with a doctor blade. Each film was allowed to stand at room temperature for 24 hr and then stripped from the glass plate. The sample films were wrapped with an aluminum foil without frame.

3. The uncured film was then placed in a vacuum oven that had been heated in advance to each imidization temperature. Four kinds of PI films obtained under the thermal imidization conditions are listed in Table 1.

Figure 3 shows the SAXS curves for PAA and PIs which were desmeared by Glatter's method [10] after correction for background and absorption effects. The SAXS curves for PAA and PI-1 first decrease monotonously with increasing scattering angle 2θ and reach nearly constant values for $2\theta > 0.4$. These behaviors are generally found in SAXS curves for randomly coiled amorphous polymers. Accordingly, the molecular aggregation of PAA and PI-1 may be concluded to be amorphous. On the other hand, diffraction maxima were observed in the SAXS curves for PI-2, 3, and 4. With increasing initial imidization temperature these maxima are shifted to lower angles and the scattering intensity is increased. The presence of diffraction maxima in the small angle region means that there exist scattering elements of some heterogeneous superstructure.

A one-dimensional model is used to successfully analyze these SAXS curves with

Table 1 Thermal Imidization Conditions

Sample	Condition
PI-1	3.5 hr at 333 K, 8 hr at 430 K, 4 hr at 470 K, 5.5 hr at 520 K
PI-2	12 hr at 530 K
PI-3	12 hr at 630 K
PI-4	2 hr at 710 K

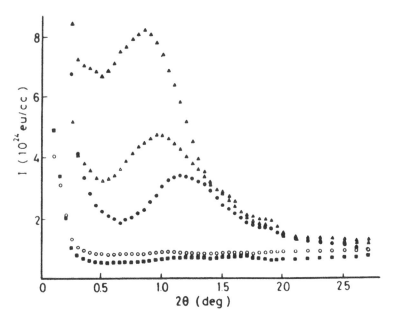

Figure 3 Desmeared SAXS curves for PAA and PI. ■, PAA; ○, PI-1; ●, PI-2; △, PI-3; and ▲, PI-4.

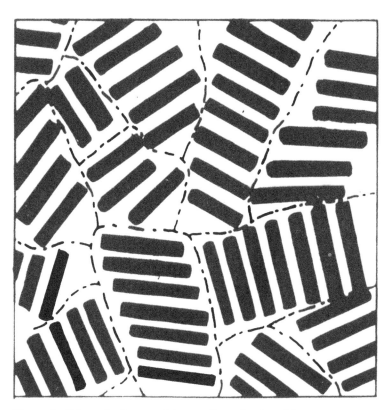

Figure 4 Schematic diagram of the one-dimensional model.

diffraction maxima [11, 12]. In this model ordered and less-ordered phases are arranged alternately in a one-dimensional array within lamellae stacks, as seen in Figure 4. The stacks are superstructure elements; it is assumed that there is no constructive interference of x-rays scattered from the different stacks.

The theoretical and experimental SAXS curves for the one-dimensional model are seen in Figure 5. The theoretical curves are calculated by Eq. A1 (see the appendix). Each result for PI-2, 3, and 4 demonstrates that the agreement between calculation and experi-

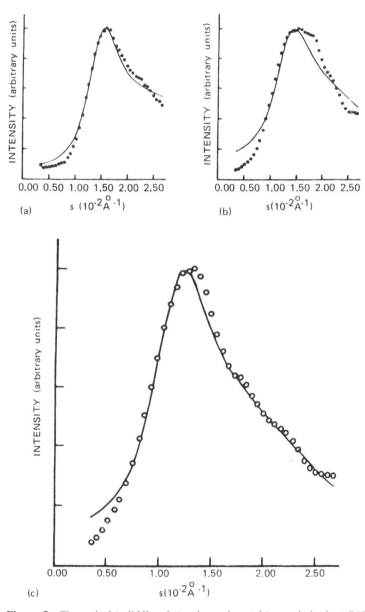

Figure 5 Theoretical (solid line, I_{th}) and experimental (open circle, I_{exp}) SAXS curves for (a) PI-2, (b) PI-3, and (c) PI-4.

ment is very good. It is concluded that the two-phase structure of PI can be explained by the one-dimensional model.

The superstructure parameters of PI obtained with the one-dimensional model are given in Table 2. Here $\bar{x}_0$ is the average thickness of the ordered lamellae, $\bar{x}_1$ is the average thickness of the less-ordered layers, $\bar{x}_b$ is the average thickness of the boundary layers, $\phi_0 = \bar{x}_0/(\bar{x}_0 + \bar{x}_1)$ is the volume fraction of the ordered phase, and Δ is the quadratic deviation of the theoretical curve from the experimental curve. Inspection of Table 2 shows that the average thickness of the ordered lamellae $\bar{x}_0$ and the volume fraction of the ordered phase ϕ_0 increase with increasing initial imidization temperature. This indicates that the molecular order in PI-2, 3, and 4 grows within molecules of PAA where segmental motions are still free. In Table 2 the average thickness of the ordered lamellae was found to have a small value of 10–25 Å. From analysis of the crystal structure of PI, the intramolecular repeat length was found to be approximately 32 Å. Considering these results, molecular chains of PI cannot be arranged in the direction normal to the lamellae; thus folded-chain structure cannot be assumed. Accordingly, it

Table 2 Superstructure Parameters for PI

Sample	$\bar{x}_0$ (Å)	$\bar{x}_1$ (Å)	$\bar{x}_b$ (Å)	Δ (%)	ϕ_0 (%)
PI-2	13	48	3	0.48	21
PI-3	18	48	4	1.57	29
PI-4	23	53	5	0.36	30

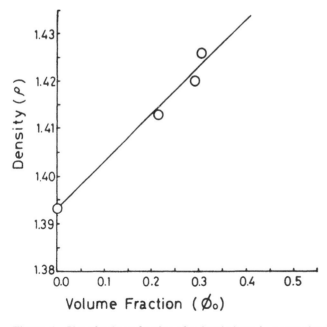

Figure 6 Plot of volume fraction of ordered phase ϕ_0 versus density ρ.

may be concluded that molecular chains in the ordered lamellae are arranged parallel to the lamellae as extended chains.

Figure 6 is a plot of the volume fraction of the ordered phase ϕ_0 in Table 2 versus the observed density ρ. The curve in Figure 6 was extrapolated to find $\rho_0 = 1.49$ g/cm³ for $\phi_0 = 100\%$ and $\rho_1 = 1.39$ g/cm³ for $\phi_0 = 0\%$, where ρ_0 and ρ_1 are the density of the ordered and the less-ordered phases, respectively. The value of ρ_0 is very small compared with that of an ideal PI crystal, $\rho_c = 1.58$ [7]. And the relative density difference $\Delta\rho/\rho$ is calculated to be approximately 7%. These results show that the ordered phases in PI are not completely crystallized and that the two-phase structure of PI is a character of low degree of inhomogeneity, differing essentially from that of ordinary crystalline polymers.

Russel [13] reported that the total SAXS intensity of PI decreased with initial imidization temperature, which contrasts with the result seen in Figure 3. The differences between these two studies may be due mainly to an effect of heating rate of PAA films on the morphology, because the molecular aggregation of PIs imidized under the conditions in Table 1 is not affected by thermal aftertreatment [9].

Takahashi et al. [14] investigated the molecular order in homogeneously condensed states of PAA and PI thermally imidized at 200°C for 1 hr by WAXD and reported that PAA chains in films exhibit a high molecular order, resembling that of a smectic liquid crystalline state, and this order is preserved and somewhat improved upon thermal imidization to form PI films. The imidization condition applied by them seems to be insufficient, because the conversion to imidization of PI imidized at 200°C for 1 hr is about 70% [15]. But their studies stimulate further investigations on the molecular order vs. chemical structure for other polymers comprising planar cyclic groups, in order to better define the boundary of disorder–order transition of semiflexible polymers [16, 17].

In-plane anisotropy can have important effects on physical properties of polyimide films [18]. Russel et al. [19] used integrated optics to investigate thin films (<8 μm) of PAA and PI as a function of molecular weight, initial imidization temperature, method of imidization, and annealing treatment. PAA films were found to exhibit a large optical anisotropy, indicating preferential alignment of the long axis of the molecule in the plane of the film. The only parameter to affect the anisotropy of the films was the method of imidization. In comparison with thermal imidization, chemical imidization was found to increase the birefringence more, indicative of a higher degree of molecular orientation parallel to the film surface.

Kapton H

The molecular aggregation of Kapton H (KH, registered trade name of PI, E. I. duPont de Nemours & Co.) is anisotropic unlike cast PI films [20]. Figure 7 shows the WAXD photographs of KH, the incident collimated x-ray beam being normal to the film plane (through direction) in Figure 7a and to the film edge (edge direction) in Figure 7b.

By considering the crystal data [7], the interplanar spacings d were calculated approximately (see Table 3). Since the spacing $d \approx 17$ Å is in fair agreement with the expected monomer length 18 Å, the (002) reflection can be reasonably attributed to the periodic repeating structure within the polymer chain. The spacing $d \approx 4.5$ Å, corresponding to the (010) reflection, may be attributed to interchain interference, i.e., to the spacing between adjacent parallel chains [8, 21]. As is seen in Figure 7b, the intensities of the intramolecular and intermolecular reflections are strongest along the equatorial and meridional directions, respectively. In addition, the meridional reflections are seen to be very broad in

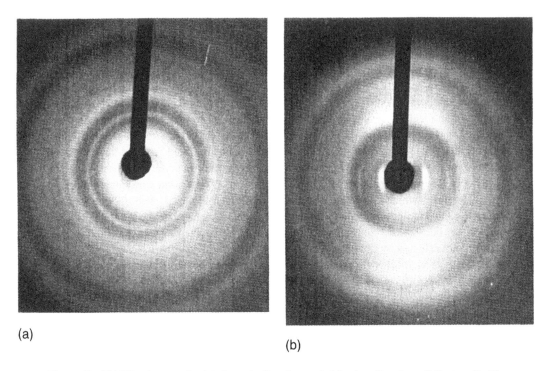

(a)

(b)

Figure 7 WAXD photographs (a) through direction and (b) edge direction of Kapton H. The primary axis of cylindrical symmetry is horizontal.

Figure 7a. These patterns indicate that molecular chains are randomly aligned within the film plane and the intermolecular order in KH is not as high as in the crystalline state.

Figures 8a and 8b show SAXS photographs of KH in the through and edge directions, respectively. The halo in the outermost region is the (002) reflection. The isotropic halo with its maximum at small scattering angle ($2\theta \approx 0.85°$) in Figure 8a suggests the existence of a two-phase structure in this film. By circular scanning densitometry of the photograph in Figure 8b, the long-range order in the small-angle region is found to be oriented along the equatorial direction, i.e., anisotropic out of the film plane.

Figure 9 plots the SAXS curve of KH, which was obtained by recording the intensity with the Kratky slit parallel to the KH film plane, i.e., perpendicular to the primary axis. The small-angle peak at $s \approx 1.0 \times 10^{-2}$ Å^{-1} is so sharp that the one-dimensional model may be adequate for analysis of the SAXS curve for KH.

Table 3 Spacings in WAXD Photographs of Kapton H

$h\ k\ l$	d (Å)	$h\ k\ l$	d (Å)
0 0 2	17.0	1 0 0	6.8
0 0 4	8.8	0 1 0	4.5
0 0 6	5.8	1 1 0	3.8

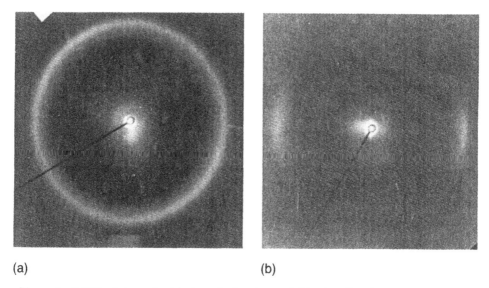

(a)   (b)

Figure 8 SAXS photographs (a) through direction and (b) edge direction of Kapton H. The primary axis is horizontal.

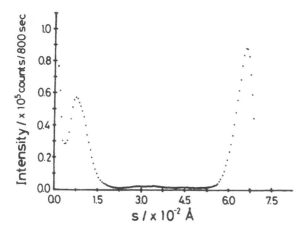

Figure 9 Desmeared SAXS curve for Kapton H; $s = 2(\sin \theta)/\lambda$ where λ is the x-ray wavelength.

The anisotropy in the SAXS from KH makes the Lorentz correction for the measured intensity complicated. For isotropic samples the Lorentz factor is s^2, but if the lamellar axis were perfectly aligned within the film plane, it would be s [22]. Table 4 lists the superstructure parameters obtained from the one-dimensional model with the anisotropic Lorentz factor s and the isotropic Lorentz factor s^2.

As is seen in Table 4, the deviation Δ is fairly large. This may be due to the fact that the SAXS peak area for KH is small and that the Lorentz factor cannot be unequivocally chosen because of the imperfect alignment within the film plane. However, since the values of $\bar{x}_0$, $\bar{x}_1$, and ϕ_0 are almost independent of the Lorentz factors used here, they

Table 4 Superstructure Parameters for Kapton H

Lorentz factor	$\bar{x}_0$ (Å)	$\bar{x}_1$ (Å)	$\bar{x}_b$ (Å)	Δ (%)	ϕ_0 (%)
s	44	43	0	2.22	51
s^2	40	44	0	2.84	48

appear to be reliable. The volume fraction ϕ_0 of the ordered phase for KH is about 50%. The PI-4 sample, which was imidized at 440°C for 2 hr in vacuum, resembles KH very closely in molecular aggregation and mechanical properties, but its volume fraction of the ordered phase is smaller than that for KH. Taking into account the fact that the density of KH is nearly the same as that of PI-4, this result indicates that the density of the ordered phase in KH is lower than that in PI-4. This is considered to result from the manufacture or subsequent processing of Kapton. The average thickness $\bar{x}_0$ of the ordered lamellae is about 40 Å, which is almost twice as large as that for PI-4. However, considering its large monomer repeat length, KH, like PI-4, appears to assume the extended-chain conformation in its ordered phase. Wellinghoff et al. [23] postulated the existence of bundles of parallel-packed PI chains to explain the yield behavior of KH. This conclusion supports their hypothesis.

Polytrimellitamideimide

Polytrimellitamideimide (PAI) is widely used in industry because of its outstanding thermal stability and solubility. The chemical structure is shown in Figure 10. The molecular aggregation and the effect on it of annealing near the glass transition temperature T_g were studied by SAXS [24]. Figure 11 shows desmeared SAXS curves for PAI films which were annealed at different temperatures in vacuum: PAI-1, as-received; PAI-2, annealed at 200°C for 1 hr; PAI-3, annealed at 250°C for 1 hr; PAI-4, annealed at 280°C for 1 hr; PAI-5, annealed at 330°C for 1 hr. The SAXS curve for PAI-1 decreases monotonically over a wide angular range beyond scattering angle $2\theta = 1.0°$. With increasing annealing temperature, the SAXS curves in Figure 11 become more intense and display a clearer scattering peak.

Assuming a density transition occurs at a phase boundary of finite width, the SAXS theory for a two-phase system gives the following equation for the scattering intensity $\bar{I}$ for infinitely long-slit collimation [25]:

$$m^3\bar{I} = K_1\left[1 - \left(\frac{2\pi^2 E^2}{3\lambda^2 R^2}\right)m^2\right] \tag{1}$$

at the high-angle tail of the scattering curve. Here $m = (2R\sin\theta)/\lambda$ is the distance between the incident and scattered x-ray in the recording plane, with R giving the distance between the sample and the recording plane and λ the wavelength of the x-ray beam; K_1 is a constant and E is the width of a transition layer.

Figure 12 shows plots of $m\bar{I}$ versus m^{-2} (Eq. 1) for PAI. The fluidlike scattering was eliminated before plotting. The plots for PAI-3, 4, and 5 are straight lines, which shows that the molecular aggregation in these samples has heterogeneous two-phase structure. The width of the density transition layer is roughly estimated to be on the order of 10 Å

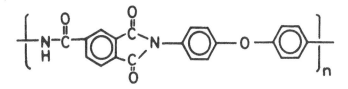

Figure 10 Chemical structure of PAI.

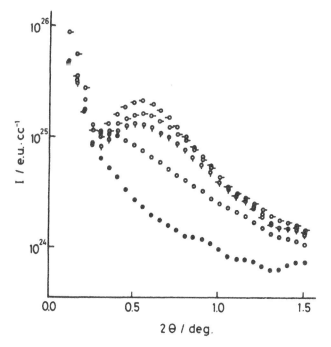

Figure 11 Desmeared SAXS curves for PAI. ●, PAI-1; ○, PAI-2; ♀, PAI-3; ○-, PAI-4; and -○, PAI-5.

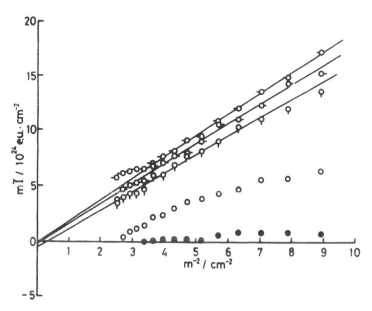

Figure 12 Plots of ml vs. m^{-2} for SAXS of PAI. ●, PAI-1; ○, PAI-2; ♀, PAI-3; ○-, PAI-4; and -○, PAI-5.

from the intercepts of these plots. On the other hand, the linearity of the plot for PAI-2 is not as good, and that for PAI-1 is very poor, which means that a two-phase model cannot be successfully applied to these samples.

The theoretical SAXS curves for PAI-3, 4, and 5 can be calculated with the equations for a one-dimensional model. The Lorentz correction for an isotropic sample is applied again. This correction factor shifts the position of the peak to higher angles by an amount that increases with the width of the peak [22]. The fitting procedure is performed to minimize the deviation of the theoretical curve from the experimental curve.

Figure 13 shows the theoretical and experimental SAXS curves for PAI-3, 4, and 5. Here $s = (2 \sin\theta)/\lambda$ denotes the reciprocal-space coordinate. It should be noted that the position of the reciprocal maxima in the experimental curves shifts to higher angles with decreasing annealing temperature when that in the desmeared curves (Figure 11) remains almost constant in scattering angle. This is caused by the Lorentz correction, as mentioned earlier. Figure 13 demonstrates that the theoretical SAXS curve fits the experimental curve and leads to the conclusion that the two-phase structure of these three samples can be elucidated by the one-dimensional model. The theoretical curves for PAI-1 and 2 do not fit experiments as well. This suggests that the ordered phases in PAI-1 and 2, if any, would not arrange so regularly.

The superstructure parameters of PAI-3, 4, and 5 determined with the one-dimensional model are listed in Table 5. Of the results shown in Table 5, the most striking is that the average thickness $\bar{x}_0$ of the ordered lamellae is only about 20–30 Å. Considering the very high rotational potential barrier for the Ph–N and Ph–C bonds [26], the main chain of PAI should have very low flexibility, and the length of the stiff unit for PAI should reasonably be estimated to equal the length of its monomer unit (about 20 Å). Hence it may be concluded that the molecular chains of PAI in the ordered lamellae are arranged parallel to the lamellar surface as extended chains. It is also noted that with increasing annealing

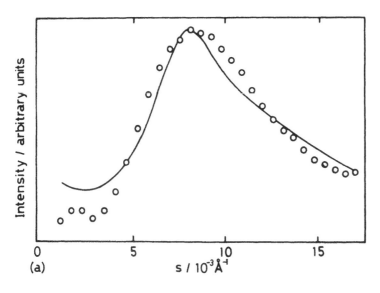

Figure 13 Experimental ($\bigcirc$, I_{exp}) and theoretical (———, I_{th}) curves for (a) PAI-3, (b) PAI-4, and (c) PAI-5.

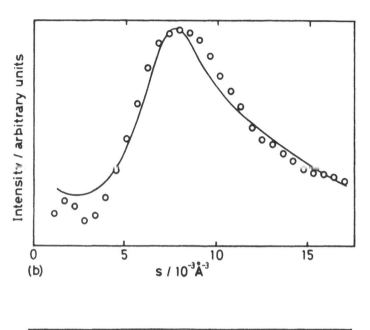

(b)

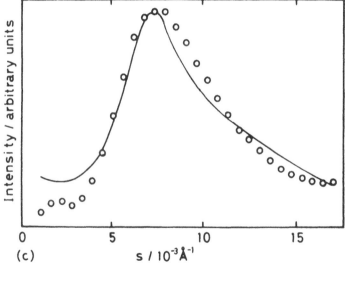

(c)

Table 5 Superstructure Parameters for PAI

Sample	$\bar{x}_0$ (Å)	$\bar{x}_1$ (Å)	$\bar{x}_b$ (Å)	Δ (%)	ϕ_0 (%)
PAI-3	20	92	14	1.97	18
PAI-4	22	94	14	1.39	19
PAI-5	26	99	15	2.59	21

Table 6 Superstructure Parameters for Two-Phase System of PAI

Sample	l_p (Å)	l_0 (Å)	l_1 (Å)	$\Delta\rho$ (g/cm³)	ρ (g/cm³)	$\Delta\rho/\rho$ (%)
PAI-3	24	29	133	0.074	1.373	5.4
PAI-4	24	30	126	0.077	1.382	5.6
PAI-5	27	34	129	0.084	1.394	6.0

temperature, the volume fraction ϕ_0 of the ordered phase shows a tendency to increase. The other important feature seen in Table 5 is that the average thickness $\bar{x}_b$ of the boundary layer has a value of about 15 Å, whereas the value $\bar{x}_b$ for PI is at most 5 Å [9]. This suggest that the molecular order in PAI does not develop as easily as in PI, because the amideimide groups in PAI are not as nearly coplanar as the imide groups in PI.

By using relationships for a generalized two-phase system, the range of inhomogeneity l_p, and the transversal length l_0 and l_1, the density difference $\Delta\rho$ and the relative density difference $\Delta\rho/\rho$ are obtained [27, 28]. These superstructure parameters for PAI-3, 4, and 5 are listed in Table 6 together with the density determined with the gradient column. Table 6 shows that the mean dimension l_0 of the ordered phase is small and the relative density difference of the two-phase structure is only ~6%. The latter value goes up to ~20% for the typical crystalline polymer polyethylene (PE) [29]. This indicates that the molecular ordering of PAI-3, 4, and 5 is intermediate between crystalline and randomly coiled amorphous states. The difference between $\bar{x}_0 + \bar{x}_1$ in Table 5 and $l_0 + l_1$ in Table 6 may be mainly due to the overestimation of invariant $\bar{Q}$.

Although the one-dimensional intensity $I(s)$ should, in principle, contain all the information on the superstructure parameters, one can get a more direct impression of the density variation in real space by deriving the one-dimensional correlation function $\gamma(x)$ [30] using the Fourier transformation

$$\gamma(x) = \frac{4\pi}{\bar{\eta}^2} \int_0^\infty s^2 I(s)\cos(2\pi sx)ds \tag{2}$$

where $\bar{\eta}^2$ is the mean square fluctuation of the electron density and x is the coordinate normal to the lamellae. If there is a periodic structure, as in a lamellae system, $\gamma(x)$ can be expected to show a maximum close to the values of x that are equivalent to the average repeat period. The experimental correlation functions for PAI-3, 4, and 5 are seen in Figure 14. All the correlation function curves have a maximum at ~120 Å; the position of the maximum for PAI-5 shows the largest value. This shows that the position of the maximum of the correlation function corresponds well with the average distance of the lamellae period $\bar{x}_0 + \bar{x}_1$.

Though SAXS on noncrystalline polymers is now one of the tools for structure investigations, the SAXS curves are more difficult to interpret than those of ordinary crystalline polymers. The method used here relies on specific model assumptions and requires parameter variation. It has been pointed out that the observation of Porod's law behavior should not be regarded as evidence for a two-phase structure [29]. Therefore, other independent proof would be helpful.

The electron micrographs and DSC measurements of PAI samples, as with those of PI samples, did not provide definitive knowledge of the superstructure [31]. This is, how-

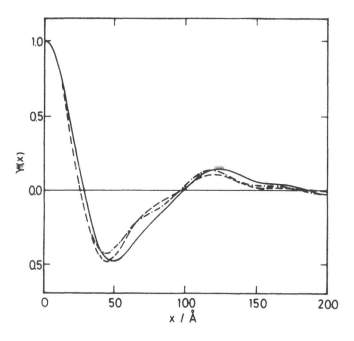

Figure 14   Correlation function $\gamma(x)$ for PAI-3 (– –), PAI-4 (– – –), and PAI-5 (———).

ever, not inconsistent with the result that the relative density difference of the two-phase structure of PAI is very small. Lamellar structures with low degrees of crystallinity usually give only one broad, first-order peak similar to that found for PAI-3, 4, and 5 in Figure 11, and in these cases there exist few interference effects [32, 33].

Heat treatment affects the glassy states of polymers in a complicated manner. Further examination of dependence of the SAXS curves on heating rate and film thickness would certainly help to determine the molecular aggregation of PAI film.

MOLECULAR AGGREGATION AND MECHANICAL PROPERTIES OF AROMATIC POLYMERS

Recently there have been some theoretical attempts to describe from a molecular viewpoint the deformation of glassy aromatic polymers with rigid chain segments [34–36]. In particular, the mechanical behavior characteristic of aromatic polyimides can be explained in terms of local molecular order consisting of bundles of parallel-packed main chains [9, 20, 34, 37].

High-Temperature Molecular Relaxations of Polypyromellitimide

Despite many mechanical and/or dielectric measurements, the glass transition temperature T_g of PI has not been clearly determined [38–40]. Isoda et al. [37] have elucidated the relation between molecular aggregation and glass transition motion of PI. Figure 15 shows the temperature dependence of dynamic tensile properties of PIs imidized under the condition in Table 1. The α dispersion is observed near 690 K. In this temperature region, PI-1, like other ordinary amorphous polymers, shows a large decrease in the storage

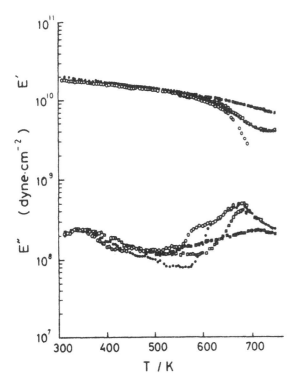

Figure 15 Temperature dependence of dynamic tensile properties of PI. ○, PI-1; ●, PI-2; □, PI-3; and ■, PI-4.

modulus E'. On the contrary, PI-4 shows a very small decrease in E' and gives a very broad loss peak in the same region. This behavior of the α dispersion in PI-4 is different from that of glass transition in ordinary amorphous polymers. It is for this reason that the molecular motion associated with the α dispersion of PI has not been definitely assigned. As seen in Figure 15, with increasing initial imidization temperature the decrease in E' in the α dispersion region became smaller and the corresponding peak in E'' became broader and smaller, shifting to higher temperature. From these results and those reported in the previous section on polypyromellitimide it is evident that the α dispersion of PI depends on molecular aggregation. It was found by the analysis of SAXS from the PIs that PI-1 is amorphous but PI-2, 3, and 4 have heterogeneous two-phase structure involving ordered phases and that these ordered phases develop with increasing initial imidization temperature. Therefore, the α dispersion of PI can be attributed to the glass transition mode, which is restrained by ordered phases. Lim et al. [41] reported that there is a steady rise in loss modulus, indicative of a large-scale micro-Brownian-type motion of the backbone chain for PI at temperatures above 600 K. Bessonov et al. [42] evaluated T_g of PI to be 690 K by using an experimental equation.

Figure 16 shows the temperature dependence of dynamic tensile properties of Kapton H both undrawn and drawn 40% [20]. For the drawn film, the dynamic storage modulus E' and the loss modulus E'' were measured in directions $\chi = 0$, 45, and 90° with respect to the drawing direction. The T_g in both directions for the drawn film is observed at about 680 K, shifted to lower temperature by about 10 K in comparison with the T_g for

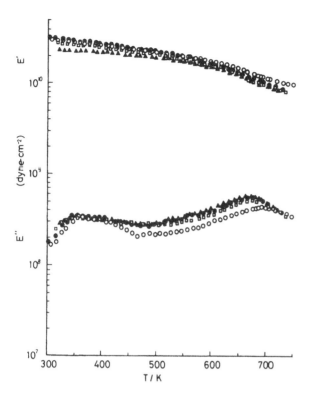

Figure 16 Temperature dependence of dynamic tensile properties of Kapton H. $\bigcirc$, undrawn; $\bullet$, drawn ($\chi = 0°$); $\square$, drawn ($\chi = 45°$); and $\blacktriangle$, drawn ($\chi = 90°$).

the undrawn film. Cold-drawing increases the density of amorphous polymers by orientation of the main chain, but decreases that of Kapton H by destruction of ordered phases existing in the virgin film [37]. Thus it may be generally concluded that a PI film having a higher degree of molecular order shows a higher T_g, because the glass transition mode in PI is restrained by ordered phases.

A broad β dispersion peak of polypyromellitimide is observed around 350 K (see Figure 15 and 16). There have been considerable differences in assignment of molecular motion responsible for this relaxation in KH. In one study it was ascribed to interplane slippage, a process analogous to that occurring in graphite, because it does not appear in dielectric measurements [6]. As can be seen in Figure 16, cold-drawing has no effect on the location and intensity of this relaxation. Furthermore, it was always observed in PI films, in which molecular aggregation leads to morphologies ranging from an amorphous state to an ordered state; the β dispersion of KH scarcely depends on molecular packing and cannot be attributed to crystal plane slippage. It is equally present in dried samples [43, 44], indicating that the idea that this peak is due to a decoupling relaxation involving loss of water should be rejected [45]. It seems more reasonable to attribute the β dispersion in KH to p-phenylene groups in the chain backbone. Butta et al. [43] pointed out that a relaxation with characteristics very similar to that shown by the β process in polyimide is exhibited by poly(p-phenylene oxide) and by many other ring-substituted poly(p-phenylene oxides). This kind of local motion of main chains has been also observed in the aromatic polyamideimide having the same p-phenylene groups in its backbone. [46].

High-Temperature Molecular Relaxations of Polytrimellitamideimide

In the previous section on this polymer it was demonstrated that the molecular ordering of PAI-1 and 2 was not very regular and that PAI-3, 4, and 5 give heterogeneous two-phase structure involving densely packed, ordered phases. Figure 17 shows the temperature dependence of the storage Young's modulus E' and the loss modulus E'' of PAI [46]. Two dispersion peaks are observed. The broad peak near 350 K is the β dispersion of PAI, which is ascribed to the local motion of *p*-phenylene groups in the chain backbone [43]. The dispersion corresponding to the glass transition of PAI appears above 450 K and shifts to higher temperature with increasing annealing temperature. The T_g of PAI-5 is higher by about 10 K than that of PAI-3 and 4. This indicates that as more highly ordered phases are developed, the micro-Brownian motions of main chains in less-ordered phases are restricted.

However, the commercial film PAI-1 does not seem to be entirely free of polar solvents, because it showed a noticeable weight loss by thermogravimetry. T_g of PAI-1 may be fairly lowered by the effect of diluent [47]. Diluents also shift β relaxations to a lower temperature [48]. As shown in Figure 18, the α dispersion for PAI-1 is about 30 K

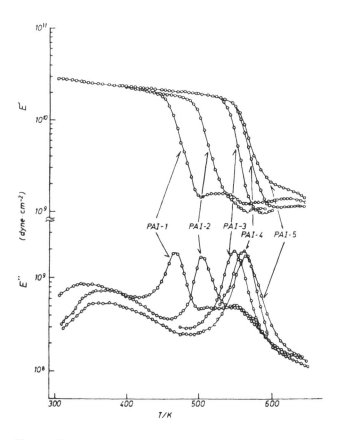

Figure 17   Temperature dependence of dynamic tensile properties of PAI.

lower than for the other PAI samples. Accordingly, the high-temperature relaxations of PAI-1 seem to be considerably affected by the solvent remaining in the film. The increase in T_g for PAI-2 may be due to a great extent to molecular packing enhanced by annealing. In fact, PAI-2 exhibited a marked increase in the small-angle x-ray scattering and the density, as compared with PAI-1.

As mentioned in the preceding section, molecular aggregation is greatly influenced by cold-drawing. Figures 18 and 19 show the temperature dependence of dynamic tensile properties of PAI-1 and 4 both undrawn and drawn 40%, respectively. For PAI-1, in which the molecular aggregation has a low-order character, the drawn and undrawn films show the α dispersion peak at near the same temperature. On the other hand, in PAI-4 the drawn film shows the α dispersion peak about 10 K lower than for the undrawn film. As is the case of Kapton H, the density and the degree of molecular order of PAI-4 decrease on cold drawing. Thus a film having a higher degree of order shows the α dispersion peak at a higher temperature. This effect of cold-drawing also reveals the correlation between dynamic mechanical properties and the state of molecular aggregation.

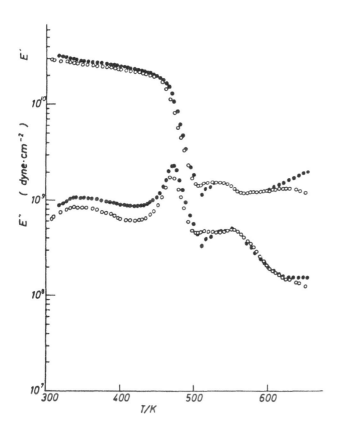

Figure 18 Effect of cold drawing on dynamic tensile properties of PAI-1. ○, PAI-1 undrawn; ●, PAI-1 drawn 40%.

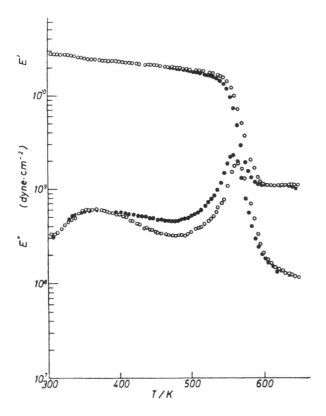

Figure 19 Effect of cold-drawing on dynamic tensile properties of PAI-4. ○, PAI-4 undrawn; ●, PAI-4 drawn 40%.

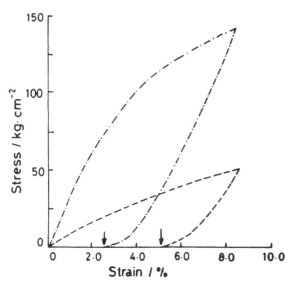

Figure 20 Stress–strain curves for PI-1 (−−) and PI-4 (−•−) at 730 K.

Thermal Distortion Temperature and Elastic Recovery of Aromatic Polymers

In aromatic polymers, static mechanical properties including thermal distortion temperature and elastic recovery are very sensitive to the change in molecular aggregation, which is often undetectable by such methods of structural analysis as SAXS and WAXD [49, 50]. Thermal distortion temperature T_d was defined as the temperature at which a load of 100 kg/cm² resulted in 3% elongation.

Figure 20 shows stress–strain curves for PI-1 and 4 measured at temperature above T_g of PI [49]. Elastic recovery was respectively determined at the point denoted by arrows to be 42% for PI-1 and 71% for PI-4. And T_d was 573 K for PI-1 and 658 K for PI-4, respectively [49]. For crystalline polymers it is well known that crystallites play the role of physical crosslinking points in mechanical properties [48]. Ordered phases can be expected to play the same role for noncrystalline polymers. The increase in T_d and elastic recovery of PI-4 is qualitatively explained in terms of the increase in the volume fraction of ordered phases, because ordered phases should restrain molecular motions of the main chains in less-ordered phases.

On the contrary, a soluble polyimide PI 2080 of the chemical structure

$$\left[\begin{array}{c} \end{array} N\text{-Ar}_{1,2} \right]_n \qquad (\text{Ar}_1 = , \text{Ar}_2 =)$$

$$(\text{Ar}_1 : \text{Ar}_2 = 20 : 80)$$

was essentially amorphous in WAXD and exhibited no heterogeneous two-phase structure having ordered phases after annealing at temperatures around T_g [50]. Accordingly, T_d and elastic recovery of PI 2080 were little affected by the annealing. The flexible carbonyl group and the statistical distribution of copolymeric repeating units in the main chain seem to prevent the formation of molecular order.

Table 7 lists the data of annealing effect on T_d and elastic recovery for the aromatic polymers poly(ether-cosulfone) (PECS), polycarbonate (PC), poly(aryl sulfone) (PASF), aromatic polyamide (PA), and poly(amide imide) (PAI). Some vinyl polymers—poly(ethyl methacrylate) (PEMA), poly(methyl methacrylate) (PMMA), and polystyrene (PS)—are included in Table 7 for comparison [49]. A state of thermodynamic equilibrium was attained by annealing for 12 hr at temperatures approximately 10 K lower than T_g. Table 7 shows that the effect of annealing on those static mechanical properties is remarkable for aromatic polymers but negligible for vinyl polymers.

This may suggest the effect of ordered phases formed in aromatic polymers by annealing. The formation of ordered phases in aromatic polymers seems to be enhanced by their planar structure originating from the rigid phenyl and/or heteroaromatic rings in the main chains and by intermolecular interactions due to those rings.

Table 7 Annealing Effect on Thermal Distortion Temperature and Elastic Recovery

Sample	Structure	T_d, K		Elastic Recovery, %	
		Original	Annealed	Original	Annealed
PEMA	$-\overset{H}{\underset{H}{C}}-\overset{CH_3}{\underset{COOC_2H_5}{C}}-$	318	318	34	39
PMMA	$-\overset{H}{\underset{H}{C}}-\overset{CH_3}{\underset{COOCH_3}{C}}-$	353	353	28	28
PS	$-\overset{H}{\underset{H}{C}}-\overset{H}{\underset{\bigcirc}{C}}-$	333	333	26	21
PECS	$-\!\bigcirc\!-SO_2\!-\!\bigcirc\!-OCH_2\underset{OH}{CH}CH_2O-$	368	423	25	65
PC	$-\bigcirc-\overset{CH_3}{\underset{CH_3}{C}}-\bigcirc-O-\overset{}{\underset{O}{C}}-O-$	413	423	62	84
PESF	$-\bigcirc-O-\bigcirc-SO_2-$	443	493	21	84
PASF	$\left(\bigcirc-O-\bigcirc-SO_2\right)_2\bigcirc-O-\bigcirc-SO_2$	508	548	44	68
PA	$\left(\overset{O}{\underset{}{C}}\bigcirc\overset{O}{\underset{}{C}}\right)_9\left(\overset{O}{\underset{}{C}}\bigcirc\overset{H}{\underset{}{N}}\right)_2\left(\overset{H}{\underset{}{N}}\bigcirc\overset{H}{\underset{}{N}}\right)_9$	453	533	23	93
PAI	$-\overset{O}{\underset{O}{N}}\bigcirc\overset{C-O}{\underset{H}{N}}\bigcirc-O-\bigcirc$	483	533	28	78

APPENDIX: ANALYSIS OF SAXS CURVES
BY THE ONE-DIMENSIONAL MODEL [11,12]

Figure A1a is a schematic presentation of the one-dimensional model. The electron density within a lamellae stack is assumed to vary only in the direction normal to the lamellae surfaces. The lateral extensions of the two phases are postulated to be much larger than the long period normal to the lamellae. Figure A1b is the electron density distribution in the direction normal to the lamellae for an ideal two-phase model, in which the mean electron density changes abruptly at each ordered–less-ordered phase boundary. The trapezium density profile is seen in Figure A1c; the density changes linearly from the ordered density ρ_0 to the less-ordered density ρ_1 over an average transition range $\bar{x}_b$. For solid polymers studied, the boundary layers in Figure A1c seem to be more realistic than those in Figure A1b. The theoretical scattering intensity $I(s)$ for an ideal two-phase structure in Figure A1b is given by

$$I(s) = I_B + I_C \tag{A1}$$

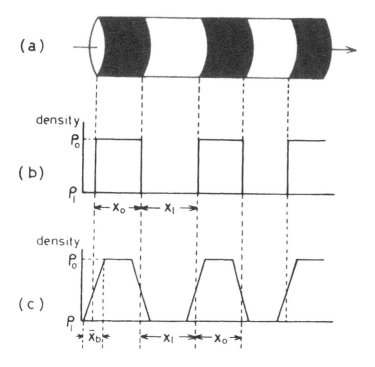

Figure A1 (a) Schematic presentation of one-dimensional model; (b) electron density distribution in direction normal to lamellae for ideal two-phase model; and (c) electron density distribution in direction normal to lamellae for pseudo two-phase model.

$$I_B = \left(\frac{N}{2\pi^2 s^2}\right) R_e \left[\frac{(1 - F_y)(1 - F_z)}{1 - F_z F_y}\right] \tag{A2}$$

$$I_C = \left(\frac{1}{2\pi^2 s^2}\right) R_e \left\{F_z \left[\frac{1 - F_y}{1 - F_z F_y}\right]^2 [1 - (F_z F_y)^N]\right\} \tag{A3}$$

$$F_y = \int_{-\infty}^{\infty} h(x_0)\exp(-2\pi i s x_0)dx_0 \tag{A4}$$

$$F_z = \int_{-\infty}^{\infty} h(x_1)\exp(-2\pi i s x_1)dx_1 \tag{A5}$$

where s is defined as $2(\sin\theta)/\lambda$, $h(x_0)$ is the thickness distribution function of the ordered phase, $h(x_1)$ is the thickness distribution function of the less-order phase, N is the number of ordered lamellae in a stack, x_0 is the thickness of a particular ordered lamellae, and x_1 is the thickness of a particular less-ordered layer.

In the case of a pseudo two-phase structure in Figure A1c, we get the correcting factor $Z(s)$ for the intensity function (Eq. A1),

$$Z(s) = \frac{1}{2\pi^2 s^2}|1 - \exp(-2\pi i s \bar{x}_b)|^2 \tag{A6}$$

where $\bar{x}_b$ is the average thickness of the boundary layer.

For analytical convenience $h(x_0)$ and $h(x_1)$ are here taken to be symmetrical Gaussian distributions defined by

$$h(x_0) = \frac{1}{(2\pi\Delta^2\bar{x}_0)^{1/2}} \exp\left[-\frac{(x_0 - \bar{x}_0)^2}{2\Delta^2\bar{x}_0}\right] \tag{A7}$$

and

$$h(x_1) = \frac{1}{(2\pi\Delta^2\bar{x}_1)^{1/2}} \exp\left[-\frac{(x_1 - \bar{x}_1)^2}{2\Delta^2\bar{x}_1}\right] \tag{A8}$$

The thickness fluctuations are designated as g_0 and g_1, related to the $\Delta^2\bar{x}_0$ and $\Delta^2\bar{x}_1$ by

$$g_0 = \left(\frac{\Delta^2\bar{x}_0}{\bar{x}_0^{\,2}}\right)^{1/2} \tag{A9}$$

and

$$g_1 = \left(\frac{\Delta^2\bar{x}_1}{\bar{x}_1^{\,2}}\right)^{1/2} \tag{A10}$$

where $\bar{x}_0$ is the average thickness of the ordered lamellae and $\bar{x}_1$ is the average thickness of the less-ordered layers.

The deviation Δ of the theoretical curve I_{th} from the experimental curve I_{exp} is given by

$$\Delta = \frac{(s_{max} - s_{min}) \int_{s_{min}}^{s_{max}} (I_{th} - I_{exp})^2 ds}{(\int_{s_{min}}^{s_{max}} I_{exp} ds)^2} \tag{A11}$$

In the one-dimensional model, the desmeared intensity I cannot be used directly as the experimental intensity I_{exp}. Instead the Lorentz correction for an isotropic sample must be applied [22]:

$$I_{exp} \propto s^2 I \tag{A12}$$

The fitting procedure has been performed such that the deviation Δ in Eq. A11 should be minimized.

REFERENCES

1. G. W. Gray and P. A. Winsor, *Liquid Crystals and Plastic Crystals,* Ellis Horwood, Sussex (1974).
2. G. C. Berry, *J. Polym. Sci. Polym. Phys. Ed., 14*: 451 (1976).
3. G. D. Wignall and G. W. Longman, *J. Mater. Sci., 8*: 1439 (1973).
4. A. P. Rudalov, M. I. Bessonov, M. M. Koton, Ye. F. Fedorova, *Dokl. Akad. Nauk. SSSR, 161*: 617 (1965).
5. R. Delasi and J. Russel, *J. Appl. Polym. Sci., 15*: 2965 (1971).
6. R. M. Ikeda, *Polym. Lett., 4*: 353 (1966).
7. L. G. Kazaryan, D. Ya. Tsvankin, B. M. Ginzburg, Sh. Tuichiev, L. N. Korzhavin, and S. Ya. Frenkel, *Vysokomol. Soedin., A14*: 1199 (1972).
8. G. Conte, L. D'Ilario, and N. V. Pavel, *J. Polym. Sci. Polym. Phys. Ed., 14*: 1553 (1976).
9. S. Isoda, H. Shimada, M. Kochi, and H. Kambe, *J. Polym. Sci. Polym. Phys. Ed., 19*: 1293 (1981).
10. O. Glatter, *J. Appl. Crystallogr., 7*: 147 (1974).
11. R. Hosemann and S. N. Baghi, *Direct Analysis of Diffraction by Matter,* North-Holland, Amsterdam (1965).

12. D. Ya. Tsvankin, *Polym. Sci. USSR, 6*: 2304 (1964).
13. T. P. Russel, *J. Polym. Sci. Polym. Phys. Ed., 22*: 1105 (1984).
14. N. Takahashi, D. Y. Yoon, and W. Parish, *Macromolecules, 17*: 2538 (1984).
15. S. Numata, K. Fujisaki, and N. Kinjo, in *Polyimides* (K. L. Mittal, ed.), Plenum, New York, p. 259 (1984).
16. P. J. Flory, *Macromolecules, 11*: 1141 (1978).
17. P. J. Flory and G. Ronca, *Mol. Cryst. Liq. Cryst., 54*: 289 (1979).
18. B. F. Blumentritt, *Polym. Eng. Sci., 18*: 1216 (1978).
19. T. P. Russel, H. Gugger, and J. D. Swalen, *J. Polym. Sci. Polym. Phys. Ed., 21*: 1745 (1983).
20. M. Kochi, H. Shimada, and H. Kambe, *J. Polym. Sci. Polym. Phys. Ed., 22*: 1919 (1984).
21. R. F. Boehme and G. S. Cargill, III, in *Polyimides* (K. L. Mittal, ed.), Plenum, New York, p. 461 (1984).
22. B. Crist and N. Morosoff, *J. Polym. Sci. Polym. Phys. Ed., 11*: 1023 (1973).
23. S. T. Wellinghoff, H. Isida, J. L. Koenig, and E. Bear, *Macromolecules, 13*: 834 (1980).
24. M. Kochi, S. Isoda, R. Yokota, and H. Kambe, *J. Polym. Sci. Polym. Phys. Ed., 24*: 1441 (1986).
25. C. G. Vonk, *J. Appl. Crystallogr., 6*: 81 (1973).
26. K. Tashiro, M. Kobayashi, and H. Tadokoro, *Macromolecules, 10*: 413 (1977).
27. G. Porod, *Kolloid Z. Z. Polym., 125*: 51 (1952).
28. G. Porod, in *Small Angle X-Ray Scattering* (O. Glatter and O. Kratky, eds.), Academic Press, London, p. 17 (1982).
29. G. R. Strobl, M. J. Shneider, and I. G. Voigt-Martin, *J. Polym. Sci. Polym. Phys. Ed., 18*: 1361 (1980).
30. C. G. Vonk and G. Kortleve, *Kolloid Z. Z. Polym., 220*: 19 (1967).
31. M. Kochi and S. Isoda, unpublished results.
32. G. R. Strobl and N. Muller, *J. Polym. Sci. Polym. Phys. Ed., 11*: 1219 (1973).
33. H. G. Killian and W. Wenig, *J. Macromol. Sci. Phys., B9*: 463 (1974).
34. A. S. Argon and M. I. Bessonov, *Philos. Mag., 35*: 917 (1977).
35. I. V. Yannas and R. R. Luise, *J. Macromol. Sci. Phys., B21*: 443 (1983).
36. N. Brown, *J. Mater. Sci., 18*: 2241 (1983).
37. S. Isoda, M. Kochi, and H. Kambe, *J. Polym. Sci. Polym. Phys. Ed., 20*: 837 (1982).
38. H. Kambe, in *Thermal Degradation and Stability of Polymers* (H. Kambe, ed.), Baifukan, Tokyo, p. 155 (1974).
39. C. E. Stroog, *J. Polym. Sci. Macromol. Rev., 11*: 161 (1976).
40. M. I. Bessonov, M. M. Koton, V. V. Kudryavtsev, and L. A. Laius, *Polyimides*, Consultants Bureau, New York (1987).
41. T. Lim, V. Frosini, V. Zaleckas, D. Morrow, and J. Sauer, *Polym. Eng. Sci., 13*: 51 (1973).
42. M. I. Bessonov and N. P. Kuznetsov, in *Polyimides* (K. L. Mittal, ed.), Plenum, New York, p. 385 (1984).
43. E. Butta, S. DePetris, and M. Pasquini, *J. Appl. Polym. Sci., 13*: 1073 (1963).
44. J. M. Perena, *Angew. Macromol. Chem., 106*: 67 (1982).
45. W. W. Wrasidlo, *J. Macromol. Sci. Phys., B6*: 559 (1972).
46. M. Kochi, S. Isoda, R. Yokota, and H. Kambe, *J. Polym. Sci. Polym. Phys., 24*: 1619 (1986).
47. J. D. Ferry, *Viscoelastic Properties of Polymers*, Wiley, New York (1960).
48. L. E. Nielsen, *Mechanical Properties of Polymers and Composites*, Marcel Dekker, New York (1975).
49. M. Kochi and H. Kambe, *Polym. Eng. Rev., 3*: 355 (1983).
50. M. Kochi, I. Mita, and R. Yokota, *Polym. Eng. Sci., 24*: 1021 (1984).

Index

About the Editor

NICHOLAS P. CHEREMISINOFF heads the product development group in the Polymers Technology Division of Exxon Chemical Company. He is responsible for directing the research and development of specialty elastomers for the consumer and various industry segment markets, and conducts research on polymer rheology/processing and development of quality control instrumentation for polymer manufacturing. Dr. Cheremisinoff has had extensive experience in the chemical and allied industries, with particular interests in the design and scale-up of multiphase reactors. The author/co-author of over thirty-five books, he received his B.S., M.S., and Ph.D. degrees in chemical engineering from Clarkson College of Technology.

Printed and bound by CPI Group (UK) Ltd, Croydon, CR0 4YY

23/10/2024

01778245-0016